ADVANCED

PHYSICS

**FIFTH
EDITION**

ADVANCED
PHYSICS

**FIFTH
EDITION**

TOM DUNCAN

JOHN MURRAY

First published in 1973
by John Murray (Publishers) Ltd
50 Albemarle Street
London W1S 4BD

Fifth edition 2000

Layouts by Fiona Webb
Artwork by Mike Humphries and Tek-Art
Typeset in 9.5/11pt Concorde and 10/12pt Franklin Gothic Demi by Wearset, Boldon, Tyne and Wear
Printed and bound in Great Britain by Butler and Tanner, Frome and London

ISBN 0 7195 7669 5

Preface to the Fifth Edition

I am indebted to my daughter Dr Heather Kennett for undertaking the major task of updating and thoroughly revising this edition so as to meet the needs of current A and AS specifications. In addition she has written a new chapter, on the increasingly popular topic *Cosmology and astrophysics*, extended the sections on particle physics and special relativity, introduced examples of the use of spreadsheets and dataloggers in physics and incorporated many examination questions from recent past papers. My grandson, Malcolm Kennett, has kindly checked the material and made many helpful suggestions for the new chapter.

Particular thanks are also due to Jonathan Ling for his detailed and very helpful comments during the early stages of the revision. At his suggestion the mathematics previously located in the introductory chapter has been more appropriately added to the *Mathematics for physics* section, terminology has been updated and material no longer required by syllabuses has been omitted. Once again Jane Roth has edited the script to her usual high standard, and prepared the material for its first publication in full colour.

T. D.

Contents

WAVES

ATOMS

Acknowledgements

AUTHOR'S ACKNOWLEDGEMENTS

I would like to thank a number of people who made many helpful comments and suggestions for earlier editions; they include Dr J. W. Warren, Mr J. Dawber, Dr J. C. Gibbings and Mr K. Munnings. Thanks also to Mr B. Baker who constructed and tested many of the objective-type questions, to my son-in-law Professor B. L. N. Kennett and my daughter Dr H. M. Kennett who checked the answers to numerical problems and to my wife who prepared the original typescript.

For permission to use questions from their examinations, grateful acknowledgement is made to the various examining boards, indicated by the following abbreviations.

AQA:
NEAB (Northern Examinations and Assessment Board)
AEB (Associated Examining Board)

OCR (previously incorporating *UCLES*)

L (London Examinations, a division of *EDEXCEL*)

IB (International Baccalaureate Organisation)

The answers given have not been provided or approved by the examining boards, who bear no responsibility for their accuracy or method of working.

T. D.

PHOTO CREDITS

The publishers are grateful to the following, who have kindly permitted the reproduction of copyright photographs or illustrations:

Figs 1, 8.20, 8.31, 18.35, 26.4, 26.9, 26.11 NASA/Science Photo Library
Fig. 2 Geoff Tompkinson/Science Photo Library
Figs 3, 1.5 John Townson, Creation
Figs 4, 1.16, 10.20, 11.4, 11.22b, p. 197, Fig. 21.7 Robert Harding Picture Library
Figs 5a, 18.3a, b, 18.17, 19.28a, b Pasco Scientific
Fig. 5b Pico Technology
p. 7 American Institute of Physics
Fig. 1.7b Auto Express

Fig. 2.1a David Scharf/Science Photo Library
Figs 2.1b, 2.9, 2.10 Omicron
Figs 2.2, 21.18 Kyocera European Office
Fig. 2.8a Astrid & Hans Freider Michler/Science Photo Library
Figs 2.17a–d, 3.12 Sir Laurence Bragg FRS & J. F. Nye, from *Proceedings, Royal Society, London A190 474-481*
Figs 2.20a, b, 3.15, 3.18b, 3.28a, b, 15.17, 22.12a, 22.30b, 24.25, 25.7b Peter Gould
Figs 3.1, 15.31 National Physical Laboratory, by permission of the Controller of the Stationery Office
Fig. 3.11 V. A. Phillips & J. A. Hugo*
Fig. 3.20 British Engine Insurance Ltd
Fig. 3.23 Vosper Thornycroft (UK) Ltd
Fig. 3.25b Clive Burnskill/Allsport
Fig. 3.25c David Cannon/Allsport
Figs 3.29, 4.8b, 7.16, 7.25, 11.9, 11.22a, 11.23, 11.25, 13.8a, b, 14.7, 14.9, 15.23a, b, 17.35, 17.40a, 18.11, 18.21a, b, c, 18.22a, b, 20.6a, b, 20.23, 23.3b, 23.9, 23.11a, 23.17a, 23.21a, 23.95, 25.7a Andrew Lambert
Figs 4.18, 14.24, 18.4, 23.25a, 23.38, 23.81a Novara, courtesy of Unilab
Figs 4.21, 4.25 RS Components
Fig. 4.24 H. Kennett and V. Edge
Fig. 5.1 Defence Science & Technology Organisation (Australia)
Fig. 5.2 Adrienne Hart-Davis/Science Photo Library
Fig. 6.34 Roger Scruton
Figs 6.39, 25.33 David Parker/Science Photo Library
Fig. 6.55a Jeff Moore, with kind permission of Peter Jones of Cranbrook, Kent
Fig. 6.84 Stephen & Donna O'Meara/Science Photo Library
Figs 6.93a, 10.10b Alex Bartel/Science Photo Library
Fig. 6.93b David Nunuk/Science Photo Library
p.111 David Ducros/Science Photo Library
Figs 7.19, 7.34a, 9.22a, b, c, A1.1 Tom Duncan
Figs 7.34b, 25.41a, b Lady Blackett*
Fig. 7.35 MIRA
Fig. 8.8 Matthew Stockman/Allsport
Fig. 8.11 Anthony Price/Ace Photo Library
Fig. 8.14 Novosti/Science Photo Library
Fig. 8.30 European Space Agency/Science Photo Library
Fig. 9.2a University of Washington Libraries Special Collections Division
Fig. 9.2b Iwasa/Rex Features
Figs 9.25, 12.14b Quadrant Picture Library

Figs 10.7, 10.8, 10.12a, 10.13 Martin Bond/Environmental Images

Fig. 10.10a Hank Morgan/Science Photo Library

Figs 10.12b, 11.2, 24.20 Martin Bond/Science Photo Library

Fig. 10.18 Mark Edwards/Still Pictures

Fig. 11.12 NHPA

Fig. 11.13 BSIP Marland/Science Photo Library

Fig. 12.15a, 19.17 J. Allan Cash

Fig. 13.3 Aerofilms

Fig. 14.35 Brookhaven National Laboratory/Science Photo Library

Figs 16.2b, 16.16 Science & Society Picture Library

Figs 16.23a, b Alstrom Electrical Machines Ltd

Fig. 16.35a Alstrom T & D Transformers Ltd

Fig. 16.35b Jeff Moore

Fig. 16.36 National Grid Company

Fig. 16.49 from *Les éditions de physique (J. de Physique,* 1951 vol. 12 p. 308)

pp. 293, 361 Dr Jeremy Burgess/Science Photo Library

Fig. 18.38 Matt Meadows/Science Photo Library

Fig. 19.33, 20.67, 25.29 Philippe Plailly/Science Photo Library

Fig. 20.3b G. D. A. Dyson*

Fig. 20.12 Bausch & Lomb Optical Co.*

Fig. 20.16 Peter Aprahamian/Science Photo Library

Fig. 20.21 C. B. Daish*

Figs 20.26a, 20.34b Last Resort Picture Library

Figs 20.40, 20.41, 20.51, 20.54 Paul Brierley

Fig. 20.58 CNES 1991 Distribution Spot Image/Science Photo Library

Fig. 21.10 from *Science Extra*, Spring 1969, with permission of BBC Publications

Figs 22.17c, d from *Nuffield Advanced Physics Student's Book Unit 1 First Edition*, Longman

Fig. 22.22 Deep Light Productions/Science Photo Library

Fig. 22.31a Prof. Henry A. Hill, University of Arizona*

Fig. 22.31b A. Mollenstedt & H. Duker*

Fig. 23.40a Alfred Pasieka/Science Photo Library

Fig. 24.14 Sheila Terry/Science Photo Library

Fig. 24.21 INTELSAT

Fig. 24.34 Scott Camazine/Science Photo Library

Figs 25.10a, b, c C. T. R. Wilson*

Fig. 25.22 CNRI/Science Photo Library

Fig. 25.24 AERE Harwell

Fig. 25.34 EFDA-JET

Fig. 25.36 Gregory Sams/Science Photo Library

Figs 26.1, 26.6 Prof. Mike Bessell

Fig. 26.12b NRAD/AUI/Science Photo Library

Figs 26.15a, b Craig Savage

*Every effort has been made to contact these copyright holders, who have given their permission for reproduction in earlier editions; the publishers apologise for any omissions and will be pleased to rectify this at the first opportunity.

About physics

STRUCTURE OF PHYSICS

Physics, like other sciences, starts with observations in the world around us or from laboratory experiments (often involving measurements) designed to obtain **facts**. The investigation of electricity, for example, began when it was noticed that amber (a glass-like fossil) attracts small light objects when it is rubbed with a cloth.

(a) Concepts

To help make sense of the facts of physics and explain the behaviour of the physical world, physicists invent terms called **concepts**. These concern quantities that can be measured. Some, such as **length**, are very basic and easily measured while others, like **potential difference** (**p.d.**) in electricity, are less 'concrete' and require more sophisticated measuring instruments.

Four of the most useful concepts are those of **atoms**, **energy**, **fields** and **waves**. They pervade the whole of physics, enabling us to build an intellectual framework that helps us to understand a wide range of phenomena.

(b) Laws

Experiments show that in many cases relationships, called **laws** or **principles**, exist between concepts. They summarize a large number of facts in an economical way, often as a mathematical equation.

For example, **Hooke's law** tells us how a spring behaves when it is stretched and relates the concepts of force and length. **Boyle's law** describes how gases respond when squeezed, using the concepts of pressure and volume. **Newton's laws of motion** deal with the action of forces on objects (often called 'bodies') and the accelerations they may produce. **Ohm's law** in electricity connects the concepts of potential difference and current.

These so-called 'laws of nature' are formulated by scientists by extracting from the facts and they do have limitations. Thus, Hooke's law is true only if the spring is not stretched too far, Newton's laws of motion do not hold for bodies moving at speeds near that of light and Ohm's law only applies to certain conductors. Nevertheless, it is by introducing concepts and discovering laws that we are able to make the physical world seem reasonable and to obtain some control over it.

(c) Theories

Frequently in physics what we are dealing with is not directly accessible to our senses and in such cases we sometimes use **theories** or 'thought-models' to help us to explain things. For instance, the **wave theory** is used to make sense of some of the properties of light and sound, and draws on our knowledge of the behaviour of 'real' waves such as surface water waves. The **field theory** enables us to deal with the invisible, action-at-a-distance (non-contact) forces occurring in electricity, magnetism and gravitation.

The **kinetic theory** gives us insight into the properties of matter in bulk, that is, it helps to relate macroscopic (large-scale) properties such as density and pressure, especially of gases, to the masses, speeds, energies, etc. of the constituent atoms and molecules that cannot be seen directly. It assumes that in some ways these sub-microscopic particles are not unlike visible particles (e.g. snooker balls) in their behaviour.

A theory connects a wide range of ideas, thus simplifying our knowledge, and from it predictions can be made and tested by new experiments. In this way the theory is further vindicated or seen to be in need of modification if it contradicts the facts. The **atomic theory** of matter has developed in this way.

It is important to remember that scientific theories are aids to understanding which, like geographical maps, are representations or analogies of reality and are not complete descriptions of it.

(d) Branches of physics and analogous laws

For the purposes of study it is often convenient to divide physics into different branches such as mechanics, heat, light, sound, electricity, etc. However, many concepts and laws cut across these artificial boundaries and are useful in more than one branch. Four such concepts have already been mentioned (i.e. atoms, energy, fields, waves) but certain laws also have analogies in other branches.

The very fundamental **conservation laws** (e.g. of mass, electric charge, momentum, energy), which state that in any changes the quantities involved are conserved, are a notable example. Similarly, the **inverse square law** describes not only how gravitational forces between masses vary with distance but also how electric forces between charges do. The mechanisms of **thermal** and **electrical conduction** have common features and the laws governing their behaviour have a similar mathematical form. The way in which changes occur in widely different phenomena such as the growth and decay of the charge on a capacitor or the decay of a radioactive material can be represented graphically by an **exponential curve**.

All these analogies and others help to create a unified structure, built on a few basic physical concepts.

PHYSICS AND TECHNOLOGY

(a) Technology the friend

Engineering and technology use our inventiveness and knowledge of physics (and other sciences) to find solutions to problems that can lead to an improvement in the material well-being of the human race. There are many examples of this.

Electrical generators, as used in power stations, are the outcome of discoveries made by Faraday about 170 years ago. So too are electric motors, the heart of so many of today's appliances.

Radio and television arose from the theoretical ideas of the physicist Clerk Maxwell concerning the connection between light, electricity and magnetism. Subsequently the efforts of Hertz, Marconi, Logie-Baird and others made possible the transmission of signals over a distance.

Predictions about the paths taken by artificial satellites and space vehicles are based on Newton's laws of motion, formulated over 300 years ago; they have contributed to the 'conquest' of space, Fig. 1, with its many beneficial spin-offs.

Nuclear power is possible because of the basic work done by Rutherford on the structure of the atom at the beginning of the twentieth century. It is one solution to our attempt to find new sources of energy.

Fig. 1 Artist's representation of the International Space Station (ISS) currently under construction in orbit

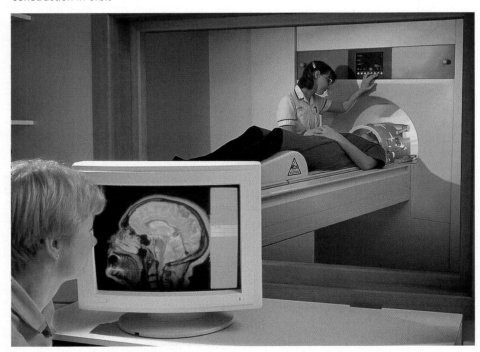

Fig. 2 MRI (magnetic resonance imaging) scan of a patient's brain. Radio wave pulses emitted by the scanner interact with atoms affected by the magnetic field; radiation absorbed or emitted is detected, digitized and used to form an image of a 'slice' of the brain on the computer screen

Modern medicine uses a host of devices such as X-ray cameras, body scanners, Fig. 2, ultrasound scanners and lasers, all developed from discoveries in physics for the diagnosis and treatment of disease.

Electronics, which arose from J. J. Thomson's 'discovery' of the electron at the end of the nineteenth century, is today being used to an ever-increasing extent in communication, control and computer systems, Fig. 3, as well as in domestic products and for medical care. Telephone exchanges now have electronic switching controlled by computers. They are linked to exchanges in other countries via large dish aerials at earth stations, Fig. 4, which send signals to earth-orbiting communication satellites for amplification and onward transmission, or via underwater optical fibres which transmit the electrical signals as pulses of laser light.

Fig. 3 Laptop computer

The laser, predicted from theoretical considerations, became a reality in 1960. At first it was regarded as a scientific curio, a light source which was 'a solution looking for a problem'. Today it has widespread applications which include, as well as optical fibre communication systems, surveying and range-finding, delicate medical operations, compact disc (CD) players, scanners at library and shop check-outs, printing and holography.

(b) Technology the foe

If not used wisely, technology can create social and environment problems such as unemployment, pollution of many types and noise problems, not to mention the ultimate folly of nuclear war.

Some scientists believe that the greenhouse effect, in which an increase of carbon dioxide in the lower levels of the atmosphere mainly from the burning of fossil fuels (e.g. coal, oil), could lead to the average temperature of the earth rising by a few °C in a few hundred years. This would cause dramatic adverse climatic and geographical changes. It is claimed that during the twentieth century the rise was 0.5 °C. As well as the desirability of reducing carbon dioxide and other 'greenhouse gas' emissions, there is also an urgent need to harness alternative, non-polluting, renewable sources of energy such as wind power, since fossil fuel resources are limited.

Damage to the ozone layer in the atmosphere, particularly above the Antarctic, by the use of chlorofluorocarbons (CFCs) in aerosols, refrigerators and air-conditioning units has also been a recent matter of concern.

Greater penetration of the earth's atmosphere by ultraviolet radiation (owing to its reduced ozone content) is believed to lead to a rise in cases of skin cancer.

It is for the human race to use technology responsibly and ensure its products are user- and environment-friendly.

(c) Interplay between physics and technology

This is a two-way process. Not only does technology depend on physics, but advances in technology are often in turn used to further the work of physicists by providing them with new techniques and instruments, e.g. the scanning tunnelling microscope (see p. 14) and the Hubble space telescope (see p. 501).

PRACTICAL WORK IN THE STUDY OF PHYSICS

(a) Types of practical work

Practical work is an essential part of any physics course and takes the following forms.

(i) *Measurement of a physical quantity* such as the acceleration of free fall (g), the end result being a numerical value and an estimate of the possible error in it.

(ii) *Verification of a well known law or principle* such as Ohm's law or the principle of conservation of momentum, which involves keeping some quantities constant while the relation between others is studied.

(iii) *Open-ended investigation* in which you do not know what the outcome will be and have to design the experiment yourself and choose the equipment required.

(iv) *Designing and constructing a system* to do a particular job. This is a popular activity in physics courses containing a section on electronics.

Experiments in categories (i) and (ii) are standard, 'bread-and-butter' types to which most time is devoted; you will find that many are outlined at appropriate points in the text. A list of suggestions for the more 'real-life' types (iii) and (iv) is given at the end of the book (p. 541); tackling two or three of these may help you to 'do' physics better.

Fig. 4 Earth station aerial for telecommunication via satellite

(b) Doing practical work

Whatever form it takes, you should

(*i*) read any instructions carefully and/or plan the procedure you will follow *before you start*,

(*ii*) record your results in a prepared table *as you make them*, to the number of significant figures the accuracy justifies, and with the units stated at the start of each column or row,

(*iii*) take *more than one measurement* of each observation if time permits,

(*iv*) take *at least eight readings* over as wide a range as possible if a graph is to be drawn, and

(*v*) *do not dismantle the apparatus* until you are certain you no longer need to use it, which may mean first plotting a graph or making calculations to ensure all your measurements are sensible.

(c) Writing up practical work

The aim should be clear from the **title** of the experiment. A clear, labelled diagram (or a circuit diagram in an electrical experiment) along with the table of results should form the basis of the **description** of what you did but any difficulties experienced or precautions taken to secure accuracy should be mentioned.

The **conclusion** is the most important part of the report and may either be a numerical value (and unit), the statement of a known law, a relationship between two quantities or a statement related to the aim.

- photoelectric effect,
- radioactive decay,
- fluid motion,
- coulomb forces,
- special relativity.

The *PEARLS* software is a comprehensive virtual physics laboratory which can be used on Macintosh or IBM-compatible PCs. It enables computer simulations to be used to model a variety of physics experiments by changing parameters and starting conditions; results can be displayed graphically.

(b) Dataloggers

These can replace a variety of standard laboratory instruments such as timers, scalers, frequency meters and storage oscilloscopes. Dataloggers can collect, store and process data.

The portable *Pasco* datalogger shown in Fig. 5*a* has a number of digital and analogue channels to which a range of compatible sensors may be connected for recording data such as temperature, sound, light, force and motion. This datalogger can be connected to the modem port on a Macintosh computer or the serial port of a PC with Windows, to display and process data with the aid of compatible software. The *Pico ADC* shown in Fig. 5*b* works as a versatile datalogger when used with PicoScope and PicoLog software on a Windows- or DOS-based PC; it plugs straight into the parallel port of the PC.

Experiments for which a datalogger is useful include

- measuring g,
- Newton's laws of motion,
- conservation of momentum,
- vibrating systems,
- potential round a sphere,
- measuring self-inductance,
- charge and discharge of a capacitor,
- radioactive decay,
- transistor characteristics.

A portable datalogging system used with a portable computer, such as the *Xemplar Pocket Book*, allows both datalogging and processing of results to be carried out in the field.

SPREADSHEETS

(a) Data manipulation

Spreadsheets can be used for a number of purposes, in particular to record and to manipulate data. A spreadsheet consists of a set of **cells** arranged in rows and columns; text, numbers and equations can be typed into the cells. For example, in an experiment to determine the Young modulus of elasticity of a wire, the values for the extension of the wire and the load (applied mass) are recorded on a sheet of a Microsoft *Excel* workbook (version 5.0) in Fig. 6*a*. Headings are typed into cells A1 and B1 in row 1, values for the extension are recorded in cells A2–A9 and the corresponding values for the applied mass in cells B2–B9.

COMPUTERS AND DATALOGGERS

(a) Computers

A computer program can be used to **simulate** an 'experiment' without using any laboratory apparatus. It should not be regarded as a substitute for the real experiment but as an aid to understanding.

Phenomena that can be studied in this way and for which programs are available include

- projectile motion,
- simple harmonic motion,
- interference and diffraction,
- capacitor discharge through a resistor,

(a) *Pasco*   **(b)** *Pico*

Fig. 5   Dataloggers

	A	B
1	Extension (mm)	Applied mass (kg)
2	0.2	5
3	0.5	10
4	0.8	15
5	1	20
6	1.3	25
7	1.5	30
8	1.8	35
9	2	40
10		
11	SLOPE (kg/mm)	m/e (kg/m)
12	19.48	19483
13		
14	E=Fl/Ae=mgl/Ae (N/m²)	
15	9.55E+10	
16		

(a) Spreadsheet of data to find the Young modulus of a wire

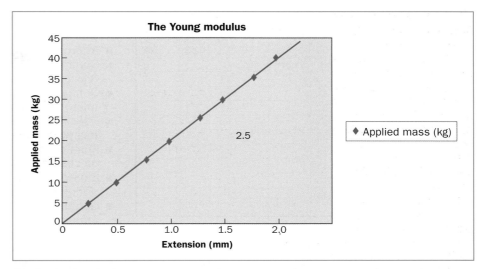

(b) Graph of applied mass versus extension

Fig. 6

The graph of the values is obtained by selecting the matrix of cells A1:A9 and B1:B9 for a graphing routine from the Chart Wizard, Fig. 6b. Lines, such as the slope, and text may be added to the graph using the drawing facility.

Text is added to the worksheet in cells A11 and B11. When a **function** or **equation** is to be entered into the worksheet it must be preceded by the = sign. To obtain the slope of the data points the function 'SLOPE' is called up from the Microsoft f_x (Function Wizard) menu by typing **=SLOPE (B2:B9,A2:A9)** into cell A12; once this function is entered the value of 19.48, the slope calculated by the computer by linear regression, appears in cell A12.

Calculation of the Young modulus of the wire can be made by means of equations entered into the worksheet. First the value of the slope (the average value of m/e) is converted from kg/mm to kg/m by entering **=A12*1000** into cell B12; the computer calculates the value to be 19483. Now the Young modulus E = stress/strain = Fl/Ae where e is the extension resulting from load $F = mg$ (see p. 28). For a wire of length $l = 1$ m, cross-section area $A = 2 \times 10^{-6}$ m² and $g = 9.8$ m s^{-2}, then $E = mg/(2 \times 10^{-6}e) = 9.8 \times$ slope/(2×10^{-6}) which is calculated on the computer by typing **=9.8*B12/0.000002** into cell A15; when this equation is entered the result of 9.55E + 10 N m^{-2} is displayed in cell A15.

Note. In equations * and / are used for multiplication and division respectively and take precedence in order of calculation over + and − signs. Brackets need to be used to change the order of operation.

Other spreadsheet programs have similar functions to *Excel*, but the specific instructions given here may need to be modified slightly.

Spreadsheet software is becoming increasingly sophisticated and allows frequently used mathematical expressions such as square roots, trigonometric functions, averages and even random numbers to be called up from the function menu and calculated by the computer. Curve-fitting facilities may be available for use on graphs; for example in 'trendline' in *Excel*, a linear, power or exponential relationship can be chosen and the equation displayed on the graph for comparison with the data. Different formulae can be entered on the spreadsheet to test for relationships between data. Large quantities of data can be automatically sorted and displayed in bar or pie charts.

(b) Simulations

Spreadsheets can also be used for simulations and simple mathematical modelling, by varying parameters in a data set or calculating changes in a system over time.

Figure 7a on p. 6 shows how a spreadsheet may be used to simulate projectile motion in the absence of air resistance (see p. 118). Values for initial velocity $u = 50$ m s^{-1} and angle of projection $\theta = 52°$ have been chosen and the numerical values entered in cells A2 and B2 respectively on a sheet of a Microsoft *Excel* workbook. The angle θ is converted to radians and the sine and cosine obtained by typing **=SIN(B2*3.142/180)** in C2 and **=COS(B2*3.142/180)** in D2. A set of times is chosen and the numerical values entered in cells A4–A12. Equation **=A\$2*A4*D\$2** is entered in B4 and **=A\$2*A4*C\$2-(E\$2*A4*A4/2)** in C4 so that the computer will calculate the horizontal distance $x = ut \cos \theta$ and the vertical distance $y = ut \sin \theta - gt^2/2$ travelled by the projectile at each chosen time.

Note. The use of the **\$** sign in these equations ensures a particular cell value (and not successive values) is used in each calculation.

Selection of the matrix of cells A3:C12 and insertion into a graphing routine allows the height of the projectile to be plotted against distance and time, Fig. 7b. By changing only the numerical values of u and θ different projectile paths may be modelled. The effect of different values for the acceleration of free fall, such as is found on other planets, could be investigated by changing the value of g in cell E2.

	A	B	C	D	E
1	Initial velocity u	Angle θ	sinθ	cosθ	g
2	50	52	0.79	0.62	9.8
3	Time	x=utcosθ	y=utsinθ−gt²/2		
4	0	0	0		
5	1	31	35		
6	2	62	59		
7	3	92	74		
8	4	123	79		
9	5	154	75		
10	6	185	60		
11	7	215	36		
12	8	246	2		
13					

(a) Spreadsheet for projectile motion simulation

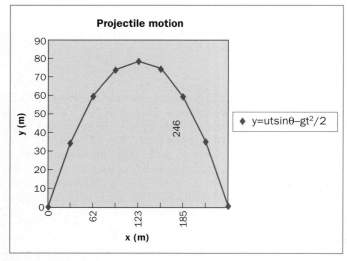

(b) Projectile trajectory

Fig. 7

WHY STUDY PHYSICS?

The answer to this question may be the very important one that you have an examination to pass which will advance your future career, but there are other reasons. Among them are that:

(a) *it can promote a knowledge and understanding of the world around us*, making it a more interesting place, for example, by accounting for the fact that a space capsule can orbit the earth at a constant speed of 8 km s⁻¹ with its engines switched off, yet to cycle at a very much smaller speed requires the bicycle pedals to be pushed;

(b) *it can encourage an appreciation of the importance of physics and its applications in technology* and perhaps enable us to use our discoveries and inventions more effectively, for example by indicating how cars can be designed to minimize damage and improve safety in a collision;

(c) *it can create an awareness of the social, economic and environmental implications* of science and hopefully help us to make well-informed judgements on such matters as nuclear power, the impact of computers on employment; and, lastly but not least,

(d) *it can be a source of enjoyment, satisfaction and intellectual stimulation.*

Some of the many areas in which physicists work today are the following.

- Alternative energy — geothermal, solar, wave, wind
- Communications — fibre optics, radar, radio, satellites, telecommunications, television
- Computing — computer-aided design, computer design, microprocessor control, robotics, system design
- Education — schools, colleges, universities, broadcasting, publishing
- Engineering — chemical, civil, control, electrical, mechanical
- Environmental science — conservation, noise control, pollution control, radiation protection
- Geophysics — mineralogy, petrology, prospecting, seismology
- Industry — aerospace, chemical, electronics, food, petroleum, semiconductors
- Materials science — metallurgy, new materials, thin films
- Medical physics — health service, instrumentation, radiology
- Meteorology — oceanography, weather forecasting
- Scientific civil service — defence, energy resources, research, patents, standards

QUESTIONS

Structure of physics

1. What is a scientific concept? Name *six* concepts.

2. What does a 'law of nature' tell us? Name *three* laws.

3. What is the purpose of a theory? Name *two* theories.

Physics and technology

4. a) Name *three* domestic appliances that contain an electric motor.

b) State *three* uses for an electric motor in a car.

5. Computers are products of physics and technology. Mention *two* ways in which they can be used

a) for the benefit,

b) to the detriment, of the community.

6. a) Why is the greenhouse effect so called?

b) Find out what it is claimed would be some of

i) the climatic effects,

ii) the geographical effects, of global warming.

c) Suggest ways of reducing 'greenhouse gas' emissions.

Spreadsheets

7. Construct a spreadsheet from the following measurements relating potential difference (V) and current (I) in a resistor (R). Use a graphing routine to plot V versus I. Decide if the graph is consistent with the relation $V = IR$ and evaluate R.

V (volts)	0.0	4.1	6.8	8.8	11.6	13.9
I (amps)	0.0	0.12	0.20	0.26	0.34	0.41

8. Design a spreadsheet to simulate projectile motion in the absence of air resistance. Using a graphing routine, plot the trajectory of a rocket launched with a velocity of 50 m s⁻² at an angle of 40° to the horizontal from the surface of the moon, where the acceleration of free fall is 1.67 m s⁻².

Materials

Polystyrene crystallites viewed through crossed polarizers

1
Materials and their uses

USE OF MATERIALS

It has been said that a scientific discovery is incomplete and immature until the technologist has found a practical application for it and 'improved the lot' of human beings. One of the essential requirements for any technological advance is the availability of the right materials. The importance of this is shown by the use of names such as Stone Age, Bronze Age and Iron Age for successive cultures in ancient times.

(a) Stone Age

In this period, dating from the earliest times recorded up to about 2500 BC, tools and weapons were made of stone. Clay was fired to make pottery, while the weaving of plant and animal fibres provided cloth, fishing nets and baskets.

When agriculture was developed, the settled existence required for tending crops and animals all the year round encouraged the building of permanent houses of wood and stone. Villages and towns grew up, requiring roads, drains, bridges and aqueducts. The resulting wealth of some communities led to envy among others and the need for town dwellers to construct fortifications for protection and to develop weapons technology.

(b) Bronze Age

This era began with the discovery around 2500 BC in eastern Europe that copper became harder and tougher when alloyed with tin to make bronze. The consequent advances in technology were, however, small compared with those of the Iron Age.

(c) Iron Age

Although iron farm implements were used in China for centuries before, it was not until about 1000 BC or so that iron came into widespread use in other parts of the world. Iron is one of the commonest metals in the earth's crust. It is extracted from its ore (iron-bearing rock) by **smelting**.

More recently, in the nineteenth century, steel (an iron alloy) became the dominant material for making tools, utensils, machinery, bridges, ships, weapons, cars and many other items.

(d) Modern era

The twentieth century saw the arrival of plastics (p. 10) and composite materials (p. 33) which have opened up a whole range of possibilities.

MATERIALS IN TENSION AND COMPRESSION

Different materials are used for different jobs, the choice depending, among other things, on the properties of the material. There are good reasons why concrete is used for constructing large buildings, wood for furniture, glass for windows, aluminium for saucepans, plastics for washing-up bowls, cotton and polyester for clothes, and rubber for elastic bands.

The use of materials in structures such as buildings and bridges depends on their **mechanical** properties. For example, it is essential to know how they behave under **tension** and **compression**. Stretching a material puts it in tension, Fig. 1.1a, while squeezing it puts it in compression, Fig. 1.1b. A material that is strong in tension can be weak in compression, and vice versa.

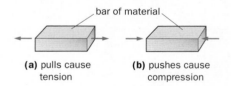

bar of material

(a) pulls cause tension **(b)** pushes cause compression

Fig. 1.1

METALS AND ALLOYS

(a) Iron and steel

Pure iron is seldom used; it is usually alloyed with other substances to form steel. **Mild steel** is iron containing a very small proportion of carbon. It is strong in both tension and compression and, being cheap, is used in large quantities for mass-produced goods like cars, cookers and refrigerators. In the construction industry scaffolding, girders, bridges, power pylons and the reinforcing for concrete are made of it.

Two disadvantages of mild steel are first that it is a heavy metal and second that it rusts. To counter rusting, an alternative to painting is to coat it with another metal which resists corrosion, such as tin, forming tinplate. **Galvanized steel** is covered by a thin layer of zinc; corrugated sheets of this are used as roofing for sheds. Chromium-plated steel is protected chiefly by a layer of nickel, on top of which a very thin layer of chromium (a hard, shiny metal) is added by electroplating.

Stainless steel contains large proportions of chromium and nickel. It is more expensive and difficult to work than mild steel but it is much stronger and harder and very resistant to corrosion. It is ideal for kitchen sinks and implements in everyday use such as cutlery.

Titanium steel has a very high melting-point and is used to make parts of jet engines, rockets, supersonic aircraft and space-shuttle nose-cones.

(b) Aluminium and duralumin

Aluminium is the most widely used metal after iron but is much more expensive to extract from its ore (bauxite); recycling of aluminium cans is a significant energy conservation measure. Its density is one-third that of iron and the thin, tough layer of oxide which forms on the surface when exposed to the air makes it very resistant to atmospheric corrosion. The pure metal tends to be weak and brittle.

Duralumin is made by alloying aluminium with small amounts of copper, manganese and magnesium. The tensile strength is then as great as that of mild steel and this, combined with its low density, makes it highly suitable for aircraft bodies.

TIMBER

There are two main types of wood. **Softwoods** like pine, besides being soft, are usually light in both weight and colour. They are used for general carpentry to make doors, window frames, floors and roof trusses in house-building. **Hardwoods** like oak and teak are stiff (see p. 26) and strong and are suitable for making good-quality furniture.

A tree grows from the centre outwards, a ring of wood being produced in the trunk each year, Fig. 1.2a. These annual rings are the **grain** marks that are seen when the trunk is cut into long planks, Fig. 1.2b. In hardwoods these marks are closer than in softwoods, which explains why the former are stiffer and stronger. In tension, the strength along the grain is greater than across it because wood consists of long tube-like fibres running up and down the tree trunk. Wood is less strong in compression.

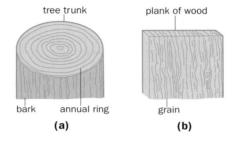

Fig. 1.2

New wood contains a great deal of moisture, most of which must be removed by **seasoning** before it is used. This is generally done naturally by stacking the freshly cut planks with spacers between, to allow air to circulate and dry them out slowly. The operation may take anything from a few weeks to several years. If a piece of wood is not properly seasoned it will gain or lose water unevenly when it is very damp or very dry. Gain of water produces expansion, loss of water causes shrinking and the result is warping.

(a) Plywood

Thin sheets of wood need less time to season than thick planks. Plywood is made by gluing thin, seasoned sheets together with the grains of alternate sheets at right angles to each other, Fig. 1.3a. An odd number of sheets is always used to give, for example, 3-ply or 5-ply. If a crack passes between the grains of one sheet it meets the next sheet across the grain, Fig. 1.3b, and does not spread. For this reason, plywood is stronger than a piece of solid wood of the same thickness. Since it consists of sheets it is called a **laminate**.

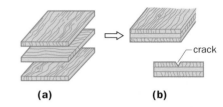

Fig. 1.3 Plywood

(b) Blockboard

This is a sandwich made by filling the space between two thin sheets of wood with strips of solid woods, as in Fig. 1.4. The grain on the outside sheets goes the same way. Like plywood, blockboard should not warp and can be fixed down firmly (e.g. as a worktop) with no fear of it moving.

Fig. 1.4 Blockboard

(c) Chipboard

This is made from wood particles and resin. It can be sawn like wood and, being reasonably strong though heavier than solid wood, it can be used for flooring, furniture, shelves, etc. It is usually sold with a thin, more attractive sheet (a **veneer**) already on one or both surfaces, Fig. 1.5. Veneers of plastic with wood grain or colour effects, or of wood itself, are available.

Fig. 1.5 Chipboard

STONE, BRICKS AND CONCRETE

A variety of materials is used to construct houses, buildings, bridges, roads and dams. Cost, climate and availability are often factors that have to be considered when deciding which to use.

(a) Stone

Deposits of stone are found in many parts of the world. They occur as granite which is hard and long-lasting, marble which is hard and attractive but does not last as long, and sandstone which is soft, easy to work and fairly long-lasting. Stone is strong in compression but weak in tension.

In cities where there is atmospheric pollution, stone buildings need cleaning periodically if they are to retain their appearance.

(b) Bricks

These are a cheap alternative to stone and have a convenient size. They are made by mixing clay with water and are then moulded into shape before being baked in ovens at a high temperature. The colour and hardness of the brick produced depends on the clay used and the baking temperature. Bricks, like stone, are weak in tension and strong in compression.

(c) Cement and mortar

Cement is a cream-coloured powder, made by heating a mixture of clay and lime to a high temperature. If mixed with sand and water, cement becomes a thick paste called **mortar**. Mortar is used to hold bricks or stones together since it becomes a hard, stone-like material when it dries.

(d) Concrete

This is now used more than any other material for building and construction work. It is made by mixing cement, sand and gravel (called 'aggregate') with water. A typical mix is 1 part cement, 2 parts sand and 4 parts aggregate, but this is varied for different purposes. If allowed to dry in a mould, any shape can be obtained and it sets as a hard, white stone.

Concrete weathers well and is strong in compression but weak in tension owing to the large number of small cracks it inevitably contains. As a result, it is brittle and unsuitable when large tensile strength is required.

In **reinforced concrete** the strength of concrete in tension is improved by inserting wires or rods of steel through the wet concrete, Fig. 1.6. As the concrete dries it sticks to the steel, giving a combination which is strong in both compression and tension.

reinforcing steel rod

concrete

Fig. 1.6 Reinforced concrete

In **prestressed concrete** even greater tensile strength is obtained, as explained in chapter 3 (p. 32). In **lightweight concrete** cinders are used as the aggregate.

POLYMERS

The properties of some of the commoner synthetic polymers are outlined here; (*a*) to (*f*) are **thermoplastics**, (*g*) and (*h*) are **thermosets** (see chapter 2, p. 23, where the molecular structure of polymers is discussed). The mechanical properties of rubber, a natural polymer, are considered in detail in chapter 3 (p. 36).

(a) Polythene

This is tough (i.e. not brittle) but flexible, and resistant to water and most solvents. It can be rolled into thin sheets and moulded into complex shapes. It is a very good electrical insulator.

(b) Polystyrene

This is more brittle than some plastics but its stiffness is taken advantage of for making small containers and toys. **Expanded polystyrene** is a solid foam containing a large number of air bubbles. Its very low density and ease of moulding to almost any shape make it a good packaging material. It is also a good heat insulator.

(c) PVC (polyvinyl chloride)

This is strong, tough, flexible and waterproof, which makes it suitable for protective sheeting and floor coverings. Being a good electrical insulator it is used to cover electric cables.

(d) Perspex

This is stiff and transparent but not so hard and brittle as glass, for which it is sometimes used as a substitute. It is easily cut and drilled, but scratches.

(e) PTFE (polytetrafluoroethylene) or Teflon

This has a much higher melting-point than most plastics and also has 'non-stick' properties. These make it useful for coating the insides of saucepans and other cooking containers and for making bearings that do not need lubricating.

(f) Nylon

Nylon fibres are used to make strong ropes and hard-wearing fabrics. Being water-resistant, clothing made from it dries quickly and requires no ironing. However, nylon shirts and blouses often feel damp to wear in hot weather because they do not absorb sweat. **Polyester** is an alternative fibre used in the textile industry.

(g) Bakelite

This is hard and brittle, but is much strengthened by the addition of, for example, sawdust. Its cheapness, low density and resistance to corrosion have made it popular for electrical fittings.

(h) Formica and melamine

These are two other thermosets with hard, smooth surfaces which make them suitable veneers for worktops and other surfaces.

OTHER MATERIALS

(a) Fibre-reinforced materials

The properties of composite materials using fibres of glass or carbon in a plastic resin (GRP and CFRP) will be considered in chapter 3 (pp. 33–35).

(b) Laminated glass

This is sometimes called 'bullet-proof' glass, and is even stronger than toughened (prestressed) glass (p. 32). It is made by fixing together layers of toughened glass with a transparent adhesive. More layers give greater strength. It is used for aircraft windscreens. If a crack gets through one layer of glass, it gets 'blunted' on meeting a layer of adhesive and is unable to penetrate the next layer, Fig. 1.7*a*. So when struck, it cracks but does not break into lots of small fragments; Fig. 1.7*b* shows the effect.

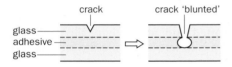

Fig. 1.7a

(c) Ceramics

These are made by mixing clay, fine sand and water into a paste which is shaped as required and fired at a high temperature. Very fine clay (e.g. china clay or kaolin) when fired at a sufficiently high temperature forms **porcelain**. Ordinary clay is unable to withstand such high temperatures and is used to make earthenware or pottery. Porcelain is usually 'glazed' by adding a layer of glass.

High-tech ceramics made from powders containing metal oxides, nitrides, carbides or borides, combine extreme hardness with a resistance to corrosion and ability to withstand high temperatures, which has seen them used in an increasing number of applications from artificial hip joints to engine rotors and cutting tools (see Figs 2.2 and 21.18).

Fig. 1.7b Laminated glass cracks but does not shatter

BEAMS

Beams or 'girders' form parts of larger structures such as bridges.

(a) Simple beams

You can see what happens when a beam is loaded by drawing lines on a piece of foam rubber, Fig. 1.8*a*, and then pushing down on its top surface. The lines at the top become shorter while those at the bottom become longer. The length of the central line is unchanged, Fig. 1.8*b*. Therefore, when a beam is loaded and bends, the top is **in compression** (squeezed), the bottom is **in tension** (stretched), and the centre, called the **neutral layer**, is neither squeezed nor stretched.

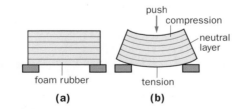

Fig. 1.8

In a solid beam, most of the material in the central region (the neutral layer) is not needed. It is wasted material whose weight simply acts as an extra load that has to be supported. If this material is removed, the much used I-beam is obtained, Fig. 1.9, which is as strong as a solid beam but much lighter. The top and bottom flanges withstand the compression and tension forces produced when the beam is loaded. Other common types of girder are L- and T-shaped.

Tubes use the same idea, the removal of unstressed material giving similar advantages. Circular tubes are most common, being equally strong in all directions at right angles to the surface.

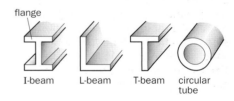

Fig. 1.9 Common types of girder

(b) Trussed beams

A simple beam can be strengthened if a **truss** is joined to it as in Fig. 1.10. If, for example, the structure is a bridge, the weight of a car on it makes CBD bend down. AB moves down too, but AC and AD hold it back so CBD does not bend so much. Loading the trussed beam therefore tends to stretch AB and puts it under tension. A beam in tension is called a **tie**.

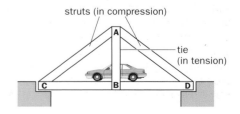

Fig. 1.10 Trussed beam

On the other hand, AC and AD are under compression and are called **struts**. They are put in this state by AB pushing down on them at A and by the bridge supports pushing up on them at C and D. The latter forces arise because the truss transfers the load to the supports.

BRIDGES

In its simplest form a bridge consists of a beam, called the **bridge-deck**, supported at the ends, Fig. 1.11.

Bridge-decks must be made of materials which can withstand both compressive and tensile stresses. They must also be fire- and water-resistant. Other important factors are cost and the amount of maintenance required.

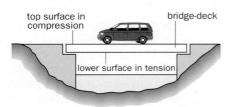

Fig. 1.11 Simple beam bridge

Early bridges were built of stone, then came steel; today reinforced, pre-stressed and lightweight concrete are most common. Where steel and concrete or stone are used, the design is such that steel bars are in parts under tension, while the stone or concrete is arranged to experience compression. Cables (of steel) are only used for parts in tension. If steel is to be under compression it is usually in the form of I- or T-shaped girders. There are many different types of bridge.

(a) Beam and pier

As well as having supports at both ends, this type has one or more pillars or piers in the middle, Fig. 1.12. The piers stop the bridge-deck from bending too much and make the bridge much stronger. A load on the bridge puts the piers under compression but, being made of stone, brick or concrete, they can withstand this.

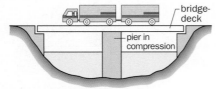

Fig. 1.12 Beam and pier bridge

(b) Arch

In an arch bridge the bridge-deck is supported by an arch either from above, as in Fig. 1.13a, or from below as in Fig. 1.13b. The arch may be of reinforced concrete or of steel girders.

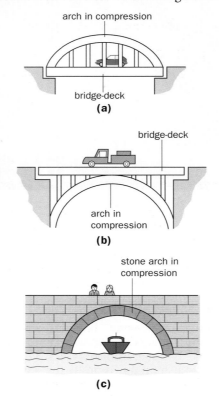

Fig. 1.13 Arch bridges

A load on the bridge-deck causes slight compression of the arch, whether it is above or below the deck. The material of the arch should therefore be strong in compression. Stone or brick bridges often have arches below the bridge-deck, Fig. 1.13c.

(c) Suspension

Most of the world's largest bridges use this construction, Fig. 1.14. They are in effect beams supported by steel cables, all of which are in tension. The main cables hang from tall towers (pylons) at each end, which must be built on rocks that can withstand the large downward forces exerted by the towers.

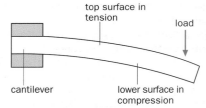

Fig. 1.14 Suspension bridge

The Forth road bridge in Scotland, in the foreground of Fig. 1.16, is of this type.

(d) Cantilever

A cantilever is a beam which is supported only at one end, Fig. 1.15. The top is in tension while the bottom is in compression.

The Forth railway bridge, built in 1890, part of which is visible in the background of Fig. 1.16, uses the cantilever principle.

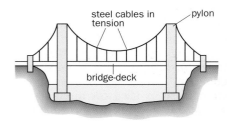

Fig. 1.15 Cantilever beam

Fig. 1.16 The Forth road bridge (suspension) and railway bridge (cantilever)

(e) Girder

Larger bridges made from steel girders are of several types. They are designed so that there is no material in the neutral layer. The top of the bridge is under compression and the bottom is in tension. Figure 1.17*a* shows a **truss girder** bridge and Fig. 1.17*b* a **lattice girder** type. Figure 1.17*c* shows how a modern **box girder** bridge is built from steel boxes.

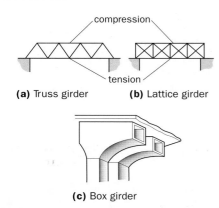

(a) Truss girder   **(b)** Lattice girder

(c) Box girder

Fig. 1.17

SOME OTHER STRUCTURES

(a) Roof truss of a house

The walls of a house with a tiled or slated roof would be pushed outwards, Fig. 1.18*a*, if the roof were not supported by a truss and tie beam, Fig. 1.18*b*. The walls are in compression, the tie beam is in tension.

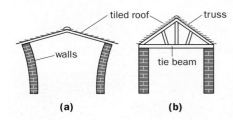

(a)   **(b)**

Fig. 1.18

(b) Balconies

The floor of a flat and its adjoining balcony are shown in Fig. 1.19. The prestressed concrete beam supports both. The steel reinforcing bar is at the bottom of the beam in the floor of the flat, because it is in tension (like a bridge-deck). However, the reinforcing steel bar is in the top of the beam forming the balcony because it is a cantilever and is in tension at the top.

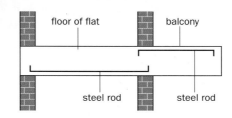

Fig. 1.19

QUESTIONS

1. a) A metal is softened or **annealed** if it is heated to a dull red heat and cooled slowly. If after heating it is plunged into cold water instead, it is hardened or **tempered** and becomes brittle. Which of the following articles have been
  i) annealed,
  ii) tempered?
A paper clip, a needle, a knife blade, a pin, a file.
b) Why does wood split more easily along the grain than across it?
c) List some of the advantages of concrete over stone.
d) Why are fabrics used for clothing often combinations of synthetics like polyester and natural fibres such as cotton?
e) Why do cracks not spread so readily through laminated glass?

2. a) Which of AB, AC or AD in Fig. 1.10 (p. 11) could be replaced by a cable?
b) A beam can also be strengthened by a truss underneath it, as in Fig. 1.20. Which of AB, AC and AD are in
  i) tension, i.e. ties,
  ii) compression, i.e. struts?

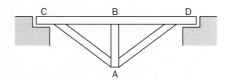

Fig. 1.20

3. a) What are
  i) the advantages,
  ii) the disadvantages,
  of wood as a material for building a bridge?
b) Why do reinforced concrete bridge-decks have steel rods at the *bottom*, Fig. 1.21?

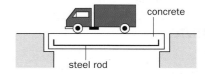

Fig. 1.21

4. a) Why does a dome not fall down?
b) What property of stone did some cathedral builders employ when they used flying buttresses to prevent the roof pushing the top of the walls outwards?

5. The crane in Fig. 1.22 is mounted on a wall and supports a load.
a) Is AB a tie or a strut?
b) Is BC a tie or a strut?

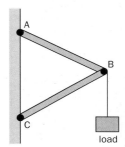

Fig. 1.22

6. In the girder bridge shown in Fig. 1.23, which girders are
a) ties,
b) struts?

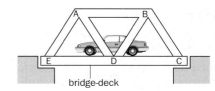

Fig. 1.23

2

Structure of materials

- Materials science
- Atoms, molecules and Brownian motion
- The Avogadro constant; the mole
- Size of a molecule
- Periodic table
- Interatomic bonds
- States of matter
- Types of solid
- Crystal structures
- Bubble raft
- X-ray crystallography
- Microwave analogue
- Polymers

MATERIALS SCIENCE

Materials technology is a long-established subject. The comparatively new subject of **materials science** is concerned with the study of materials as a whole and not just with their physical, chemical or engineering properties. As well as asking *how* materials behave, the materials scientist also wants to know *why* they behave as they do. Why is steel strong, glass brittle and rubber extensible? To begin to find answers to such questions has required the drawing together of ideas from physics, chemistry, metallurgy and other disciplines.

The deeper understanding of materials that we now have has come from realising that the properties of matter in bulk depend largely on the way the atoms are arranged when they are close together. Progress has been possible because of the invention of instruments for 'seeing' finer and finer details. The electron microscope, which uses beams of electrons instead of beams of light as in the optical microscope, reveals structure just above the atomic level. The scanning tunnelling microscope and X-ray apparatus allow investigation at the atomic level.

The scanning tunnelling microscope (STM), developed in 1981, has a charged metal tip a few atoms wide, Fig. 2.1*a*, which is brought very close to a conducting or semiconducting surface. An image of the surface is obtained as the tip moves across the

sample, revealing the position of individual atoms and molecules. This enables scientists to study the fundamental behaviour of atoms on surfaces. Figure 2.1*b* is an STM image showing surface atoms of silicon; such images aid in designing computer circuitry on thin films or ultra-small chips of silicon.

Fig. 2.1b STM image of the surface of silicon

Fig. 2.1a Scanning electron micrograph of the tip of a scanning tunnelling microscope (STM) (x2500)

The importance of materials science lies in the help it can give with the selection of materials for particular applications, the design of new materials and the improvement of existing ones. For example, high-tech ceramics are now being used as cutting tools, as Fig. 2.2 shows.

Fig. 2.2 Knives and scissors with ceramic blades

ATOMS, MOLECULES AND BROWNIAN MOTION

The modern atomic theory was proposed in 1803 by John Dalton, an English schoolteacher. He thought of atoms as tiny, indivisible particles, all the atoms of a given element being exactly alike and different from those of other elements in behaviour and mass. By making simple assumptions he explained the gravimetric (i.e. by weight) laws of chemical combination but failed to account satisfactorily for the volume relationships which exist between combining gases. This required the introduction in 1811 by the Italian scientist, Amedeo Avogadro, of the molecule as the smallest particle of an element or compound capable of existing independently and consisting of two or more atoms, not necessarily identical. So, while we could only have atoms of elements, molecules of both elements and compounds were possible.

At the end of the nineteenth century some scientists felt that evidence, more direct than that provided by the chemist, was needed to justify the basic assumption that atoms and molecules exist. In 1827 the Scottish botanist, Robert Brown, discovered that fine pollen grains suspended in water were in a state of constant movement, describing small, irregular paths but never stopping. The effect, which has been observed with many kinds of small particles suspended in both liquids and gases, is called **Brownian motion**. It is now considered to be due to the unequal bombardment of the suspended particles by the molecules of the surrounding medium.

Very small particles are essential. If the particle is very large compared with the size of a molecule, the impacts, occurring on every side and irregularly, will cancel out and there will be no average resultant force on the particle. However, if the particle is small enough to suffer impacts with only a few hundred molecules at any instant, the chances of these cancelling out are proportionately less. It is then likely that for a short time most of the impacts will be in one direction; shortly afterwards the direction will have changed. The phenomenon can be observed in smoke in a small glass cell which is illuminated strongly from one side and viewed from above with a low-power microscope, Fig. 2.3. How would the random motion be affected by (*i*) cooling the air to a low temperature, (*ii*) using smaller smoke particles?

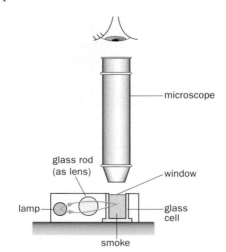

Fig. 2.3 Viewing Brownian motion

The effect, on its own, does not offer conclusive proof for molecules but it clearly reveals that on the microscopic scale there is great activity in matter which macroscopically (on a large scale) appears to be at rest. The theory of the motion was worked out by Einstein and found to correspond closely with observation. His basic assumption was that the suspended particles have the same mean kinetic energy as the molecules of the fluid and so behave just like very large molecules. Their motions should therefore be similar to those of the fluid molecules.

THE AVOGADRO CONSTANT; THE MOLE

Atomic and **molecular masses** give the masses of atoms and molecules compared with the mass of another kind of atom. Originally the hydrogen atom was taken as the standard, with atomic mass 1, since it has the smallest mass. In 1960 it was agreed internationally to base atomic and molecular masses on the carbon-12 isotope $^{12}_{6}C$. The atomic mass of carbon-12 is taken as exactly 12, making that of hydrogen 1.008 and of oxygen 16.00. Atomic masses are found very accurately using a **mass spectrometer** (see p. 400).

From the definition of atomic mass, any number of atoms of carbon will have 12 times the mass of the *same* number of atoms of hydrogen; 1 g of hydrogen will contain the same number of atoms as 12 g of carbon. In general, the atomic mass of any element, expressed in grams, contains the same number of atoms as 12 g of carbon. This number is, by definition, a constant. It is called the **Avogadro constant** and is denoted by L. Its accepted experimental value is 6.02×10^{23}.

The number of molecules in the molecular mass in grams of a substance is also the same for all substances and equal to the Avogadro constant. There are, therefore, 6.02×10^{23} molecules in 2 g of hydrogen (molecular mass 2) and in 18 g of water (molecular mass 18). The Avogadro constant is useful when dealing with other particles besides atoms and molecules, and a quantity which contains 6.02×10^{23} particles is called a **mole**. We can have a mole of atoms, a mole of molecules, a mole of ions, or a mole of electrons; all contain 6.02×10^{23} particles.

$$L = 6.02 \times 10^{23} \text{ particles per mole}$$

Sometimes quantities are expressed in terms of kilogram-moles; the number of particles per kilogram-mole is 6.02×10^{26}.

The Avogadro constant has been measured in various ways. In an early method alpha particles emitted by a radioactive source were counted by allowing those within a small known angle to strike a fluorescent screen. Each particle produced one scintillation on the screen and if it is assumed that one particle is emitted by each radioactive atom an approximate value for L can be obtained (see question 6, p. 24). Other methods give more reliable results; one involves X-ray crystallography (p. 22).

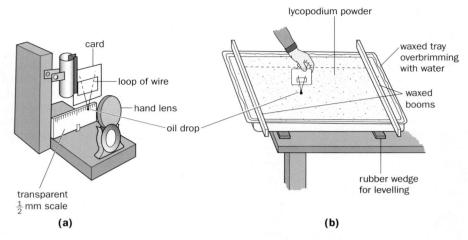

Fig. 2.4   Determining the size of an oil molecule

SIZE OF A MOLECULE

(a) Monolayer experiments

An experimental determination of the size of a molecule was made by Lord Rayleigh in 1899. He used the fact that certain organic substances, such as olive oil, spread out over a clean water surface to form very thin films.

A simple procedure for performing the experiment is to obtain a drop of olive oil by dipping the end of a loop of thin wire, mounted on a card, into olive oil, quickly withdrawing it and then estimating the diameter of the drop by holding it against a $\frac{1}{2}$ mm scale and viewing the drop and scale though a lens, Fig. 2.4a. If the drop is then transferred to the centre of a waxed tray over-brimming with water, the surface of which has been previously cleaned by drawing two waxed booms across it and then lightly dusted with lycopodium powder, Fig. 2.4b, it spreads out into a circular film pushing the powder before it. Assuming the drop is spherical, the thickness of the film can be calculated if its diameter is measured. It is found to be about 2×10^{-9} metre, i.e. 2 nanometres (2 nm).

Oil-film experiments do not necessarily prove that matter is particulate but from them we can infer that if molecules exist and if the film is one molecule thick, i.e. a monolayer, then in the case of olive oil one dimension of its molecule is 2 nm.

(b) Predictions from the kinetic theory of gases

Information about the molecular world can sometimes be obtained from observations of the behaviour of matter in bulk, i.e. from macroscopic observations. Thus with the help of the kinetic theory of gases, expressions can be derived relating such properties as rate of diffusion with the size of the gas molecules involved.

(c) Using the Avogadro constant

Consider copper which has atomic mass 64 and density 9.0 g cm^{-3}. One mole of copper atoms, therefore, has mass 64 g and volume 64/9 cm³; it contains 6.0×10^{23} atoms. The volume available to each atom is $64/(9 \times 6 \times 10^{23})$ cm³ and the radius r of a sphere having this volume is given by

$$\frac{4}{3}\pi r^3 = \frac{64}{9 \times 6 \times 10^{23}} \text{ cm}^3$$

$$\therefore \quad r = 0.14 \times 10^{-7} \text{ cm}$$
$$= 0.14 \times 10^{-9} \text{ m}$$
$$= 0.14 \text{ nm}$$

If copper atoms are spherical, would their radius be larger or smaller than this even if they were packed tightly? Why? A more accurate way of calculating the size of a copper atom is indicated in questions 10 to 13 on p. 25.

PERIODIC TABLE

If the elements are arranged in order of increasing atomic mass then, at certain repeating intervals, elements occur with similar chemical properties. Sometimes it is necessary to place an element of larger atomic mass before one of slightly smaller atomic mass to preserve the pattern. The first eighteen elements of this arrangement, called the **periodic table**, are shown in Table 2.1. The third and eleventh (3 + 8) elements are the alkali metals lithium and sodium; the ninth and seventeenth (9 + 8) are the halogens fluorine and chlorine — here the repeating interval is eight. The serial number of an element in the table is called its **atomic number**.

Table 2.1

Group 1	Group 2	Group 3	Group 4	Group 5	Group 6	Group 7	Group 0
1 hydrogen							2 helium
3 lithium	4 beryllium	5 boron	6 carbon	7 nitrogen	8 oxygen	9 fluorine	10 neon
11 sodium	12 magnesium	13 aluminium	14 silicon	15 phosphorus	16 sulphur	17 chlorine	18 argon

The periodic table suggests that the atoms of the elements may not be simple entities but are somehow related. There must be similarities between the atoms of similar elements and it would seem that the similarity might be due to the way they are built up.

We now believe that atoms are composed of three types of particle — protons, neutrons and electrons. Protons and neutrons are packed together into a very small nucleus which is surrounded by a 'cloud' of electrons, the diameter of the atom as a whole being at least 10 000 times greater than that of the nucleus. The comparative masses and charges of the three basic particles are given in Table 2.2. The nucleus is positively charged and the electron cloud negatively charged but the number of protons equals the number of electrons so that the atom is electrically neutral.

Table 2.2

Particle	Mass	Charge
electron	1	$-e$
proton	1836	$+e$
neutron	1839	0

e = electronic charge

The number of protons in the nucleus of an atom has been found to be the same as its atomic number which therefore means that each element in the periodic table has one more proton and one more electron in its atom than the previous element. Hydrogen, the first element, has one proton and one electron. Helium, the second element, has two protons and two electrons. Lithium, with atomic number three, has three protons and three electrons. Neutrons are present in all nuclei except that of hydrogen.

INTERATOMIC BONDS

Materials consist of atoms held together by the attractive forces they exert on each other. These forces are electrical in nature and create interatomic **bonds** of various types. The type formed in any case depends on the outer electrons in the electron clouds of the atoms involved.

(a) Ionic bond

This is formed between the atoms of elements at opposite sides of the periodic table, for example between sodium (Group 1) and chlorine (Group 7) when they are brought together to form sodium chloride, or common salt. A sodium atom has a loosely held outer electron which is readily accepted by a chlorine atom. The sodium atom becomes a **positive ion**, i.e. an atom deficient of an electron, and the chlorine atom becomes a **negative ion**, i.e. an atom with a surplus electron. The two ions are then bonded by the electrostatic attraction between their unlike charges.

A sodium ion attracts *all* neighbouring chloride ions in other pairs of bonded ions and vice versa. Each ion becomes surrounded by ions of opposite sign and the resulting structure depends among other things on the relative sizes of the two kinds of ion.

The ionic bond is strong. Ionic compounds are usually solid at room temperature and have high melting-points. They are good electrical insulators in the solid state since the electrons are nearly all firmly bound to particular ions and few are available for conduction. In the molten state some ionic bonds are broken and conduction is possible.

(b) Covalent bond

In ionic bonding electron **transfer** occurs from one atom to another. In covalent bonding electron **sharing** occurs between two or more atoms. Thus the atoms of carbon can form covalent bonds with other carbon atoms. Each carbon atom has four outer electrons, Fig. 2.5a, and all can be shared with four other carbon atoms to make four bonds, Fig. 2.5b, each consisting of two interlocking electron clouds.

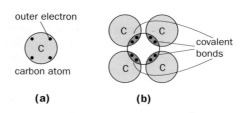

Fig. 2.5

Covalent bonds are also strong and many covalent compounds have similar mechanical properties to ionic compounds, but they do not conduct electricity when molten.

(c) Metallic bond

Metal atoms have one or two outer electrons that are in general loosely held and are readily lost. In a metal we picture many free electrons drifting around randomly, not attached to any particular atom as they are in covalent bonding. All atoms share *all* the free electrons. The atoms thus exist as positive ions in a 'sea' of free electrons, Fig. 2.6; the strong attraction between the ions and electrons constitutes the metallic bond.

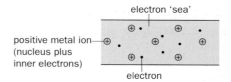

Fig. 2.6

The nature of the metallic bond has a profound influence on the various properties of metals.

(d) Van der Waals bond

Van der Waals forces are very weak and are present in all atoms and molecules. They arise because, although the centres of negative and positive charges in an atom coincide over a period of time, they do not coincide at any instant. There is a little more of the electron cloud on one side of the nucleus than the other. A weak **electric dipole** is produced giving rise to an attractive force between opposite ends of such dipoles in neighbouring atoms.

The condensation and solidification at low temperatures of oxygen, hydrogen and other gases is caused by van der Waals forces binding their molecules together. (In the molecules of such gases the atoms are held together covalently.) Van der Waals forces are also important when considering polymers (p. 23).

Two further points: first, sometimes more than one of the above four types of bonding is involved in a given case; second, information about the strength of interatomic bonds in solids is obtained from heat of sublimation measurements, in which solid is converted directly to vapour and all atoms separated from their neighbours (see question 8, p. 24); for liquids latent heat of vaporization measurements provide the information (see p. 69).

STATES OF MATTER

The existence of three states or **phases** of matter is due to a struggle between interatomic (intermolecular) forces and the motion which atoms (molecules) have because of their internal energy (see p. 65).

(a) Solids

In addition to attractive interatomic forces there must also be interatomic repulsion, otherwise matter would collapse. Evidence suggests that at distances greater than one atomic diameter the attractive force exceeds the repulsive one, while for small distances, i.e. less than one atomic diameter, the reverse is true. In Fig. 2.7a the dotted graphs show how the short-range attractive force and the very short-range repulsive force between two atoms vary with the separation of the atoms; the total or resultant force is shown by the continuous graph.

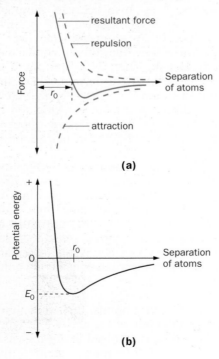

(a)

(b)

Fig. 2.7

It can be seen that for one value of the separation, r_0, the resultant interatomic force is zero. This is the situation that normally exists in a solid; if the atoms come closer together — for example, when the solid is compressed — they repel each other; they attract when they are pulled farther apart.

In an ionic bond the short-range attractive part of the interatomic force arises from the attraction between positive and negative ions which pulls them together until their electron clouds start to overlap, thus creating a very short-range repulsive force.

Now consider the motion of the atoms, the other contestant. In a solid the atoms vibrate about their equilibrium positions, alternately attracting and repelling one another, but the interatomic forces have the upper hand. The atoms are more or less locked in position and so solids have shape and appreciable stiffness.

The corresponding potential energy (p.e.) versus separation curve for two atoms (or molecules) is shown in Fig. 2.7b. At the equilibrium separation r_0 when the resultant force is zero, the p.e. must have its minimum value E_0. This is so because any attempt to change the separation involves overcoming an opposing force — an attractive one if the separation increases and a repulsive one if it decreases. E_0 is called the **bonding energy**; it is the energy needed to pull the atoms apart so that their p.e. increases to zero. They are then quite free from one another's influence.

Bonding energy will be considered later — see latent heat (p. 69).

(b) Liquids

As the temperature is increased the atoms have larger amplitudes of vibration and eventually they are able partly to overcome the interatomic forces of their immediate neighbours. For short spells they are within range of the forces exerted by other atoms not quite so near. There is less order and the solid melts. The atoms or molecules of a liquid are not much farther apart than in a solid but they have great speeds, due to the increased temperature, and move randomly in the liquid while continuing to vibrate. The difference between solids and liquids is a difference of structure rather than a difference of distance between atoms or molecules.

Although the forces between the molecules in a liquid do not enable it to have a definite shape, they must still exist otherwise the liquid would not hold together, or exhibit surface tension, viscosity, or latent heat of vaporization.

(c) Gases

In a gas or vapour the atoms and molecules move randomly with high speeds through all the space available and are comparatively far apart. On average their spacing at s.t.p. (standard temperature and pressure, i.e. 0 °C and atmospheric pressure) is about 10 molecular diameters and their mean free path (the distance travelled between collisions) is roughly 300 molecular diameters. Molecular interaction only occurs for those brief spells when molecules collide and large repulsive forces operate between them.

TYPES OF SOLID

There are three main types of solid.

(a) Crystalline

Most solids, including all metals and many minerals, are crystalline. In substances such as sugar the crystal form is evident but less so in the case of metals, although large crystals of zinc are often visible on a freshly galvanized iron surface.

The crystalline structure of a metal can be revealed by polishing the surface, treating it with an etching chemical, sometimes a dilute acid, and then viewing it under an optical microscope. The metal is seen to consist of a mass of tiny crystals, called 'grains', at various angles to one another; it is said to be **polycrystalline**. Grain sizes are generally small, often about 0.25 mm across. Figure 2.8a shows crystal grains in a cross-section of an aluminium–titanium alloy. The grains show up on the surface after etching because 'steps' are formed on each grain due to the rate of chemical action differing with different grain orientations. Light is then reflected in various directions by the different grains so that some appear light and others dark, Fig. 2.8b.

Fig. 2.8a Crystal grains in a metal alloy

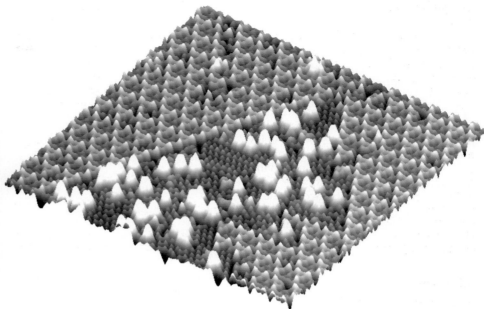

Fig. 2.9 STM image of a crystal grain boundary

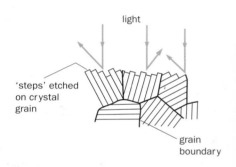

Fig. 2.8b The grains reflect light in different directions

A grain boundary is revealed as an abrupt change in the arrangement of the atoms in the STM image shown in Fig. 2.9.

The essence of the structure of a crystal, whether a large single crystal or a tiny grain in a polycrystalline specimen, is that the arrangement of atoms, ions or molecules repeats itself regularly many times, i.e. there is a long-range order.

A combination of analysis techniques is often used to study materials; the equipment shown in Fig. 2.10 incorporates a scanning tunnelling microscope (STM), scanning electron microscope (SEM) and chemical analysis techniques.

Fig. 2.10 Multi-technique equipment for materials research

(b) Amorphous or glassy

Here the particles are assembled in a more disordered way and only show order over short distances; there is no long-range order. The structure of an amorphous solid has been likened to that of an instantaneous photograph of a liquid. It is much more difficult to unravel but is the subject of considerable research. The many types of glass are the commonest of the amorphous solids; we can think of them as having a structure of groups of atoms (e.g. of silicon and oxygen) that would have been crystalline had it not been distorted. The **glass transition temperature** is the temperature at which a material changes from a liquid to an amorphous glassy solid.

(c) Polymers

These are considered later (p. 23).

CRYSTAL STRUCTURES

The structure adopted by a crystalline solid depends on various factors including the kind of bond(s) formed and the size and shape of the particles involved. For example, in metals where all the positive ions attract all electrons (p. 17), the bonding pulls equally in all directions, i.e. is non-directional, and every ion tends to surround itself by as many other ions as is geometrically possible. A **close-packed structure** results. On the other hand, in covalent solids the bonding is directional, i.e. every shared electron is localized between only two atoms. This does not encourage close-packing since the number of atoms immediately surrounding each atom is limited to the number of covalent bonds it forms. Would you describe the ionic bond as directional or non-directional?

Some common crystal structures will now be described.

(a) Face-centred cubic (FCC)

This arrangement of close-packing is shown in Fig. 2.11a; there is a particle at the centre of each of the six faces of the cube in addition to the eight at the corners. Copper and aluminium have this structure. The sodium chloride crystal can be regarded as two interpenetrating FCC structures, one of

sodium ions and the other of chloride ions, Fig. 2.11b; each sodium ion is surrounded by six chloride ions and vice versa.

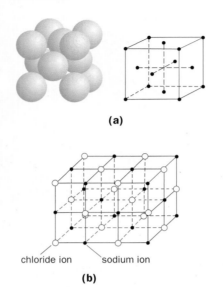

(a)

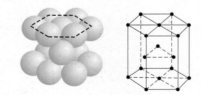

chloride ion   sodium ion

(b)

Fig. 2.11   FCC structure

(b) Hexagonal close-packing (HCP)

This is represented in Fig. 2.12; it is built up from layers of hexagons. Zinc and magnesium form HCP crystals.

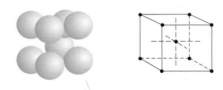

Fig. 2.12   HCP structure

These two structures give the closest possible packing and account for 60% of all metals. They are not very dissimilar if we consider how they can be assembled from successive layers. Figure 2.13 shows a layer A of hexagonal close-packed spheres in which each sphere touches six others — this is the closest packing possible for spheres. A second hexagonal close-packed layer B can be placed on top and the packing of these two layers will be closest when the spheres of B sit in the hollows formed by three neighbouring spheres of A. A third hexagonal close-packed layer can be placed on top of B in two ways. If it rests in the hollows of B so that its spheres are directly above the *spheres* in A, then an HCP crystal results,

Fig. 2.13 (bottom), and the layer stacking is ABAB. However, if the third layer rests in other hollows in B, its spheres can be above *hollows* in A and the structure is FCC with layer stacking ABCABC (top).

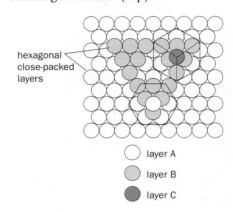

hexagonal close-packed layers

○ layer A

◔ layer B

● layer C

Fig. 2.13   Close-packing

(c) Body-centred cubic (BCC)

This arrangement of packing has a particle at the centre of the cube and one at each corner, Fig. 2.14. Alkali metals have this less closely packed structure.

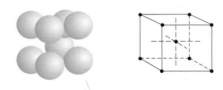

Fig. 2.14   BCC structure

(d) Tetrahedral

Tetrahedral structures have a particle at the centre of a regular tetrahedron and one at each of the four corners, Fig. 2.15a. This more open arrangement is found in carbon (as diamond), silicon and germanium — all substances which form covalent bonds. The hardness of diamond is partly due to the fact that its atoms are not in layers and so cannot slide over each other as they can in graphite, the other crystalline form of carbon. Graphite forms layers of six-membered rings of carbon atoms that are about two-and-a-half times farther apart than are the carbon atoms in the layers, Fig. 2.15b. The forces between the layers are weak, thus explaining why graphite flakes easily and is soft, and suitable for use in pencils and as a lubricant.

Graphite and diamond provide a good example of the importance of structure in determining properties.

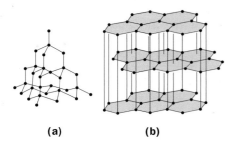

(a)   **(b)**

Fig. 2.15 Carbon structures

Two further points: first, there is in every crystal structure a typical cell, called the **unit cell**, which is repeated over and over again — Figs 2.11*a*, 2.12 and 2.14 are examples of unit cells; second, the structures described are those of perfect crystals. In practice there are imperfections in crystals and these are important in determining the properties of a material, as we shall see later.

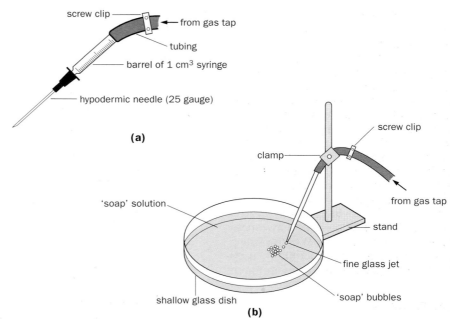

Fig. 2.16 Making a bubble raft

BUBBLE RAFT

Soap bubbles pack together in an orderly manner and provide a good representation, in two dimensions, of how atoms are packed in a crystal.

A bubble raft is made by attaching a glass jet (about 1 mm bore) or a 25 gauge hypodermic needle on a 1 cm³ syringe barrel, to the gas tap, via a length of rubber tubing and a screw clip, Fig. 2.16*a*. The jet is held below the surface of a 'soap' solution (1 Teepol, 8 glycerol and 32 water is satisfactory) in a shallow glass dish, at a constant depth which gives bubbles of about 2 mm diameter, Fig. 2.16*b*. If the dish is placed on an overhead projector a magnified image of the raft can be viewed. What pulls the bubbles together and what keeps them from getting too close?

A perfect, hexagonally close-packed array is shown in Fig. 2.17*a*. Grain boundaries are readily seen in Fig. 2.17*b*, 'vacancies' in Fig. 2.17*c* and a bubble of different size in Fig. 2.17*d*. What might (*c*) and (*d*) represent in a real crystal and what effect do they have on the structure?

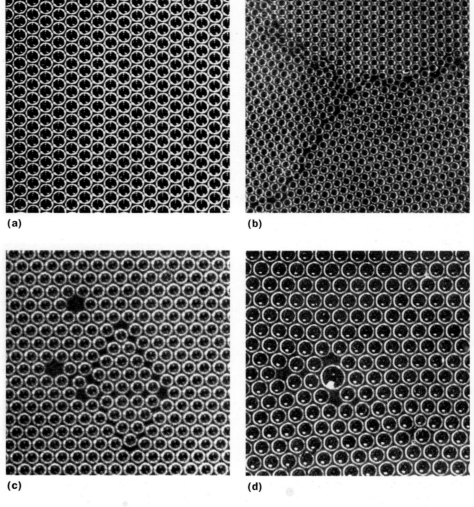

Fig. 2.17 Bubble-raft models of crystal structures

X-RAY CRYSTALLOGRAPHY

Before the discovery that enabled the arrangement of atoms in a crystal to be determined experimentally, crystallographers had simply assumed that the regular external shapes of crystals were due to the atoms being arranged in regular, repeating patterns.

In 1912 three German physicists, Max von Laue, W. Friedrich and P. Knipping, found that a beam of X-rays, on passing through a crystal, formed a pattern of spots on a photographic plate (see Fig. 2.20*b*). Shortly afterwards W. L. (later Sir Lawrence) Bragg and his father Sir William Bragg showed how the pattern could be used to reveal the positions of the atoms in a crystal. Together they proceeded to unravel the atomic structures of many substances and started the science of X-ray crystallography. The structures of many complex organic molecules, including some like DNA (deoxyribonucleic acid) that play a vital part in the life process, have been discovered by this technique.

X-rays, like light, have a wave-like nature and when they fall on a crystal they are scattered by the atoms. In some directions the scattered beams reinforce each other while in others they cancel each other. X-rays are used because their wavelengths are of the same order as the atomic spacings in crystals — about 10 000 times less than those of light.

When X-rays fall on a crystal made up of regularly spaced layers of atoms, each layer produces a weak 'reflected' beam, in the same way that light is reflected by a mirror. In Fig. 2.18*a* reflection by a single layer of atoms is shown: most of the beam passes through. Figure 2.18*b* shows a beam of X-rays falling on a set of parallel layers of atoms. If all the reflected waves combine to produce a strong reflected beam (and give an intense spot on a photograph) then they must all emerge in step. For this to happen the path difference between successive layers must be a whole number of wavelengths of the X-rays — as they are in this case. Otherwise crests of one wave may coincide with troughs from another and the two tend to cancel out.

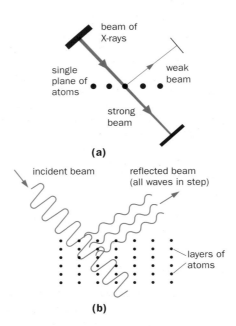

Fig. 2.18

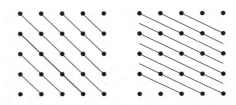

Fig. 2.19

(a) Polycrystalline sample

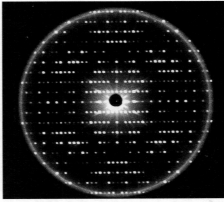

(b) Single crystal

Fig. 2.20   X-ray photographs of crystals

The atoms in a crystal lie in several different sets of parallel planes, from all of which strong reflections may be obtained to give a pattern of spots characteristic of the particular structure. Two other possible sets of planes are shown for the array in Fig. 2.19. In a polycrystalline sample many planes are involved at once and thousands of spots are produced resulting in circles or circular arcs on the photograph. Figure 2.20*a* is due to a polycrystalline sample and 2.20*b* to a single crystal of the same material.

MICROWAVE ANALOGUE

Microwaves are very short radio waves with wavelengths extending from 1 cm to about 1 m and are used for radar and satellite communication. Regularly spaced polystyrene tiles or layers of spheres give strong reflections of microwaves at certain angles depending on the tile or layer separation. This can be explained as being due to interference between waves reflected from successive layers just as with X-rays and layers of atoms.

The apparatus is shown in Fig. 2.21*a*; wax lens A produces a parallel beam from the 3 cm microwave transmitter and lens B focuses it on the detector. First it should be shown, with the transmitter and detector in line, that microwaves can largely penetrate a polystyrene ceiling tile but not a metal sheet.

(a) One tile

A single tile held vertically on the turntable reflects (partially) the beam at *any* angle. Through what angle from the straight-through position must the detector be turned when the **glancing angle** (i.e. the angle between the tile and the incident beam) is θ? (See p. 77.)

(b) Two tiles

When a second tile is brought up behind and parallel to the first (already positioned for the reflected beam to be detected), the intensity of the reflection rises and falls as the extra path to the second tile varies between an even and an odd number of half-wavelengths. Interference is occurring between the waves reflected from each tile when they come together. Will this occur for all glancing angles?

(c) Ten tiles at 3 cm centre-to-centre spacing

If the detector is swung round as the array of tiles revolves on the turntable, a strong reflection is obtained only when the detector makes an angle of 60° with the straight-through position. The tiles then bisect the angle between the detector and the straight-through position. They make an angle of 30° with the straight-through position, and give a glancing angle of 30°, Fig. 2.21b.

(d) Polystyrene ball 'crystal'

This is made from seven hexagonal close-packed layers of 5 cm diameter spheres glued together to give a face-centred cubic (FCC) structure (see Appendix 1). As the crystal rotates on the turntable, *two pairs* of strong reflections are obtained per revolution when the detector is at 44° from the straight-through position ($\theta = 22°$), the time between each pair being greater than that between the two signals in each. At 50° ($\theta = 25°$) there are *two single* strong reflections per revolution and similarly at 74° ($\theta = 37°$).

The first pair of 44° reflections is produced when two easily identifiable sets of vertical 'hexagonal' layers (with the packing of spheres in each layer the closest possible) bisect *in turn* the angle between the straight-through position and the detector; the second pair arises half a revolution later when the 'backs' of the two sets of the same layers are in the bisecting position. At 50°, a set of 'square' layers (with less closely packed spheres — see question 9, p. 24) is responsible for the strong signals when, twice in each revolution, it bisects the angle between the beam and the detector.

POLYMERS

Polymers are materials with giant molecules, each containing anything from 1000 to 100 000 atoms, and are usually carbon (organic) compounds. An example of a natural polymer is cellulose whose long, tough fibres give strength and stiffness to the roots, stems and leaves of plants and trees. Rubber, wool, proteins, resins and silk are others. Synthetic polymers include plastics such as polythene, Perspex and polystyrene, fibres like nylon and polyester, synthetic rubbers and the epoxy resins which are well known for their strong bonding properties and toughness.

The unravelling of the intricacies of nature's polymers required X-ray apparatus, the electron microscope and other instruments. Their molecules were found to consist of a large number of repeating units, called **monomers**, arranged in a long flexible chain. Thus every molecule of cellulose comprises a long chain of from a few hundred to several thousand glucose sugar ($C_{16}H_{12}O_6$) molecules.

Synthetic polymers are made by a chemical reaction known as **polymerization**, in which large numbers of small molecules join together to form a large one. Polythene, or polyethylene to give it its full name, is made by polymerizing ethylene (C_2H_4), a gas obtained when petroleum is 'cracked'. In one process the ethylene molecules, heated to 100–300 °C under a pressure several thousand times greater than atmospheric, link with one another to give the long chain molecules of polythene, Fig. 2.22.

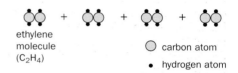

ethylene molecule (C_2H_4)

○ carbon atom

• hydrogen atom

Fig. 2.22 Polymerization of ethylene

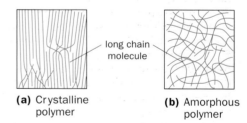

long chain molecule

(a) Crystalline polymer

(b) Amorphous polymer

(c) Polymer with crystalline and amorphous regions

Fig. 2.23

If the chains run parallel to each other, like wires in a cable, the structure shows a certain amount of order and is said to be 'crystalline', Fig. 2.23a. This contrasts with the disorder of tangled chains in an 'amorphous' structure, Fig. 2.23b. Many polymers have both crystalline and amorphous regions, Fig. 2.23c. If crystallinity predominates an X-ray photograph shows sharp spots (but the pattern is never as sharp as for wholly

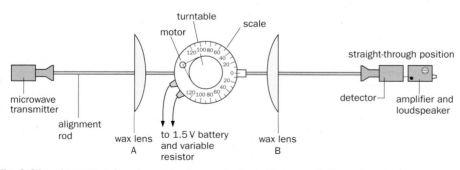

microwave transmitter

alignment rod

wax lens A

motor

turntable

to 1.5 V battery and variable resistor

scale

wax lens B

straight-through position

detector

amplifier and loudspeaker

Fig. 2.21a Apparatus for microwave demonstration of X-ray crystallography

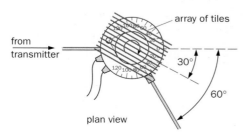

from transmitter

array of tiles

30°

60°

plan view

Fig. 2.21b An array of tiles simulates layers of atoms

crystalline materials) and the polymer is fairly strong and rigid. A polymer with a largely amorphous structure is soft and flexible and gives diffuse rings in an X-ray photograph. The proportion of crystalline to amorphous regions in a polymer depends on its chemical composition, molecular arrangement and how it has been processed. The intermolecular forces between chains are of the weak van der Waals type, but in crystalline structures the chains are close together over comparatively large distances and so the total effect of these forces is to produce a stiff material.

When an amorphous (glassy) polymer is stretched the chains become less coiled and tangled. They line up giving a more ordered ('crystalline') structure, making the material stiffer and able to take strains of 100% or more. Rubber behaves similarly when stretched (p. 36) and the polymer is said to change from the **glassy** to the **rubbery** state if stressed.

Crystallization is one of two principles that have been used to produce strong, stiff polymers (e.g. polythene, nylon); the other is the formation of strong covalent bonds between chains — a process known as cross-linking. In vulcanizing raw rubber, i.e. heating it with a controlled amount of sulphur, a certain number of sulphur atoms form cross-links between adjacent rubber molecules to give a more solid material than raw rubber which is too soft for use, Fig. 2.24. As more cross-links are added to rubber it stiffens and ultimately becomes the hard material called ebonite.

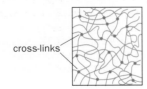

cross-links

Fig. 2.24 Vulcanized rubber

Polymers such as ebonite and Bakelite (the first plastic to be made) with many strong cross-links do not soften with increased temperature but set once and for all after their initial moulding. They are called **thermosetting plastics** or **thermosets** and remain comparatively strong until excessive heating leads to breakdown of the cross-links and chemical decomposition. By contrast, in **ther-** **moplastic** polymers only the weak van der Waals forces hold the chains together and these materials can be softened by heating and if necessary remoulded. On cooling they recover their original properties and retain any new shape. This treatment can be repeated almost indefinitely so long as temperatures are below those causing decomposition, i.e. breakdown of the covalent bonds that hold together the atoms in the long chain.

The possibility of using synthetic polymers in the future as load-bearing structural materials for houses, buildings, cars, boats and aircraft depends largely on how far their strength and stiffness can be increased. As we have seen, two methods are used to do this at present. One is by having 'crystallized' long chains — a physical feature — and the other requires cross-links to be formed between chains — a chemical feature. Current research is directed towards producing molecular chains which are themselves stiff (most existing synthetic polymer chains are inherently flexible) by polymerizing monomers which have a ring-shaped structure. It is now possible to produce polymers with the strength of steel by crystallizing and cross-linking such chains.

QUESTIONS

1. Experiment shows that 3 g of carbon combine with 8 g of oxygen to form 11 g of carbon dioxide. One atom of carbon reacts with two atoms of oxygen to give one molecule of carbon dioxide (i.e. $C + O_2 = CO_2$).
   a) Compare the mass of an oxygen atom with that of a carbon atom. (*Hint:* start by supposing 1 g of C contains x atoms.)
   b) What is the atomic mass of oxygen on the carbon-12 scale?
   c) What mass of oxygen contains the same number of atoms as 12 g of C?

2. **a)** If the atomic mass of nitrogen is 14, what mass of nitrogen contains the same number of atoms as 12 g of carbon?
   b) What mass of chlorine contains the same number of atoms as 32 g of oxygen? (Atomic masses of chlorine and oxygen are 35.5 and 16 respectively.)

3. Taking the value of the Avogadro constant as 6.0×10^{23}, how many atoms are there in

   a) 14 g of iron (at. mass 56),
   b) 81 g of aluminium (at. mass 27),
   c) 6.0 g of carbon (at. mass 12)?

4. What is the mass of
   a) one atom of magnesium (at. mass 24),
   b) three atoms of uranium (at. mass 238),
   c) one molecule of water (mol. mass 18)?
   Take the value of the Avogadro constant as 6.0×10^{23}.

5. **a)** A **mole** is the name given to the *quantity of substance* which contains a certain number of particles. What is this number?
   b) What is the mass of 1 mole of hydrogen molecules?
   c) If the density of hydrogen at s.t.p. is 9.0×10^{-5} g cm^{-3} what volume does 1 mole of hydrogen molecules occupy at s.t.p.?
   d) How many molecules are there in 1 cm^3 of hydrogen at s.t.p.?

6. By counting scintillations it is found that 1.00 mg of polonium in decaying completely emits approximately 2.90×10^{18} alpha particles. If one particle is emitted by each atom and the atomic mass of polonium is 210, what is the Avogadro constant?

7. Estimate
   a) the mass, and
   b) the diameter
   of a water molecule (assumed spherical) if water has molecular mass 18 and the Avogadro constant is 6.0×10^{23} per mole.

8. **a)** Suggest an *approximate* but reasonable value for the heat of sublimation of copper (in J g^{-1}) from the following data.
   Specific latent heat of fusion $= 2.0 \times 10^2$ J g^{-1}
   Specific latent heat of vaporization $= 4.8 \times 10^3$ J g^{-1}
   What additional information would enable a better estimate to be made?
   b) If the Avogadro constant is 6.0×10^{23} per mole and the atomic mass of copper is 64, what is the heat of sublimation of copper in J/atom? Why is a knowledge of this quantity useful?

9. **a)** 'Square' and 'hexagonal' methods of packing spheres are shown in Figs 2.25a and b respectively. How many other spheres are touched by
   i) A,
   ii) B?
   In which arrangement is the packing closest?

b) Figure 2.25*c* is a pyramid of spheres in which the second and successive layers are formed by placing balls in the hollows of the layer below it. How are the balls packed in
   i) the sloping sides of the pyramid,
   ii) the horizontal layers?

(a) Square packing

(b) Hexagonal packing

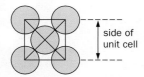

(c) Pyramid

Fig. 2.25

10. One face of the unit cell of an FCC crystal is shown in Fig. 2.26, atoms being represented by spheres. If *r* is the atomic radius in cm, calculate
   a) the length of a side of the unit cell,
   b) the volume of a unit cell,
   c) the number of unit cells in 1 cm³.

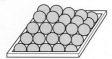

side of unit cell

Fig. 2.26

11. In a crystal built up from a large number of similar unit cells the atoms at the corners and on the faces of individual cells are shared with neighbouring cells. In an FCC unit cell
   a) how many corner atoms are there?
   b) how many neighbouring cells share each corner atom?
   c) what is the effective number of corner atoms per cell?
   d) how many *face* atoms are there?
   e) how many neighbouring cells share each face atom?
   f) what is the effective number of face atoms per cell?
   g) what is the total effective number of atoms per cell?
   h) what is the total effective number of atoms in 1 cm³ of unit cells? (Use your answer to question 10c).)

12. X-ray diffraction shows that copper, atomic mass 64 and density 9.0 g cm⁻³, has an FCC structure. If the Avogadro constant is 6.0×10^{23} per mole, how many atoms are there in 1.0 cm³ of copper?

13. Using your answers to 11h) and 12, calculate the atomic radius of copper.

14. For crystalline sodium chloride, draw the unit cell which is repeated throughout the lattice. Label precisely the two kinds of particle at the lattice sites. What are the forces maintaining them in their relative positions?

Calculate the distance between adjacent particles in crystalline sodium chloride, given that its formula weight is 58.5 and its density is 2.16 g cm⁻³ (2.16×10^3 kg m⁻³). (Avogadro constant $= 6.03 \times 10^{23}$ mole⁻¹)

Discuss the effect of a small stress on a crystalline lattice.

15. In an experiment to estimate the size of an atom, a spherical oil droplet of diameter 0.46 mm is placed on a clean smooth surface of water. The oil spreads over part of the water surface to form a circular patch of uniform thickness and diameter 78 mm. The thickness of the patch is calculated by using the volume of the droplet and the diameter of the circular patch. The thickness of the patch is taken as an estimate of the size of an atom.
   a) i) Using the information above, obtain an estimate of the size of an atom.
      ii) What is the commonly accepted approximate size of atoms?
      iii) State *two* reasons why the estimate obtained in part a) i) differs from the accepted approximate size of atoms.
   b) Name *one* other method of obtaining a value for the size of atoms.
      (*NEAB, AS/A PH05, June 1998*)

16. Parcels are often sealed by using plastic packing tape. Such tape is difficult to tear across its width but is easily torn along its length.
   a) Describe the microstructure of the tape, making reference to the two main types of bonding between molecules.
   b) By referring to your description of the microstructure explain why the tape is easily torn along its length, but difficult to tear across its width.
      (*AQA: NEAB, AS/A PH05, March 1999*)

17. A single crystal of quartz (silica: SiO_2) can be transformed into glass. The process involves crushing the crystal to a powder, melting the powder and then cooling the melt.
   a) Describe the nature of the microstructure at each stage of this process:
   single crystal, powder stage, liquid stage, solid glass stage.
   b) Describe the photographs which would arise from X-ray crystallographic analysis of each of the four structures you have outlined in part a).
      (*NEAB, AS/A PH05, March 1998*)

3

Mechanical properties

- ■ Stress and strain
- ■ The Young modulus
- ■ Stretching experiments
- ■ Deformation and dislocations
- ■ Strengthening metals
- ■ Cracks and fracture
- ■ Fatigue and creep
- ■ Composite materials
- ■ Strain energy
- ■ Rubber
- ■ Elastic moduli

STRESS AND STRAIN

The mechanical properties of a material are concerned with its behaviour under the action of external forces — a matter of importance to engineers when selecting a material for a particular job. Four important mechanical properties are **strength**, **stiffness**, **ductility** and **toughness**.

Strength deals with how great an applied force a material can withstand before breaking. **Stiffness** tells us about the opposition a material sets up to being distorted by having its shape or size, or both, changed. A stiff material is not very flexible. There is no such thing as a perfectly stiff or rigid (unyielding) material; all 'give' in some degree although the deformation may be very small. **Ductility** or workability relates to the ability of the material to be hammered, pressed, bent, rolled, cut or stretched into useful shapes. A **tough** material is one that is not brittle, i.e. it does not crack readily. Steel has all four properties, putty has none of them. Glass is strong and stiff but not tough or ductile. Which properties would you ascribe to rubber, nylon and diamond?

Information about mechanical properties may be obtained by observing the behaviour of a wire or strip of material when it is stretched. The stretching of short rods or 'test-pieces' is done using a machine like that in Fig. 3.1.

Fig. 3.1 Testing a material by stretching a test-piece

The extension produced in a sample of material depends on (*i*) the nature of the material, (*ii*) the stretching force, (*iii*) the cross-section area of the sample and (*iv*) its original length. What effect would you expect (*iii*) and (*iv*) to have? To enable fair comparisons to be made between samples having different sizes the terms 'stress' and 'strain' are used when referring to the deforming force and the deformation it produces.

Stress σ **is the force** (in N) **acting on unit cross-section area** (1 m²). For a force F and area A (Fig. 3.2) we can write

$$\text{stress} = \frac{\text{force}}{\text{area}} \quad \text{or} \quad \sigma = \frac{F}{A}$$

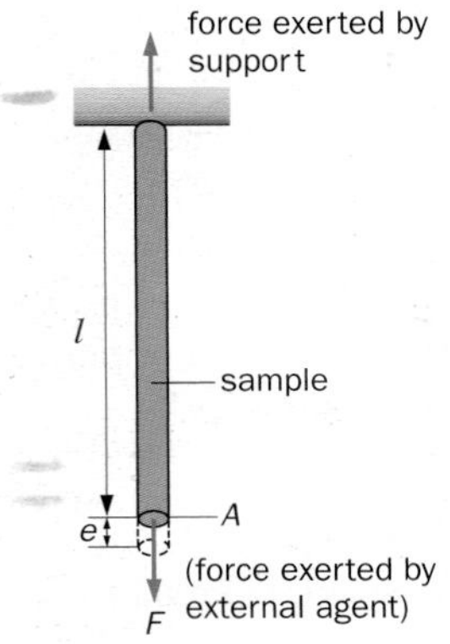

The unit of stress is the **pascal** (Pa) which equals one newton per square metre (N m⁻²).

Strain ϵ **is the extension of unit length** (1 m). If e = extension and l = original length (Fig. 3.2) then

$$\text{strain} = \frac{\text{extension}}{\text{original length}} \quad \text{or} \quad \epsilon = \frac{e}{l}$$

Strain is a ratio and has no unit.

A stress that causes an increase of length puts the sample in tension, and so we talk about a **tensile stress** and a **tensile strain**.

The shape of the stress–strain graph for the stretching of a sample (e.g. a wire) depends not only on the material but also on its previous treatment and method of manufacture. For a ductile material, i.e. a metal, it has the *general* form shown by OEPAD in Fig. 3.3. There are two main parts, as follows.

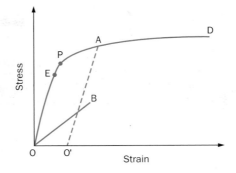

Fig. 3.3

(a) Elastic deformation

The first part of the graph from O to E is a straight line through the origin showing that strain is directly proportional to stress, i.e. doubling the stress doubles the strain. Over this range the material suffers **elastic** deformation, i.e. it returns to its original length when the stress is removed and none of the extension remains.

(b) Plastic deformation

As the stress is increased the graph becomes non-linear but the deformation remains elastic until at a certain stress corresponding to point P, and called the **yield point**, permanent or **plastic** deformation starts. The material then behaves rather like Plasticine and retains some of its extension if the stress is removed. Recovery is no longer complete and on reducing the stress at A, for example, the specimen recovers along AO′ where AO′ is almost parallel to OE. OO′ is the permanent plastic extension produced. If the stress is reapplied, the curve O′AD is followed. At D the specimen develops one or more 'waists' and **ductile fracture** occurs at one of them. The stress at D is the greatest the material can bear and is called the **breaking stress** or **ultimate tensile strength**; it is a useful measure of the strength of a material.

The specimen appears to 'give' at P, and over the plastic region a given stress increase produces a greater increase of strain than previously. None the less it still opposes deformation and any increase of strain requires increased stress. Beyond P the material is said to **work-harden** or **strain-harden**.

Few ductile materials behave elastically for strains as great as ½% (e.g. for an extension of ½ cm in a 100 cm long wire) but they may bear large plastic strains, up to 50%, before fracture. The breaking stress for steels may occur at stresses half as great again as that at the yield point, while for very ductile metals with low yield points it may be several times greater than the stress at the yield point. It is desirable that metals used in engineering structures should carry loads which only deform them elastically. On the other hand, the fabrication of metals into objects of various shapes requires them to withstand considerable plastic deformation before fracture, and so be very ductile. The dominant position of metals in modern technology arises from their strength and ductility.

A brittle material such as a glass may give a curve like OB in Fig. 3.3 and fractures almost immediately after the elastic stage; little or no plastic deformation occurs and the glass is non-ductile.

THE YOUNG MODULUS

The stress–strain curve for the stretching of metals and some other materials (e.g. glasses), over almost all the elastic region, is a straight line through the origin. That is, **tensile strain is directly proportional to tensile stress during elastic deformation**. This statement is known as Hooke's law and in more elementary work it is often stated in the form 'extension varies as the load'. In mathematical terms it can be written

$$\text{tensile strain} \propto \text{tensile stress}$$

or

$$\frac{\text{tensile stress}}{\text{tensile strain}} = \text{a constant}$$

This constant is called **the Young modulus** and is denoted by E. Its value is given by the slope of the straight part of the stress–strain graph and depends on the nature of the material and *not* on the dimensions of the sample. If a material has large E, it resists elastic deformation strongly and a large stress is required to produce a small strain. E is thus a measure of the opposition of a material to change-of-length strains such as occur when a wire or rod is stretched elastically, i.e. it measures **elastic stiffness**.

Fig. 3.2

In the early days of iron railway bridge construction, engineers relied heavily on 'rule of thumb' methods. It required a series of disasters like that of the Tay Bridge in Scotland in 1879 and a reputed collapse rate of 25 bridges per year at about the same time in the USA before it was accepted that reliable strength calculations were necessary for safety and the economical use of materials. The value of E is one of the pieces of information that must be known in order to calculate the deformation (deflections) that will occur in a loaded structure and its parts.

If a stretching force F acting on a wire of cross-section area A and original length l causes an extension e we can write

$$E = \frac{\text{tensile stress}}{\text{tensile strain}} = \frac{F/A}{e/l} = \frac{Fl}{Ae}$$

Like stress, E is expressed in pascals (Pa) since strain is a ratio.

Suppose a load of 1.5 kg attached to the end of a wire 3.0 m long of diameter 0.46 mm stretches it by 2.0 mm then $F = ma = mg = 1.5 \times 9.8$ N ($g = 9.8$ m s^{-2}).

$$E = \frac{Fl}{Ae}$$

$$= \frac{(1.5 \times 9.8 \text{ N})(3.0 \text{ m})}{(\pi \times 0.23^2 \times 10^{-6} \text{ m}^2)(2.0 \times 10^{-3} \text{ m})}$$

$$= \frac{1.5 \times 9.8 \times 3.0}{\pi \times 0.23^2 \times 10^{-6} \times 2.0 \times 10^{-3}} \frac{\text{Nm}}{\text{m}^2\text{m}}$$

$$= 1.3 \times 10^{11} \text{ Pa}$$

Approximate values of E for some common materials are given in Table 3.1.

Table 3.1

Material	The Young modulus $E/10^{10}$ Pa
steel	21
copper	13
glasses	7
polythene	about 0.5
rubber	about 0.005

Glasses are surprisingly stiff (and strong). The high elasticity of rubber (it has the ability to regain its original shape after a very large deformation) is not to be confused with its low elastic modulus. For steel a large stress gives a small strain while the same stress applied to rubber will give a very much larger strain; steel has a greater modulus of elasticity than rubber.

STRETCHING EXPERIMENTS

1. Copper

Using an arrangement like that in Fig. 3.4, the extensions produced in a 2 metre length of copper wire (SWG 32) are found as it is loaded to breaking with 100 gram slotted weights.

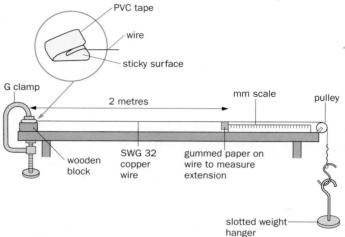

Fig. 3.4 Stretching experiment

A load–extension graph is plotted to see if Hooke's law is obeyed and to find the percentage strain copper can withstand before its elastic limit is exceeded.

The breaking stress of copper can be determined and, if wire of another gauge is used (say SWG 26), whether the breaking stress depends on wire thickness. The percentage plastic strain borne by copper before it breaks may also be found.

2. Steel

If experiment 1 is repeated for 2 metres of steel wire (SWG 44) the same kind of information can be obtained.

3. Rubber

A load–extension graph can be plotted for a strip of rubber (5 cm long and 2 mm wide cut from a rubber band), suspended vertically and loaded with 100 g slotted weights and then unloaded. Information may be obtained about adherence to Hooke's law, and also the number of times its original length that rubber can be extended. The breaking stress of rubber should be found.

4. Polythene

A strip 15 cm long and 1 cm wide, cut *cleanly* from a piece of polythene, can be investigated as in experiment 3.

5. Glass

The breaking stress of glass may be found by hanging weights from a glass thread which has been freshly drawn from a length of 3 mm diameter soda glass rod. If parts of the rod are left at the top and bottom, the thread can be supported by clamping one end and a hook made at the other end for the weights.

6. The Young modulus for a wire

Using the apparatus of Fig. 3.5*a* or *b* the extensions of a wire can be measured with greater accuracy.

In Fig. 3.5*a* the right-hand wire is under test and carries a vernier scale (see p. 551) which, when the right-hand wire is loaded, moves over a millimetre scale attached to the left-hand wire and enables the extension to be measured.

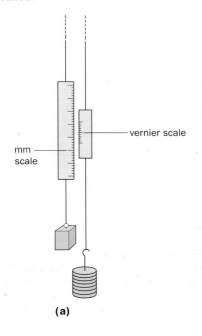

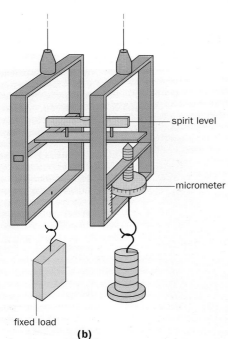

(a)

(b)

Fig. 3.5 Accurate determination of extensions

The alternative and more accurate arrangement in Fig. 3.5*b* is known as Searle's apparatus. In this case the micrometer screw (see p. 551) is adjusted, after the addition of a load to the right-hand wire, so that the bubble of the spirit level is centralized. The extension is then found from the scale readings. By having two wires of the same material suspended from the same support, errors are eliminated if there is a change of temperature or if the support yields, since both wires will be affected equally.

Initially both wires should have loads that keep them taut and free from kinks. Readings are then taken as the load on the right-hand wire is increased by equal steps, without exceeding the elastic limit. The strain should therefore not be more than 0.1%, i.e. the wire should not be stretched much beyond 1/1000th of its original length. The length l of the wire to the top of the vernier or micrometer is measured with a metre rule and the diameter ($2r$) found at various points along its length with a micrometer screw gauge.

From a graph of load against extension, an average value of load/extension in kg mm^{-1} is given by PQ/OQ, Fig. 3.6. The Young modulus can then be calculated from

$$E = \frac{Fl}{Ae}$$

where F/e is expressed in N m^{-1}, i.e. $F/e = \text{PQ} \times g/(\text{OQ} \times 10^{-3})$, A ($= \pi r^2$) in m^2 and l in m. It may be helpful to use a spreadsheet; see p. 4.

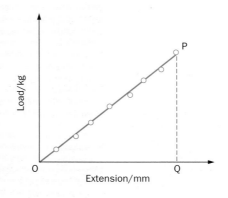

Fig. 3.6

DEFORMATION AND DISLOCATIONS

The deformation behaviour of materials can be explained at the atomic level.

(a) Elastic strain

This is due to the stretching of the interatomic bonds that hold atoms together. The atoms are pulled apart very slightly; each is displaced a tiny distance from its equilibrium position and the material lengthens. Hooke's law is a result of the fact that the 'interatomic force–separation' graph, Fig. 3.7*a*, is a straight line for atomic separations close to the equilibrium separation r_0, Fig. 3.7*b*.

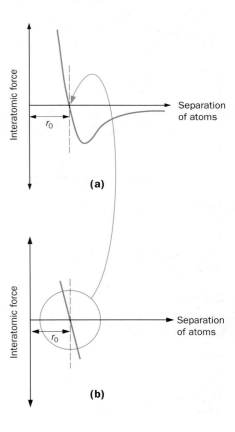

Fig. 3.7

The Young modulus E is high for materials with strong interatomic bonds. Covalent and ionic solids, and to a lesser extent metals, are in this category. Diamond (pure carbon), the hardest known natural substance, has a large number of very strong covalent bonds per unit volume and a very high value of E.

As well as determining the stiffness of a material, the Young modulus also governs its strength since this too depends on the forces between atoms. However, there are other factors which prevent solids from displaying their theoretical strengths (see p. 31).

(b) Plastic strain

The ability to undergo plastic strain (and be ductile) is a property of crystalline materials. The yielding that occurs could therefore be attributed to the slipping of layers of atoms (or ions) over one another. With close-packed layers like those in Fig. 3.8a, the atoms would have to be moved farther apart, Fig. 3.8b. This would be resisted by the inter-atomic bonds, many of which will have to be broken simultaneously.

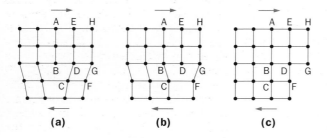

(a)   **(b)**

Fig. 3.8

Calculations based on the known strength of bonds show that the stresses needed to produce slip in this way are many times greater than those that cause plastic strain. The problem of explaining the weakness of metals led to a search for defects in crystal structures and in 1934 G. I. Taylor of Cambridge University proposed the **dislocation** as one such defect.

Occasionally, due perhaps to growth faults during crystallization, there is an incomplete plane of atoms (or ions) in the crystal lattice, for example AB in Fig. 3.9a. The movement of the dislocation produces the same effect as a plane of atoms slipping over other planes, but much more easily.

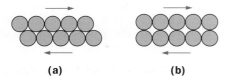

(a)   **(b)**   **(c)**

Fig. 3.9   Movement of a dislocation

If a stress is applied as shown by the arrows, atom B, whose bonds have already been weakened by the distortion of the structure, moves a small distance to the right and forms a bond with atom C, Fig. 3.9b. Plane of atoms DE is now incomplete. Then D flicks over and joins with F leaving GH as the incomplete plane, Fig. 3.9c. The result is just the same as if half-plane AB had slipped over the planes to the right, to the surface of the crystal. This process would have involved breaking a great many bonds at the same time. Instead, the dislocation, by moving a single line at a time, has broken many fewer bonds and

required a much smaller stress to do it. No atom has moved more than a small fraction of the atomic spacing. Plastic deformation by this mechanism is clearly only possible in the well-ordered structure of a crystalline material.

The passage of a dislocation in a crystal is like the movement of a ruck in a carpet. A greater force is needed to drag one carpet over another by pulling one end of it, than to make a ruck in the carpet and kick it along, Fig. 3.10.

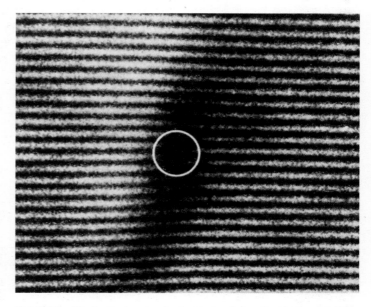

ruck

Fig. 3.10

Calculations confirm that the stresses to make dislocations move in metals are in good agreement with their measured plastic flow stresses.

The first and most direct evidence for the existence of dislocations was obtained in 1956 by J. W. Menter, also of Cambridge University, using an electron microscope. An electron micrograph (an electron microscope photograph) is shown in Fig. 3.11 of an aluminium–copper alloy in which the planes of atoms are spaced about 0.20 nm apart. A dislocation can be seen; the extra plane of atoms (on the left-hand side) ends in the white circle and distorts the arrangement of the surrounding planes.

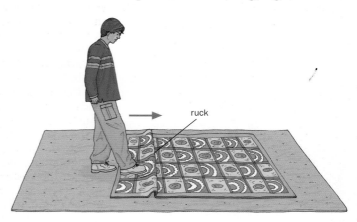

Fig. 3.11   Electron micrograph showing a dislocation in the crystal planes

Dislocations can be obtained in a bubble raft (p. 21) and made to move if the raft is squeezed between two glass slides dipping into the 'soap' solution. One can be seen in Fig. 3.12.

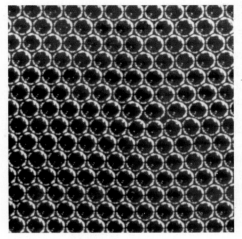

Fig. 3.12 Bubble-raft model of a crystal dislocation

STRENGTHENING METALS

Pure metals produced commercially are generally too weak or too soft to be of much mechanical use — a rod of pure copper the thickness of a pencil is easily bent by hand. Their weakness can be attributed to the fact that they contain a moderate number of dislocations which can move about easily in the orderly crystal structure, thus allowing deformation under relatively small stresses. The traditional methods of making metals stronger and stiffer all involve obstructing dislocation movement by 'barriers', i.e. pockets of disorder in the lattice. Three barriers will be considered.

(a) 'Foreign' atoms

In an alloy such as steel 'foreign' atoms (e.g. carbon) are introduced into the lattice of iron, disturbing its perfection and opposing dislocation motion. This makes for greater strength and stiffness.

(b) Other dislocations

One problem with the dislocation model is that when the dislocations have slipped out of the crystal, as in Fig. 3.9c, the crystal is then perfect and should have its theoretical strength. In general this is not observed and it would seem that further dislocations are generated while slip is occurring. As the metal is submitted to further stress, dislocations are created, move, meet and thereby obstruct each other's progress. A 'traffic jam' of dislocations builds up. Such stress occurs when a metal is hammered, stretched, bent, etc. and work-hardens.

On the other hand, in the process of annealing, in which a metal is heated and cooled slowly so that it is softer and easier to work, some dislocations can disappear. This may be due either to dislocations reaching the surface of the metal and forming a 'step' there, or to different types of dislocation meeting, cancelling each other out to give a perfect lattice.

(c) Grain boundaries

In practice most metal samples are polycrystalline, consisting of many small crystals or grains at different angles to each other (see Fig. 2.8a, p. 19). The boundary between two grains is imperfect and can act as an obstacle to dislocation movement. In general, the smaller the grains the more difficult it is to deform the metal. Why?

An obvious way of strengthening metals would be to eliminate dislocations altogether and produce in effect perfect crystals. So far this has only been possible for tiny, hairlike single-crystal specimens called 'whiskers' that are only a few micrometres thick and are seldom more than a few millimetres long. They can withstand elastic strains of 4 or 5% (compared with $\frac{1}{2}$% or less for most common materials). Unfortunately, perhaps owing to surface oxidation, dislocations soon develop and the 'whisker' weakens.

CRACKS AND FRACTURE

Cracks, both external and internal and however small, play an important part in the fracture of a material and prevent it having maximum strength. Different types of fracture usually occur in brittle and ductile materials.

(a) Brittle fracture

This happens after little or no plastic deformation by the very rapid propagation of a crack. It takes place, for example, when a glass rod is cut by making a small but sharp notch on it with a glass knife or file and then 'bending' it, as in Fig. 3.13 — with the notch on the far side of the rod. Why?

notch in glass opposite thumbs

Fig. 3.13

Around a scratch, notch or crack there is a concentration of stress which in general is greater the smaller the radius of curvature of the tip of the crack. Such stress concentrations may be seen by viewing a lamp through two 'crossed' Polaroid squares (one square is rotated to cut off most of the light coming from the other to give a dark field of view) with a strip of polythene between them, Fig. 3.14. Stretching the strip causes colours to appear and is an indication that the polythene is under stress. The phenomenon is called **photoelasticity** and is used to study stresses in plastic models of structures (see p. 348). When the strip is cut half-way across and again pulled, colours are seen at the tip of the cut, showing the stress is high there. Figure 3.15 shows the stress pattern around two holes in a plastic block under compression. If the stress at the tip of a crack is sufficiently high, interatomic bonds are broken, the crack spreads and breaks a few more bonds at the new tip. Eventually complete fracture occurs.

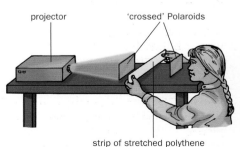

projector   'crossed' Polaroids

strip of stretched polythene

Fig. 3.14

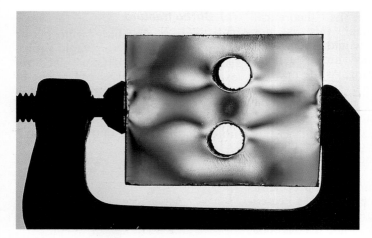

Fig. 3.15 Photoelastic stress pattern

Even tiny surface scratches, which seem to arise inevitably on all materials, can lead to fracture in brittle materials. For example, a freshly drawn ½-metre-long glass fibre can be bent into an arc but it fractures if 'scratched' at A, Fig. 3.16, by gently stroking a few times with another glass fibre. It is less likely to break if scratched at B. Why?

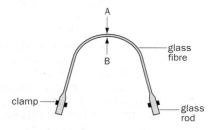

Fig. 3.16

In brittle materials like concrete and glass, cracks spread more readily when the specimen is stretched or bent, i.e. is in tension, Fig. 3.17. Crack propagation is much more difficult if such materials are used in compression (i.e. squeezed) so that any cracks close up. Prestressed concrete contains steel rods that are in tension because they were stretched while the concrete was poured on them and set. As well as providing extra tensile strength these keep the concrete in compression even if the whole prestressed structure is in tension.

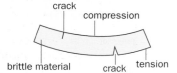

Fig. 3.17

Glass can also be prestressed by its surface being compressed, making it more resistant to crack propagation. In thermal toughening, jets of air are used to cool the hot glass and cause the outside to harden and contract whilst the inside is still soft. Later the inside contracts and pulls on the now reluctant-to-yield, compressed surface. As we saw previously, stresses in transparent materials are revealed by polarized light. The pattern of the air jets used for cooling the prestressed glass of a car windscreen can be seen through polarizing spectacles or sometimes by sunlight that has been partially polarized by reflection from the car bonnet. Prestressing glass in this way can increase its toughness 3000 times.

(b) Ductile fracture

In this case fracture follows appreciable plastic deformation, by *slow* crack propagation. After thinning uniformly along its length during the plastic stage, the specimen develops a 'waist' or 'neck' in which cavities form. These join up into a crack and this travels out to the surface of the specimen, Fig. 3.18a, to give 'cup and cone' shaped, dull fracture surfaces, like those in Fig. 3.18b for a metal rod. It is possible that the internal cavities are formed during the later stages of plastic strain when stress concentrations arise in regions having a large number of interlocking dislocations.

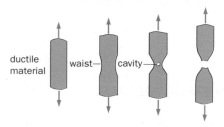

Fig. 3.18a Ductile fracture

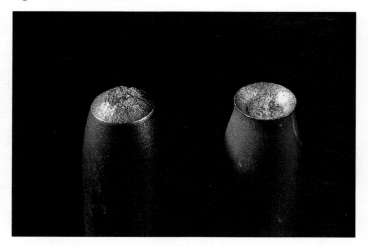

Fig. 3.18b Fractured metal rod

Cracks propagate with much greater difficulty in metals owing to the action of dislocations. These can move to places of high stress such as the tip of a crack, thus reducing the effective stress by causing it to be shared among a greater number of interatomic bonds, Fig. 3.19. The crack tip is thereby deformed plastically, blunted by the dislocation and further cracking possibly stopped. For the same reason surface scratches on metals have practically no effect. In brittle materials dislocation movement is impossible (why?) and high local stresses can build up at cracks under an applied force.

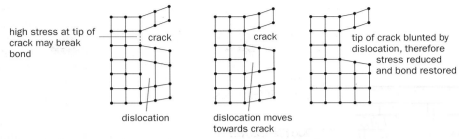

high stress at tip of crack may break bond

crack

crack

tip of crack blunted by dislocation, therefore stress reduced and bond restored

dislocation

dislocation moves towards crack

Fig. 3.19 Dislocations in a metal can prevent cracks propagating

FATIGUE AND CREEP

These are two other important aspects of the mechanical behaviour of metals that are conveniently considered here.

(a) Fatigue

This may cause fracture, often with little or no warning, and happens when a metal is subjected to a large number of cycles of *varying* stress, even if the maximum value of the stress could be applied *steadily* with complete safety. It is estimated that about 90% of all metal failures are due to fatigue; it occurs in aircraft parts, in engine connecting rods, axles, etc.

A typical fatigue fracture in a steel shaft is shown in Fig. 3.20; starting as a fine crack, probably at a point of high stress, it has spread slowly, producing a smooth surface (as on right of photograph), until it was halfway across the shaft which then broke suddenly. Stress concentrations may be due to bad design (e.g. rapid changes of diameter) or bad workmanship (e.g. a tool mark) and are common at holes.

For many ferrous (iron-based) metals there is a safe stress variation below which failure will not occur even for an infinite number of cycles. With other materials 'limited-life' design only is possible. As yet no fully comprehensive model of fatigue has been developed.

(b) Creep

In general this occurs at high temperature and results in the metal continuing to deform as time passes, even under constant stress. The effect is thought to be associated with dislocation motion due to the vibratory motion of atoms, and in the creep-resistant alloys that have been developed this motion is restricted. Such alloys are used, for example, to make the turbine blades of jet engines, where high stress at high temperature has to be withstood without change of dimensions.

Some low-melting-point metals can creep at room temperature, for example unsupported lead pipes gradually sag and the lead sheeting on church roofs has to be replaced periodically.

COMPOSITE MATERIALS

Composites are produced by combining materials so that the combination has the most desirable features of the components. The idea is not new. Wattle and daub (interlaced twigs and mud) have been used to build homes for a long time; straw and clay are the ingredients of bricks; Inuit people freeze moss into ice to give a less brittle material for igloo construction; reinforced concrete contains steel rods or steel mesh. In all cases the composite has better mechanical properties than any of its components.

The production of composites is an attempt to copy nature. Wood is a composite of cellulose fibres cemented together with lignin. Bone is another composite material. Many modern technological applications require materials that are strong and stiff but light and heat-resistant. The development of composites to meet these requirements is at present a major concern of materials scientists throughout the world and offers exciting possibilities to engineers in the future.

The highest **strength-to-weight** and **stiffness-to-weight ratios** are possessed not by metals but by materials such as glass, carbon and boron whose atoms are linked by many strong covalent bonds. (The strength of covalent bonds and their number per unit volume in these covalent solids accounts for the high strength and stiffness; the directional nature of the bond explains the non-close-packed structure and consequently small density, see p. 20). Unfortunately these materials are brittle, partly because their structures make dislocation motion difficult under an applied stress and partly because they usually have small surface scratches that develop into cracks.

In modern composite materials the desirable properties of covalent solids are exploited by incorporating them as **fibres** in a weaker, yielding material called the **matrix**. Freshly drawn fibres are fairly scratch-free and are therefore strong. The matrix has three functions: first, it has to bond with and hold the fibres together so that the applied load is transmitted to them; second, it must protect the surface of the fibres from scratches; third, if cracks do appear it should prevent them from spreading from one fibre to another — it can do this by acting as a barrier to the crack and deflecting it harmlessly along the interface it forms with the fibre, Fig. 3.21. A plastic resin or a ductile metal makes a suitable matrix. Fibre-reinforced composites are strong to stresses applied along the fibres.

Fig. 3.20 Fatigue fracture

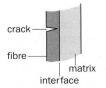

crack

fibre

matrix

interface

crack deflected along interface

Fig. 3.21

Fig. 3.22 HMS *Penzance*, a Sandown-class minehunter built of glass-reinforced plastic

(a) Glass-reinforced plastic (GRP)

'Fibreglass' was the first of the successful modern composites. It consists of high-strength glass fibres in a plastic resin and is used for making boat hulls, storage tanks, pipes, and car components. Figure 3.22 shows a Sandown-class minehunter vessel, one of the largest ships built of GRP afloat. The main reason for using GRP for these vessels is that it is non-magnetic (unlike steel). Its other advantages are that it does not corrode, rot, warp, shrink or split, and is resistant to marine borers. Also, it is fire-resistant and needs three times less maintenance than steel or wooden hulls.

Lightweight lift-jet engines for vertical take-off also use GRP extensively for low-temperature parts; they can produce a thrust sixteen times their own weight. The best all-metal engines in commercial service today produce a thrust less than five times their weight.

(b) Carbon-fibre-reinforced plastic (CFRP)

This is similar to GRP but carbon fibres (about 6×10^5 per cm^2 of cross-section) of greater strength and stiffness replace glass fibres (Fig. 3.23). Carbon fibres are stronger and stiffer than steel and much lighter. CFRPs are particularly attractive to the aircraft industry. They are not subject to

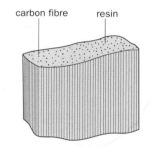

carbon fibre   resin

Fig. 3.23 A section through CFRP

fatigue failure or to high-stress concentrations around holes and cracks as metals are. They do not, however, flow plastically like metals but are elastic until failure, when extensive damage may occur. Their resistance to corrosion is also poor and prohibits certain applications.

Today CFRP-type materials are used extensively to make parts of Formula One racing cars, including the driver's 'safety cell', the nose cone, the front and rear 'wings' (in effect these are aircraft wings turned upside down to force the car on to the track and improve cornering), Fig. 3.24. Many racing drivers owe their survival in recent high-speed crashes to carbon fibres. The lightness and strength of CFRPs makes them ideal for tennis racquet frames, Fig. 3.25*a*. More controversially, CFRP was used for the frame and wheels of the *Lotus Sport* pursuit bicycle, Fig. 3.25*b*, which was of revolutionary design at the 1992 Barcelona Olympic Games and won a gold medal in the 4 km race.

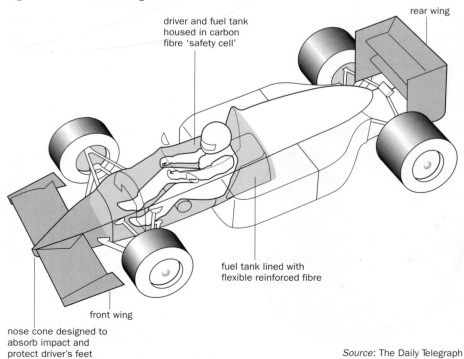

rear wing

driver and fuel tank housed in carbon fibre 'safety cell'

fuel tank lined with flexible reinforced fibre

front wing

nose cone designed to absorb impact and protect driver's feet

Source: The Daily Telegraph

Fig. 3.24 Use of CFRP (blue areas) in a Formula One racing car

(c) Fibre-reinforced metals

Metals mixed with fibres of silicon carbide and boron are being developed. These 'metal matrix composites', as well as being lighter and stronger than steel, can be 'engineered' by using the fibres to strengthen specific points of important components. They are also more environmentally friendly, since the advanced manufacturing techniques for fabricating them use laser and water-jet systems that are less energy-hungry than steel furnaces.

Much research is taking place on composites. The interface between a fibre and its matrix holds the key to the formation of a successful composite; at present a great deal is not understood about the properties of the interface. If the fibres are not covered uniformly with the matrix, small holes form at the interface. In service these cause high-stress concentrations and premature failure. They may also allow liquids and gases to penetrate the composite and attack the fibres. The effect of coating the fibre with some other material to protect its surface from scratches, corrosion, etc., before combining it with the matrix is being studied.

STRAIN ENERGY

Energy has to be supplied to stretch a wire. If the stretching force is provided by hanging weights, there is a loss of potential energy, some of which is stored in the stretched wire as **strain energy**. Provided the elastic limit is not exceeded this energy can usually be recovered completely (rubber is a notable exception; see the next section). If it is exceeded, the part of the energy used to cause crystal slip (plastic strain) is retained by the wire.

To determine the strain energy stored in a wire stretched by a known amount, consider a material with a force–extension graph like that in Fig. 3.26. (This is of the same form as its stress–strain graph.)

Suppose the wire is already extended by e_1 and then suffers a further extension δe_1 which is so small that the shaded area is near enough a rectangle. If F_1 is the average but *nearly* constant value of the stretching force during the extension δe_1, then the

Fig. 3.25a Martina Hingis using a CFRP racquet

Fig. 3.25b The Lotus Sport – a recent design of racing bicycle which used CFRP for the frame and wheels

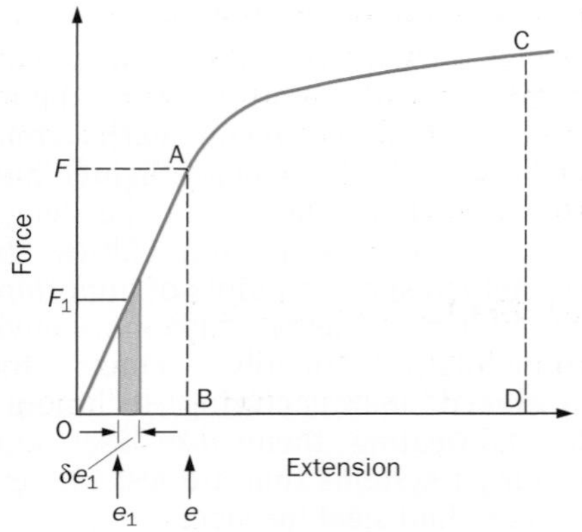

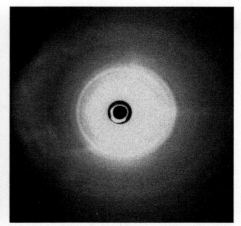

Fig. 3.26

energy transferred to the wire, i.e. the work done δW_1, is given by

$$\delta W_1 = \text{force} \times \text{distance}$$
$$= F_1 \times \delta e_1$$
$$= \text{area of shaded strip}$$

∴   total work done during
   whole extension e = area OAB

i.e. strain energy stored in wire
   = area OAB

If Hooke's law is obeyed up to A, OA is a straight line (as shown) and OAB is a triangle.

Therefore the strain energy in the wire for the whole extension e and final stretching force F, is given by

$$\text{strain energy} = \text{area } \triangle\text{OAB}$$
$$= \tfrac{1}{2}\text{AB} \times \text{OB}$$
$$= \tfrac{1}{2}Fe$$

The expression $\tfrac{1}{2}Fe$ gives the strain energy in joules if F is in newtons and e in metres.

If l is the original length of the wire and A its cross-section area, then volume of wire = Al.

∴   strain energy per unit volume

$$= \tfrac{1}{2}Fe/(Al)$$

$$= \tfrac{1}{2}\left(\frac{F}{A} \times \frac{e}{l}\right)$$

But F/A = stress and e/l = strain, so

strain energy per unit volume	$= \tfrac{1}{2}(\text{stress} \times \text{strain})$

This is the area under the stress–strain graph.

If the wire suffers plastic deformation in an extension OD, the total strain energy stored is again the area under the force–extension graph, i.e. area OCD, and can be calculated by counting the squares on the graph paper, knowing the 'energy value' of one square.

The **specific energy** is the energy stored in the wire per unit mass.

RUBBER

The two most striking mechanical properties of rubber are (a) its range of elasticity is great — some rubbers can be stretched to more than ten times their original length (i.e. 1000% strain) before the elastic limit is reached, and (b) its value of the Young modulus is about 10^4 times smaller than most solids and *increases* as the temperature rises, an effect not shown by any other material.

Rubber is a polymer consisting of up to 10^4 isoprene molecules (C_5H_8) joined end-to-end into a long chain of carbon atoms, Fig. 3.27. The enormous extensibility and low value of E cannot be due to the stretching of the strong covalent bonds between atoms in the carbon chain.

If a sample of stretched rubber is 'photographed' by a beam of high-energy electrons, sharp spots are obtained, Fig. 3.28a, similar to those produced by X-rays and a crystal. This suggests there is some order among the molecules in such a sample. Figure 3.28b is a similar photograph of unstretched rubber. A plausible explanation of the behaviour of rubber might be that its long-chain molecules are intertwined and jumbled up like cooked spaghetti. Figure 3.29 is a model of one rubber molecule. A stretching force would tend to make the chains uncoil and straighten out into more or less orderly lines alongside each other. When the force is removed they coil up again. There is also some cross-bonding between chains, achieved during manufacture by vulcanizing raw rubber (see p. 24); as well as causing stiffening this cross-linking also greatly increases the reversible strain possible by anchoring together the long molecules. Fully extended rubber is strong because the bonds are then stretched directly.

The rise in value of E with temperature can be attributed to the greater disorder among the chains when the material is heated; their resistance to alignment by a stretching force therefore increases.

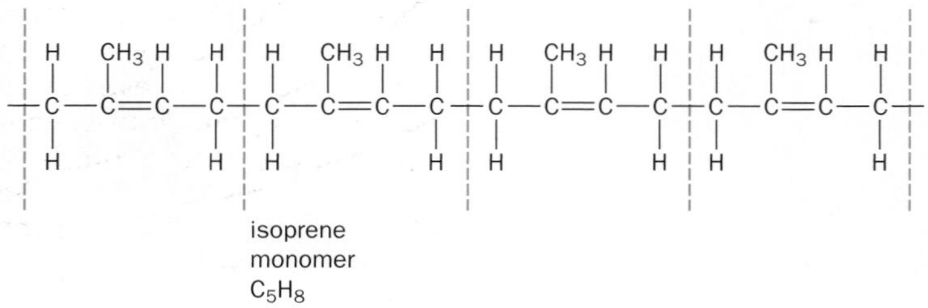

isoprene
monomer
C_5H_8

Fig. 3.27 The chemical structure of rubber

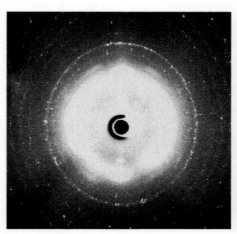

(a) Stretched rubber

(b) Unstretched rubber

Fig. 3.28 Electron micrographs of rubber

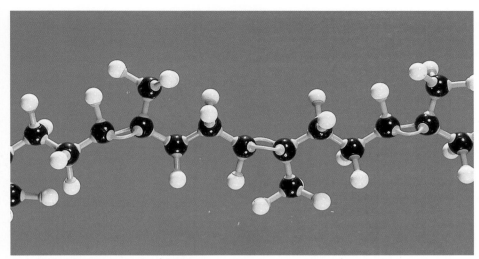

Fig. 3.29   Model of a rubber molecule

If a stress–strain curve is plotted for the loading and unloading of a piece of rubber the two parts do not coincide, Fig. 3.30. OABC is for stretching and CDEO for contracting. The strain for a given stress is greater when unloading than when loading. The unloading strain can be considered to 'lag behind' the loading strain; the effect is called **elastic hysteresis**. It occurs with other substances, noticeably with polythene and glasses and to a small extent with metals.

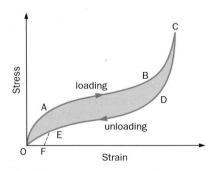

Fig. 3.30   Hysteresis loop

Note that Hooke's law is not obeyed. This is typical of non-crystalline polymers and contrasts with the usually linear behaviour of crystalline materials, e.g. metals which obey the law during elastic strain. It is also evident that rubber stretches easily at first but is stiffer at large extensions. It has been shown (p. 36) that the area enclosed by OABC and the strain axis represents the energy supplied to cause stretching; similarly the area under CDEO represents the energy given up by the rubber during contraction. The shaded area is called

a **hysteresis loop**; it is a measure of the energy 'lost' as heat during one expansion–contraction cycle. The changes in temperature can be felt by placing a wide rubber band on the lips; if it is stretched quickly the temperature rises because of the transfer of mechanical energy to the rubber. When released under control the temperature falls. (The temperature change on free contraction is small.)

Rubber with a hysteresis loop of small area is said to have **resilience**. This is an important property where the rubber undergoes continual compression and relaxation, as each part of a car tyre does when it touches the road and rotates on. If the rubber used in tyres does not have high resilience there is appreciable loss of energy resulting in increased petrol consumption or lower maximum speed. Should the heat build-up be large the tyre may disintegrate.

When rubber is stretched and released there may be a small permanent 'set' as shown by the dotted line EF in Fig. 3.30.

ELASTIC MODULI

All deformations of a body, whether stretches, compressions, bends or twists, can be regarded as consisting of one or more of three basic types of strain. For many materials experiment shows that *provided the elastic limit is not exceeded*

$$\frac{\text{stress}}{\text{strain}} = \text{a constant}$$

This is a more general statement of Hooke's law. The constant is called an **elastic modulus** of the material for the type of strain under consideration. There are three moduli, one for each kind of strain.

(a) The Young modulus E

This has already been considered (p. 27) and is concerned with change-of-length strains. It is defined by

$$E = \frac{\text{tensile stress}}{\text{tensile strain}}$$

where stress is force per unit area (F/A) and strain is change of length per unit length (e/l).

(b) Shear modulus G

In this case the strain involves a change of shape without change of volume. Thus if a tangential force F is applied along the top surface of area A of a rectangular block of material fixed to the bench, the block suffers a change of shape and is deformed so that the front and rear faces become parallelograms, Fig. 3.31.

The shear stress is F/A and angle α is taken as a measure of the strain produced. (The force F on the bottom surface of the block is exerted by the bench.) The shear modulus is defined as

$$G = \frac{\text{shear stress}}{\text{shear strain}} = \frac{F}{A\alpha}$$

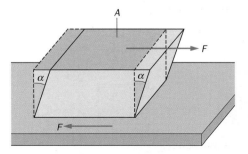

Fig. 3.31

When a wire is twisted, a small square on the surface becomes a rhombus, Fig. 3.32, and is an example of a shear strain. G can be found from experiments on the twisting of wires. If a spiral spring is stretched, the wire itself is not extended but is twisted, i.e. sheared. The extension thus depends on the shear modulus of the material as well as on the dimensions of the spring.

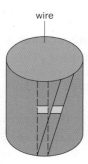

Fig. 3.32

(c) Bulk modulus K

If a body of volume V is subjected to a small *increase* of external pressure δp which changes its volume by the small amount δV, Fig. 3.33, the deformation is a change of volume without a change of shape. The bulk stress is δp, i.e. increase in force per unit area, and the bulk strain is $\delta V/V$, i.e. change of volume/original volume; the bulk modulus K is defined by

$$K = \frac{\text{bulk stress}}{\text{bulk strain}} = \frac{-\delta p}{\delta V/V}$$

$$= -V\frac{\delta p}{\delta V}$$

The negative sign is introduced to make K positive since δV, being a decrease, is negative.

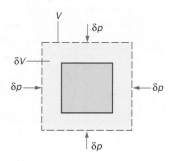

Fig. 3.33

Solids have all three moduli, liquids and gases only K. All moduli have the same units — pascals (Pa).

QUESTIONS

1. a) Why are stresses and strains rather than forces and extensions generally considered when describing the deformation behaviour of solids?
b) A length of copper of square cross-section measuring 1.0 mm by 1.0 mm is stretched by a tension of 40 N. What is the tensile stress in Pa?
c) If the breaking stress of steel is 1.0×10^9 Pa will a wire of this material of cross-section area 4.0×10^{-4} cm² break when a 10 kg mass is hung from it? (Take $g = 10$ m s⁻².)
d) A strip of rubber 6 cm long is stretched until it is 9 cm long. What is the tensile strain in the rubber as
    i) a ratio,
    ii) a percentage?
e) A wire originally 2 m long suffers a 0.1% strain. What is its stretched length?

2. a) The Young modulus for steel is greater than that for brass. Which would stretch more easily? Which is stiffer?
b) How does a deformed body behave when the deforming force is removed if the strain is
    i) elastic,
    ii) plastic?
c) A brass wire 2.5 m long of cross-section area 1.0×10^{-3} cm² is stretched 1.0 mm by a load of 0.40 kg. Calculate the Young modulus for brass. (Take $g = 10$ m s⁻².)
What percentage strain does the wire suffer? Use the value of E to calculate the force required to produce a 4.0% strain in the same wire. Is your answer for the force reliable? If it isn't, would it be greater or smaller than your answer? Explain.

3. a) A 0.50 kg mass is hung from the end of a wire 1.5 m long of diameter 0.30 mm. If the Young modulus for the material of the wire is 1.0×10^{11} Pa, calculate the extension produced. (Take $g = 10$ m s⁻².)
b) Two wires, one of steel and one of phosphor bronze, each 1.5 metres long and of diameter 0.20 cm, are joined end to end to form a composite wire of length 3.0 metres. What tension in this wire will produce a total extension of 0.064 cm? (The Young modulus for steel $= 2.0 \times 10^{11}$ Pa and for phosphor bronze $= 1.2 \times 10^{11}$ Pa.)

4. Figure 3.34 shows tensile stress–strain curves for three different materials X, Y and Z.

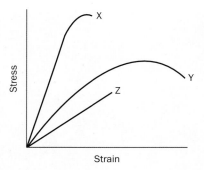

Fig. 3.34

a) For each material named below, state which curve is typical of the material, giving the reasoning behind your choice:
    i) copper,
    ii) glass,
    iii) hard steel.
b) Choose one of the materials named in part a).
    i) Name the material.
    ii) Describe, in terms of microstructure, the stages leading to the fracture of a cylindrical specimen of the material.
    iii) What is the name given to the type of fracture you have described in part ii)?
    (*AQA: NEAB, AS/A PH05, March 1999*)

5. Define **stress**, **strain**, **the Young modulus**.
Describe in detail how the Young modulus for a steel wire may be determined by experiment.
A vertical steel wire 350 cm long, diameter 0.100 cm, has a load of 8.50 kg applied at its lower end. Find
a) the extension,
b) the energy stored in the wire. (Take the Young modulus for steel as 2.00×10^{11} Pa and $g = 9.81$ m s⁻².)

6. In an experiment to determine the Young modulus of the material of a wire the following data was obtained.

Applied load/kg	Extension/mm
0.0	0.0
1.0	2.8
2.0	6.2
3.0	8.7
4.0	12.1
5.0	15.0

The length of the wire was measured as 2.0 m and the diameter as 0.2 mm. By using a spreadsheet and graphing routine (see p. 4), or otherwise,
a) plot a graph of load against extension,
b) determine the slope of the graph,

c) calculate the Young modulus of the wire.

7. a) Classify each of the following polymers as amorphous or as semicrystalline: Nylon, Perspex, Polythene.

The graph in Fig. 3.35 shows the variation of the Young modulus E with temperature T for an amorphous polymer.

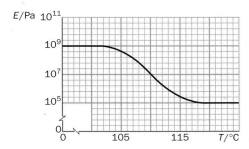

Fig. 3.35

Estimate the glass transition temperature of this polymer.

Explain in microscopic terms why the Young modulus of an amorphous polymer is large at temperatures below 105 °C but is much smaller at temperatures above 115 °C.

In what way would the Young modulus–temperature graph for a semicrystalline polymer differ from that for an amorphous polymer?

b) Figure 3.36 shows the arrangement of the atoms in a crystal in the region of an edge dislocation.

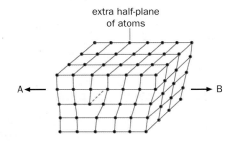

Fig. 3.36

Using labelled diagrams, show how the dislocation moves when stress is applied in the directions A and B.

Explain why the stress required to deform a perfect crystal is greater than that required to deform one containing dislocations.

In some circumstances the presence of dislocations can result in an *increase* in the strength of a metal. Explain how this can occur.

(L, A PH4 part qn, June 1998)

8. A wire is stretched by a tensile force. Figure 3.37 shows the force–extension graph for the wire in two idealised straight-line sections (PQ and QR).

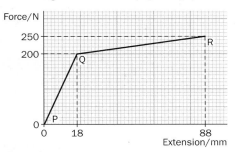

Fig. 3.37

The unextended length of the wire is 2.00 m.

a) Calculate the strain in the wire
  i) at Q,
  ii) at R.

b) Calculate the strain energy in the wire at Q.

c) Calculate the work done in stretching the wire from P to R.

(OCR, Basic 1, June 1999)

9. The graph in Fig. 3.38 shows the behaviour of a material when subjected to stress. Give an example of a material which behaves in this way when put under stress.

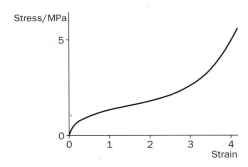

Fig. 3.38

Copy Fig. 3.38 and add a second line which shows the behaviour of a brittle material with a Young modulus of approximately 10 MPa.

Describe with the aid of diagrams the difference in structure between a crystalline solid and a polymeric one.

(L, AS/A PH2, June 1999)

10. a) A materials scientist needs to be able to determine several different properties of a new material. Some of these properties are
  i) density,
  ii) the Young modulus,
  iii) the ultimate tensile stress.
Explain the meaning of each of these terms and outline how the Young modulus and the ultimate tensile stress may be determined.

b) Figure 3.39 shows the force–extension graphs for identically-shaped pieces of steel and a new material.

Compare the behaviour of the two specimens.

Suggest, with two reasons, which of the two materials would be the more suitable for use in a car bumper.

(OCR, Basic 1, March 1999)

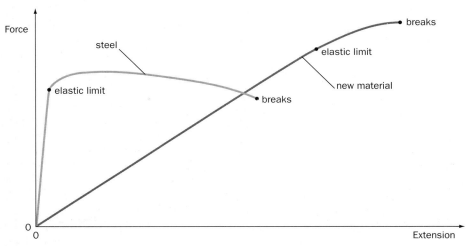

Fig. 3.39

4

Electrical properties

- Conduction in solids
- Current and charge
- Drift velocity of electrons
- Potential difference
- Resistance
- Current–p.d. relationships
- Types of resistor
- Resistor networks
- Potential divider

- Using ammeters and voltmeters
- Shunts, multipliers and multimeters
- Resistivity
- Electrical strain gauge
- Effects of temperature on resistance
- Electromotive force
- Internal resistance

- Kirchhoff's laws
- Power and heating effect
- Wheatstone bridge
- Potentiometer
- Electrolysis
- Electric cells
- Thermoelectric effect
- Datalogging

CONDUCTION IN SOLIDS

Materials exhibit a very wide range of electrical conductivities. The best conductors (silver and copper) are over 10^{23} times better than the worst conductors, i.e. the best insulators (e.g. polythene). Between these extreme cases is the important group of semiconductors (e.g. germanium and silicon).

The first requirement for conduction is a supply of charge carriers that can wander freely through the material. In most solid conductors, notably metals, we believe that the carriers are loosely held outer electrons. With copper, for example, every atom has one 'free' electron (the one that helps to form the metallic bond, p. 17) which is not attached to any particular atom and so can participate in conduction. On the other hand, if all electrons are required to form the bonds (covalent or ionic) that bind the atoms of the material together then the material will be an insulator. In semiconductors only a small proportion of the charge carriers are free to move.

The free electrons in a solid conductor are in a state of rapid motion, moving within the crystal lattice at speeds calculated to be about 1/1000 of the speed of light. This motion is normally completely haphazard (like that of gas molecules) and as many

electrons with a given speed move in one direction as in the opposite direction with the same speed. There is, therefore, no net flow of charge and so no current.

If a battery is connected across the ends of the conductor, an electric field is created in the conductor which causes the electrons to accelerate and gain kinetic energy. Collisions occur between the accelerating electrons and atoms (really positive ions) of the conductor that are vibrating about their mean position in the crystal lattice but are not free to undergo translational motion. As a result the electrons lose kinetic energy and slow down whilst the ions gain vibrational energy. The net effect is to transfer chemical energy from the battery, via the electrons, to internal energy (see p. 65) of the ions. This shows itself on the macroscopic scale as a temperature rise in the conductor and subsequently energy may pass from the conductor to the surroundings as heat.

The electrons are again accelerated and the process is repeated. The *overall* acceleration of the electrons, however, is zero on account of their collisions. They acquire a constant **average drift velocity** in the direction from negative to positive of the battery and it is this resultant drift of charge that constitutes an electric current. An analogous situation may arise when a

ball rolls down a long flight of steps. The acceleration caused by the earth's gravitational field when the ball drops can be cancelled by the force it experiences on 'colliding' with the steps. The ball may roll down the stairs with zero average acceleration, i.e. at constant average speed.

The 'free electron theory' is able to account in a general way for many of the facts of conduction and, although in more advanced work it has been extended by the 'band theory', it will be adequate for our present purposes.

CURRENT AND CHARGE

In metals, current is the movement of negative charge, i.e. electrons; in gases and electrolytes both positive and negative charges may be involved. Under the action of a battery, charges of opposite sign move in opposite directions and so a convention for current direction has to be chosen. As far as most external effects are concerned, positive charge moving in one direction is the same as negative charge moving in the opposite direction. By agreement all current is assumed to be due to the motion of positive charges and so when current arrows are marked on circuits they are directed from the positive to the negative of the supply. If the charge carriers are negative they move in the opposite direction to that of the arrow.

The basic electrical unit is the unit of current — the **ampere** (abbreviated to A); it is defined in terms of the magnetic effect of a current. The unit of electric charge, the **coulomb** (C), is defined in terms of the ampere.

One coulomb is the quantity of electric charge carried past a given point in a circuit when a steady current of 1 ampere flows for 1 second.

If 2 amperes flow for 1 second, 2×1 coulombs (ampere-seconds) pass; if 2 amperes flow for 3 seconds then 2×3 coulombs pass. In general if a steady current I (in amperes) flows for time t (in seconds) then the quantity Q (in coulombs) of charge that passes is given by

$$Q = It$$

The flow of charge in a conductor is often compared with the flow of water in a pipe. The flow of water in litres per second, say, corresponds to the flow of charge in coulombs per second, i.e. amperes.

The charge on an electron, i.e. the electronic charge, is 1.60×10^{-19} C and is much too small as a practical unit. In 1 C there are therefore $1/(1.60 \times 10^{-19})$, i.e. 6.24×10^{18}, electronic charges. A current of 1 A is thus equivalent to a drift of 6.24×10^{18} electrons past each point in a conductor every second.

Smaller units of current are the **milliampere** (10^{-3} A), abbreviated to mA, and the **microampere** (10^{-6} A), abbreviated to µA.

DRIFT VELOCITY OF ELECTRONS

On the basis of the free electron theory an expression can be derived for the drift velocity of electrons in a current and an estimate made of its value. The results are surprising.

Consider a conductor of length l and cross-section area A having n 'free' electrons per unit volume each carrying a charge e, Fig. 4.1.

Volume of conductor	$= Al$
Number of 'free' electrons	$= nAl$
Total charge Q of 'free' electrons	$= nAle$

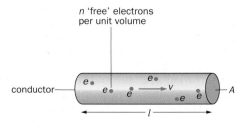

n 'free' electrons per unit volume

conductor

Fig. 4.1

Suppose that a battery across the ends of the conductor causes the charge Q to pass through length l in time t with average drift velocity v. The resulting steady current I is given by

$$I = \frac{Q}{t}$$
$$= \frac{nAle}{t}$$

But $v = l/t$, therefore $t = l/v$.

$$\therefore \quad I = \frac{nAle}{l/v} = nAev \quad \text{and so}$$

$$v = \frac{I}{nAe}$$

To obtain a value for v consider a current of 1.0 ampere in SWG 28 copper wire of cross-section area 1.1×10^{-7} square metre. If we assume that each copper atom contributes one 'free' electron, it can be shown (see question 2a, p. 59) that $n \approx 10^{29}$ electrons per cubic metre. Then, since $e = 1.6 \times 10^{-19}$ coulomb (the charge on an electron),

$$v = \frac{I}{nAe}$$
$$= \frac{1.0 \text{ A}}{(10^{29} \text{ m}^{-3})(1.1 \times 10^{-7} \text{ m}^2)(1.6 \times 10^{-19} \text{ C})}$$
$$= \frac{1.0}{1.1 \times 1.6 \times 10^3} \cdot \frac{\text{A}}{\text{m}^{-1} \text{ C}}$$
$$\approx 6 \times 10^{-4} \frac{\text{C s}^{-1}}{\text{m}^{-1} \text{ C}} \quad (1 \text{ A} = 1 \text{ C s}^{-1})$$
$$\approx 6 \times 10^{-4} \text{ m s}^{-1}$$
$$\approx 0.6 \text{ mm s}^{-1}$$

This is a remarkably small velocity and means that it takes electrons about half an hour to drift 1 m when a current of 1 A flows in this wire. The tiny drift velocity of electrons contrasts with their random speeds due to their thermal motion (about 1/1000 of the speed of light) and is not to be confused with the speed at which the electric field causing their drift motion travels along a conductor. This is close to the speed of light, i.e. 3×10^8 m s^{-1} (see Appendix 2). Current therefore starts to flow almost simultaneously at all points in a circuit.

The same expression for drift velocity holds for charge carriers other than electrons (in fact, it holds for the transport of other things as well as electric charge). In an electrolyte, conduction is due to ions and using the arrangement of Fig. 4.2 information can be obtained about their motion. On applying an electric field (from a 250 volt d.c. supply), the purple stains from the permanganate crystals travel very slowly towards the positive of the supply, and if we make the not unreasonable assumption that the stain travels with the charge carriers then ions too would appear to have tiny drift velocities, of a similar value to those calculated for electrons.

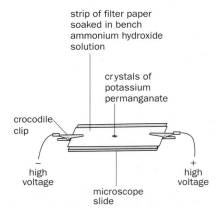

strip of filter paper soaked in bench ammonium hydroxide solution

crystals of potassium permanganate

crocodile clip

− high voltage

microscope slide

+ high voltage

Fig. 4.2

Despite the slow movement of the carriers in conductors and their very small charge, large currents are possible. Why?

The term **current density** J is sometimes used and is defined as the current per unit cross-section area at right angles to the direction of flow. It is given by

$$J = \frac{I}{A} = nev$$

POTENTIAL DIFFERENCE

In an electric circuit electrical energy is transformed into other forms of energy. A lamp transforms electrical energy into heat and light, and an electric motor transforms electrical energy into mechanical energy. Such energy transfers, produced by suitable devices, are a useful feature of electric circuits and form the basis of the definition of the term **potential difference (p.d.)** — an idea that helps us to make sense of circuits.

The potential difference V between two points in a circuit is the amount of electrical energy transformed into other forms of energy when unit charge passes from one point to the other.

The unit of potential difference is the **volt** (V) and equals the p.d. between two points in a circuit in which 1 joule of electrical energy is transformed when 1 coulomb passes from one point to the other. If 2 joules are transformed per coulomb then the p.d. is 2 volts. If the passage of 3 coulombs is accompanied by the transfer of 9 joules of energy, the p.d. is 9/3 joules per coulomb, i.e. 3 volts.

It therefore follows that if the p.d. between two points A and B (we more commonly talk about the p.d. *across* AB) is 5 volts then when 4 coulombs pass from A to B, the energy transferred will be 5 joules per coulomb, i.e. 5×4 joules. In general, if a charge of Q (in coulombs) flows in a part of a circuit across which there is a p.d. of V (in volts) then the energy change W (in joules) is given by

$$W = QV$$

If Q is in the form of a steady current I (in amperes) flowing for time t (in seconds) then $Q = It$ and

$$W = ItV$$

Although it is always the p.d. between two points that is important in electric circuits, there are some occasions when it is helpful to consider what is called the **potential at a point**. This involves selecting a convenient point in the circuit and saying it has zero potential. The potentials of all other points are then stated with reference to it, i.e. the potential at any point is then the p.d. between the point and the point of zero potential. In practice one part of a piece of electrical equipment (e.g. a power supply) is often connected to earth; the earth and all points in the circuit joined to it are then taken as having zero potential.

If positive charge moves (i.e. conventional current flows) from a point A to a point B then A is regarded as being at a higher potential than B. Negative charge flow is therefore from a lower to a higher potential, i.e. from B to A. We can look upon p.d. as a kind of electrical 'pressure' that drives conventional current from a point at a higher potential to one at a lower potential.

RESISTANCE

When the same p.d. is applied across different conductors different currents flow. Some conductors offer more opposition or **resistance** to the passage of current than others.

The resistance R of a conductor is defined as the ratio of the potential difference V across it to the current I flowing through it.

$$R = \frac{V}{I}$$

The unit of resistance is the **ohm** (symbol Ω, the Greek letter omega) and is the resistance of a conductor in which the current is 1 ampere when a p.d. of 1 volt is applied across it. Larger units are the **kilohm** (10^3 ohm), symbol kΩ, and the **megohm** (10^6 ohm), symbol MΩ.

The smaller I is for a given V, the greater the resistance R of the conductor.

The resistance of a metal can be regarded as arising from the interaction which occurs between the crystal lattice of the metal and the 'free' electrons as they drift through it under an applied p.d. This interaction is due mainly to collisions between electrons and the vibrating ions of the metal, but collisions between defects in the crystal lattice (e.g. impurity atoms and dislocations) also play a part, especially at very low temperatures.

The **conductance** (G) of a specimen is the reciprocal of its resistance and is measured in **siemens** (S = Ω^{-1}).

$$G = \frac{1}{R}$$

CURRENT–P.D. RELATIONSHIPS

Using one of the ammeter–voltmeter circuits described later (Fig. 4.16), the p.d. V across a component can be varied and the corresponding current I measured. A graph of I against V shows the relationship between these two quantities and is called the **characteristic** of the component. It summarizes pictorially how the component behaves.

(a) Metals and alloys

These give *I–V* graphs which are straight lines through the origin, Fig. 4.3*a*, and are called **linear** or **ohmic** conductors. For them $I \propto V$ and it follows that V/I = a constant (equal to the reciprocal of the slope of the graph, i.e. by OB/AB). They obey **Ohm's law**, which states that **the resistance of a metallic conductor does not change with p.d., provided the temperature is constant**.

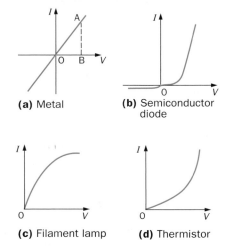

Fig. 4.3 Characteristics of components

(b) Semiconductor diodes

The typical *I–V* graph in Fig. 4.3*b* shows that current passes when the p.d. is applied in one direction but is almost zero when it acts in the opposite direction. A diode therefore has a small resistance if the p.d. is applied one way round but a very large resistance when the p.d. is reversed. It conducts in one direction only and is a **non-ohmic** or **non-linear** conductor. This one-way property makes it useful as a **rectifier** for changing alternating current (a.c.) to direct current (d.c.).

(c) Filament lamps

The *I–V* graph of, for example, a torch bulb, Fig. 4.3*c*, bends over as *V* and *I* increase, indicating that a given change of *V* causes a smaller change in *I* at larger values of *V*. That is, the resistance (*V/I*) of the tungsten wire filament increases as the current raises its temperature and makes it white-hot. In general, the resistance of metals and alloys *increases* with temperature rise.

(d) Thermistors

These are made of semiconductor materials. The *I–V* graph of the commonest type bends upwards, Fig. 4.3*d*, i.e. its resistance *decreases* sharply as its temperature rises.

TYPES OF RESISTOR

Conductors specially constructed to have resistance are called **resistors**, denoted by ⎓⎓⎓ ; they are required for many purposes in electric circuits. Several types exist.

(a) Carbon composition resistors

These are made from mixtures of carbon black (a conductor), clay and resin binder (non-conductors) which are pressed and moulded into rods by heating. The resistivity of the mixture depends on the proportion of carbon. The stability of such resistors is poor and their values are usually only accurate to within ±10%. They have largely been replaced by other types for general use.

Values are sometimes shown by colour markings as in Fig. 4.4. The tolerance colours are gold ±5%, silver ±10%, no colour ±20%.

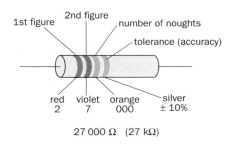

Fig. 4.4 Colour markings of resistors

The alternative printed code of letters and numbers is simpler and is marked on the resistor and used on diagrams.

Value	0.27 Ω	1 Ω	3.3 Ω	10 Ω	220 Ω
Mark	R27	1R0	3R3	10R	K22

Value	1 kΩ	68 kΩ	100 kΩ	1 MΩ	6.8 MΩ
Mark	1K0	68K	M10	1M0	6M8

Tolerances are indicated by adding a letter: F = ±1%, G = ±2%, J = ±5%, K = ±10%, M = ±20%. Thus 5K6K = 5.6 kΩ ± 10%.

(b) Carbon film resistors

Ceramic rods are heated to about 1000 °C in methane vapour which decomposes and deposits a uniform film of carbon on the rod. The resistance of the film depends on its thickness and can be increased by cutting a spiral groove in it, Fig. 4.5. The film is protected by an epoxy resin coating. The stability and hence accuracy of this type of resistor is commonly ±2% and the power rating $\frac{1}{8}$ to $\frac{1}{2}$ watt.

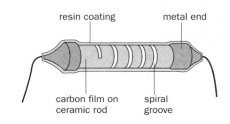

Fig. 4.5 Carbon film resistor

(c) Wire-wound resistors

High-accuracy, high-stability resistors are always wire-wound, as are those required to have a large power rating (i.e. over 2 watts). They use the fact that the resistance of a wire increases with its length. Manganin (a copper, manganese and nickel alloy) wire is used for high-precision standard resistors because of its low temperature coefficient of resistance (p. 49); constantan or Eureka (copper–nickel) wire is used for several purposes and Nichrome (nickel–chromium) wire for commercial resistors.

Adjustable known resistances, called resistance boxes, are used for electrical measurements in the laboratory.

They consist of a number of constantan coils which can be connected in series by plugs or switches to give the required value, Figs 4.6 and 4.7. It is especially important with resistance boxes not to exceed the maximum safe currents, since overheating may change the resistance value or even burn out the coils. The power limit is about 1 watt per coil and so a 1 ohm coil should not carry more than 1 ampere and a 100 ohm coil not more than 0.1 ampere (from power = I^2R, see p. 53).

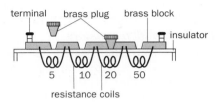

reading=65 Ω

Fig. 4.6 Plug-type resistance box

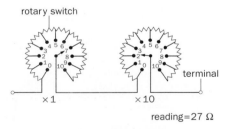

reading=27 Ω

Fig. 4.7 Switch-type resistance box

(d) Variable resistors

Those used in electronic circuits, often as sound volume or other controls and sometimes called potentiometers, consist of an incomplete circular track of carbon composition or wire-wound card, with fixed connections to each end and a rotating arm contact which can slide over the track. Figures 4.8*a* and *b* show the outside and inside respectively of a wire-wound variable resistor. If the track is 'linear', the resistance tapped off is proportional to the distance moved by the sliding contact; if it is 'logarithmic' it is proportional to the log of the distance, and at the end of the track a small movement of the sliding contact causes a larger increase of resistance than at the start.

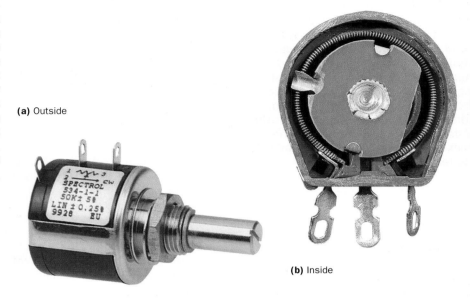

(a) Outside

(b) Inside

Fig. 4.8 Wire-wound variable resistor

Larger current versions, as used in many electrical experiments, consist of constantan wire wound on a straight ceramic tube with the sliding contact carried on a metal bar above the tube.

There are two ways of using a variable resistor. It may be used as a **rheostat** for controlling the current in a circuit, when only one end connection and the sliding contact are required, Fig. 4.9. It can also act as **potential divider** (see p. 45) for controlling the p.d. applied to a device, all three connections then being used. In Fig. 4.10 any fraction of the total p.d. from the battery can be tapped off by varying the sliding contact.

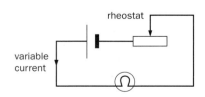

Fig. 4.9

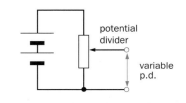

Fig. 4.10

RESISTOR NETWORKS

A network of resistors like that in Fig. 4.11 has a combined or equivalent resistance which can be found experimentally from the ratio of the voltmeter reading to the ammeter reading. Its value may also be calculated.

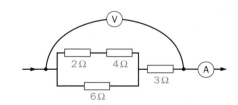

Fig. 4.11

(a) Resistors in series

Resistors are in series if the same current passes through each in turn. In Fig. 4.12 if the total p.d. across all three resistors is V and the current is I, the combined resistance R is given by

$$R = \frac{V}{I}$$

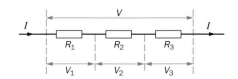

Fig. 4.12 Resistors in series

The electrical energy transformed per coulomb in passing through all the resistors equals the sum of that transformed in each resistor. Therefore if V_1, V_2 and V_3 are the p.ds across R_1, R_2 and R_3 respectively then

$$V = V_1 + V_2 + V_3$$

By the definition of resistance, $R_1 = V_1/I$, $R_2 = V_2/I$ and $R_3 = V_3/I$, therefore $V_1 = IR_1$, $V_2 = IR_2$, $V_3 = IR_3$ and since $V = IR$ we have

$$IR = IR_1 + IR_2 + IR_3$$

$$R = R_1 + R_2 + R_3$$

(b) Resistors in parallel

Here alternative routes are provided to the current, which splits, and we would expect the combined resistance to be less than the smallest individual resistance. In Fig. 4.13 if I is the total current through the network and I_1, I_2 and I_3 are the currents in the separate branches then since current is not used up in a circuit

$$I = I_1 + I_2 + I_3$$

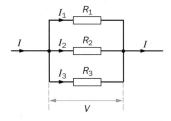

Fig. 4.13 Resistors in parallel

In a parallel circuit the p.d. across each parallel branch is the same. If this is V then by the definition of resistance $R_1 = V/I_1$, $R_2 = V/I_2$, $R_3 = V/I_3$ and if R is the combined resistance then $R = V/I$. Therefore $I = V/R$, $I_1 = V/R_1$, $I_2 = V/R_2$ and $I_3 = V/R_3$. Hence

$$\frac{V}{R} = \frac{V}{R_1} + \frac{V}{R_2} + \frac{V}{R_3}$$

$$\frac{1}{R} = \frac{1}{R_1} + \frac{1}{R_2} + \frac{1}{R_3}$$

The single resistance R which would have the same resistance as the whole network can be calculated.

For the special case of two equal resistors in parallel we have $R_1 = R_2$.

Then

$$\frac{1}{R} = \frac{1}{R_1} + \frac{1}{R_1} = \frac{2}{R_1}$$

$$\therefore \quad R = \frac{R_1}{2}$$

In general for n equal resistances R_1 in parallel, the combined resistance is R_1/n.

The combined resistance of the network in Fig. 4.11 is 6 Ω; do you agree?

POTENTIAL DIVIDER

A potential divider provides a convenient way of getting a variable p.d. from a fixed p.d. and in its simplest form consists of two resistors in series.

(a) Using two fixed resistors

In Fig. 4.14a the two fixed resistors $R_1 = 10$ Ω and $R_2 = 20$ Ω are connected in series across a p.d. $V = 6.0$ V. The same current I passes through each resistor, therefore

$$I = V/(R_1 + R_2) = 6.0 \text{ V}/30 \text{ Ω}$$
$$= \tfrac{1}{5} \text{A} = 0.20 \text{ A}$$

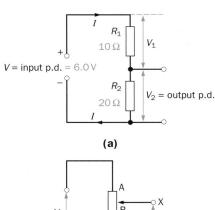

(a)

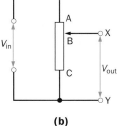

(b)

Fig. 4.14 Potential divider circuits

Hence

$$V_1 = IR_1 = 0.20 \text{ A} \times 10 \text{ Ω} = 2.0 \text{ V}$$

and

$$V_2 = IR_2 = 0.20 \text{ A} \times 20 \text{ Ω} = 4.0 \text{ V}$$

The p.ds appearing across the resistors are therefore in the ratio of their resistances, i.e. $V_1/V_2 = R_1/R_2 = \tfrac{1}{2}$. The p.d. is divided by the resistors into *three* equal parts, each of 2.0 V; one part, i.e. 2.0 V, appears across R_1 and two parts, i.e. 4.0 V, appear across R_2. If V is regarded as the **input p.d.** to the potential divider and V_2 as the **output p.d.**, then V_2 can be tapped off and applied to another circuit.

If the resistors are changed so that $R_1 = 5.0$ Ω and $R_2 = 25$ Ω, then $R_1/R_2 = 5.0/25 = \tfrac{1}{5}$ and V_1/V_2 would also equal $\tfrac{1}{5}$. The input p.d. of 6.0 V would be divided in effect into *six* equal parts with

$$V_1 = \tfrac{1}{6} \times 6.0 \text{ V} = 1.0 \text{ V}$$

and $$V_2 = \tfrac{5}{6} \times 6.0 \text{ V} = 5.0 \text{ V}$$

Different voltages are therefore obtained across R_1 and R_2 by changing their ratio.

In general, for an input p.d. V we can write, since $I = V/(R_1 + R_2)$,

$$V_1 = IR_1 = \left(\frac{R_1}{R_1 + R_2}\right)V$$

and $$V_2 = IR_2 = \left(\frac{R_2}{R_1 + R_2}\right)V$$

Therefore if $V = 5.0$ V, $R_1 = 30$ kΩ and $R_2 = 20$ kΩ, then

$$V_1 = \frac{30 \text{ kΩ}}{50 \text{ kΩ}} \times 5.0 \text{ V} = 3.0 \text{ V}$$

and $$V_2 = \frac{20 \text{ kΩ}}{50 \text{ kΩ}} \times 5.0 \text{ V} = 2.0 \text{ V}$$

Note that $V_1 + V_2$ must equal V.

(b) Using a variable resistor

A variable resistor connected as in Fig. 4.14b gives an easier way of changing the ratio R_1/R_2 and getting an output p.d. that varies more 'smoothly'. B is the position of the sliding contact on resistor AC. The resistance between A and B represents R_1 and that between B and C represents R_2. A continuously variable output p.d. V_{out}, from 0 to V_{in}, is available between X and Y, depending on the position of the sliding contact B. $V_{out} = \tfrac{1}{2}V_{in}$ when $R_1 = R_2$, which, in a linear variable resistor (p. 44), occurs when AB = BC.

(c) Effect of load on output p.d.

When the output p.d. from a potential divider drives current through another circuit, this current 'loads' the potential divider and the output p.d. is less than the calculated value because it is now developed across the smaller resistance of R_2 in *parallel* with the load resistance of the external circuit.

It can be shown that the output p.d. does not change much so long as the resistance of the load is always at least *ten* times greater than R_2. Putting it another way, the current drawn by the load should not exceed $\frac{1}{10}$ of the current through the potential divider.

Other devices apart from resistors can be used in potential dividers (see p. 426).

USING AMMETERS AND VOLTMETERS

Most ammeters and voltmeters are basically galvanometers (i.e. current detectors capable of measuring currents of the order of milliamperes or microamperes) of the moving-coil type which have been modified by connecting suitable resistors in parallel or in series with them as described in the next sections. Moving-coil instruments are accurate, sensitive and reasonably cheap and robust.

Connecting an ammeter or voltmeter should cause the minimum disturbance to the current or p.d. it has to measure. The current in a circuit is measured by breaking the circuit and inserting an ammeter **in series** so that the current passes through the meter. The resistance of an ammeter must therefore be *small* compared with the resistance of the rest of the circuit. Otherwise, inserting the ammeter changes the current to be measured. The perfect ammeter would have zero resistance, the p.d. across it would be zero and no energy would be absorbed by it.

The p.d. between two points A and B in a circuit is most readily found by connecting a voltmeter across the points, i.e. **in parallel** with AB. The resistance of the voltmeter must be *large* compared with the resistance of AB, otherwise the current drawn from the main circuit by the voltmeter (which is required to make it operate)

becomes an appreciable fraction of the main current and the p.d. across AB changes. A voltmeter can be treated as a resistor which automatically records the p.d. between its terminals. The perfect voltmeter would have infinite resistance, take no current and absorb no energy.

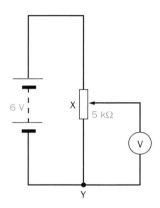

Fig. 4.15

It is instructive to set up the circuit in Fig. 4.15 for measuring the p.d. tapped off by the potential divider between X and Y, using first a high-resistance voltmeter (e.g. one with a 100 μA movement) for Ⓥ and then a low-resistance voltmeter (e.g. one with a 10 mA movement). The reading in the second case is much lower. What will it be if both voltmeters are connected across XY?

The most straightforward method of measuring resistance uses an ammeter and a voltmeter as in Fig. 4.16*a*. The voltmeter records the p.d. across R but the ammeter gives the sum of the currents in R and in the voltmeter. If the voltmeter has a much higher resistance than R, the current through it will be small by comparison and the error in calculating R can be neglected. However, if the resistance of the voltmeter is not sufficiently high, perhaps because R is very high, the voltmeter should be connected across both R and the ammeter as in Fig. 4.16*b*. The ammeter now gives the true current in R. The voltmeter indicates the p.d. across R and the ammeter together, but the resistance of the latter is usually negligible compared with that of R and so the p.d. across it will be so small as to make the error in calculating R negligible.

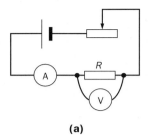

(a)

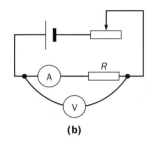

(b)

Fig. 4.16

SHUNTS, MULTIPLIERS AND MULTIMETERS

(a) Conversion of a microammeter into an ammeter

Consider a moving-coil meter which has a resistance (due largely to the coil) of 1000 Ω and which gives a full-scale deflection (f.s.d.) when 100 μA (0.0001 A) passes through it. If we wish to convert it to an ammeter reading 0–1 A this can be done by connecting a resistor (perhaps a misnomer) of very low value in parallel with it. Such a resistor is called a **shunt** and it must be chosen so that only 0.0001 A passes through the meter and the rest of the 1 A, namely 0.9999 A, passes through the shunt, Fig. 4.17. A full-scale deflection of the meter will then indicate a current of 1 A.

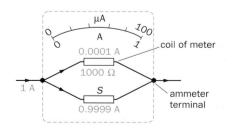

Fig. 4.17   Connection of a shunt

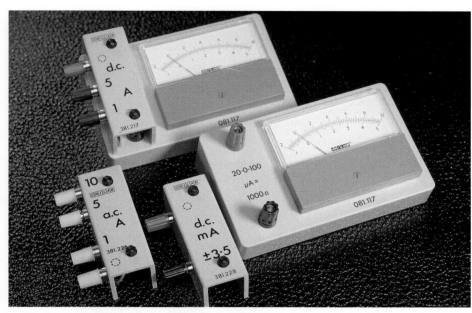

Fig. 4.18 A microammeter together with shunts for conversion to an ammeter

To obtain the value S of the shunt, we use the fact that the meter and the shunt are in parallel. Therefore

p.d. across meter = p.d. across shunt

Applying Ohm's law to both meter and shunt,

$$0.0001\,A \times 1000\,\Omega = 0.9999\,A \times S$$
$$\text{(from } V = IR\text{)}$$
$$\therefore \quad S = \left(\frac{0.0001 \times 1000}{0.9999}\right)\frac{A\,\Omega}{A}$$
$$= 0.1\,\Omega$$

The combined resistance of the meter and the shunt in parallel will now be very small (less than $0.1\,\Omega$) and the current in a circuit will be virtually undisturbed when the ammeter is inserted.

In Fig. 4.18 a microammeter (20–0–100 μA) with its matching shunts is shown.

(b) Conversion of a microammeter into a voltmeter

To convert the same moving-coil meter of resistance $1000\,\Omega$ and f.s.d. 100 μA to a voltmeter reading 0–1 V, a resistor of high value must be connected in series with the meter. The resistor is called a **multiplier** and it must be chosen so that when a p.d. of 1 V is applied across the meter and resistor in series, only 0.0001 A flows through the meter and a full-scale deflection results, Fig. 4.19.

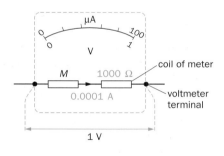

Fig. 4.19 Connection of a multiplier

To obtain the value M of the multiplier, we apply Ohm's law when there is an f.s.d. of 0.0001 A. Hence

p.d. across multiplier and meter in series = 0.0001 A $(M + 1000\,\Omega)$

But the meter is to give an f.s.d. when the p.d. across it and the multiplier in series is 1 V. Therefore

$$0.0001\,A(M + 1000\,\Omega) = 1\,V$$
$$\therefore \quad M + 1000\,\Omega = \frac{1\,V}{0.0001\,A}$$
$$= 10\,000\,V\,A^{-1}$$
$$\therefore \quad M = 9000\,\Omega$$

Voltmeters are often graded according to their 'resistance per volt' at f.s.d. For the above voltmeter, 1 V applied across its terminals produces a full-scale deflection, i.e. a current of 100 μA, and so the resistance of the meter (coil + multiplier) must be $10\,000\,\Omega$ (since $R = V/I = 1/0.0001\,\Omega = 10\,000\,\Omega$). The 'resistance per volt' of the meter is thus $10\,000\,\Omega/V$. To be used as a voltmeter with an f.s.d. of 10 V it would need to have a total resistance of $100\,000\,\Omega$, i.e. a multiplier of $99\,000\,\Omega$, to limit the full-scale current to 100 μA — but its resistance for every volt of deflection is still $10\,000\,\Omega$. A $100\,\Omega/V$ voltmeter has a resistance of $100\,\Omega$ for an f.s.d. of 1 V and draws a full-scale current of 10 mA ($I = V/R = \frac{1}{100}\,A = 0.01\,A = 10\,mA$). Hence a voltmeter with a higher 'resistance per volt' will draw a smaller current and the circuit to which it is connected will be less disturbed. A good voltmeter should have a resistance of at least $1000\,\Omega/V$.

In electronic circuits resistances of $1\,M\Omega$ or higher are encountered and electronic voltmeters which have very high resistances have to be used.

(c) Multimeters

A multi-range instrument or multimeter is a moving-coil galvanometer adapted to measure current, p.d. and resistance. There is a tapped shunt S across the meter and a tapped multiplier M in series with it, Fig. 4.20. A rotary switch allows the various ranges to be chosen.

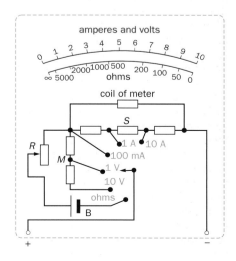

Fig 4.20 Connections in a multimeter

One other position of the switch is marked 'ohms' and puts a dry cell B (usually 1.5 volts) and a rheostat R in series with the meter. To measure resistance the terminals are short-circuited and R adjusted until the pointer gives a full-scale deflection, i.e. is on the zero of the ohms scale. The unknown resistance then replaces the short circuit across the terminals. The current falls and the pointer indicates the value of ohms. Figure 4.21 shows a multimeter.

Digital electronic meters are now in use.

(d) Electronic meters

Electronic meters constructed from transistors or integrated circuits have very high input resistance (e.g. 10 MΩ). They affect most circuits very little and give highly accurate readings. They often have a digital readout and are available as voltmeters (DVM) (see p. 445) and multimeters (DMM).

Fig. 4.21   An analogue multimeter

RESISTIVITY

The resistance of a conductor depends on its size as well as on the material of which it is made. To make fair comparisons of the abilities of different materials to conduct, the resistance of specimens of the same size must be considered.

Experiment shows that the resistance R of a uniform conductor of a given material is directly proportional to its length l and inversely proportional to its cross-section area A. So

$$R \propto \frac{l}{A}$$

or

$$R = \frac{\rho l}{A}$$

where ρ is a constant (for fixed temperature and other physical conditions), called the **resistivity** of the material of the conductor.

Since $\rho = AR/l$ we can say that **the resistivity of a material is numerically the resistance of a sample of unit length and unit cross-section area, at a certain temperature**. The unit of ρ is ohm metre (Ω m) since those of AR/l are metre^2 × ohm/metre, i.e. ohm metre.

Knowing the resistivity of a material the resistance of *any* specimen of that material may be calculated.

For example, if the cross-section area of the live rail of an electric railway is 50 cm^2 and the resistivity of steel is 1.0×10^{-7} Ω m then, neglecting the effect of joints, the resistance per kilometre of rail R follows —

$$R = \frac{\rho l}{A}$$

$$= \frac{(1.0 \times 10^{-7}\ \Omega\ \text{m}) \times (10^3\ \text{m})}{(50 \times 10^{-4}\ \text{m}^2)}$$

$$= \left(\frac{1.0 \times 10^{-7} \times 10^3}{50 \times 10^{-4}}\right) \frac{\Omega\ \text{m m}}{\text{m}^2}$$

$$= 2.0 \times 10^{-2}\ \Omega$$

Table 4.1   Resistivities at 20 °C

Material		Resistivity /Ω m	Use
CONDUCTORS			
Metal	Silver	1.6×10^{-8}	Contacts on small switches
	Copper	1.7×10^{-8}	Connecting wires
	Aluminium	2.7×10^{-8}	Power cables
	Tungsten	5.5×10^{-8}	Lamp filaments
Alloys	Manganin (Cu–Mn–Ni)	44×10^{-8}	High-precision standard resistors
	Constantan/Eureka (Cu–Ni)	49×10^{-8}	Resistance boxes, variable resistors
	Nichrome (Ni–Cr)	110×10^{-8}	Heating elements
	Carbon	3000×10^{-8}	Radio resistors
SEMICONDUCTORS			
	Germanium	0.6	Transistors
	Silicon	2300	Transistors
INSULATORS			
	Glass	10^{10}–10^{14}	
	Polystyrene	10^{15}	

The resistivities at 20 °C of various materials are given in Table 4.1; their experimental determination is briefly described on p. 54.

The **conductivity** (σ) of a material is the reciprocal of its resistivity (ρ), i.e. $\sigma = 1/\rho$, and has unit ohm^{-1} metre^{-1} ($\Omega^{-1}\,\text{m}^{-1}$).

Silver is the best conductor, i.e. has the lowest resistivity, and is followed closely by copper which, being much less expensive, is used for electrical connecting wire. Although the resistivity of aluminium is nearly twice that of copper, its density is only about one-third of copper's. The ratio of current-carrying-capacity to weight of aluminium is therefore greater than that of copper. This accounts for its use in the overhead power cables of the Grid, where aluminium strands are wrapped round a core of steel wires (Fig. 4.22). The cable then has the strength it requires for suspension in long spans between pylons.

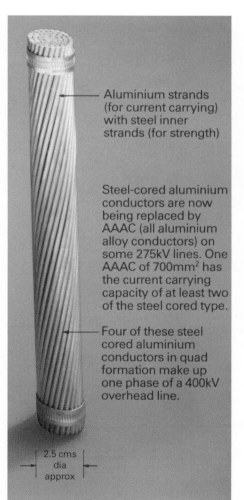

Aluminium strands (for current carrying) with steel inner strands (for strength)

Steel-cored aluminium conductors are now being replaced by AAAC (all aluminium alloy conductors) on some 275kV lines. One AAAC of 700mm² has the current carrying capacity of at least two of the steel cored type.

Four of these steel cored aluminium conductors in quad formation make up one phase of a 400kV overhead line.

2.5 cms dia approx

Fig. 4.22 An overhead power cable

The resistivity of a pure metal is increased by small amounts of 'impurity' and alloys have resistivities appreciably greater than those of any of their constituents. On the other hand, the addition of tiny traces of impurities to pure semiconductors, a process known as 'doping' the semiconductor (see p. 416), reduces their resistivity. Impurity atoms in a crystal lattice act as 'defects' and restrict the movement of charge carriers. When a semiconductor is doped this restriction of movement is more than offset by the production of extra 'free' charges.

ELECTRICAL STRAIN GAUGE

One device that engineers employ to obtain information about the size and distribution of strains in structures such as buildings, bridges and aircraft is the electrical strain gauge. It converts mechanical strain into a resistance change in itself by using the fact that the resistance of a wire depends on its length and cross-section area.

One type of gauge consists of a very fine wire (of an alloy containing mostly nickel, iron and chromium) cemented to a piece of thin paper as in Fig. 4.23. In use it is securely attached with a very strong adhesive to the component under test so that it experiences the same strain as the component when loaded. If, for example, an increase of length strain occurs, the gauge wire gets longer and thinner and on both counts its resistance increases. Thick leads connect the gauge to a resistance-measuring circuit (e.g. a Wheatstone bridge, p. 54) and previous calibration of the gauge enables the strain to be measured directly. What is the advantage of using a parallel-wire arrangement for the strain gauge?

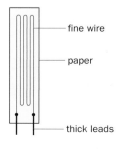

fine wire

paper

thick leads

Fig. 4.23 Electrical strain gauge

EFFECTS OF TEMPERATURE ON RESISTANCE

(a) Temperature coefficient

The resistance of a material varies with temperature and the variation can be expressed by its **temperature coefficient of resistance** α. If a material has resistance R_0 at 0 °C and its resistance increases by δR due to a temperature rise $\delta\theta$ then α for the material is defined by the equation

$$\alpha = \frac{\delta R}{R_0} \cdot \frac{1}{\delta\theta}$$

α is **the fractional increase in the resistance at 0 °C per unit rise in temperature**. The unit of α is °C^{-1}, since δR and R_0 have the same units (ohms) and $\delta R/R_0$ is a ratio. For copper $\alpha \approx 4 \times 10^{-3}$ °C^{-1}, which means that a copper wire having a resistance of 1 ohm at 0 °C increases in resistance by 4×10^{-3} ohm for every °C temperature rise.

Experiment shows that the value of α varies with the temperature at which $\delta\theta$ occurs but, to a good approximation, for metals and alloys we can generally assume it is constant in the range 0 to 100 °C. So if a specimen has resistances R_θ and R_0 at temperatures θ and 0 °C respectively then, replacing δR by $R_\theta - R_0$ and $\delta\theta$ by θ in the expression for α, we obtain

$$\alpha = \frac{R_\theta - R_0}{R_0\theta}$$

Rearranging gives $R_\theta - R_0 = R_0\alpha\theta$ and

$$R_\theta = R_0(1 + \alpha\theta)$$

When using the equation where accuracy is important, R_0 should be the resistance at 0 °C. A calculation shows the procedure when R_0 is not known.

Suppose a copper coil has a resistance of 30 Ω at 20 °C and its resistance at 60 °C is required. Taking α for copper as 4.0×10^{-3} °C^{-1} we have

$$R_{20} = R_0(1 + 20\alpha)$$

and

$$R_{60} = R_0(1 + 60\alpha)$$

Dividing,

$$\frac{R_{60}}{R_{20}} = \frac{1 + 60\alpha}{1 + 20\alpha}$$

$$\therefore \quad R_{60} = \frac{30 \ \Omega(1 + 60 \ °C \times 4.0 \times 10^{-3} \ °C^{-1})}{(1 + 20 \ °C \times 4.0 \times 10^{-3} \ °C^{-1})}$$

$$= 34.5 \ \Omega$$

If the calculation had *not* been based on the resistance at 0 °C and we had taken the original resistance (i.e. R_{20}) as R_0 then using $R_{60} = R_0(1 + \alpha\theta)$ where $\theta = (60 - 20) \ °C = 40 \ °C$ and $R_0 = 30 \ \Omega$, we get

$$R_{60} = 30 \ \Omega(1 + 4.0 \times 10^{-3} \times 40) = 34.8 \ \Omega$$

This approximate method is quicker but in this example introduces an error of 0.3 in 34.5, i.e. about 1%, which is acceptable for many purposes.

The experimental determination of α is outlined on p. 54. Metals and alloys have **p**ositive **t**emperature **c**oefficients (they are p.t.c. materials), i.e. their resistance *increases* with temperature rise. The values for pure metals are of the order of 4×10^{-3} per °C or roughly 1/273 per °C, the same as the cubic expansivity of a gas. In a tungsten-filament electric lamp the current raises the temperature of the filament to over 2730 °C when lit. The 'hot' resistance of the filament is, therefore, more than ten times the 'cold' resistance. Why doesn't a fuse blow every time lights are switched on? Alloys have much lower temperature coefficients of resistance than pure metals; that for Manganin is about 2×10^{-5} per °C and a small temperature change has little effect on its resistance.

Graphite, semiconductors and most non-metals have **n**egative **t**emperature **c**oefficients (they are n.t.c. materials), i.e. their resistance *decreases* with temperature rise.

(b) Superconductors

When certain metals (e.g. tin, lead) and alloys are cooled to near −273 °C with liquid helium an *abrupt* decrease of resistance occurs. Below a definite temperature known as the transition temperature, which is different for each material, the resistance vanishes and a current once started seems to flow for ever. Such materials are called **superconductors**.

Recently superconductors have been made from the oxides of copper, barium and yttrium, which need only be cooled with the plentiful and cheaper liquid nitrogen (at −196 °C).

The use of superconductors in electrical power engineering, low-noise electronic devices and energy storage systems is being investigated. For example, powerful electromagnets can be made more energy-efficient by using coils made from superconducting wires, since these have zero resistance and so do not generate internal energy. A novel high-speed transportation system which uses superconducting electromagnets to levitate vehicles above a magnetic track, leading to much-reduced energy losses from friction, is already on trial in Japan.

Superconductors expel magnetic fields from their interior and this can give rise to magnetic levitation in the 'Meissner effect', as shown in Fig. 4.24 where a magnet floats above superconducting material.

Fig. 4.24 Magnetic levitation above superconductors

(c) Thermistors

These are devices whose resistance varies quite markedly with temperature. (Their name is derived from **therm**al res**istors**.) Depending on their composition they can have either n.t.c. or p.t.c. characteristics. The n.t.c. type consists of a mixture of oxides of iron, nickel and cobalt with small amounts of other substances and is used in electronic circuits to compensate for resistance increase in other components when the temperature rises and also as a thermometer for temperature measurement. The p.t.c. type, which is based on barium titanate, can show a resistance increase of 50 to 200 times for a temperature rise of a few degrees. It is useful as a temperature-controlled switch. Why? Figure 4.25 shows a selection of thermistors.

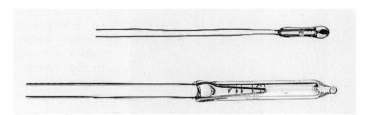

(a) Bead-in-glass

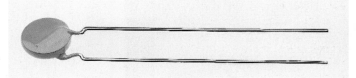

(b) Disc

Fig. 4.25 Types of thermistor

(d) Electrons, resistance and temperature

The 'free electron theory' can account qualitatively for the variation of resistance with temperature of different materials. The increased average separation of the ions in a metal which accompanies a temperature rise (see p. 71) causes local distortion of the crystal lattice. As a result there is increased interaction between the lattice and the 'free' electrons when they drift under an applied p.d. The average drift speed is reduced and the resistance therefore increases. In semiconductors this is more than compensated for when greater vibration of the atoms breaks bonds, 'freeing' more charge carriers (an insignificant effect in metals) and thereby producing a marked decrease of resistance with temperature rise. Heavily doped semiconductors (see p. 416) can acquire metallic properties, a temperature rise *increasing* their resistance.

ELECTROMOTIVE FORCE

Batteries and generators are able to maintain one terminal positive (i.e. deficient in electrons) and the other negative (i.e. with an excess of electrons). If we consider the motion of positive charges, then a battery, for example, moves positive charges from a place of low potential (the negative terminal) through the battery to a place of high potential (the positive terminal). The action may be compared with that of a pump causing water to move from a point of low gravitational potential to one of high potential.

A battery or generator therefore **does work** on charges and so energy must be transferred within it. (Work is a measure of energy transfer.) In a battery chemical energy is transferred to electrical energy which we consider to be stored in the electric and magnetic fields produced. When current flows in an external circuit this stored electrical energy is transformed, for example, into heat, but it is replenished at the same rate at which it is transferred. The electric and magnetic fields thus act as a temporary storage reservoir of electrical energy in the transfer of chemical energy to heat. A battery or generator is said to produce an **electromotive force (e.m.f.)**.

The e.m.f. is defined in terms of energy transfer:

> The electromotive force E of a source (a battery, generator, etc.) is the energy (chemical, mechanical, etc.) transferred to electrical energy when unit charge passes through it.

The unit of e.m.f., like the unit of p.d., is the **volt** and equals the e.m.f. of a source which transfers 1 joule of chemical, mechanical or other form of energy to electrical energy when 1 coulomb passes through it. A car battery with an e.m.f. of 12 volts supplies 12 joules per coulomb passing through it; a power station generator with an e.m.f. of 25 000 volts is a much greater source of energy and supplies 25 000 joules per coulomb — 2 coulombs would receive 50 000 joules and so on. In general, if a charge Q (in coulombs) passes through a source of e.m.f. E (in volts), the electrical energy supplied by the source W (in joules) is

$$W = QE$$

It should be noted that although e.m.f. and p.d. have the same unit, they deal with different aspects of an electric circuit. Whilst e.m.f. applies to a source supplying electrical energy, p.d. refers to the transfer of electrical energy out of a circuit. The term electromotive force is misleading to some extent, since it measures energy per unit charge and not force. It is true, however, that the source of e.m.f. is responsible for moving charges round the circuit. Both e.m.f. and p.d. are often called **voltage**.

A voltmeter measures p.d. and one connected across the terminals of an electrical supply such as a battery records what is called the **terminal p.d.** of the battery. If the battery is not connected to an external circuit and the voltmeter has a very high resistance then the current through the battery will be negligible. We can regard the voltmeter as measuring the number of joules of electrical energy the battery supplies per coulomb, i.e. its e.m.f. A working but less basic definition of e.m.f. is to say that it equals the terminal p.d. of a battery or generator **on open circuit**, i.e. when not maintaining current.

INTERNAL RESISTANCE

A high-resistance voltmeter connected across a cell on open circuit records its e.m.f. (very nearly), Fig. 4.26a. Let this be E. If the cell is now connected to an external circuit in the form of a resistor R and maintains a steady current I in the circuit, the voltmeter reading falls; let it be V, Fig. 4.26b. V is the terminal p.d. of the cell (but not on open circuit) and it is also the p.d. across R (assuming the connecting leads have zero resistance). Since V is less than E, then not all the energy supplied per coulomb by the cell (i.e. E) is transformed in the external circuit into other forms of energy (often heat). What has happened to the 'lost' energy per coulomb?

(a)

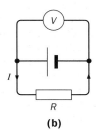

(b)

Fig. 4.26

The deficiency is due to the cell itself having some resistance. A certain amount of electrical energy per coulomb is wasted in getting through the cell and so less is available for the external circuit. The resistance of a cell is called its **internal** or **source resistance** (r). Taking stock of the energy transfers in the complete circuit including the cell, we can say, assuming conservation of energy:

energy supplied = per coulomb by cell

| energy transferred per coulomb to external circuit | + energy wasted per coulomb in internal resistance of battery |

Or, from the definitions of e.m.f. and p.d.,

e.m.f. = p.d. across R + p.d. across r

In symbols

E	=	V	+	v
e.m.f.		useful volts		'lost' volts

where v is the p.d. across the internal resistance of the cell, a quantity which cannot be measured directly but obtained only by subtracting V from E. From the equation $E = V + v$ we see that **the sum of the p.ds across all the resistance in a circuit (external and internal) equals the e.m.f.**

Since $V = IR$ and $v = Ir$ we can rewrite the previous equation as

$$E = IR + Ir$$

or

$$E = I(R + r)$$

Suppose a high-resistance voltmeter reads 1.5 V when connected across a dry battery on open circuit, and 1.2 V when the same battery is supplying a current of 0.30 A through a lamp of resistance R. What is (a) the e.m.f. of the battery, (b) the internal resistance of the battery and (c) the value of R?

Using symbols with their previous meanings:

(a) Since the terminal p.d. on open circuit equals the e.m.f., we have $E = 1.5$ V.

(b) $E = V + v = V + Ir$ where $V = 1.2$ V and $I = 0.30$ A. Therefore

$$Ir = E - V$$

and

$$r = \frac{E - V}{I} = \frac{1.5\text{ V} - 1.2\text{ V}}{0.30\text{ A}} = 1.0\ \Omega$$

(c) From $V = IR$,

$$R = \frac{V}{I} = \frac{1.2\text{ V}}{0.30\text{ A}} = 4.0\ \Omega$$

The internal resistance of an electrical supply depends on several factors and is seldom constant as is often assumed in calculations. However, it is sometimes useful to know its rough

value. An estimate can be made by taking several p.d. (V) and current (I) measurements and plotting a graph of V against I to get an average value for r from its slope (since the equation $V = E - Ir$ is of the form $y = c + xm$ where the slope $m = r$ and the intercept on the $V(y)$ axis is $c = E$). The circuit in Fig. 4.26b could be used if R is a variable resistor and an ammeter is connected in series with it to measure I.

Sources such as low-voltage supply units and car batteries from which large currents are required must have very low internal resistances. On the other hand, if a 5000 V power supply does not have an internal resistance of the order of megohms to limit the current it supplies, it will be dangerous.

The effect of internal resistance can be seen when a bus or car starts with the lights on. Suppose the starter motor requires a current of 100 A from the battery of e.m.f. 12 V and internal resistance 0.04 Ω to start the engine. How many volts are 'lost'? What is the terminal p.d. of the battery with the starter motor working? Why do the lights dim if they are designed to operate on a 12 V supply?

The effect of a load (i.e. an external circuit drawing current) on a source is to reduce its terminal p.d. The equation $V = E - Ir$ shows that if either I or r increases, V decreases. Also note that the maximum current that can be supplied by a source is E/r and occurs when $R = 0$; it is called the 'short-circuit' current and would damage most sources.

The terminal p.d. of a battery on open circuit as measured by even a very-high-resistance voltmeter is not quite equal to the e.m.f. because the voltmeter must take some current, however small, to give a reading. A small part of the e.m.f. is, therefore, 'lost' in driving current through the internal resistance of the battery. A potentiometer is used to measure e.m.f. to a very high accuracy (p. 55).

KIRCHHOFF'S LAWS

The statements that have been made about steady currents and p.ds in series and parallel circuits are summarized by Kirchhoff's two laws.

(a) First law

At a junction in a circuit, the current arriving equals the current leaving.

If currents arriving are given, say, a positive sign and those leaving a negative sign, then the law may be stated in symbols as

$$\Sigma I = 0$$

Σ (pronounced 'sigma') stands for 'the algebraic sum of' — in this case the currents I. For example in Fig. 4.27a, at junction A,

current arriving = I

current leaving = $I_1 + I_2 + I_3$

$\therefore I = I_1 + I_2 + I_3$ or $I - I_1 - I_2 - I_3 = 0$

i.e. $\qquad \Sigma I = 0$

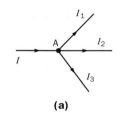

(a)

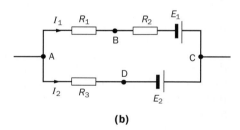

(b)

Fig. 4.27

The law is a statement of our belief that when we are dealing with steady currents, **charge is conserved** and flows in a circuit without being destroyed or accumulating at any point.

(b) Second law

Round any closed circuit or loop the algebraic sum of the e.m.f. E is equal to the algebraic sum of the products of current I and resistance R.

In symbols

$$\Sigma E = \Sigma IR$$

If we adopt the sign conventions that, in going round a circuit or loop,

(*i*) e.m.fs are positive if we pass from the positive terminal of the supply round the rest of the circuit or loop to the negative terminal, and negative if the opposite happens, and

(*ii*) *IR* products are positive when there is a drop of potential and negative when there is a rise, then in Fig. 4.27*b* we can say for a clockwise journey round ABCDA, and with the current directions shown,

$$\sum E = E_1 - E_2$$
$$\sum IR = I_1R_1 + I_1R_2 - I_2R_3$$
$$\therefore E_1 - E_2 = I_1R_1 + I_1R_2 - I_2R_3$$

The law is a statement of the **conservation of energy** using electrical quantities.

POWER AND HEATING EFFECT

Current flow is accompanied by the transfer of electrical energy and it is often necessary to know the **rate** at which a device brings about this transfer.

The **power** of a device is the rate at which it transfers energy.

If the p.d. across a device is *V* and the current through it is *I*, the electrical energy *W* transferred from it in time *t* is (from the definition of p.d., p. 42)

$$W = ItV$$

The power *P* of the device will be

$$P = \frac{W}{t} = \frac{ItV}{t}$$

$$P = IV$$

The unit of power is the **watt** (W) and equals an energy transfer rate of 1 joule per second, i.e. $1 \text{ W} = 1 \text{ J s}^{-1}$. In the expression $P = IV$, *P* will be in watts if *I* is in amperes and *V* in volts. A larger unit is the **kilowatt** (kW) which equals 1000 watts.

If *all* the electrical energy is transformed into heat by the device it is called a 'passive' resistor and the rate of production of heat will also be *IV*. If its resistance is *R*, Fig. 4.28, then since $R = V/I$ we have

$$P = IV$$
$$= \frac{V}{R} \cdot V = \frac{V^2}{R}$$
$$= I \cdot IR = I^2R$$

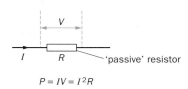

$$P = IV = I^2R$$

Fig. 4.28

There are thus three alternative expressions for power *but* the last two are only true when all the electrical energy is transformed into heat. The first, $P = IV$, gives the rate of production of all forms of energy. For example, if the current in an electric motor is 5 A when the applied p.d. is 10 V then 50 W of electric power is supplied to it. However, it may only produce 40 W of mechanical power, the other 10 W being the rate of production of heat by the motor windings due to their resistance.

(a) Heating elements and lamp filaments

The expression $P = V^2/R$ shows that for a **fixed supply p.d.** of *V*, the rate of heat production by a resistor increases as *R* decreases. Now $R = \rho l/A$, therefore $P = V^2A/\rho l$ and so where a high rate of heat production at constant p.d. is required, as in an electric fire on the mains, the heating element should have a large cross-section area *A*, a small resistivity ρ and a short length *l*. It must also be able to withstand high temperatures without oxidizing in air (and becoming brittle). Nichrome is the material that best satisfies all these requirements.

Electric lamp filaments have to operate at even higher temperatures in order to emit light. In this case, tungsten, which has a very high melting-point (3400 °C), is used either in a vacuum or more often in an inert gas (nitrogen or argon). The gas reduces evaporation of the tungsten (why?) and prevents the vapour condensing on the inside of the bulb and blackening it. In projector lamps there is a little iodine which forms tungsten iodide with the tungsten

vapour and remains as vapour when the lamp is working, thereby preventing blackening.

(b) Fuses

When current flows in a wire its temperature rises until the rate of loss of heat to the surroundings equals the rate at which heat is produced. If this temperature exceeds the melting-point of the material of the wire, the wire melts. A fuse is a short length of wire, often tinned copper, selected to melt when the current through it exceeds a certain value. It thereby protects a circuit from excessive currents.

It can be shown (see question 21, p. 61) that

(*i*) the temperature reached by a given wire depends only on the current through it and is independent of its length (provided it is not so short for significant heat loss from the ends where it is supported); and

(*ii*) the current required to reach the melting-point of the wire increases as the radius of the wire increases.

Fuses which melt at progressively higher temperatures can thus be made from the same material by using wires of increasing radius.

(c) The kilowatt-hour (kWh)

For commercial purposes the kilowatt-hour is a more convenient unit of electrical energy than the joule.

The kilowatt-hour is the quantity of energy transformed into other forms of energy by a device of power 1 kilowatt in 1 hour.

The energy transferred by a device in kilowatt-hours is therefore calculated by multiplying the power of the device in kilowatts by the time in hours for which it is used. Hence a 3 kW electric radiator working for 4 hours uses 12 kWh of electrical energy — often called 12 'units'. How many joules are there in 1 kWh?

(d) Maximum power theorem

The power delivered to a load of resistance *R* is a maximum when *R* equals the source resistance *r*. The maximum power *P* is given by the following.

$$P = I^2R = E^2R/(R + r)^2$$
$$= E^2R/4R^2 = E^2/4R$$

The power wasted in the source is $E^2/4r$, i.e. the same as that in R since $r = R$. The efficiency of the power-transfer process is thus 50%.

WHEATSTONE BRIDGE

(a) Theory

The Wheatstone bridge circuit enables resistance to be measured more accurately than by the ammeter–voltmeter method (p. 46). It involves making adjustments until a galvanometer is undeflected and so, being a 'null' method, it does not depend on the accuracy of an instrument. Other known resistors are, however, required.

Four resistors P, Q, R, S are joined as in Fig. 4.29a. If P is the unknown resistor, Q must be known as must the values of R and S or their ratio. A sensitive galvanometer G and a cell (dry or Leclanché) are connected as shown. One or more of Q, R and S are adjusted until there is no deflection on G. The bridge is then said to be balanced and it can be shown that

$$\frac{P}{Q} = \frac{R}{S}$$

from which P can be found. The proof of this expression follows.

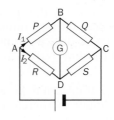

(a) Principle

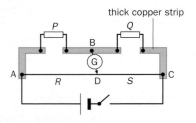

(b) Practical form — the metre bridge

Fig. 4.29 The Wheatstone bridge

At balance, no current flows through G, therefore the p.d. across BD is zero and so

p.d. across AB = p.d. across AD

Also, current through P = current through $Q = I_1$, and current through R = current through $S = I_2$. Therefore

$$I_1 \times P = I_2 \times R$$

Similarly, p.d. across BC = p.d. across DC. Therefore

$$I_1 \times Q = I_2 \times S$$

Dividing, $\quad \dfrac{P}{Q} = \dfrac{R}{S}$

It can be shown that the same condition holds if the cell and G are interchanged.

(b) Metre bridge

This is the simplest practical form of the Wheatstone bridge, Fig. 4.29b. The resistors R and S consist of a wire AC of uniform cross-section and 1 m long, made of an alloy such as constantan, with a resistance of several ohms. The ratio of R to S is altered by changing the position on the wire of the movable contact or 'jockey' D. The other arm of the bridge contains the unknown resistor P and a known resistor Q. Thick copper strips of low resistance connect the various parts. Figures 4.29a and b have identical lettering to show their similarity.

The position of D is adjusted until there is no deflection on G, then

$$\frac{P}{Q} = \frac{R}{S} = \frac{\text{resistance of AD}}{\text{resistance of DC}}$$

Since the wire is uniform, resistance will be proportional to length and therefore

$$\frac{P}{Q} = \frac{AD}{DC}$$

(c) Measurement of resistivity

The resistivity ρ of a material can be determined using a metre bridge by measuring the resistance R of a known length l of wire and also its average diameter d using a micrometer screw gauge. Then $\rho = AR/l$ where $A = \pi(d/2)^2$.

(d) Measurement of temperature coefficient of resistance

This may be found for, say, copper by measuring its resistance at different temperatures with the apparatus shown in Fig. 4.30a. A graph of resistance against temperature is plotted. Over small temperature ranges it is a straight line and from it the temperature coefficient α is calculated using $\alpha = (R_\theta - R_0)/(R_0\theta)$, Fig. 4.30$b$.

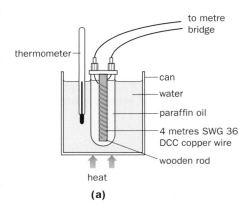

(a)

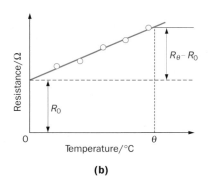

(b)

Fig. 4.30 Measurement of temperature coefficient of resistance

POTENTIOMETER

(a) Theory

A potentiometer is an arrangement which measures p.d. accurately. It can be adapted to measure current and resistance.

In its simplest form it consists of a length of resistance wire AB of **uniform cross-section area**, lying alongside a millimetre scale, and through which a steady current is maintained by a cell, called the driver cell, Fig. 4.31. (This is usually an accumulator because this gives a steady current for a long time.) As a result there is a p.d. between any two points on the wire which is proportional to their distance apart. Basically it is a potential divider. Part of the p.d. across AB is tapped off and used to counter-balance the p.d. to be measured. If the p.d. across AB is 2 volts and if the wire is uniform and the current steady, what p.ds can be tapped off between (*i*) A and P, (*ii*) A and Q, (*iii*) A and R, (*iv*) Q and B?

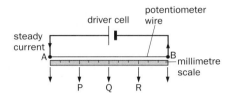

Fig. 4.31   Principle of the potentiometer

In practice the unknown p.d. V_1 is connected with its positive side to X in Fig. 4.32 if the positive terminal of the driver cell is joined to A, as shown. The negative side of the unknown p.d. goes to a galvanometer G and a jockey C. In AX and YGC there are thus two p.ds trying to cause current flow. The one tapped off between A and C from the potentiometer wire tends to drive current in an anticlockwise direction whilst the unknown p.d. V_1 acts in a clockwise direction. The direction of current flow through G therefore depends on whether V_1 is greater or less than the p.d. across AC. When the position of the jockey on the wire is such that there is **no current** through G, these two p.ds are equal and the potentiometer is said to be **balanced**. The balance length l_1 is then measured.

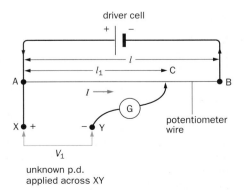

Fig. 4.32   The potentiometer in practice

If the resistance of AB per cm is r then the resistance of the balance length l_1 (in cm) is $l_1 r$ and if the steady current through AB from the driver cell is I, we have

$$\text{unknown p.d. } V_1 = \text{p.d. tapped off at balance}$$

$$\therefore V_1 = I \times l_1 r$$

Since I and r are constants

$$V_1 \propto l_1$$

Let the p.d. across the whole potentiometer wire AB be V, then if $l = $ AB

$$V = I \times lr$$

Therefore

$$\frac{V_1}{V} = \frac{Il_1 r}{Ilr} = \frac{l_1}{l}$$

$$\therefore V_1 = \frac{l_1}{l} \cdot V$$

Knowing V, l and l_1, we can find V_1. When the driver cell is an accumulator of low internal resistance, V may be taken as its e.m.f. If this is 2.0 volts and if $l = 100$ cm and $l_1 = 80$ cm then $V_1 = (80/100) \times 2 = 1.6$ volts.

A potentiometer is a kind of voltmeter but is much more accurate than the best dial instrument since its 'scale' (i.e. the wire) may be made as long as we wish and its adjustment, being a 'null' method, does not depend on the calibration of the galvanometer. It has the further advantage of not altering the p.d. to be measured since at balance no current is drawn by it from the unknown p.d.; it behaves like a voltmeter of infinite resistance. On the other hand, the wire form considered here is bulky and slow to use compared with an ordinary voltmeter.

(b) Measuring internal resistance of a cell

The balance length l is found first with the cell on open circuit, Fig. 4.33 (solid black lines). The p.d. across XY therefore equals the p.d. at the terminals of the cell on open circuit, i.e. its e.m.f. E; therefore $E \propto l$. A known resistance R is then connected across the cell (blue dotted lines) and if l_1 is the new balance length, the p.d. across XY falls and equals the p.d. V across the cell when it maintains current through R; therefore $V \propto l_1$. Hence

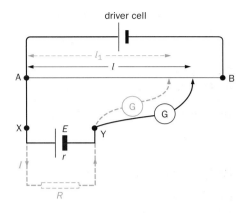

Fig. 4.33

$$\frac{E}{V} = \frac{l}{l_1}$$

If the current through R (at balance) is I and r is the internal resistance of the cell, Ohm's law applied first to the whole circuit and then to R alone gives

$$E = I(R + r) \quad \text{and} \quad V = IR$$

$$\therefore \frac{E}{V} = \frac{R + r}{R} = \frac{l}{l_1}$$

$$\therefore 1 + \frac{r}{R} = \frac{l}{l_1}$$

$$\frac{r}{R} = \frac{l}{l_1} - 1 = \frac{l - l_1}{l_1}$$

Hence r can be calculated.

ELECTROLYSIS

(a) Ionic theory

Liquids that undergo chemical change when a current passes through them are called **electrolytes** and the process is known as **electrolysis**. Solutions in water of acids, bases and salts are electrolytes. Liquids that conduct without suffering chemical decomposition are non-electrolytes; molten metals such as mercury are examples.

Conduction in an electrolyte is due to the movement of positive and negative ions. There is evidence from X-ray crystallography that, in the solid state, compounds such as sodium chloride consist of regular structures of positive and negative ions (p. 20) held together by electrostatic forces. When such substances are dissolved in water (and some other solvents) the inter-ionic forces are weakened so that the ions can separate and move about easily in the solution. **Ionization** or **dissociation** is said to have occurred as a result of solution; the ionization of other salts, bases and acids is similarly explained.

In Fig. 4.34 when a p.d. is applied to the plates that dip into the electrolyte, the electrodes create an electric field which causes positive ions to move towards one electrode (the **cathode**), while negative ions are attracted to the other (the **anode**). The two streams of oppositely charged ions, drifting slowly in opposite directions (see p. 41), constitute the current in the electrolyte. At the electrodes, for conduction to continue, either (*i*) the ions must be discharged, i.e. give up their excess electrons if they are negative or accept electrons if they are positive, or (*ii*) fresh ions must be formed from the electrode and pass into solution. In any event the anode must gain electrons and the cathode lose them to maintain electron flow in the external circuit. After being discharged the ions usually come out of solution and are liberated as uncharged matter, being either deposited on the electrodes or released at them as bubbles of gas. (The electrodes must be made of metal or carbon. Why?)

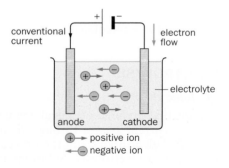

Fig. 4.34 Electrolysis

Consider the electrolysis of copper sulphate solution with copper electrodes. The solution contains copper ions with a double positive charge (Cu^{2+}) and sulphate ions with a double negative charge (SO_4^{2-}). Under an applied p.d. the copper ions drift to the cathode where each receives two electrons ($2e^-$) and forms a copper atom that is deposited on the cathode.

$$Cu^{2+} + 2e^- \text{ from cathode} \rightarrow Cu$$

Sulphate ions collect round the anode and the most likely reaction to occur there, because it involves less energy than any other, is the formation of fresh copper ions by copper atoms of the anode going into solution. The anode is thus able to acquire electrons because every copper atom must lose two electrons to form a copper ion. Also, the fresh copper ions neutralize the negatively charged sulphate ions tending to gather round the anode.

$$Cu \rightarrow Cu^{2+} + 2e^- \text{ to anode}$$

The net result is that copper is deposited on the cathode and goes into solution from the anode. In general, metals and hydrogen are liberated at the cathode and non-metals at the anode.

Electrolysis is used in many industrial processes. By allowing chemical reactions to occur at different places in the same solution, it keeps the products separate and makes feasible reactions that are otherwise impossible.

(b) Specific charge of an ion

Faraday's first law of electrolysis states that:

> The mass of a substance liberated or deposited in electrolysis is directly proportional to the total charge passed.

This is confirmed by experiment and implies that the same kind of ion always has the same charge and mass. If m is the mass liberated and Q the charge passed for a particular substance, Q/m is a constant for the ions of that substance. It is called the **specific charge of the ion** and is expressed in coulombs per kilogram ($C\,kg^{-1}$).

For example, if in the electrolysis of copper sulphate solution a steady current $I = 1.0\,A$ deposits $1.2\,g$ of copper on the cathode in a time $t = 1$ hour, the specific charge of the copper ion is given by

$$\text{specific charge} = \frac{Q}{m} = \frac{It}{m}$$

$$= \frac{(1.0\,A) \times (3600\,s)}{(1.2 \times 10^{-3}\,kg)}$$

$$= 3.0 \times 10^6\,C\,kg^{-1}$$

Knowing that a copper ion carries a double electronic charge, i.e. $2e$, where $e = 1.6 \times 10^{-19}$ C, the mass m of a copper ion is then given by

$$\text{specific charge} = \frac{\text{charge on ion}}{\text{mass of ion}} = \frac{2e}{m}$$

that is, $$m = \frac{2e}{\text{specific charge}}$$

hence, $$m = \frac{2 \times 1.6 \times 10^{-19}\,C}{3.0 \times 10^6\,C\,kg^{-1}}$$

$$= 1.1 \times 10^{-25}\,kg$$

Large-scale measurements on matter in bulk thus enable us, with the help of theory, to obtain information about atomic-sized particles. The results agree well with those from more direct methods (e.g. mass spectrometry, p. 399).

(c) The Faraday constant F

This is the quantity of electric charge which liberates **one mole** of any singly charged ion. Experiment gives its value as

$$F = 9.65 \times 10^4\,C\,mol^{-1}$$

If e is the charge on a hydrogen ion and L is the number of ions in 1 mole of hydrogen ions, i.e. the Avogadro constant (p. 15), then

$$F = Le$$

since 1 mole of hydrogen ions is liberated by 9.65×10^4 coulombs. X-ray crystallography measurements give $L = 6.02 \times 10^{23}$ per mole and so

$$e = \frac{F}{L} = \frac{9.65 \times 10^4\,C}{6.02 \times 10^{23}} = 1.60 \times 10^{-19}\,C$$

This is the charge on a singly charged ion and is found to be the same as that on an electron. The above expression gives one of the most accurate ways of obtaining the electronic charge e.

(d) Ohm's law and electrolytes

The variation of current with p.d. for an electrolyte may be investigated using the circuit of Fig. 4.35. The p.d. is varied from 0 to 4 volts by the potential divider and measured with a high-resistance voltmeter.

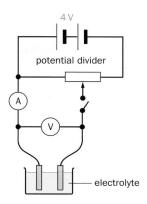

Fig. 4.35

With copper sulphate solution and copper electrodes, the graph of current against p.d. is a straight line through the origin, Fig. 4.36a, and Ohm's law is fairly well obeyed. The smallest p.d. causes current to flow and supports the ionic theory assumption that electrolytes, as soon as they dissolve, split into ions which are immediately available for conduction.

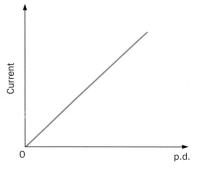

(a) Copper sulphate solution and copper electrodes

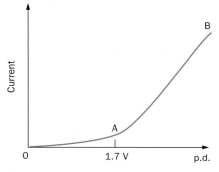

(b) Acidified water and platinum electrodes

Fig. 4.36

Using water (acidified) and platinum electrodes there is no appreciable current flow until the p.d. exceeds 1.7 volts. Thereafter increases in p.d. cause proportionate increases of current, and hydrogen and oxygen are evolved at the cathode and anode respectively. The current–p.d. graph is shown in Fig. 4.36b. The virtual absence of current for p.ds below 1.7 volts is attributed to the existence of a **back e.m.f.** of maximum value 1.7 volts, which the applied p.d. must exceed before the electrolyte conducts. The back e.m.f. is due to polarization, i.e. the accumulation at the electrodes of products of electrolysis formed when the circuit is first made. In this case hydrogen at the cathode and oxygen at the anode effectively replace the platinum electrodes by gas electrodes and act as a chemical cell with an e.m.f. which opposes the applied p.d. (If the switch in Fig. 4.35 is opened the back e.m.f. is recorded on the voltmeter and falls rapidly.)

AB in the graph of Fig. 4.36b is a straight line showing that, if allowance is made for the back e.m.f., acidified water obeys Ohm's law. The equation of AB is

$$V - E = IR$$

where R is the resistance of the electrolyte, I is the current when the applied p.d. is V, and E is the back e.m.f.

ELECTRIC CELLS

These transfer chemical energy to electrical energy and consist of two different metals (or a metal and carbon) separated from each other by an electrolyte. Their e.m.f. depends on the nature and concentration of the chemicals used; their size affects the internal resistance and the amount of electrical energy they can supply, i.e. their **capacity**.

Many different cells have been invented since the first was made by Volta at the end of the eighteenth century. Volta's **simple cell** consisted of plates of copper and zinc in dilute sulphuric acid and had an e.m.f. of about 1 V.

(a) Primary cells

In general these have to be discarded after use and are popularly called 'dry' batteries though this description is not strictly correct. Some types used today are listed in Table 4.2.

The graphs in Fig. 4.37 show roughly how the e.m.f. of different cells of size AA vary with time when supplying moderate currents (e.g. 30 mA).

Table 4.2 Types of primary cell

Cell	e.m.f.	Properties	Use
Carbon–zinc	1.5 V	Relatively cheap; most popular type; e.m.f. falls as current increases; best for low currents or occasional use	Torches Radios
Alkaline–manganese	1.5 V	Medium price; e.m.f. does not fall so much in use; long 'shelf' life; better for higher continuous currents; lasts about four times longer than same size carbon–zinc cell	Radios Calculators
Mercury	1.4 V	Large capacity for their size; made as 'buttons'; e.m.f. almost constant till discharged; best for low current use; expensive	Hearing aids Cameras Watches Calculators
Silver oxide	1.5 V	Similar to mercury cell	As for mercury cell
Weston standard	1.0186 V at 20 °C	e.m.f. constant if current does not exceed 10 μA	Laboratory standard of e.m.f.

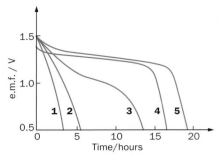

1 Carbon–zinc
2 High-power carbon–zinc
3 Alkaline–manganese
4 Mercury
5 Silver oxide

Fig. 4.37

(b) Secondary cells

These can be charged and discharged repeatedly (but not indefinitely). They supply 'high' continuous currents depending on their capacity, which is expressed in ampere-hours (A h) for a particular discharge rate. For example, a cell with a capacity of 30 A at the 10-hour rate will sustain a current of 3 A for 10 hours, but whilst 1 A would be supplied for more than 30 hours, 6 A would not flow for 5 hours.

The **lead–acid** type is generally called an accumulator. It has an e.m.f. of 2.0 V which it maintains until it is nearly discharged. A 12 V car battery consists of six in series and may supply a current of several hundred amperes for a few seconds to the starter motor. The internal resistance of one cell is of the order of 0.01 Ω; consequently for ordinary currents the 'lost' volts are negligible.

Three compact rechargeable types, commonly used for example in portable computers and mobile phones, are the **Nickel Cadmium** (NiCad) cell and the newer **Nickel Metal Hydride** (NiMH) and **Lithium ion** cells. The first is cheap but should be fully discharged before recharging from a constant current source to prolong its life. The second has a greater capacity and no 'memory effect' like some NiCads so can be fully charged each time. The third is the most efficient and most expensive. 'Button' types are available.

THERMOELECTRIC EFFECT

If two different metals such as copper and iron are joined in a circuit and their junctions are kept at different temperatures, a small e.m.f. is produced and current flows, Fig. 4.38. The effect is known as the **thermoelectric** or **Seebeck effect** and the pair of junctions is called a **thermocouple**.

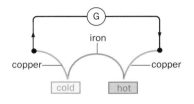

Fig. 4.38 A thermocouple circuit

The value of the thermo-e.m.f. depends on the metals used and the temperature difference between the junctions; the e.m.f.–temperature-difference curve is always approximately a parabola. Figure 4.39 shows the curves for (*a*) copper–iron and (*b*) iron–constantan. The iron–constantan curve, although part of a parabola, is almost linear over a large range and produces about 10 times the e.m.f. of a copper–iron couple. The temperature of the hot junction at which the e.m.f. is a maximum is called the **neutral temperature**.

Thermocouples are used as thermometers, particularly for measuring varying temperatures or the temperature at a point.

The direct transformation of heat into electricity by metal thermocouples is a very inefficient process but better couples are now available based on semiconductors such as iron

disilicide. On account of their reliability, long life and cheapness, these are suitable as small power supply units in space probes, weather buoys and weatherships. Radiation from a radioactive source (e.g. strontium-90) in the unit falls on the hot junction and produces the necessary temperature rise in it.

DATALOGGING

When parameters such as potential difference, resistance or current are changing quickly in a circuit it is advantageous to replace a multimeter by a datalogger. To record measurements using a datalogger the quantity to be measured must be converted into a potential difference; datalogging of a changing p.d. can thus be done directly. For example, if changing temperatures are to be monitored using a thermocouple, the meter in Fig. 4.38 could be replaced by a datalogger allowing a continuous record of the variation of the thermocouple e.m.f. with time to be displayed on a computer screen.

The easiest way to display varying currents on the computer is to pass the current through a 1 Ω resistor and then monitor the potential difference across the resistor as the current changes. Since $V = IR$, the potential difference recorded by the datalogger will be numerically equal to the current in the circuit.

There are many different types of sensor available for the measurement of temperature, current, motion, light intensity, etc. but the function of each type is to convert the input into a potential difference, which can then be fed to a datalogger.

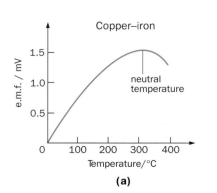

(a)

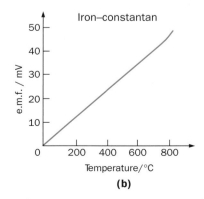

(b)

Fig. 4.39

QUESTIONS

Current and charge

1. If the heating element of an electric radiator takes a current of 4.0 A, what charge passes each point every minute? If the charge on an electron is 1.6×10^{-19} C how many electrons pass a given point in this time?

2. **a)** If the density of copper is 9.0×10^3 kg m^{-3} and 63.5 kg of copper contains 6.0×10^{26} atoms, find the number of free electrons per cubic metre of copper assuming that each copper atom has one free electron.
   b) How many free electrons will there be in a 1.0 m length of copper wire of cross-section area 1.0×10^{-6} m^2 (i.e. 1.0 mm^2)?
   c) Taking the charge on an electron as 1.6×10^{-19} C, what is the total charge of the free electrons per metre of wire?
   d) Assuming that the free electrons are responsible for conduction, how long will the charge in c take to travel 1 m when a current of 2.0 A flows?
   e) What is the drift velocity of the free electrons?

3. Explain in terms of the motion of free electrons what happens when an electric current flows through a metallic conductor.
   A metal wire contains 5.0×10^{22} electrons per cm^3 and has a cross-sectional area of 1.0 mm^2. If the electrons move along the wire with a mean drift velocity of 1.0 mm s^{-1}, calculate the current in amperes in the wire if the electronic charge is 1.6×10^{-19} C.

Potential difference; resistance; meters

4. **a)** What is the p.d. between two points in a circuit if 200 joules of electrical energy is changed to other forms of energy when 25 coulombs of electric charge pass? If the charge flows in 10 seconds, what is the current?
   b) What is the p.d. across an immersion heater which transforms 3.6×10^3 joules of electrical energy into heat every second and takes a current of 15 amperes?

5. Three voltmeters $\widehat{V}$, $\widehat{V_1}$ and $\widehat{V_2}$ are connected as in Fig. 4.40.

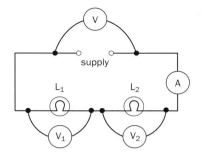

Fig. 4.40

a) If $\widehat{V}$ reads 12 volts and $\widehat{V_1}$ reads 8.0 volts, what does $\widehat{V_2}$ read?
b) If the ammeter $\widehat{A}$ reads 0.50 ampere, how much electrical energy is transformed into heat and light by L_1 in 1 minute?
c) Copy the diagram and mark with a $+$ the positive terminals of the voltmeters and ammeter for correct connection.

6. A p.d. of V drives current through two resistors of 2 ohms and 3 ohms joined in series, Fig. 4.41.

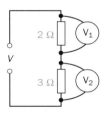

Fig. 4.41

a) If voltmeter $\widehat{V_1}$ reads 4 volts, what is the current in the 2 ohm resistor?
b) What is the current in the 3 ohm resistor?
c) What does voltmeter $\widehat{V_2}$ read?
d) What is the value of V?
e) Find the value of the single equivalent resistor which, if it replaced the 2 ohm and 3 ohm resistors in series, would allow the same current to flow when joined to the same p.d. V.

7. Two resistors of 3 ohms and 6 ohms are connected in parallel across a p.d. of 6 volts, Fig. 4.42a.
   a) How are I, I_1 and I_2 related?
   b) What is the p.d. across each resistor?
   c) Find the values of I, I_1 and I_2 in amperes.
   d) Find the value of the single equivalent resistor R which, if it replaced the 3 ohm and 6 ohm resistors in parallel, would allow the value of I found in c) to flow when the p.d. across it is 6 volts, Fig. 4.42b.

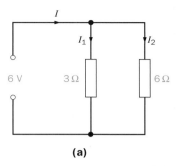

(a)

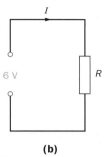

(b)

Fig. 4.42

8. **a)** In the circuit of Fig. 4.43a what is the p.d. across i) AB, ii) BC?
   b) What do these p.ds become when the circuit is altered as in Fig. 4.43b?

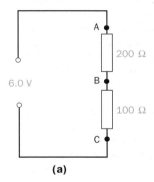

(a)

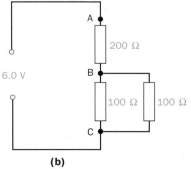

(b)

Fig. 4.43

9. A resistor of 500 ohms and one of 2000 ohms are placed in series with a 60 volt supply. What will be the reading on a voltmeter of internal resistance 2000 ohms when placed across
   a) the 500 ohm resistor,
   b) the 2000 ohm resistor?

10. The circuit diagram in Fig. 4.44 shows a 12 V power supply connected across a potential divider R by the sliding contact P. The potential divider is linked to a resistance wire XY through an ammeter. A voltmeter is connected across the wire XY.

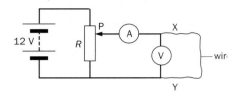

Fig. 4.44

Explain, with reference to this circuit, the term **potential divider**.

The circuit has been set up to measure the resistance of the wire XY. A set of voltage and current measurements is recorded and used to draw the graph in Fig. 4.45 below. Explain why the curve deviates from a straight line at higher current values.

Calculate the resistance of the wire for low current values.

To determine the resistivity of the material of the wire, two more quantities would have to be measured. What are they?

Explain which of these two measurements you would expect to have the greater influence on the error in a calculated value for the resistivity. How would you minimise this error?

(L, AS/A PH1, June 1998)

11. Measurements of p.d. and current on five different 'devices' A, B, C, D, E gave the graphs in Fig. 4.46. Suggest, with reasons, what each might be.

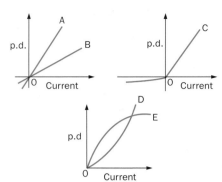

Fig. 4.46

12. A number of measurements were made of the p.d. across a light bulb for different currents flowing through it:

p.d./V	0.0	0.27	0.54	0.81	1.08	1.40	1.90	2.80
current/mA	0.0	10	20	30	40	50	60	70

a) Using a spreadsheet and graphing routine or otherwise, plot a graph of the variation of the p.d. with the current.
b) Explain why the graph is non-linear at large currents.
c) Estimate the resistance of the bulb at a low and a high current value.
d) For what current does the bulb shine brightest?

13. If a moving-coil ammeter gives a full-scale deflection for a current of 15 mA and has a resistance of 5.0 ohms, how would you adapt it so that it could be used
   i) as a voltmeter reading to 1.5 V,
   ii) as an ammeter reading to 1.5 A?

Resistivity; temperature coefficient

14. Assuming that the resistivity of copper is half that of aluminium and that the density of copper is three times that of aluminium, find the ratio of the masses of copper and aluminium cables of equal resistance and length.

15. A wire has a resistance of 10.0 ohms at 20.0 °C and 13.1 ohms at 100 °C. Obtain a value for its temperature coefficient of resistance.

16. When the current passing through the Nichrome element of an electric fire is very small the resistance is found to be 50.9 Ω, room temperature being 20.0 °C. In use the current is 4.17 A on a 240 V supply. Calculate
a) the rate of energy transfer by the element,
b) the steady temperature reached by it.
(The temperature coefficient of resistance of Nichrome may be taken to have the constant value 1.70×10^{-4} °C^{-1} over the temperature range involved.)

Electrical energy; e.m.f.; internal resistance; Kirchhoff's laws

17. How much electrical energy does a battery of e.m.f. 12 volts supply when
a) a charge of 1 coulomb passes through it,
b) a charge of 3 coulombs passes through it,
c) a current of 4 amperes flows through it for 5 seconds?

18. In the circuit of Fig. 4.47, each cell has an e.m.f. of 1.5 V and zero internal resistance. Each resistor has a resistance of 10 Ω. There are currents I_1 and I_2 in the branches, as shown.

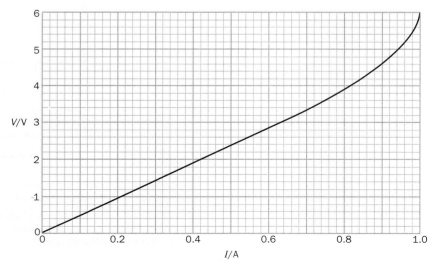

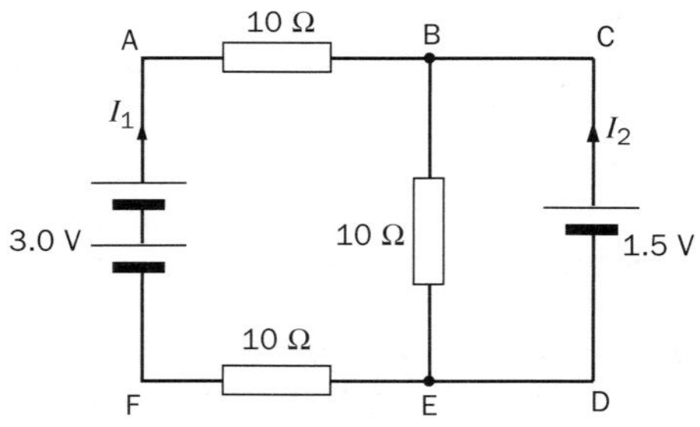

Fig. 4.47

Fig. 4.45

a) Use Kirchhoff's first law to write down an expression for the current in BE, in terms of I_1 and I_2.
b) Use Kirchhoff's second law to write down equations for the circuit loops
 i) ABEFA,
 ii) CBEDC.
You are not required to solve these equations.
(*UCLES,* Further Physics, March 1998)

19. a) A flashlamp bulb is marked '2.5 V, 0.30 A' and has to be operated from a dry battery of e.m.f. 3.0 V for the correct p.d. of 2.5 V to be produced across it. Why?
b) How much heat and light energy is produced by a 100 W electric lamp in 5 minutes?
c) What is the resistance of a 240 V, 60 W bulb?

20. The p.d. across the terminals XY of a box is measured by a very-high-resistance voltmeter V. In arrangement **A**, Fig. 4.48, the voltmeter reads 105 volts. In arrangement **B**, with the same box, the reading of the voltmeter is 100 volts. The inside of the box is not altered between the two arrangements.

Explain what you think may be in the box. You may, if you want to, draw in extra features on copies of the diagrams to use in your explanation.

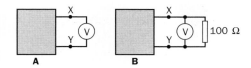

Fig. 4.48

21. By considering a wire of radius r, length l and resistivity ρ, through which a current I flows, show that
a) the rate of production of heat by it is $I^2\rho l/\pi r^2$,
b) the rate of loss of heat from its surface is $2\pi rlh$ where h is the heat lost per unit area of surface per second,
c) the steady temperature it reaches is independent of its length and depends only on I.

Wheatstone bridge; potentiometer

22. Two resistance coils, P and Q, are placed in the gaps of a metre bridge. A balance point is found when the movable contact touches the bridge wire at a distance of 35.5 cm from the end joined to P. When the coil Q is

shunted with a resistance of 10 ohms, the balance point is moved through a distance of 15.5 cm. Find the values of the resistances P and Q.

23. How would you investigate the way in which the current through a metal wire depends on the potential difference between its ends? What conditions should be fulfilled if Ohm's law is to hold?

Explain the theory of the Wheatstone bridge method of comparing resistances.

In an experiment with a simple metre bridge, the unknown X is kept in the left-hand gap and there is a fixed resistance in the right-hand gap. X is heated gradually, and when its temperature is 30 °C, the balance point on the bridge is found to be 51.5 cm from the left-hand end of the slide wire. When its temperature is 100 °C the balance point is 54.6 cm from that end. Find the temperature coefficient of resistance for the material of X, and calculate where the balance point would be if X were cooled to 0 °C.

24. A two-metre potentiometer wire is used in an experiment to determine the internal resistance of a voltaic cell. The e.m.f. of the cell is balanced by the fall of potential along 90.6 cm of wire. When a standard resistor of 10.0 ohms is connected across the cell the balance length is found to be 75.5 cm. Draw a labelled circuit diagram and calculate, from first principles, the internal resistance of the cell. How may the accuracy of this determination be improved? Assume that other electrical components are available if required.

25. In the circuit of Fig. 4.49, the internal resistance of the cell is negligible.

The distance of the slider from the left-hand end of the slide wire is l. The graph of Fig. 4.50 shows the variation with l of the current I in the cell.

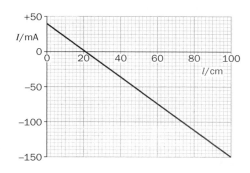

Fig. 4.50

a) Explain why, for different values of l, the current I can be positive, zero or negative.
b) Show that the e.m.f. of the cell is 1.26 V.
(*OCR,* Further Physics, March 1999)

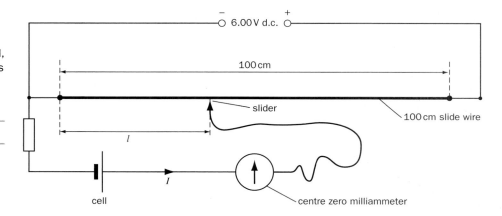

Fig. 4.49

5

Thermal properties

- ■ Temperature and thermometers
- ■ Heat and internal energy
- ■ Specific heat capacity
- ■ Measuring specific heat capacities

- ■ Heat loss and cooling corrections
- ■ Cooling laws and temperature fall
- ■ Latent heat
- ■ Heat calculations
- ■ Expansion of solids

- ■ Thermal stress
- ■ Thermal conductivity

TEMPERATURE AND THERMOMETERS

A knowledge of the thermal properties of materials is desirable when deciding, for example, what material to use for an electric storage heater, or for insulation in a refrigerator, or for specialist clothing, Fig. 5.1. Before studying some of these properties, certain basic ideas will first be considered.

(a) Defining a temperature scale

Temperature is sometimes called the 'degree of hotness' and is a quantity which is such that when two bodies are placed in contact, heat flows from the body at the higher temperature to the one at the lower temperature. To measure temperatures, a **temperature scale** has to be established as follows.

(*i*) Some property of matter is selected whose value varies continuously with the degree of hotness. Suitable properties must be accurately measurable over a wide range of temperature with fairly simple apparatus and must vary in a similar way to many other physical properties.

(*ii*) Two standard degrees of hotness are chosen — called the **fixed points** — and numbers assigned to them. On the Celsius method of numbering (previously known as the centigrade method) the lower fixed point is the **ice point**, i.e. the temperature of pure ice in equilibrium with air-saturated water at standard atmospheric pressure[1], and is designated as 0 degrees

Fig. 5.1 A thermal imaging camera was used to record this soldier's 'hot spots' when wearing a protective vest

Celsius (0 °C). The upper fixed point is the **steam point**, i.e. the temperature at which steam and pure boiling water are in equilibrium at standard atmospheric pressure, and is taken as 100 °C.

(*iii*) The values X_{100} and X_0 of the temperature-measuring property are found at the steam and ice points respectively; then $(X_{100} - X_0)$ gives the **fundamental interval** of the scale. If X_θ is the value of the property at some other temperature θ which we wish to know then the value of θ in °C is given by the following equation.

$$\frac{\theta}{100} = \frac{X_\theta - X_0}{X_{100} - X_0}$$

This equation *defines* temperature θ in °C on the scale based on this particular temperature-measuring property. Note that it has been defined so that equal increases in the value of the property represent equal increases of temperature, i.e. the temperature scale is *defined* so that the property varies uniformly or linearly with temperature measured on its own scale.

[1] Standard atmospheric pressure is defined to be 1.013×10^5 Pa (1 Pa = 1 pascal = 1 N m^{-2}) and equals the pressure at the foot of a column of mercury 760 mm high of specified density and subject to a particular value of *g*.

Some thermometers using different temperature-measuring properties will now be considered briefly.

(b) Mercury-in-glass thermometer

The change in length of a column of mercury in a glass capillary tube was one of the first thermometric properties to be chosen. If l_0 and l_{100} are the lengths of a mercury column at 0 °C and 100 °C respectively and if l_θ is the length at some other temperature θ, then θ in °C is *defined* on the mercury-in-glass scale by the equation

$$\frac{\theta}{100} = \frac{l_\theta - l_0}{l_{100} - l_0}$$

For example, if a mercury thread has lengths 5.0 cm and 20 cm at the ice and steam points and is 8.0 cm long at another temperature θ, then

$$\frac{\theta}{100\,°C} = \frac{(8.0 - 5.0)\,cm}{(20 - 5.0)\,cm} = \frac{3.0\,cm}{15\,cm}$$

$$\therefore \quad \theta = \left(\frac{3.0}{15} \times 100\right)\left(\frac{cm\,°C}{cm}\right)$$

$$= 20\,°C$$

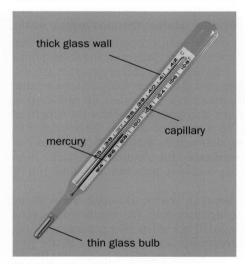

Fig. 5.2 Mercury-in-glass thermometer

Inaccuracies arise in mercury thermometers from (*i*) non-uniformity of the bore of the capillary tube, (*ii*) the gradual change in the zero owing to the bulb shrinking for a number of years after manufacture, and (*iii*) the mercury in the stem not being at the same temperature as that in the bulb.

Properties of this and other types of thermometer are given in Table 5.1.

Table 5.1 Types of thermometer

Thermometer	Range/°C	Comments
Mercury-in-glass	−39 to 500	Simple, cheap, portable, direct reading but not very accurate. Everyday use, clinical work, and weather recording
Constant-volume gas	−270 to 1500	Very wide range, very accurate, very sensitive but bulky, slow to respond and not direct reading. Used as a standard to calibrate other more practical types
Platinum resistance	−200 to 1200	Wide range, very accurate but unsuitable for rapidly changing temperatures because of large heat capacity. Best for small steady temperature differences
Thermocouple	−250 to 1500	Wide range, fairly accurate, robust and compact. Widely used in industry for rapidly changing temperatures and temperatures at a 'point'
Pyrometer	above 1000	See chapter 20
'Modern' radiation	−50 to 3500	Automatic non-contact measurement (see chapter 20)

(c) Constant-volume gas thermometer

If the volume of a fixed mass of gas is kept constant, its pressure changes appreciably when the temperature changes. A temperature θ in °C is *defined* on the constant volume gas scale by the equation

$$\frac{\theta}{100} = \frac{p_\theta - p_0}{p_{100} - p_0}$$

where p_0, p_θ and p_{100} are the pressures at the ice point, the required temperature θ and the steam point.

A simple constant-volume thermometer is shown in Fig. 5.3. The gas (air in school models, but hydrogen, helium or nitrogen in more accurate versions) is in bulb A, which is at the temperature to be measured. As the temperature increases the gas expands, pushing the mercury down in B and up in tube C. By raising C the mercury level in B is restored to the reference mark R and the volume of gas thus kept constant. The gas pressure is then $h + H$ where H is atmospheric pressure.

In accurate work corrections are made for (*i*) the gas in the 'dead-space' D not being at A's temperature, (*ii*) thermal expansion of A, and (*iii*) capillary effects at the mercury surfaces (see p. 183).

(d) Platinum resistance thermometer

The electrical resistance of a pure platinum wire increases with temperature (by about 40% between the ice and steam points) and since resistance can be found very accurately it is a good property on which to base a temperature scale. A temperature θ in °C on the platinum resistance scale is *defined* by the equation

$$\frac{\theta}{100} = \frac{R_\theta - R_0}{R_{100} - R_0}$$

where R_0 and R_{100} are the resistances of the platinum wire at the ice and steam points respectively and R_θ is the resistance at the temperature required.

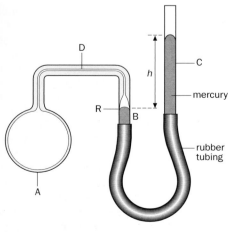

Fig. 5.3 Constant-volume gas thermometer

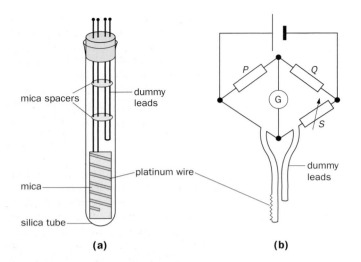

Fig. 5.4 Platinum resistance thermometer and its circuit

A platinum resistance thermometer, Fig. 5.4*a*, consists of a fine platinum wire wound on a strip of mica (an electrical insulator) and connected to thick copper leads. A pair of identical short-circuited dummy leads are enclosed in the same silica tube (which can withstand high temperatures) and compensate exactly for changes in resistance with temperature of the leads to the platinum wire. The thermometer is connected to a Wheatstone bridge circuit (p. 54) as in Fig. 5.4*b* and if $P = Q$, the resistance of the platinum wire equals that of S.

(e) Thermistor thermometer

A thermistor (e.g. TH3) connected to a digital multimeter on its lowest ohm range (e.g. 0–1 kΩ) can be used as a thermometer since its resistance changes with temperature. It is not sufficiently accurate, however, nor does it have a large enough range, for its variation in resistance to define a temperature scale.

The thermistor needs to be calibrated using the arrangement shown in Fig. 5.5. The multimeter reading is noted at various temperatures, measured by the mercury-in-glass thermometer, as the water is heated. Before each reading is taken, the burner is removed and the water is stirred thoroughly. A graph of resistance (*y*-axis) against temperature (*x*-axis) is then plotted.

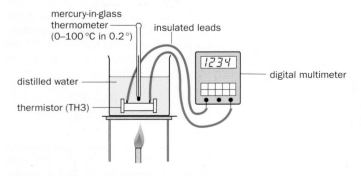

Fig. 5.5 Calibration of a thermistor thermometer

To find an unknown temperature, such as the melting-point of a solid (e.g. octadecan-1-ol), a boiling tube is half-filled with the solid and placed in boiling water. When the solid has melted the thermistor and leads are placed in it and left until the resistance reading becomes steady. The boiling tube and contents are then removed from the boiling water and allowed to cool. Resistance readings are taken at regular intervals, say every 2 minutes, until the substance has solidified completely. By plotting a graph of resistance (*y*-axis) against time (*x*-axis), the resistance at the melting-point (i.e. when the change from liquid to solid occurs and the temperature levels off) can be found. An estimate of the melting-point is then read from the calibration graph of resistance against temperature.

A thermistor can also be used to control temperature. An alarm circuit using a thermistor in a potential divider circuit which operates a transistor is described on p. 426.

(f) Thermocouple thermometer

The thermoelectric effect, considered on p. 58, is widely used to measure temperature. If great accuracy is not required, especially at high temperatures, the thermocouple can be connected across a meter marked to read temperatures directly (calibrated using the known melting-points of metals). When rapidly changing temperatures need to be monitored, the meter connected to the thermocouple may be replaced by a datalogger (see p. 4).

It is not usual to define a thermoelectric scale of temperature, but what would be the shape of a graph of thermocouple e.m.f. against temperature measured on such a scale?

(g) Disagreement between scales; thermodynamic scale

Thermometers based on different properties given different values for the same temperature, except at the fixed points where they must agree by definition. All are correct according to their own scales and the discrepancy arises because, not unexpectedly, thermometric properties do not keep in step as the temperature changes. Thus when the length of the mercury column in a mercury-in-glass thermometer is, for example, mid-way between its 0 and 100 °C values (i.e. reading 50 °C) the resistance of a platinum resistance thermometer is not exactly mid-way between its 0 and 100 °C values.

The disagreement between scales, although small in the range 0 to 100 °C, is inconvenient. We could always state the temperature scale involved when giving a temperature, e.g. 50 °C on the mercury-in-glass scale, but a better procedure is to take one scale as a standard in which all temperatures are expressed, however they are measured. The one chosen is called the **absolute thermodynamic scale**. At this stage it is enough to say that it is the fundamental temperature scale in science and that the SI unit of temperature, the **kelvin** (denoted by K *not* °K) is defined in terms of it. The zero of this scale is called **absolute zero** and it is thought that temperatures below this do not exist; certainly so far all attempts to reach it have been unsuccessful, although it has been approached very closely.

On the thermodynamic scale 0 °C = 273.15 K (273 K for most purposes) and 100 °C = 373.15 K, hence a temperature interval of one Celsius degree equals one kelvin.

HEAT AND INTERNAL ENERGY

Temperature is a useful idea when describing some aspects of the behaviour of matter in bulk. It is a quantity that is measurable in the laboratory as we have just seen and is capable of perception by the sense of touch. One of the aims of modern science is to relate macroscopic (i.e. large-scale) properties, such as temperature, to the masses, speeds, energies, etc., of the constituent atoms and molecules. That is, to explain the macroscopic in terms of the microscopic.

The kinetic theory regards the atoms of a solid as vibrating to and fro about their equilibrium positions, alternately attracting and repelling one another. Their energy, called **internal energy**, is considered to be partly kinetic and partly potential. The kinetic component is due to the vibratory motion of the atoms and according to the theory depends on the temperature; the potential component is stored in the interatomic bonds that are continuously stretched and compressed as the atoms vibrate, and it depends on the forces between the atoms and their separation. In a solid both forms of energy are present in roughly equal amounts and there is continual interchange between them. In a gas, where the intermolecular forces are weak, the internal energy is almost entirely kinetic. The kinetic theory thus links **temperature** with the **kinetic energy of atoms and molecules**.

Heat, in science, is defined as **the energy that is transferred from a body at a higher temperature to one at a lower temperature, by conduction, convection or radiation**. Like other forms of energy it is measured in joules. When a transfer of heat occurs the internal energy of the body receiving the heat increases and if the kinetic component increases, the temperature of the body rises. Heat was previously regarded as a fluid called 'caloric', which all bodies were supposed to contain. It was measured in calories — a unit now little used – one thousand of which equal the dietician's Calorie.

The internal energy of a body can also be increased by doing work, i.e. by a force undergoing a displacement in its own (or a parallel) direction. Thus the temperature of the air in a bicycle pump rises when it is compressed, i.e. it becomes hotter. Work done by the compressing force has become internal energy of the air and its temperature rises, as it would by heat transfer. It is impossible to tell whether the temperature rise of a given sample of hot air is due to compression (i.e. work done) or to heat flow from a hotter body.

The expression 'heat in a body', although often used, is misleading, for it may be that the body has become hot yet no heat flow has occurred. We should talk about the 'internal energy' of the body. It is sometimes said that 'the quantity of heat contained *in* a cup of boiling water is greater than *in* a spark of white-hot metal'. What is really meant is that the boiling water has more *internal energy* and more heat *can be obtained* from it than from the spark.

The internal energy of a body may be changed in two ways: by doing **work** or by transferring **heat**. Work and heat are both concerned with energy **in the process of transfer** and when the transfer is over, neither term is relevant. Work is energy being transferred by a force moving its point of application, and the force may arise from a mechanical, gravitational, electrical or magnetic source; heat flow arises from a temperature difference.

In a wire carrying a current, electrical energy is transformed into internal energy (i.e. more vigorous vibration of the atoms of the wire) and a temperature rise occurs. Subsequently this energy may be given out by the wire to the surroundings as heat. We sum up the whole process by saying that an electric current has a 'heating effect'.

SPECIFIC HEAT CAPACITY

(a) Definition

Materials differ from one another in the quantity of heat needed to produce a certain rise of temperature in a given mass. The **specific heat capacity** c enables comparisons to be made. Thus if a quantity of heat δQ raises the temperature of a mass m of a material by $\delta \theta$ then c is defined by the equation

$$c = \frac{\delta Q}{m \delta \theta}$$

In words, we can say that c is **the quantity of heat required to produce unit rise of temperature in unit mass**. (The word 'specific' before a quantity means per unit mass.)

The unit of c is joule per kilogram kelvin ($J\,kg^{-1}\,K^{-1}$), since in the above expression δQ is in joules, m in kilograms and $\delta \theta$ in kelvin. Sometimes it is more convenient to consider mass in grams, when the unit is $J\,g^{-1}\,K^{-1}$.

In the expression for c, as the temperature rise $\delta \theta$ tends to zero, c approaches the specific heat capacity at a particular temperature and experiment shows that its value for a given material is not constant but varies slightly with temperature. Mean values are thus obtained over a temperature range and to be strictly accurate this range should be stated. For ordinary purposes, however, it is often assumed constant.

The approximate specific heat capacity of water at room temperature is $4.2 \times 10^3\ J\,kg^{-1}\,K^{-1}$ (or $4.2\ J\,g^{-1}\,K^{-1}$) and is large compared with the values for most substances. At temperatures approaching absolute zero (0 K) all values of c tend to zero. Values for some other materials at ordinary temperatures are shown in Table 5.2.

High specific heat capacity is desirable in a material if only a small temperature rise is required for a given heat input. This accounts for the efficiency of water as a coolant in a car radiator and of hydrogen gas in enclosed electric generators (the latter also because of its comparatively good thermal conductivity).

Table 5.2 Mean specific heat capacities/$J\,kg^{-1}\,K^{-1}$

aluminium	9.1×10^2
brass	3.8×10^2
copper	3.9×10^2
glass (ordinary)	6.7×10^2
iron	4.7×10^2
mercury	1.4×10^2
lead	1.3×10^2
ice	2.1×10^3
rubber	1.7×10^3
wood	1.7×10^3
alcohol	2.5×10^3
glycerine	2.5×10^3
paraffin oil	2.1×10^3
turpentine	1.8×10^3

(b) Molar heat capacity

If heat capacities are referred to one mole (p. 15) of the material instead of to unit mass, the quantity obtained by multiplying the specific heat capacity by the atomic or molecular mass is called the **atomic** or **molar heat capacity**. It is very nearly 25 J mol^{-1} K^{-1} for many solids. This fact is known as Dulong and Petit's law. Since 1 mole of any substance contains the same number of atoms or molecules, the heat required per atom or molecule to raise the temperature of many solids by a given amount is about the same. The implication is that the heat capacity of a solid depends on the *number* of atoms or molecules present, not on their mass, and is further evidence for the atomic theory of matter.

(c) Useful equation

The equation defining specific heat capacity may be written

$$Q = mc(\theta_2 - \theta_1)$$

This expression is useful in heat calculations and gives the quantity of heat Q taken in by a body of mass m and mean specific heat capacity c when its temperature rises from θ_1 to θ_2. It also gives the heat lost by the body when its temperature falls from θ_2 to θ_1. In words, we can say

$$\begin{array}{c}\text{heat given out} \\ \text{(or taken in)}\end{array} = \text{mass} \times \begin{array}{c}\text{specific heat} \\ \text{capacity}\end{array} \times \begin{array}{c}\text{temperature} \\ \text{change}\end{array}$$

If the temperature of a body of mass 0.5 kg and specific heat capacity 400 J kg^{-1} K^{-1} rises from 15 °C to 20 °C (288 K to 293 K) the heat taken in is

$$Q = (0.5 \text{ kg}) \times (400 \text{ J kg}^{-1}\text{K}^{-1}) \times (5 \text{ K})$$
$$= (0.5 \times 400 \times 5) \text{ kg J kg}^{-1}\text{K}^{-1}\text{K}$$
$$= 1000 \text{ J}$$

(d) Heat capacity

The **heat capacity** or **thermal capacity** of a body is a term in common use and is defined as **the quantity of heat needed to produce unit rise of temperature in the body**. It is measured in joules per kelvin (J K^{-1}) and from the definition of specific heat capacity it follows that

$$\text{heat capacity} = \text{mass} \times \text{specific heat capacity}$$

The heat capacity of a copper vessel of mass 0.1 kg and specific heat capacity 390 J kg^{-1} K^{-1} is 39 J K^{-1}.

MEASURING SPECIFIC HEAT CAPACITIES

1. Electrical method

(a) *Solids.* The method is suitable for metals, such as copper and aluminium, which are good thermal conductors. A cylindrical block of the material is used, having holes for an electric heater (12 V, 2–4 A) and a thermometer, Fig. 5.6. The mass m of the block is found and its initial temperature θ_1 recorded. The block is lagged with expanded polystyrene and a suitable steady current switched on as a stop-clock is started. The voltmeter and ammeter readings V and I are noted. When the temperature has risen by about 10 K, the current is stopped and the time t taken for which it passed. The highest reading θ_2 on the thermometer is noted.

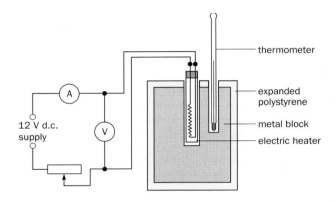

Fig. 5.6 Measuring specific heat capacity of a metal

Assuming that no energy loss occurs,

electrical energy supplied by heater
$$= \text{heat received by block}$$
$$ItV = mc(\theta_2 - \theta_1)$$

where c is the specific heat capacity of the metal. So

$$c = \frac{ItV}{m(\theta_2 - \theta_1)}$$

Note that:
- if I is in amperes, t in seconds, V in volts, m in g, θ_1 and θ_2 in K, then c is in J g^{-1} K^{-1};
- the small amount of heat received by the thermometer and heater has been neglected.

(b) *Liquids.* The apparatus is shown in Fig. 5.7, a calorimeter being a vessel in which heat measurements are made. The procedure is similar to that for solids except that the liquid is stirred continuously during the heating. If m is the mass of liquid, c its specific heat capacity, m_c the mass of the calorimeter and stirrer, c_c the known specific heat capacity of the material of the calorimeter and stirrer, and if θ_1, θ_2, I, V, t have their previous meanings, then assuming

$$\begin{array}{c}\text{energy supplied} \\ \text{by heater}\end{array} = \begin{array}{c}\text{energy received} \\ \text{by liquid}\end{array} + \begin{array}{c}\text{energy received} \\ \text{by calorimeter} \\ \text{and stirrer}\end{array}$$

we have

$$ItV = mc(\theta_2 - \theta_1) + m_c c_c(\theta_2 - \theta_1)$$
$$= (mc + m_c c_c)(\theta_2 - \theta_1)$$

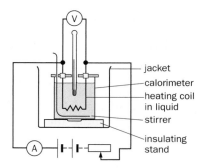

Fig. 5.7 Measuring specific heat capacity of a liquid

2. Method of mixtures

(a) *Solids.* The solid is weighed to find its mass m, heated in boiling water at temperature θ_3 for 10 minutes, Fig. 5.8a, and then quickly transferred to a calorimeter of mass m_c containing a mass of water m_w at temperature θ_1, Fig. 5.8b. The water is stirred and the highest reading θ_2 on the thermometer noted.

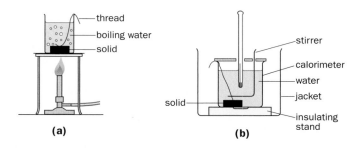

Fig. 5.8 Method of mixtures

Assuming no heat loss from the calorimeter when the hot solid is dropped into it, we have

| heat given out by solid cooling from θ_3 to θ_2 | = | heat received by water warming from θ_1 to θ_2 | + | heat received by calorimeter warming from θ_1 to θ_2 |

If c is the specific heat capacity of the solid, c_w that of water and c_c that of the calorimeter, then

$$mc(\theta_3 - \theta_2) = m_w c_w(\theta_2 - \theta_1) + m_c c_c(\theta_2 - \theta_1)$$
$$= (m_w c_w + m_c c_c)(\theta_2 - \theta_1)$$
$$\therefore \quad c = \frac{(m_w c_w + m_c c_c)(\theta_2 - \theta_1)}{m(\theta_3 - \theta_2)}$$

Hence c can be found knowing c_w and c_c.

(b) *Liquids.* In this case a hot solid of known specific heat capacity is dropped into the liquid whose specific heat capacity is required; the procedure and calculation are the same as in (a).

HEAT LOSS AND COOLING CORRECTIONS

In experiments with calorimeters certain precautions can be taken to minimize heat losses. These include (*i*) polishing the calorimeter to reduce radiation loss, (*ii*) surrounding it by an outer container or a jacket of a poor heat conductor to reduce convection and conduction loss, and (*iii*) supporting it on an insulating stand or supports to minimize conduction.

When the losses, despite all precautions, are not small, or where great accuracy is required, an estimate can be made of the temperature that would have been reached, i.e. a 'cooling correction' is made which when added to the observed maximum temperature gives the estimated maximum temperature had no heat been lost. Alternatively the need to make a cooling correction can be eliminated; one way of doing this is explained in section (**b**).

(a) Graphical method

As well as being suitable for electrical heating experiments, this method is convenient when finding the specific heat capacity of a bad thermal conductor (e.g. glass or rubber) by the method of mixtures. In the latter case the hot solid is slow to transfer heat to the calorimeter and water and some time elapses before the mixture reaches its maximum temperature. During this time appreciable cooling occurs even if the calorimeter is lagged.

To make the cooling correction, the temperature is taken at half-minute intervals starting *just before* the hot solid is added to the calorimeter and ending when the temperature has fallen by at least 1 K from its observed maximum value. A graph of temperature against time is plotted. In that shown in Fig. 5.9, θ_1 is the initial temperature of the calorimeter and contents (i.e. room temperature) and θ_2 is the observed maximum temperature. The dotted line shows how the temperature might have risen if no heat were lost.

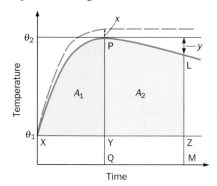

Fig. 5.9

The cooling correction required is x. To obtain it, PQ is drawn through the top of the curve parallel to the temperature axis and similarly LM further along the curve so that y is 1 K. XYZ is then drawn through θ_1, parallel to the time axis. The areas A_1 and A_2 are found by counting the squares on the graph paper and it can be shown that the cooling correction is given by

$$x = \frac{A_1}{A_2} \times y$$

where $y = 1 \text{ K}$ for convenience. The estimated maximum temperature is then $\theta_2 + x$.

This method is based on the assumption that the rate of loss of heat is directly proportional to the temperature difference between the body (e.g. calorimeter) and its surroundings. This is true for heat loss by (*i*) conduction (see p. 72), (*ii*) convection so long as it is forced (i.e. a draught) or, if natural, provided the temperature difference is small (see below) and (*iii*) radiation if the temperature excess is small.

In electrical heating experiments, temperature–time readings are taken during and immediately after the heating, and the cooling correction obtained from a graph as explained above.

If a heat sensor and datalogger are available, the temperature in this type of experiment can be monitored continuously and a cooling curve like that shown in Fig. 5.9 can be displayed on a computer screen in real time.

(b) Initial cooling method

If the calorimeter and its contents are cooled to about 5 K below room temperature and then heated steadily during the experiment to about 5 K above, the heat gained from the surroundings during the first half of the time will be nearly equal to that lost to the surroundings during the second half. No cooling correction is then necessary. The method is suitable when finding the specific heat capacity of a liquid by electrical heating.

COOLING LAWS AND TEMPERATURE FALL

(a) Five-fourths power law

For cooling in still air by natural convection the five-fourths power law holds. It states that

$$\text{rate of loss of heat} \propto (\theta - \theta_0)^{5/4}$$

where θ is the temperature of the body in surroundings at temperature θ_0. If the temperature excess $(\theta - \theta_0)$ is small, the relation becomes approximately linear.

(b) Newton's law of cooling

Under conditions of forced convection, i.e. in a steady draught, Newton's law applies. It states that

$$\text{rate of loss of heat} \propto (\theta - \theta_0)$$

This is true for quite large temperature excesses.

(c) Rate of fall of temperature

As well as the temperature excess, the rate of loss of heat from a body depends on the area and nature of its surface (i.e. whether it is dull or shiny). Hence for a body having a uniform temperature θ and a surface area A we can say, if Newton's law holds,

$$\text{rate of loss of heat} = kA(\theta - \theta_0)$$

where k is a constant depending on the nature of the surface.

If the temperature θ of the body falls we can also write

rate of loss of heat
$$= mc \times \text{rate of fall of temperature}$$

where m is the mass of the body and c is its specific heat capacity. Hence

rate of fall of temperature
$$= kA(\theta - \theta_0)/(mc)$$

The mass of a body is proportional to its volume and so the rate of fall of temperature of a body is proportional to the ratio of its surface area to its volume, i.e. is inversely proportional to a linear dimension. A small body, therefore, cools faster than a large one (its temperature falls faster), as everyday experience confirms. In calorimeter experiments the use of large apparatus, etc., minimizes the effect of errors due to heat loss.

LATENT HEAT

The heat that a body absorbs in melting, evaporating or sublimating, and gives out in freezing or condensing is called **latent** (hidden) **heat** because it does not produce a change of temperature in the body — it causes a change of state or phase. When water is boiling, for example, its temperature remains steady at 100 °C (at standard atmospheric pressure) although heat, called **latent heat of vaporization**, is being supplied to it. Similarly the temperature of liquid

naphthalene stays at 80 °C while it is freezing; there is no fall of temperature until all the liquid has solidified, but heat, called **latent heat of fusion**, is still being given out by the liquid.

The kinetic theory sees the supply of latent heat to a melting solid as enabling the molecules to overcome sufficiently the forces between them for the regular crystalline structure of the solid to be broken down. The molecules then have the greater degree of freedom and disorder that characterize the liquid state. So, whereas heat that increases the kinetic energy component of molecular internal energy causes a temperature rise, the supply of latent heat is regarded as increasing the potential energy component since it allows the molecules to move both closer together and farther apart.

When vaporization of a liquid occurs a large amount of energy is needed to separate the molecules and allow them to move around independently as gas molecules. In addition some energy is required to enable the vapour to expand against the atmospheric pressure. The energy for both these operations is supplied by the latent heat of vaporization and, like latent heat of fusion, we regard it as increasing the potential energy of the molecules.

(a) Specific latent heat of fusion

The specific latent heat of fusion is the quantity of heat required to change unit mass of a substance from solid to liquid without change of temperature.

It is denoted by the symbol l_m and is measured in J kg^{-1} or J g^{-1}.

The specific latent heat of fusion of ice can be determined by the method of mixtures. A calorimeter of mass m_c is two-thirds filled with a mass m_w of water warmed to about 5 K above room temperature. The temperature θ_1 of the water is noted, then a sufficient number of small pieces of ice, carefully dried on blotting paper, are added one at a time and the mixture stirred, until the temperature is about 5 K below room temperature. The lowest temperature θ_2 is noted. The calorimeter and contents are then weighed to find the mass of ice m.

The heat given out by the calorimeter and warm water in cooling from θ_1 to θ_2 does two things. First it supplies the latent heat needed to melt the ice at 0 °C to water at 0 °C and second it provides the heat to raise the now melted ice from 0 °C to the final temperature of the mixture θ_2. Hence

heat given out		heat used to		heat used to warm
by calorimeter	=	melt ice at	+	melted ice from
and water cooling		0 °C		0 °C to θ_2

If c_c and c_w are the specific heat capacities of the calorimeter and water respectively and l_m is the specific latent heat of fusion of ice then

$$m_c c_c(\theta_1 - \theta_2) + m_w c_w(\theta_1 - \theta_2) = m l_m + m c_w(\theta_2 - 0)$$

Simplifying,

$$(m_c c_c + m_w c_w)(\theta_1 - \theta_2) = m(l_m + c_w \theta_2)$$

$$\therefore \ l_m + c_w \theta_2 = \frac{(m_c c_c + m_w c_w)(\theta_1 - \theta_2)}{m}$$

$$\text{and } l_m = \frac{(m_c c_c + m_w c_w)(\theta_1 - \theta_2)}{m} - c_w \theta_2$$

For ice the accepted value of l_m is 334 J g^{-1}. No cooling correction is necessary (see *Initial cooling method*, p. 68) but the temperature of the mixture must not be taken more than 5 K below room temperature otherwise water vapour in the air may condense to form dew on the calorimeter and give up latent heat to it.

(b) Specific latent heat of vaporization

The specific latent heat of vaporization is the quantity of heat required to change unit mass of a substance from liquid to vapour without change of temperature.

It is denoted by l_v and measured in J kg^{-1} or J g^{-1}.

A value can be found for l_v by a continuous-flow method using the apparatus of Fig. 5.10. The liquid is heated electrically by a coil carrying a steady current I and having a p.d. V across it. Vapour passes down the inner tube of a condenser where it is changed back to liquid by cold water flowing through the outer tube.

After the liquid has been boiling for some time it becomes surrounded by a 'jacket' of vapour at its boiling-point and a steady state is reached when the rate of vaporization equals the rate of condensation. All the electrical energy supplied is then used to supply latent heat to the liquid (and none to raise its temperature) and to make good any heat loss from the 'jacket'. If a mass m of liquid is now collected in time t from the condenser, we have

$$ItV = m l_v + h$$

where l_v is the specific latent heat of vaporization of the liquid and h is the heat lost from the 'jacket' in time t. The 'jacket' of vapour makes h small and if it is neglected l_v can be found. Alternatively it may be eliminated by a second determination with a different power input.

The specific latent heat of vaporization of water is 2.3×10^3 J g^{-1}.

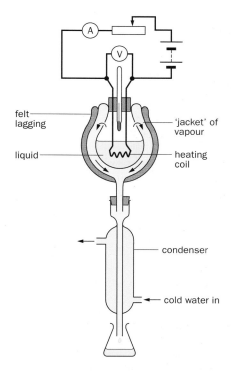

Fig. 5.10 Continuous-flow method for finding specific latent heat of vaporization

(c) Bonding energy and latent heat

The bonding energy for two atoms or molecules is the amount of energy that has to be supplied to pull them apart and make their potential energy zero. In Fig. 2.7*b* (p. 18) it equals E_0, the minimum value of the p.e. at the equilibrium separation r_0.

If a molecule has n neighbours, the total bonding energy for each molecule with its neighbours is nE_0. For a mole, containing L molecules (the Avogadro constant), the energy needed to break all the bonds would be $\frac{1}{2}nLE_0$ (the $\frac{1}{2}$ is necessary because, for any pair of molecules A and B, A is regarded as a neighbour of B and then B as a neighbour of A — otherwise each bond would be considered twice).

Taking the conversion of water to steam as an example, 2.3×10^6 J are needed to vaporize 1 kg of water (at 373 K), so, since 1 mole of water has mass 0.018 kg, the energy to be supplied is given by

$$\frac{1}{2}nLE_0 = (2.3 \times 10^6 \text{ J kg}^{-1}) \times (0.018 \text{ kg mol}^{-1})$$
$$= 4.14 \times 10^4 \text{ J mol}^{-1}$$

But $L = 6.02 \times 10^{23}$ mol^{-1} and for a liquid $n \approx 10$, therefore

$$E_0 = \frac{2 \times 4.14 \times 10^4 \text{ J mol}^{-1}}{10 \times 6.02 \times 10^{23} \text{ mol}^{-1}}$$

$$= 1.4 \times 10^{-20} \text{ J}$$

This is a rough estimate of the bonding energy of water.

HEAT CALCULATIONS

Example 1. A piece of copper of mass 100 g is heated to 100 °C and then transferred to a well-lagged copper can of mass 50.0 g containing 200 g of water at 10.0 °C. Neglecting heat loss, calculate the final steady temperature of the water after it has been well stirred. Take the specific heat capacities of copper and water as 4.00×10^2 J kg^{-1} K^{-1} and 4.20×10^3 J kg^{-1} K^{-1} respectively.

Let the final steady temperature $= \theta$
Fall in temperature of piece of copper $= (100 °C − \theta)$
Rise in temperature of can and water $= (\theta − 10 °C)$

Expressing masses in kg,

heat given out by copper $= 0.1 \times 400 \times (100 − \theta)$ J
heat received by copper can $= 0.05 \times 400 \times (\theta − 10)$ J
heat received by water $= 0.2 \times 4200 \times (\theta − 10)$ J

Heat given out $=$ heat received

Therefore

$$40(100 − \theta) = 20(\theta − 10) + 840(\theta − 10)$$
$$= (20 + 840)(\theta − 10)$$
$$4000 − 40\theta = 860\theta − 8600$$
$$900\theta = 12\,600$$
$$\theta = 14.0 °C$$

Example 2. When a current of 2.0 A is passed through a coil of constant resistance 15 Ω immersed in 0.5 kg of water at 0 °C in a vacuum flask, the temperature of the water rises to 8 °C in 5 minutes. If instead the flask originally contained 0.25 kg of ice and 0.25 kg of water, what current must be passed through the coil if this mixture is to be heated to the same temperature in the same time? (Specific heat capacity of water $= 4.2 \times 10^3$ J kg^{-1} K^{-1}; specific latent heat of fusion of ice $= 3.3 \times 10^5$ J kg^{-1}.)

Assuming no heat is lost from the vacuum flask, then in time t

$$\begin{matrix} \text{electrical} & \text{heat} & \text{heat received} \\ \text{energy} & = \text{received} + \text{by vacuum} \\ \text{supplied} & \text{by water} & \text{flask} \end{matrix} \qquad (1)$$

But, electrical energy supplied $= I^2Rt$ joules where $I = 2$ A, $R = 15$ Ω and $t = 5 \times 60 = 300$ s. Also, heat received by water $= mc(\theta_1 − \theta_2)$ where $m = 0.5$ kg, $c = 4.2 \times 10^3$ J kg^{-1} K^{-1} and $\theta_1 − \theta_2 = 8 °C − 0 °C = 8$ K. If h is the heat received by the vacuum flask in time t, then from (1)

$$(2 \text{ A})^2(15 \text{ Ω})(300 \text{ s})$$
$$= (0.5 \text{ kg})(4.2 \times 10^3 \text{ J kg}^{-1} \text{ K}^{-1})(8 \text{ K}) + h$$
$$(4 \times 15 \times 300) \text{ J} = (0.5 \times 4.2 \times 10^3 \times 8) \text{ kg J kg}^{-1}\text{K}^{-1}\text{K} + h$$
$$\therefore \quad h = (18\,000 − 16\,800) \text{ J} = 1200 \text{ J}$$

In the second part, in the same time t,

$$\begin{matrix} \text{electrical} & \text{heat to} & \text{heat to} & \text{heat received} \\ \text{energy} & = \text{melt } 0.25 \text{ kg} + \text{warm } 0.5 \text{ kg} + \text{by vacuum} \\ \text{supplied} & \text{ice at } 0 °C & \text{water from} & \text{flask} \\ & & 0 °C \text{ to } 8 °C & (\text{i.e. } h) \end{matrix}$$

$$\therefore \quad I^2(15 \text{ Ω})(300 \text{ s}) = (0.25 \text{ kg})(3.3 \times 10^5 \text{ J kg}^{-1})$$
$$+ (0.5 \text{ kg})(4.2 \times 10^3 \text{ J kg}^{-1} \text{ K}^{-1})(8 \text{ K})$$
$$+ 1200 \text{ J}$$

where I is the current required to warm the ice and water to 8 °C in 300 s.

$$\therefore \quad I^2(15 \times 300) \text{ Ω s} = (0.25 \times 3.3 \times 10^5) \text{ J}$$
$$+ (0.5 \times 4.2 \times 10^3 \times 8) \text{ J} + 1200 \text{ J}$$
$$\therefore \quad I^2(4500) \text{ Ω s} = 82\,500 \text{ J} + 16\,800 \text{ J} + 1200 \text{ J} = 100\,500 \text{ J}$$
$$I^2 = \frac{100\,500}{4500} \frac{\text{J}}{\text{Ω s}} = 22.3 \text{ V C Ω}^{-1} \text{ s}^{-1} = 22.3 \text{ A}^2$$
$$(\text{since A} = \text{C s}^{-1} = \text{V Ω}^{-1})$$
$$\therefore \quad I = 4.7 \text{ A}$$

EXPANSION OF SOLIDS

(a) Linear expansion

The change of length that occurs with temperature change in a solid has to be allowed for in the design of many devices. The variation is described by the **linear expansivity** α. If a solid of length l increases in length by δl owing to a temperature rise $\delta \theta$, α for the material is defined by the equation

$$\alpha = \frac{\delta l}{l} \cdot \frac{1}{\delta \theta}$$

In words, α is **the fractional increase of length** (i.e. $\delta l/l$) **per unit rise of temperature**. The unit of α is K^{-1} since δl and l have the same units (metres) and so $\delta l/l$ is a ratio.

As the temperature rise $\delta \theta$ tends to zero, α approaches the linear expansivity at a particular temperature and experiment shows that its value for a given material is not constant but varies slightly with temperature. Mean values are therefore obtained over a temperature range and in accurate work this range is stated. For ordinary purposes α can be assumed constant in the range 0 to 100 °C. The mean value for copper (at room temperature) is 1.7×10^{-5} K^{-1}; a copper rod 1 metre long therefore increases in length by 1.7×10^{-5} metre for every 1 K temperature rise.

A useful expression is obtained if we consider a solid of original length l_0 which increases to l_θ for a temperature *rise* of θ. Replacing δl by $l_\theta − l_0$ and $\delta \theta$ by θ in the expression for α, we get

$$\alpha = \frac{l_\theta − l_0}{l_0 \theta}$$

Rearranging,

$$l_\theta − l_0 = l_0 \alpha \theta \quad \text{and}$$

$$l_\theta = l_0(1 + \alpha \theta)$$

Note that l_0 is the *original* length and, since the values of α are very small, it need not be the length at 0 °C (unlike R_0 in the temperature coefficient of resistance formula $R_\theta = R_0(1 + \alpha \theta)$ on p. 49, which is generally taken to be the resistance at 0 °C).

If the temperature of a 2 metre long copper rod rises from 15 °C to 25 °C then $l_0 = 2$ m, $\theta = (25 - 15)$ K = 10 K, $\alpha = 1.7 \times 10^{-5}$ K^{-1} and

$$
\begin{aligned}
l_\theta - l_0 &= l_0 \alpha \theta \\
&= (2 \text{ m})(1.7 \times 10^{-5} \text{ K}^{-1})(10 \text{ K}) \\
&= 2 \times 1.7 \times 10^{-5} \times 10 \text{ m K}^{-1} \text{ K} \\
&= 3.4 \times 10^{-4} \text{ m} = 0.34 \text{ mm}
\end{aligned}
$$

Thermal expansion of a solid can be explained on the atomic scale with the help of Fig. 5.11a, which shows that the repelling forces between atoms increases more rapidly than the attractive forces as the atomic separation varies. A rise of temperature increases the amplitude of vibration of the atoms about their equilibrium position but this will be greater on the extension side of that position, for the reason just given, and the average separation of the atoms increases. At low temperatures the amplitudes of oscillation are small and symmetrical, Fig. 5.11b, about the equilibrium position and so we would not expect expansion with increase of temperature — a fact confirmed by experiment. A further conclusion from this argument is that linear expansivities should increase with rising temperature, as they do.

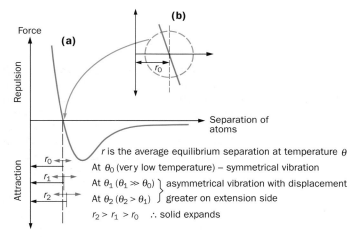

r is the average equilibrium separation at temperature θ
At θ_0 (very low temperature) – symmetrical vibration
At θ_1 ($\theta_1 \gg \theta_0$) ⎫ asymmetrical vibration with displacement
At θ_2 ($\theta_2 > \theta_1$) ⎭ greater on extension side
$r_2 > r_1 > r_0$ ∴ solid expands

Fig. 5.11

(b) Area expansion

The change of area of a surface with temperature change is described by the **area** or **superficial expansivity** β. If an area A increases by δA owing to a temperature rise $\delta\theta$ then β is given by

$$
\beta = \frac{\delta A}{A} \cdot \frac{1}{\delta\theta}
$$

In words, β is **the fractional increase of area** (i.e. $\delta A/A$) **per unit rise of temperature**.

The variation of area with temperature is given by an equation similar to that for linear expansion, that is

$$
A_\theta = A_0(1 + \beta\theta)
$$

where A_0 and A_θ are the original and new areas respectively and θ is the temperature rise.

It can be shown (see below) that, for a given material, $\beta \approx 2\alpha$.

Consider a square plate of side l_0, Fig. 5.12. We have $A_0 = l_0^2$.

A temperature rise of θ causes the length of each side to become $l_0(1 + \alpha\theta)$ if the material is isotropic, i.e. has the same properties in all directions. The new area A_θ is

$$
\begin{aligned}
A_\theta &= l_0^2(1 + \alpha\theta)^2 \\
&= l_0^2(1 + 2\alpha\theta + \alpha^2\theta^2)
\end{aligned}
$$

Now $\alpha^2\theta^2$ is very small compared with $2\alpha\theta$ and since $A_0 = l_0^2$ we have

$$
A_\theta \approx A_0(1 + 2\alpha\theta)
$$

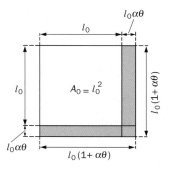

Fig. 5.12

Comparing this with $A_\theta = A_0(1 + \beta\theta)$, it follows that

$$
\beta \approx 2\alpha
$$

(c) Volume expansion

Changes of volume of a material with temperature are expressed by the **cubic expansivity** γ. If a volume V increases by δV for a temperature rise $\delta\theta$ then γ is given by

$$
\gamma = \frac{\delta V}{V} \cdot \frac{1}{\delta\theta}
$$

In words, γ is **the fractional increase of volume** (i.e. $\delta V/V$) **per unit rise of temperature**.

The equation

$$
V_\theta = V_0(1 + \gamma\theta)
$$

is also useful and it can be shown that, for a given material, $\gamma \approx 3\alpha$. The proof involves calculating the volume change of a cube in terms of the linear expansion of each side, in a similar manner to that adopted for areas. Cubic and area expansivities for solids are not given in tables of physical constants since they are readily calculated from linear expansivities. The comment on the constancy of α (p. 70) also applies to β and γ.

A hollow body such as a bottle expands as if it were solid throughout, otherwise it would not retain the same shape when heated.

THERMAL STRESS

Forces are created in a structure when thermal expansion or contraction is resisted. An idea of the size of such forces can be obtained by considering a metal rod of initial length l_0, cross-section area A and linear expansivity α, supported between two fixed end plates, Fig. 5.13. If the temperature of the rod is raised by $\theta\,°C$, it tries to expand but is prevented by the plates and a compressive stress arises in it.

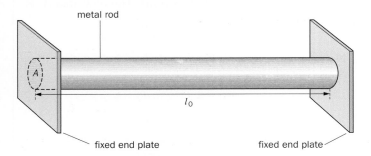

metal rod

A

l_0

fixed end plate   fixed end plate

Fig. 5.13

Removing one of the plates would allow the rod to expand freely and its new length l_θ would be $l_0 + l_0\alpha\theta$. We can therefore consider the plate, when fixed in position, as exerting force F on the rod and reducing its length from l_θ to l_0. Then

$$\text{compressive strain} = \frac{\text{change in length}}{\text{original length}}$$

$$= \frac{l_\theta - l_0}{l_\theta} = \frac{l_0\alpha\theta}{l_\theta}$$

Also,

$$\text{compressive stress} = \frac{F}{A}$$

If the Young modulus for the material of the rod is E (the compressive modulus is the same as the tensile one for small compressions) then

$$E = \frac{\text{stress}}{\text{strain}} = \frac{Fl_\theta}{Al_0\alpha\theta}$$

The difference between l_θ and l_0 would be small compared with either and so to a good approximation $l_\theta = l_0$. Therefore

$$E = \frac{F}{A\alpha\theta} \quad \text{and} \quad F = EA\alpha\theta$$

For a steel girder with $E = 2.0 \times 10^{11}$ Pa (N m^{-2}), $A = 100$ cm$^2 = 10^{-2}$ m^2, $\alpha = 1.2 \times 10^{-5}$ K^{-1} and a temperature rise $\theta = 20$ K,

$$\begin{aligned} F &= (2 \times 10^{11}\,\text{N m}^{-2}) \times (10^{-2}\,\text{m}^2) \\ &\quad \times (1.2 \times 10^{-5}\,\text{K}^{-1}) \times (20\,\text{K}) \\ &= 2 \times 10^{11} \times 10^{-2} \times 1.2 \times 10^{-5} \times 20\,\text{N m}^{-2}\,\text{m}^2\,\text{K}^{-1}\,\text{K} \\ &= 4.8 \times 10^{5}\,\text{N} \end{aligned}$$

This is a sizeable force.

The original length of the rod does not affect the force but long rods tend to buckle at lower compressive stresses than short ones.

Thermal stress is put to good use in the technique of shrink-fitting in which, for example, a large gear wheel is fitted on to a shaft of the same material. The diameter of the central hole in the wheel is smaller at room temperature than the outside diameter of the shaft. If the shaft is cooled with solid carbon dioxide ('dry ice') at $-78\,°C$ it can be fitted into the wheel. At room temperature the shaft is under compression and the wheel under tension and a tight-fitting joint results.

THERMAL CONDUCTIVITY

Heat transfer, i.e. the passage of energy from a body at a higher temperature to one at a lower temperature, occurs by the three processes of conduction, convection and radiation, although evaporation and condensation may often play an important part. In some cases the aim of the heat engineer is to encourage heat flow (as in a boiler) while in others it is to minimize it (e.g. lagging a house). Here we shall discuss **conduction**, i.e. the transfer of energy due to the temperature difference between neighbouring parts of the same body.

(a) Definition

Consider a *thin* slab of material of thickness δx and uniform cross-section area A between whose faces a *small* temperature difference $\delta\theta$ is maintained, Fig. 5.14. If a quantity of heat δQ passes through the slab by conduction in time δt, the **thermal conductivity** k of the material is defined by the equation

$$\frac{\delta Q}{\delta t} = -kA\frac{\delta\theta}{\delta x}$$

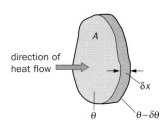

A

direction of heat flow

δx

θ $\theta - \delta\theta$

Fig. 5.14

The negative sign indicates that heat flows towards the lower temperature, i.e. as x increases, θ decreases, making $\delta\theta/\delta x$ — called the **temperature gradient** — negative. Inserting the negative sign ensures that $\delta Q/\delta t$ and k will be positive. In words, we may define k as **the rate of flow of heat through a material per unit area, per unit temperature gradient**.

The unit of k is J s^{-1} m^{-1} K^{-1} or W m^{-1} K^{-1} as can be seen by inserting units for the various quantities in the previous expression and remembering that 1 watt is 1 joule per second. Its value for copper is 390 W m⁻¹ K⁻¹ and for polystyrene 0.08 W m⁻¹ K⁻¹ (both at room temperature).

When $\delta\theta \to 0$, k approaches the thermal conductivity at a particular temperature and experiment shows that its value for a given material varies slightly with temperature. If measurements are not made over too great a temperature range, a constant mean value for k is usually assumed.

In the limiting case when $\delta x \to 0$, a cross-section is then being considered and the equation defining k can be more precisely written in calculus notation as

$$\frac{dQ}{dt} = -kA\frac{d\theta}{dx}$$

(b) Temperature gradients

When heat has been passing along a conductor for some time from a source of fixed high temperature, a steady state may be reached with the temperature at each point of the conductor becoming constant.

In the *unlagged* bar of Fig. 5.15*a* the quantity of heat passing in a given time through successive cross-sections decreases because of heat loss from the sides. The lines of heat flow are divergent and the temperature falls faster near the hotter end. For steady-state conditions a graph of temperature θ against distance x from the hot end is as shown. The temperature gradient at any point is given by the slope of the tangent at that point (in calculus notation by $d\theta/dx$).

In a *lagged* bar whose sides are well wrapped with a good insulator, Fig. 5.15*b*, heat loss from the sides is negligible and the rate of flow of heat is the same all along the bar. The lines of heat flow are parallel and, in the steady state, the temperature falls at a constant rate as shown. The temperature gradient in this case is the slope of the graph, i.e. $\delta\theta/\delta x$.

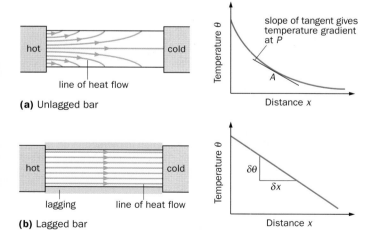

(a) Unlagged bar

(b) Lagged bar

Fig. 5.15 Conduction of heat down an unlagged and a lagged bar

There is a useful expression applicable to many simple problems in which the *lines of heat flow are parallel*, for example in a lagged bar and in a plate of large cross-section area. Why the latter? Consider a conductor of length x, cross-section area A and thermal conductivity k whose opposite ends are maintained at temperatures θ_2 and θ_1 ($\theta_2 > \theta_1$). From what has been said in the previous paragraph it follows that the quantity of heat Q passing any point in time t when the **steady state** has been reached is given by

$$\frac{Q}{t} = kA\left(\frac{\theta_2 - \theta_1}{x}\right)$$

We will now use this expression, which is sometimes called **Fourier's law**.

(c) Composite slab problem

Suppose we wish to find the rate of flow of heat through a plaster ceiling which measures $5\,m \times 3\,m \times 15\,mm$, (*i*) without and (*ii*) with a 45 mm thick layer of insulating fibreglass if the inside and outside surfaces are at the surrounding air temperatures of 15 °C and 5 °C respectively.

($k_{plaster} = 0.60$ W m⁻¹ K⁻¹ and $k_{fibreglass} = 0.040$ W m⁻¹ K⁻¹)

Assuming steady states are reached and lines of heat flow are parallel we can use rate of flow of heat = $Q/t = kA(\theta_2 - \theta_1)/x$. We have $A = 5\,m \times 3\,m = 15\,m^2$.

(*i*) Without fibreglass, Fig. 5.16*a*, $x = 15\,mm = 0.015\,m$,

$$\frac{Q}{t} = \frac{(0.60\ \text{W m}^{-1}\ \text{K}^{-1})(15\ \text{m}^2) \times (10\ \text{K})}{(0.015\ \text{m})}$$

$$= \left(\frac{0.60 \times 15 \times 10}{0.015}\right)\frac{\text{W m}^{-1}\ \text{K}^{-1}\ \text{m}^2\ \text{K}}{\text{m}}$$

$$= 6.0 \times 10^3\ \text{W}$$

(In practice it will be very much less than this, see later.)

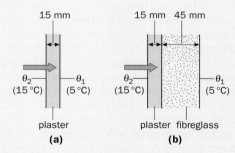

Fig. 5.16

(*ii*) With fibreglass, Fig. 5.16*b*. Let the temperature of the plaster–fibreglass boundary be θ. **The rate of flow of heat is the same through both materials.**

$$\therefore \frac{Q}{t} = \frac{(0.60\ \text{W m}^{-1}\ \text{K}^{-1})(15\ \text{m}^2)(15\ °C - \theta)}{(0.015\ \text{m})}$$

$$= \frac{(0.04\ \text{W m}^{-1}\ \text{K}^{-1})(15\ \text{m}^2)(\theta - 5\ °C)}{(0.045\ \text{m})}$$

Solving for θ we get

$$\theta = 14.8\ °C$$

Substituting θ in the first equation,

$$\frac{Q}{t} = \frac{0.60 \text{ W m}^{-1} \text{ K}^{-1} \times 15 \text{ m}^2 \times (15 - 14.8) \text{ K}}{0.015 \text{ m}}$$

$$= 1.2 \times 10^2 \text{ W}$$

The example above shows that when heat flows through a composite slab the temperature fall per mm is greater across the poorer conductor. This is of practical importance where a good conductor is in contact with a layer of a bad conductor; the latter controls the rate of conduction of heat through the good conductor. If, in calculations, it is assumed that the surface of a good conductor is at the same temperature as that of the surrounding air, values of heat flow will be obtained that are of the order of one hundred times too large. In Fig. 5.17, for example, most of the temperature drop occurs in the layer of gas between the flame and the boiler plate and in any scale deposited on the plate by the water.

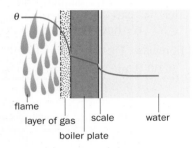

flame
layer of gas | scale | water
boiler plate

Fig. 5.17 Temperature drop across a boiler plate

(d) Mechanisms of thermal conduction

In solids (and liquids) two processes seem to be involved. The first concerns atoms and the second free electrons.

Atoms at a higher temperature vibrate more vigorously about their equilibrium positions in the lattice than their colder neighbours. But because they are coupled to them by interatomic bonds they pass on some of their vibratory energy and cause them to vibrate more energetically as well. These in turn affect other atoms and thermal conduction occurs. However, the process is generally *slow* because atoms, compared with electrons, are massive and the increases in vibratory motion are therefore fairly small. Consequently materials that are electrical insulators, in which this is the main conduction mechanism, are usually poor thermal conductors.

This transfer of vibratory energy is often regarded as resembling the passage of elastic waves through the material. As light waves are considered to have a dual nature, sometimes behaving as particles called photons, so too are these elastic waves considered to have particle-like forms called **phonons**, and thermal conduction is said to be due to phonons having collisions with, and transferring energy to, atoms in the lattice.

The second process concerns materials with a supply of free electrons. In these the electrons share in any gain of energy due to temperature rise of the material and their velocities increase much more than those of the atoms in the lattice, since they are considerably lighter. They are able to move over larger distances and pass on energy *quickly* to cooler parts. Electrical conductors in which this mechanism predominates are therefore good thermal conductors; free electrons are largely responsible for both properties.

It should not be thought that only metals are good thermal conductors: phonons can be a very effective means of heat transfer, especially at low temperatures. At about $-180\ ^\circ\text{C}$ synthetic sapphire (Al_2O_3) is a better thermal conductor than copper.

(e) Heating buildings

This is considered in Chapter 10, where the terms **U-value** and **thermal resistance**, used by heating engineers, are explained.

QUESTIONS

Temperature; thermometers

1. Explain what is meant by a **scale of temperature** and how a temperature is defined in terms of a specified property.

   When a particular temperature is measured on scales based on different properties it has a different numerical value on each scale except at certain points. Explain why this is so and state
   a) at what points the values agree, and
   b) what scale of temperature is used as a standard.

   Explain the principles of two different types of thermometer, one of which is suitable for measuring a rapidly varying temperature and the other for measuring a steady temperature whose value is required to a high degree of accuracy. Give reasons for your choice of thermometer in each case. Experimental details are not required.

2. The temperature of the water in a hot bath is measured with a thermometer. For each of the statements in the table below, indicate whether the statement is true or false.

   A thermocouple has the cold junction immersed in an ice–water mix at 0 °C. When the hot junction is in boiling water, the e.m.f. is 1.65 mV. Estimate the temperature of the hot junction when the e.m.f. is 1.47 mV.

   (L, A PH3, Jan 1996)

	True	False
The thermometer will *not* give the correct temperature unless it is in thermal contact with the water.		
The thermometer will *not* give the correct temperature unless it is in thermal equilibrium with the water.		
The thermometer will *not* give the correct temperature unless it, itself, is in a thermal equilibrium state.		

Specific heat capacity; latent heat

3. A current of 2.50 A passing through a heating coil immersed in 180 g of paraffin (specific heat capacity 2.00 J g^{-1} K^{-1}) contained in a 100 g calorimeter (specific heat capacity 0.400 J g^{-1} K^{-1}) raises the temperature from 5 °C below room temperature to 5 °C above room temperature in 100 s. What should be the reading of a voltmeter connected across the heating coil?

4. A student pours 500 g of water into an aluminium saucepan of mass 1.20 kg, heats it over a steady flame and records the temperature as it heats up. The temperatures are plotted as shown in Fig. 5.18.

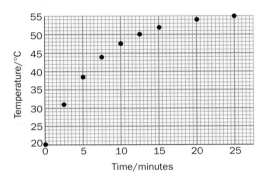

Fig. 5.18

Calculate the total heat capacity of the saucepan and water.
Specific heat capacity of water = 4200 J kg^{-1} K^{-1}.
Specific heat capacity of aluminium = 900 J kg^{-1} K^{-1}.

Find the rate of rise of water temperature at the beginning of the heating process.

Hence find the rate at which energy is supplied to the saucepan and water.

Explain why the rate at which the temperature rises slows down progressively as the heating process continues.

(*L, A PH3, June 1996*)

5. A student seals 200 g of ice-cold water in a glass vacuum (Thermos) flask and finds that it warms up by 3.5 K in one hour. The specific heat capacity of water is 4200 J kg^{-1} K^{-1}.

Calculate the average rate of heat flow into the flask in watts.

To check this result over a longer period, the student fills the flask with equal amounts of ice and water, all at 0 °C, and leaves it for four hours. The specific latent heat (enthalpy) of fusion of ice is 0.33 MJ kg^{-1}. How much ice would you expect to have melted at the end of the four hours?

The student found instead that the glass flask had collapsed into small pieces. Suggest a reason for the pressure inside the glass flask to drop sufficiently for the collapse to occur.

(*L, A PH3, June 1998*)

Expansion

6. The steel cylinder of a car engine has an aluminium alloy piston. At 15 °C the internal diameter of the cylinder is exactly 8.0 cm and there is an all-round clearance between the piston and the cylinder wall of 0.050 mm. At what temperature will they fit perfectly? (Linear expansivities of steel and the aluminium alloy are 1.2×10^{-5} and 1.6×10^{-5} K^{-1} respectively.)

7. The height of the mercury column in a barometer is 76.46 cm as read at 15 °C by a brass scale which was calibrated at 0 °C. Calculate the error caused by the expansion of the scale and hence find the true height of the column. (Linear expansivity of brass is 1.900×10^{-5} K^{-1}.)

8. Calculate the minimum tension with which platinum wire of diameter 0.10 mm must be mounted between two points in a stout Invar frame if the wire is to remain taut when the temperature rises 100 °C. Platinum has linear expansivity 9.0×10^{-6} K^{-1} and the Young modulus 1.7×10^{11} Pa. The thermal expansion of Invar may be neglected.

Thermal conductivity

9. A cubical container full of hot water at a temperature of 90 °C is completely lagged with an insulating material of thermal conductivity 6.4×10^{-4} W cm^{-1} K^{-1} (6.4×10^{-2} W m^{-1} K^{-1}). The edges of the container are 1.0 m long and the thickness of the lagging is 1.0 cm. Estimate the rate of flow of heat through the lagging if the external temperature of the lagging is 40 °C. Mention any assumptions you make in deriving your result.

Discuss qualitatively how your result will be affected if the thickness of the lagging is increased considerably, assuming that the temperature of the surrounding air is 18 °C.

10. Explain what is meant by **temperature gradient**.

An ideally lagged compound bar 25 cm long consists of a copper bar 15 cm long joined to an aluminium bar 10 cm long and of equal cross-section area. The free end of the copper is maintained at 100 °C and the free end of the aluminium at 0 °C. Calculate the temperature gradient in each bar when steady state conditions have been reached.
Thermal conductivity of copper = 3.9 W cm^{-1} K^{-1}.
Thermal conductivity of aluminium = 2.1 W cm^{-1} K^{-1}.

11. The walls of a container used for keeping objects cool consist of two thicknesses of wood 0.50 cm thick separated by a space 1.0 cm wide packed with a poorly conducting material. Calculate the rate of flow of heat per unit area into the container if the temperature difference between the internal and external surfaces is 20 °C. (Thermal conductivity of wood = 2.4×10^{-3} W cm^{-1} K^{-1}; of the poorly conducting material = 2.4×10^{-4} W cm^{-1} K^{-1}.)

12. The ends of a straight uniform metal rod are maintained at temperatures of 100 °C and 20 °C, the room temperature being below 20 °C. Draw sketch-graphs of the variation of the temperature of the rod along its length when the surface of the rod is
a) lagged,
b) coated with soot,
c) polished.

Give a qualitative explanation of the form of the graphs.

A liquid in a glass vessel of wall area 595 cm^2 and thickness 2.0 mm is agitated by a stirrer driven at a uniform rate by an electric motor rated at 100 W. The efficiency of conversion of electrical to mechanical energy in the motor is 75%. The temperature of the outer surface of the glass is maintained at 15.0 °C. Estimate the equilibrium temperature of the liquid, stating any assumptions you make.
Thermal conductivity of glass = 0.840 W m^{-1} K^{-1}.

6

Optical properties

INTRODUCTION

The scientific study of light and optical materials is involved in the making of spectacles, cameras, projectors, binoculars, microscopes and telescopes. The most important optical materials are the various kinds of glass, but many others such as plastics, Polaroid, synthetic and natural crystals have increasingly useful applications.

In this chapter we shall consider the behaviour of certain optical components and instruments. Light will be treated as a form of energy which travels in straight lines called **rays**, a collection of rays being termed a **beam**. The ray treatment of light is known as geometrical optics and is developed from

- rectilinear propagation, i.e. straight-line travel;
- the laws of reflection;
- the laws of refraction.

When light comes to be regarded as waves it will be seen that shadows cast by objects are not as sharp as rectilinear propagation suggests. However, at this stage it will be sufficiently accurate to assume that light does travel in straight lines so long as we exclude very small objects and apertures (those with diameters less than about 10^{-2} mm).

REFLECTION AT PLANE SURFACES

(a) Laws of reflection

When light falls on a surface it is partly reflected, partly transmitted and partly absorbed. Considering the part reflected, experiments with rays of light and mirrors show that two laws hold.

1. The angle of reflection equals the angle of incidence, i.e.

$$i_1 = i_2$$

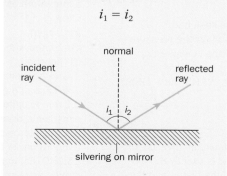

Fig. 6.1

2. The reflected ray is in the same plane as the incident ray and the normal to the mirror at the point of incidence. (The reflected ray is not turned to either side of the normal as seen from the incident ray.)

Note that the angles of incidence and reflection are measured to the normal to the surface and not to the surface itself.

(b) Regular and diffuse reflection

A mirror in the form of a highly polished metal surface or a piece of glass with a deposit of silver on its back surface reflects a high percentage of the light falling on it. If a parallel beam of light falls on a plane (i.e. flat) mirror in the direction IO, Fig. 6.2*a*, it is reflected as a parallel beam in the direction OR and **regular reflection** is said to occur. Most objects, however, reflect light diffusely and the rays in an incident parallel beam are reflected in many directions as in Fig. 6.2*b*. Diffuse reflection is due to the surface of the object not being perfectly smooth like a mirror and although at each point on the surface the laws of reflection are observed, the angle of

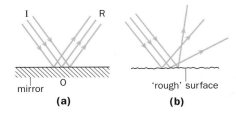

Fig. 6.2 Regular and diffuse reflection

incidence and therefore the angle of reflection varies from point to point.

(c) Rotation of a mirror

When a mirror is rotated through a certain angle, the reflected ray turns through *twice* that angle. This is a useful fact which can be easily proved (see below).

Consider the plane mirror of Fig. 6.3*a*. When it is in position MM, the ray IO, incident at angle *i*, is reflected along OR so that $\angle$ RON = *i*.

The mirror is now rotated through an angle θ to position M'M' and the *direction of the incident ray kept constant*, Fig. 6.3*b*. The angle of incidence $\angle$ ION' becomes $(i + \theta)$ since the angle between the first and second positions of the normals, i.e. $\angle$ NON', is also θ. Let OR' be the new direction of the reflected ray; then $\angle$ R'ON' = $(i + \theta)$. The reflected ray is thus turned through $\angle$ ROR' and

$$\begin{aligned}
\angle \text{ROR'} &= \angle \text{R'ON} - \angle \text{RON} \\
&= \angle \text{R'ON'} + \angle \text{N'ON} \\
&\quad - \angle \text{RON} \\
&= (\theta + i) + \theta - i \\
&= 2\theta \\
&= \text{twice the angle of} \\
&\quad \text{rotation of the mirror}
\end{aligned}$$

(d) Optical lever and light-beam galvanometers

Some sensitive galvanometers use a beam of light in conjunction with a small mirror as a pointer. The arrangement is called an **optical lever**. It uses the 'rotation of a mirror' principle and increases the ability of the meter to detect small currents, i.e. makes it more sensitive.

A tiny mirror is attached to the part (e.g. a coil of wire) of the meter which rotates when a current flows in it, Fig. 6.4. A beam of light from a fixed lamp falls on the mirror and is reflected onto a scale. For a given current, the longer the pointer (i.e. the reflected beam) the greater the deflection observed on the scale.

Besides being almost weightless the arrangement has the additional advantage of doubling the rotation of the moving part since the angle the reflected beam turns through is twice the angle of rotation of the mirror, the direction of the incident ray remaining fixed.

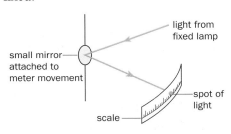

Fig. 6.4 An optical lever

IMAGES IN PLANE MIRRORS

(a) Point object

The way in which the image of a point object is seen in a plane mirror is shown in Fig. 6.5. Rays from the object at O are reflected according to the laws of reflection so that they *appear* to come from a point I behind

the mirror and this is where the observer imagines the image to be. The image at I is called an unreal or **virtual image** because the rays of light do not actually pass through it, they only seem to come from it. It would not be obtained on a screen placed at I as would a **real image**, which is one where rays really do meet. (The image produced on a cinema screen by a projector is a real image.) Rays OA and AE are real rays, but IA is a virtual ray that appears to have travelled a certain path but in fact has not; it gives rise to a virtual image.

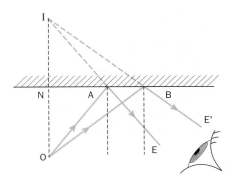

Fig. 6.5 Virtual image of a point object

Everyday observation suggests and experiment shows that **the image in a plane mirror is as far behind the mirror as the object is in front and that the line joining the object to the image is perpendicular to the mirror**, i.e. in Fig. 6.5 ON = NI and OI is at right angles to the mirror. It is possible to show, using the first law of reflection and congruent triangles, that this is so and also that *all* rays from O, after reflection, appear to intersect at I. A perfect image is thus obtained, i.e. all rays from the point object appear to come from *one* point on the image; a plane mirror is one of the few optical devices achieving this.

It is possible for a plane mirror to give a real image. In Fig. 6.6*a* a convergent beam is reflected so that the reflected rays actually pass through a point I in front of the mirror. There is a real image at I which can be picked up on a screen. At the point O, towards which the incident beam was converging before it was intercepted by the mirror, there is considered to be a **virtual object**. Later we will find it useful on occasion to treat a convergent beam in this way. Comparing Figs 6.6*a* and *b*, we see that in the

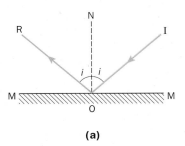

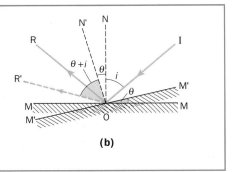

(a)   **(b)**

Fig. 6.3

first, a **convergent beam** regarded as a virtual point object gives a real point image, whilst in the second, a **divergent beam** from a real point object gives a virtual point image. In both cases object and image are equidistant from the mirror.

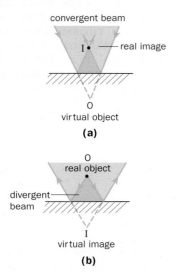

Fig. 6.6

(b) Extended object

Each point on an extended (finite-sized) object produces a corresponding point image. In Fig. 6.7 the image of a point A on the object is at A', the two points being equidistant from the mirror. The image of point B is at B'. If an eye at E views the object directly it sees A on the right of B, but if it observes the image in the mirror A' is on the left. The right-hand side of the object therefore becomes the left-hand side of the image and vice versa. The image is said to be **laterally inverted**, i.e. the wrong way round, as you can check by looking at yourself in a mirror.

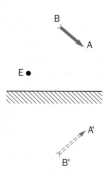

Fig. 6.7

(c) 'No parallax' method of locating images

Suppose the object is a small pin O placed in front of a plane mirror M. To find the position of its virtual image I, a large locating pin P is placed behind M, Fig. 6.8*a*, and moved towards or away from M until P and the image of O always appear to move together when the observer moves his/her head from side to side. P and I are then **coincident**: P is at the position of the image of O. When P and I do not coincide there is relative movement, called **parallax**, between them when the observer's head is moved sideways. The location of I can be achieved more quickly by remembering that if P is farther from M than I, then P appears to move in the *same* direction as the observer, Fig. 6.8*b*.

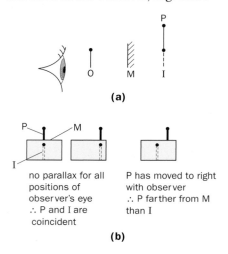

Fig. 6.8

The 'no parallax' method is used to find real as well as virtual images, as we shall see later when curved mirrors and lenses are considered.

(d) Inclined mirrors

Two mirrors M_1 and M_2 at 90° form *three* images of an object P placed between them, Fig. 6.9*a*. I_1 is formed by a single reflection at M_1, I_2 by a single reflection at M_2, and I_3 by reflections at M_1 and M_2. The line joining each image and its object is perpendicularly bisected by the mirror involved (we can think of I_3 as being either the image of I_1 acting as an object for mirror M_2 extended to the left or the image of I_2 as an object for mirror M_1 extended downwards) and so it follows that $OP = OI_1 = OI_2 = OI_3$.

Hence, P, I_1, I_2 and I_3 lie on a circle, centre O.

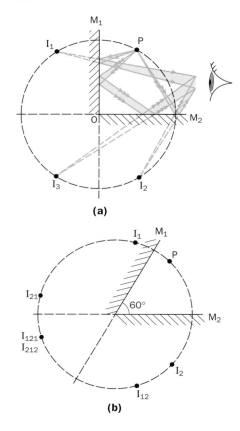

Fig. 6.9

Two mirrors at 60° to each other form *five* images, Fig. 6.9*b*. As the angle between the mirrors decreases the number of images increases and in general for an angle θ (which is such that $360°/\theta$ is an integer) it can be shown that $[(360°/\theta) - 1]$ images are formed. When the mirrors are parallel $\theta = 0°$ and in theory an infinite number of images should be obtained, all lying on a straight line passing through the object and perpendicular to the mirrors, Fig. 6.10. In practice some light is lost at each reflection and only a limited number of images are seen. If the distances of P from M_1 and M_2 are *a* and *b* respectively, can you prove that the separation of the images is successively 2*a*, 2*b*, 2*a*, 2*b*, etc.?

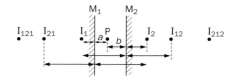

Fig. 6.10

CURVED MIRRORS

We shall consider mainly spherical mirrors, i.e. those which are part of a spherical surface.

(a) Terms and definitions

There are two types of spherical mirror, concave and convex, Figs 6.11*a* and *b*. For a concave mirror the centre C of the sphere of which the mirror is a part is in front of the reflecting surface; for a convex mirror it is behind. C is the **centre of curvature** of the mirror, and P, the centre of the mirror surface, is called the **pole**. The line CP produced is the **principal axis**. AB is the **aperture** of the mirror.

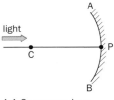

(a) Concave mirror

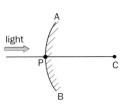

(b) Convex mirror

Fig. 6.11

Observation shows that a *narrow* beam of rays, parallel and near to the principal axis, is reflected from a concave mirror so that all rays converge to a point F on the principal axis, Fig. 6.12. F is called the **principal focus** of the mirror and it is a **real** focus since light actually passes through it. Concave mirrors are also known as converging mirrors because of their action on a parallel beam of light. They are used as reflectors in car headlamps and searchlights, and are an essential component of the largest telescopes.

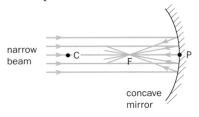

Fig. 6.12 A converging mirror

A narrow beam of rays, parallel and near to the principal axis, falling on a convex mirror is reflected to form a divergent beam which *appears* to come from a point F behind the mirror, Fig. 6.13. A convex mirror therefore has a **virtual** principal focus; it is also called a diverging mirror.

Rays that are close to the principal axis and make small angles with it, i.e. are nearly parallel to the axis, are called **paraxial** rays. Our treatment of spherical mirrors will be restricted here to such rays, which, in effect, means we shall consider only mirrors of small aperture. In diagrams, however, they will be made larger for clarity.

Spherical mirrors form a point image of *all* paraxial rays from a point object, as well as bringing paraxial rays that are parallel to the principal axis to a point focus F.

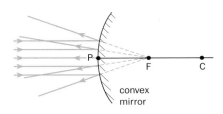

Fig. 6.13 A diverging mirror

(b) Relation between f and r

The distance PC from the pole to the centre of curvature of a spherical mirror is called its **radius of curvature** (*r*); the distance PF from the pole to the principal focus is its **focal length** (*f*), Fig. 6.14.

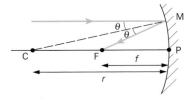

Fig. 6.14

A simple relation exists between *f* and *r*. For small values of θ (that is for paraxial rays):

$$f = \frac{r}{2}$$

The focal length of a spherical mirror is half the radius of curvature.

This relation also holds for a convex mirror, Fig. 6.15.

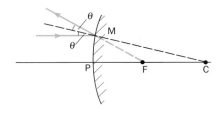

Fig. 6.15

IMAGES IN SPHERICAL MIRRORS

In general the *position* of the image formed by a spherical mirror and its *nature* (i.e. whether it is real or virtual, inverted or upright, magnified or diminished) depend on the distance of the object from the mirror. Information about the image in any case can be obtained either by drawing a ray diagram or by calculation using formulae.

(a) Ray diagrams

We shall assume that small objects on the principal axes of mirrors of small aperture are being considered so that all rays are paraxial. Point images will therefore be formed of points on the object.

To construct the image, *two* of the following three rays are drawn from the *top of the object*.

1. A ray parallel to the principal axis which after reflection actually passes through the principal focus or appears to diverge from it.

2. A ray through the centre of curvature which strikes the mirror normally and is reflected back along the same path.

3. A ray through the principal focus which is reflected parallel to the principal axis, i.e. a ray travelling the reverse path to that in (1).

The diagrams for a concave mirror are shown in Fig. 6.16 and for a convex mirror in Fig. 6.17. In the latter case no matter where the object is, the image is always virtual, upright and diminished.

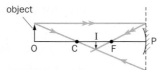

(a) Object beyond C
Image between C and F, real, inverted, diminished

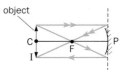

(b) Object at C
Image at C, real, inverted, same size

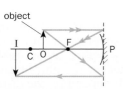

(c) Object between C and F
Image beyond C, real, inverted, magnified

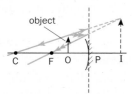

(d) Object between F and P
Image behind mirror, virtual, upright, magnified

Fig. 6.16 Ray diagrams for a concave mirror[1]

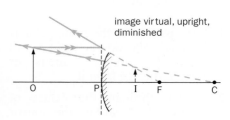

Fig. 6.17 Ray diagram for a convex mirror

Fig. 6.18 shows that a convex mirror gives a wider **field of view** than a plane mirror. Convex mirrors are therefore used as car wing-mirrors, for safety on bends of roads, and for security in shops and on the stairs of double-decker buses. They make the

estimation of distances more difficult, however, because there is only a small movement of the image for a large movement of the object.

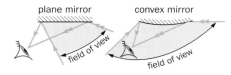

Fig. 6.18 Convex mirrors give a wide field of view

(b) The mirror formula

In Figs 6.19a and b a ray OM from a point object O on the principal axis is reflected at M so that the angles θ, made by the incident and reflected rays with the normal CM, are equal. A ray OP strikes the mirror normally and is reflected back along PO. The intersection I of the reflected rays MI and PO in (a) gives a **real** point image of O, and that of MI and PO both produced backwards in (b) gives a **virtual** point image of O.

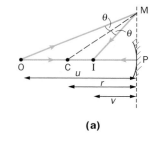

(a)

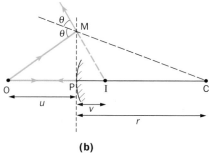

(b)

Fig. 6.19

We now introduce a **sign convention** so that distances are given a positive or a negative sign. We shall adopt the 'real is positive' rule, as follows.

A **real** object or image distance is **positive**.

A **virtual** object or image distance is **negative**.

The focal length of a concave mirror is thus positive (since its principal focus is real) and of a convex mirror negative. The radius of curvature takes the same sign as the focal length.

In Figs 6.19a and b let u, v and r stand for the *numerical values* and *signs* of the object and image distance and radius of curvature respectively, then for both concave and convex mirrors it can be shown that, for paraxial rays,

$$\frac{1}{v} + \frac{1}{u} = \frac{2}{r}$$

Also, since r = 2f, we have

$$\frac{1}{v} + \frac{1}{u} = \frac{1}{f}$$

Notes. (*i*) *All* paraxial rays from a point object O must, after reflection, pass through I to give a point image.

(*ii*) When the numerical values for u, v, r or f are substituted in the formula, the appropriate sign *must* also be included; the sign (as well as the value) of the distance to be found comes out in the answer and so even if the sign is known from other information it must *not* be inserted in the equation.

(*iii*) Only two cases have been considered but the same formula holds for others, e.g. a concave mirror forming a virtual image of a real object, and a convex mirror giving a real image of a virtual object (i.e. of converging light).

(c) Magnification

The lateral, transverse or linear magnification m (abbreviated to magnification) produced by a mirror is defined by

$$m = \frac{\text{height of image}}{\text{height of object}}$$

---

[1] **Notes.** (*i*) In (a) and (c) O and I are interchangeable; such positions of object and image are called **conjugate points**.
  (*ii*) C is a **self-conjugate point** — as (b) shows, object and image are coincident at C.
  (*iii*) If the object is at infinity (i.e. a long way off), a real image is formed at F. Conversely an object at F gives a real image at infinity.
  (*iv*) In all cases the foot of the object is on the principal axis and its image also lies on this line.

In Fig. 6.20, II' is the real image formed by a concave mirror of a finite object OO'. Triangles O'PO and I'PI are similar and so

$$\frac{\text{height of image}}{\text{height of object}} = \frac{\text{I'I}}{\text{O'O}} = \frac{\text{IP}}{\text{OP}}$$

$$m = \frac{v}{u}$$

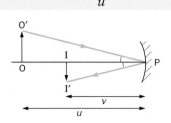

Fig. 6.20

For example, if the image distance (*v*) is twice the object distance (*u*), the image is twice the height of the object.

Note that no signs need be inserted in this formula for *m*, i.e. it is a numerical and not an algebraic formula.

The same result can be derived for other cases.

MIRROR CALCULATIONS

Example 1. An object is placed 15 cm from (*a*) a concave mirror, (*b*) a convex mirror, of radius of curvature 20 cm. Calculate the image position and magnification in each case.

(*a*) *Concave mirror*
The object is real, therefore $u = +15$ cm.

Since the mirror is concave $r = +20$ cm, therefore $f = +10$ cm.

Substituting values and signs in $1/v + 1/u = 1/f$,

$$\frac{1}{v} + \frac{1}{(+15)} = \frac{1}{(+10)}$$

$$\therefore \frac{1}{v} = \frac{1}{10} - \frac{1}{15} = \frac{1}{30}$$

$$\therefore v = +30 \text{ cm}$$

The image is real since *v* is positive and it is 30 cm in front of the mirror. Also,

$$\text{magnification } m = \frac{v}{u} \text{ (numerically)}$$

$$= \frac{30}{15} = 2.0$$

The image is twice as high as the object (see Fig. 6.16*c*).

(*b*) *Convex mirror*
We have $u = +15$ cm but $r = -20$ cm and $f = -10$ cm since the mirror is convex.

Substituting as before in $1/v + 1/u = 1/f$,

$$\frac{1}{v} + \frac{1}{(+15)} = \frac{1}{(-10)}$$

$$\therefore \frac{1}{v} = -\frac{1}{10} - \frac{1}{15} = -\frac{5}{30}$$

$$\therefore v = -\frac{30}{5} = -6.0 \text{ cm}$$

The image is virtual since *v* is negative and it is 6.0 cm behind the mirror. Also

$$m = \frac{v}{u} \text{ (numerically)}$$

$$= \frac{6}{15} = \frac{2}{5}$$

The image is two-fifths as high as the object (see Fig. 6.17).

Example 2. When an object is placed 20 cm from a concave mirror, a real image magnified three times is formed. Find (*a*) the focal length of the mirror, (*b*) where the object must be placed to give a virtual image three times the height of the object.

(*a*) The object is real, therefore $u = +20$ cm.

Also,

$$m = 3 = \frac{v}{u} = \frac{v}{20} \text{ (numerically)}$$

The image is real,

$$\therefore v = +3 \times 20 = +60 \text{ cm}$$

Substituting in $1/v + 1/u = 1/f$,

$$\frac{1}{(+60)} + \frac{1}{(+20)} = \frac{1}{f}$$

$$\therefore \frac{1}{f} = \frac{4}{60}$$

$$\therefore f = +15 \text{ cm}$$

(*b*) Let the *numerical* value of the object distance $= x$.

Therefore, $u = +x$ since the object is real and $v = -3x$ since the image is virtual; $f = +15$ cm.

Using the mirror formula,

$$\frac{1}{(+x)} + \frac{1}{(-3x)} = \frac{1}{(+15)}$$

$$\therefore \frac{3}{3x} - \frac{1}{3x} = \frac{1}{15}$$

$$\therefore \frac{2}{3x} = \frac{1}{15}$$

$$\therefore x = 10 \text{ cm}$$

The object should be 10 cm in front of the mirror.

Note. By letting *x* be the numerical value of *u* we are able to substitute for *u* since we know its sign — a useful dodge.

METHODS OF MEASURING *f* FOR SPHERICAL MIRRORS

(*a*) *Concave mirror*

1. Rough method. The image formed by the mirror of a distant window (several metres away) is focused sharply on a screen, Fig. 6.21*a*. The distance between the mirror and the screen is *f* since rays of light from a point on such an object are approximately parallel, Fig. 6.21*b*.

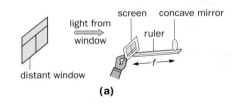

(a)

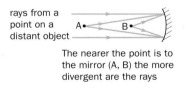

The nearer the point is to the mirror (A, B) the more divergent are the rays

(b)

Fig. 6.21

2. Self-conjugate point method. The position of an object is adjusted until it coincides with its own image. This occurs when the object is at the centre of curvature and distance *r*, i.e. 2*f*, from the mirror. A point at which an object and its image coincide is said to be 'self-conjugate'.

The object can be a pin moved up and down above the mirror until there is no parallax between it and its real, inverted image, Fig. 6.22*a*. If an illuminated object is used, it is moved to and from the mirror until a clear image is obtained on a screen beside the object, Fig. 6.22*b*.

The pin/no parallax method generally gives more accurate results.

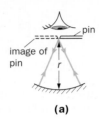

(a)

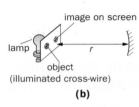

(b)

Fig. 6.22

3. *Mirror formula method.* Values of the image distance *v* corresponding to different values of the object distance *u* are found using either the pin/no parallax method or an illuminated object and screen. For each pair of values, *f* is calculated from $1/f = 1/v + 1/u$ and the average taken.

(b) Convex mirror

Auxiliary converging lens method. A convex mirror usually forms a virtual image of a real object. Such an image cannot be located by a screen and is not easy to find by a pin/no parallax method. With the help of a converging lens, however, a real image can be obtained.

In Fig. 6.23 the lens L forms a real image at C of an object O when the convex mirror is absent. This image is located and the distance LC noted. The mirror is then placed between L and C and moved until O coincides in position with its own image.

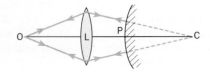

Fig. 6.23

The light from L is then falling normally on the mirror and is retracing its path to form a real inverted image at O. If produced, the rays from L must pass through the centre of curvature of the mirror since they are normal to the mirror. C, the position of the image of O formed by L alone, must therefore also be the centre of curvature of the mirror and so PC = *r*. Distance LP is measured and then $r = 2f = PC = LC - LP$.

REFRACTION AT PLANE SURFACES

(a) Laws of refraction

When light passes from one medium, say air, to another, say glass, Fig. 6.24, part is reflected back into the first medium and the rest passes into the second medium with its direction of travel changed. The light is said to be bent or **refracted** on entering the second medium and the angle of refraction is the angle made by the refracted ray OB with the normal ON. There are two laws of refraction.

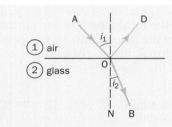

Fig. 6.24

1. For two particular media, the ratio of the sine of the angle of incidence to the sine of the angle of refraction is constant, i.e.

$$\frac{\sin i_1}{\sin i_2} = \text{a constant}$$

This is known as **Snell's law** after its discoverer.

2. The refracted ray is in the same plane as the incident ray and the normal to the medium at the point of incidence but on the opposite side of the normal from the incident ray.

The constant ratio $\sin i_1/\sin i_2$ is dependent on the **refractive index** of each of the media. If the media containing the incident and refracted rays are denoted by 1 and 2 respectively, the corresponding refractive indices are n_1 and n_2, and Snell's law becomes

$$\frac{\sin i_1}{\sin i_2} = \frac{n_2}{n_1}$$

or, writing it more symmetrically:

$$n_1 \sin i_1 = n_2 \sin i_2$$

The refractive index of a medium depends on the colour of the light and is usually stated for yellow light. The refractive index of water is 1.33, of crown glass about 1.5 and of air at normal pressure about 1.0003 — which is 1 near enough, the same as for a vacuum.

The greater the refractive index of a medium the greater is the change in direction suffered by a ray of light when it passes from air to the medium. Refraction is therefore greater from air to crown glass than from air to water. In both cases the refracted ray is bent *towards* the normal, i.e. towards ON in Fig. 6.25*a*; the light is travelling into an 'optically denser' medium. A ray travelling from glass or water to air is bent *away from* the normal, Fig. 6.25*b*.

Refraction can be attributed to the fact that light has different speeds in different media (see p. 297).

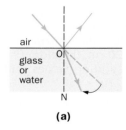

(a)

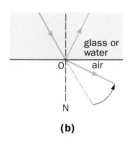

(b)

Fig. 6.25

(b) Refractive index relationships

The symmetrical form of Snell's law is useful in calculations.

> For example, if a ray of light is incident on a water–glass boundary at 30° then $i_1 = i_w = 30°$ and if $n_1 = n_w = \frac{4}{3}$ and $n_2 = n_g = \frac{3}{2}$, the angle of refraction $i_2 = i_g$ is given by
>
> $$n_w \sin i_w = n_g \sin i_g$$
>
> i.e. $\quad \frac{4}{3}\sin 30° = \frac{3}{2}\sin i_g$
>
> $\therefore \quad \sin i_g = \frac{4}{3}\times\frac{1}{2}\times\frac{2}{3}$ $(\sin 30° = \frac{1}{2})$
>
> $\qquad\qquad = \frac{5}{9}$
>
> $\therefore \quad i_g = 26°$

Suppose a ray AB travels from air (medium 1), to glass (medium 2), to water (medium 3), to air (medium 1) as in Fig. 6.26.

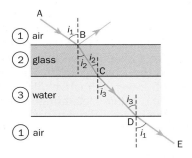

Fig. 6.26

Applying Snell's law to each parallel boundary,

$$n_1 \sin i_1 = n_2 \sin i_2$$
$$n_2 \sin i_2 = n_3 \sin i_3$$
so $\quad n_1 \sin i_1 = n_2 \sin i_2 = n_3 \sin i_3 \ldots$

When Snell's law is applied to the lowest boundary then we find that the emergent angle must be i_1, equal to the incident angle. Experiment shows that the emergent ray DE, although laterally displaced, is parallel to the incident ray AB as predicted.

(c) Real and apparent depth

One effect of refraction is that the apparent depth of a pool of clear water, when viewed from above the surface, is less than its real depth; also, an object under water is not where it seems to be to an outside observer.

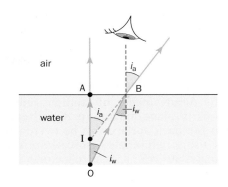

Fig. 6.27

In Fig. 6.27 rays from a point O underwater are bent away from the normal at the water–air boundary and appear to come from I, the image of O. For refraction at B from water to air,

$$n_w \sin i_w = n_a \sin i_a = \sin i_a \quad (n_a = 1)$$

$$\therefore \quad n_w = \frac{\sin i_a}{\sin i_w}$$

$$= \frac{AB/IB}{AB/OB} = \frac{OB}{IB}$$

If the observer is directly above O, i_w and i_a are small, rays OB and IB are close to OA, thus making OB $\approx$ OA and IB $\approx$ IA. Then

$$n_w = \frac{OA}{IA} \quad \text{(approximately)}$$

> $$n_w = \frac{\text{real depth}}{\text{apparent depth}}$$

A pool of water appears even shallower when viewed obliquely rather than from vertically above. As the observer moves, the image of a point O traces out a curve, called a **caustic**, whose apex is at I_1, Fig. 6.28.

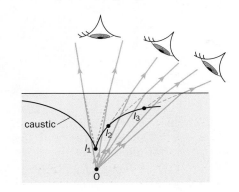

Fig. 6.28

(d) Multiple images in mirrors

Several images are seen when an object is viewed obliquely in a thick glass mirror with silvering on the back surface. In Fig. 6.29, I_1 is a faint image of object O formed by the weak reflected ray AB from the front surface of the mirror. I_2, the main image, is bright and is due to the refracted ray AC being reflected at the back (silvered) surface and again refracted at the front surface. I_3 and other weaker images are formed as shown. The net effect of these multiple reflections and refractions is to reduce the sharpness of the primary image I_2.

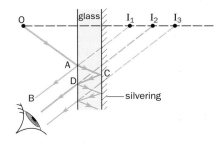

Fig. 6.29

(e) Mirages

These are often seen as small, distant pools of water on a hot tarmac road, particularly when vision is very oblique as it is for anyone at ground level. They are caused by refraction in the atmosphere.

The air near a road heated by the sun is hot; higher up the air is cool and its density greater. Consequently rays from the sky travelling towards the road are gradually refracted away from the normal as they pass from denser to less dense air. Upward bending of the light occurs, Fig. 6.30, and the blue light from the sky then seems to an observer to have been reflected from the road and gives the appearance of puddles.

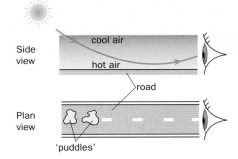

Fig. 6.30 Formation of mirages

TOTAL INTERNAL REFLECTION

(a) Critical angle

For small angles of incidence a ray of light travelling from one medium to another of smaller refractive index, say from glass to air, is refracted away from the normal, Fig. 6.31*a*; a weak internally reflected ray is also formed. Increasing the angle of incidence increases the angle of refraction and at a certain angle of incidence *c*, called the **critical angle**, the refracted ray just emerges along the surface of the glass and the angle of refraction is 90°, Fig. 6.31*b*. At this stage the internally reflected ray is still weak but just as *c* is exceeded it suddenly becomes bright and the refracted ray disappears, Fig. 6.31*c*. **Total internal reflection** is now said to be occurring since all the incident light is reflected inside the optically denser medium.

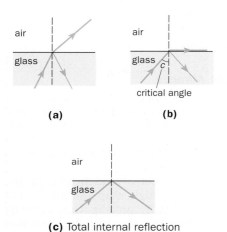

(a)   **(b)**

(c) Total internal reflection

Fig. 6.31

Applying Snell's law in the form $n_1 \sin i_1 = n_2 \sin i_2$ to the critical ray at a glass–air boundary, we have

n_1 = refractive index of glass = n_g
i_1 = critical angle for glass = c
n_2 = refractive index of air = 1
i_2 = angle of refraction in air = 90°

$$\therefore \quad n_g \sin c = 1 \sin 90° = 1$$

$$n_g = \frac{1}{\sin c}$$

Taking $n_g = \frac{3}{2}$ (crown glass), $\sin c = \frac{2}{3}$ and $c \approx 42°$. So if the incident angle in the crown glass exceeds 42°, total internal reflection occurs. Can it occur when a ray of light in glass ($n_g = \frac{3}{2}$ say) is incident on a boundary with water ($n_w = \frac{4}{3}$)?

(b) Totally reflecting prisms

The disadvantages of multiple reflections in plane mirrors (p. 83) silvered on the back surface can be overcome by using right-angled isosceles prisms (angles 90°, 45°, 45°) as reflectors.

The critical angle of crown glass is about 42° and a ray OA incident normally on face PQ of such a prism, Fig. 6.32*a*, suffers total internal reflection at face PR since the angle of incidence in the optically denser medium is 45°. A bright ray AB emerges at right angles to face QR since the angle of reflection at PR is also 45°. The prism thus reflects the ray through 90°.

Light can be reflected through 180° and an erect image obtained of an inverted one (as in prism binoculars, p. 101) if the prism is arranged as in Fig. 6.32*b*.

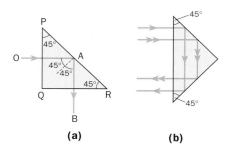

(a)   **(b)**

Fig. 6.32 Prism reflectors

(c) Optical fibres

Light can be confined within a bent glass rod by total internal reflection and so 'piped' along a twisted path, as in Fig. 6.33. The beam is reflected from side to side practically without loss (except for that due to absorption in the glass) and emerges only at the end of the rod, where it strikes the surface almost normally, i.e. at an angle less than the critical angle.

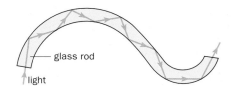

— glass rod

light

Fig. 6.33 Principle of an optical fibre

A single, very thin, solid glass fibre behaves in the same way and if several thousand are taped together a flexible light pipe is obtained that can be used, for example in medicine and engineering, to illuminate some otherwise inaccessible spot. Leakage of light at places of contact between the fibres is reduced by coating each fibre with glass of lower refractive index than its own, called 'cladding', thereby encouraging total internal reflection.

If the aim is to transport an *image* and not simply to transport *light*, the fibres must occupy the same position in the bundle relative to one another. Such bundles are more difficult to make and cost more. In Fig. 6.34 the

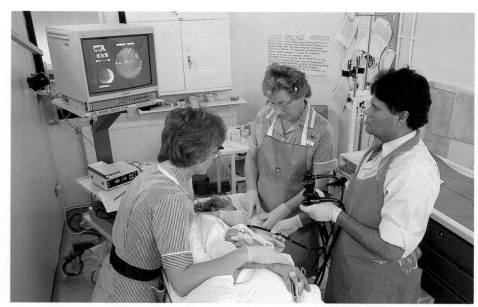

Fig. 6.34 An endoscope in use

doctor is viewing an image of the patient's inside via flexible optical fibres (an 'endoscope').

Optical fibres are now used to carry telephone, television and computer signals as pulses of light (see p. 458).

REFRACTION THROUGH PRISMS

A prism has two plane surfaces inclined to each other, such as LMQP and LNRP in Fig. 6.35. Angle MLN is called the **refracting angle** of the prism, LP is the **refracting edge** and any plane such as XYZ which is perpendicular to LP is a **principal plane**.

The importance of the prism really depends on the fact that the angle of deviation suffered by light at the first refracting surface, say LMQP, is not cancelled out by the deviation at the second surface LNRP (as it is in a parallel-sided block), but is added to it. This is why it can be used in a spectrometer, an instrument for analysing light into its component colours. In what follows, expressions for the angle of deviation will be obtained and subsequently used.

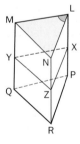

Fig. 6.35

(a) General formulae

In Fig. 6.36, EFGH is a ray lying in a principal plane XYZ of a prism of refracting angle A and passing from air, through the prism and back to air again. KF and KG are normals at the points of incidence and emergence of the ray.

Equations (1) and (2) in the following derivations are true for any prism. (The position and shape of the third side of the prism does not affect the refraction under consideration and so is shown as an irregular line in Fig. 6.36.)

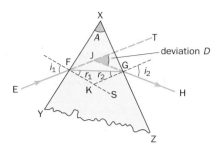

Fig. 6.36 Deviation by a prism

For refraction at XY

angle of deviation = angle JFG
$$= i_1 - r_1$$

For refraction at XZ

angle of deviation = angle JGF
$$= i_2 - r_2$$

Since both deviations are in the same direction, the **total deviation** D is given by

angle TJH = angle JFG + angle JGF

i.e. $\quad D = (i_1 - r_1) + (i_2 - r_2) \quad$ (1)

Another expression arises from the geometry of Fig. 6.36. In quadrilateral XFKG

angle XFK + angle XGK = 180°

$\therefore \quad A + $ angle FKG = 180°

But since FKS is a straight line

angle GKS + angle FKG = 180°

$\therefore \quad$ angle GKS = A

In triangle KFG, angle GKS is an exterior angle.

$\therefore \quad$ angle GKS = $r_1 + r_2$

$\therefore \quad A = r_1 + r_2 \quad$ (2)

(b) Minimum deviation

The angle of deviation D varies with the angle of incidence i_1 of the ray incident on the first refracting face of the prism. The variation is shown in Fig. 6.37a; for one angle of incidence it has a minimum value D_{min}. At this value *the ray passes symmetrically through the prism*, i.e. the angle of emergence of the ray from the second face equals the angle of incidence of the ray on the first face: $i_2 = i_1 = i$, Fig. 6.37b. It therefore follows that $r_1 = r_2 = r$.

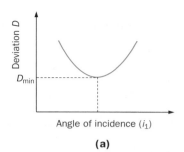

(a)

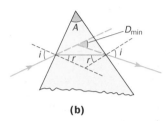

(b)

Fig. 6.37

From equation (1) of the previous section, the angle of minimum deviation D_{min} is thus given by

$$D_{min} = (i - r) + (i - r)$$

i.e. $\quad D_{min} = 2(i - r) \quad$ (3)

Also, from equation (2),

$$A = r + r = 2r$$

$$\therefore \quad r = \frac{A}{2}$$

Substituting for r in (3),

$$D_{min} = 2i - A$$

$$\therefore \quad i = \frac{A + D_{min}}{2}$$

If n is the refractive index of the material of the prism then

$$n = \frac{\sin i}{\sin r}$$

and we have

$$n = \frac{\sin [(A + D_{min})/2]}{\sin (A/2)}$$

If, for example, $A = 60°$ and $D_{min} = 40°$, then $(A + D_{min})/2 = 50°$ and so $n = \sin 50°/ \sin 30° = 1.5$.

Two points should be considered. First, no values of D are shown on the graph of Fig. 6.37a for small values of i (less than about 30° for a crown glass prism of refracting angle 60° and $n = 1.5$). Why? Second, the above formula for minimum deviation only holds for a prism of angle A less than twice the critical angle. Why?

(c) Small-angle prism

The expression for the deviation in this case will be used later for developing lens theory.

Consider a ray falling almost normally in air on a prism of small angle A (less than about 6° or 0.1 radian) so that angle i_1 in Fig. 6.38 is small.

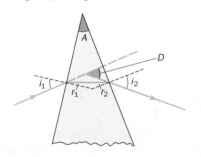

angles exaggerated for clarity

Fig. 6.38

Now $n = \sin i_1/\sin r_1$ where n is the refractive index of the material of the prism, therefore r_1 will also be small. Hence, since the sine of a small angle (like the tangent) is nearly equal to the angle in radians, we have

$$i_1 = nr_1$$

Also, $A = r_1 + r_2$ (equation (2), p. 85), and so if A and r_1 are small, r_2 and i_2 will also be small. From $n = \sin i_2/\sin r_2$ we can say

$$i_2 = nr_2$$

The deviation D of a ray passing through any prism is given by

$$D = (i_1 - r_1) + (i_2 - r_2)$$

(equation (1), p. 85). Substituting for i_1 and i_2,

$$D = nr_1 - r_1 + nr_2 - r_2$$
$$= n(r_1 + r_2) - (r_1 + r_2)$$
$$= (n - 1)(r_1 + r_2)$$

The deviation D is thus given by

$$D = (n - 1)A$$

This expression shows that for a given angle A *all* rays entering a *small-angle* prism at *small angles of incidence* suffer the *same* deviation.

(d) Dispersion

Newton found that when a beam of white light (e.g. sunlight) passes through a prism it is spread out by the prism into a band of all the colours of the rainbow from red to violet, Fig. 6.39. The band of colours is called a **spectrum** and the separation of the colours by the prism is known as **dispersion**. He concluded that white light is a mixture of light of various colours and identified red, orange, yellow, green, blue, indigo and violet.

Fig. 6.39 Dispersion by a prism

Red is deviated least by the prism and violet most as shown by the exaggerated diagram of Fig. 6.40a. The refractive index of the material of the prism for violet light is greater than for red light.

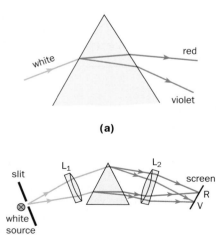

(a)

(b)

Fig. 6.40

A method of producing a **pure** spectrum, i.e. one in which the different colours do not overlap (as they do when a prism is used on its own), is shown in Fig. 6.40b. A diverging beam of white light, emerging from a very narrow slit, is made parallel by lens L_1 and then dispersed by the prism into a number of different coloured parallel beams, each travelling in a slightly different direction. Lens L_2 brings each colour to a separate focus on a screen. The spectrum is a series of monochromatic images of the slit and the narrower this is the purer the spectrum. (L_1 and L_2 are achromatic doublets, see p. 94).

MEASUREMENT OF n

(a) Real and apparent depth methods (solids and liquids)

A travelling microscope is focused on a pencil dot O on a sheet of white paper lying on the bench and the reading on the microscope scale noted, Fig. 6.41. Let it be x. If the refractive index of glass is required, a block of the material is placed over the dot and the microscope refocused on the image I of O as seen through the block. Let the reading be y. Finally the microscope is focused on the top T of the block, made visible by a sprinkling of lycopodium powder.

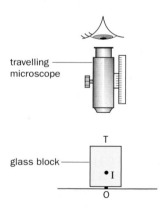

Fig. 6.41

Suppose the reading is now z, then

$$\text{real depth of O} = OT = z - x$$
$$\text{apparent depth of O} = IT = z - y$$
$$\therefore \quad n = \frac{\text{real depth}}{\text{apparent depth}} = \frac{z - x}{z - y}$$

The refractive index of a liquid in a beaker can be found by a similar procedure.

The method satisfies the assumption made in deducing this expression for n (p. 83) because the microscope collects only rays very close to the normal OT; accuracy of $\pm 1\%$ is possible if the microscope has a small depth of focus.

(b) Minimum deviation method (solids and liquids)

A solid prism of the material is placed on the table of a spectrometer (p. 101), A and $D_{\min}$ are measured and n calculated from

$$n = \frac{\sin[(A + D_{\min})/2]}{\sin(A/2)}$$

The method is suitable for liquids if a hollow prism with perfectly parallel, thin walls is used. Accuracy of $\pm 0.1\%$ is possible.

(c) Concave mirror method (liquids)

The centre of curvature C of the mirror is first found by moving an object pin up and down above the mirror until it coincides in position with its image (see *Method 2*, p. 81). Some liquid is then poured into the mirror and the object pin moved until point O is found where it again coincides with its image. In Fig. 6.42 ray ONB must be retracing its own path after striking the mirror normally at B and if BN is produced it will pass through C.

For the refraction at N

i_1 = angle of incidence

   $= \angle$ ONA $= \angle$ NOM (alt. angles)

i_2 = angle of refraction

   $= \angle$ BND $= \angle$ NCM (corr. angles)

The refractive index n of the liquid is given by

$$n = \frac{\sin i_1}{\sin i_2} = \frac{\sin \text{NOM}}{\sin \text{NCM}}$$
$$= \frac{\text{NM/NO}}{\text{NM/NC}} = \frac{\text{NC}}{\text{NO}}$$

If ray ONB is close to the principal axis CP of the mirror then to a good approximation NC = MC and NO = MO,

$$\therefore \quad n = \frac{\text{MC}}{\text{MO}}$$

Both distances can be measured and n found. The method is useful when only a small quantity of liquid is available.

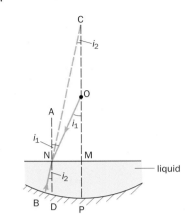

Fig. 6.42

THIN LENSES

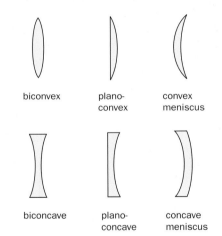

Fig. 6.43   Types of lens

Lenses are of two basic types, **convex** which are thicker in the middle than at the edges and **concave** for which the reverse holds. Figure 6.43 shows examples of both types, bounded by spherical or plane surfaces.

The **principal axis** of a spherical lens is the line joining the centres of curvature of its two surfaces, Fig. 6.44. Our treatment here will be confined to **paraxial** rays, i.e. rays close to the axis and making very small angles with it. In effect this means we shall only consider lenses of small aperture but in diagrams both angles and lenses will be made larger for clarity. The case of wide-angle beams will be considered briefly later (pp. 92–3).

The **principal focus F** of a thin lens is the point on the principal axis towards which all paraxial rays (parallel to the principal axis) converge in the case of a convex lens, or from which they appear to diverge in the case of a concave lens, after refraction, Figs 6.44*a* and *b*. Since light can fall on either surface, a lens has two principal foci, one on each side, and these are equidistant from its centre P (if the lens is thin and has the same medium on both sides, e.g. air). The distance FP is the **focal length** f of the lens. A convex lens is a **converging** lens[2] and has real foci. A concave lens is a **diverging** lens and has virtual foci.

A parallel beam at a small angle to the axis of a lens is refracted to converge to, or to appear to diverge from, a point in the plane containing F, perpendicular to the axis and known as the **focal plane**, Figs 6.45*a* and *b*.

As we shall see shortly, the important property of a lens is that it focuses *all* paraxial rays from a point object (and not just parallel paraxial rays) to form a point image.

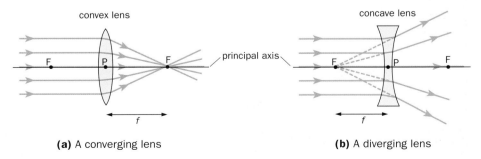

(a) A converging lens   **(b)** A diverging lens

Fig. 6.44

[2] This is only true if the convex lens has a greater refractive index than the surrounding medium. In water a biconvex air lens diverges light.

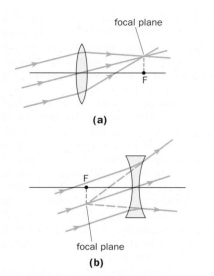

Fig. 6.45

IMAGES FORMED BY THIN LENSES

Information about the position and nature of the image in any case can be obtained either from a ray diagram or by calculation.

(a) Ray diagrams

To construct the image of a small object perpendicular to the axis of a lens, *two* of the following three rays are drawn from the top of the object.

1. A ray parallel to the principal axis which after refraction passes through the principal focus or appears to diverge from it.

2. A ray through the centre of the lens (called the **optical centre**) which continues straight on undeviated (it is only slightly displaced laterally because the middle of the lens acts like a thin parallel-sided block), Fig. 6.46.

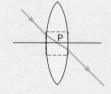

Fig. 6.46

3. A ray through the principal focus which is refracted parallel to the principal axis, i.e. a ray travelling the reverse path to that in (1).

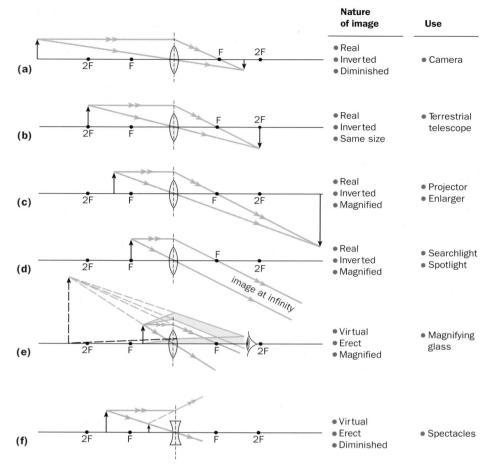

Fig. 6.47 Ray diagrams for a convex lens (*a–e*) and a concave lens (*f*)

The diagrams for a converging lens are shown in Figs 6.47*a* to *e* and for a diverging lens in Fig. 6.47*f*. The latter, like a convex mirror, always forms a virtual upright and diminished image whatever the object position. Note that a thin lens is represented by a straight line at which all the refraction is considered to occur; in practice light is usually refracted both on entering and leaving the lens.

It must also be emphasized that the lines drawn are *constructional* ones; two narrow cones of rays that actually enter the eye of an observer from the top and bottom of an object are shown shaded in Fig. 6.47*e*. They are obtained by working back from the eye, from right to left here.

(b) Simple formula for a thin lens

We can regard a thin lens as made up of a large number of small-angle prisms whose angles increase from zero at the middle of the lens to a small value at its edge.

Consider one such prism at distance h from the optical centre P of a lens, Figs 6.48*a* and *b*. If a paraxial ray parallel to the axis is incident on this prism it suffers small deviation D (since the prism is small-angled) and is refracted through the principal focus F. Hence, since the tangent of a small angle equals the small angle in radians,

$$D = \frac{h}{FP} \qquad (1)$$

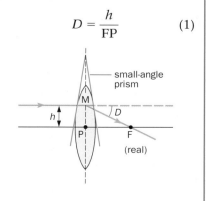

Fig. 6.48a Converging

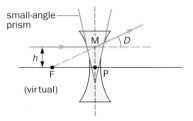

Fig. 6.48b Diverging

Now consider a point object O on the axis which gives rise to a point image I, Figs 6.49*a* and *b*. If a paraxial ray from O is incident on the small-angle prism at distance *h* from the axis, it must also suffer deviation *D* (since all rays entering a small-angle prism at small angles of incidence suffer the same deviation (p. 86)).

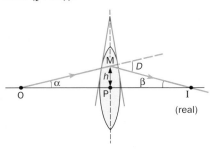

(a) Converging

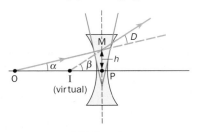

(b) Diverging

Fig. 6.49

Let angles α and β be as shown. In triangle IOM, since the exterior angle of a triangle equals the sum of the interior opposite angles,

Converging	*Diverging*
$D = \alpha + \beta$	$D = \beta - \alpha$
$= \dfrac{h}{OP} + \dfrac{h}{IP}$	$= \dfrac{h}{IP} - \dfrac{h}{OP}$

Therefore from (1)

$$\frac{h}{FP} = \frac{h}{OP} + \frac{h}{IP} \qquad (2)$$

and
$$\frac{h}{FP} = \frac{h}{IP} - \frac{h}{OP} \qquad (2)'$$

If we introduce the 'real is positive' sign convention given on p. 80, **the focal length of a converging lens is positive** and **of a diverging lens negative**. If *u*, *v* and *f* stand for the *numerical values* and *signs* of the object and image distances and focal length respectively then for *both* cases we have the *algebraic relationship*

$$\frac{1}{v} + \frac{1}{u} = \frac{1}{f}$$

Notes. (*i*) The formula is independent of the angle the incident ray makes with the axis, therefore *all* paraxial rays from point object O must, after refraction, pass through I to give a point image, i.e. the small-angle prisms to which the lens is equivalent deviate the various rays from O by varying amounts depending on the angle of the prism but always so that they all pass through I.

(*ii*) When numerical values for *u*, *v* and *f* are inserted in the formula, the appropriate sign *must* also be included.

(c) Magnification

The lateral, transverse or linear magnification *m* (abbreviated to magnification) produced by a lens is defined by

$$m = \frac{\text{height of image}}{\text{height of object}}$$

In Fig. 6.50 I'I' is the real image formed by a converging lens of a finite object OO'. Triangles O'PO and I'PI are similar, therefore

$$\frac{\text{height of image}}{\text{height of object}} = \frac{I'I}{O'O} = \frac{IP}{OP}$$

and so

$$m = \frac{v}{u}$$

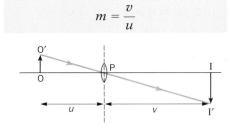

Fig. 6.50

This is a numerical formula and no signs need be inserted.

LENS CALCULATIONS

Example 1. An object is placed 20 cm from (*a*) a converging lens, (*b*) a diverging lens, of focal length 15 cm. Calculate the image position and magnification in each case.

(*a*) *Converging lens*
The object is real, therefore
$$u = +20 \text{ cm}$$
Since the lens converges,
$$f = +15 \text{ cm}$$
Substituting values and signs in
$$\frac{1}{v} + \frac{1}{u} = \frac{1}{f}$$
gives $\quad \dfrac{1}{v} + \dfrac{1}{(+20)} = \dfrac{1}{(+15)}$
$$\therefore \frac{1}{v} = \frac{1}{15} - \frac{1}{20} = \frac{4}{60} - \frac{3}{60} = \frac{1}{60}$$
$$\therefore v = +60 \text{ cm}$$

The image is real since *v* is positive and it is 60 cm from the lens.
Also,
$$\text{magnification } m = \frac{v}{u} \text{ (numerically)}$$
$$= \frac{60}{20} = 3.0$$

The image is three times as high as the object (see Fig. 6.47*c*).

(*b*) *Diverging lens*
We have
$$u = +20 \text{ cm} \qquad f = -15 \text{ cm}$$
Substituting as before in $1/v + 1/u = 1/f$,
$$\frac{1}{v} + \frac{1}{(+20)} = \frac{1}{(-15)}$$
$$\therefore \frac{1}{v} = -\frac{1}{15} - \frac{1}{20} = -\frac{7}{60}$$
$$\therefore v = -\frac{60}{7} = -8.6 \text{ cm}$$

The image is virtual since *v* is negative and it is 8.6 cm from the lens.
Also,
$$m = \frac{v}{u} \text{ (numerically)}$$
$$= \frac{60/7}{20} = \frac{3}{7}$$

The image is three-sevenths as high as the object (see Fig. 6.47*f*).

Example 2. An object is placed 6.0 cm from a thin converging lens A of focal length 5.0 cm. Another thin converging lens B of focal length 15 cm is placed coaxially with A and 20 cm from it on the side away from the object. Find the position, nature and magnification of the final image.

For lens A,

$$u = +6.0 \text{ cm} \qquad f = +5.0 \text{ cm}$$

Substituting in $1/v + 1/u = 1/f$,

$$\frac{1}{v} + \frac{1}{(+6)} = \frac{1}{(+5)}$$

$$\therefore \quad \frac{1}{v} = \frac{1}{5} - \frac{1}{6} = \frac{6}{30} - \frac{5}{30} = \frac{1}{30}$$

$$\therefore \quad v = +30 \text{ cm}$$

Image I_1 in Fig. 6.51 is real, therefore *converging* light falls on lens B, and I_1 acts as a virtual object for B. Applying $1/v + 1/u = 1/f$ to B we have

$$u = -(AI_1 - AB) = -(30 - 20)$$
$$= -10 \text{ cm}$$

$$f = +15 \text{ cm}$$

$$\therefore \quad \frac{1}{v} + \frac{1}{(-10)} = \frac{1}{(+15)}$$

$$\therefore \quad \frac{1}{v} = \frac{1}{15} + \frac{1}{10} = \frac{2}{30} + \frac{3}{30} = \frac{5}{30}$$

$$\therefore \quad v = +6.0 \text{ cm}$$

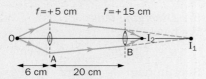

Fig. 6.51

Image I_2 is real and is formed 6.0 cm beyond B.

Magnification by A

$$= m_1 = \frac{v}{u} = \frac{30}{6} = 5.0$$

Magnification by B

$$= m_2 = \frac{v}{u} = \frac{6}{10} = \frac{3}{5}$$

Total magnification $m = m_1 \times m_2$

$$= 5.0 \times \frac{3}{5} = 3.0$$

The final image is three times the size of the object.

(Note that m_1 and m_2 are multiplied and *not* added. Why?)

OTHER THIN LENS FORMULAE

(a) Full formula for a thin lens

Here we will find a relationship, sometimes called the 'lens-maker's formula', between the focal length of a thin lens, the radii of curvature of its surfaces and the refractive index of the lens material. It will be assumed that (*i*) the lens can be replaced by a system of small-angle prisms and (*ii*) all rays falling on the lens are paraxial, i.e. the lens has a small aperture and all objects are near the axis.

Consider the prism of small angle A which is formed by the tangents XL and XM to the lens surfaces at L and M, Fig. 6.52. XL and XM are perpendicular to the radii of curvature C_1L and C_2M respectively, C_1 and C_2 being the centres of curvature of the surfaces. Therefore angle LXM (i.e. A) between the tangents equals angle MYC_1 between the radii,

$$\therefore \quad \angle MYC_1 = A = \theta_1 + \theta_2$$

(ext. angle of $\triangle YC_2C_1$ = sum of int. opp. $\angle$s)

But since θ_1 and θ_2 are small we can say $\theta_1 = \tan \theta_1$ and $\theta_2 = \tan \theta_2$,

$$\therefore \quad A = \frac{h}{C_1P} + \frac{h}{C_2P} \qquad (h = YP)$$

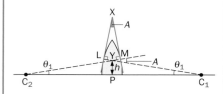

Fig. 6.52

The deviation D produced in *any* ray incident at a small angle on a prism of small angle A and refractive index n is (p. 86)

$$D = (n-1)A$$

$$= (n-1)\left(\frac{h}{C_1P} + \frac{h}{C_2P}\right)$$

If we now consider a ray parallel to the axis and at height h above it, it suffers the same deviation D as any other paraxial ray and since it is refracted through the principal focus F, $D = h/FP$ (from equation (1), p. 88).

Hence

$$\frac{h}{FP} = (n-1)\left(\frac{h}{C_1P} + \frac{h}{C_2P}\right)$$

$$\therefore \quad \frac{1}{FP} = (n-1)\left(\frac{1}{C_1P} + \frac{1}{C_2P}\right)$$

Introducing a sign convention for distances converts this numerical relationship to an algebraic one, which is applicable to all lenses and all cases.

If f, r_1 and r_2 stand for the numerical values *and signs* of the focal length and radii of curvature respectively of the lens we have

$$\frac{1}{f} = (n-1)\left(\frac{1}{r_1} + \frac{1}{r_2}\right)$$

In the 'real is positive' convention the rule for the sign of a radius of curvature is that **a surface convex to the less dense medium has a positive radius while a surface concave to the less dense medium has a negative radius**. A positive surface thus converges light; a negative one diverges it.

For the convex meniscus lens of Fig. 6.53*a* we have $n = 1.5$, $r_1 = +10$ cm (since it is convex to the air on its left), $r_2 = -15$ cm (since it is concave to the air on its right).

$$\therefore \quad \frac{1}{f} = (1.5 - 1)\left(\frac{1}{(+10)} + \frac{1}{(-15)}\right)$$

$$= \frac{1}{2}\left(\frac{3}{30} - \frac{2}{30}\right) = \frac{1}{60}$$

$$\therefore \quad f = +60 \text{ cm}$$

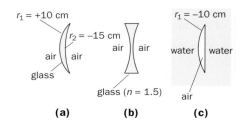

Fig. 6.53

Calculate f for the biconcave lens in Fig. 6.53*b* whose radii of curvature are each 20 cm. (*Answer:* -20 cm)

A more general form of the formula is

$$\frac{1}{f} = \left(\frac{n_2}{n_1} \sim 1\right)\left(\frac{1}{r_1} + \frac{1}{r_2}\right)$$

where n_2 is the refractive index of the lens material and n_1 that of the surrounding medium. The term $(n_2/n_1 \sim 1)$ *is always taken to be positive* since refractive indices do not have signs. ($\sim$ means the 'difference between'.)

For a plano-convex *air* lens of radius 10 cm, in *water* of refractive index $\frac{4}{3}$, we have $n_2 = 1$, $n_1 = \frac{4}{3}$, $r_1 = -10$ cm (since it is concave to the air), $r_2 = \infty$, Fig. 6.53c.

$$\therefore \frac{1}{f} = \left(\frac{1}{4/3} \sim 1\right)\left(\frac{1}{(-10)} + \frac{1}{\infty}\right)$$

$$= \frac{1}{4} \times \left(\frac{-1}{10} + 0\right)$$

$$\therefore f = -40 \text{ cm} \quad \text{(a diverging lens)}$$

(b) Focal length of two thin lenses in contact

Combinations of lenses in contact are used in many optical instruments to improve their performance. A and B in Fig. 6.54 are two thin lenses in contact, of focal lengths f_1 and f_2.

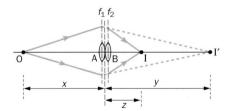

Fig. 6.54

Paraxial rays from point object O on the principal axis are refracted through A and would, in the absence of B, give a real image of O at I'. Hence for A, $u = +x$ and $v = +y$. From the simple formula for a thin lens

$$\frac{1}{(+x)} + \frac{1}{(+y)} = \frac{1}{f_1}$$

For B, I' acts as a *virtual object* (i.e. converging light falls on B from A) giving a real image of O at I and so for B, $u = -y$ and $v = +z$,

$$\therefore \frac{1}{(-y)} + \frac{1}{(+z)} = \frac{1}{f_2}$$

Adding, $\quad \dfrac{1}{(+x)} + \dfrac{1}{(+z)} = \dfrac{1}{f_1} + \dfrac{1}{f_2}$

Considering the combination, I is the real image formed of O by both lenses, therefore $u = +x$ and $v = +z$,

$$\therefore \frac{1}{(+x)} + \frac{1}{(+z)} = \frac{1}{f}$$

where f is the combined focal length, i.e. the focal length of the single lens that would be exactly equivalent to the two in contact.

The combined focal length f is then given by

$$\frac{1}{f} = \frac{1}{f_1} + \frac{1}{f_2}$$

For example, if a converging lens of 5.0 cm focal length is in contact with a diverging lens of 10 cm focal length, then $f_1 = +5.0$ cm, $f_2 = -10$ cm and the combined focal length f is given by

$$\frac{1}{f} = \frac{1}{(+5)} + \frac{1}{(-10)} = \frac{1}{5} - \frac{1}{10}$$

$$= +\frac{1}{10}$$

$$\therefore f = +10 \text{ cm}$$
$$\text{(a converging combination)}$$

(c) Power of a lens

The shorter the focal length of a lens, the more it converges or diverges light. The **power** F of a lens is defined as the reciprocal of its focal length f in metres.

$$\text{power } F = \frac{1}{f}$$

The unit of power is the radian per metre (rad m^{-1}) since $f = h/D$ (equation (1), p. 88), where the distance h is in metres and the angle of deviation D is in radians. (Previously the unit was called the dioptre; 1 dioptre = 1 radian per metre.)

The power of a lens of focal length
(a) 1 metre is 1 rad m^{-1},
(b) 25 cm (0.25 m) is 4.0 rad m^{-1}.

The sign of F is the same as f, i.e. positive for a converging lens and negative for a diverging one.

Opticians obtain the power of a lens using a lens measurer, Fig. 6.55a. This has three legs, the centre one being spring-loaded and connected to a pointer moving over a scale, Fig. 6.55b. By measuring the surface curvature, the power may be obtained quickly and accurately of any lens made of material of a certain refractive index. Lenses of materials of other refractive indices are catered for by using a scale of refractive indices along with the instrument.

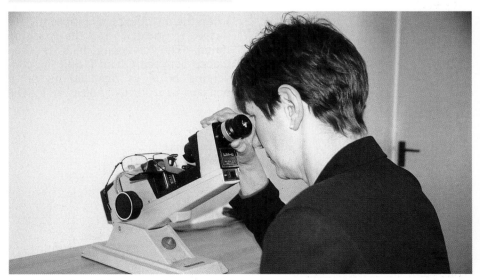

Fig. 6.55a An optician measuring a spectacle lens

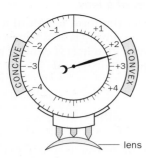

Fig. 6.55b   A lens measurer

	A	B	C	D	E	F
1	u (cm)	v (cm)	1/u	1/v	1/f = 1/u + 1/v	f (cm)
2			= 1/A2	= 1/B2	= C2+D2	= 1/E2
3						
4						
5						
6						
7						= AVERAGE (F2:F6)

Fig. 6.57   Spreadsheet for calculating the focal length of a lens

Although f is useful for constructing ray diagrams, it is more convenient to use F when calculating the combined effect of several optical parts. Thus the combined power F of three thin lenses of power F_1, F_2 and F_3 in contact is

$$F = F_1 + F_2 + F_3$$

METHODS OF MEASURING *f* FOR LENSES

(a) Converging lens

1. Rough method. The image formed by the lens of a *distant* window is focused sharply on a screen. The distance between the lens and the screen is f. Why?

2. Plane mirror method. Using the arrangement shown in Fig. 6.56a or b, a pin or illuminated object is adjusted until it coincides in position with its image, located by 'no parallax' or by a screen. The rays from the object must emerge from the lens and fall on the plane mirror normally to retrace their path. The object is therefore at the principal focus.

3. Lens formula method. Several values of the image distance v, corresponding to different values of the object distance u, are found using either the pin/no parallax method or an illuminated object and screen. For each pair of values, f is calculated from $1/f = 1/v + 1/u$ and the average taken.

A spreadsheet such as that shown in Fig. 6.57 is useful for calculating f; if the first pair of values for u and v are entered into cells A2 and B2, the calculated value for f will appear in cell F2. If five pairs of u and v readings are entered into cells A2 to A6 and B2 to B6, the average value for f is displayed in cell F7 (see question 18, p. 106).

(b) Diverging lens

Auxiliary converging lens method. A diverging lens usually forms a virtual image of a real object. Such an image cannot be located by a screen and is not easily found by a pin/no parallax method. However, with the help of a converging lens, a real image can be obtained.

In Fig. 6.58 the converging lens forms a real image at I' of an object O when the diverging lens is absent. This image is located and its position noted. The diverging lens is then placed between C and I' and the converging beam of light falling on it behaves as a virtual object at I'. The

diverging lens forms a real image of I' at I, which is located. For the diverging lens $u = -$I'D and $v = +$ID, from which f can be calculated.

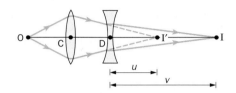

Fig. 6.58

DEFECTS IN IMAGES

So far our discussion of the formation of images by spherical mirrors and lenses has been confined to paraxial rays; we have assumed that the mirror or lens had a small aperture and that object points were on or near the principal axis. In such cases it is more or less true to say that point images are formed of point objects. However, when objects are extended and/or mirrors and lenses are of large aperture, rays are non-paraxial and the image can differ in shape, sharpness and colour from the object. Two of the most important image defects or **aberrations** will be considered.

(a) Spherical aberration

This arises with mirrors and lenses of large aperture and results in the image of an object point not being a point. The defect is due to the fact that the focal length of the mirror or lens for marginal rays is *less* than for paraxial rays — a property of a spherical surface.

Consider a point object at infinity (i.e. a long distance off) on the principal axis of a mirror or lens whose aperture is not small. The incident rays are parallel to the axis and are reflected or refracted so that the marginal rays farthest from the axis come

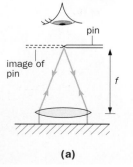

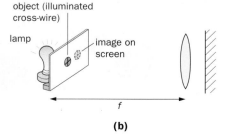

Fig. 6.56

to a focus at F_m while the paraxial rays give a point focus at F_p, Figs 6.59*a* and *b*. (All the reflected or refracted rays are tangents to a surface, called a **caustic surface**, which has an apex at F_p. A caustic curve may be seen on the surface of a cup of tea in bright light, the inside of the cup acting as the mirror.) The nearest approach to a sharp image is the **circle of least confusion**, i.e. the smallest circle through which all the reflected or refracted rays pass.

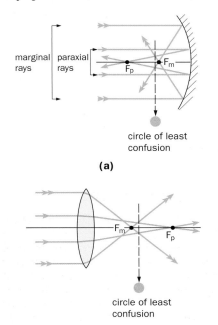

(a)

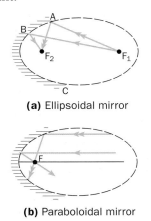

(b)

Fig. 6.59 Spherical aberration

In general, the image of any object point, on or off the axis, is a circular 'blur' and not a point. The distance F_mF_p in Fig. 6.59 is the longitudinal spherical aberration of the mirror or lens for the particular object distance.

It is not possible to construct a mirror which always forms a point image of a point object on the axis, but an ellipsoidal mirror achieves this for one definite point on the axis for both paraxial and marginal rays. In Fig. 6.60*a*, ABC represents an ellipsoidal mirror with foci F_1 and F_2; *all* rays from F_1 are reflected through F_2.

A parabola is an ellipse with one focus at infinity and so a paraboloidal mirror brings all rays from an object point *on* the axis at infinity to a point focus, Fig. 6.60*b*, thus accounting for its use as the objective in an astronomical telescope. (It should be

noted, however, that it does not form a point image of a point object *off* the axis.) Searchlights and car headlamps have paraboloidal reflectors which produce a roughly parallel beam from a small light source at the focus. A perfectly parallel beam does not spread out as the distance from the reflector increases and its intensity does not therefore decrease on this account.

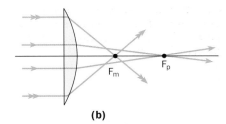

(a) Ellipsoidal mirror

(b) Paraboloidal mirror

Fig. 6.60

For a lens, spherical aberration can be *minimized* if the angles of incidence at each refracting surface are kept small, thus, in effect, making all rays paraxial. This is achieved by sharing the deviation of the light as equally as possible between the surfaces. Figure 6.61 shows parallel light falling on a plano-convex lens; spherical aberration is smaller in (*a*) than in (*b*). Why? Why would it be better to have the convex side towards the object if the lens was used as a tele-

scope objective but the other way around for a microscope objective?

Spherical aberration can also be reduced by placing a stop in front of the lens or mirror to cut off marginal rays but this has the disadvantage of making the image less bright.

(b) Chromatic aberration

This defect occurs only with lenses and causes the image of a white object to be blurred with coloured edges. A lens has a greater focal length for red light than for violet light, as shown in the exaggerated diagram of Fig. 6.62*a*. (This follows from $1/f = (n - 1) \times (1/r_1 + 1/r_2)$ bearing in mind that $n_{violet} > n_{red}$.) Thus a converging lens produces a series of coloured images of an extended white object, of slightly different sizes and at different distances from the lens, Fig. 6.62*b*. The eye, being most sensitive to yellow–green light, would focus the image of this colour on a screen but superimposed on it would be the other images, all out of focus.

Chromatic aberration can be eliminated for *two* colours (and reduced for all) by an **achromatic doublet**. This consists of a converging lens of crown glass combined with a diverging lens of flint glass. One surface of each lens has the same radius of curvature to allow them to be cemented together with Canada balsam, Fig. 6.63, and thereby reduce light loss by reflection. The flint glass of the diverging lens produces the same dispersion as the crown glass of the converging

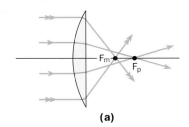

(a)

Fig. 6.61 Spherical aberration for a plano-convex lens

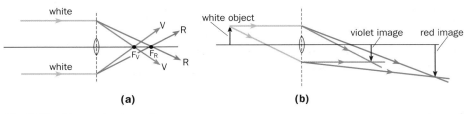

(a)

(b)

Fig. 6.62 Chromatic aberration

lens but in the opposite direction and with less deviation of the light, so that overall the combination is converging. In Fig. 6.64, the dispersions (exaggerated) θ_1 and θ_2 are equal and opposite; for the deviations, $D_1 > D_2$.

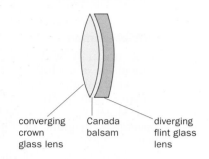

converging crown glass lens   Canada balsam   diverging flint glass lens

Fig. 6.63 An achromatic doublet

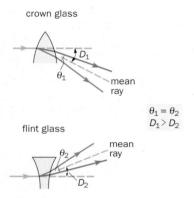

crown glass

D_1

θ_1

mean ray

$\theta_1 = \theta_2$
$D_1 > D_2$

flint glass

θ_2

mean ray

D_2

Fig. 6.64

THE EYE AND ITS DEFECTS

The construction of the human eye is shown in Fig. 6.65. The image on the retina is formed by successive refraction at the surfaces between the air, the cornea, the aqueous humour, the lens and the vitreous humour. The retina consists of a complex array of nerves and receptors known as rods and cones which transform light into electrical signals. Rods are particularly sensitive to dim light, while cones are responsible for acute and colour vision. The brain interprets the information transmitted to it as electrical impulses from the retinal image and appreciates by experience that an inverted image means an upright object, Fig. 6.66.

In good light the eye automatically focuses the image of an object on a very small region towards the centre

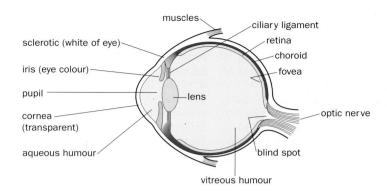

muscles   ciliary ligament
sclerotic (white of eye)   retina
    choroid
iris (eye colour)   fovea
pupil   lens
cornea (transparent)   optic nerve
    blind spot
aqueous humour
vitreous humour

Fig. 6.65 Structure of the eye

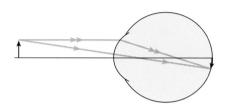

Fig. 6.66 An upright object gives an inverted image on the retina

of the retina called the fovea, where the density of cones is high. The fovea permits the best observation of detail — to about two minutes of arc, i.e. to $\frac{1}{10}$ mm at about 20 cm. The periphery of the retina can only detect much coarser detail but it is more sensitive to dim light. The iris automatically controls the amount of light entering the eye by changing the size of the pupil.

Light intensity and colour can influence a person's perception of their surroundings. Cool, light colours such as blue, green, grey and white appear to enlarge space, while warm, dark colours such as red, orange, brown and black, diminish space. Contrasting colours and tones are used to attract attention, for example in advertising; strong or subtle lighting is used in interior design to create atmosphere.

Objects at different distances are focused by the ciliary ligaments changing the shape of the lens — a process known as **accommodation**. The lens becomes more convex to view closer objects.

The farthest point which can be seen distinctly by the unaided eye is called the **far point** — infinity for the normal eye; the nearest point that can be focused distinctly by the unaided eye is called the **near point** — 25 cm

for a normal adult eye but less for younger people. The distance of 25 cm is known as the **distance of most distinct vision**. The range of accommodation of the normal eye is thus from 25 cm to infinity and when relaxed it is focused on infinity.

(a) Short sight (myopia)

A short-sighted person sees near objects clearly but the far point is closer than infinity. The image of a distant object is focused in front of the retina because the focal length of the eye is too short for the length of the eyeball, Fig. 6.67a. The defect is corrected by a *diverging* spectacle lens whose focal length f is such that it produces a virtual image at the far point of the eye of an object at infinity, Fig. 6.67b. For example, if the far point is 200 cm, then $v = -200$ cm, $u = \infty$ and from $1/f = 1/v + 1/u$ we get $1/f = -1/200 + 1/\infty$. Therefore $f = -200$ cm.

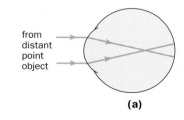

from distant point object

(a)

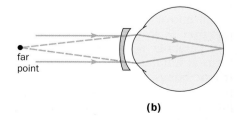

far point

(b)

Fig. 6.67 Short sight and its correction

(b) Long sight (hypermetropia)

A long-sighted person sees distant objects clearly but the near point is more than 25 cm from the eye. The image of a near object is focused behind the retina because the focal length of the eye is too long for the length of the eyeball, Fig. 6.68*a*. The defect is corrected by a *converging* spectacle lens of focal length *f* which gives a virtual image at the near point of the eye for an object at 25 cm, Fig. 6.68*b*. For example, if the near point is 50 cm, then $u = +25$ cm, $v = -50$ cm and $1/f = -1/50 + 1/25 = +1/50$. Therefore $f = +50$ cm.

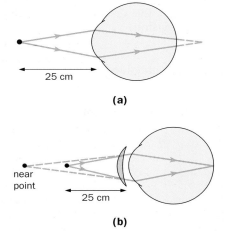

(a)

(b)

Fig. 6.68 Long sight and its correction

(c) Presbyopia

With this defect, which often develops with age, the eye loses its power of accommodation. Two pairs of spectacles may be needed, one for distant objects and the other for reading. Sometimes 'bifocals' are used which have a diverging top part to correct for distant vision and a converging lower part for reading, Fig. 6.69.

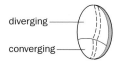

diverging —

converging —

Fig. 6.69 Bifocal lens

(d) Astigmatism

If the curvature of the cornea varies in different directions, rays in different planes from an object are focused in different positions by the eye and the image is distorted. The defect is called astigmatism and anyone suffering from it will see one set of lines in Fig. 6.70 more sharply than the others. It may be possible to correct it with a non-spherical spectacle lens whose curvature increases the effect of that of the cornea in its direction of minimum curvature or decreases it in the maximum curvature direction.

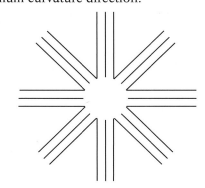

Fig. 6.70

(e) Contact lenses

These consist of tiny, unbreakable plastic lenses held to the cornea by the surface tension (p. 180) of eye fluid. Early 'hard' contact lenses were made of Perspex and many people experienced discomfort in wearing them. Most modern rigid lenses are made from plastics which are permeable to oxygen, making them more comfortable. 'Soft' contact lenses are flexible, being made from a jelly-like plastic which contains water.

As well as being safer for sports, contact lenses may help certain eye defects which spectacles cannot. If the cornea is conical-shaped, for example, vision is very distorted but if a contact lens is fitted and the space between the lens and the cornea filled with a saline solution of the same refractive index as the cornea, normal vision results, Fig. 6.71.

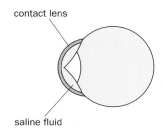

contact lens

saline fluid

Fig. 6.71 Correction of an eye defect by a contact lens

MAGNIFYING POWER OF OPTICAL INSTRUMENTS

Previously, when considering the magnification produced by mirrors and lenses, we used the idea of **linear magnification** *m* and showed that it was given by $m = v/u$. However, if the image is formed at infinity, as it can be with some optical instruments, then *m* should be infinite! The difficulty is that we cannot get to the image to view it and in such cases *m* is therefore not a very helpful indication of the improvement produced by the instrument. A more satisfactory term is clearly necessary to measure this.

The *apparent* size of an object depends on the size of its image on the retina and, as Fig. 6.72 shows, this depends not so much on the actual size of the object as on the angle it subtends at the eye, i.e. on the **visual angle**. Thus, AB is larger than CD but because it subtends the same visual angle as CD, it *appears* to be of equal size. The **angular magnification** or **magnifying power** *M* of an optical instrument is defined by the equation

$$M = \frac{\beta}{\alpha}$$

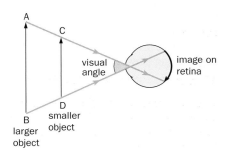

A
C
visual angle
image on retina
B larger object
D smaller object

Fig. 6.72

where β = angle subtended at the eye by the image formed when using the instrument, and α = angle subtended at the unaided eye (i.e. without the instrument) by the object at some 'stated distance'.

In the case of a telescope the 'stated distance' must be where the object (e.g. the moon) is; for a microscope it is usually taken to be the 'distance of most distinct vision' (i.e. 25 cm away) since it is at that distance the object is seen most distinctly by the normal, unaided eye.

The difference between the magnifying power M and the magnification m should be noted. M is the ratio of the *apparent* sizes of image and object and involves a comparison of visual angles; m is the ratio of the *actual* sizes of image and object. They do not necessarily have the same value but in some cases they do.

MAGNIFYING GLASS

This is also called the 'simple microscope' and consists of a converging lens forming a virtual, upright, magnified image of an object placed inside its principal focus, Fig. 6.73. The image appears largest and clearest when it is at the near point.

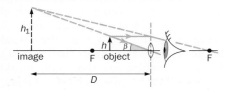

Fig. 6.73   The simple microscope

Assuming rays are paraxial and that the eye is close to the lens, we can say $\beta = h_1/D$ where h_1 is the height of the image and D is the *magnitude* of the distance of most distinct vision (usually 25 cm). If the object is viewed at the near point by the unaided eye, Fig. 6.74, we have $\alpha = h/D$ where h is the height of the object. Hence, the magnifying power M is given by

$$M = \frac{\beta}{\alpha} = \frac{h_1/D}{h/D} = \frac{h_1}{h}$$

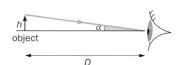

Fig. 6.74

So in this case $M = m$, where the linear magnification $m = v/u$, v and u being the image and object distances respectively.

$$\therefore \quad M = \frac{v}{u}$$

If $1/v + 1/u = 1/f$ is multiplied throughout by v, we get $v/u = v/f - 1$.

$$\therefore \quad M = \frac{v}{f} - 1$$

It follows that a lens of short focal length has a large magnifying power. For example, if $f = +5.0$ cm and $v = -D = -25$ cm (since the image is virtual) then

$$M = \frac{v}{f} - 1 = \frac{-D}{f} - 1 = \frac{-25}{+5} - 1$$
$$= -6.0$$

The magnifying power is 6.0, i.e. $(D/f + 1)$; since M is a number the negative sign can be omitted.

Draw a ray diagram for a magnifying glass forming a virtual image at *infinity* (where must the object be placed?) and use it to show that in this case $M = D/f$ numerically, i.e. M is one less than when the image is at the near point.

What is the effect, if any, on M if the eye is moved back from the lens when the image is at (*i*) the near point, (*ii*) infinity?

COMPOUND MICROSCOPE

The focal length of a lens can be decreased and its magnifying power thereby increased by making its surfaces more curved. However, serious distortion of the image results from excessive curvature and to obtain greater magnifying power a 'compound microscope' is used, consisting of two separated, converging lenses of short focal lengths.

The lens L_1 nearer to the object, called the **objective**, forms a real, magnified, inverted image I_1 of an object O placed just outside its principal focus F_o. I_1 is just inside the principal focus F_e of the second lens L_2, called the **eyepiece**, which acts as a magnifying glass and produces a magnified, virtual

image I_2 of I_1. The microscope is said to be in 'normal adjustment', when I_2 is at the near point. Fig. 6.75 shows the usual constructional rays (see p. 88) drawn to locate I_1 and I_2; note that the object is seen inverted.

(a) Magnifying power

We shall assume that (*i*) all rays are paraxial, (*ii*) the eye is close to the eyepiece and (*iii*) the microscope is in normal adjustment. Then $M = \beta/\alpha$ where, in this case,

β = angle subtended at the eye by I_2 *at the near point*
  = h_2/D (h_2 being the height of I_2 and D the distance of most distinct vision)

and

α = angle subtended at the eye by O at the near point, without the microscope
  = h/D (h being the height of O)

So
$$M = \frac{\beta}{\alpha} = \frac{h_2/D}{h/D} = \frac{h_2}{h}$$
$$= \frac{h_2}{h_1} \times \frac{h_1}{h}$$

(where h_1 is the height of I_1). Now h_2/h_1 is the linear magnification m_e produced by the eyepiece and h_1/h is the linear magnification m_o due to the objective. Therefore

$$M = m_e \times m_o$$

i.e. when the microscope is in normal adjustment with the final image at the near point, the magnifying power equals the linear magnification (as it does for a magnifying glass with the image at the near point). It follows that M will be large if f_o and f_e are small.

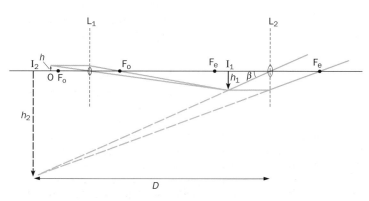

Fig. 6.75   Constructional ray diagram for a compound microscope

Many school-type microscopes have a low-power objective $(f_o \approx 16 \text{ mm})$ magnifying 10 times and a high-power objective $(f_o \approx 4 \text{ mm})$ magnifying 40 times. When used with a ×10 eyepiece, the overall magnifying power is therefore 100 on low power and 400 on high power.

If prolonged observation is to be made it is more restful for the eye to view the final image I_2 at infinity instead of at the near point. The intermediate image I_1 must then be at the principal focus F_e of the eyepiece so that the emergent rays from the eyepiece are parallel. It can be shown that the magnifying power is then slightly less than for normal adjustment.

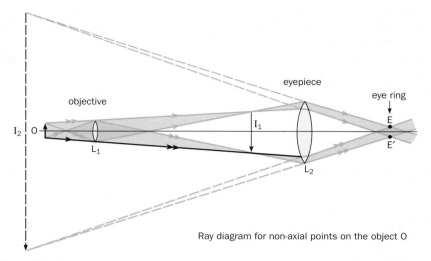

Ray diagram for non-axial points on the object O

Fig. 6.77   The position of the eye ring

(b) Resolving power

This is the ability of an optical instrument to reveal detail, i.e. to form separate images of objects that are very close together. To increase the magnifying power without also increasing the resolving power can be likened to stretching an elastic sheet on which a picture has been painted — the picture gets bigger but no more detail is seen.

It can be shown that the resolving power of a microscope is greater (i) the greater the angle θ subtended at the objective by a point in the object, Fig. 6.76, and (ii) the shorter the wavelength of the light used. Both these factors impose a definite limit on the resolving power and, having regard to this limit, the maximum useful magnifying power is about 600. In practice, in the interests of eye comfort, this value is often exceeded (over ×2000 is attainable in the best instruments).

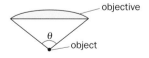

Fig. 6.76

(c) The eye ring

The best position for an observer to place the eye when using a microscope is where it gathers most light from that passing through the objective. The image is then brightest and the field of view greatest.

In Fig. 6.77 the paths of two cones of rays are shown coming from the top and bottom of an object, filling the whole aperture of the objective and passing through the microscope. (Constructional rays, not shown here, are first drawn as in Fig. 6.75 to locate I_1 and I_2.) The only position at which the eye would receive both these cones of light (and also those falling on the objective from all other object points) is at EE' where they cross. All light from the objective refracted by the eyepiece will pass through a small circle of diameter EE', which must therefore be the image of the objective formed by the eyepiece. This image at EE' is called the **eye ring** (or exit-pupil) and it is the best position for an observer's eye.

Ideally EE' should equal the diameter of the average eye pupil and in a microscope a circular opening of this size is often fixed just beyond the eyepiece to indicate the eye ring position. If, for example, the objective is 15 cm from the eyepiece of focal length 1 cm then the distance v of the eye ring from the eyepiece is given by $1/v + 1/(+15) = 1/(+1)$, i.e. $v = 1.1$ cm.

(d) A calculation

The objective and the eyepiece of a microscope may be treated as thin lenses with focal lengths of 2.0 cm and 5.0 cm respectively. If the distance between them is 15 cm and the final image is formed 25 cm from the eyepiece, calculate (i) the position of the object and (ii) the magnifying power of the microscope.

Let the positions of the object O, the first image I_1 and the final image I_2 be as in Fig. 6.78. A ray from the top of O through the optical centre P_1 of the objective passes through the top of I_1 and a ray from the top of I_1 through the optical centre P_2 of the eyepiece passes through the top of I_2 when produced backwards. Let h and h_1 be the heights of O and I_1 respectively.

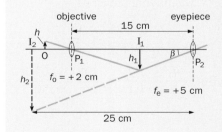

Fig. 6.78

(i) Consider the eyepiece. I_1 acts as the object and the final image I_2 is *virtual*. We have $v = -25$ cm and $f_e = +5.0$ cm,

$$\therefore \quad \frac{1}{(-25)} + \frac{1}{u} = \frac{1}{(+5)}$$

$$\therefore \quad \frac{1}{u} = \frac{1}{5} + \frac{1}{25} = \frac{6}{25}$$

$$\therefore \quad u = I_1 P_2 = +4\tfrac{1}{6} \text{ cm}$$

Now consider the objective. Image I_1 is *real* and at a distance P_1I_1 from the objective where $P_1I_1 = P_1P_2 - I_1P_2 = 15 - 4\frac{1}{6} = 10\frac{5}{6}$ cm. Hence $v = P_1I_1 = +10\frac{5}{6} = +65/6$ cm. Also $f_o = +2.0$ cm,

$$\therefore \quad \frac{1}{(+65/6)} + \frac{1}{u} = \frac{1}{(+2)}$$

$$\therefore \quad \frac{1}{u} = \frac{1}{2} - \frac{6}{65} = \frac{53}{130}$$

$$\therefore \quad u = OP_1 = +\frac{130}{53} \text{ cm}$$

$$= +2\frac{24}{53} \text{ cm} \quad (\approx 2.5 \text{ cm})$$

The object O is about 2.5 cm from the objective.

(*ii*) Assuming the observer is close to the eyepiece, the angle subtended at the eye is given by

$$\beta = \frac{h_1}{I_1P_2} = \frac{h_1}{(25/6)} = \frac{6h_1}{25}$$

The angle α subtended at the observer's eye when the object is viewed at the near point (assumed to be 25 cm away) without the microscope is given by $\alpha = h/25$, Fig. 6.79.

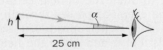

Fig. 6.79

$$\therefore \quad \text{Magnifying power } M = \frac{\beta}{\alpha}$$

$$= \frac{6h_1/25}{h/25} = \frac{6h_1}{h}$$

But $\dfrac{h_1}{h} = \dfrac{P_1I_1}{P_1O} = \dfrac{10\frac{5}{6}}{2\frac{24}{53}} = \dfrac{65/6}{130/53}$

$$= \frac{53}{12}$$

$$\therefore \quad M = 6 \times \frac{53}{12} = 27$$

REFRACTING ASTRONOMICAL TELESCOPE

A lens-type astronomical telescope consists of two converging lenses; one is an objective of long focal length and the other an eyepiece of short focal length. The objective L_1 forms a real, diminished, inverted image I_1 of a distant object at its principal focus F_o since the rays incident on L_1 from a *point* on such an object can be assumed parallel. The eyepiece L_2 acts as a magnifying glass and forms a magnified, virtual image of I_1 and, when the telescope is in normal adjustment, this image is at infinity. I_1 must therefore be at the principal focus F_e of L_2, hence F_o and F_e coincide.

In Fig. 6.80 three *actual* rays are shown coming from the top of a distant object and passing through the top of I_1 in the focal plane of L_1. They must emerge parallel from L_2 to appear to come from the top of the final image at infinity. They must also be parallel to the line joining the top of I_1 to the optical centre of L_2.

(a) Magnifying power

It will be assumed that (*i*) all rays are paraxial, (*ii*) the eye is close to the eyepiece and (*iii*) the telescope is in normal adjustment. Then $M = \beta/\alpha$ and, in this case,

β = angle subtended at the eye by the final image at infinity

= angle subtended at the eye by I_1

= h_1/f_e (h_1 being the height of I_1)

and

α = angle subtended at the eye by the object without the telescope

= angle subtended at the objective by the object

(since the distance between L_1 and L_2 is very small compared with the distance of the object from L_1). From the diagram (Fig. 6.80)

$$\alpha = h_1/f_o$$

Hence $\quad M = \dfrac{\beta}{\alpha} = \dfrac{h_1/f_e}{h_1/f_o}$

$$M = \frac{f_o}{f_e}$$

Notes. (*i*) The above expression for M is true only for normal adjustment; the separation of the objective and eyepiece is then $f_o + f_e$.

(*ii*) A telescope is in normal adjustment when the final image is formed at infinity; a microscope is in normal adjustment with the final image at the near point.

For a high magnifying power the objective should have a large focal length and the eyepiece a small one. The largest lens telescope in the world is at the Yerkes Observatory, USA; the objective has a focal length of about 20 metres and the most powerful eyepiece has a focal length of about 6.5 mm. The maximum value of M is therefore $20 \times 10^3/6.5 \approx 3000$.

If it is desired to form the final image at the near point, i.e. telescope not in normal adjustment, the eyepiece must be moved so that I_1 is closer to it than F_e. The magnifying power is then slightly greater than f_o/f_e.

(b) Resolving power

It can be shown that the ability of a telescope to reveal detail increases as the diameter of the objective increases. However, large lenses are not

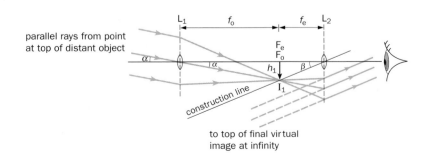

Fig. 6.80 The refracting astronomical telescope

only difficult to make but they tend to sag under their own weight. The objective of the Yerkes telescope has a diameter of 1 metre which is about the maximum possible. There is no point in increasing the magnifying power of a telescope further if the resolving power cannot also be increased.

(c) The eye ring

As in the case of the microscope, the eye ring is in the best position for the eye and is the circular image of the objective formed by the eyepiece. All rays incident on the objective which leave the telescope pass through it. In Fig. 6.81a two cones of rays are shown coming from the top and bottom of a distant object and crossing at the eye ring EE'.

If the telescope is in normal adjustment, the separation of the lenses is $f_o + f_e$ and, from similar triangles in Fig. 6.81b,

$$\frac{AB}{EE'} = \frac{f_o}{f_e}$$

But the magnifying power M for a telescope in normal adjustment is f_o/f_e, hence

$$M = \frac{\text{diameter of objective}}{\text{diameter of eye ring}}$$

This expression enables M to be found simply by illuminating the objective with a sheet of frosted glass and a lamp, locating the image of the objective formed by the eyepiece (i.e. the eye ring) on a screen and measuring the diameter.

(d) Brightness of image

A telescope increases the light-gathering power of the eye and in the case of a point object, such as a star, forms a brighter image. Thus, when the diameter of the objective is doubled, the telescope collects four times more light from a given star (why?) and since a point image is formed of a point object, whatever the magnifying power, the star appears brighter. Many more faint stars become visible when the diameter of the objective lens is increased.

The brightness of the background is not similarly increased because it acts as an extended object and, as we will now see, a telescope does not increase the brightness of such an object.

When the diameter of the eye ring equals the diameter of the pupil of the eye, almost all the light entering the telescope enters the eye. If M is the magnifying power of the telescope, the diameter of the objective is M times the diameter of the eye ring and the area of the objective is M^2 times greater. M^2 times more light enters the eye via the telescope than would enter it from the object directly. However, the image of an extended object has an area M^2 times that of the object, since the telescope makes the object appear M times as high and M times as wide as it does to the unaided eye at the same distance. The brightness of the image cannot therefore exceed that of the object and in fact is less, owing to loss of light in the instrument. The contrast between a star and its background is therefore increased by a telescope.

(e) A calculation

An astronomical telescope has an objective of focal length 100 cm and an eyepiece of focal length 5.0 cm. Calculate the power when the final image of a distant object is formed (i) at infinity, (ii) 25.0 cm from the eyepiece.

(i) When the final image is at infinity, the telescope is in normal adjustment and the magnifying power is given by

$$M = \frac{f_o}{f_e} = \frac{100}{5.0} = 20$$

(ii) Let the position of I_1 be as shown in Fig. 6.82. A ray from the top of the distant object through the optical centre P_1 of the objective passes through the top of I_1. Also a ray from the top of I_1 through the optical centre P_2 of the eyepiece passes through the top of the final image when produced backwards.

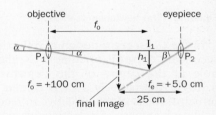

objective   eyepiece

Fig. 6.82

For the eyepiece we can say $v = -25.0$ cm (final image *virtual*) and $f_e = +5.0$ cm.

$$\therefore \quad \frac{1}{(-25)} + \frac{1}{u} = \frac{1}{(+5)}$$

$$\therefore \quad \frac{1}{u} = \frac{1}{5} + \frac{1}{25} = \frac{6}{25}$$

$$\therefore \quad u = I_1 P_2 = 4\tfrac{1}{6}\ \text{cm}$$

If the eye is close to the eyepiece the angle β subtended at the eye is given by $\beta = h_1/I_1 P_2 = h_1/(25/6) = 6h_1/25$. The angle α subtended at the unaided eye by the object = angle subtended at the objective by the object (see p. 98) = $h_1/f_o = h_1/100$.

Hence $M = \dfrac{\beta}{\alpha} = \dfrac{6h_1/25}{h_1/100} = \dfrac{6 \times 100}{25}$

$$= 24$$

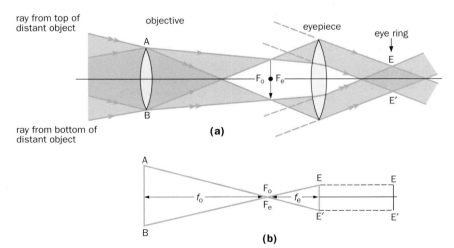

ray from top of distant object   objective   eyepiece   eye ring

ray from bottom of distant object   **(a)**

(b)

Fig. 6.81

REFLECTING ASTRONOMICAL TELESCOPE

The largest modern astronomical telescopes use a concave mirror of long focal length as the objective instead of a converging lens but the principle is the same as for the refracting telescope. One arrangement, called the **Newtonian** form after the inventor of the reflecting telescope, is shown in Fig. 6.83. Parallel rays from a distant point object on the axis are reflected first at the objective and then at a small plane mirror to form a real image I_1 which can be magnified by an eyepiece or photographed by having a film at I_1. The plane mirror, whose area is negligible compared with that of the concave mirror, deflects the light sideways without altering the effective focal length f_o of the objective. In normal adjustment the magnifying power is f_o/f_e where f_e is the focal length of the eyepiece.

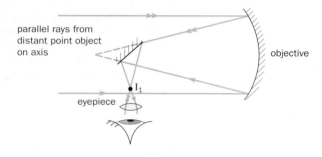

Fig. 6.83   The Newtonian telescope

In the *Cassegrain* design of reflecting telescope the small secondary mirror is orientated to reflect the light through a hole at the centre of the objective (see question 30b, p. 107). Alternatively a CCD camera can be placed at the principal focus of the objective to record the image.

The advantages of reflecting telescopes are

- no chromatic aberration since no refraction occurs at the objective;
- no spherical aberration for a point object on the axis at infinity if a paraboloidal mirror is used (see p. 93);
- a mirror can have a much larger diameter than a lens (since it can be supported at the back) thereby giving greater resolving power and a brighter image of a point object;
- only one surface requires to be ground (compared with two for a lens), thus reducing costs.

The large reflecting optical telescope on Mauna Kea, Hawaii, Fig. 6.84, has a concave paraboloidal mirror of diameter 3.6 metres. It is used in conjunction with spectrometers, cameras and other instruments in temperature-controlled, air-conditioned surroundings.

Blurring of images from ground-based telescopes occurs because of variations in the refractive index of the earth's atmosphere. By placing the telescope above the atmosphere, as with the Hubble space telescope (see chapter 26), this effect can be eliminated. Very sharp images have been obtained from the Hubble telescope, which has a mirror diameter of 2.4 m.

For general astronomical work lens telescopes are more easily handled than large mirror telescopes; the latter are used only where high resolving power is required.

Fig. 6.84   Large reflecting optical telescope, Mauna Kea, Hawaii

OTHER TELESCOPES

(a) Terrestrial telescope

The final image in an astronomical telescope is inverted and although this is not a handicap for looking at a star, it is when viewing objects on earth.

A terrestrial telescope is a refracting astronomical telescope with an intermediate 'erecting' lens arranged as in Fig. 6.85 to be at a distance of $2f$ (where f is the focal length of the erecting lens) from the inverted image I_1 formed by the objective. An erect image I_2 of the same size as I_1 is formed at $2f$ beyond the erecting lens and acts as an 'object' for the eyepiece in the usual way. A disadvantage of this arrangement is the increase in length of the telescope by $4f$.

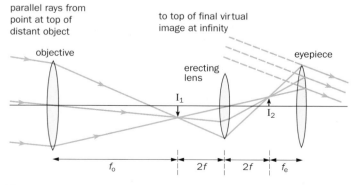

Fig. 6.85   The terrestrial telescope

(b) Prism binoculars

These consist of a pair of refracting astronomical telescopes with two totally reflecting prisms (angles 90°, 45° and 45°) between each objective and eyepiece, as in Fig. 6.86. Prism A causes lateral inversion and prism B inverts vertically so that the final image is the same way round and the same way up as the object. Each prism reflects the light through 180°, making the effective length of each telescope three times the distance between the objective and the eyepiece. Good magnifying power is thus obtained with compactness.

Prism binoculars marked '7 × 50' have a magnifying power of 7 and objectives of diameter 50 mm.

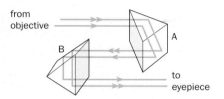

Fig. 6.86 Arrangement in prism binoculars

(c) Galilean telescope

A final erect image is obtained in the Galilean telescope using only two lenses — a converging objective of large focal length f_o and a diverging eyepiece of small focal length f_e.

The image of a distant object would, in the absence of the eyepiece, be formed by the objective at I_1, where $P_1I_1 = f_o$, Fig. 6.87. With the eyepiece in position at a distance f_e from I_1, the separation of the lenses is $f_o - f_e$ (numerically) and rays falling on the eyepiece emerge parallel, so that to the eye the top of the final image is *above* the axis of the telescope. An upright image at infinity is thereby obtained. The converging light falling on the eyepiece behaves like a virtual object at I_1 and a virtual image of it is formed.

If the telescope is in normal adjustment, i.e. final image at infinity, the magnifying power is f_o/f_e, as for an astronomical telescope. In Fig. 6.87 the ray from the top of I_1 passing through the centre P_2 of the eyepiece goes to the top of the final image at infinity. It must therefore be parallel to the three parallel rays emerging from the eyepiece. The angle β subtended at

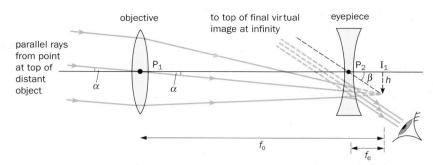

Fig. 6.87 The Galilean telescope

the eye (close to the telescope) is thus given by $\beta = h/I_1P_2 = h/f_e$ where h is the height of I_1. Also, the angle α subtended at the unaided eye by the object is very nearly equal to the angle subtended at the objective by the object, hence $\alpha = h/I_1P_1 = h/f_o$. So the magnifying power M is given by

$$M = \frac{\beta}{\alpha} = \frac{h/f_e}{h/f_o} = \frac{f_o}{f_e}$$

The Galilean telescope is shorter than the terrestrial telescope but the field of view is very limited because the eye ring is between the lenses (why?) and so inaccessible to the eye. Opera glasses consist of two telescopes of this type.

SPECTROMETER

The spectrometer is designed primarily to produce and make measurements on the spectra of light sources and is generally used with a diffraction grating but a prism can be employed. It also provides a very accurate method of measuring refractive index.

The instrument consists of (*i*) a fixed collimator with a movable slit of adjustable width (to produce a parallel beam of light from the source illuminating the slit), (*ii*) a turntable (having a circular scale) on which the grating or prism is placed and (*iii*) a telescope (with a vernier scale) rotatable about the same vertical axis as the turntable, Fig. 6.88. The converging lenses in the collimator and telescope are achromatic.

CAMERA

A typical arrangement is shown in Fig. 6.89. The lens system has to have a field of view of about 50° (compared with 1° or so for an average microscope objective) and so the reduction of aberrations is a major consideration. Very large apertures give blurred images because of aberrations. So do very small apertures, in this case due to the phenomenon called diffraction. The best images are therefore generally obtained with intermediate apertures. For some types of optical system, e.g. eyes, cameras and enlargers, aberrations are more significant than diffraction and the aperture has to be reduced to obtain clear images. For others, e.g. telescopes, diffraction is usually more significant and apertures have to be made as large as is practicable.

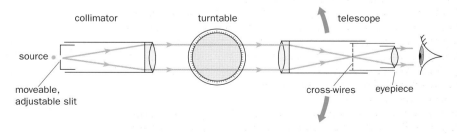

Fig. 6.88 Spectrometer

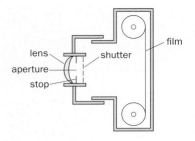

Fig. 6.89 Camera

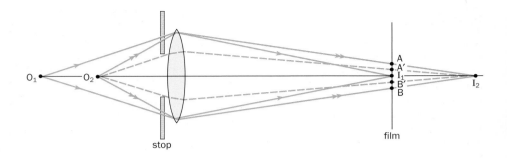

Fig. 6.90 Stopping down a camera lens increases the depth of field

Inexpensive cameras use a meniscus lens, which is usually an achromatic doublet, and a stop to restrict the aperture. More expensive cameras have a lens system of several components designed to minimize the various aberrations. Focusing of objects at different distances is achieved by slightly altering the separation of the lens from the film.

In many cameras the amount of light passing through the lens can be altered by an aperture control or stop of variable width. This has a scale of **f-numbers** with all or some of the following settings — 1.4, 2, 2.8, 4, 5.6, 8, 11, 16, 22, 32. These are such that reducing the f-number by one setting, say from 8 to 5.6, *doubles* the area of the aperture, i.e. the *smaller* the f-number the *larger* the aperture. An f-number of 4 means the diameter d of the aperture is $\frac{1}{4}$ the focal length f of the lens, i.e. $d = f/4$.

The aperture affects (*i*) the exposure time and (*ii*) the depth of field. Consider (*i*). Using the next lower f-number halves the exposure time needed to produce the same illumination on the film (since the area of the aperture has been doubled). The exposure required depends on the lighting conditions and must be brief if the object is moving.

The **depth of field** is the range of distances in which the camera can more or less focus objects simultaneously. A landscape photograph needs a large depth of field, whereas for a family group it may be desirable to have the background out of focus. The depth of field is increased by reducing the lens aperture, as can be seen from Fig. 6.90, in which the images formed by a lens of point objects O_1 and O_2 are at I_1 and I_2 respectively. The diameter of the circular patch of light on a film in focus for I_1 is, for the out-of-focus I_2, AB if the whole aperture is

used but only A'B' if the lens is stopped down. **Depth of focus** is the tiny distance the film plane can be moved to or from the lens without defocusing the image.

TELEPHOTO LENS

To obtain a larger image of a distant object on the film of a camera, the focal length of the lens needs to be increased. This is achieved without making the camera too long by using a **telephoto** lens.

It consists of a convex lens in front of a concave lens which converges the light from the convex lens less rapidly, Fig. 6.91. The combination acts as a convex lens of greater focal length.

The distance from the concave lens to the film (the **back focal length**) is appreciably less than the effective focal length of both lenses together. The camera is thus shorter than it would be if just one convex lens had been used to give the required magnification.

The **telephoto ratio** of the combination is defined by

$$\text{telephoto ratio} = \frac{\text{equivalent focal length}}{\text{back focal length}}$$

Typically, the ratio is 2, making the camera around half the size it would be with an equivalent single lens. (See question 26, p. 106.)

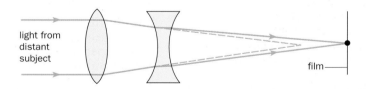

Fig. 6.91 Telephoto lens system

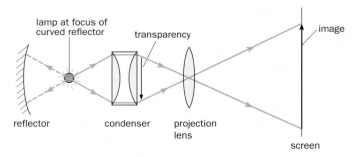

Fig. 6.92 Projector

PROJECTOR

A projector is designed to throw on a screen a magnified image of a film or transparency. It consists of an illumination system and a projection lens, Fig. 6.92.

The image on the screen is usually so highly magnified that very strong but uniform illumination of the film with white light is necessary if the image is also to be bright. This is achieved by directing light from a high-power, intense, tungsten–iodine filament lamp on to the film by means of a curved reflector and a condenser lens system arranged as shown. Since the screen is generally a considerable distance away, the film, inverted, must be just outside the principal focus of the projection lens. Any chance of the image of the lamp appearing on the screen is also removed. To keep the projector as short as possible, the condenser has a short focal length f and the lamp is placed at a distance of $2f$ from it. What will be, approximately, (*i*) the separation of the condenser and the projection lens and (*ii*) the focal length of the projection lens if the film is close to the condenser?

RADIO TELESCOPES

Certain 'objects' in space, among them the sun, emit radio signals which can be picked up by huge aerials called radio telescopes. These do not give visual pictures as do optical telescopes, but produce electrical signals which are recorded graphically and can be converted into visual images by computer. There are various types of

Fig. 6.93b Very Large Array (VLA) of radio aerials in New Mexico

radio telescope: the 'Mark IA' one at Jodrell Bank, Cheshire, shown on the right in Fig. 6.93*a*, consists of a steerable metal reflector or 'dish', 75 metres in diameter; others have isolated aerials distributed over a large area, Fig. 6.93*b*.

The resolving power of any telescope depends on the diameter of the objective (lens, mirror or aerial) and the wavelength of the radiation from the object. The larger the diameter and the shorter the wavelength, the closer together two distant points can be and still be separated by the telescope. The Mount Palomar optical telescope has an objective (concave mirror) of diameter 5 metres but its resolving power is more than 1000 times that of the Jodrell Bank radio telescope with an aerial of diameter 75 metres. This is because the shortest wavelength of the radio signals from outer space that can penetrate the ionized layers in the upper atmosphere is about 1 cm, whereas the mean wavelength of light is 6×10^{-5} cm. (The longest wavelength that can pass through the earth's radio 'window' is about 30 metres.) Large arrays of radio aerials (Fig. 6.93*b*) can, with computer-aided deciphering of their combined signals, provide a much greater resolving power.

One of the strongest radio sources in our own star system (**galaxy**) is the Crab nebula, a mass of luminous gas in the constellation of the Bull (see Fig. 26.6). This nebula is believed to be the remains of a star which underwent a tremendous explosion, becoming a **supernova** and shining so brightly that it was observed in 1054, according to Chinese records, in broad daylight for several months. The radio signals arise from the highly excited gas which is still expanding outwards from the explosion centre.

ELECTRON MICROSCOPES

The electron microscope is analogous in principle to the optical microscope but its performance is far superior. The maximum magnifying power attainable with the best optical microscope is about 2000; for an electron microscope up to 100 000 is typical. An electron microscope can resolve detail about 10^{-9} m across, as compared with 10^{-6} m for the best optical microscope. Since atomic dimensions are of the order of 10^{-9} m (1 nm), this means that in some cases an electron microscope can reveal separate molecules.

The similarity between the paths of light in an optical microscope and the paths of electrons in an electron microscope can be seen from Fig. 6.94. In the electron microscope, electrons are produced by an electron gun and the 'lenses' are electromagnets designed so that their fields focus the electron beam to give an image on a fluorescent screen or photographic plate. The focal lengths of the 'lenses' are variable and determined by the current through the electromagnet coils.

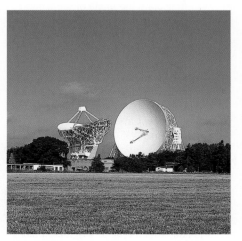

Fig. 6.93a Radio telescopes at Jodrell Bank

When the object is struck by electrons, more penetrate in some parts than in others, depending on the thickness and density of the part. The image is brightest where most electrons have been transmitted. The object must be very thin, otherwise too much electron scattering occurs and no image forms. Also, the whole arrangement is highly evacuated. Why? An air-lock device permits objects to be inserted and removed without loss of vacuum.

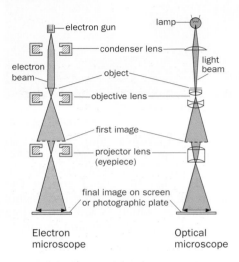

Electron microscope   Optical microscope

Fig. 6.94

The high resolving power of an electron microscope arises from the fact that, just as light can be considered to have both wave-like and particle-like properties, so *moving* electrons seem to have the characteristics of both particles and waves. (Electrons as well as light can give interference and diffraction effects.) The wavelength associated with a moving electron depends on its speed which in turn depends on the p.d. accelerating it in the electron gun. It can be shown that for a p.d. of 100 kV, common in many electron microscopes, the wavelength is about 3.5×10^{-12} m. The resolving power of a microscope increases as the wavelength of the radiation falling on the object decreases and therefore, if we compare the above very small wavelength with that of light (say 6×10^{-5} m), we see why the electron microscope shows very much greater detail.

Two recent developments in electron microscopy are the high-voltage electron microscope and the scanning tunnelling microscope. The first oper-

ates at 1 million volts and enables thicker specimens to be studied. In a scanning tunnelling microscope (STM) a metal tip is scanned very close to the surface of a flat conducting or semiconducting specimen, Fig. 6.95. A high potential difference between tip and specimen enables electrons to tunnel between the two and produce a current, the size of which increases as the tip gets closer to the surface of the specimen. By maintaining a constant current, i.e. a constant separation between tip and surface during a scan, the topography of the surface can be mapped on the atomic scale. The positions of individual atoms and molecules are revealed at resolutions of 10^{-10} m. Photographs taken with this instrument are shown on pp. 14 and 19.

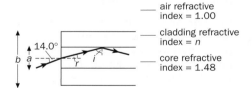

Fig. 6.95

QUESTIONS

Spherical mirrors

1. If a concave mirror has a focal length of 10 cm, find the two positions where an object can be placed to give, in each case, an image twice the height of the object.

2. A convex mirror of radius of curvature 40 cm forms an image which is half the height of the object. Find the object and image positions.

3. A concave mirror of radius of curvature 25 cm faces a convex mirror of radius of curvature 20 cm and is 30 cm from it. If an object is placed midway between the mirrors, find the nature and position of the image formed by reflection first at the concave mirror and then at the convex mirror.

4. i) Give two similarities between an astronomical reflecting telescope and a single dish radio telescope.
   ii) State how the resolving powers of the two telescopes compare. Justify your answer.
   (*NEAB*, AS/A PH04 part qn, June 1998)

Refraction at plane surfaces

5. A ray of light in air passes successively through parallel-sided layers of water and glass. If the angle of incidence in air is 60° and the refractive indices of water and glass are $\frac{4}{3}$ and $\frac{3}{2}$ respectively, calculate
   a) the angle of refraction in the water,
   b) the angle of incidence at the water–glass boundary and
   c) the angle of refraction in the glass.

6. The refractive indices of crown glass and of a certain liquid are 1.51 and 1.63 respectively. Determine the conditions under which total internal reflection can occur at a surface separating them.

7. a) A ray of light is incident at 45° on one face of a 60° prism of refractive index 1.5. Calculate the total deviation of the ray.
   b) A ray of light just undergoes total internal reflection at the second face of a prism of refracting angle 60° and refractive index 1.5. What is its angle of incidence on the first face?

8. Figure 6.96 (not to scale) shows a ray of monochromatic light entering a multimode optical fibre at such an angle that it *just* undergoes total internal reflection at the boundary between the core and the cladding.

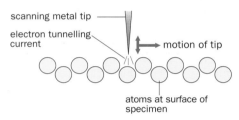

air refractive index = 1.00
cladding refractive index = n
core refractive index = 1.48

Fig. 6.96

Suggest appropriate magnitudes for the diameters labelled a and b.
   Calculate the angle of refraction r.
   Find the angle of incidence i at the boundary between the core and the cladding.
   Hence estimate the refractive index n of the cladding.
   Copy the diagram and show what would happen to the ray of light if it were incident at an angle slightly greater than 14°.
   (*L*, AS/A PH2, June 1997)

9. Figure 6.97a shows a glass, part filled with water. A narrow beam of light is shone vertically down into the water and passes straight through. Figure 6.97b shows the glass tilted until the angle α is such that the light is refracted along the lower surface of the glass.

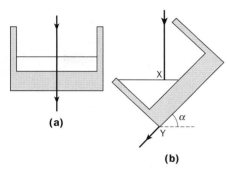

(a)

(b)

Fig. 6.97

Copy and complete Fig. 6.97*b* to show the path of the ray of light from when it enters the water at X to when it leaves the glass at Y.

The refractive indexes of air, glass and water are listed below:

refractive index of air $n_a = 1.00$
refractive index of glass $n_g = 1.50$
refractive index of water $n_w = 1.33$

Calculate the critical angle at the glass/air surface.

Calculate the value of α.

(L, AS/A PH2, June 1996)

10. A 60.0° prism is made of glass whose refractive index for a certain light is 1.65. At what angle of incidence will minimum deviation occur? Between what limits must the angle of incidence lie, if light is to pass through the prism by refraction at adjacent faces?

11. Figures 6.98*a* and *b* show a device for detecting when the level of liquid in a tank is above a certain level. The device uses a solid glass rod bent into a U-shape. The refractive indices *n* are shown on the diagram.

Calculate the critical angle for the glass/air interface.

Calculate the critical angle for the glass/liquid interface.

Copy and complete the diagrams to show the path of the ray of light when
i) the liquid level is low,
ii) the liquid level is high.
Explain how the liquid level detector works.

(L, AS/A PH2, June 1999)

Lenses; the eye

12. **a)** The filament of a lamp is 80 cm from a screen and a converging lens forms an image of it on the screen, magnified three times. Find the distance of the lens from the filament and the focal length of the lens.
b) An erect image 2.0 cm high is formed 12 cm from a lens, the object being 0.5 cm high. Find the focal length of the lens.

13. **a)** A small object is placed on the principal axis of, and 150 mm away from, a diverging lens of focal length 100 mm.
i) Copy Fig. 6.99 and draw rays to show how an image is formed by the lens.

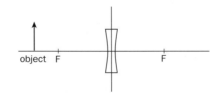

object F   F

Fig. 6.99

ii) Calculate the distance of the image from the lens.
b) The diverging lens in part a) is replaced by a converging lens, also of focal length 100 mm. The object remains in the same position and an image is formed by the converging lens.

Compare *two* properties of this image with those of the image formed by the diverging lens in part a).

(AQA: NEAB, AS/A PH04, March 1999)

14. A lens forms the image of a distant object on a screen 30 cm away. Where should a second lens, of focal length 30 cm, be placed so that the screen has to be moved 8.0 cm towards the first lens for the new image to be in focus?

15. The radii of curvature of the faces of a thin converging meniscus lens of glass of refractive index $\frac{3}{2}$ are 15 cm and 30 cm. What is the focal length of the lens
a) in air,
b) when completely surrounded by water of refractive index $\frac{4}{3}$?

16. Write an essay on the physics of sight. Your account should include
a) an explanation of the formation of a clear image by a healthy eye,
b) an explanation of the terms **accommodation** and **depth of focus**,
c) a description of an experiment to demonstrate how a converging lens may be used to correct long sight,
d) an outline of the social implications of a person's perception of light intensity and of colour advertising, giving examples where appropriate.

(UCLES, Health Physics, March 1998)

17. A student writes the following sentences in an attempt to explain the eye defect known as short sight.

'People suffering from myopia (short sight) are unable to see clearly things which are far away but can see clearly things which are close to them. The lens in the eye, which does all of the refracting, is too powerful and always focuses light to form an image in front of the retina.'

a) Comment on the passage above, identifying and explaining two mistakes (in physics) which the student has made.
b) State the type of lens required to correct short sight.
c) The furthest distance from the eye at which a person can see clearly is 1.2 m and the cornea to retina distance for this person is 1.7 cm.
i) Calculate the focal length of the refracting system of this eye. State any assumptions made.
ii) Show that the power of the refracting system which would allow this eye to view distant objects clearly is 58.8 D.

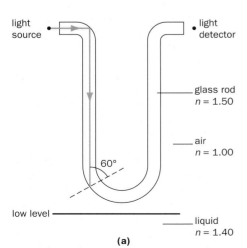

light source

light detector

glass rod
$n = 1.50$

air
$n = 1.00$

60°

low level

liquid
$n = 1.40$

(a)

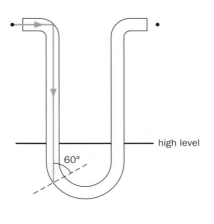

60°

high level

(b)

Fig. 6.98

iii) Calculate the power of the lens required to correct this defect.
(*UCLES,* Health Physics, June 1998)

18. **a)** In an experiment to find the focal length of a converging lens the following values were recorded for object (*u*) and image (*v*) distances.

u/cm	20	24	28	32	36
v/cm	60	40	32	28	26

Using a spreadsheet such as that shown in Fig. 6.57 (p. 92), or otherwise, find the focal length of the lens.
b) How could you modify the spreadsheet used in a) so that it can be used to calculate the image positions (for object distances as in a)) for lenses of different focal lengths?

19. What spectacles are required by a person whose near and far points are 40.0 cm and 200 cm away respectively to bring her near point to a distance of 25.0 cm? Find her new range of vision.

Optical instruments

---

20. Explain the difference between the terms **magnifying power** and **magnification**, as used about optical systems. Illustrate this, by calculating both, in the case of an object placed 5.0 cm from a simple magnifying glass of focal length 6.0 cm, assuming that the minimum distance of distinct vision for the observer is 25 cm.

21. **a)** Explain the terms **magnifying power** and **resolving power** in connection with a microscope.
b) A compound microscope is formed from two lenses of focal lengths 1.0 and 5.0 cm. A small object is placed 1.1 cm from the objective and the microscope adjusted so that the final image is formed 30 cm from the eyepiece. Calculate the angular magnification of the instrument. (Assume that the nearest distance of distinct vision is 25 cm.)

22. Describe, with the help of diagrams, how
a) a single biconvex lens can be used as a magnifying glass,
b) two biconvex lenses can be arranged to form a microscope. State
    i) one advantage,
    ii) one disadvantage,
of setting the microscope so that the final image is at infinity rather than at the near point of the eye.
A centimetre scale is set up 5.0 cm in front of a biconvex lens whose focal length is 4.0 cm. A second biconvex

lens is placed behind the first, on the same axis, at such a distance that the final image formed by the system coincides with the scale itself and that 1.0 mm in the image covers 2.4 cm in the scale. Calculate the position and focal length of the second lens.

23. State what is meant by **normal adjustment** in the case of an astronomical telescope.
Trace the paths of three rays from a distant non-axial point source through an astronomical telescope in normal adjustment.
Define the **magnifying power** of the instrument, and, by reference to your diagram, derive an expression for its magnitude.
A telescope consists of two thin converging lenses of focal lengths 100 cm and 10.0 cm respectively. It is used to view an object 2.00×10^3 cm from the objective. What is the separation of the lenses if the final image is 25.0 cm from the eyepiece? Determine the magnifying power for an observer whose eye is close to the eyepiece.

24. **a)** A particular camera has a single converging lens. Objects at infinity are focused on to the film when the film is 10 cm from the lens. The distance between the film and the lens can be changed by 1.0 cm. Determine the range of object distances which can be brought to focus on the film.
b) i) Using this camera a photograph is taken of a small object which lies within the range of distances obtained in a). With the aid of a ray diagram explain how, without altering the distance between the lens and the film, other objects within the **depth of field** are considered to be in focus.
ii) State how the depth of field obtained in b) i) may be increased without altering the distance between the lens and the film.
iii) Indicate on your ray diagram in b) i) the **depth of focus**.
(*NEAB, AS/A PH04, June 1997*)

25. Two identical single-lens cameras are used to photograph the launch of a space shuttle. The *f*-numbers on camera A and camera B are set at 2.8 and 8, respectively. The two lenses have the same focal length.
    i) Define *f*-number.
    ii) The exposure time for camera A is set at 1/125 s. Calculate the exposure time for camera B if the

same amount of light is to enter each camera.
(*AQA: NEAB, AS/A PH04, March 1999*)

26. The telephoto lens in a camera consists of a convex lens of focal length 10.0 cm with a concave lens of focal length 5.0 cm placed 7.0 cm behind it. The subject to be photographed is 5.0 m in front of the convex lens.
a) What is the total length of the camera from the convex lens to the film when the image is in focus on the film?
b) What magnification is produced?
c) If the telephoto lens is replaced by a single convex lens of a focal length which gives the same magnification as the telephoto lens (the subject still being 5.0 m away), what is the total length of the camera now?
d) What is the focal length of the single equivalent lens?

27. Which pair of lenses with the focal lengths (in cm) listed below could be used to make a telescope which gives an upright, magnified final image?

	A	**B**	**C**	**D**	**E**
Objective	+50	+5.0	+50	+50	+5.0
Eyepiece	−50	−5.0	+5.0	−5.0	−50

28. An astronomical telescope consisting of two convex lenses is used to view a distant piece of paper with two arrows drawn on it as in Fig. 6.100*a*. Which of **A** to **E** in Fig. 6.100*b* represents the image seen?

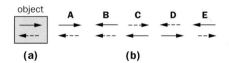

Fig. 6.100

29. A sharp, magnified image of a slide is produced on a screen by a projector. To obtain a larger image that is still sharp, how would you move
a) the screen and
b) the projector lens,
in relation to the slide?

30. **a)** A telescope is made from two converging lenses of focal lengths 2.50 m and 0.020 m.
    i) Show, with the aid of a labelled diagram, how the lenses would be placed for normal adjustment. Show, on the diagram, the principal focus of each lens.
    ii) The telescope is used to observe a planet which subtends an angle of 5.0×10^{-5} rad at the objective.

Calculate the angle subtended at the eye by the final image.

b) Figure 6.101 (not drawn to scale) shows an incomplete Cassegrain reflecting telescope. F_1 is the principal focus of the concave mirror.

  i) Copy the diagram and add the second necessary mirror, M.

  ii) Complete the path of the two rays through the telescope when it is in normal adjustment. Show your reasoning in drawing these rays, either on the diagram or below. Label the principal focus, F_2, of the eye lens, and the position of C, the centre of curvature of M.

c) i) State what is meant by chromatic aberration and explain the effect it would have on the image in an uncorrected refracting telescope.

  ii) Explain why the Cassegrain telescope would be almost free of chromatic aberration.

(AQA: NEAB, AS/A PH01, June 1999)

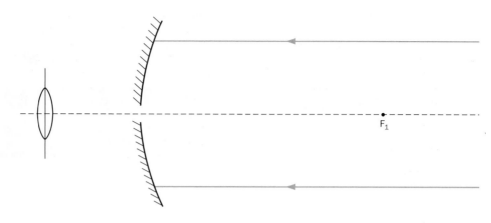

Fig. 6.101

Objective-type revision questions 1

The first figure of a question number gives the relevant chapter, e.g. **2.2** is the second question for chapter 2.

Multiple choice

Select the response that you think is correct.

2.1 The density of aluminium is 2.7 g cm^{-3}, its atomic mass is 27 and the Avogadro constant is 6.0×10^{23} atoms per mole. If aluminium atoms are assumed to be spheres, packed so that they occupy three-quarters of the total volume, the volume of an aluminium atom in cm^3 is
A $4 \times 27/(3 \times 2.7 \times 6.0 \times 10^{23})$
B $3 \times 27/(4 \times 2.7 \times 6.0 \times 10^{23})$
C $3 \times 2.7/(4 \times 27 \times 6.0 \times 10^{23})$
D $4 \times 27 \times 6.0 \times 10^{23}/(3 \times 2.7)$
E $3 \times 2.7 \times 6.0 \times 10^{23}/(4 \times 2.7)$

2.2 The weakest form of bonding in materials is
A van der Waals **B** ionic **C** covalent **D** metallic

3.1 Which one of the following is the Young modulus (in Pa) for the wire having the stress–strain curve of Fig. R1?
A 36×10^{11} **B** 8.0×10^{11} **C** 2.0×10^{11}
D 0.50×10^{11} **E** 16×10^{11}

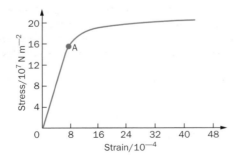

Fig. R1

3.2 Which one of the statements A–E is a correct statement about the evidence provided by Fig. R1?
A The wire only obeys Hooke's law between O and A and after A it becomes much more difficult to stretch it.
B The wire does not obey Hooke's law between O and A and after A it becomes much more difficult to stretch it.
C The wire only obeys Hooke's law between O and A and after A it becomes much easier to stretch it.

D The wire does not obey Hooke's law between O and A and after A it becomes much easier to stretch it.
E The wire does not obey Hooke's law at all and is no harder or easier to stretch before A than after A.

3.3 Wires X and Y are made from the same material. X has twice the diameter and three times the length of Y. If the elastic limits are not reached when each is stretched by the same tension, the ratio of energy stored in X to that in Y is
A $2:3$ **B** $3:4$ **C** $3:2$ **D** $6:1$ **E** $12:1$

3.4 Reasons for the good stiffness and strength of three different materials are given below. Select from the list of five materials the *one* to which *each* statement best applies (three answers).
   i) It has 'foreign' atoms in the lattice which oppose dislocation movement.
   ii) It has high covalent bond density.
   iii) It has long-chain molecules lying more or less parallel along their length.
A steel **B** rubber **C** copper **D** polythene
E glass

4.1 The drift velocity of the free electrons in a conductor is independent of one of the following. Which is it?
A the length of the conductor
B the number of free electrons per unit volume
C the cross-section area of the conductor
D the electronic charge
E the current

4.2 The value of *X* in ohms which gives zero deflection on the galvanometer in Fig. R2 is
A 3 **B** 6 **C** 15 **D** 18 **E** 27

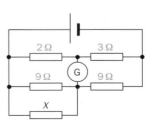

Fig. R2

4.3 If X, Y and Z in Fig. R3 are identical lamps, which of the following changes to the brightnesses of the lamps occur when switch S is closed?
  A X stays the same, Y decreases
  B X increases, Y stays the same
  C X increases, Y decreases
  D X decreases, Y increases
  E X decreases, Y decreases

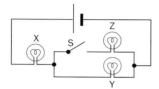

Fig. R3

4.4 A moving-coil galvanometer has a resistance of 10 Ω and gives a full scale deflection for a current of 0.01 A. It could be converted into a voltmeter reading up to 10 V by connecting a resistor of value
  A 0.10 Ω in parallel with it
  B 90 Ω in series with it
  C 0.10 Ω in series with it
  D 990 Ω in parallel with it
  E 990 Ω in series with it

4.5 Which of the graphs in Fig. R4 best shows the variation of current with time in a tungsten filament lamp, from the moment current flows?

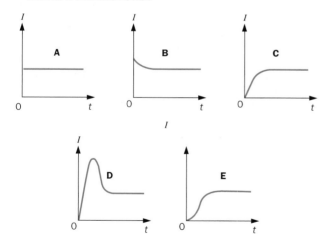

Fig. R4

4.6 Two resistors are connected in parallel as shown in Fig. R5.

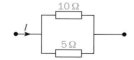

Fig. R5

A current passes through the parallel combination. The power dissipated in the 5.0 Ω resistor is 40 W. Which one of the following is the power dissipated in watts in the 10 Ω resistor?
  A 10 **B** 20 **C** 40 **D** 80

4.7 If each resistor in Fig. R6 is 2 Ω, the effective resistance in ohms between X and Y is
  A $\frac{2}{5}$ **B** 1 **C** 2 **D** $2\frac{2}{3}$ **E** $3\frac{1}{2}$

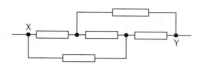

Fig. R6

4.8 A cylindrical copper rod is re-formed to twice its original length. Which one of the following statements describes the way in which the resistance is changed?
  A The resistance remains constant.
  B The resistance increases by a factor of two.
  C The resistance increases by a factor of four.
  D The resistance increases by a factor of eight.

4.9 A thermocouple thermometer is to be designed using the circuit of Fig. R7. AB is a potentiometer wire of resistance 5.0 Ω and ED is a thermocouple whose e.m.f. is 20 mV at 400 °C and zero at 0 °C. For a temperature measurement range from 0 °C to 400 °C, the required value for resistor R in ohms is
  A 195 **B** 295 **C** 395 **D** 495

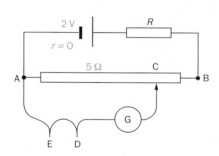

Fig. R7

4.10 A wire of length l and cross-section area A has a conductance G. Another wire, made from the same material, has length $l/2$ and cross-section area $2A$. What is the resulting conductance when the two wires are connected in parallel?
  A $G/4$ **B** $2G$ **C** $5G/4$ **D** $4G$

4.11

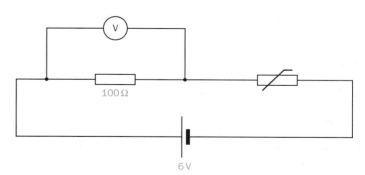

Fig. R8

A thermistor is connected in series to a resistance of 100 Ω and a 6 V battery in the potential divider circuit shown in Fig. R8. When the temperature of the thermistor increases from 20 °C to 40 °C, its resistance drops from 300 Ω to 200 Ω. What will be the corresponding increase in the reading shown on a voltmeter connected across the fixed 100 Ω resistor?

A 0.5 V **B** 1.0 V **C** 1.5 V **D** 2.0 V

5.1 Choose from the following statements one which does *not* apply to the platinum resistance thermometer.
  A It can give high accuracy.
  B It is suitable for measuring the temperature in a small object.
  C It has high heat capacity.
  D It can cover a wide range of temperature.
  E It can only be used for steady temperatures.

5.2 Spheres P and Q are uniformly constructed from the same material, which is a good conductor of heat, and the radius of Q is twice the radius of P. The rate of fall of temperature of P is x times that of Q when both are at the same surface temperature. The value of x is
  A $\frac{1}{4}$ **B** $\frac{1}{2}$ **C** 2 **D** 4 **E** 8

5.3 Heat flows through the bar XYZ in Fig. R9a, the ends X and Z being maintained at fixed temperatures (temperature at X > temperature at Z). If only the part YZ is lagged, which graph in Fig. R9b shows the variation of temperature (θ) with distance along XZ for steady-state conditions?

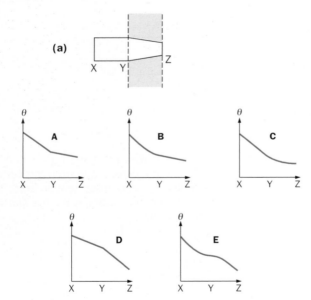

Fig. R9

6.1 A lens of focal length 12 cm forms an upright image three times the size of a real object. The distance in cm between the object and image is
  A 8.0 **B** 16 **C** 24 **D** 32 **E** 48

6.2 When a lens is inserted between an object and a screen which are a fixed distance apart the size of the image is either 6 cm or $\frac{2}{3}$ cm. The size of the object in cm is
  A 2 **B** 3 **C** 4 **D** $4\frac{1}{3}$ **E** 9

6.3 Which *one* of the following combinations of lenses is used as a compound microscope? (The objective is listed first.)
  A long focus converging and shorter focus converging
  B long focus converging and shorter focus diverging
  C long focus converging and long focus converging
  D short focus converging and longer focus converging
  E short focus converging and longer focus diverging

6.4 A converging lens is used to produce an image on a screen of an object. What change is needed for the image to be formed nearer to the lens?
  A Increase the focal length of the lens.
  B Insert a diverging lens between the lens and the screen.
  C Increase the distance of the object from the lens.
  D Move the object closer to the lens.

Multiple selection

In each question one or more of the responses may be correct. Choose one letter from the answer code given.

*Answer **A** if i), ii) and iii) are correct.*
*Answer **B** if only i), ii) are correct.*
*Answer **C** if only ii) and iii) are correct.*
*Answer **D** if i) only is correct.*
*Answer **E** if iii) only is correct.*

4.12 In the potentiometer circuit of Fig. R10 the galvanometer reveals a current in the direction shown wherever the sliding contact touches the wire. This could be caused by
  i) E_1 being too low
  ii) 3.0 Ω being too high
  iii) a break in PQ.

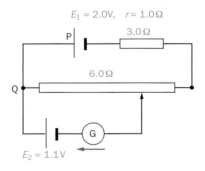

Fig. R10

6.5 A microscope with a short-focus objective
  i) allows more light to be collected
  ii) keeps the distance between objective and eyepiece small
  iii) gives high magnifying power.

Mechanics

Communications satellite in geostationary orbit above the earth

7
Statics and dynamics

- Mechanics
- Composition and resolution of forces
- Moments and couples
- Equilibrium of coplanar forces
- Laws of friction
- Nature of friction
- Triangle and polygon of forces
- Velocity and acceleration
- Equations for uniform acceleration

- Velocity–time graphs
- Motion under gravity
- Projectiles
- Relative velocity
- Newton's laws of motion
- *F* = *ma* calculations
- Air resistance: terminal velocity
- Momentum
- Conservation of momentum
- Rocket and jet propulsion
- Momentum calculations

- Work, energy and power
- Kinetic and potential energy
- Conservation of energy
- Energy calculations
- Elastic and inelastic collisions
- Collisions in two dimensions
- Car collisions and safety

MECHANICS

Mechanics is concerned with the action of forces on a body. If the forces balance the body is said to be **in equilibrium** and the branch of mechanics which deals with such cases is called **statics** — the subject reviewed in the first part of this chapter. In the second part of the chapter we will consider the effects of forces which are not in equilibrium — a study known as **dynamics**.

Many engineering problems, such as designing buildings, bridges, roads, reservoirs, jet engines and aircraft, require the application of the principles of mechanics so that structures with the necessary strength are obtained using the minimum of material. Not only are these principles useful for dealing with ordinary experience but, suitably supplemented, they enable us to deal with the physics of the atom on the one hand, and astronomy and space travel on the other.

COMPOSITION AND RESOLUTION OF FORCES

(a) Scalar and vector quantities

A scalar quantity has magnitude only and is completely described by a certain number of appropriate units. A vector quantity has both magnitude and direction; it can be represented by a straight line whose length represents the magnitude of the quantity on a particular scale and whose direction (shown by an arrow) indicates the direction of the quantity.

For example, if the points X and Y in Fig. 7.1*a* are 2 metres apart, the statement that XY = 2 m fully describes the distance between them; distance is a scalar. However, the **displacement XY** between the points is 2 m in a direction of 30° east of north, Fig. 7.1*b*; displacement is a vector (like other vectors it is often printed in bold type). Other examples of scalars are mass, time, density, speed and energy. Force, velocity (displacement per unit time) and momentum are vectors.

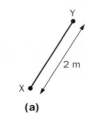

(a)   (b)

Fig. 7.1

(b) Parallelogram of forces

Scalars and vectors require different mathematical treatment. Scalars are added arithmetically but vectors are added geometrically by the **parallelogram law** which ensures their directions as well as their magnitudes are taken into account. The law will be illustrated by the addition of two displacements.

Suppose we walk from A to B and then from B to C as in Fig. 7.2*a* so that we suffer successive displacements **AB** and **BC**. The resultant displacement is given by **AC** in magnitude and direction. The same resultant displacement (i.e. **AC**) would be obtained if we started from the same point A and drew AD equal to BC in magnitude and direction and then drew DC equal to AB, Fig. 7.2*b*. The sum of two vectors therefore equals the diagonal of the parallelogram of which the vectors are adjacent sides. How would you subtract two vectors?

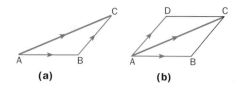

Fig. 7.2 Parallelogram law

The parallelogram law for the addition (composition) of forces is stated as follows.

If two forces acting at a point are represented in magnitude and direction by the sides of a parallelogram drawn from the point, their resultant is represented by the diagonal of the parallelogram drawn from the point.

(c) Resolution of forces

The reverse process to the addition of two vectors by the parallelogram law, is the splitting or **resolving** of one vector into two components. It is particularly useful in the case of forces when the components are taken at right angles to each other.

Suppose the force F is represented by OA in Fig. 7.3a and that we wish to find its components along OX and OY ($\angle$ XOY = 90°). A perpendicular AB is dropped from A on to OX and another AC from A on to OY, to give rectangle (parallelogram) OCAB. OB and OC are the required components of resolved parts. If $\angle$ AOB = θ then

$$\cos \theta = OB/OA = OB/F$$
$$\therefore \quad OB = F \cos \theta$$

and
$$\sin \theta = AB/OA = OC/F$$
$$\therefore \quad OC = F \sin \theta$$

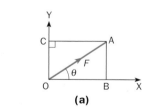

(a)

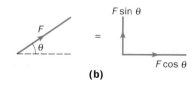

(b)

Fig. 7.3

The two mutually perpendicular forces $F \cos \theta$ and $F \sin \theta$ are thus equivalent to F, Fig. 7.3b. The total effect of F along OX is represented by $F \cos \theta$. Also note that if $\theta = 0$, $F \cos \theta = F$ and $F \sin \theta = 0$, hence a force has no effect in a perpendicular direction. Resolving a force (or any vector) gives two quite independent forces and is a process we shall use frequently.

For example, the component of a force of 10 N in a direction making an angle of 60° with it is $10 \cos 60°$ N, i.e. 5 N; in a direction perpendicular to this component the effective value of the force is $10 \sin 60°$ N, i.e. $5\sqrt{3}$ N.

MOMENTS AND COUPLES

(a) Moment of a force

A force applied to a hinged or pivoted body changes its rotation about the hinge or pivot. Experience shows that the turning effect or **moment** of the force is greater the greater the magnitude of the force and the greater the distance of its point of application from the pivot.

The moment of a force about a point is measured by the product of the force and the perpendicular distance from the line of action of the force to the point.

For example, in Fig. 7.4a if OAB is a trap-door hinged at O and acted on by forces P and Q as shown then

$$\text{moment of } P \text{ about O} = P \times OA$$

and

$$\text{moment of } Q \text{ about O} = Q \times OC$$

Note that the perpendicular distance must be taken. Alternatively, we can resolve Q into component $Q \cos \theta$ perpendicular to OB and $Q \sin \theta$ along OB, Fig. 7.4b. The moment of the latter about O is zero since its line of action passes through O. For the former we have

$$\text{moment of } Q \cos \theta \text{ about O}$$
$$= Q \cos \theta \times OB$$
$$= Q \times OC \quad (\text{since } \cos \theta = OC/OB)$$

which is the same as before. Moments are measured in newton metres (N m) and are given a positive sign if they tend to produce anticlockwise rotation.

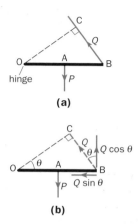

(a)

(b)

Fig. 7.4

(b) Couple

A **couple** consists of two equal and opposite parallel forces whose lines of action do not coincide; it always tends to change rotation. A couple is applied to a water tap to open it. From Fig. 7.5 we can say that the magnitude of the moment or **torque** of the couple P–P about O

$$= P \times OA + P \times OB$$
$$\quad \text{(both are clockwise)}$$
$$= P \times AB$$

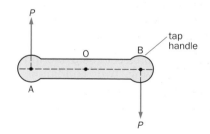

Fig. 7.5 A couple

Hence

torque of couple = one force × perpendicular distance between forces

(c) Useful theorem

When dealing with rotating bodies (chapter 8) it is useful to note that **several forces acting on a body can be reduced to a single force through the centre of mass (gravity) and a couple.**

That this is so can be seen from Fig. 7.6. By obtaining the resultant F of all the forces acting on the body (e.g. by using the polygon of forces, see p. 116, or by repeated use of the parallelogram of forces), diagram (b) can be derived from diagram (a). In diagram (c) the system is not altered if we add two equal and opposite forces F, parallel to the resultant and displaced from it by a perpendicular distance d, acting through the centre of mass O. We can then regard the three forces as a single force F acting through O and an anticlockwise couple T of moment $F \times d$.

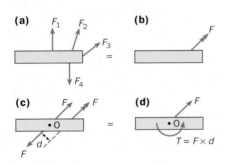

Fig. 7.6

EQUILIBRIUM OF COPLANAR FORCES

(a) General conditions for equilibrium

If a body is acted on by a number of coplanar forces (i.e. forces in the same plane) and is in equilibrium (i.e. there is rest or unaccelerated motion) then the following two conditions must apply.

1. The components of the forces in both of any two directions (usually taken at right angles) must balance.

2. The sum of the clockwise moments about any point equals the sum of the anticlockwise moments about the same point.

The first statement is a consequence of there being no translational motion in any direction and the second follows since there is no rotation of the body. In brief, if a body is in equilibrium the forces and the moments must *both* balance.

(b) Worked example

A sign of mass 5.0 kg is hung from the end B of a uniform bar AB of mass 2.0 kg. The bar is hinged to a wall at A and held horizontal by a wire joining B to a point C which is on the wall vertically above A. If angle ABC = 30°, find the force in the wire and that exerted by the hinge. $(g = 10 \text{ m s}^{-2})$

The weight of the sign will be 50 N and of the bar 20 N (since $W = mg$). The arrangement is shown in Fig. 7.7a. Let P be the force in the wire and suppose Q, the force exerted by the hinge, makes angle θ with the bar. The bar is uniform and so its weight acts vertically downwards at its centre G. Let the length of the bar be $2l$.

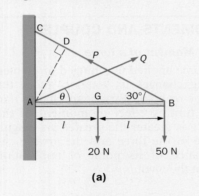

(a)

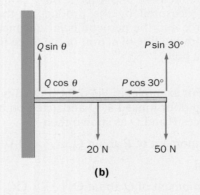

(b)

Fig. 7.7

(*i*) *There is no rotational acceleration*, therefore taking moments about A we have

clockwise moments
$\qquad$ = anticlockwise moment

i.e.

$20 \times l + 50 \times 2l = P \times \text{AD}$
$\qquad$ (AD is perpendicular to BC)

$\therefore 120l = P \times \text{AB} \sin 30°$
$\qquad$ (sin 30° = AD/AB)

$\qquad = P \times 2l \times 0.5$

$\therefore P = 1.2 \times 10^2 \text{ N}$

Note. By taking moments about A there is no need to consider Q since it passes through A and so has zero moment.

(*ii*) *There is no translational acceleration*, therefore the vertical components (and forces) must balance, likewise the horizontal components. Hence resolving Q and P into vertical and horizontal components (which now replace them, Fig. 7.7b) we have:

Vertically

$Q \sin \theta + P \sin 30° = 20 + 50$

$\therefore \quad Q \sin \theta = 70 - 120(1/2)$

$\qquad Q \sin \theta = 10 \qquad (1)$

Horizontally

$Q \cos \theta = P \cos 30° = 120(\sqrt{3}/2)$

$\therefore \quad Q \cos \theta = 60\sqrt{3} \qquad (2)$

Dividing (1) by (2),

$\qquad \tan \theta = 10/(60\sqrt{3})$

$\therefore \quad \theta = 5.5°$

Squaring (1) and (2) and adding,

$Q^2(\sin^2 \theta + \cos^2 \theta) = 100 + 10\,800$

$\therefore Q^2 = 10\,900 \ (\sin^2 \theta + \cos^2 \theta = 1)$

and $\qquad Q = 1.04 \times 10^2 \text{ N}$

(c) Structures

Forces act at a joint in many structures and if these are in equilibrium then so too are the joints. The joint O in the bridge structure of Fig. 7.8 is in equilibrium under the action of forces P and Q exerted by the girders and the normal force S exerted by the bridge support at O. The components of the forces in two perpendicular directions at the joint must balance. Therefore

$\qquad S = Q \sin \theta \quad \text{and} \quad P = Q \cos \theta$

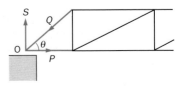

Fig. 7.8

If θ and S are known (the latter from the weight and loading of the bridge) then P and Q (which the bridge designer may wish to know) can be found. Other joints may be treated similarly.

LAWS OF FRICTION

Frictional forces act along the surface between two bodies whenever one moves or tries to move over the other, and in a direction so as to oppose relative motion of the surfaces. Sometimes it is desirable to reduce friction to a minimum but in other cases its presence is essential. For example, it is the frictional push of the ground on the soles of our shoes that enables us to walk. Otherwise our feet would slip backwards as they do when we try to walk on an icy road.

(a) Coefficients of friction

Experiment shows that when a body is at rest, the frictional force to be overcome before it moves, called **limiting friction**, is greater than that which acts once it is moving, called **sliding**, **kinetic** or **dynamic friction** (all terms are used).

The laws of friction, which hold approximately, can be summarized as follows.

1. The limiting frictional force F is directly proportional to the normal force N exerted by the surface on the body (which equals the weight of the body), i.e. $F \propto N$ or $F/N = $ constant.

2. The dynamic frictional force F' is directly proportional to the normal force N, i.e. $F' \propto N$ or $F'/N = $ constant, and is reasonably independent of the speed of motion.

3. The frictional force does not depend on the area of contact of the surfaces if the normal reaction is constant.

The coefficients of limiting and dynamic friction are denoted by μ and μ' respectively and are defined by the equations

$$\mu = F/N \quad \text{and} \quad \mu' = F'/N$$

For two given surfaces μ' is usually less than μ but they are often assumed equal. For wood on wood μ is about 0.2 to 0.5.

In general a surface exerts a frictional force, and the resultant force on a body on the surface has two components — a normal force N perpendicular to the surface and a frictional force F along the surface, Fig. 7.9. If the surface is smooth, as is sometimes assumed in mechanics calculations, $\mu = 0$ and so $F = 0$. Therefore a smooth surface only exerts a force at right angles to itself, i.e. a normal force N.

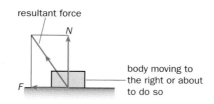

resultant force

body moving to the right or about to do so

Fig. 7.9

The coefficient of limiting friction, μ, can also be found by placing the block on the surface and tilting the latter to the angle θ at which the block is just about to slip, Fig. 7.10a. The three forces acting on the block are its weight mg, the normal force N of the surface and the limiting frictional force F ($= \mu N$). They are in equilibrium and if mg is resolved into components $mg \sin \theta$ along the surface and $mg \cos \theta$ perpendicular to the surface, Fig. 7.10b, then

$$F = \mu N = mg \sin \theta$$

and $$N = mg \cos \theta$$

Dividing, $$\mu = \tan \theta$$

Hence μ can be found by measuring θ, called the **angle of friction**.

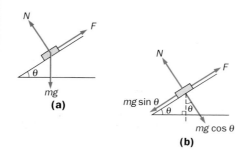

mg

(a)

$mg \sin \theta$

$mg \cos \theta$

(b)

Fig. 7.10

(b) Worked example

A uniform ladder 4.0 m long, of mass 25 kg, rests with its upper end against a smooth vertical wall and with its lower end on rough ground. What must be the least coefficient of friction between the ground and the ladder for it to be inclined at 60° with the horizontal without slipping? ($g = 10 \text{ m s}^{-2}$)

The weight (mg) of the ladder is 250 N and the forces acting on it are shown in Fig. 7.11. The wall is smooth and so the force S of the wall on the ladder is normal to the wall. Since the ladder is uniform its weight W can be taken to act at its mid-point G. If it is about to slip there will be a force exerted on it by the ground which can be resolved into a normal force N and a limiting frictional force $F = \mu N$, where μ is the required coefficient of friction.

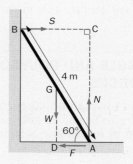

Fig. 7.11

The forces are in equilibrium.

Resolving vertically,

$$N = W = 250 \text{ newtons}$$

Resolving horizontally,

$$F = \mu N = S$$

Taking moments about A,

$$S \times AC = W \times AD$$

$$S \times 4.0 \cos 30° = 250 \times 2.0 \sin 30°$$

$$= 250$$

$$\therefore \quad S = 125/\sqrt{3} \text{ newtons}$$

Hence $$\mu = \frac{S}{N} = \frac{125}{250\sqrt{3}}$$

$$= 0.29$$

NATURE OF FRICTION

Close examination of the flattest and most highly polished surfaces shows that they have hollows and humps more than one hundred atoms high. When one solid is placed on another, contact therefore only occurs at a few places of small area. From electrical resistance measurements of two metals in contact it is estimated that, in the case of steel, the actual area that is touching may be no more than 1/10 000th of the apparent area.

The pressures at the points of contact are extremely high and cause the humps to flatten out (undergoing plastic deformation) until the increased area of contact enables the upper solid to be supported. It is thought that at the points of contact small, cold-welded 'joints' are formed by the strong adhesive forces between molecules which are very close together. These have to be broken before one surface can move over the other.

TRIANGLE AND POLYGON OF FORCES

These are two laws that enable statics problems to be solved graphically. The **triangle of forces** law is stated as follows.

> If three forces acting at a point can be represented in size and direction by the sides of a closed triangle, taken in order, then the forces are in equilibrium.

'Taken in order' means that the arrows showing the force directions follow each other in the *same direction* round the triangle. If the triangle is not closed the forces have a resultant, represented by the line which closes the shape.

In Fig. 7.12*a* three forces are shown acting at a point. In the scale diagram of Fig. 7.12*b*, their sizes and directions are represented by the sides of a closed triangle taken in order. The forces are therefore in equilibrium.

The converse of the triangle of forces is often more useful; it states that if three forces acting at a point are in equilibrium, they can be represented in size and direction by the sides of a triangle taken in order. For example, it could be applied to the three forces P, Q and S acting at O in the structure of Fig. 7.8.

With more than three forces a polygon is drawn; if it is closed, the forces are in equilibrium, Fig. 7.12*c*, *d*. Otherwise, the line closing the polygon is the resultant.

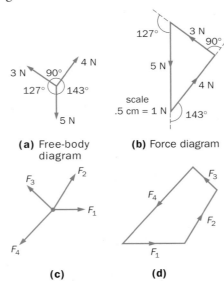

(a) Free-body diagram   **(b)** Force diagram

(c)   **(d)**

Fig. 7.12   Triangle and polygon of forces

In the case of *three non-parallel forces in equilibrium* their lines of action must all pass through the *same* point. Otherwise, if the third force does not act through the intersection of the other two, there is a resultant couple about that point. The worked example on p. 114 and questions 2 and 3 at the end of the chapter can be solved graphically using this fact.

VELOCITY AND ACCELERATION

(a) Speed and velocity

If a car travels from A to B along the route shown in Fig. 7.13*a*, its **average speed** is defined as **the actual distance travelled**, i.e. AXYZB, **divided by the time taken**. The speed at any instant is found by considering a very short time interval. Speed has magnitude only and is a scalar quantity.

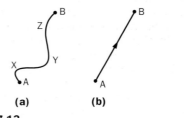

(a)   **(b)**

Fig. 7.13

Velocity is defined as **the distance travelled in a particular direction divided by the time taken**. The average velocity of the car in the direction AB is the distance between A and B, i.e. the length of the straight line AB, Fig. 7.13*b*, divided by the time actually taken for the journey from A to B. AB is the displacement of the car; in this case it is not the actual path followed from A to B, although of course it could be in other cases. Velocity can therefore also be defined as **the change of displacement in unit time**; it has both magnitude and direction and is a vector quantity.

The SI unit of velocity is 1 metre per second (1 m s^{-1}); $10 \text{ m s}^{-1} = 36 \text{ km h}^{-1}$. In Fig. 7.13 if the route AXYZB is 40 kilometres and the car takes 1 hour for the journey, its average **speed** is 40 km h^{-1}. If B is 25 km north-east of A as the crow flies, the car's average **velocity** is only 25 km h^{-1} towards the north-east.

The velocity v of a body which undergoes a very small displacement δs in the very small time δt is given by the equation

$$v = \frac{\delta s}{\delta t}$$

More strictly, in calculus notation, the velocity v at an instant is defined by

$$v = \lim_{\delta t \to 0} \left(\frac{\delta s}{\delta t} \right) = \frac{ds}{dt}$$

Velocity is therefore the **rate of change of displacement**.

A body that covers equal distances in the same straight line in equal time intervals, no matter how short these are, is said to be moving with constant or **uniform velocity**. Only a body moving in a straight line can have uniform velocity. The direction of motion of a body travelling in a curved path is continually changing and so it cannot have uniform velocity even though its speed may be constant.

(b) Acceleration

A body is said to accelerate when its velocity changes. If a very small velocity change δv occurs in a very small time interval δt, the acceleration a of the body is given by

$$a = \frac{\text{change in velocity}}{\text{time taken for change}} = \frac{\delta v}{\delta t}$$

More correctly, in calculus notation, the instantaneous acceleration a is defined by

$$a = \lim_{\delta t \to 0} \left(\frac{\delta v}{\delta t} \right) = \frac{dv}{dt} = \frac{d^2 s}{dt^2}$$

In words, **acceleration is the rate of change of velocity**.

For a car accelerating towards the north from 10 m s^{-1} to 20 m s^{-1} in 5.0 seconds we can say

average acceleration

$$= \frac{(20 - 10) \text{ m s}^{-1}}{5.0 \text{ s}}$$

$$= 2.0 \text{ m s}^{-2} \text{ towards the north}$$

That is, on average, the velocity of the car increases by 2.0 m s^{-1} every second. Since 10 m s^{-1} = 36 km h^{-1} and 20 m s^{-1} = 72 km h^{-1}, we could also say the average acceleration is $(72 - 36)$ km h^{-1}/5 s = 7.2 km h^{-1} per second = 7.2 km h^{-1} s^{-1}.

EQUATIONS FOR UNIFORM ACCELERATION

The acceleration of a body is uniform if its velocity changes by equal amounts in equal times. We will now derive four useful equations for a body moving in a straight line with uniform acceleration.

Suppose the velocity of the body increases steadily from u to v in time t then the uniform acceleration a is given by

$$a = \frac{\text{change of velocity}}{\text{time taken}}$$

$$= \frac{v - u}{t}$$

Therefore

$$v = u + at \qquad (1)$$

Since the velocity is increasing steadily, the average velocity is the mean of the initial and final velocities, i.e.

$$\text{average velocity} = \frac{u + v}{2}$$

If s is the displacement of the body in time t, then since average velocity = displacement/time = s/t, we can say

$$\frac{s}{t} = \frac{u + v}{2} \qquad (2)$$

$$\therefore \quad s = \tfrac{1}{2}(u + v)t$$

But $\qquad v = u + at$

$$\therefore \quad s = \tfrac{1}{2}(u + u + at)t$$

$$s = ut + \tfrac{1}{2}at^2 \qquad (3)$$

If we eliminate t from (3) by substituting $t = (v - u)/a$ from (1), we get on simplifying

$$v^2 = u^2 + 2as \qquad (4)$$

Knowing any three of u, v, a, s and t the others can be found.

VELOCITY–TIME GRAPHS

Acceleration is rate of change of velocity (in calculus notation dv/dt) and equals at any instant the slope of the velocity–time graph. In Fig. 7.14, curve 1 has zero slope and represents uniform velocity, curve 2 is a straight line of constant slope and represents uniform acceleration, while curve 3 is for variable acceleration since its slope varies.

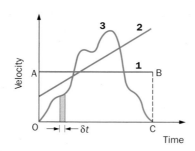

Fig. 7.14

The distance travelled by a body during any interval of time can also be found from a velocity–time graph, a fact which is especially useful in cases of non-uniform acceleration since the three equations of motion do not then apply. For the constant velocity case, curve 1, the distance travelled in time OC = velocity × time = OA × OC = area OABC. For curve 3, if we consider a small enough time interval δt, the velocity is almost constant and the

distance travelled in δt will be the area of the very thin shaded strip. By dividing up the whole area under curve 3 into such strips it follows that the **total distance travelled in time OC equals the area between the velocity–time graph and the time-axis**.

MOTION UNDER GRAVITY

(a) Free fall

Experiments show that at a particular place all bodies falling freely under gravity, in a vacuum or when air resistance is negligible, have the *same* constant acceleration irrespective of their masses. This acceleration towards the surface of the earth, known as the **acceleration of free fall**, is denoted by g. Its magnitude varies slightly from place to place on the earth's surface and is approximately 9.8 m s^{-2}. The velocity of a freely falling body therefore increases by 9.8 m s^{-1} every second; in the equations of motion g replaces a.

A direct determination of g may be made using the apparatus of Fig. 7.15, in which the time for a steel ball-bearing to fall a known distance from rest is measured (to about 0.005 second) by an electric stop-clock. When the two-way switch is changed to the 'down' position, the electromagnet releases the ball and simultaneously the clock starts. At the end of its fall the ball opens the 'trap door' of the impact switch and the clock stops. Air resistance is negligible for a small dense object such as a ball-bearing falling a short distance. The result is found from $s = ut + \tfrac{1}{2}at^2$ with $u = 0$ and $a = g$.

The measurement of g using a simple pendulum is described on p. 153.

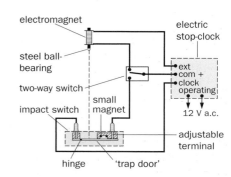

Fig. 7.15 Determination of g

(b) Vertical projection

The velocity of a body projected upwards from the ground decreases by 9.8 m s^{-1} (near enough to 10 m s^{-1}) every second, neglecting the effect of air resistance. Hence if a ball is projected straight upward with an initial velocity of 30 m s^{-1} then in just over 3 seconds it will have zero velocity and be at its highest point.

PROJECTILES

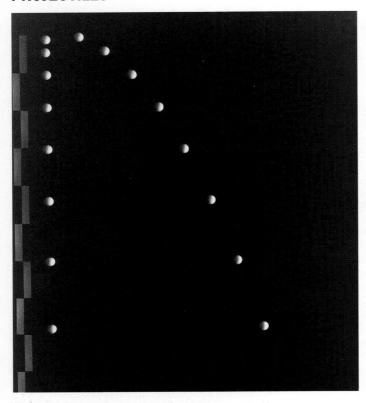

Fig. 7.16   Free fall and projectile motion compared

Figure 7.16 is a multiflash photograph of the motion of two balls, one released from rest and the other projected simultaneously with a horizontal velocity. It is clear that the vertical motion of the projected ball (a constant acceleration = g) is unaffected by its horizontal motion (a constant velocity). The two motions are quite independent of each other.

Consider a body projected obliquely from O with velocity u at an angle θ to the horizontal, Fig. 7.17. Suppose we wish to know the height attained by the body and its horizontal range. If we resolve u into horizontal and vertical components u cos θ and u sin θ respectively, each component can be considered independently of the other.

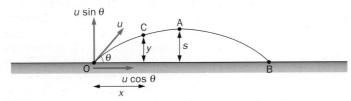

Fig. 7.17

Vertical motion

The body is subject to a constant acceleration $a = -g$. (Here it is convenient to take downward directed vectors as negative, which explains why g has a negative sign.) If s is the height attained then, since the initial velocity is $u \sin \theta$ and the final velocity zero, we have from equation (4) on p. 117

$$0 = u^2 \sin^2 \theta - 2gs$$

$$\therefore \ s = \frac{u^2 \sin^2 \theta}{2g}$$

Also, if t is the time to reach the highest point A, it follows from equation (1) that

$$0 = u \sin \theta - gt$$

$$\therefore \ t = \frac{u \sin \theta}{g}$$

The time taken by the body to fall to the horizontal level of O is also t. Therefore

$$\text{time of flight} = 2t = 2 \ \frac{u \sin \theta}{g}$$

Horizontal motion

Neglecting air resistance, the horizontal component $u \cos \theta$ remains constant during the flight since g has no effect in a horizontal direction. The horizontal distance travelled, OB,

$$= \text{horizontal velocity} \times \text{time of flight}$$

$$= \frac{u \cos \theta \times 2u \sin \theta}{g}$$

$$= \frac{2u^2 \sin \theta \cos \theta}{g} = \frac{u^2 \sin 2\theta}{g}$$

$$(\text{since } \sin 2\theta = 2 \sin \theta \cos \theta)$$

For a given velocity of projection the range is a maximum when sin 2θ = 1, i.e. when θ = 45°, and has the value u^2/g.

Trajectory

From equation (3) on p. 117, after time t the horizontal distance travelled $x = ut \cos \theta$ (since a = 0) and the corresponding vertical distance $y = ut \sin \theta - gt^2/2$ (since $a = -g$). The spreadsheet represented in Fig. 7 (p. 6) shows that when y is plotted against x the trajectory of the projectile is a parabola (neglecting the effect of air resistance).

RELATIVE VELOCITY

Suppose that a car A travelling on a straight road at 80 km h^{-1} passes a car B going in the *same* direction at 50 km h^{-1}, Fig. 7.18a. Then, velocity of A relative to B is given by

$$v_A - v_B = 80 \text{ km h}^{-1} - 50 \text{ km h}^{-1}$$

$$= 30 \text{ km h}^{-1} \text{ (to the right in Fig. 7.18}b)$$

If A and B are travelling in *opposite* directions, Fig. 7.18*c*, we show this by giving one velocity a + sign, say v_A, and the other a − sign. Hence we can write

$$v_A - v_B = +80 \text{ km h}^{-1} - (-50 \text{ km h}^{-1})$$
$$= (80 + 50) \text{ km h}^{-1}$$
$$= 130 \text{ km h}^{-1}$$

(to the right in Fig. 7.18*d*)

In effect, in both cases, to find the velocity of A relative to B, we have applied B's velocity *reversed* to both cars. It is then just as if B is at rest and A has two velocities v_A and v_B, which are subtracted when v_A and v_B are in the same direction and added when they are in opposite directions.

When the velocities are not in the same straight line, they must be added by the parallelogram law.

NEWTON'S LAWS OF MOTION

Newton (1642–1727) studied and developed Galileo's (1564–1642) ideas about motion and subsequently stated the three laws which now bear his name. He established the subject of dynamics. His laws are a set of statements and definitions that we believe to be true because the results they predict are found to be in very exact agreement with experiment over a wide range of conditions. We do not regard them as absolutely true and more exact laws are required for certain extreme cases (see p. 504).

(a) First law

If a body is at rest it remains at rest or if it is in motion it moves with uniform velocity (i.e. constant speed in a straight line) until it is acted on by a resultant force.

The second part of the law appears to disagree with certain everyday experiences which suggest that a steady effort has to be exerted on a body, e.g. a bicycle, even to keep it moving with constant velocity (let alone to accelerate it), otherwise it comes to rest. The law on the other hand states that a moving body retains its motion naturally and if any change occurs (i.e. if it is accelerated) some outside agent — a force — must be responsible. It seems

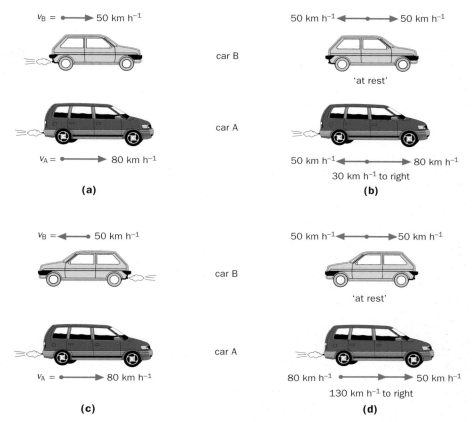

Fig. 7.18

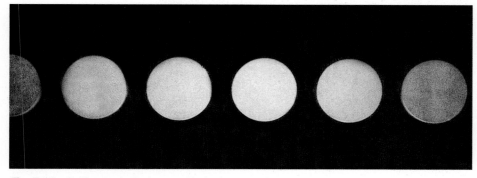

Fig. 7.19 Uniform velocity in near-frictionless conditions

that the question to be asked about a moving body is not 'what keeps it moving' but 'what changes or stops its motion'. Frequently it is friction and if a body does move under near-frictionless conditions, its velocity is in fact almost uniform. This is shown in Fig. 7.19, which is a photograph of a puck, illuminated at equal time intervals by a flashing xenon lamp, moving on a cushion of carbon dioxide gas across a clean, level glass plate.

This law really defines a force as something that changes the state of rest or uniform motion of a body. Contact may be necessary, as when we push a body with our hands, or it may not be, as in the case of gravitational, electric and magnetic forces.

(b) Mass

The first law implies that matter has a built-in reluctance to change its state of rest or motion. This property, possessed by all bodies, is called **inertia**. Its effects are evident when a vehicle suddenly stops, causing the passengers to lurch forward (tending to keep moving), or starts, jerking the passengers backwards (since they tend to remain at rest).

The **mass** of a body is a measure of its inertia; a large mass requires a large force to produce a particular acceleration. The unit of mass is the **kilogram** (kg) and is the mass of a piece of platinum–iridium carefully preserved at Sèvres, near Paris.

In principle, the mass m of a body can be measured by comparing the accelerations a and a_0 produced by the same force in the body and in the standard kilogram (m_0) respectively. The ratio of the two masses is then defined by

$$\frac{m}{m_0} = \frac{a_0}{a}$$

from which m can be calculated. *In practice*, this method is neither quick nor accurate and mass is most readily found using a beam balance to compare the **weight** of the body with that of a standard. It can be shown (see section *(d)* below) that the mass of a body is proportional to its weight and so a beam balance also compares masses.

The second law indicates how forces can be measured.

(c) Second law

> The rate of change of momentum of a body is proportional to the resultant force and occurs in the direction of the force.

The momentum p of a body of constant mass m moving with velocity u is, by definition, mu. That is,

momentum = mass × velocity

Suppose a force F acts on the body for time t and changes its velocity from u to v, then

change of momentum = $mv - mu$

∴ rate of change of momentum

$$= \frac{m(v - u)}{t}$$

Hence, by the second law,

$$F \propto \frac{m(v - u)}{t}$$

If a is the acceleration of the body then

$$a = \frac{v - u}{t}$$

$$\therefore \quad F \propto ma$$

or

$$F = kma$$

where k is a constant. One **newton is defined as the force which gives a mass of 1 kilogram an acceleration of 1 metre per second per second**. So if $m = 1$ kg and $a = 1$ m s^{-2} then $F = 1$ N and substituting these values in $F = kma$ we obtain $k = 1$. So *with these units $k = 1$* and

$$F = ma$$

This expression is one form of Newton's second law and it indicates that a force can be measured by finding the acceleration it produces in a known mass. It can be verified experimentally using, for example, photogate timers and vehicles on an inclined air track. Two points should be noted when using $F = ma$ to solve numerical problems. First, F is the *resultant* (or unbalanced) force causing acceleration a in a particular direction, and, second, F must be in newtons, m in kilograms and a in metres per second squared.

(d) Weight

The weight W of a body is the force of gravity acting on it towards the centre of the earth. Weight is thus a **force**, not to be confused with mass which is independent of the presence or absence of the earth. If g is the acceleration of the body towards the centre of the earth then we can substitute F (force accelerating the body) = W and $a = g$ in $F = ma$, giving

$$W = mg$$

If $g = 9.8$ m s^{-2}, a body of mass 1 kg has a weight of 9.8 N (roughly 10 N). The mass m of a body is constant but its weight mg varies with position on the earth's surface since g varies from place to place. Weight can be measured by a calibrated spring balance.

If two bodies of masses m_1 and m_2 have weights W_1 and W_2 at the same place then

$$W_1 = m_1 g \quad \text{and} \quad W_2 = m_2 g$$

$$\therefore \quad \frac{W_1}{W_2} = \frac{m_1}{m_2}$$

That is, the **weight of a body is proportional to its mass**, a fact we use when finding the mass of a body by comparing its weight with that of standard masses on a beam balance.

(e) Third law

> If body A exerts a force on body B, then body B exerts an equal but opposite force on body A.

The law is stating that forces never occur singly but always in pairs, as a result of the interaction between two bodies. For example, when you step forward from rest your foot pushes backwards on the earth and the earth exerts an equal and opposite force forward on you. Two bodies and two forces are involved. The comparatively small force you exert on the large mass of the earth produces no noticeable acceleration of the earth, but the equal force it exerts on your very much smaller mass causes you to accelerate. It is important to note that the equal and opposite forces *do not act on the same body*; if they did, there could never be any resultant forces and all acceleration would be impossible.

If you pull a string attached to a block with a force P to the right, Fig. 7.20, the string pulls you with an equal force P to the left. Generally we can assume the string transmits the force unchanged and so there is another pair of equal and opposite forces at the block. The string exerts a pull P to the right *on the block* and the block exerts an equal pull to the left *on the string* — one force acts on the block and the other on the string. The string is pulled outwards at both ends and is in a state of tension.

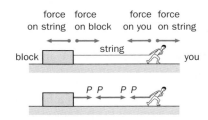

Fig. 7.20

In calculations it is often useful to draw **free-body diagrams** which show all the forces acting on each body.

F = ma CALCULATIONS

Example 1. A rocket develops an initial thrust of 3.3×10^7 N and has a lift-off mass of 2.8×10^6 kg. Find the initial acceleration of the rocket at lift-off. (Take $g = 10$ m s^{-1}.)

Let T be the initial thrust on the rocket, m its mass and W its weight, Fig. 7.21. Then

$$W = mg = 2.8 \times 10^6 \text{ kg} \times 10 \text{ m s}^{-2}$$
$$= 2.8 \times 10^7 \text{ N}$$

Fig. 7.21

The *resultant* upwards force *on the rocket* is $(T - W)$ and if a is the initial vertical acceleration, then from $F = ma$ we have

$$(T - W) = ma$$
$$\therefore \quad a = \frac{T - W}{m}$$
$$= \frac{3.3 \times 10^7 - 2.8 \times 10^7}{2.8 \times 10^6} \frac{\text{N}}{\text{kg}}$$
$$= 1.8 \text{ m s}^{-2} \quad (\text{N kg}^{-1})$$

Note. We can apply $F = ma$ here since the rocket is instantaneously at rest. In general this is not possible because the mass of the rocket changes.

Example 2. Two blocks A and B are connected as in Fig. 7.22 on a horizontal frictionless floor and pulled to the right with an acceleration of 2.0 m s^{-2} by a force P. If $m_1 = 50$ kg and $m_2 = 10$ kg, what are the values of T and P?

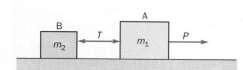

Fig. 7.22

The forces acting *on the blocks* are shown. Apply $F = ma$ to each block.

For B,
$$T = m_2 a = 10 \text{ kg} \times 2 \text{ m s}^{-2}$$
$$= 20 \text{ N}$$

For A,
$$P - T = m_1 a = 50 \text{ kg} \times 2 \text{ m s}^{-2}$$
$$= 100 \text{ N}$$
$$\therefore \quad P = 120 \text{ N}$$

Example 3. A helicopter of mass M and weight W rises with vertical acceleration, a, due to the upward thrust U generated by its rotor. The crew and passengers of total mass m and total weight w exert a combined force R on the floor of the helicopter. Write an equation for the motion of (a) the helicopter, (b) the crew and passengers.

(a) The forces acting *on the helicopter* are the upwards force U due to the action of the rotor on the surrounding air, its weight W downwards due to the earth and the force R downwards exerted on the floor by the crew and passengers, Fig. 7.23a.

$\therefore$ *Resultant* upwards force on helicopter $= U - W - R$

Hence, by the second law

$$U - W - R = Ma \qquad (1)$$

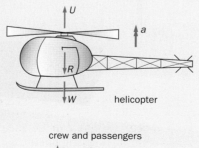

helicopter

crew and passengers

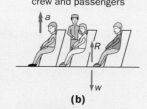

(b)

Fig. 7.23

(b) The forces acting *on the crew and passengers* are the upwards push of the floor of the helicopter (which by the third law must equal the downwards push R of the crew and passengers on the floor) and their weight w downwards, Fig. 7.23b.

$\therefore$ *Resultant* upwards force on crew and passengers $= R - w$

Hence, by the second law

$$R - w = ma \qquad (2)$$

The required equations are (1) and (2).

AIR RESISTANCE : TERMINAL VELOCITY

When an object falls in air, the air resistance opposing its motion increases as its speed rises, so reducing its acceleration. Eventually, air resistance acting upwards equals the weight of the object acting downwards. The resultant force on the object is then zero (since the two opposing forces balance) and the object falls with a constant velocity, called its **terminal velocity**. The value of the terminal velocity depends on the size, shape and weight of the object.

A small dense object such as a steel ball-bearing has a high terminal velocity and falls a considerable distance with a constant acceleration of about 9.8 m s^{-2} before air resistance equals its weight. A light object such as a raindrop, or one with a large surface area such as a parachute, has a low terminal velocity and accelerates over a comparatively short distance before air resistance equals its weight. A sky diver in free fall has a terminal velocity of more than 50 m s^{-1} (100 miles per hour).

MOMENTUM

The **momentum** of a body is the mass of the body multiplied by its velocity. If SI units are used, Newton's second law may be written

force = rate of change of momentum

In symbols

$$F = \frac{mv - mu}{t}$$

where F is the force acting on a body of mass m which increases its velocity from u to v in time t. Therefore

$$Ft = mv - mu$$

The quantity Ft is called the **impulse** of the force on the body. It is a vector and, like linear momenta, impulses in opposite directions must be given positive and negative signs. In words, the impulse–momentum equation is

impulse = change of momentum

The equation shows that impulse and momentum have the same unit, i.e. N s or kg m s^{-1}.

These ideas are important in sport. The good cricketer or tennis player 'follows through' with the bat or racquet when striking the ball. The force applied then acts for a longer time, the impulse is greater and so also is the change of momentum (and velocity) of the ball. On the other hand, when a cricket ball is caught its momentum is reduced to zero. This is achieved by an impulse in the form of an opposing force acting for a particular time, and, while any number of combinations of force and time will give a particular impulse, the 'sting' can be removed from the catch by drawing back the hands as the ball is caught. A smaller force is then applied for a longer time.

Using a force sensor we find that in collisions of this and other types, the force is not constant but builds up to a maximum value as the deformation of the colliding bodies increases. It does, however, have an average value.

CONSERVATION OF MOMENTUM

(a) Principle

Suppose a body A of mass m_1 and velocity u_1 collides with another body B of mass m_2 and velocity u_2 moving in the same direction, Fig. 7.24a. If A exerts a force F to the *right* on B for time t then by Newton's third law, B will exert an equal but opposite force F on A, also for time t (since the time of contact is the same for each) but to the *left*.

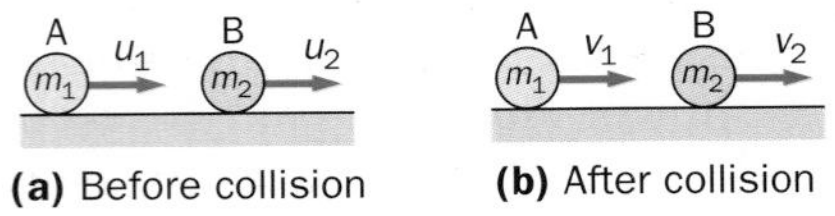

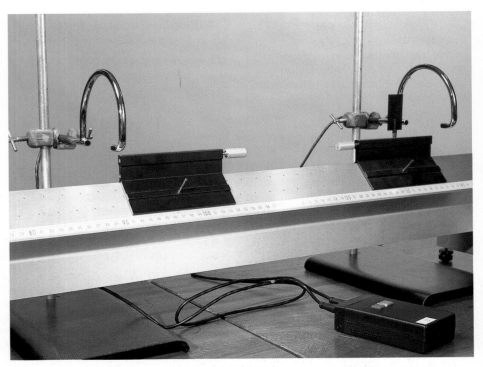

A u_1 B u_2
m_1 m_2

(a) Before collision

A v_1 B v_2
m_1 m_2

(b) After collision

Fig. 7.24

The bodies thus receive equal but opposite impulses Ft and so it follows from the impulse–momentum equation that the changes of momentum must be equal and opposite. The total momentum change of A and B is therefore zero, or in other words the **total momentum of A and B together remains constant** despite the collision. So if A has a reduced velocity v_1 after the collision and B has an increased velocity v_2, both in the same direction as before, Fig. 7.24b, then

$$m_1u_1 + m_2u_2 = m_1v_1 + m_2v_2$$

This important result, known as the **principle of conservation of momentum**, has been deduced from Newton's second and third laws and is a universal rule of the physical world which still holds even in some extreme (relativistic) conditions where Newton's laws fail. It applies not only to collisions but to any interaction between two or more bodies. When a gun is fired, for example, the backward momentum component of the gun in a horizontal direction equals the component of the forward momentum of the shell and propellant gases so that the *total* momentum of the gun–shell system remains zero even though the momentum of each part changes.

No *external* agent must act on the interacting bodies otherwise momentum may be added to the system. Sometimes momentum does appear to be gained (or lost). For example, a body falling towards the earth increases its downward momentum but the body is interacting with the earth (the external agent) which gains an equal amount of upward momentum from the attraction of the body on the earth. The complete system consists of the body *and* the earth and their total momentum remains constant. Similarly, when a car comes to rest we believe that all the momentum it loses is transferred by the action of friction to the earth, although we cannot easily prove this.

The general statement of the principle is as follows.

When bodies in a system interact the total momentum remains constant provided no external force acts on the system.

Fig. 7.25   Testing the principle of conservation of momentum using a linear air track

(b) Experimental test

The principle can be investigated experimentally using a linear air track on which two gliders move with negligible friction, Fig. 7.25. Velocity is determined from the time it takes for a glider to move through a photogate. This is recorded by an electronic timer with memory (so that successive passes through the gate are recorded) or by a datalogger and computer; a separate photogate is needed to monitor the motion of each glider. Velocities before and after the collision are calculated for each glider and used to confirm that momentum is conserved in the collision, if the friction on the track is negligible.

ROCKET AND JET PROPULSION

The principle of rocket and jet propulsion is illustrated by the behaviour of an inflated balloon when released with its neck open. With the neck closed there is a state of balance inside the balloon with equal pressure at all points, Fig. 7.26a. When the neck is opened the pressure on the surface opposite the neck is now unbalanced and the balloon is forced to move in the opposite direction to that of the escaping air, Fig. 7.26b. According to the principle of conservation of momentum, the air and the balloon have equal but opposite amounts of momentum, that is

$$m_{\text{air}} \times v_{\text{air}} = m_{\text{balloon}} \times v_{\text{balloon}}$$

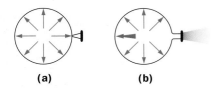

(a)   **(b)**

Fig. 7.26

In a rocket and a jet engine a stream of gas is produced at very high temperature and pressure which then escapes at high velocity through an exhaust nozzle. The thrust arises from the large increase in momentum of the exhaust gases.

A rocket carries its own supplies of oxygen (liquid) and fuel (e.g. kerosene or liquid hydrogen), Fig. 7.27a. The mass of a rocket is not constant but decreases appreciably as it uses

fuel (often at a rate of over 3000 kg s^{-1}). The acceleration consequently increases.

A jet engine uses the surrounding air for its oxygen supply and so is unsuitable for space travel. Figure 7.27b is a simplified drawing of one type of jet engine (gas turbine). The compressor draws in air at the front, compresses it, fuel is injected and the mixture burns to produce hot exhaust gases which escape at high speed from the rear of the engine. These cause forward propulsion and drive the turbine which in turn rotates the compressor.

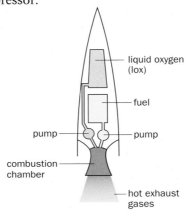

liquid oxygen (lox)

fuel

pump   pump

combustion chamber

hot exhaust gases

(a)   Rocket propulsion

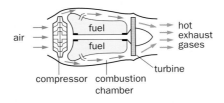

air

fuel

fuel

hot exhaust gases

turbine

compressor   combustion chamber

(b)   Jet engine

Fig. 7.27

MOMENTUM CALCULATIONS

The concepts of impulse and momentum are useful when considering collisions and explosions, i.e. situations in which forces (called impulsive forces) act for a short time.

Example 1. A jet of water emerges from a hose-pipe of cross-section area 5.0×10^{-3} m^2 with a velocity of 3.0 m s^{-1} and strikes a wall at right angles. Calculate the force on the wall assuming the water is brought to rest and does not rebound. (Density of water $= 1.0 \times 10^3$ kg m^{-3}.)

The water arrives with a velocity of 3.0 m s^{-1} and hits an area of wall 5.0×10^{-3} m^2. Hence volume of water striking wall per second

$$= (3.0 \text{ m s}^{-1})(5.0 \times 10^{-3} \text{ m}^2)$$
$$= 1.5 \times 10^{-2} \text{ m}^3 \text{ s}^{-1}$$

Therefore mass of water striking wall per second

$$= 1.5 \times 10^{-2} \times 1.0 \times 10^3 \text{ kg s}^{-1}$$
$$= 15 \text{ kg s}^{-1}$$

Velocity change of water on striking wall

$$= 3.0 - 0 = 3.0 \text{ m s}^{-1}$$

Therefore momentum change per second of water on striking wall

$$= (15 \text{ kg s}^{-1})(3.0 \text{ m s}^{-1})$$
$$= 45 \text{ kg m s}^{-2}$$

But

$$\text{force} = \text{momentum change per second}$$
$$= 45 \text{ N}$$

(In practice the horizontal momentum of the water is seldom completely destroyed and so the answer is only approximate.)

Example 2. A railway truck A of mass 2×10^4 kg travelling at 0.5 m s^{-1} collides with another truck B of half its mass moving in the opposite direction with a velocity of 0.4 m s^{-1}. If the trucks couple automatically on collision, find the common velocity with which they move, Fig. 7.28.

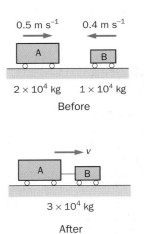

0.5 m s^{-1} 0.4 m s^{-1}

A   B

2×10^4 kg 1×10^4 kg

Before

v

A   B

3×10^4 kg

After

Fig. 7.28

Total momentum *to the right* of A and B before collision

$$= 2 \times 10^4 \times 0.5$$
$$- 1 \times 10^4 \times 0.4 \text{ kg m s}^{-1}$$
$$= 0.6 \times 10^4 \text{ kg m s}^{-1}$$

(If the momentum of A is taken as positive, that of B must be negative.)

Total momentum *to the right* of A and B after collision

$$= (3 \times 10^4) \text{ kg} \times v$$

By the principle of conservation of momentum

$$(3 \times 10^4) \text{ kg} \times v$$
$$= 0.6 \times 10^4 \text{ kg m s}^{-1}$$
$$\therefore \quad v = \frac{0.6 \times 10^4 \text{ kg m s}^{-1}}{3 \times 10^4 \text{ kg}}$$
$$= 0.2 \text{ m s}^{-1}$$

Example 3. A jet engine on a test bed takes in 20.0 kg of air per second at a velocity of 100 m s^{-1} and burns 0.80 kg of fuel per second. After compression and heating the exhaust gases are ejected at 500 m s^{-1} relative to the aircraft. Calculate the thrust of the engine.

Velocity change of 20 kg of air

$$= (500 - 100) \text{ m s}^{-1} = 400 \text{ m s}^{-1}$$

$\therefore$ Momentum change per second of 20 kg of air

$$= 20 \text{ kg s}^{-1} \times 400 \text{ m s}^{-1}$$

Initial velocity of fuel is zero so its velocity change is 500 m s^{-1}.

$\therefore$ Momentum change per second of 0.80 kg of fuel

$$= 0.80 \text{ kg s}^{-1} \times 500 \text{ m s}^{-1}$$

$\therefore$ Total momentum change per second of air and fuel

$$= (20 \times 400 + 0.80 \times 500) \text{ kg m s}^{-2}$$
$$= 8.4 \times 10^3 \text{ kg m s}^{-2}$$

But

force (thrust) = total change of momentum per second

$\therefore$ thrust of engine $= 8.40 \times 10^3$ N
(1 kg m s^{-2} = 1 N)

Note. If the engine is in an aircraft flying at 100 m s^{-1}, taking in air at this speed, the thrust would be about the same.

WORK, ENERGY AND POWER

(a) Work

In science the term 'work' has a definite meaning which differs from its everyday one. For example, someone holding a heavy weight at rest may say and feel he is doing hard work but in fact none is being done on the weight in the scientific sense.

Work is done when a force moves its point of application along the direction of its line of action.

In the simple case of Fig. 7.29a, the constant force F and the displacement s are in the same direction and we define the work W done by the force on the body by

$$W = Fs$$

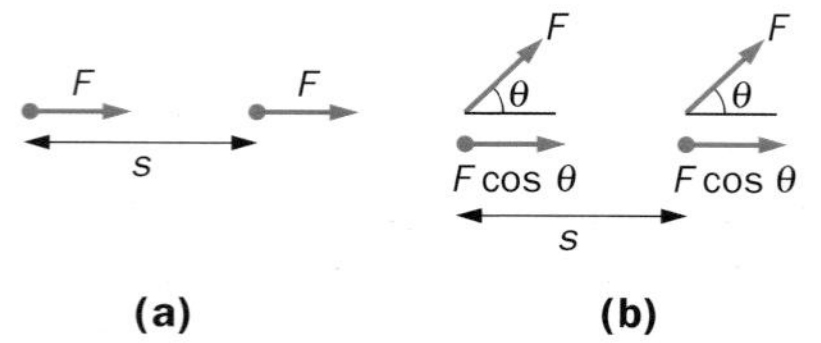

Fig. 7.29

If the force does not act in the direction in which motion occurs but at an angle θ to it as in Fig. 7.29b, then the work done is defined as the product of the component of the force in the direction of motion and the displacement in that direction. That is,

$$W = (F \cos \theta)s$$

When $\theta = 0$, $\cos \theta = 1$ and so $W = Fs$, in agreement with the first equation. When $\theta = 90°$, $\cos \theta = 0$ and F has no component in the direction of motion and so no work is done. Thus the work done by the force of gravity when a body is moved horizontally is zero.

If the force varies, the work done can be obtained from a force–displacement graph in which the component of the force in the direction of the displacement is plotted, Fig. 7.30. Suppose the force is F when the displacement is x, then the work done during a further, very small displacement δx is $F \delta x$, i.e. the shaded area. By dividing up the whole area under the curve into narrow strips we see that the total work done during displacement s is represented by area OABC.

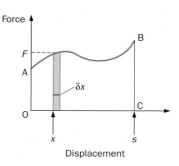

Fig. 7.30

Work can be either positive or negative. It is positive if the force acts in the same direction as the displacement (Figs 7.29a and b), but negative if it is oppositely directed (θ is then $> 90°$ and $\cos \theta$ is negative). The work done by friction when it opposes one body sliding over another is negative.

One joule is the work done by a force of 1 newton when its point of application moves through a distance of 1 metre in the direction of the force. Thus

1 joule (J) = 1 newton metre (N m)

Work is a scalar although force and displacement are both vectors.

(b) Energy

When a body A does work by exerting a force on another body B, the body A is said to lose energy, equal in amount to the work it performs. **Energy** is therefore often defined as **that which enables a body to do work**; it is measured in joules, like work. When an interchange of energy occurs between two bodies we can consider **the work done as measuring the quantity of energy transferred between them**. So if body A does 5 joules of work on body B then the energy transfer from A to B is 5 joules.

(c) Power

The **power** of a machine is **the rate at which it does work**, i.e. the rate at which it transfers energy. The unit of power is the **watt** (W) and equals a rate of transfer of 1 joule per second, i.e. $1\text{ W} = 1\text{ J s}^{-1}$.

In calculus notation, since power P is the rate at which work W is done or energy transferred, we can write

$$P = \frac{\mathrm{d}W}{\mathrm{d}t}$$

If δW is the energy transferred by a *constant force* F when a body moves with *constant velocity* v through a small displacement δs in time δt, then $\delta W = F \times \delta s$ and P is given by

$$P = F\frac{\mathrm{d}s}{\mathrm{d}t} = Fv$$

(see p. 116). If F is in N and v in m s^{-1}, the unit of P, i.e. Fv, is N m s^{-1}, or J s^{-1}, or W.

KINETIC AND POTENTIAL ENERGY

(a) Kinetic energy

Kinetic energy (k.e.) is the energy a body has because of its motion. For example, a moving hammer does work against the resistance of the wood into which a nail is being driven. An expression for kinetic energy can be obtained by calculating the amount of energy transferred from the body while it is being brought to rest.

Consider a body of constant mass m moving with velocity u. Let a constant force F act on it and bring it to rest in a distance s, Fig. 7.31. Since the final velocity v is zero, from $v^2 = u^2 + 2as$ we have

$$0 = u^2 + 2as$$
$$\therefore \quad a = -\frac{u^2}{2s}$$

The negative sign shows that the acceleration a is opposite in direction to u (as we would expect). The acceleration in the direction of F is there-

fore $+u^2/2s$. The original kinetic energy of the body equals the work W it does against F, that is, the energy transferred in being brought to rest.

$$\text{kinetic energy of body} = W = Fs$$
$$= mas \qquad \text{(since } F = ma\text{)}$$
$$= ms\frac{u^2}{2s} \qquad \left(\text{since } a = \frac{u^2}{2s}\right)$$

Therefore

$$\text{k.e.} = \tfrac{1}{2}mu^2$$

Conversely, if work is done on a body the gain of kinetic energy when its velocity increases from zero to u can be shown to be $\tfrac{1}{2}mu^2$.

In general, if the velocity of a body of mass m increases from u to v when work is done on it by a force F acting over a distance s, then

$$Fs = \tfrac{1}{2}mv^2 - \tfrac{1}{2}mu^2$$

This is called the work–energy equation and may be stated as

work done by forces = change in k.e. acting on the body   of the body

(b) Potential energy

Potential energy (p.e.) is the energy a system of bodies has because of the relative positions of its parts, i.e. owing to its configuration. It arises when a body experiences a force in a field such as the earth's gravitational field. In that case the body occupies a position with respect to the earth and the potential energy is regarded as a joint property of the body–earth system and not of either body separately. The relative positions of the parts of the system, i.e. of the body and earth, determine its potential energy; the greater the separation the greater the potential energy.

Normally we are only concerned with differences of potential energy. In the gravitational case it is convenient to consider that the potential energy is zero when the body is at the surface of the earth. The potential energy when a body of mass m is at height h above ground level equals the work which must be done against the downward pull of gravity to raise the body to this height. A force, equal and opposite to mg, has to be exerted on the body

over displacement h (assuming g is constant near the earth's surface). Therefore

work done by external force against gravity

$$= \text{force} \times \text{displacement}$$
$$= mgh$$

and so

$$\text{p.e.} = mgh$$

On returning to ground level an amount of potential energy equal to mgh would be transferred. A good example of this occurs when the water in a mountain reservoir falls to a lower level and does work by driving a power station turbine.

A stretched or compressed spring is also considered to have potential energy.

CONSERVATION OF ENERGY

If a body of mass m is thrown vertically upwards with velocity u at A, it has to do work against the constant force of gravity, Fig. 7.32. When it has risen to B let its reduced velocity be v.

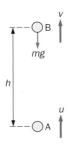

Fig. 7.32

By the definition of kinetic energy (k.e.),

loss of k.e. between A and B
      = work done against gravity

By the definition of potential energy (p.e.),

gain of p.e. between A and B
      = work done against gravity

$$\therefore \quad \text{loss of k.e.} = \text{gain of p.e.}$$
$$\therefore \quad \tfrac{1}{2}mu^2 - \tfrac{1}{2}mv^2 = mgh$$

Fig. 7.31

This is called the **principle of conservation of mechanical energy** and may be stated as follows.

> The total amount of mechanical energy (k.e. + p.e.) which the bodies in an isolated system possess is constant.

It applies only to frictionless motion, i.e. to **conservative systems**. Otherwise, in the case of a rising body, work has to be done against friction as well as against gravity and the body gains less p.e. than when friction is absent. Furthermore, the gain of p.e. would depend on the path taken; it does not in a conservative system.

Work done against frictional forces is generally accompanied by a temperature rise. This suggests that we might include in our energy accountancy what is called **internal energy**. This would then extend the energy conservation principle to non-conservative systems and we can then say, for example,

loss of k.e. = gain of p.e.
+ gain of internal energy

The mechanics of a body seen to be in motion has thus been related to a phenomenon which is apparently not mechanical and in which motion is not directly detected. (However, we believe that internal energy is composed of random molecular kinetic and potential energy, see p. 65.) In a similar way, the idea of energy has been extended to other areas of physics and is now a unifying theme. In fact, physics is sometimes said to be the study of energy transformations, measured in terms of the work done by the forces involved in the transformation.

The principle of conservation of mechanical energy is a special case of the more general **principle of conservation of energy** — one of the fundamental laws of science.

> Energy may be transformed from one form into another, but it cannot be created or destroyed, i.e. the total energy of a system is constant.

Energy uses and the background to energy supply and demand are considered in chapter 10.

ENERGY CALCULATIONS

The work–energy equation

$$Fs = \tfrac{1}{2}mv^2 - \tfrac{1}{2}mu^2$$

is useful for solving problems when the distance over which a force acts is known.

Example 1. A car of mass 1.0×10^3 kg travelling at 72 km h^{-1} on a horizontal road is brought to rest in a distance of 40 m by the action of the brakes and frictional forces. Find (*a*) the average stopping force, (*b*) the time taken to stop the car.

72 km h^{-1} = 72 $\times$ 10^3 m/3600 s = 20 m s^{-1}

(*a*) If the car has mass m and initial speed u, then

kinetic energy lost by car = $\tfrac{1}{2}mu^2$

If F is the average stopping force and s the distance over which it acts, then

work done by car against $F = Fs$

But $Fs = \tfrac{1}{2}mu^2$

$\therefore\ F \times 40\ \text{m} = \tfrac{1}{2} \times (1.0 \times 10^3\ \text{kg}) \times (20\ \text{m s}^{-1})^2$

$\therefore\ F = \dfrac{1.0 \times 10^3 \times 400}{2 \times 40}\ \dfrac{\text{kg m}^2\ \text{s}^{-2}}{\text{m}}$

= 5.0×10^3 N

(*b*) Assuming constant acceleration and substituting $v = 0$, $u = 20$ m s^{-1} and $s = 40$ m in $v^2 = u^2 + 2as$ we have

$0 = 20^2\ \text{m}^2\ \text{s}^{-2} + 2a \times 40\ \text{m}$

$\therefore\ a = -\dfrac{400\ \text{m}^2\ \text{s}^{-2}}{80\ \text{m}} = -5.0\ \text{m s}^{-2}$

(The negative sign indicates the acceleration is in the opposite direction to the displacement.)

Using $v = u + at$

$0 = 20\ \text{m s}^{-1} + t \times (-5\ \text{m s}^{-2})$

$\therefore\ t = \dfrac{-20\ \text{m s}^{-1}}{-5\ \text{m s}^{-2}} = 4.0\ \text{s}$

Example 2. A bullet of mass 10 g travelling horizontally at a speed of 1.0×10^2 m s^{-1} embeds itself in a block of wood of mass 9.9×10^2 g suspended by strings so that it can swing freely. Find (*a*) the vertical height through which the block rises,

(*b*) how much of the bullet's energy becomes internal energy. ($g = 10$ m s^{-2})

(*a*) The bullet is brought to rest very quickly because of the resistance offered by the block and we assume that the block (with the bullet embedded) hardly moves until the bullet is at rest. Momentum is conserved in the collision, so

$$mu = (M + m)v$$

where m and M are the masses of the bullet and block respectively, u is the velocity of the bullet before impact and v is the velocity of the block + bullet as they move off.

$\therefore\ 10 \times 10^{-3}$ kg $\times 1.0 \times 10^2$ m s^{-1}
$= (990 + 10) \times 10^{-3}$ kg $\times v$

$\therefore\ v = 1.0$ m s^{-1}

When the block has swung to its maximum height h, all its kinetic energy has become potential energy, if frictional forces are neglected. Conservation of energy therefore holds and we can say

$$\tfrac{1}{2}(M + m)v^2 = (M + m)gh$$

$\therefore\ h = \dfrac{v^2}{2g}$

$= \dfrac{(1\ \text{m s}^{-1})^2}{2 \times 10\ \text{m s}^{-2}}$

$= \dfrac{1}{2 \times 10}\ \dfrac{\text{m}^2\ \text{s}^{-2}}{\text{m s}^{-2}}$

$\therefore\ h = 5.0 \times 10^{-2}$ m

(*b*) Original kinetic energy of bullet = $\tfrac{1}{2}mu^2$

$= \tfrac{1}{2}(10 \times 10^{-3}\ \text{kg})(1.0 \times 10^2\ \text{m s}^{-1})^2$

$= \tfrac{1}{2} \times 10 \times 10^{-3} \times 1.0 \times 10^4\ \text{kg m}^2\ \text{s}^{-2}$

$= 50$ N m = 50 J

Kinetic energy of block + bullet after impact

$= \tfrac{1}{2}(M + m)v^2$

$= \tfrac{1}{2}(1000 \times 10^{-3}\ \text{kg})(1\ \text{m s}^{-1})^2$

$= 0.50$ J

Therefore

internal energy = loss of kinetic
produced   energy

$= (50 - 0.50)$ J
$= 49.5$ J

ELASTIC AND INELASTIC COLLISIONS

In all collisions momentum is conserved but there is generally a loss of k.e., usually to internal energy and to a very small extent to sound energy. When there is a loss of k.e. (see *Example 2* on p. 126), the collision is called an **inelastic** (no-bounce) one. In a **perfectly elastic** collision k.e. is conserved (as well as momentum). Different collisions can be investigated using trolleys and tickertape or gliders on a linear air track.

(a) Relative velocity rule

This rule states that *in a perfectly elastic collision*

$$\frac{\text{relative velocity}}{\text{before collision}} = -\left(\frac{\text{relative velocity}}{\text{after collision}}\right)$$

It can be proved using the principles of conservation of momentum and energy. Suppose bodies of masses m_1 and m_2 are moving in the same direction with velocities u_1 and u_2 before collision, Fig. 7.33a. After the collision their velocities are v_1 and v_2, still to the right, Fig. 7.33b.

(a) Before collision   **(a)** After collision

Fig. 7.33

By the principle of conservation of momentum

$$m_1u_1 + m_2u_2 = m_1v_1 + m_2v_2$$
$$\therefore \quad m_1(u_1 - v_1) = m_2(v_2 - u_2) \quad (1)$$

By the principle of conservation of energy

$$\tfrac{1}{2}m_1u_1{}^2 + \tfrac{1}{2}m_2u_2{}^2 = \tfrac{1}{2}m_1v_1{}^2 + \tfrac{1}{2}m_2v_2{}^2$$
$$\therefore \quad m_1(u_1{}^2 - v_1{}^2) = m_2(v_2{}^2 - u_2{}^2)$$
$$\therefore \quad m_1(u_1 - v_1)(u_1 + v_1) = m_2(v_2 - u_2)(v_2 + u_2) \quad (2)$$

Substituting (1) in (2)

$$m_2(v_2 - u_2)(u_1 + v_1) = m_2(v_2 - u_2)(v_2 + u_2)$$
$$\therefore \quad u_1 + v_1 = v_2 + u_2$$

or

$$u_1 - u_2 = -(v_1 - v_2) \quad (3)$$

That is, the velocity of m_1 relative to m_2 before collision equals minus the velocity of m_1 relative to m_2 after collision.

(b) Energy transfer in collisions

If a moving body has a perfectly elastic collision with a *stationary body of the same mass*, we can show that the moving body comes to rest and the stationary one travels on with the velocity of the moving one, i.e. there is a complete transfer of k.e. from the moving to the stationary body. Therefore, since $m_1 = m_2$ and $u_2 = 0$, from equation (1)

$$u_1 - v_1 = v_2 \quad \text{or} \quad u_1 = v_1 + v_2 \quad (4)$$

From the relative velocity rule, i.e. equation (3),

$$u_1 = v_2 - v_1 \quad (5)$$

Adding (4) and (5) gives

$$2u_1 = 2v_2 \quad \text{or} \quad u_1 = v_2$$

From (4) it follows that $v_1 = 0$.

The energy transfer is therefore a maximum when the colliding masses are equal. This is also true for interacting electric circuits. For example, the maximum energy (and power) transfer from an amplifier to a loudspeaker occurs when their impedances ('inertias': see p. 444) are equal.

In the case of a *very small mass colliding elastically with a very large one at rest*, it can similarly be proved that the small one rebounds with the *same* velocity while the large one *remains at rest*. In the opposite case of a moving body of *very large mass* colliding elastically with a stationary body of *very small mass*, the greatest velocity the latter can have is *twice* that of the large mass. In both these extreme cases the energy transfer during the collision is negligible.

COLLISIONS IN TWO DIMENSIONS

In the last section *head-on*, *one-dimensional* collisions were considered. *Oblique, two-dimensional* collisions can be studied using pucks moving on a cushion of carbon dioxide gas over a glass plate. Such an event is shown by the multiflash photograph of Fig. 7.34a between a moving puck and a stationary one of *equal mass*.

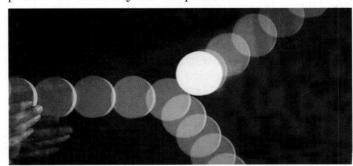

(a) Pucks of equal mass

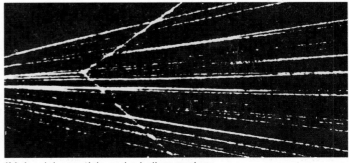

(b) An alpha particle and a helium nucleus

Fig. 7.34   Oblique, elastic collisions

Momentum can be shown to be conserved by making measurements on successive images before and after the collision. Then, either use is made of the parallelogram law or the velocities are resolved into components at right angles to each other and momentum equated separately for each direction (since momentum and velocity are vectors).

The kinetic energy of the system before and after the collision can also be compared to confirm that it is perfectly elastic. Note that the angle between the directions of motion of the pucks after the collision is 90°. This is always true for an oblique elastic collision between bodies of equal mass, one of which is at rest initially.

Collisions between atoms and other atomic particles were first studied in a cloud chamber (p. 468); many such collisions are perfectly elastic. Figure 7.34b shows a cloud chamber photograph of an alpha particle colliding with a helium nucleus. Compare this with Fig. 7.34a. Assuming atomic particles and magnetic pucks behave similarly, what can be concluded about (i) the mass of an alpha particle compared with that of a helium nucleus and (ii) the type of collision that has occurred? (The plane of the collision tracks in Fig. 7.34b is not exactly in the plane of the photograph but calculations from other photographs taken at the same time from different angles show that the actual angle between the tracks after the collision is 90°.)

Electrons can have elastic or inelastic collisions with the atoms of a gas and the latter give information about the electronic structure of atoms.

CAR COLLISIONS AND SAFETY

When a car stops rapidly in a collision, large forces are produced on the car and its passengers, Fig. 7.35. The equations we met earlier connecting force to kinetic energy and momentum are helpful when considering how to minimize damage and improve safety. They are

$$Fs = \tfrac{1}{2}mv^2 \quad \text{and} \quad Ft = mv$$

where v is the speed of the car, s and t are the stopping distance and stopping

Fig. 7.35   New car designs have to undergo impact tests before the cars are marketed. This car has crashed into a 120-ton concrete block fronted by a metal structure (to simulate another car) at 63 km h^{-1}

time respectively. Increasing both s and t reduces F; doubling v quadruples s. There are four aspects to the problem.

(a) Car design

This should be such that the front and rear of the car collapse in a way that enables them to absorb the kinetic energy of the collision and preferably recover their previous shape. Shock-absorbing bumpers are one solution for minor crashes. In the one shown in Fig. 7.36, when the bumper hits something the piston moves to the left and fluid is forced through small holes in the 'dashpot', so absorbing the energy gradually.

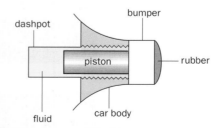

Fig. 7.36   Shock-absorbing car bumper

At the same time the passengers need to be protected by a strong central 'cell' which stops more slowly than the front and rear 'crumple' zones. Some cars are now being designed with special collapsible features to increase the safety of drivers and passengers in *side* impacts.

(b) Seat belts

If a car travelling at 15 m s^{-1} (about 30 mph) hits something, the effect felt by anyone not using a seat belt is roughly the same as that produced by jumping off a building 12 m high!

A person of mass 45 kg moving at 15 m s^{-1} has k.e. of about 5000 J. In a collision a seat belt exerts a backwards force F through the stopping distance s and does work Fs in reducing the k.e. to zero, i.e. $Fs = 5000$ J. A typical value for s might be 0.5 m (the distance between the person and the car windscreen or seat in front).

If F were constant throughout the impact, it would have a steady value of 5000 J/0.50 m = 10 000 N. This force would cause pain but could be stood for a short time if the seat belt was fairly tight and spread the force over the correct parts of the body. The time t it would last is given by $Ft = mv$, i.e. $t = mv/F = 45$ kg × 15 m s^{-1}/10^4 N = 0.068 s ≈ $\frac{1}{15}$ s.

If s were 0.05 m, F would be a lethal 100 000 N. Figure 7.37a shows how each of these two *steady* forces can reduce a k.e. of 5000 J to zero. In each case the area under the F–s graph represents the work done, i.e. $Fs = \tfrac{1}{2}mv^2$.

The force exerted by a real seat belt increases as it stretches and is not constant, Fig. 7.37b, but the area under the graph shows 5000 J of k.e. is still absorbed before F becomes too large. In a head-on collision the front

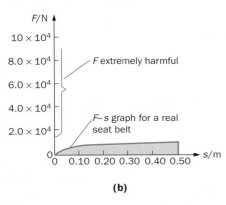

(a)

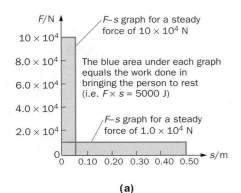

(b)

Fig. 7.37

of a car crumples through 0.5 m or so, giving an occupant a total stopping distance of about 1.0 m before hitting the windscreen, Fig. 7.38.

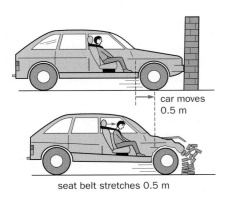

car moves 0.5 m

seat belt stretches 0.5 m

Fig. 7.38

(c) Head restraints

If you are in a car hit from behind, your head tends to stay in the same place while your body moves forward, causing severe damage to the top of the spine. Head restraints can prevent this by pushing your head forwards so that your whole body moves forward as one.

The chance of being killed in an accident is about five times less if seat belts are worn and head restraints are installed.

(d) Air bags

These inflate on collision and stop the steering wheel injuring the driver. They are fitted in many new cars.

QUESTIONS

Assume $g = 10$ m s^{-2} unless stated otherwise.

Statics

1. State the principle of moments.
   Figure 7.39 shows a lorry on a light bridge. Calculate the downwards force on each of the bridge supports.

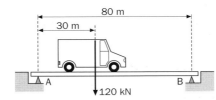

Fig. 7.39

Copy the axes in Fig. 7.40 and sketch a graph to show how the force on the support at A changes as the centre of gravity of the lorry moves from A to B. Mark the value of the intercept on the force axis.

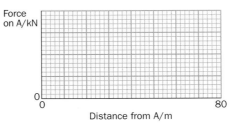

Fig. 7.40

(*L*, AS/A PH1, June 1998)

2. State the conditions of equilibrium of a body acted on by a system of coplanar forces.
   An aerial attached to the top of a radio mast 20 m high exerts a horizontal force on it of 3.0×10^2 N. A stay-wire from the mid-point of the mast to the ground is inclined at 60° to the horizontal. Assuming the action of

the ground on the mast can be regarded as a single force, find
   a) the force exerted on the mast by the stay-wire,
   b) the magnitude and direction of the action of the ground.

3. A uniform ladder 5.0 m long and having mass 40 kg rests with its upper end against a smooth vertical wall and with its lower end 3.0 m from the wall on rough ground. Find the magnitude and direction of the force exerted at the bottom of the ladder.

4. a) Explain what is meant by a vector quantity.

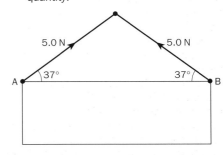

Fig. 7.41

   b) Figure 7.41 shows a picture supported by two wires. The left hand wire exerts a force of 5.0 N on the picture at A. The right hand wire exerts a force of 5.0 N on the picture at B. Resolve the force acting at B to calculate
      i) its vertical component,
      ii) its horizontal component.
   c) Calculate the weight of the picture assuming it hangs freely.
      (*NEAB*, AS/A PH01, June 1997)

Dynamics

5. Figure 7.42 shows a mass attached by a piece of string to a glider which is free to glide along an air track. A student finds that the glider takes 1.13 s to move a distance of 90 cm starting from rest.

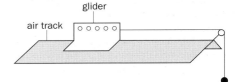

Fig. 7.42

Calculate the speed of the glider after 1.13 s.
   Calculate its average acceleration during this time.

How would you test whether or not the acceleration of the glider is constant? (*L, AS/A PH1, Jan 1997*)

6. A projectile is fired from ground level with a velocity of 500 m s^{-1} at 30° to the horizontal. Calculate its horizontal range, the greatest height it reaches and the time taken to rise to that height. (Neglect air resistance.)

7. A body slides, with constant velocity, down a plane inclined at 30° with the horizontal. Show in a diagram the forces acting on the body, and find the coefficient of kinetic friction between the body and the plane.

If the plane were now tilted so as to make an angle of 60° with the horizontal, with what acceleration would the body slide down the plane? What force, applied parallel to this plane, would be required to cause the body to move up the plane with a constant velocity?

8. An object of mass *m* rests on the floor of a lift which is ascending with acceleration *a*. Draw a diagram to show the external forces acting on the object, and write down its equation of motion. How do these forces arise? Show graphically how their magnitudes vary with the acceleration of the lift. What force constitutes the second member of the action–reaction pair in the case of each of these external forces?

9. Five identical cubes, each of mass *m*, lie in a straight line, with their adjacent faces in contact, on a horizontal surface, as shown in Fig. 7.43.

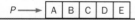

Fig. 7.43

Suppose the surface is frictionless and that a constant force *P* is applied from left to right to the end face of A.

What is the acceleration of the system and what is the resultant force acting on each cube? What force does cube C exert on cube D?

If friction is present between the cubes and the surface, draw a graph to illustrate how the total frictional force varies as *P* increases uniformly from zero.

10. A train of mass 1.4×10^5 kg accelerates uniformly from rest along a level track. It travels 100 m in the first 26 s. Calculate
i) the acceleration of the train,

ii) the speed reached after 26 s,
iii) the resultant force required to produce this acceleration,
iv) the average power required.
(*NEAB, AS/A PH01, Feb 1997*)

11. A stationary ball of mass 6.0×10^{-2} kg is hit horizontally with a tennis racket. The ball is in contact with the racket for 30 ms and leaves the racquet with a speed of 27 m s^{-1}.
a) Calculate
i) the change in the momentum of the ball,
ii) the average force which the racquet exerts on the ball.
b) Calculate the horizontal distance travelled by the ball before it hits the ground, if it leaves the racquet at a vertical height of 2.5 m.
c) i) Explain what is meant by an **inelastic collision**.
ii) Suggest a reason why the collision between the ball and the racquet is inelastic.
(*NEAB, AS/A PH01, June 1998*)

12. The Saturn V rockets which launched the Apollo space missions had the following specifications:
mass at lift-off = 3.0×10^6 kg
velocity of exhaust gases
= 1.1×10^4 m s^{-1}
initial rate of fuel consumption at lift-off
= 3.0×10^3 kg s^{-1}
a) Calculate
i) the force (thrust) produced at lift off,
ii) the resultant force acting on the rocket at lift-off.
b) If the thrust of the engines were constant, give one reason why the acceleration increased as the flight progressed.
(*NEAB, AS/A PH01, Feb 1997*)

13. a) An empty railway truck of mass 10 000 kg is travelling horizontally at a speed of 0.50 m s^{-1}. For this truck calculate the
i) momentum,
ii) kinetic energy.
b) Sand falls vertically into the truck at a constant rate of 40 kg s^{-1}. Calculate the additional horizontal force which must be applied to the truck, if it is to maintain a steady speed of 0.50 m s^{-1}.
c) Had no additional force been applied to the truck while sand continued to fall, explain without calculation what would happen to the truck's
i) momentum,
ii) speed.
(*NEAB, AS/A PH01, June 1997*)

14. An aeroplane is flying horizontally at a steady speed of 67 m s^{-1}. A parachutist falls from the aeroplane and falls freely for 80 m before the parachute opens. For the purposes of calculation, you may assume that air resistance is negligible before the parachute opens.
a) i) Show that the vertical component of the velocity is approximately 40 m s^{-1} when the parachutist has fallen 80 m. The acceleration of free fall *g* is 9.8 m s^{-2}.
ii) Determine the magnitude and direction of the resultant velocity of the parachutist at this point.
iii) State and explain how the magnitudes of the horizontal and vertical components of the parachutist's velocity are actually affected by the air resistance before the parachute opens.
b) The parachutist lands with a vertical velocity of 7.0 m s^{-1} and no horizontal velocity. The parachutist, of mass 85 kg, lands in one single movement, taking 0.25 s to come to rest.
i) Calculate the average retarding force on the parachutist during the landing.
ii) Explain how the parachutist's loss of momentum on landing is consistent with the principle of conservation of momentum.
(*AEB, 0635/1, Summer 1998*)

15. Write down an expression for the kinetic energy of a body.
a) A car of mass 1.00×10^3 kg travelling at 20 m s^{-1} on a horizontal road is brought to rest by the action of its brakes in a distance of 25 m. Find the average retarding force.
b) If the same car travels up an incline of 1 in 20 at a constant speed of 20 m s^{-1}, what power does the engine develop if the frictional resistance is 100 N?

16. What is the vertical upward force acting on the axle of the fixed, smooth, weightless pulley in Fig. 7.44, when the masses move under gravity?

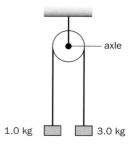

Fig. 7.44

17. What is the average force exerted on a vertical steel wall by steel ball-bearings, each of mass m, fired at the rate of n per second with a horizontal velocity u, assuming they rebound with the same speed?

18. A ball P moving with velocity v on a smooth horizontal surface has a head-on, perfectly elastic collision with another ball Q of the same mass which is at rest. What happens in this collision to
    a) kinetic energy,
    b) linear momentum, and
    c) the velocities of P and Q?

19. Design a spreadsheet to simulate projectile motion in the absence of air resistance (see pp. 5–6).

    Using a graphing routine, plot the trajectory of a rocket launched from earth with a velocity of 60 m s^{-1} at an angle of 45° to the horizontal.

    Use your spreadsheet to plot the trajectories of a rocket launched with the same initial conditions if the acceleration of free fall is
      i) 1.67 m s^{-2},
      ii) 15 m s^{-2}.

20. **a)** Starting with the definition of work, deduce the change in the gravitational potential energy of a mass m, when moved a distance h upwards against a gravitational field of field strength g.
    b) By using the equations of motion, show that the kinetic energy E_K of an object of mass m travelling with speed v is given by

    $$E_K = \tfrac{1}{2}mv^2$$

    c) A cyclist, together with his bicycle, has a total mass of 90 kg and is travelling with a constant speed of 15 m s^{-1} on a flat road at A, as illustrated in Fig. 7.45. He then goes down a small slope to B so descending 4.0 m.

Calculate
  i) the kinetic energy at A,
  ii) the loss of potential energy between A and B,
  iii) the speed at B, assuming that all the lost potential energy is transformed into kinetic energy of the cyclist and bicycle.

d) i) A cyclist travelling at a constant speed of 15 m s^{-1} on a level road provides a power of 240 W. Calculate the total resistive force.
  ii) The cyclist now travels at a higher constant speed. Explain why the cyclist needs to provide a greater power.

e) It is often stated that many forms of transport transform chemical energy into kinetic energy. Explain why a cyclist travelling at constant speed is not making this transformation. Explain what transformations of energy are taking place.

(OCR, 9244/3, Nov 1999)

21. **a)** i) Give an equation showing how the principle of conservation of momentum applies to the colliding snooker balls shown in Fig. 7.46.

before collision

after collision

Fig. 7.46

  ii) State the condition under which the principle of conservation of momentum applies.

b) A trolley, A, of mass 0.25 kg and a second trolley, B, of mass 0.50 kg are held in contact on a smooth horizontal surface. A compressed spring inside one of the trolleys is released and they then move apart. The speed of A is 2.2 m s^{-1}.
  i) Calculate the speed of B.
  ii) Calculate a minimum value for the energy stored in the spring when compressed.

c) The rotor blades of a helicopter sweep out a cross-sectional area, A. The motion of the blades helps the helicopter to hover by giving a downward velocity, v, to a cylinder of air, density ρ. The cylinder of air has the same cross-sectional area as that swept out by the rotor blades.

  Explaining your reasoning,
    i) derive an expression for the mass of air flowing downwards per second, and
    ii) derive an expression for the momentum given per second to this air.
    iii) Hence show that the motion of the air results in an upward force, F, on the helicopter given by

    $$F = \rho A v^2$$

d) A loaded helicopter has a mass of 2500 kg. The area swept out by its rotor blades is 180 m^2. If the downward flow of air supports 50% of the weight of the helicopter, what speed must be given to the air by the motion of the rotor blades when the helicopter is hovering? Take the density of air to be 1.3 kg m^{-3}.

(AQA: NEAB, AS/A PH01, June 1999)

4.0 m

A

B

Fig. 7.45

8

Circular motion and gravitation

- ■ Motion in a circle
- ■ Two useful expressions
- ■ Deriving $a = v^2/r$
- ■ Centripetal force
- ■ Rounding a bend
- ■ Other examples of circular motion
- ■ Moment of inertia

- ■ Kinetic energy of a rotating body
- ■ Equations for uniform angular acceleration
- ■ Work done by a couple
- ■ Angular momentum
- ■ Kepler's laws
- ■ Gravity and the moon
- ■ Law of universal gravitation

- ■ Testing gravitation
- ■ Masses of the sun and planets
- ■ Newton's work
- ■ Earth's gravitational field
- ■ Acceleration of free fall
- ■ Artificial satellites
- ■ Weightlessness
- ■ Speed of escape

MOTION IN A CIRCLE

In everyday life, in atomic physics and in astronomy and space travel there are many examples of bodies moving in paths which, if not exactly circular, are nearly so. In this chapter we will see how ideas developed for dealing with straight-line motion enable us to tackle circular motion.

A body which travels equal distances in equal times along a circular path has constant speed but *not* constant velocity. This is because of the way we have defined speed and velocity; speed is a scalar quantity, velocity is a vector quantity. Figure 8.1 shows a ball attached to a string being whirled round in a horizontal circle. The velocity of the ball at P is directed along the tangent at P; when it reaches Q its velocity is directed along the tangent at Q. If the speed is constant the **magnitudes** of the velocities at P and Q are the same but their **directions** are different and so the velocity of the ball has changed. A change of velocity is an acceleration and a body moving uniformly in a circular path or arc is therefore accelerating.

In everyday language acceleration usually means going faster and faster, i.e. involves a change of speed. However, in physics it means a change of velocity and since the velocity changes not only when the speed changes, but also when the direction of motion

changes, then, for example, a car rounding a bend (even at constant speed) is accelerating.

Fig. 8.1

TWO USEFUL EXPRESSIONS

We will use these from time to time when dealing with circular motion.

(a) Angles in radians: $s = r\theta$

Angles can be measured in radians as well as in degrees. In Fig. 8.2 the angle θ, in radians, is defined by the equation

$$\theta = \frac{s}{r}$$

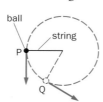

Fig. 8.2

If $s = r$ then $\theta = 1$ radian (rad). Therefore 1 radian is the angle subtended at the centre of a circle by an arc equal in length to the radius. When $s = 2\pi r$ (the circumference of a circle of radius r) then $\theta = 2\pi$ radians $= 360°$.

$$\therefore \quad 1 \text{ radian} = 360°/2\pi \approx 57°$$

From the definition of a radian it follows that the length s of an arc which subtends an angle θ at the centre of a circle of radius r, is

$$s = r\theta$$

where θ is in radians.

(b) Angular velocity: $v = r\omega$

The speed of a body moving in a circle can be specified either by its speed along the tangent at any instant, i.e. by its linear speed, or by its angular velocity. This is the angle swept out in unit time by the radius joining the body to the centre of the circle. It is measured in radians per second (rad s^{-1}).

We can derive an expression connecting angular velocity and linear speed. Consider a body moving uniformly from A to B in time t so that radius OA rotates through an angle θ, Fig. 8.2. The angular velocity ω of the body about O is

$$\omega = \frac{\theta}{t}$$

If arc AB has length s and if v is the constant speed of the body then

$$v = \frac{s}{t}$$

But from the previous section, $s = r\theta$ where r is the radius of the circle,

$$\therefore \quad v = \frac{r\theta}{t}$$

But $\omega = \theta/t$, therefore

$$v = r\omega$$

If $r = 3$ m and $\omega = 1$ revolution per second $= 2\pi$ rad s^{-1} then the linear speed $v = 6\pi$ m s^{-1}.

DERIVING $a = v^2/r$

To obtain an expression for the acceleration of a small body (i.e. a particle) describing circular motion, consider such a body moving with *constant speed* v in a circle of radius r, Fig. 8.3a. If it travels from A to B in a short interval of time δt then, since distance = speed × time,

$$\text{arc AB} = v\,\delta t$$

Also, by the definition of an angle in radians

$$\text{arc AB} = r\,\delta\theta \qquad (\delta\theta = \angle\,\text{AOB})$$

$$\therefore \quad r\,\delta\theta = v\,\delta t$$

$$\therefore \quad \delta\theta = \frac{v\,\delta t}{r} \qquad (1)$$

The vectors $\mathbf{v}_A$ and $\mathbf{v}_B$ drawn tangentially at A and B represent the velocities at these points. The *change* of velocity between A and B is obtained by subtracting $\mathbf{v}_A$ from $\mathbf{v}_B$. That is

$$\text{change of velocity} = \mathbf{v}_B - \mathbf{v}_A$$

But $\qquad \mathbf{v}_B - \mathbf{v}_A = \mathbf{v}_B + (-\mathbf{v}_A)$

So to subtract vector $\mathbf{v}_A$ from vector $\mathbf{v}_B$ we *add* vectors $\mathbf{v}_B$ and $(-\mathbf{v}_A)$ by the parallelogram law.

In Fig. 8.3b, XY represents $\mathbf{v}_B$ in magnitude (v) and direction (BD); YZ represents $(-\mathbf{v}_A)$ in magnitude (v) and direction (CA). The resultant, which gives the change of velocity, is then seen from the figure to be, in effect, a vector represented by XZ.

Since one vector $(-\mathbf{v}_A)$ is perpendicular to OA and the other $(\mathbf{v}_B)$ is perpendicular to OB, $\angle\,\text{XYZ} = \angle\,\text{AOB} = \delta\theta$. If δt is very small, $\delta\theta$ will also be small and XZ in Fig. 8.3b will have almost the same length as arc XZ in Fig. 8.3c which subtends angle $\delta\theta$ at the centre of a circle of radius v. Arc XZ $= v\,\delta\theta$ (from definition of radian) and so

$$\text{XZ} = v\,\delta\theta$$

But from (1) $\qquad \delta\theta = \frac{v\,\delta t}{r}$

$$\therefore \quad \text{XZ} = \frac{v^2}{r}\delta t$$

The *magnitude* of the acceleration a between A and B is

$$a = \frac{\text{change of velocity}}{\text{time interval}} = \frac{\text{XZ}}{\delta t}$$

Therefore

$$a = \frac{v^2}{r}$$

If ω is the angular velocity of the body, $v = r\omega$ and we can also write

$$a = \omega^2 r$$

The *direction* of the acceleration is *towards the centre* O of the circle, as can be seen if δt is made so small that A and B all but coincide; XZ is then perpendicular to $\mathbf{v}_A$ (or $\mathbf{v}_B$), i.e. along AO (or BO). We say the body has a **centripetal acceleration** (i.e. centre-seeking).

Does a body moving uniformly in a circle have *constant* acceleration? (Remember that acceleration is a vector.)

CENTRIPETAL FORCE

Since a body moving in a circle (or a circular arc) is accelerating, it follows from Newton's first law of motion that there must be a force acting on it to cause the acceleration. This force, like the acceleration, will also be directed towards the centre and is called the **centripetal force**. It causes the body to deviate from the straight-line motion which it would naturally follow if the force were absent. The value F of the centripetal force is given by Newton's second law, that is

$$F = ma = \frac{mv^2}{r}$$

where m is the mass of the body and v is its speed in the circular path of radius r. If the angular velocity of the body is ω we can also say, since $v = r\omega$,

$$F = m\omega^2 r$$

When a ball attached to a string is swung round in a horizontal circle, the centripetal force which keeps it in a circular orbit arises from the tension in the string. We can think of the tension as tugging continually on the body and 'turning it in' so that it remains at a fixed distance from the centre. If the ball is swung round faster, a larger force is needed and if this is greater than the tension the string can bear, the string breaks and the ball continues to travel along a tangent to the circle at the point of breaking, Fig. 8.4.

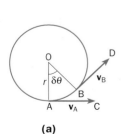
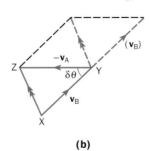
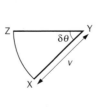

(a)   **(b)**   **(c)**

Fig. 8.3

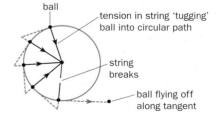

Fig. 8.4

Other examples of circular motion will be discussed in the following sections. In all cases it is important to appreciate that the forces acting on the body must provide a resultant force of magnitude mv^2/r towards the centre. What is the nature of the centripetal force for (a) a car rounding a bend, (b) a satellite circling the earth?

One arrangement for testing $F = mv^2/r$ experimentally is shown in Fig. 8.5a. The turntable, driven by the electric motor, is *gradually* speeded up and the spring extends until the truck just reaches the stop at the end of the track. The speed v of the *truck* in orbit is found by measuring the time for one revolution of the turntable and then, with the turntable at rest, the radius r of the circle described by the truck (i.e. the distance from the centre of the turntable to the centre of the truck). Knowing the mass m of the truck, mv^2/r can be calculated.

The tension in the stretched spring is the centripetal force and this can be found using a force sensor or by measuring with a spring balance the tension required to extend the spring *by the same amount* as it is when the truck is at the end stop, Fig. 8.5b. The value obtained should agree with the value of mv^2/r to within a few per cent.

If motion and force sensors are available, v and F can be monitored continuously using a datalogger and computer. The mass of the truck can be altered by loading it with lead plates and the experiment repeated for each mass.

ROUNDING A BEND

If a car is travelling round a circular bend with uniform speed on a horizontal road, the resultant force acting on it must be directed to the centre of its circular path, i.e. it must be the centripetal force. This force arises from the interaction of the car with the air and the ground. The direction of the force exerted by the air on the car will be more or less opposite to the instantaneous direction of motion. The other and more important horizontal force is the frictional force exerted inwards by the ground on the tyres of the car, Fig. 8.6. The resultant of these two forces is the centripetal force.

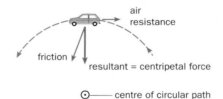

Fig. 8.6

The successful negotiation of a bend on a flat road therefore depends on the tyres and the road surface being in a condition that enables them to provide a sufficiently high frictional force, otherwise skidding occurs. Safe cornering that does not rely on friction is achieved by 'banking' the road.

The problem is to find the angle θ at which a bend should be banked so that the centripetal force acting on the car arises entirely from a component of the normal force N of the road, Fig. 8.7a. Treating the car as a particle

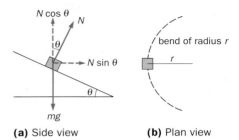

(a) Side view   **(b)** Plan view

Fig. 8.7

and resolving N vertically and horizontally we have, since $N \sin \theta$ is the centripetal force,

$$N \sin \theta = \frac{mv^2}{r}$$

where m and v are the mass and speed respectively of the car and r is the radius of the bend, Fig. 8.7b. Also, the car is assumed to remain in the same horizontal plane and so has no vertical acceleration, therefore

$$N \cos \theta = mg$$

Hence, by division,

$$\tan \theta = \frac{v^2}{gr}$$

The equation shows that for a given radius of bend, the angle of banking is only correct for one speed. In a velodrome the banking becomes steeper towards the outside of the track and the cyclist selects a position according to his speed, Fig. 8.8. In an attempt to reduce speeds on cornering, banking is not now used on most car racing tracks. A car's suspension tilts the car automatically.

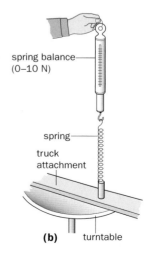

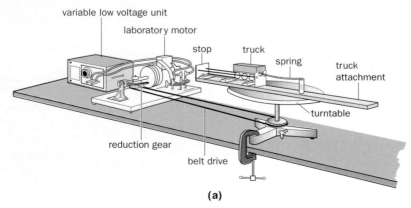

(a)

Fig. 8.5 Experiment to calculate centripetal force

Fig. 8.8 Banked pursuit track in Italy

A bend in a railway track is banked, in this case so that at a certain speed no lateral thrust has to be exerted by the outer rail on the flanges of the wheels of the train, otherwise the rails are strained. The horizontal component of the normal force of the rails on the train then provides the centripetal force.

An aircraft in straight, level flight experiences a lifting force at right angles to the surface of its wings, which balances its weight. To turn, the ailerons are operated so that the aircraft banks and the horizontal component of the lift supplies the necessary centripetal force, Fig. 8.9. The aircraft's weight is now opposed only by the vertical component of the lift, and height will be lost unless the lift is increased by, for example, increasing the speed.

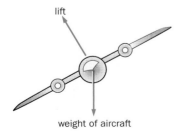

Fig. 8.9 An aircraft banking to provide centripetal force

OTHER EXAMPLES OF CIRCULAR MOTION

(a) The rotor

This device is sometimes present in amusement parks. It consists of an upright drum of diameter about 4 metres, inside which people stand with their backs against the wall. The drum is spun at increasing speed about its central vertical axis and at a certain speed the floor is pulled downwards. The occupants do not fall but remain 'pinned' against the wall of the rotor.

Fig. 8.10 The rotor

The forces acting on a passenger of mass m are shown in Fig. 8.10. N is the normal force of the wall on the passenger and is the centripetal force needed to keep him moving in a circle. Hence if r is the radius of the rotor and v the speed of the passenger then

$$N = \frac{mv^2}{r}$$

F is the frictional force acting upwards between the passenger and the rotor wall and since there is no vertical motion of the passenger

$$F = mg$$

If μ is the coefficient of limiting friction between passenger and wall, we have $F = \mu N$

$$\therefore \quad \mu N = mg$$

$$\therefore \quad \mu = \frac{mg}{N} = \frac{mg}{mv^2/r}$$

$$\therefore \quad \mu = \frac{gr}{v^2}$$

This equation gives the minimum coefficient of friction required to prevent the passenger slipping; it does not depend on the passenger's weight. A typical value of μ between clothing and a rotor wall (of canvas) is about 0.40 and so if $r = 2$ m, v must be about 7 m s^{-1} (or more). What will be the angular velocity of the drum? How many revolutions will it make per minute?

(b) Looping the loop

Fig. 8.11 Looping the loop at an amusement park

A pilot who is not strapped into the aircraft can loop the loop without falling downwards at the top of the loop. A bucket of water can be swung round in a vertical circle without spilling. A ball-bearing can loop the loop on a length of curtain rail in a vertical plane. All these effects have similar explanations.

Consider the bucket of water when it is at the top of the loop, A, in Fig. 8.12. If the weight mg of the water is *less than* mv^2/r, the normal force N of the bottom of the bucket on the water provides the rest of the force required to maintain the water in its circular path. However, if the bucket is swung more slowly then mg will be greater than mv^2/r and the 'unused' part of the weight causes the water to leave the bucket. What provides the centripetal force for the *water* when the bucket is at (*i*) B, (*ii*) C and (*iii*) D?

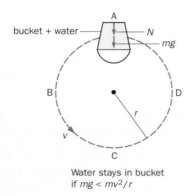

Fig. 8.12

(c) Centrifuges

Centrifuges separate solids suspended in liquids, or liquids of different densities. The mixture is in a tube, Fig. 8.13*a*, and when it is rotated at high speed in a horizontal circle the

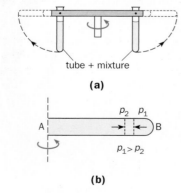

Fig. 8.13 The centrifuge

Fig. 8.14 Human centrifuge used for training Russian cosmonauts. It is spun in a horizontal plane at high speed

less dense matter moves towards the centre of rotation. On stopping the rotation, the tube returns to the vertical position with the less dense matter at the top. Cream is separated from milk in this way.

The action uses the fact that if a horizontal tube of liquid is rotated, the force exerted by the closed end must be greater than when the tube was at rest so that it can provide the necessary centripetal force acting radially inwards. In Fig. 8.13*b* the liquid pressure at B is greater than at A and a pressure gradient exists along the tube. For any part of the liquid the force due to the pressure difference supplies exactly the centripetal force required. If this part of the liquid is replaced by matter of smaller density (and thus of smaller mass), the force is too large and the matter moves inwards.

During the launching and re-entry of space vehicles, accelerations of about 8*g* occur and the resulting large forces which act on the surface of the astronaut's body cause blood to drain from some parts and congest others. If the brain is deprived, loss of vision and unconsciousness may follow. Tests with large centrifuges in which passengers are subjected to high centripetal accelerations, Fig. 8.14, show that a person will tolerate 15*g* for a few minutes when the body is perpendicular to the direction of the acceleration but only 6*g* when in the direction of acceleration. What will be the best position for an astronaut to adopt at lift-off and re-entry?

MOMENT OF INERTIA

In most of the cases of circular motion considered so far we have treated the body as a 'particle' so that all of it, in effect, revolves in a circle of the same radius. When this is not realistic we have to regard the rotating body as a system of connected 'particles' moving in circles of different radii. The way in which the mass of the body is distributed then affects its behaviour.

This may be shown by someone who is sitting on a freely rotating stool with a heavy weight in each hand, Fig. 8.15. When she extends her arms the speed of rotation decreases but increases again when she brings them in. The angular velocity of the system clearly depends on how the mass is distributed about the axis of rotation. A concept is needed to express this property.

Fig. 8.15

The mass of a body is a measure of its in-built opposition to any change of linear motion, i.e. mass measures inertia. The corresponding property for rotational motion is called the **moment of inertia**. The more difficult it is to change the angular velocity of a body rotating about a particular axis, the greater is its moment of inertia about that axis. Experiment shows that a wheel with most of its mass in the rim (e.g. a flywheel) is more difficult to start and stop than a uniform disc of equal mass rotating about the same axis; the former has a greater moment of inertia. Similarly, the moment of inertia of the person on the rotating stool is greater when her arms are extended. It should be noted that moment of inertia is a property of a body rotating about a particular axis; if the axis changes so does the moment of inertia.

We now require a measure of moment of inertia that takes into account the distribution of mass about the axis of rotation, and which plays a role in rotational motion similar to that played by mass in linear, i.e. straight-line, motion.

KINETIC ENERGY OF A ROTATING BODY

Suppose the body of Fig. 8.16 is rotating about an axis through O with constant angular velocity ω. A particle A, of mass m_1, at a distance r_1 from O, describes its own circular path and if v_1 is its linear velocity along the tangent to the path at the instant shown, then $v_1 = r_1\omega$ and

$$\text{kinetic energy of A} = \tfrac{1}{2}m_1 v_1^2$$
$$= \tfrac{1}{2}m_1 r_1^2 \omega^2$$

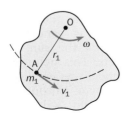

Fig. 8.16

The kinetic energy of the whole body is the sum of the kinetic energies of its component particles. If these have masses m_1, m_2, m_3, etc., and are at dis-tances r_1, r_2, r_3, etc., from O then, since all particles have the same angular velocity ω (the body being rigid), we have

kinetic energy of whole body

$$= \tfrac{1}{2}m_1 r_1^2 \omega^2 + \tfrac{1}{2}m_2 r_2^2 \omega^2 + \tfrac{1}{2}m_3 r_3^2 \omega^2 + \ldots$$
$$= \tfrac{1}{2}\omega^2(\Sigma\, mr^2)$$

where $\Sigma\, mr^2$ represents the sum of the mr^2 values for all the particles of the body. The quantity $\Sigma\, mr^2$ depends on the mass and its distribution and is taken as a measure of the moment of inertia of the body about the axis in question. It is denoted by the symbol I and so

$$I = \Sigma\, mr^2$$

Therefore

$$\text{rotational k.e. of body} = \tfrac{1}{2}I\omega^2$$

Comparing this with the expression $\tfrac{1}{2}mv^2$ for linear kinetic energy we see that the mass m is replaced by the moment of inertia I and the velocity v is replaced by the angular velocity ω. The unit of I is kg m².

Values of I for regular bodies can be found using calculus; that for a uniform rod of mass m and length l about an axis through its centre is $ml^2/12$. About an axis through its end it is $ml^2/3$.

It must be emphasized that rotational kinetic energy ($\tfrac{1}{2}I\omega^2$) is not a new type of energy but is simply the sum of the linear kinetic energies of all the particles of the body. It is a convenient way of stating the kinetic energy of a rotating rigid body.

The mass of a flywheel is concentrated in the rim, thereby giving it a large moment of inertia. When rotating, its kinetic energy is therefore large and explains why it is able to keep an engine (e.g. in a car) running at a fairly steady speed even though energy is supplied intermittently to it. Some toy cars have a small flywheel which is set into rapid rotation by a brief push across a solid surface. The kinetic energy of the flywheel will then keep the car in motion for some distance.

EQUATIONS FOR UNIFORM ANGULAR ACCELERATION

The angular acceleration α of a rotating body is its rate of change of angular velocity ω. If a very small angular velocity change $\delta\omega$ occurs in a very small time interval δt, the angular acceleration α is given by

$$\alpha = \frac{\text{change in angular velocity}}{\text{time taken for change}} = \frac{\delta\omega}{\delta t}$$

and is measured in rad s⁻². In calculus notation, the instantaneous angular acceleration α is defined by

$$\alpha = \lim_{\delta t \to 0}\left(\frac{\delta\omega}{\delta t}\right) = \frac{d\omega}{dt}$$

The equations for uniform linear acceleration (p. 117) have rotational counterparts which are, by analogy:

$$\omega = \omega_0 + \alpha t \qquad (1)$$
$$\frac{\theta}{t} = \frac{\omega + \omega_0}{2} \qquad (2)$$
$$\omega = \omega_0 t + \tfrac{1}{2}\alpha t^2 \qquad (3)$$
$$\omega^2 = \omega_0^2 + 2\alpha\theta \qquad (4)$$

where ω_0 is the initial angular velocity and ω is the final angular velocity (both in rad s⁻¹) after the body has rotated through angular displacement θ (in rad) with constant angular acceleration α (in rad s⁻²) in a time interval t (in s).

WORK DONE BY A COUPLE

Rotation is changed by a couple, that is, by two equal and opposite parallel forces whose lines of action do not coincide. It is often necessary to find the work done by a couple so that the energy transfer occurring as a result of its action on a body is known.

Consider the wheel in Fig. 8.17 of radius r on which the two equal and opposite forces P act tangentially and rotation through angle θ (in radians) occurs.

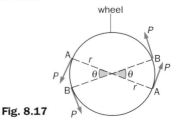

Fig. 8.17

Work done by each force
$$= \text{force} \times \text{distance}$$
$$= P \times \text{arc AB} = P \times r\theta$$

∴ total work done by couple
$$= Pr\theta + Pr\theta = 2Pr\theta$$

But, torque (or moment) of couple
$$= P \times 2r = 2Pr \qquad \text{(p. 113)}$$

Therefore

work done by couple
$$= \text{torque} \times \text{angle of rotation}$$
$$= T\theta$$

For example, if $P = 2.0$ N, $r = 0.50$ m and the wheel makes 10 revolutions then $\theta = 10 \times 2\pi$ rad and $T = P \times 2r = 2.0$ N $\times 2 \times 0.50$ m $= 2$ N m. Hence, work done by couple $= T\theta = 2 \times 20\pi = 1.3 \times 10^2$ J.

In general, if a couple of torque T about a certain axis acts on a body of moment of inertia I through an angle θ about the same axis and its angular velocity increases from 0 to ω, then

work done by couple
$$= \text{kinetic energy of rotation}$$
$$T\theta = \tfrac{1}{2}I\omega^2$$

ANGULAR MOMENTUM

(a) Definition

In linear motion it is often useful to consider the (linear) momentum of a body. In rotational motion, **angular momentum** is important.

Consider a rigid body rotating about an axis O and having angular velocity ω at some instant, Fig. 8.16 (p. 137). Let A be a particle of this body, distance r_1 from O and having linear velocity v_1 as shown, then the linear momentum of A $= m_1 v_1 = m_1 \omega r_1$ (since $v_1 = \omega r_1$).

The angular momentum of A about O is defined as the **moment of its momentum** about O. Hence

angular momentum of A
$$= r_1 \times m_1 \omega r_1 = \omega m_1 r_1^2$$

∴ total angular momentum of rigid body
$$= \Sigma \, \omega m r^2$$
$$= \omega \Sigma \, mr^2$$
$$= I\omega$$

where I is the moment of inertia of the body about O. Angular momentum is the equivalent of linear momentum (mv), with I replacing m and ω replacing v.

(b) Newton's second law

A body rotates when it is acted on by a couple. The rotational form of Newton's second law of motion may be written (by analogy with $F = ma$)

$$T = I\alpha$$

where T is the torque or moment of the couple causing rotational acceleration α.

(c) Conservation

A similar argument to that used to deduce the principle of conservation of linear momentum from Newton's third law can be employed to derive the principle of conservation of angular momentum.

The total angular momentum of a system remains constant provided no external torque acts on the system.

Fig. 8.18

Ice skaters, ballet dancers, acrobats and divers use the principle. The diver in Fig. 8.18 leaves the high-diving board with outstretched arms and legs and some initial angular velocity about her centre of gravity. Her angular momentum ($I\omega$) remains constant since no external torques act on her (gravity exerts no torque about her centre of gravity). To make a somersault she must increase her angular velocity. She does this by pulling in her legs and arms so that I decreases

and ω therefore increases. By extending her arms and legs again, her angular velocity falls to its original value. Similarly, a skater can whirl faster on ice by folding his or her arms.

The principle of conservation of angular momentum is useful for dealing with large rotating bodies such as the earth, as well as tiny, spinning particles such as electrons.

(d) Worked example

A shaft rotating at 3.0×10^3 revolutions per minute is transmitting a power of 10 kilowatts. Find the magnitude of the driving couple.

Work done per second by driving couple = power transmitted by shaft

Hence, since 1 W = 1 J s^{-1},
$$T\theta = 10 \times 10^3 \text{ J s}^{-1}$$

where T is the moment of the couple and θ is the angle through which the shaft rotates in 1 second. Now

3.0×10^3 revs per minute
$$= 3.0 \times 10^3 / 60 \text{ revs per second}$$
$$= 50 \text{ revs per second}$$

∴ $\theta = 50 \times 2\pi$ rad s^{-1}
(since 2π rad $= 360° = 1$ rev)

and so $T = \dfrac{10 \times 10^3}{50 \times 2\pi}$ N m
$$= 32 \text{ N m}$$

KEPLER'S LAWS

The Polish monk Copernicus (1473–1543) first proposed that the earth, rather than being the centre of the universe as was generally thought, revolved round the sun, as did the other planets. This heliocentric (sun-centred) model was greatly developed by the German astronomer Kepler (1571–1630) who, following on prolonged study of accurate observations made by his tutor the Danish astronomer Tycho Brahe over a period of twenty years, arrived at a very complete description of planetary motion. Kepler announced his first two laws in 1609 and the third in 1619.

1. Each planet moves in an ellipse which has the sun at one focus.

2. The line joining the sun to the moving planet sweeps out equal areas in equal times.

3. The squares of the times of revolution (T) of the planets about the sun are proportional to the cubes of their mean distances (r) from it:

$$r^3/T^2 \text{ is a constant}$$

In Fig. 8.19, if planet P takes the same time to travel from A to B as from C to D then the green areas are equal. Strictly speaking the distances in law 3 should be the semi-major axes of the ellipses but the orbits are sufficiently circular for the mean radius to be taken. The third column of Table 8.1 shows the constancy of r^3/T^2. Note that not all these planets were known of in Kepler's time.

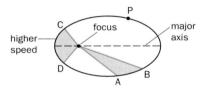

Fig. 8.19

Kepler's three laws enabled planetary positions, both past and future, to be determined accurately without the complex array of geometrical constructions used previously which were due to the Greeks. His work was also important because by stating his empirical laws (i.e. laws based on observation, not on theory) in mathematical terms he helped to establish the equation as a form of scientific shorthand.

Table 8.1

Planet	Mean radius of orbit r (metres)	Period of revolution T (seconds)	r^3/T^2
Mercury	5.79×10^{10}	7.60×10^6	3.36×10^{18}
Venus	1.08×10^{11}	1.94×10^7	3.35×10^{18}
Earth	1.49×10^{11}	3.16×10^7 (1 year)	3.31×10^{18}
Mars	2.28×10^{11}	5.94×10^7 (1.9 years)	3.36×10^{18}
Jupiter	7.78×10^{11}	3.74×10^8 (11.9 years)	3.36×10^{18}
Saturn	1.43×10^{12}	9.30×10^8 (29.5 years)	3.37×10^{18}
Uranus	2.87×10^{12}	2.66×10^9 (84.0 years)	3.34×10^{18}
Neptune	4.50×10^{12}	5.20×10^9 (165 years)	3.37×10^{18}
Pluto	5.90×10^{12}	7.82×10^9 (248 years)	3.36×10^{18}

Fig. 8.20 Jupiter and its four planet-sized moons

GRAVITY AND THE MOON

Kepler's laws summed up neatly *how* the planets of the solar system behaved without indicating *why* they did so. One of the problems was to find the centripetal force that kept a planet in its orbit round the sun, or a moon round its planet (Fig. 8.20), in a way which agreed with Kepler's laws.

Much later Newton realised (perhaps in his garden when the apple fell) that the earth exerts an inward pull on nearby objects causing them to fall. He then speculated whether this same force of gravity might extend out farther to pull on the moon and keep it in orbit. If it did, perhaps the sun also pulled on the planets in the same way with the same kind of force. He decided to test the idea first on the moon's motion — as we will do now.

If r is the radius of the moon's orbit round the earth and T is the time it takes to complete one orbit, i.e. its period, Fig. 8.21, then using accepted values we have

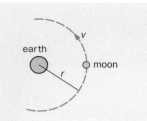

Fig. 8.21

$$r = 3.84 \times 10^8 \text{ m}$$

$$T = 27.3 \text{ days}$$

$$= 27.3 \times 24 \times 3600 \text{ s}$$

(The time between full moons is 29.5 days but this is because of the earth also moving round the sun. The moon has therefore to travel a little farther to reach the same position relative to the sun. Judged against the background of the stars, the moon takes 27.3 days to make one complete orbit of the earth, which is its true period *T*.)

The speed *v* of the moon along its orbit (assumed circular) is

$$v = \frac{\text{circumference of orbit}}{\text{period}} = \frac{2\pi r}{T}$$

$$= \frac{2\pi \times 3.84 \times 10^8}{27.3 \times 24 \times 3600} \text{ m s}^{-1}$$

$$= 1.02 \times 10^3 \text{ m s}^{-1}$$

The moon's centripetal acceleration *a* will be

$$a = \frac{v^2}{r} = \frac{(1.02 \times 10^3 \text{ m s}^{-1})^2}{3.84 \times 10^8 \text{ m}}$$

$$= 2.72 \times 10^{-3} \text{ m s}^{-2}$$

The acceleration of free fall at the earth's surface is 9.81 m s^{-2} and so if gravity is the centripetal force for the moon it must weaken between the earth and the moon. The simplest assumption would be that gravity halves when the distance doubles and at the moon it would be 1/60 of 9.81 m s^{-2} since the moon is 60 earth-radii from the centre of the earth and an object at the earth's surface is 1 earth-radius from the centre. But 9.81/60 = 1.64 $\times 10^{-1}$ m s^{-2}, which is still too large.

The next relation to try would be an inverse square law by which gravity is one-quarter when the distance doubles, one-ninth when it trebles and so on. At the moon it would be 1/60^2 of 9.81 m s^{-2}, i.e. 9.81/3600 = 2.72 $\times$ 10^{-3} m s^{-2} — the value of the moon's centripetal acceleration.

LAW OF UNIVERSAL GRAVITATION

Having successfully tested the idea of 'inverse square law' gravity for the motion of the moon round the earth, Newton turned his attention to the solar system.

His proposal, first published in 1687 in his great work the *Principia* ('Mathematical principles of natural knowledge'), was that the centripetal force which keeps the planets in orbit round the sun is provided by the gravitational attraction of the sun for the planets. This, according to Newton, was the same kind of attraction as that of the earth for an apple. Gravity — the attraction of the earth for an object — was thus a particular case of gravitation. In fact, Newton asserted that every object in the universe attracted every other object with a gravitational force and that this force was responsible for the orbital motion of celestial (heavenly) bodies.

Newton's hypothesis, now established as a theory and known as the **law of universal gravitation**, may be stated quantitatively as follows.

> Every particle of matter in the universe attracts every other particle with a force which is directly proportional to the product of their masses and inversely proportional to the square of their distances apart.

The **gravitational force** *F* between two particles of masses m_1 and m_2, distance *r* apart is therefore given by

$$F \propto \frac{m_1 m_2}{r^2} \quad \text{or} \quad F = G \frac{m_1 m_2}{r^2}$$

where *G* is a constant, called the **universal gravitational constant**, and assumed to have the same value everywhere for all matter.

Newton believed the force was directly proportional to the mass of each particle because the force on a falling body is proportional to its mass ($F = ma = mg = m \times$ constant, therefore $F \propto m$), i.e. to the mass of the *attracted* body. Hence, from the third law of motion, he argued that since the falling body also attracts the earth with an equal and opposite force that is proportional to the mass of the earth, then the gravitational force between the bodies must also be proportional to the mass of the *attracting* body. The moon test justified the use of an inverse square relation between force and distance.

The law applies to *particles* (i.e. bodies whose dimensions are very small compared with other distances involved), but Newton showed that the attraction exerted at an external point by a sphere of uniform density (or a sphere composed of uniform concentric shells) was the same as if its whole mass were concentrated at its centre. We tacitly assumed this for the earth in the previous section and will use it in the following.

The gravitational force between two ordinary objects (say two 1 kg masses 1 metre apart) is extremely small and therefore difficult to detect. What does this indicate about the value of *G* in SI units? What will be the SI unit of *G*?

TESTING GRAVITATION

To test $F = Gm_1m_2/r^2$ for the sun and planets the numerical values of all quantities on both sides of the equation need to be known. Newton neither had reliable information about the masses of the sun and planets nor did he know the value of *G* and so he could not adopt this procedure. There are alternatives however.

(a) Deriving Kepler's laws

The behaviour of the solar system is summarized by Kepler's laws and any theory that predicts these would, for a start, be in agreement with the facts.

Suppose a planet of mass *m* moves with speed *v* in a circle of radius *r* round the sun of mass *M*, Fig. 8.22.

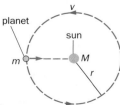

Fig. 8.22

Then

gravitational attraction of sun for planet

$$= G\frac{Mm}{r^2}$$

If this is the centripetal force keeping the planet in orbit, then

$$G\frac{Mm}{r^2} = \frac{mv^2}{r}$$

$$\therefore \quad \frac{GM}{r} = v^2$$

If T is the time for the planet to make one orbit,

$$v = \frac{2\pi r}{T}$$

$$\therefore \quad \frac{GM}{r} = \frac{4\pi^2 r^2}{T^2}$$

$$\therefore \quad GM = \frac{4\pi^2 r^3}{T^2}$$

Hence

$$\frac{r^3}{T^2} = \frac{GM}{4\pi^2}$$

Since GM is constant for any planet, r^3/T^2 is constant, which is Kepler's third law. We have considered a circular orbit but the same result holds for an elliptical one.

The first law can be derived by showing that if an inverse square law holds for gravitation, a planet moves in an orbit that is a conic section (i.e. a circle, ellipse, parabola or hyperbola) with the sun at one focus. Also, when a planet is acted on by *any* force, not just one governed by an inverse square law, directed from the planet towards the sun, the radius covers equal areas in equal times — which is the second law.

(b) Discovery of other planets

Theories can never be proved correct, they are only disproved by making predictions which conflict with observations. A good theory should lead to new discoveries. Newton's theory of gravitation has not only enabled us to work out problems connected with space travel, leading to new knowledge about the solar system, but it has also resulted in the discovery of planets not known in his day.

The planets exert gravitational pulls on one another but, except in the case of the larger planets like Jupiter and Saturn, the effect is only slight. The French scientist Laplace showed after Newton's time how to predict the effect of these disturbances (called perturbations) on Kepler's simple elliptical orbits.

The planet Uranus, discovered in 1781, showed small deviations from its expected orbit even after allowance had been made for the effects of known neighbouring planets. Two astronomers, Adams in England and Leverrier in France, working quite independently, predicted from the law of gravitation the position, size and orbit of an unknown planet that could cause the observed perturbations. A search was made and the new planet located in 1846 in the predicted position by the Berlin Observatory. Neptune had been discovered.

In 1930, history was repeated when American astronomers discovered Pluto from perturbations of the orbit of Neptune.

MASSES OF THE SUN AND PLANETS

The theory of gravitation enables us to obtain information about the mass of any celestial body having a satellite. If the value of the gravitational constant G is known, the actual mass can be calculated. Otherwise only a comparison is possible. A determination of G was not made until after Newton's death.

The principle is simply to measure all the quantities in $F = Gm_1m_2/r^2$ except G which can then be calculated. The earliest determinations used a measured mountain as the 'attracting' mass and a pendulum as the 'attracted' one. The first laboratory experiment was performed by Cavendish in 1798. He measured the very small gravitational forces exerted on two small lead balls (m_1 and m_2) by two larger ones (M_1 and M_2) using a torsion balance, Fig. 8.23. In this, the force twists a calibrated wire. Modern measurements give the value

$$G = 6.7 \times 10^{-11} \text{ N m}^2 \text{ kg}^{-2}$$
$$\text{(or m}^3 \text{ s}^{-2} \text{ kg}^{-1})$$

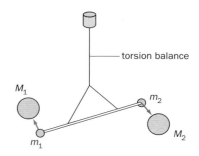

Fig. 8.23 Cavendish's determination of G

(a) Mass of the sun

Consider the earth of mass m_e moving with speed v_e round the sun of mass m_s in a circular orbit of radius r_e, Fig. 8.24*a*. The gravitational attraction of the sun for the earth is the centripetal force.

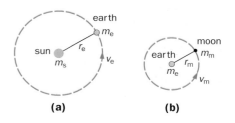

Fig. 8.24

$$G\frac{m_s m_e}{r_e^2} = \frac{m_e v_e^2}{r_e}$$

$$\therefore \quad m_s = \frac{v_e^2 r_e}{G}$$

If T_e is the time for the earth to make one orbit, then

$$v_e = \frac{2\pi r_e}{T_e} \quad \text{and} \quad m_s = \frac{4\pi^2}{G} \cdot \frac{r_e^3}{T_e^2}$$

Substituting for G, $r_e = 1.5 \times 10^{11}$ m and $T_e = 3.0 \times 10^7$ s (1 year), we get

$$m_s = 2.0 \times 10^{30} \text{ kg}$$

(b) Mass of the earth

Considering the moon of mass m_m moving with speed v_m round the earth of mass m_e in a circular orbit of radius r_m, Fig. 8.24*b*, we similarly obtain

$$G\frac{m_e m_m}{r_m^2} = \frac{m_m v_m^2}{r_m}$$

$$\therefore \quad m_e = \frac{v_m^2 r_m}{G}$$

If T_m is the period of the moon then $v_m = 2\pi r_m/T_m$ and so

$$m_e = \frac{4\pi^2}{G} \cdot \frac{r_m^3}{T_m^2}$$

Substituting for G, $r_m = 4.0 \times 10^8$ m and $T_m = 2.4 \times 10^6$ s (1 month) we find that

$$m_e = 6.0 \times 10^{24} \text{ kg}$$

The ratio of the mass of the sun to that of the earth is $2.0 \times 10^{30} : 6.0 \times 10^{24}$, i.e. 330 000 : 1. Table 8.2 gives the relative masses and densities of bodies in the solar system.

Table 8.2

	Mass (earth = 1)	Density (water = 1)
Sun	330 000	1.4
Moon	0.012	3.3
Mercury	0.056	6.1
Venus	0.82	5.1
Earth	1.0	5.5
Mars	0.11	4.1
Jupiter	320	1.4
Saturn	95	0.7
Uranus	15	1.6
Neptune	17	2.3
Pluto	6×10^{-4}	0.6

NEWTON'S WORK

(a) Scientific explanation

The accusation is sometimes made that science does not get down to underlying causes and give the 'true' reasons. In many cases this is so. Newton's work raises the question of what is meant by scientific explanation.

Consider gravitation. Newton did not really explain why a body falls or why the planets move round the sun. He attributed these effects to something called 'gravitation' and this, like other basic scientific ideas, seems by its very nature to defy explanation in any simpler terms. It appears that we must accept it as a fundamental concept of science, which is very useful because it enables us to regard apparently different phenomena — the falling of an apple and the motion of the planets — as having the same 'cause'.

A scientific explanation is very often an idea or concept that provides a connecting link between effects and so simplifies our knowledge. Explanations in terms of such concepts as energy, momentum, molecules, atoms and electrons fall into this category. Concepts that do not cast their net wide are of little value in science.

(b) Influence of Newton's work

Starting from the laws of motion and gravitation, Newton created a model of the universe which explained known facts, led to new discoveries and produced a unified body of knowledge. He united the physics of 'heaven and earth' by the same set of laws and so brought to a grand climax the work begun by Copernicus, Kepler and Galileo.

The success of Newtonian mechanics had a profound influence on both scientific and philosophical thought for 200 years. There arose a widespread belief that by using scientific laws the future of the whole universe could be predicted if the positions, velocities and accelerations of all the particles in it were known at a certain time. This 'mechanistic' outlook regarded the universe as a giant piece of clockwork, wound up initially by the 'divine power' and now ticking over according to strict mathematical laws.

Today scientists are humbler and probability has replaced certainty. Although Newtonian mechanics is still perfectly satisfactory for the world of ordinary experience, it has been supplemented by two other theories. The **theory of relativity** (see p. 504) has joined it for situations in which bodies are moving at very high speeds, and **quantum mechanics** enables us to deal with the physics of the atom.

(c) Chaos theory

More recently a different view of the universe has emerged in the form of 'chaos theory'. It suggests that it may not always be possible to predict the future of a system, be it simple or complex. Some scientists believe it heralds a new approach, not only to physics but also to other disciplines such as weather forecasting, engineering, economics, chemistry, electronics, astronomy and biology. It will be considered more fully in chapter 25.

EARTH'S GRAVITATIONAL FIELD

An action-at-a-distance effect, in which one body A exerts a force on another body B not in contact with it, can be regarded as due to a 'field of force' in the region around A.

We can think of the sun and all other celestial and terrestrial bodies as each having a gravitational field which exerts a force on any other body in the field. The **strength of a gravitational field is defined as the force acting on unit mass placed in the field**. If a body of mass m experiences a force F when in the earth's field, the strength of the earth's field is F/m (in newtons per kilogram). Measurement shows that if $m = 1$ kg, then $F = 9.8$ N (at the earth's surface); the strength of the earth's field is therefore 9.8 N kg^{-1}. However, if a mass m falls freely under gravity its acceleration g would be $F/m = 9.8$ m s^{-2} (since $F = ma = mg$).

We seem to have two ways of looking at g. When considering bodies falling freely we can think of it as an acceleration (of 9.8 m s^{-2}), but when a body of known mass is *at rest* or is *unaccelerated* in the earth's field and we wish to know the gravitational force (in newtons) acting on it we regard g as the earth's gravitational field strength (of 9.8 N kg^{-1}).

ACCELERATION OF FREE FALL

(a) Relation between g and G

A body of mass m at a place on the earth's surface where the acceleration of free fall is g experiences a force $F = mg$ (i.e. its weight) because of its attraction by the earth, Fig. 8.25. Assuming the earth behaves as if its whole mass M were concentrated at its centre O, then, by the law of gravitation, we can also say that F is the gravitational pull of the earth on the body. So

$$F = G\frac{Mm}{r^2}$$

where r is the radius of the earth.

$$\therefore \quad mg = G\frac{Mm}{r^2}$$

$$g = \frac{GM}{r^2}$$

Fig. 8.25

It is worth noting that the mass m in $F = ma = mg$ is called the **inertial mass** of the body; it measures the opposition of the body to change of motion, i.e. its inertia. The mass of the same body when considering the law of gravitation is known as the **gravitational mass**. Experiments show that to a high degree of accuracy these two masses are equal for a given body and so we can, as we have done here, represent each by m.

(b) Variation of g with height

If g' is the acceleration of free fall at a distance a from the centre of the earth where $a > r$, r being the earth's radius, then from section **(a)**,

$$g' = \frac{GM}{a^2} \quad \text{and} \quad g = \frac{GM}{r^2}$$

Dividing, $\quad \dfrac{g'}{g} = \dfrac{r^2}{a^2}$

or $\qquad g' = \dfrac{r^2}{a^2} g$

Above the earth's surface, the acceleration of free fall g' therefore varies inversely as the square of the distance a from the centre of the earth (since r and g are constant), i.e. it decreases with height as shown in Fig. 8.26.

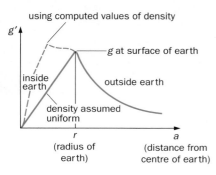

Fig. 8.26

At height h above the surface, $a = r + h$,

$$\therefore \quad g' = \frac{r^2}{(r + h)^2} g = \frac{1}{(1 + h/r)^2} g$$

$$= \left(1 + \frac{h}{r}\right)^{-2} g$$

If h is very small compared with r (6400 km) we can neglect powers of (h/r) higher than the first. Then

$$g' = \left(1 - \frac{2h}{r}\right) g$$

(c) Variation of g with depth

At a point such as P below the surface of the earth, it can be shown that if the shaded spherical shell in Fig. 8.27 has uniform density, it produces no gravitational field inside itself. The gravitational acceleration g_1 at P is then due entirely to the sphere of radius b and if this is assumed to be of uniform density, then from section **(a)**,

$$g_1 = \frac{GM_1}{b^2} \quad \text{and} \quad g = \frac{GM}{r^2}$$

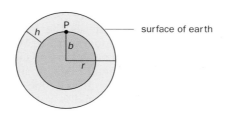

Fig. 8.27

where M_1 is the mass of the sphere of radius b. The mass of a uniform sphere is proportional to its radius cubed, hence

$$\frac{M_1}{M} = \frac{b^3}{r^3}$$

But $\qquad \dfrac{g_1}{g} = \dfrac{M_1}{M} \cdot \dfrac{r^2}{b^2}$

$$\therefore \quad \frac{g_1}{g} = \frac{b}{r} \quad \text{or} \quad g_1 = \frac{b}{r} g$$

Thus, assuming the earth has uniform density, the acceleration g_1 is directly proportional to the distance b from the centre, i.e. it decreases linearly with depth, Fig. 8.26. At depth h below the earth's surface, $b = r - h$,

$$\therefore \quad g_1 = \left(\frac{r - h}{r}\right) g = \left(1 - \frac{h}{r}\right) g$$

In fact, because the earth's density is not constant, g_1 actually *increases* for all depths now attainable, as shown by part of the dotted curve in Fig. 8.26.

(d) Variation of g with latitude

The observed variation of g over the earth's surface is largely due to (*i*) the equatorial radius of the earth exceeding its polar radius by about 21 km and thereby making g greater at the poles than at the equator, where a body is farther from the centre of the earth, and (*ii*) the effect of the earth's rotation which we will now consider.

A body of mass m at any point of the earth's surface (except at the poles) must have a centripetal force acting on it. This force is supplied by part of the earth's gravitational attraction for it. On a stationary earth the gravitational pull of the earth on m would be mg where g is the acceleration of free fall under such conditions. However, because of the earth's rotation, the observed gravitational pull is less than this and equals mg_0 where g_0 is the *observed* acceleration of free fall. Therefore

centripetal force on body $= mg - mg_0$

At the equator, the body is moving in a circle of radius r, where r is the earth's radius, and it has the same angular velocity ω as the earth. The centripetal force is then $m\omega^2 r$ and so

$$mg - mg_0 = m\omega^2 r$$
$$\therefore \quad g - g_0 = \omega^2 r$$

Substituting

$$r = 6.4 \times 10^6 \text{ m}$$

and $\quad \omega = 1$ revolution in 24 hours

$$= 2\pi/(24 \times 3600) \text{ rad s}^{-1}$$

we get

$$g - g_0 = 3.4 \times 10^{-2} \text{ m s}^{-2}$$

Assuming the earth is perfectly spherical, this is also the difference between the polar and equatorial values of the acceleration of free fall. (At the poles $\omega = 0$ and so $g = g_0$.) The observed difference is 5.2×10^{-2} m s^{-2}, of which 1.8×10^{-2} m s^{-2} arises from the nonsphericity of the earth.

At latitude θ on an assumed spherical earth, the body describes a circle of radius $r \cos \theta$, Fig. 8.28a. The magnitude of the centripetal force required is thus $m\omega^2 r \cos \theta$ and is smaller than at the equator since ω has the same value. However, its direction is along PQ whereas mg acts along PO towards the centre of the earth. The observed gravitational pull mg_0 is therefore less than mg by a force

$m\omega^2r\cos\theta$ along PQ and will be in a different direction from mg. The value and direction of mg_o must be such that when it is compounded by the parallelogram law with $m\omega^2r\cos\theta$ along PQ, it gives mg along PO, Fig. 8.28*b*. The direction of g_o as shown by a falling body or a plumb line is not exactly towards the centre of the earth except at the poles and the equator.

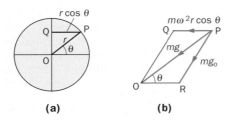

(a)   **(b)**

Fig. 8.28

ARTIFICIAL SATELLITES

(a) Satellite orbits

The centripetal force which keeps an artificial satellite in orbit round the earth is the gravitational attraction of the earth for it. For a satellite of mass m travelling with speed v in a circular orbit of radius R (measured from the centre of the earth), we have

$$\frac{mv^2}{R} = \frac{GMm}{R^2}$$

where M is the mass of the earth.

$$\therefore \quad v^2 = \frac{GM}{R}$$

But

$$g = \frac{GM}{r^2} \quad \text{(see p. 142)}$$

where r is the radius of the earth and g is the acceleration of free fall at the earth's surface.

$$\therefore \quad v^2 = \frac{gr^2}{R}$$

If the satellite is close to the earth, say at a height of 100–200 km, then $R \approx r$ and

$$v^2 = gr$$

Substituting values $r = 6.4 \times 10^6$ m (6400 km) and $g = 9.8$ m s^{-2},

$$v = \surd(gr)$$
$$= \surd(9.8 \text{ m s}^{-2} \times 6.4 \times 10^6 \text{ m})$$
$$= 7.9 \times 10^3 \text{ m s}^{-1}$$

$$\therefore \quad v \approx 8 \text{ km s}^{-1}$$

The time for the satellite to make one complete orbit of the earth, i.e. its period T, is

$$T = \frac{\text{circumference of earth}}{\text{speed}}$$
$$= \frac{2\pi r}{v}$$
$$= \frac{2\pi \times 6.4 \times 10^6 \text{ m}}{7.9 \times 10^3 \text{ m s}^{-1}}$$

$$\therefore \quad T \approx 5000 \text{ s} \approx 83 \text{ minutes}$$

A satellite in orbit is being continually pulled in by gravity from a straight-line tangent path to a circular path, Fig. 8.29. It 'falls' again and again from the tangents instead of continuing along them; its horizontal speed is such that it 'falls' by the correct distance to keep it in a circle. Although the satellite has an acceleration towards the centre of the earth, it has no vertical velocity because it 'falls' at the same rate as the earth's surface falls away underneath it. With respect to the earth's surface its velocity in a vertical direction is zero since the distance between the satellite and the earth's surface remains constant. In practice it is very difficult to achieve an exactly circular orbit.

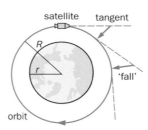

Fig. 8.29

(b) Launching a satellite

To be placed in orbit a satellite must be raised to the desired height and given the correct speed and direction by the launching rocket. A typical launching sequence using a two-stage rocket might be as follows.

At lift-off, the rocket, with a manned capsule or an unmanned payload such as a satellite on top, is held down by clamps on the launching pad for a few seconds until the exhaust gases have built up an upward thrust which exceeds the rocket's

weight. The clamps are then removed by remote control and the rocket accelerates upwards. To penetrate the dense lower part of the atmosphere by the shortest possible route, the rocket rises vertically initially and, after this, is gradually tilted by the guidance system. The first-stage rocket, which may burn for about 2 minutes producing a speed of 3 km s^{-1} or so, lifts the vehicle to a height of around 60 km, then separates and falls back to earth, landing many kilometres from the launch site.

The vehicle now coasts in free flight (unpowered) to its orbital height, say 160 km, where it is momentarily moving horizontally (i.e. parallel to the earth's surface immediately below). The second-stage rocket then fires and increases the speed to that required for a circular orbit at this height (about 8 km s^{-1}). By firing small rockets, the payload is separated from the second stage which follows behind, also in orbit.

Figure 8.30 shows the lift-off of a *three*-stage rocket, Ariane 4.

Fig. 8.30 Lift-off of an Ariane 4 rocket with a payload of an Intelsat VII communications satellite

The equation $v^2 = gr^2/R$ for circular orbits shows that each orbit requires a certain speed and the greater the orbit radius R the smaller the speed v.

'Synchronous' or 'geostationary' satellites are 35 800 km above the equator and have a period of 24 hours, so remain in the same position above the earth, apparently stationary. By acting as relay stations, they make continuous, world-wide communications (e.g. of telephone calls and television programmes) possible (see chapter 24).

Some notable space flights are given in Table 8.3. Re-usable space shuttles now launch satellites into low-earth orbits, and recover them for repair.

WEIGHTLESSNESS

An astronaut orbiting the earth in a space vehicle with its rocket motors off is said to be 'weightless'. If weight means the pull of the earth on a body, then the statement, although commonly used, is misleading. A body is not truly weightless unless it is outside the earth's (or any other) gravitational field, i.e. at a place where $g = 0$. In fact it is gravity which keeps an astronaut and his vehicle in orbit. To appreciate what 'experiencing the sensation of weightlessness' means we will consider similar situations on earth.

We are aware of our weight because the ground (or whatever supports us) exerts an *upward* push on us as a result of the *downward* push our feet exert on the ground. It is this upward push which makes us 'feel' the force of gravity. When a lift suddenly starts upwards the push of the floor on our feet increases and we feel heavier. On the other hand if the support is reduced we seem to be lighter. In fact **we judge our weight from the upward push exerted on us by the floor**. If our feet are completely unsupported we experience weightlessness. Passengers in a lift that has a continuous downward acceleration equal to g would get no support from the floor since they, too, would be falling with the same acceleration as the lift. There is no upward push on them and so no sensation of weight is felt. The condition is experienced briefly when we jump off a wall or dive into a swimming pool, as we are then briefly in 'free fall'.

An astronaut in an orbiting space vehicle is not unlike a passenger in a freely falling lift. The astronaut is moving with *constant speed* along the orbit, but since he is travelling in a circle he has a centripetal acceleration — of the same value as that of the space vehicle and equal to g at that height. The walls of the vehicle exert no force on the astronaut; he is unsupported, the physiological sensation of weight disappears and he floats about 'weightless'. Similarly, any object released in the vehicle does not 'fall' (Fig. 8.31, p. 146). Anything not in use must be firmly fixed and liquids will not pour.

Summing up, to be strictly correct we should not use the term 'weightless' unless by weight we mean the force exerted on (or by) a body by (or on) its support, and generally we do not.

Also, it is important to appreciate that although 'weightless' a body still has mass and it would be just as difficult to push it in space as on earth. An astronaut floating in an orbiting vehicle could still be injured by hitting a hard but 'weightless' object.

Table 8.3 Some notable space flights

Name	Launch date	Descent date or lifetime	Period (min)	Height (km)	Notes
Sputnik 1	4 Oct 1957	10 years	96.2	215–939	First artificial earth satellite. Mass 83.6 kg
Vostok 1	12 Apr 1961	12 Apr 1961	89.3	169–315	First manned space flight. One orbit by Yuri Gagarin
Mariner 2	27 Aug 1962	Last contact 3 Jan 1963	—	—	First successful interplanetary space probe. Flew past Venus in December 1962
Vostok 6	16 June 1963	19 June 1963	88	210	First woman in space. Valentina Tereshkova made 48 orbits of the earth
Early Bird	6 Apr 1965	Now superseded by INTELSAT models	1437	35 000–36 000	First commercial synchronous communications satellite: 'stationary' between Africa and South America
Apollo 11	16 July 1969	24 July 1969	—	—	First men (Armstrong and Aldrin) to land on the moon and return to earth with 'moon samples'
Salyut 1	19 April 1971	11 Oct 1971	88.5	212–280	First manned orbiting space station. Two cosmonauts spent 22 days in the station in July 1971 but died on their return journey
Columbia	12 April 1981	14 April 1981	57	300	First re-usable space shuttle to orbit and return to earth
Galileo	18 Oct 1989	10 years	—	—	Particularly successful NASA interplanetary space probe launched from shuttle Atlantis. Detailed photos sent back of Jupiter and its moons (1995–1999)
Hubble space telescope	April 1990	15 years	—	600	US space telescope with 2.4 m mirror commissioned for astronomical research
Mars Global	7 Nov 1996	—	118	378	Orbiter and lander to survey Mars

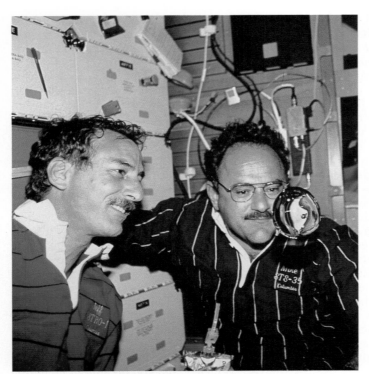

Fig. 8.31 In an orbiting space shuttle a water bubble floats 'weightless'

SPEED OF ESCAPE

The faster a ball is thrown upwards, the higher it rises before it is stopped and pulled back by gravity. We will show that to escape from the earth into outer space an object must have a speed of just over 11 km s^{-1} — called the **escape speed**.

The escape speed is obtained from the fact that the potential energy gained by the body equals its loss of kinetic energy, if air resistance is neglected. The work done measures the energy transferred. Let m be the mass of the escaping body and M the mass of the earth. The force F exerted on the object by the earth when it is distance x from the centre of the earth is

$$F = G\frac{Mm}{x^2}$$

Therefore work done δW by gravity when the body moves a further short distance δx upwards is

$$\delta W = -F\delta x = -G\frac{Mm}{x^2}\delta x$$

(the negative sign shows the force acts in the opposite direction to the displacement, see p. 124). Therefore

total work done while body escapes

$$= \int_r^\infty -G\frac{Mm}{x^2}\,dx \quad (r = \text{radius of earth})$$

$$= -GMm\left[-\frac{1}{x}\right]_r^\infty = GMm\left[\frac{1}{x}\right]_r^\infty$$

$$= -\frac{GMm}{r}$$

If the body leaves the earth with speed v and just escapes from its gravitational field,

$$\tfrac{1}{2}mv^2 = \frac{GMm}{r}$$

$$\therefore v = \sqrt{\left(\frac{2GM}{r}\right)}$$

But $$g = \frac{GM}{r^2} \qquad\qquad \text{(p. 142)}$$

$$\therefore v = \sqrt{(2gr)}$$

Substituting $r = 6.4 \times 10^6$ m and $g = 9.8$ m s^{-2}, we get $v \approx 11$ km s^{-1}.

Possible paths for a body projected at different speeds from the earth are shown in Fig. 8.32.

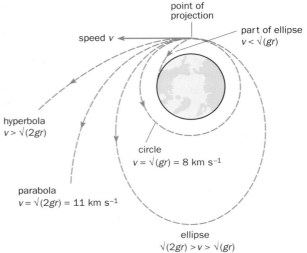

Fig. 8.32

The multi-stage rockets in use at present burn their fuel in a comparatively short time to obtain the best performance. They behave rather like objects thrown upwards, i.e. like projectiles. For a journey to, say, the moon, they therefore have to attain the escape speed. In many cases this is done by first putting the final vehicle into a 'parking' orbit round the earth with a speed of 8 km s^{-1} and then firing the final rocket again to reach escape speed in the appropriate direction.

The attainment of the escape speed is not a necessary condition however. The essential thing is that a certain amount of *energy* is required to escape from the earth and if rockets were available which could develop large power over a long time, escape would still be possible without ever achieving escape speed. In fact if we had a long enough ladder and the necessary time and energy we could walk to the moon!

Air molecules at 0 °C have an average speed of about 0.5 km s^{-1} which, being much less than the escape speed, ensures that the earth's gravitational field is able to maintain an atmosphere of air round the earth. The average speed of hydrogen molecules at 0 °C is more than three times that of air molecules and explains their rarity in the earth's atmosphere. The moon has no atmosphere. Can you suggest a possible reason?

QUESTIONS

Assume $g = 10$ m s^{-2} unless stated otherwise.

Circular motion

1. A particle moves in a semicircular path AB of radius 5.0 m with constant speed 11 m s^{-1}, Fig. 8.33. Calculate
   a) the time taken to travel from A to B (take $\pi = 22/7$),
   b) the average velocity,
   c) the average acceleration.

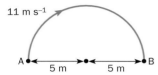

Fig. 8.33

2. A turntable makes 33 revolutions per minute. Calculate
   a) its angular velocity in rad s^{-1},
   b) the linear velocity of a point 0.12 m from the centre.
3. A grinding wheel of diameter 0.12 m spins horizontally about a vertical axis, as shown in Fig. 8.34. P is a typical grinding particle bonded to the edge of the wheel.

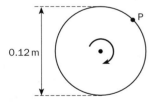

Fig. 8.34

   a) If the rate of rotation is 1200 revolutions per minute, calculate
      i) the angular velocity,
      ii) the acceleration of P,
      iii) the magnitude of the force acting on P if its mass is 1.0×10^{-4} kg.
   b) The maximum radial force at which P remains bonded to the wheel is 2.5 N.
      i) Calculate the angular velocity at which P will leave the wheel if its rate of rotation is increased.
      ii) If the wheel exceeds this maximum rate of rotation, what will be the speed and direction of motion of particle P immediately after it leaves the wheel?
      (NEAB, AS/A PH01, June 1997)

4. What is meant by a **centripetal force**? Why does such a force do no work in a circular orbit?
   a) An object of mass 0.50 kg on the end of a string is whirled round in a horizontal circle of radius 2.0 m with a constant speed of 10 m s^{-1}. Find its angular velocity and the tension in the string.
   b) If the same object is now whirled in a vertical circle of the same radius with the same speed, what are the maximum and minimum tensions in the string?
5. A car is travelling round a bend in a road at a constant speed of 22 m s^{-1}. The driver moves along a circular path of radius 25 m.
   a) Explain why, although the speed of the driver is constant, his velocity is not constant.
   b) Figure 8.35 illustrates the position of the car and the driver at an instant during this motion.

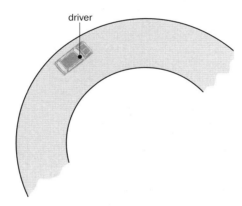

Fig. 8.35

   Copy the diagram and show the directions of the velocity and of the acceleration of the driver at this instant.
   c) Calculate the magnitude of the driver's acceleration.
   d) Suggest what provides a force to cause this acceleration.
      (UCLES, Basic 1, June 1998)
6. A racing car goes around a circular curve as fast as it can without skidding. The radius of the curve is 50 m and the road is banked at 20° to allow faster speeds. The coefficient of static friction between the road and the tyres is 0.80.

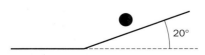

Fig. 8.36

   a) Copy Fig. 8.36 and draw in vectors to represent all the forces acting on the car while it is travelling at this maximum speed.
   b) Resolve the forces into horizontal and vertical components and apply Newton's laws of motion in these two directions.
   c) Use these equations and that for the maximum frictional force to determine a value for the maximum speed that the car can have.
   d) Explain why the coefficient of **static** friction can be used in this problem, despite the fact that the car is moving.
      (IB, Higher Paper 2, May 1998)

Moment of inertia

7. As an alternative to a petrol engine, a prototype car is fitted with a flywheel. This is spun to a high speed at the start of a journey. To start the car the flywheel is connected to the drive wheels by a system of gears. The rotational energy stored in the flywheel is then used to propel the car.
   The flywheel has a moment of inertia of 90 kg m^2. Before each journey the flywheel is accelerated until it has an angular speed of 300 rad s^{-1}.
   a) State the factors which affect the moment of inertia of the flywheel.
   b) Assuming the angular acceleration to be uniform, determine the torque required to accelerate the flywheel to its maximum speed in 2.5 minutes.
   c) Calculate the energy stored in the flywheel at the start of a journey.
   d) The average resistive force on the car during one journey is 400 N. Calculate the maximum distance the car can travel during this journey.
      (AEB, 0635/2, Summer 1998)
8. Derive an expression for the kinetic energy of a rotating rigid body.
   An electric motor supplies a power of 5×10^2 W to drive an unloaded flywheel of moment of inertia 2 kg m^2 at a steady speed of 6×10^2 revolutions per minute. How long will it be before the flywheel comes to rest after the power is switched off assuming the frictional couple remains constant?
9. a) Define **angular momentum**. State the principle of conservation of angular momentum.
   An ice dancer is spinning about a vertical axis with his arms extended vertically upwards. Will he spin faster or slower when he allows his arms to fall until they are horizontal? Has his kinetic

energy been increased or decreased? How do you account for the change?

b) A horizontal disc rotating freely about a vertical axis makes 90 revolutions per minute. A small piece of putty of mass 2.0×10^{-2} kg falls vertically on to the disc and sticks to it at a distance of 5.0×10^{-2} m from the axis. If the number of revolutions per minute is thereby reduced to 80, calculate the moment of inertia of the disc.

Gravitation

10. The mass of the earth is 5.98×10^{24} kg and the gravitational constant is 6.67×10^{-11} m^3 kg^{-1} s^{-2}. Assuming the earth is a uniform sphere of radius 6.37×10^6 m, find the gravitational force on a mass of 1.00 kg at the earth's surface.

11. State the period of the earth about the sun. Use this value to calculate the angular speed of the earth about the sun in rad s^{-1}.

    The mass of the earth is 5.98×10^{24} kg and its average distance from the sun is 1.50×10^{11} m. Calculate the centripetal force acting on the earth.

    What provides this centripetal force?
    (*L*, AS/A PH1, June 1997)

12. A satellite orbits the earth once every 120 minutes. Calculate the satellite's angular speed.

    Draw a free-body force diagram for the satellite.

    The satellite is in a state of free fall. What is meant by the term **free fall**? How can the height of the satellite stay constant if the satellite is in free fall?
    (*L*, AS/A PH1, June 1996)

13. It is proposed to place a communic - ations satellite in a circular orbit round the equator at a height of 3.59×10^7 m above the earth's surface. Find the period of revolution of the satellite in hours and comment on the result. (Use the values given in question 10 for the radius and mass of the earth and the gravitational constant.)

14. **a)** State Newton's law of gravitation. If the acceleration of free fall, g_m, at the moon's surface is 1.70 m s^{-2} and its radius is 1.74×10^6 m, calculate the mass of the moon.

    b) To what height would a signal rocket rise on the moon, if an identical one fired on earth could reach 200 m? (Ignore atmospheric resistance.) Explain your reasoning.

c) Explain, using algebraic symbols and stating which quantity each represents, how you could calculate the distance D of the moon from the earth (mass M_e) if the moon takes t seconds to move once round the earth.

d) What is meant by 'weightlessness', experienced by an astronaut orbiting the earth, and how is it caused? Explain also whether the astronaut would have the same experience when falling freely back to earth in the capsule just prior to re-entry in the earth's atmosphere. (Gravitational constant = 6.67×10^{-11} m^3 kg^{-1} s^{-2}; acceleration of free fall = 9.81 m s^{-2}.)

15. **a)** Select a method for measuring g, the gravitational field strength at the surface of the earth. For your chosen method:
    i) sketch the apparatus you would use,
    ii) state the measurements that you would need to make,
    iii) state how you would make your measurements,
    iv) show how you would use your measurements to determine the gravitational field strength.

    b) A white dwarf star has a mass of 1.4×10^{30} kg and a radius of 1.2×10^6 m. Show that the gravitational field strength at its surface is 6.5×10^7 N kg^{-1}. (Universal gravitational constant, $G = 6.7 \times 10^{-11}$ N m^2 kg^{-2}.)

    c) The escape speed for a body is the speed that it must be given at the surface of the body for it to be able to escape from the gravitational field without a further input of energy. Calculate the escape speed for the white dwarf star described in b).
    (*AEB*, 0635/2, Summer 1997)

16. Ganymede is one of the moons of Jupiter. Ganymede has a mass M_G and orbits Jupiter, mass M_J. The radius of the orbit is r.

    Write down Newton's law of gravitation as applied to the Jupiter–Ganymede system.

    Write down an expression for the centripetal force F required to cause Ganymede to orbit Jupiter with an angular speed of ω.

    Show that $r^3 \omega^2 = GM_J$ where G is the universal gravitational constant.

    Ganymede orbits Jupiter once every 7.16 days and the radius of its orbit is 1.07×10^9 m. Calculate the mass of Jupiter.
    (*L*, A PH4, June 1999)

17. Newton's law of gravitation may be expressed as
$$F = \frac{GmM}{d^2}$$

    a) State what each of these five symbols represents.

    b) State what is meant by the **period** of a planet in its orbit.

    c) Use Newton's law of gravitation to derive an expression for the radius of the circular orbit of a planet around the Sun in terms of its period.

    d) i) The average radius of Pluto's orbit is 39.4 AU and its period is 9.04×10^4 days. However, irregularity in Pluto's orbit has been observed. There has been speculation that, in order to explain such irregularity, a tenth planet X exists.
    If X were to exist and if its period were twice that of Pluto, calculate at what average distance X would orbit around the Sun.
    ii) The diameter of Pluto is about 6.0×10^6 m. Imagine that Pluto is observed from a point O on planet X. Assuming that both orbits are circular, calculate the maximum value of angle β as shown in Fig. 8.37.

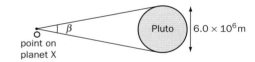

Fig. 8.37

[*Note:* 1 AU = 1.5×10^{11} m]
(*OCR*, Cosmology, June 1999)

9

Mechanical oscillations

- ■ Introduction
- ■ Simple harmonic motion
- ■ Equations of s.h.m.
- ■ Expression for ω

- ■ Mass on a spring
- ■ Simple pendulum
- ■ Some calculations
- ■ Energy of s.h.m.

- ■ Damped oscillations
- ■ Forced oscillation and resonance
- ■ A mathematical model

INTRODUCTION

In previous chapters linear and circular motion were considered. Another common type of motion is the to-and-fro repeating movement called a **vibration** or **oscillation**.

Examples of oscillatory motion are provided by a swinging pendulum, a mass on the end of a vibrating spring, and the strings and air columns of musical instruments when producing a note. Sound waves are transmitted by the oscillation of the particles of the medium in which the sound is travelling. Atoms in a solid vibrate about fixed positions in their lattice.

Vibrations can occur in turbines, aircraft, cars and tall buildings, and were responsible for the collapse of the Tacoma Narrows suspension bridge when a moderate gale set the bridge oscillating until the main span broke up, Fig. 9.2a. In metal structures vibrations can cause fatigue failure. Vibrations of the ground due to seismic waves during an earthquake can cause great destruction, Fig. 9.2b.

In a mechanical oscillation there is a continual interchange of potential and kinetic energy because of the system having (*i*) **elasticity** (or springiness) which allows it to store p.e. and (*ii*) **mass** (or inertia) which enables it to have k.e. So when a body on the lower end of a spiral spring, Fig. 9.1, is pulled down and released, the elastic restoring force pulls the body up and it accelerates towards its equilibrium position O with increasing velocity. The accelerating force decreases as the body approaches O (since the spring is stretched less) and so the *rate of change* of velocity (the acceleration) decreases.

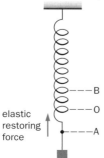

elastic
restoring
force

- - - B
- - - O
- - - A

Fig. 9.1

Fig. 9.2a Collapsed Tacoma Narrows bridge, America, 1940

Fig. 9.2b Earthquake damage in Kobe, Japan, 1995

At O the restoring force is zero but because the body has inertia it overshoots the equilibrium position and continues to move upwards. The spring is now compressed and the elastic restoring force acts again, but downwards towards O this time. The body therefore slows down and at an increasing rate due to the restoring force increasing at greater distances from O. The body eventually comes to rest above O and repeats its motion in the opposite direction, p.e. stored as elastic energy of the spring being continually changed to k.e. of the moving body and vice versa. The motion would continue indefinitely if no energy loss occurred, but energy *is* lost. Why?

The time for a complete oscillation from A to B and back to A, or from O to A to O to B and back to O again is the **period** *T* of the motion. The **frequency** *f* is the number of complete oscillations per unit time and a little thought (perhaps using numbers) will indicate that

$$f = \frac{1}{T}$$

An oscillation (or cycle) per second is a **hertz** (Hz). The maximum displacement OA or OB is called the **amplitude** of the oscillation.

Some other simple oscillatory systems are shown in Fig. 9.3. It is worth trying to discover experimentally (*i*) which have a constant period (compared with a watch), (*ii*) what factors determine the period (or frequency) of the oscillation, and (*iii*) whether 'time-traces' of their motions can be obtained and what they look like.

SIMPLE HARMONIC MOTION

In Fig. 9.4 N is a body oscillating in a straight line about O, between A and B; N could be a mass hanging from a spiral spring. In linear motion we considered accelerations that were constant in magnitude and direction, and in circular motion the accelerations (centripetal) were constant in magnitude if not in direction. In oscillatory motion, the accelerations, like the displacements and velocities, change periodically in both magnitude and direction.

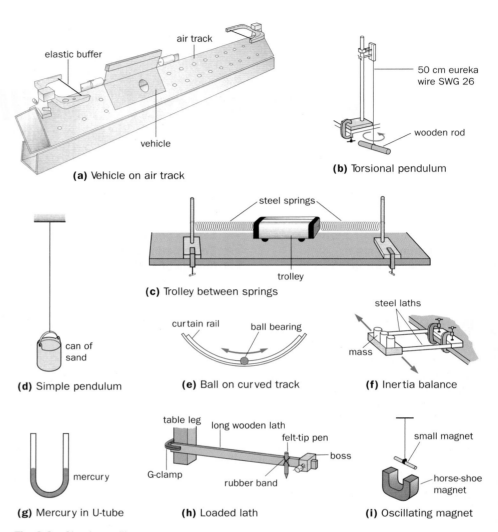

Fig. 9.3 Simple oscillatory systems

(a) Vehicle on air track
(b) Torsional pendulum
(c) Trolley between springs
(d) Simple pendulum
(e) Ball on curved track
(f) Inertia balance
(g) Mercury in U-tube
(h) Loaded lath
(i) Oscillating magnet

Consider first displacements and velocities. When N is below O, the displacement (measured from O) is downwards; the velocity is directed downwards when N is moving away from O but upwards when it moves towards O and is zero at A and B. When N is above O, the displacement is upwards and the velocity upwards or downwards according to whether N is moving away from or towards O.

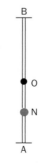

Fig. 9.4

The variation of acceleration can be seen by considering a body oscillating on a spiral spring. The magnitude of the elastic restoring force increases with displacement but always acts towards the equilibrium position (i.e. O); the resulting acceleration must therefore behave likewise, increasing with displacement but being directed to O whatever the displacement. So if N is below O, the displacement is downwards and the acceleration upwards, but if the displacement is upwards the acceleration is downwards. If we adopt the sign convention that quantities acting downwards are positive and those acting upwards are negative then acceleration and displacement always have opposite signs in an oscillation. Figure 9.5 summarizes these facts and should be studied carefully.

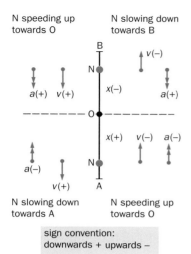

Fig. 9.5

The simplest relationship between the magnitudes of the acceleration and the displacement would be one in which the acceleration a of the body is directly proportional to its displacement x. Such an oscillation is said to be a **simple harmonic motion (s.h.m.)** and is defined as follows.

If the acceleration of a body is directly proportional to its distance from a fixed point and is always directed towards that point, the motion is simple harmonic.

The equation relating acceleration and displacement can be written

$$a \propto -x$$

or $\quad a = -(\text{positive constant}) . x$

The negative sign indicates that although the acceleration is larger at larger displacements it is always in the opposite direction to the displacement, i.e. towards O. What kind of motion would be represented by the above equation if a positive sign replaced the negative sign? (It would not be an oscillation.)

In practice many mechanical oscillations are nearly simple harmonic, especially at small amplitudes, or are combinations of such oscillations. In fact any system which obeys Hooke's law will exhibit this type of motion when vibrating. The equation for s.h.m. turns up in many problems in sound, optics, electrical circuits and even in atomic physics. In calculus notation it is written

$$\frac{\mathrm{d}^2x}{\mathrm{d}t^2} = -\text{constant} . x$$

where $a = \mathrm{d}v/\mathrm{d}t = \mathrm{d}^2x/\mathrm{d}t^2$. Using calculus this second-order differential equation can be solved to give expressions for displacement and velocity. However, we shall use a simple geometrical method which links circular motion and simple harmonic motion.

EQUATIONS OF S.H.M.

Suppose a point P moves round a circle of radius r and centre O with uniform angular velocity ω; its speed v round the circumference will be constant and equal to ωr, Fig. 9.6a. As P revolves, N, the foot of the perpendicular from P on the diameter AOB, moves from A to O to B and returns through O to A as P completes each revolution. Let P and N be in the positions shown at time t after leaving A, with radius OP making angle θ with OA and distance ON being x. We will now show that N describes s.h.m. about O.

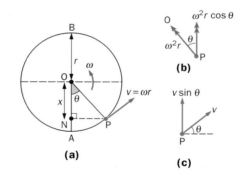

Fig. 9.6

(a) Acceleration

The motion of N is due to that of P, therefore the acceleration of N is the *component* of the acceleration of P parallel to AB. The acceleration of P is $\omega^2 r$ (or v^2/r) along PO and so the component of this parallel to AB is $\omega^2 r \cos \theta$, Fig. 9.6b. So the acceleration a of N is

$$a = -\omega^2 r \cos \theta$$

The negative sign, as explained before, indicates mathematically that a is always directed towards O. Now $x = r \cos \theta$, therefore

$$a = -\omega^2 x$$

Since ω^2 is a positive constant, this equation states that the acceleration of N towards O is directly proportional to its distance from O. N therefore describes s.h.m. about O as P moves round the circle — called the **auxiliary circle** — with constant speed.

The table below gives values of a for different values of x and we see that a is zero at O and a maximum at the limits A and B of the oscillation where the direction of motion changes.

x	0	$+r$	$-r$
a	0	$-\omega^2 r$	$+\omega^2 r$

Using the arrangement of Fig. 9.7 the shadow of a ball moving steadily in a circle can be viewed on a screen. The shadow moves with s.h.m. and represents the *projection* of the ball on the screen.

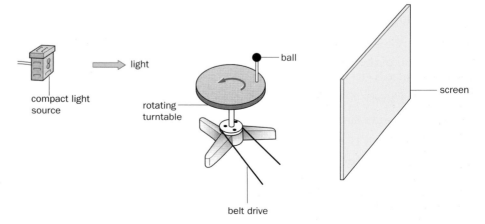

Fig. 9.7 Viewing s.h.m.

(b) Period

The period T of N is the time for N to make one complete oscillation from A to B and back again. In the same time P will travel once round the auxiliary circle, so

$$T = \frac{\text{circumference of auxiliary circle}}{\text{speed of P}}$$

$$= \frac{2\pi r}{v}$$

Therefore

$$T = \frac{2\pi}{\omega}$$

since $v = \omega r$. For a particular s.h.m. ω is constant and so T is constant and independent of the amplitude r of the oscillation. If the amplitude increases, the body travels faster and so T remains unchanged. A motion which has a constant period whatever the amplitude is said to be **isochronous** and this property is an important characteristic of s.h.m.

(c) Velocity

The velocity of N is the *component* of P's velocity parallel to AB, i.e.

$$-v \sin \theta \quad \text{(see Fig. 9.6c)}$$

$$\text{or} \quad -\omega r \sin \theta \quad \text{(since } v = \omega r\text{)}$$

Since $\sin \theta$ is positive when $0° < \theta < 180°$, i.e. N moving upwards, and negative when $180° < \theta < 360°$, i.e. N moving downwards, the negative sign ensures that the velocity is negative when acting upwards and positive when acting downwards (see Fig. 9.5). The variation of the velocity of N with time t (assuming P, and so N, start from A at zero time) is given by

$$\text{velocity} = -\omega r \sin \omega t$$

$$\text{(since } \theta = \omega t\text{)}$$

The variation of the velocity of N with displacement x is given by

$$\text{velocity} = \pm \omega r \sqrt{(1 - \cos^2 \theta)}$$

$$\text{(since } \sin^2 \theta + \cos^2 \theta = 1\text{)}$$

$$= \pm \omega r \sqrt{\{1 - (x/r)^2\}}$$

$$\text{(since } x = n \cos \theta, \text{ see Fig. 9.6a)}$$

$$= \pm \omega \sqrt{(r^2 - x^2)}$$

The velocity of N is $\pm \omega r$ (a maximum) when $x = 0$, and zero when $x = \pm r$.

(d) Displacement

This is given by

$$x = r \cos \theta = r \cos \omega t$$

The graph of the variation of the displacement of N with time (i.e. its 'time-trace') is shown in Fig. 9.8a and, like those for velocity and acceleration in Figs 9.8b and c, it is sinusoidal. Note that when the velocity is zero the acceleration is a maximum and vice versa. We say there is a **phase difference** of a quarter of a period (i.e. $T/4$) between the velocity and the acceleration. What is the phase difference between the displacement and the acceleration?

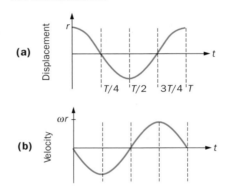

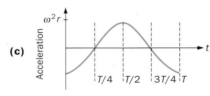

(a)

(b)

(c)

Fig. 9.8 Simple harmonic motion

Graphs such as those shown in Fig. 9.8 can be obtained experimentally for the arrangement shown in Fig. 9.9 by using a motion sensor, datalogger and computer.

(e) Summary

$$\text{acceleration} = -\omega^2 x$$

$$= -\omega^2 r \cos \omega t$$

$$\text{velocity} = \pm \omega \sqrt{(r^2 - x^2)}$$

$$= -\omega r \sin \omega t$$

$$\text{displacement} = r \cos \omega t$$

$$\text{period} = 2\pi/\omega$$

$$\text{frequency} = \omega/2\pi$$

These equations are true for *any* s.h.m.

EXPRESSION FOR ω

Consider the equation for simple harmonic motion $a = -\omega^2 x$. We can write (ignoring signs)

$$\omega^2 = \frac{a}{x} = \frac{ma}{mx} = \frac{ma/x}{m}$$

where m is the mass of the system. The force causing the acceleration a at displacement x is ma, therefore ma/x is the force per unit displacement. So

$$\omega = \sqrt{\left(\frac{\text{force per unit displacement}}{\text{mass of oscillating system}}\right)}$$

The period T of the s.h.m. is given by

$$T = \frac{2\pi}{\omega}$$

$$= 2\pi \sqrt{\left(\frac{\text{mass of oscillating system}}{\text{force per unit displacement}}\right)}$$

This expression shows that T increases if (*i*) the *mass* of the oscillating system *increases* and/or (*ii*) the *force per unit displacement decreases*, i.e. if the elasticity factor decreases.

MASS ON A SPRING

(a) Period of oscillations

The extension of a spiral spring which obeys Hooke's law is directly proportional to the extending tension. A mass m attached to the end of a spring exerts a downward tension mg on it and if it stretches it by an amount l as in Fig. 9.9a, then if k is the tension required to produce unit extension (called the **spring constant** and measured in N m^{-1}) the stretching tension is kl.

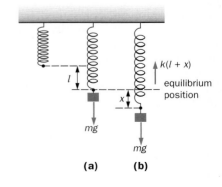

(a) **(b)**

Fig. 9.9

$$\therefore \quad mg = kl$$

Suppose the mass is now pulled down a further distance x below its equilibrium position, the stretching tension acting downwards is $k(l + x)$ which is also the tension in the spring acting upwards, Fig. 9.9*b*. Hence the resultant restoring force *upwards* on the mass

$$= k(l + x) - mg$$
$$= kl + kx - kl \quad \text{(since } mg = kl)$$
$$= kx$$

When the mass is released it oscillates up and down. If it has an acceleration a at extension x then by Newton's second law

$$-kx = ma$$

The negative sign indicates that at the instant shown a is upwards (negative by our sign convention) while the displacement x is downwards (i.e. positive).

$$\therefore \quad a = -\frac{k}{m}x = -\omega^2 x$$

where $\omega^2 = k/m =$ a positive constant since k and m are fixed. The motion is therefore simple harmonic about the equilibrium position so long as Hooke's law is obeyed. The period T is given by $2\pi/\omega$, therefore

$$T = 2\pi\sqrt{\frac{m}{k}}$$

It follows that $T^2 = 4\pi^2 m/k$. If the mass m is varied and the corresponding periods T found, a graph of T^2 against m is a straight line but it does not pass through the origin as we might expect from the above equation. This is because of the mass of the spring itself being neglected in the above derivation. Its effective mass and a value of g can be found experimentally.

(b) Measurement of g and effective mass of spring

Let m_s be the effective mass of the spring, then

$$T = 2\pi\sqrt{\left(\frac{m + m_s}{k}\right)}$$

But $\qquad\qquad mg = kl$

Substituting for m in the first equation and squaring, we get

$$T^2 = \frac{4\pi^2}{k}\left(\frac{kl}{g} + m_s\right)$$
$$\therefore \quad l = \frac{g}{4\pi^2} \cdot T^2 - \frac{gm_s}{k}$$

By measuring (*i*) the static extension l and (*ii*) the corresponding period T, using several different masses in turn, a graph of l against T^2 can be drawn. It is a straight line of slope $g/4\pi^2$ and intercept gm_s/k on the l axis, Fig. 9.10. This enables g and m_s to be found. Theory suggests that the effective mass of a spring is about one-third of its actual mass.

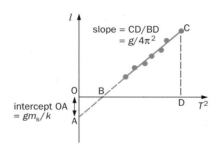

Fig. 9.10

SIMPLE PENDULUM

(a) Period of oscillations

The simple pendulum consists of a small bob (in theory a 'particle') of mass m suspended by a light inextensible thread of length l from a fixed point B, Fig. 9.11. If the bob is drawn aside slightly and released, it oscillates to and fro in a vertical plane along the arc of a circle. We shall show that it describes s.h.m. about its equilibrium position O.

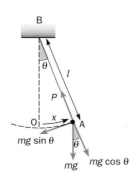

Fig. 9.11 The simple pendulum

Suppose at some instant the bob is at A where arc OA = x and $\angle$ OBA = θ. The forces on the bob are P and the weight mg of the bob acting vertically downwards. Resolving mg radially and tangentially at A we see that the tangential component $mg \sin \theta$ is the unbalanced restoring force acting towards O. If a is the acceleration of the bob along the arc at A due to $mg \sin \theta$ then the equation of motion of the bob is

$$-mg \sin \theta = ma$$

The negative sign indicates that the force is towards O while the displacement x is measured along the arc from O in the opposite direction.

When θ is small, $\sin \theta = \theta$ in radians and $x = l\theta$ (see p. 132). Then

$$-mg\theta = -mg\frac{x}{l} = ma$$
$$\therefore \quad a = -\frac{g}{l}x = -\omega^2 x \quad \text{(where } \omega^2 = g/l)$$

The motion of the bob is thus simple harmonic *if the oscillations are of small amplitude*, i.e. θ does not exceed 10°. The period T is given by $2\pi/\omega$, therefore

$$T = 2\pi\sqrt{\frac{l}{g}}$$

T is therefore independent of the amplitude of the oscillations and at a given place on the earth's surface where g is constant, it depends only on the length l of the pendulum.

(b) Measurement of g

A fairly accurate determination of g can be made by measuring T for different values of l and plotting a graph of l against T^2. A straight line AB is then drawn so that the points are evenly distributed about it, Fig. 9.12. It should pass through the origin and its slope BC/CA gives an average value of l/T^2 from which g can be calculated since

$$T = 2\pi\sqrt{\frac{l}{g}}$$
$$\therefore \quad T^2 = 4\pi^2\frac{l}{g}$$
$$\therefore \quad g = 4\pi^2\frac{l}{T^2} = 4\pi^2\frac{\text{BC}}{\text{CA}}$$

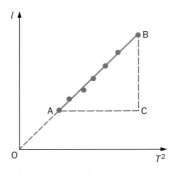

Fig. 9.12

The experiment requires (*i*) 100 oscillations to be timed, (*ii*) an angle of swing less than 10°, (*iii*) the length *l* to be measured to the centre of the bob, (*iv*) the oscillations to be counted as the bob passes the equilibrium position O. Why?

SOME CALCULATIONS

Example 1. A particle moving with s.h.m. has velocities of 4 cm s^{-1} and 3 cm s^{-1} at distances of 3 cm and 4 cm respectively from its equilibrium position. Find (*a*) the amplitude of the oscillation, (*b*) the period, (*c*) the velocity of the particle as it passes through the equilibrium position.

(*a*) Using the previous notation and taking the case shown in Fig. 9.13, the equation for the velocity is

$$\text{velocity} = -\omega\sqrt{(r^2 - x^2)} \quad \text{(p. 152)}$$

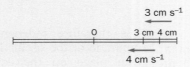

Fig. 9.13

We are taking velocities and displacements to the left as negative and those to the right positive.

When $x = +3 \text{ cm}$, velocity $= -4 \text{ cm s}^{-1}$, therefore (units implicit for clarity)

$$-4 = -\omega\sqrt{(r^2 - 9)}$$

When $x = +4 \text{ cm}$, velocity $= -3 \text{ cm s}^{-1}$, therefore

$$-3 = -\omega\sqrt{(r^2 - 16)}$$

Squaring and dividing these equations we get

$$\frac{16}{9} = \frac{r^2 - 9}{r^2 - 16}$$

Hence $r = \pm 5 \text{ cm}$.

(*b*) Substituting for *r* in one of the velocity equations we find

$$\omega = 1 \text{ s}^{-1}$$

$$\therefore \; T = \frac{2\pi}{\omega} = 2\pi \text{ s}$$

(*c*) At the equilibrium position

$$x = 0$$

$$\therefore \quad \text{velocity} = \pm\omega\sqrt{(r^2 - x^2)}$$

$$= \pm\omega r$$

$$= \pm 5 \text{ cm s}^{-1}$$

Example 2. A light spiral spring is loaded with a mass of 50 g and it extends by 10 cm. Calculate the period of small vertical oscillations. ($g = 10 \text{ m s}^{-2}$)

The period *T* of the oscillations is given by

$$T = 2\pi\sqrt{\frac{m}{k}}$$

where $m = 50 \times 10^{-3} \text{ kg}$

and $k = $ force per unit displacement

$$= \frac{50 \times 10^{-3} \times 10 \text{ N}}{10 \times 10^{-2} \text{ m}} = 5.0 \text{ N m}^{-1}$$

$$\therefore \; T = 2\pi\sqrt{\left(\frac{50 \times 10^{-3}}{5}\right)} \text{ s}$$

$$= 2\pi\sqrt{10^{-2}} \text{ s}$$

$$= 2\pi \times 10^{-1} \text{ s}$$

$$= 0.63 \text{ s}$$

Example 3. A simple pendulum has a period of 2.0 s and an amplitude of swing 5.0 cm. Calculate the maximum magnitudes of (*a*) the velocity of the bob, (*b*) the acceleration of the bob.

(*a*) $T = 2\pi/\omega$, therefore $\omega = 2\pi/T = 2\pi/2 \text{ s}^{-1} = \pi \text{ s}^{-1}$. Velocity is a maximum at the equilibrium position where $x = 0$.

$$v = \pm\omega\sqrt{(r^2 - x^2)}$$

$$= \pm\pi\sqrt{25} \text{ cm s}^{-1} \quad \text{(since } r = \pm 5 \text{ cm)}$$

$$= \pm 5\pi \text{ cm s}^{-1}$$

$$= \pm 16 \text{ cm s}^{-1}$$

(*b*) Acceleration is a maximum at the limits of the swing where $x = r = \pm 5.0 \text{ cm}$.

$$a = -\omega^2 r$$

$$= -\pi^2 \times 5 \text{ cm s}^{-2}$$

$$= -50 \text{ cm s}^{-2}$$

ENERGY OF S.H.M.

In an oscillation there is a constant interchange of energy between the kinetic and potential forms, and if the system does no work against resistive forces (is undamped) its total energy is constant.

(a) Kinetic energy

The velocity of a particle N of mass m at a distance x from its centre of oscillation O, Fig. 9.14, is

$$v = +\omega\sqrt{(r^2 - x^2)}$$

$\therefore$ k.e. at displacement x

$$= \tfrac{1}{2}m\omega^2(r^2 - x^2)$$

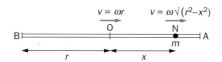

Fig. 9.14

(b) Potential energy

As N moves out from O towards A (or B) work is done against the force trying to restore it to O. So N loses k.e. and gains p.e. When $x = 0$, the restoring force is zero; at displacement x, the force is $m\omega^2x$ (since the acceleration has magnitude ω^2x). Therefore

average force on N while moving to displacement x

$$= \tfrac{1}{2}m\omega^2x$$

$\therefore$ work done = average force $\times$ displacement in direction of force

$$= \tfrac{1}{2}m\omega^2x \times x = \tfrac{1}{2}m\omega^2x^2$$

$\therefore$ p.e. at displacement $x = \tfrac{1}{2}m\omega^2x^2$

(c) Total energy

At displacement x we have

total energy = k.e. + p.e.

$$= \tfrac{1}{2}m\omega^2(r^2 - x^2) + \tfrac{1}{2}m\omega^2x^2$$

$$\boxed{\text{total energy} = \tfrac{1}{2}m\omega^2r^2}$$

This is constant, does not depend on x and is directly proportional to the product of (i) the mass, (ii) the square of the frequency and (iii) the square of

the amplitude. Figure 9.15 shows the variation of k.e., p.e. and total energy with displacement.

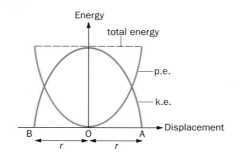

Fig. 9.15

In a simple pendulum all the energy is kinetic as the bob passes through the centre of oscillation and at the top of the swing it is all potential.

(d) Variation of k.e. and p.e. with time

These vary with time as shown by the graphs in Fig. 9.16, since

$$\text{k.e.} = \tfrac{1}{2}mv^2 = \tfrac{1}{2}m\omega^2r^2\sin^2\omega t$$
$$(v = -\omega r\sin\omega t)$$

$$\text{p.e.} = \tfrac{1}{2}m\omega^2x^2 = \tfrac{1}{2}m\omega^2r^2\cos^2\omega t$$
$$(x = r\cos\omega t)$$

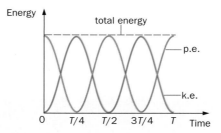

Fig. 9.16

DAMPED OSCILLATIONS

The amplitude of the oscillations of, for example, a pendulum gradually decreases to zero owing to the resistive force that arises from the air. The motion is said to be **damped** by air resistance; its energy becomes internal energy of the surrounding air.

The behaviour of a mechanical system depends on the extent of the damping. The damping of the mass on the spring in Fig. 9.17 is greater than when it is in air. Undamped oscillations are said to be **free**, Fig. 9.18a. If

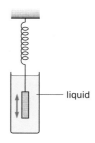

Fig. 9.17

a system is slightly damped, oscillations of decreasing amplitude occur, Fig. 9.18b. When heavily damped no oscillations occur and the system returns very slowly to its equilibrium position, Fig. 9.18c. When the time taken for the displacement to become zero is a minimum, the system is said to be **critically damped**, Fig. 9.18d. This minimum time is $T/4$, where T is the period of free oscillations.

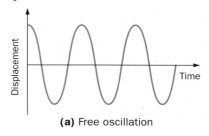

(a) Free oscillation

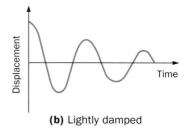

(b) Lightly damped

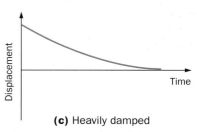

(c) Heavily damped

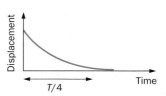

(d) Critically damped

Fig. 9.18 The effect of damping on an oscillating system

The motion of many devices is critically damped on purpose. Thus the shock absorbers on a car critically damp the suspension of the vehicle and so resist the setting up of vibrations which could make control difficult or cause damage. In the shock absorber of Fig. 9.19 the motion of the suspension up or down is opposed by viscous forces when the liquid passes through the transfer tube from one side of the piston to the other. The damping of a car can be tested by applying your weight to the suspension momentarily; the car should rapidly return to its original position without vibrating.

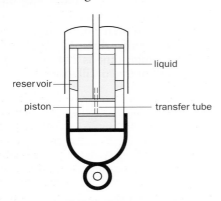

Fig. 9.19   Shock-absorbing car suspension

Instruments such as balances and electrical meters are critically damped (i.e. 'dead-beat') so that the pointer moves quickly to the correct position without oscillating. The damping is often produced by electromagnetic forces.

FORCED OSCILLATION AND RESONANCE

(a) Barton's pendulums

The assembly consists of a number of paper cone pendulums (made by folding paper circles of about 5 cm diameter) of lengths varying from $\frac{1}{4}$ m to $\frac{3}{4}$ m, each loaded with a plastic curtain ring. All are suspended from the same string as a 'driver' pendulum which has a heavy bob and a length of $\frac{1}{2}$ m, Fig. 9.20.

The driver pendulum is pulled well aside and released so that it oscillates in a plane perpendicular to that of the diagram. The motion settles down after a short time and all the pendulums oscillate with very nearly the

same frequency as the driver but with *different* amplitudes. This is a case of **forced oscillation**.

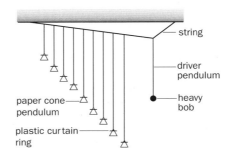

Fig. 9.20   Barton's pendulums

The pendulum whose length equals that of the driver has the greatest amplitude; its natural frequency of oscillation is the same as the frequency of the driving pendulum. This is an example of **resonance** and the driving oscillator then transfers its energy most easily to the other system, i.e. the paper cone pendulum of the same length.

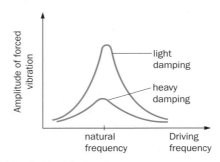

Fig. 9.21   Resonance

The amplitudes of oscillations also depend on the extent to which the system is damped. For example, removing the rings from the paper cone pendulums reduces their mass and so increases the damping. All amplitudes are then found to be reduced and that of the resonant pendulum is less pronounced. These results are summarized by the graphs in Fig. 9.21 which indicate that the sharpest resonance occurs in a lightly damped system. Figures 9.22a and b are time-exposure photographs taken with the camera looking along the line of swinging pendulums towards and at the same level as the bob of the driver. Is (a) more or less damped than (b)?

Careful observation shows that the resonant pendulum is always a quarter of an oscillation behind the driver pendulum, i.e. there is a phase difference of a quarter of a period. The shorter pendulums are nearly in phase with the driver, while those that are longer than the driver are almost half a period behind it. This is evident from the instantaneous photograph of Fig. 9.22c, taken when the driver is at maximum displacement to the left.

(b) Hacksaw blade oscillator

The arrangement, shown in Fig. 9.23, provides another way of finding out what happens when one oscillator (a loaded hacksaw blade) is driven by another (a heavy pendulum), as often occurs in practice. The positions of

(a)

(b)
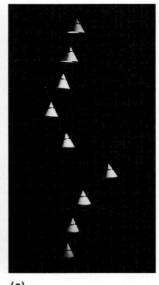
(c)

Fig. 9.22   Barton's pendulums in action: (a), (b) time-exposure photographs, (c) instantaneous photograph

the mass on the blade and the pendulum bob can both be adjusted to alter the natural frequencies. By using different rubber bands the degree of coupling may be varied, as can the damping, by turning the card. The motion of the driver is maintained by gentle, timely taps just below its support.

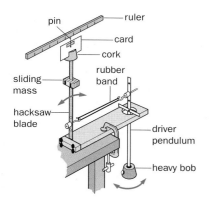

Fig. 9.23 Hacksaw blade oscillator

There is scope for investigating (*i*) the transient oscillations that occur as the motion starts and before the onset of steady conditions, (*ii*) resonance, (*iii*) phase relationships, (*iv*) damping and (*v*) coupling.

(c) Examples of resonance

These are common throughout science and are generally useful. In the production of musical sounds from air columns in wind instruments resonance occurs, in many cases, between the vibrations of air columns and of small vibrating reeds. Electrical resonance occurs when a radio circuit is tuned by making its natural frequency for electrical oscillations equal to that of the incoming radio signal.

Information about the strength of chemical bonds between ions can be obtained by a resonance effect. We regard electromagnetic radiation (light, infrared, etc.) as a kind of oscillating electrical disturbance which, when incident on a crystal, subjects the ions to an oscillating electrical force. With radiation of the correct frequency, the ions can be set into oscillation by resonance, Fig. 9.24. Energy is absorbed from the radiation and the absorbed frequency can be found using a suitable spectrometer. With sodium chloride, absorption of infrared radiation occurs.

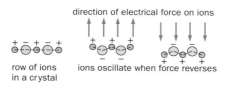

Fig. 9.24 Resonance in crystals

Resonance in mechanical systems is a source of trouble to engineers. An oscillation of large amplitude can build up and destroy a structure. Figure 9.25 shows a wind-tunnel model of an aircraft being tested for resonance. It is important that the natural frequencies of vibration of an aircraft do not equal any that may be produced by the forces experienced in flight. Otherwise resonance might occur and undue stress result.

(d) Quality factor Q

This is approximately equal to the number of natural (free) oscillations that occur before the oscillator loses all its energy. Lightly damped oscillations that resonate sharply and die away slowly have high Q values. Heavily damped ones have low values. Q for a car suspension is 1, for a simple pendulum and a guitar string 10^3 and for the quartz crystal of a watch 10^5. If middle C (256 Hz) is sounded on a guitar string it should last for $1000/256 \approx 4$ s.

A MATHEMATICAL MODEL

Real oscillators, such as a motor cycle on its suspension, a tall chimney swaying in the wind and atoms (or ions) vibrating in a crystal, only approximate to the ideal type of motion we call s.h.m.

Simple harmonic motion is the description in mathematical terms of a kind of motion which is not fully realised in practice. It is a mathematical model and is useful because it represents well enough many real oscillations. This is due to its simplicity; complications such as damping, variable mass and variable stiffness (elastic modulus) are omitted and the only conditions imposed on the system are that the restoring force should be directed towards the centre of the motion and be proportional to the displacement.

A more complex model might, for example, take damping into account and as a result be a better description of a particular oscillator, but it would probably not be so widely applicable. On the other hand, if a model is too simple it may be of little use for dealing with real systems. A model must have just the correct degree of complexity. The mathematical s.h.m. model has this and so is useful in practice.

Fig. 9.25 Model aircraft can be tested for resonance in a wind tunnel

QUESTIONS

1. Write a short account of **simple harmonic motion** explaining the terms **amplitude**, **time period** and **frequency**.

   A particle of mass m moves such that its displacement from the equilibrium position is given by $y = a \sin \omega t$ where a and ω are constants. Derive an expression for the kinetic energy of the particle at a time t and show that its value is a maximum as the particle passes through the equilibrium position.

   A steel strip clamped at one end vibrates with a frequency of 30 Hz, and an amplitude of 4.0 mm at the free end. Find

   a) the velocity of the free end as it passes through the equilibrium position, and

   b) the acceleration at the maximum displacement.

2. A simple pendulum has a period of 2.0 s and oscillates with an amplitude of 10 cm. What is the frequency of the oscillations?

   At what point of the swing is the speed of the pendulum bob a maximum? Calculate this maximum speed.

   At what points of the swing is the acceleration of the pendulum bob a maximum? Calculate this acceleration.

   (L, A PH4, Jan 1996)

3. Figure 9.26 shows a mass of 0.51 kg suspended at the lower end of a spring. The graph shows how the tension, F, in the spring varies with the extension, Δx, of the spring.

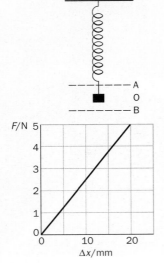

Fig. 9.26

   Use the graph to find a value for the spring constant k.

   The mass, originally at point O, is set into small vertical oscillations between the points A and B. Choose A, B or O to complete the following sentences.
   The speed of the mass is a maximum when the mass is at ...
   The velocity and acceleration are both in the same direction when the mass is moving from ... to ...

   Calculate the period of oscillation T of the mass.

   What energy transformations take place while the mass moves from B to O?

   (L, AS/A PH2, Jan 1996)

4. A trolley of mass 0.40 kg is attached to two fixed points by springs as shown in Fig. 9.27. When it is displaced horizontally 0.20 m from its equilibrium position and released, it moves with simple harmonic motion with a period of 4.0 s.

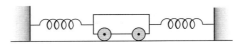

Fig. 9.27

   a) Ignoring any frictional effects, calculate

   i) the maximum speed of the trolley,
   ii) the maximum kinetic energy of the trolley,
   iii) the effective stiffness constant, k, of the spring system.

   b) In practice, frictional forces gradually reduce the kinetic energy of the trolley. If the trolley loses one-tenth of its kinetic energy in every complete oscillation, calculate the kinetic energy of the trolley at the equilibrium position when it has made four complete oscillations. Assume that the kinetic energy when it first passes the equilibrium position is the value which you have calculated in a) ii).

   (NEAB, AS/A PH02, June 1997)

5. **a)** Describe an experiment to demonstrate the effects of damping on the oscillatory motion of a vibrating system undergoing

   i) free, and
   ii) forced harmonic oscillations.
   Draw labelled diagrams to illustrate the results you would expect to obtain.

   b) What is the physical origin of the damping mechanisms in the case of
   i) the oscillations of a simple pendulum in air,

   ii) the vibrations of a bell sounding in air, and
   iii) oscillatory currents in an electrical circuit?

   c) A spring is supported at its upper end. When a mass of 1.0 kg is hung on the lower end the new equilibrium position is 5.0 cm lower. The mass is then raised 5.0 cm to its original position and released. Discuss as fully as you can the subsequent motion of the system.

6. A mass of 2.0 kg is hung from the lower end of a spiral spring and extends it by 0.40 m. When the mass is displaced a further short distance x and released, it oscillates with acceleration a towards the rest position. If $a = -kx$ and if the tension in the spring is always directly proportional to its extension, what is the value of the constant k? (Earth's gravitational field strength is 9.8 N kg^{-1}.)

7. **a)** If a mass hanging on a vertical spring which obeys Hooke's law is given a small vertical displacement it will oscillate vertically, above and below its equilibrium position. Why is the motion of the mass simple harmonic?

   b) Explain why the oscillations of a simple pendulum, consisting of a mass m hanging from a thread of length l, are almost perfectly simple harmonic, and why the deviations from perfect simple harmonic motion become greater if the amplitude of the swing is increased.

   c) A metre ruler is clamped to the top of a table so that most of its length overhangs and is free to vibrate with vertical simple harmonic motion. Calculate the maximum velocity of the tip of the ruler if the amplitude of the vibration is 5.0 cm and the frequency is 4.0 Hz.

   d) What is the maximum possible amplitude of vibration of the ruler in c) if a small object placed on the ruler at the vibrating end is to maintain contact with it throughout the vibration? If this amplitude were very slightly exceeded, at what stage in the vibration would contact between the object and the ruler be broken? Explain clearly why it would happen at this point.
   ($g = 10$ m s^{-2})

8. Figure 9.28 shows the displacement–time graph for the centre of a loudspeaker cone when it is displaced manually and then allowed to vibrate naturally.

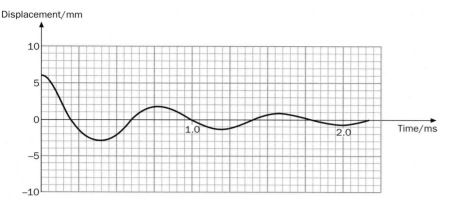

Fig. 9.28

a) i) State the feature of the graph (Fig. 9.28) which shows that the oscillations are simple harmonic.
ii) State the feature which shows that the oscillations are damped.

b) i) Calculate the natural frequency of oscillation of the cone.
ii) Calculate the maximum acceleration of the cone during the time interval shown in Fig. 9.28.
iii) Draw a graph to show how the acceleration of the body changes with time in this time interval.

c) The graph in Fig. 9.29 shows how the amplitude of the cone varies with frequency (the frequency response graph) when a signal generator provides alternating current of different frequencies but similar amplitude to the loudspeaker.

Sketch this graph, then on the same axes draw graphs to show the effect of using:

i) a coil of greater mass (label this graph C);

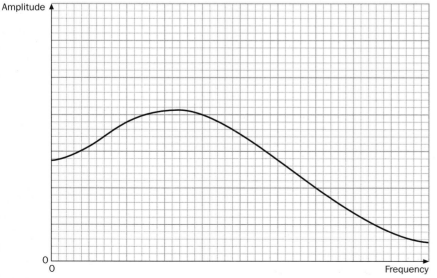

Fig. 9.29

ii) a system of the same stiffness and mass, but less damping (label this graph D).

(AEB, 0635/1, Summer 1997)

9. a) Explain the meanings of the following terms: free vibration, forced vibration, resonance.

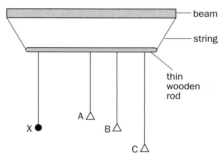

Fig. 9.30

b) In Fig. 9.30, A, B and C are three pendulums, each with a light paper cone for a bob. X is another pendulum with a heavy metal bob. Pendulum X is displaced perpendicular to the plane containing the bobs at rest and then released.

Compare the motions of pendulums A, B and C to that of X, with reference to period and amplitude.

(NEAB, AS/A PH02, June 1997)

10. One simple model of the hydrogen molecule assumes that it is composed of two oscillating hydrogen atoms joined by two springs as shown in Fig. 9.31.

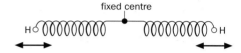

Fig. 9.31

If the spring constant of each spring is 1.13×10^3 N m^{-1}, and the mass of a hydrogen atom is 1.67×10^{-27} kg, show that the frequency of oscillation of a hydrogen atom is 1.31×10^{14} Hz.

Using this spring model, discuss why light of wavelength 2.29×10^{-6} m would be strongly absorbed by the hydrogen molecule.

(L, A PH4, June 1997)

11. Investigate the effect of changing the values of r and ω in the expression $r\cos \omega t$ using a spreadsheet such as that shown in Fig. 9.32a.

The value chosen for r is typed into cell C2 and that for ω into cell D2; **=C\$2*COS(D\$2*A2)** is entered into cell B2 to enable $r\cos \omega t$ to be calculated by the computer for the times entered in column A.

Choose more times and a graphing routine to obtain a plot like that shown in Fig. 9.32b.

	A	**B**	**C**	**D**
1	t	rcosωt	r	ω
2	0	= C\$2*COS(D\$2*A2)	1	3
3	0.2			
4	0.4			

(a)

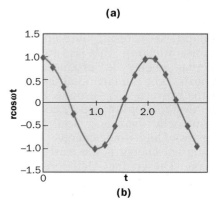

(b)

Fig. 9.32

10
Energy and its uses

ABOUT ENERGY

(a) Conservation

One of the basic laws of physics is the principle of conservation of energy, that is, energy cannot be destroyed (p. 126). However, energy is continually being transferred in many different processes all around us.

Some forms of energy are more useful than others; they are more suitable for doing work and being transformed into other forms of energy. Electrical and chemical energy are in this category and are called **high-grade** energy; internal energy (i.e. kinetic energy of the molecules due to their random motion *and* potential energy due to the forces between molecules) is **low-grade** energy that is not easily transformed into anything else.

(b) Degradation

In the end, all energy transfers result in the surroundings being heated (for example as a result of doing work against friction), that is, low-grade internal energy is produced. This appears to be the fate of all the energy in the universe. The process is called the 'degradation' of energy and is the reason why there is a need for new sources of high-grade energy.

Internal energy is the least ordered energy. The gradual 'deterioration' of all energy into this state of maximum disorder is described by saying that the **entropy** of the universe is increas-ing. Decreasing the entropy is impossible; it would be like a hot drink that had gone cold warming itself up again from the colder surroundings.

(c) Units

In physics all forms of energy are measured in joules (J). In the energy industry, the kilowatt-hour (kWh: see p. 53), the therm and the British thermal unit (Btu) are also used. Table 10.1 shows how they are related.

The megatonne of coal equivalent (Mtce) is also in common use. It is defined as the energy obtained by burning completely 1 million tonnes of coal (1 tonne = 1000 kg), but since the energy content of coal varies from about 20×10^9 J to 30×10^9 J per tonne it is not an exact unit like the others.

Three matters of great practical importance to the future of the energy industry are:

■ energy sources;
■ energy transfer; and
■ energy consumption.

Table 10.1 Units of energy

	J	kWh	therm	Btu
1 J =	1	0.28×10^{-6}	9.5×10^{-9}	950×10^{-6}
1 kWh =	3.6×10^6	1	34×10^{-3}	3.4×10^3
1 therm =	110×10^6	29	1	100×10^3
1 Btu =	1.1×10^3	290×10^{-6}	10×10^{-6}	1

ENERGY SOURCES

Sources of energy that are used in the form in which they occur naturally are called **primary** energy sources. They fall into two groups — finite sources and renewable sources.

(a) Finite sources

These include the fossil fuels **coal**, **oil** and **natural gas** which are formed from the fossilized remains of plants (and the tiny organisms that fed on them) buried 600 million years ago and are not renewable. They are complex hydrocarbon compounds containing stores of chemical energy (derived originally by the action of solar energy on plants), which is released when they are burnt.

Predictions about how long coal, oil and gas will last are full of uncertainties. First, the exact size of the reserves is unknown and 'recoverable' reserves are less than 'proved' reserves; but if the price of fuel rises, deposits previously considered to be uneconomic to extract can become economic. Second, future demand is very difficult to forecast and depends, among other things, on the growth of the world's population (over 6000 million at present and expected to reach about 7000 million by the year 2013), the standard of living and the way industry develops in different countries. However, it is clear that even at present rates of consumption, the world's recoverable deposits of oil and gas will be running low early in the 21st century; coal should last for about 200 years or so (see question 3, p. 173).

Nuclear fuels such as uranium are also finite resources. At present we have enough of these to last at least 100 years.

(b) Renewable sources

These cannot be exhausted. **Solar energy**, **biofuels**, **hydroelectric power**, **wind** and **wave power** are solar in origin; **tidal** and **geothermal energy** are not (although the sun does have some influence on the tides). Each will be discussed later.

(c) Energy density

In an energy source such as coal, which has a high energy density, the energy is concentrated and the converter needed to release it (i.e. a furnace) is relatively small. By contrast, the energy density of solar energy is low (despite the huge amounts of energy falling upon the earth) and as a result a solar converter must be larger for the same power output.

(d) Availability

An energy source should be available when it is needed and capable of meeting increased demand. In the case of, for example, solar and wind sources, this may not be possible.

ENERGY TRANSFER

Engines, machines and many other appliances invented by the human race are used to do work in industry, agriculture, the home and for transport. Such devices transfer energy, not necessarily from a primary energy source, to do the job required.

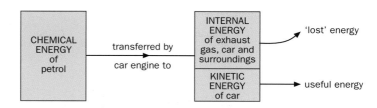

Fig. 10.1   Energy transfer in a car engine

A car engine is designed to transfer the chemical energy of petrol into mechanical energy (k.e. of the car) but it also produces unwanted internal (low-grade) energy which heats the exhaust gases, the car and the surrounding air. The energy transfer process is represented by the flow diagram in Fig. 10.1.

An electric motor transfers electrical energy to mechanical energy of its moving parts but again internal energy is also produced.

The **efficiency** of an energy transfer process is defined by

$$\text{efficiency} = \frac{\text{useful energy output}}{\text{total energy input}}$$

and is expressed as a fraction or as a percentage. The useful energy is energy produced for the job in hand, for example, mechanical energy (not internal energy) in the case of a car engine.

Table 10.2 lists the percentage efficiencies of some devices and the energy transfers involved.

One of the guiding principles of energy-efficient design is that, whenever possible, the grade of the energy supplied for a job should match the end-use need. Only in that way can the energy losses be reduced to a minimum and efficiency increased.

ENERGY CONSUMPTION

(a) Patterns of consumption

The percentage contribution of the various forms of primary energy to the world's consumption at the present time are given in Table 10.3. (Biofuels, although important in many countries, are not included because their use is not well documented.) In 1998 the total consumption was about 3.7×10^{20} J.

Table 10.3   Primary energy use

Oil	Coal	Gas	Nuclear	Hydro-electric
41%	26%	24%	7%	2%

The pattern for the UK is similar (see Fig. 10.3, p. 162). The great dependence on fossil fuels is evident, and in view of the predictions about how soon these are likely to be used up, it is clear that the world has an energy problem. The need to find alternative and preferably renewable primary sources is urgent.

Consumption varies from one country to another; North America and Europe are responsible for about two-thirds of the world's energy consumption each year. Table 10.4 (p. 162) shows approximate values for the annual consumption per head of population for various areas. For the world as a whole the average figure is 61×10^9 J per head per year.

Table 10.2   Efficiencies of some devices

Device	% Efficiency	Energy transfer
Large electric motor	90	Electrical to mechanical
Large electric generator	90	Mechanical to electrical
Domestic gas boiler	75	Chemical to internal energy
Steam turbine	45	Internal energy to mechanical
Car engine	25	Chemical to mechanical

Table 10.4 Energy consumption per head per year/J × 10⁹

N. America	UK	Japan	S. America	China	Africa
270	135	120	40	25	15

The UK figure is the equivalent of about 100 kWh per head per day but includes the primary energy used by industry and transport and the transfer losses that occur before the energy reaches the consumer.

(b) Growth of demand

The demand for primary energy has escalated dramatically in the last half-century. The consumption of oil doubled every eight or nine years during the 1950s and 60s, while as much coal was burnt in the 30 years following World War II as had been used since it was first mined hundreds of years ago. The growth of demand is shown by the graph in Fig. 10.2. In recent years the rate of growth has slowed down owing to world economic recession, but demand is still enormous.

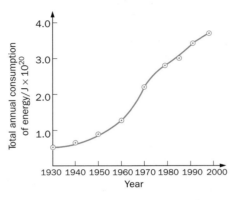

Fig. 10.2 Growth in energy consumption

ENERGY USE IN THE UK

The flow diagram in Fig. 10.3 shows, in round numbers, how the various sources of energy are currently used in the UK. It brings out the following points:

(*i*) About one-third of the **primary energy** is lost as internal energy (i.e. heating of the surroundings) during transfer to **delivered (secondary) energy**. (*Boxes 1 and 2*)

(*ii*) Approaching one-half of the oil used as delivered energy is consumed by cars, lorries, aircraft, etc. (*Box 3*)

(*iii*) Industrial and domestic users each take about one-third of the delivered energy, with transport accounting for one-quarter and others (e.g. agriculture and commerce) consuming the rest. (*Box 4*)

(*iv*) Three-quarters of all the energy supplied to homes is used for space heating; the rest is required for water heating, electrical appliances and cooking. (*Box 5*)

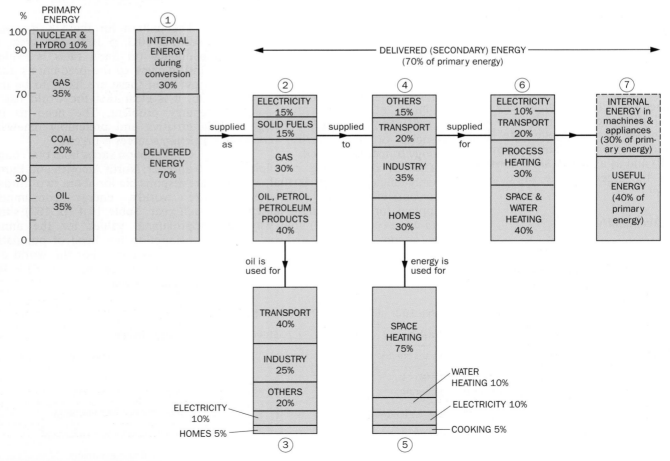

Fig. 10.3 Use of energy in the UK

(*v*) Over two-thirds of the delivered energy is used for space, water and industrial process heating and is mostly low-grade energy. One-fifth or so is required as liquid fuel for transport and the remainder for lighting, electrical devices and processes, and in telecommunications. (*Box 6*)

(*vi*) About one-third of the total primary energy consumption is again lost when delivered energy is transformed into **useful energy** (end-use energy), giving an overall efficiency for the complete transfer process of roughly 40%. (*Box 7*)

THERMAL POWER STATIONS

The electrical energy produced at a power station is a very convenient form of high-grade secondary (delivered) energy. It has the added advantage of being easily distributed by cables and for many uses there is no alternative to it.

A thermal power station uses coal, oil, natural gas or uranium as the primary energy source. The layout of a fossil-fuel/nuclear station and the energy flow diagram are given in Fig. 10.4*a,b*.

(a) Action

In a fossil-fuel station, the coal, oil or gas is burnt in a **furnace** and produces hot gases; in a nuclear station, carbon dioxide gas is heated by the fission of uranium in a **reactor** (see chapter 25). In the first case, the hot gases convert water into high-pressure steam, with a temperature of 560 °C, in a **boiler**; in the second case energy is transferred in a **heat exchanger**.

The steam passes through pipes to the high-pressure cylinder of a **steam turbine** consisting of a ring of stationary blades fixed to the cylinder wall.

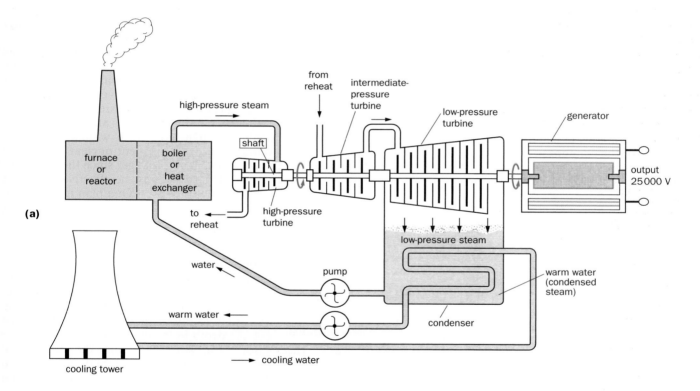

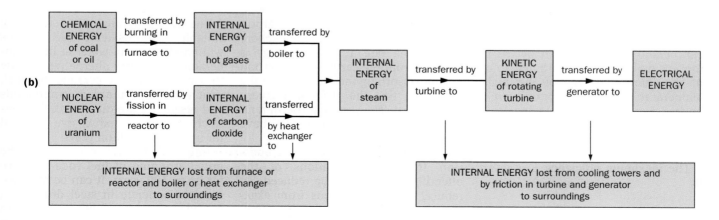

Fig. 10.4   A thermal power station and its energy transfer diagram

These direct the steam on to a second ring of blades, secured to the turbine shaft, which turns under the force of the steam. Some of the internal energy of the steam is transferred to mechanical (kinetic) energy by the turbine. When it leaves the high-pressure cylinder, the steam returns to the boiler or heat exchanger for reheating before it enters the intermediate and low-pressure cylinders.

In its passage through the turbine, the steam's temperature and pressure fall, and it expands. Because of this expansion, the blades are much larger towards the low-pressure end of the turbine.

In the **condenser** below the turbine the steam is condensed back to water for re-use in the boiler or heat exchanger. A partial vacuum is created in the condenser (owing to the greater density of water) and this increases the efficiency of the process allowing more energy to be extracted from the steam. Large amounts of cooling water are needed for the condenser; after use this water is cooled in large **cooling towers** and re-used.

The turbine drives a **generator** which transfers mechanical (kinetic) energy to electrical energy in the form of alternating current (of frequency 50 Hz and voltage 25 000 V in the UK).

(b) Efficiency

The efficiency of a thermal power station in transforming primary energy into electrical energy is about 30%; i.e. 70% of the energy input is 'lost', about half of the loss being to the surroundings via the cooling towers. The transfer of internal energy to mechanical energy is a very inefficient process and has a theoretical upper limit. It can be shown that when it occurs in for example a steam turbine, some of the internal energy of the high-temperature fuel (i.e. steam) *must* be transferred to the low-temperature surroundings, the amount depending on the two temperatures (see chapter 21).

The maximum efficiency E is given by

$$E = \frac{T_1 - T_2}{T_1}$$

where T_1 and T_2 are the high and low temperatures respectively, in K. In a modern power station steam enters the turbines at about 830 K (560 °C) and if it is rejected at say 300 K (27 °C) then $E = (830 - 300)/830 = 530/830 \approx 64\%$. In practice, other factors reduce this further.

In recently constructed 'combined cycle' gas turbine (CCGT) power stations, natural gas is burnt in a gas turbine linked directly to an electricity generator. The hot exhaust gases from the turbine are not released into the atmosphere but used to produce steam in a boiler. The steam is used to generate more electricity from a steam turbine driving another generator. Over 50% efficiency is claimed, with less pollution.

(c) Combined heat and power (CHP)

In this system the internal energy that would otherwise be lost from the cooling towers is used to provide space heating and hot water to homes and factories near the power station.

FUELS AND POLLUTION

The use of fuels, fossil or nuclear, causes pollution, i.e. has an effect on the environment that is generally considered to be adverse.

(a) Fossil-fuel power stations

These discharge waste gases (for example sulphur dioxide from burning coal and oil, and carbon dioxide from all fossil fuels) into the atmosphere. Sulphur oxides cause 'acid rain' when they combine with water in the atmosphere. Any increase in the carbon dioxide content of the atmosphere reduces the rate at which energy is radiated from the earth, due to carbon dioxide absorbing the radiation (in the same way that the glass of a greenhouse does). The resulting rise in temperature of the earth could have dramatic climatic effects.

To tackle these problems, the emission from power stations of the offending gases has to be reduced. To reduce sulphur emissions from existing coal-fired power stations, low-sulphur coal can be burnt or flue-gas desulphurization processes employed.

Efficient, cleaner CCGTs (combined cycle gas turbines) are preferred for new stations. To reduce carbon dioxide emissions, the only really effective ways are by consuming less electric power and by using non-polluting, renewable energy sources instead of fossil fuels.

Polluting emissions from cars also need to be reduced. Fitting 'catalytic converters' to car exhausts gives some reduction.

(b) Nuclear power stations

These discharge radioactive materials in the form of gases into the atmosphere and liquids into the sea, but discharges are limited to what are believed to be safe levels. Problems also arise from the need to store safely — sometimes for many years — other forms of radioactive waste product. These fall into three categories — high-, intermediate- and low-level.

High-level waste, which results from reprocessing spent nuclear fuel, poses the greatest danger since the treatment it receives must be effective for about a million years. One plan is to incorporate it in a very stable solid, which would be buried deep in the earth. This immobilizing solid must do two things.

First, it must withstand contact with water without disintegrating and allowing the radioactive material to escape, even at high temperatures (resulting from the increase of temperature with depth in the earth, about 20–30 °C per km, and also from the radioactive heating effect). Second, its structure must survive bombardment from the alpha particles emitted by the radioactive material within it, for the necessary length of time.

The solid generally considered to be suitable is a borosilicate glass, but this will break down in water at high temperatures. A better alternative may be to use the synthetic ceramic 'Synroc', which contains titanium compounds and resembles a natural rock that has safely contained radioactive material for up to 2000 million years.

Intermediate-level waste from the initial stripping of fuel rods is active for up to 300 years. It can be incorporated in concrete in steel drums and buried in the ground, but there remain some reservations about the safety and desirability of this method of disposal.

Low-level waste, such as paper towels, gloves and other protective clothing, is large in volume but its radioactivity is weak, and it can be compacted and stored safely in steel containers in concrete vaults.

The safe transportation, reprocessing and storage of radioactive waste are significant problems which the nuclear industry has to deal with and which are a continual source of public concern and debate.

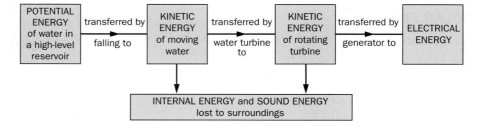

Fig. 10.5 Energy transfer in a hydroelectric power station

HYDROELECTRIC POWER STATIONS

(a) Action

In hydroelectric power stations the turbines are driven by fast-flowing water, obtained when water stored behind a dam in a high-level reservoir falls to a lower level. The energy flow diagram is given in Fig. 10.5.

The power available depends on the head of water (i.e. the vertical height through which it falls) and the rate of flow. Doubling either roughly doubles the power, doubling both gives approximately four times the power.

(b) Efficiency

The efficiency of a large installation can be as high as 85–90% since many of the causes of energy losses in thermal power stations are absent. In some cases the electrical energy is produced at less than half the cost per unit compared with a coal-fired plant.

(c) Pumped storage

Electrical energy cannot be stored; it must be used as it is generated, but the demand for it varies with the time of day and season, as the typical graphs in Fig. 10.6 show.

In a hydroelectric power station with pumped storage, like that at Ffestiniog, Fig. 10.7, electricity generated at 'off-peak' periods is used to pump water back up from the low-level reservoir to the high-level reservoir. It is easier to do this than to reduce the output of the generators. At 'peak' demand times, the potential energy of the water in the high-level reservoir is converted back into electrical energy. Three-quarters of a unit is produced for every unit used to pump the water.

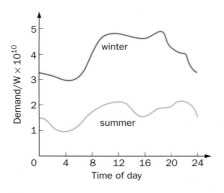

Fig. 10.6 Variation in power demand

(d) Further points

In the UK, hydroelectric power stations generate about 2% of the electricity supply. Most are located in Scotland and Wales where the average rainfall is higher than in other areas. In some countries there are very large hydroelectric undertakings: examples include the Snowy Mountains scheme in Australia, the Kariba Dam project in Zimbabwe, and in Norway where mountain reservoirs abound.

Hydroelectric power is a renewable source which is reliable and easily maintained. There are risks connected with the construction of dams but these and the impact on the environment are minimal with good planning and management.

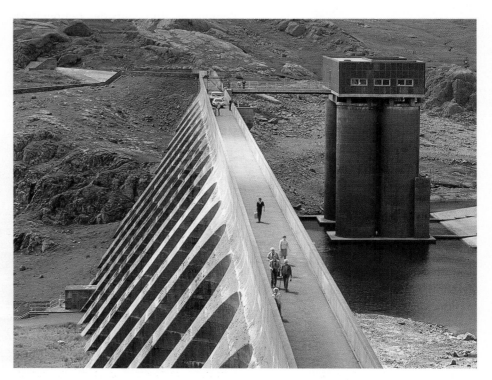

Fig. 10.7 Ffestiniog hydroelectric power station, North Wales

SOLAR ENERGY

(a) Nature

The sun's energy is produced by thermonuclear fusion (see chapter 26). It consists of electromagnetic radiation ranging from short-wavelength X-rays to long-wavelength radio waves, with about 99% of it in the form of light, infrared and ultraviolet radiation.

Not all of the solar radiation arriving at the edge of the earth's atmosphere reaches the earth's surface; about 30% is reflected back into space by atmospheric dust and by the polar ice caps. Another 47% is absorbed during the day by land and sea and becomes internal energy, i.e. it heats the earth. At night this is radiated back into space (as infrared) by the earth as it cools down.

A further 23% causes evaporation from the oceans to form water vapour that is carried upwards by convection currents as clouds. When it rains some of this energy becomes gravitational potential energy of the water stored in mountain lakes and reservoirs, i.e. the source of **hydroelectric power**.

About 0.2% causes convection currents in the air, creating **wind power** which in turn causes **wave power**. Finally, 0.02% is absorbed by plants during photosynthesis and is stored in them as chemical energy (e.g. of sugar). As we shall see later, plants are a source of **biofuels**.

(b) Solar constant

This is defined as the solar energy falling per second on a square metre placed normal to the sun's rays at the edge of the earth's atmosphere, when the earth is at its mean distance from the sun. Its value is $1.35 \, \text{kW m}^{-2}$ or $1.35 \times 10^3 \, \text{J s}^{-1} \text{m}^{-2}$.

The amount of solar radiation received at any point on the earth's surface depends on:

- the **geographical location**; for example, the amount falling annually on each square metre of the UK is about half that in sunny regions like California which are nearer the equator;
- the **season**, especially in northern latitudes where the variations from summer to winter are considerable;
- the **time of day**; the lower the sun is in the sky the greater is the atmospheric absorption; and

- the **altitude**; the greater the height above sea level the less is the absorption by the atmosphere, clouds and pollution.

In the UK solar energy is received at an average rate of about $200 \, \text{W m}^{-2}$.

SOLAR DEVICES

The solar energy falling on the earth in an *hour* equals the total energy used by the world in a *year*. Unfortunately its low **energy density** requires large collecting devices and its **availability** is variable. Its greatest potential use is as a source for low-temperature space and water heating, but it can also produce high-temperature heating, at temperatures of over $3000 \, °\text{C}$, or be converted directly into electricity.

(a) Passive solar heating

This is achieved by capturing the sun's rays directly (in the northern hemisphere) through large south-facing windows. The internal energy so acquired is distributed through the building naturally by conduction, convection and radiation. The use of building materials with a high specific heat capacity, which warm up during the day and gradually release their energy as the outside temperature falls, helps to smooth out variations.

Double-glazed windows (and small north-facing ones) and well-insulated walls and roofs are also essential.

(b) Active solar heating

Solar collectors are used, their shape and size depending on the temperature range they cover. Two of the many types will be considered.

Flat plate collectors (solar panels) are low-temperature devices used to heat swimming-pools and to produce domestic hot water up to about $70 \, °\text{C}$. Some for the latter use are shown in Fig. 10.8. The construction of one is shown in Fig. 10.9.

Thin copper tubes containing the water to be heated are partly embedded in a copper collector plate (both blackened), which is mounted on a good thermal insulator in a metal frame. Solar radiation falls on the tubes and plate through a glass cover where it is trapped, as in a greenhouse. The warmed water is pumped to a heat-exchange coil in an insulated hot-water tank connected to the house hot-water system. In the UK the panel should face south roughly and be tilted at an angle of between 30° and 40° for maximum annual output.

Parabolic dish collectors capable of creating very high temperatures use mirrors to concentrate the sun's rays on to a small area at their focus, Fig. 10.10*a*. In some arrangements, a

Fig. 10.8 Flat plate solar collectors on the roofs of London houses

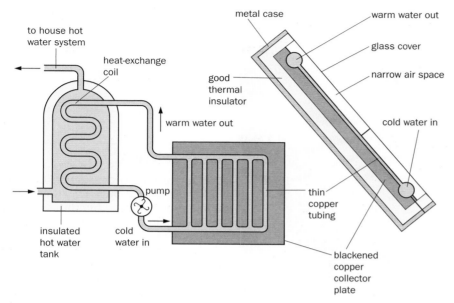

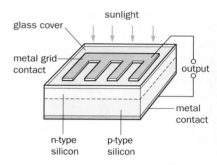

(c) Photovoltaic devices (solar cells)

These transform the energy in sunlight directly into electricity. They are made from semiconducting p- and n-type silicon (see chapter 23). One is shown in Fig. 10.11.

Fig. 10.11 A photovoltaic cell

Fig. 10.9 Construction of a flat plate solar collector

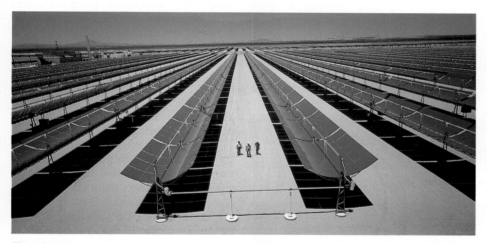

Fig. 10.10a Array of computer-controlled parabolic solar collectors which track the sun in the Mojave Desert, California

Fig. 10.10b Solar furnace at Odeillo in the French Pyrenees

furnace housed in a tower at the focus of the array produces steam to drive a turbine coupled to an electrical generator. Several prototype solar power stations are working or are under construction in the USA and in various Mediterranean countries, but the capital costs are high.

The solar furnace at Odeillo in the French Pyrenees, Fig. 10.10*b*, generates 1 MW of internal energy and temperatures greater than 3000 °C. The 45 m diameter reflector consists of about 10 000 small plane mirrors in a parabolic shape and receives the solar radiation reflected from over 60 large mirrors which, under computer control, follow the sun.

When light falls on the surface it frees loosely held electrons in the material and an e.m.f. of about 0.5 V is generated between two metal contacts on its top and bottom surfaces, the former being a grid. Protection is given by a glass or clear plastic cover.

Currently the best-designed cells can generate 240 W m^{-2} in bright sunlight, with a maximum transfer efficiency of 24%. Panels of connected cells have greater power outputs. They are used, for example, on satellites to supply the electronic equipment. Figure 10.12*a* shows a solar-powered car whose back-up batteries are being recharged at a solar 'filling' station. Panels of photovoltaic cells are used for small-scale electricity generation in remote areas of developing countries where there is sufficient sun but no electricity grid. There is also increasing use in less sunny countries, for power supply (with back-up) to large industrial and commercial buildings, Fig. 10.12*b*.

Recent developments in photovoltaic cell technology have made large-scale power generation more cost-effective. There is already a 10 MW solar power plant in California, but the capital costs involved in building such a solar power station are very high. This is due to the large surface area of cells required for generating high power outputs and the need to convert the d.c. voltages generated to a.c. for transmission.

Fig. 10.12a Charging up an electric car at a solar-powered charging station, California

Fig. 10.12b Panels of photovoltaic cells on the roof of the Union Bank of Switzerland, Lugano, Switzerland

A novel idea is for a solar power station to be assembled in space, in a geostationary orbit above an earth receiving station to which the solar energy, collected by the satellite's panels, would be beamed down as microwaves.

WIND ENERGY

Winds are due to convection currents in the air caused by uneven heating of the earth's surface by the sun. The main winds over the earth arise from cold air from the polar regions forcing the warm air over the tropics to rise; the path they follow is affected by the earth's rotation.

Wind speeds and directions change with the seasons but at any given place the pattern is fairly constant over the years. Speed increases with height, being greatest in hilly areas; it is also greater over the sea and coastal areas where there is less surface drag.

(a) Power of the wind

Wind has kinetic energy due to its motion. A device such as a wind turbine (the modern version of a windmill), which is able to slow it down, can extract part of the energy. Not all the energy can be extracted, otherwise the wind would stop and no more air could pass. In practice, a well-designed turbine will transfer about 35% of the available energy in the wind.

It can be shown (see question 14, p. 174) that the power P developed by a wind turbine (with a horizontal axis) is directly proportional to:

- the *cube of the wind speed v*, i.e. if the wind speed doubles, eight times more power is available; and
- the *area A swept out* by the turbine blades, so that doubling the diameter d of the blades quadruples the power (since $A \propto d^2$).

In symbols

$$P \propto Av^3 \quad \text{or} \quad P = kAv^3$$

where k is a constant. Therefore for maximum power a wind turbine should have long blades and be sited where there are high wind speeds.

(b) Wind turbines

In most machines the axis carrying the blades is horizontal like that in the foreground of Fig. 10.13. The whole assembly of axis, blades and generator is mounted well above the ground on a tower. There may be any number of blades, made from wood, metal or composites such as GRP (p. 34), sweeping an area of diameter up to 50 m or so.

Fig. 10.13 Horizontal-axis and vertical-axis wind turbines in Carmarthen Bay, Wales

Wind turbines are designed to work between certain wind speeds. The 'cut-in' speed at which the generator starts to produce electricity is typically 4 m s^{-1} (about 10 mph). The 'cut-out' wind speed at which the machine is shut down to prevent damage is around 25 m s^{-1} (about 60 mph).

Vertical-axis machines such as that in the background of Fig. 10.13 are less common but they have the advantage of responding to wind from any direction and requiring less support since the generator can be on the ground.

Conditions in the UK favour the use of wind energy. The prevailing winds are strong, especially in winter when the energy demand is greatest, and the coastline is long. There are currently 52 operating wind projects connected to the National Grid, with a total output of 353 MW — equivalent to the annual needs of about 234 000 homes. Turbines are usually grouped in arrays, or 'wind farms', often of 10 to 20 but in some cases many more. Two of the largest wind farms are in Powys, Wales — with 56 and 103 turbines and outputs of over 30 MW.

At the present time, the cost of producing electricity from either solar energy or from wind power is substantially higher than for a thermal power station, and generators are less reliable since they require sunshine or wind to function; however, they can provide useful 'top-ups' to the National Grid. Currently 1% of the state of California's electricity supply comes from 17 000 windmills, with the equivalent output of one fossil-fuel plant!

WATER POWER

(a) Wave energy

Waves form when the wind blows over the sea. The stronger the wind, the greater the waves, and in a gale they can be 20–25 m from crest to trough. Measurements show that the average power over a year of waves well out in the Atlantic is about 100 kW per metre of wavefront, falling to around half this value nearer the coasts of the British Isles. The energy increases with the *square* of the wave height and varies with time, its peak coinciding with the peak energy demand in winter. The best wave-energy sites around the British Isles are shown in Fig. 10.14.

Fig. 10.14 Best sites for wave-energy converters

Many ingenious wave-energy converters have been proposed. The best can extract about *half* of the wave energy available and convert around *half of that* into electrical energy for delivery by cable to the shore. For example, if the power level is 60 kW per metre, the output would be about 15 kW per metre of wavefront intercepted. In practice, the slow oscillations of the waves have to be converted by some form of turbine into the rotary motion required to drive a generator. The principle of the **rocking-boom converter** is shown in Fig. 10.15.

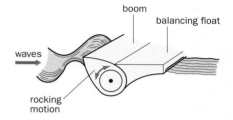

Fig. 10.15 The rocking-boom wave-energy converter

Recent research has shown that difficulties in construction, maintenance, energy transfer and transmission make large-scale production of electricity from wave-energy converters unlikely in the near future. Small systems are being developed, however, to supply small coastal or island communities with power. A prototype shore-based converter on Islay in the Inner Hebrides generates 75 kW. It uses the rise and fall of the waves inside a chamber to force air through a turbine which can extract power from the airstream when it flows in either direction.

(b) Tidal energy

Tides are caused by the gravitational pull of the moon, and to a lesser extent the sun, on the oceans. There is a high tide at places nearest the moon and also opposite on the far side, Fig. 10.16. As the earth rotates on its axis, the positions of high tide move over its surface, giving two high and two low tides daily or, more exactly, every 24 hours 50 minutes. The extra 50 minutes is due to the moon travelling round the earth in the same direction as the earth's daily spin, so the earth has to make just over one revolution before a given place is again opposite the moon.

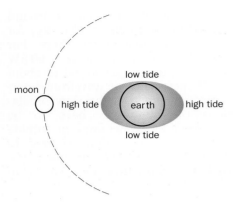

Fig. 10.16

The tidal range is a maximum when there is a new or a full moon. The sun, moon and earth are then in line, Fig. 10.17*a* with the lunar and solar pulls reinforcing to produce extra high (spring) tides. The lowest (neap) tides occur when there is a half moon and the sun and moon are pulling at right angles to each other, Fig. 10.17*b*.

Tidal energy can be harnessed by building a barrage (barrier), containing water turbines and sluice gates, across the mouth of a river. Large gates are opened during the incoming (flood) tide, allowing the water to pass until high tide, when they are closed. On the outgoing (ebb) tide, when a sufficient head of water has built up, small gates are opened, letting the potential energy of the trapped water drive the turbines and generate electricity.

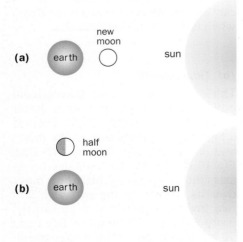

Fig. 10.17 Positions of earth, moon and sun which produce (*a*) spring and (*b*) neap tides

One of the largest working tidal schemes in the world is in France, at the estuary of the River Rance. It has 24 turbines and a peak output of 240 MW. In the UK a feasibility study of a barrage across the Severn Estuary, where the tide can rise and fall by 11 m, has been undertaken and has confirmed it to be possible technologically. The proposed barrage would be about 10 miles long and would have 216 turbines generating a peak output of 8640 MW. It could produce about 7% of present electricity consumption in England and Wales. Further studies are under way to assess its potential environmental effects, the implications for shipping, and the financing of the scheme. A barrage across the River Mersey is also subject to major studies. This could supply three-quarters of the electricity demand of the city of Liverpool. Smaller schemes are also being investigated.

'Tidemills' — offshore underwater turbines using similar technology to wind turbines to capture the energy of tidal currents — are also the subject of research. As with wave-energy converters, difficulties with maintenance of machinery and with electricity transmission (often across landscapes of great natural beauty) mean that large-scale exploitation is unlikely in the short term.

BIOFUELS

(a) Biomass

This is the name for organic (and other) material from which biofuels (e.g. methane gas and alcohol) can be obtained by processing. Much of it, like fossil fuel, is stored solar energy, in the form of starch and cellulose made under the sun's action in green plants by photosynthesis (with a conversion efficiency of about 2%). It includes the following:

- cultivated (energy) crops, e.g. oil-seed rape, sugar beet, red clover, artichokes, rye grass, maize;
- crop residues, e.g. vegetable stems and leaves, cereal straw, pea vines;
- waste from food crops, e.g. animal manure;
- natural vegetation, e.g. gorse, reeds, bracken;
- trees grown for their wood, e.g. pine, spruce;

- domestic and industrial refuse, e.g. paper, cardboard, plastics, glass, metal, clothing;
- sewage.

Biomass, especially wood and animal dung, is the main source of fuel in the developing world. Industrialized countries are now recognizing and investigating its potential. Its widespread adoption requires control of land use, e.g. by farming subsidies, to ensure that energy crops grown for commercial reasons do not replace essential food crops. Uncontrolled deforestation to obtain wood could lead to soil erosion and flash-flood damage. Transport, storage and conversion costs also need to be considered.

(b) Biofuel production

Each type of biomass may be treated in many different ways to give a range of useful biofuels. The fuel may be a gas, e.g. methane, a liquid, e.g. alcohol (ethanol), or a solid, e.g. wood, straw.

In **anaerobic digestion** (Fig. 10.18) green plants, sewage sludge or animal dung decompose by bacterial action in the absence of air to produce **biogas** — a mixture of methane and carbon dioxide. The process occurs naturally but is accelerated if it happens in a thermally insulated, air-tight tank, heated to 35 °C, and stirred. The gas forms in a week or two and can be piped off to a storage tank. If wood shavings, straw or treated refuse are used, the conversion is much slower. When bacterial action is complete, what remains in the tank is rich in nitrogen and makes a good fertilizer.

Fig. 10.18 Feeding a small biogas digester in rural India

In **fermentation** crushed biomass containing starch (e.g. maize, potatoes) or sugar (e.g. sugar cane or beet) is broken down by the action of enzymes in yeast to give **ethanol**. After about 30 hours, the 'brew' contains around 10% ethanol which is then removed by distillation.

(c) Biofuels and their uses

Solid biofuels such as wood and straw are burnt for domestic, industrial and agricultural heating. Refuse, if dried, sorted to recover metals and glass, shredded and pressed into pellets can also be used, alone or with other fuels, e.g. coal.

Liquid biofuels like the alcohols methanol and ethanol are now replacing petrol for vehicles in the developing world. They contain up to 50% less energy per litre than petrol but they are free of lead and sulphur and so cleaner, and burn more efficiently at lower temperatures. They can be blended with unleaded petrol and engine modifications are needed only if the alcohol content exceeds 20%.

Vegetable oils from crushed seeds and nuts (e.g. rape seed, peanuts) can be a direct replacement for diesel fuel or blended with it.

Biogas is a mixture of methane (50–70%) and carbon dioxide with an energy content about two-thirds that of natural gas. It can be burnt in stationary engines to produce mechanical energy or used for heating and cooking. Removing the carbon dioxide improves its quality.

Methane is a clean gas with a high **heating value** (heat production per unit mass of fuel) which can be fed straight into the gas mains (natural gas is 95% methane). It can also be converted into methanol.

The heating values of some fuels are given in Table 10.5.

GEOTHERMAL ENERGY

The rocky crust enclosing the earth's core is around 3 km thick and its temperature increases rapidly with depth. In volcanic and earthquake regions it rises by over 80 degrees C per km, in other areas by about 30 degrees C per km. Steam geysers and hot-water springs are evidence of geothermal energy.

Table 10.5 Heating values of fuels/kJ g^{-1}

Solids	Value	Liquids	Value	Gases	Value
Wood	17	Ethanol	30	Biogas	34
Coal	25–33	Crude oil	50	Methane	55

Strictly speaking, geothermal energy is not renewable in quite the way that solar, wind and water power are. As internal energy is taken from the rocks, their temperature falls and eventually the available energy is exhausted (in perhaps several decades). However, after a few thousands of years they would have been warmed again by the surrounding hot rocks, albeit very slowly since rock is a poor thermal conductor. So in the very long term geothermal energy can be thought of as renewable.

Two methods of extraction are possible, depending on the nature of the rocks.

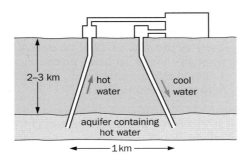

Fig. 10.19 Extracting geothermal energy from a hot aquifer

(a) Hot aquifers

These are layers of permeable (porous) rock such as sandstone or limestone at a depth of 2–3 km which *contain* hot water at temperatures of 60–100 °C. A shaft is drilled to the aquifer and the hot water pumped up it to the surface where it is used for district space- and water-heating schemes or to generate electricity. Sometimes a second shaft is drilled and the cool spent geothermal water returned by it to rocks at least 1 km from the bottom of the aquifer shaft (why?), Fig. 10.19.

This is the method used by most of the countries that have exploited geothermal energy to date. Figure 10.20 shows a geothermal plant in Iceland, and its surface lagoon of warm water.

(b) Hot dry rocks

These are impermeable (non-porous) rocks found throughout the world, which at depths of 5–6 km have temperatures of 200 °C or higher. In this case two shafts are drilled so that they terminate at different levels in the hot rock (e.g. 300 m apart), Fig. 10.21. The rocks near the ends of the shafts are fractured by explosion or by other

Fig. 10.20 A geothermal power station making use of hot aquifers in Iceland

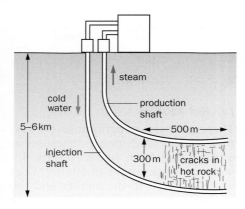

Fig. 10.21 Extracting geothermal energy from hot dry rock

methods, to reduce the resistance to the flow of cold water which is pumped, under very high pressure (300 atmospheres), down the injection shaft and emerges as steam from the top of the shallower (production) shaft.

The method has been used successfully in the USA — electricity has been produced from a small turbogenerator driven by geothermal steam, but very inefficiently. An experimental project was undertaken in Cornwall, where radioactivity in the granite causes twice the average heating, but the technical difficulties experienced and the costs involved prevented commercial development.

ENERGY LOSSES FROM BUILDINGS: *U*-VALUES

The inside of a building can only be kept at a *steady* temperature above that outside by heating it at a rate which equals the rate it is losing energy. The loss occurs mainly by conduction through the walls, roof, floors and windows. For a typical house in the UK where no special precautions have been taken, the contribution each of these makes to the total loss is as shown in Table 10.6*a*. The substantial reduction of this loss which can be achieved, especially by wall and roof insulation, is shown in Table 10.6*b*. These factors need to be fully quantified in designing buildings.

Table 10.6a

Percentage of total energy loss due to

walls	roof	floors	windows	draughts
35	25	15	10	15

Table 10.6b

Percentage of each loss saved by

Insulating

walls	roof	carpets	double glazing	draught excluders
65	80	≈30	50	≈60

Percentage of total loss saved = 60

(a) U-value

In practice, heating engineers find the *U*-value of a material to be a more convenient quantity to work with than thermal conductivity *k* (p. 72). It is defined as follows.

> The *U*-value of a specified thermal conductor (e.g. a single-glazed window) is the rate of flow of energy through it per square metre for a temperature difference of 1 K between its two surfaces.

It is measured in W m^{-2} K^{-1} and given by the equation

> U-value = $\dfrac{\text{rate of loss of energy}}{\text{surface area} \times \text{temperature difference}}$

The rate of flow of energy through a thermal conductor can therefore be calculated, knowing its *U*-value, from

> $\dfrac{\text{rate of}}{\text{loss of energy}} = \dfrac{U\text{-value}}{\times \text{ surface area}}$
> $\times \text{ temp. difference}$

Table 10.7

Thermal conductor	U-value /W m^{-2} K^{-1}
Brick wall with air cavity	1.7
Brick wall with 75 mm thick cavity insulation	0.6
Tiled roof without insulation	2.2
Tiled roof with 75 mm thick roof insulation	0.45
Single-glazed 6 mm thick glass window	5.6
Double-glazed window with 20 mm air gap	2.9

Some *U*-values are given in Table 10.7; note that good insulators have low *U*-values.

(b) Thermal resistance coefficient

This term can be employed instead of *U*-value and arises from the analogy between energy flow and electric charge flow. We can write

> energy flow rate $= \dfrac{\text{temperature difference}}{\text{resistance to flow}}$

In symbols

$$\phi = \frac{\Delta\theta}{R_e} \quad \text{or} \quad \Delta\theta = \phi R_e \qquad (1)$$
$$(\text{cf. } V = IR)$$

where $\Delta\theta$ is the temperature difference in K, ϕ is the energy flow rate in W and R_e is the **thermal resistance** of the sample in K W^{-1}. Comparison with the thermal conductivity equation (p. 73) gives

$$R_e = \frac{x}{kA} \qquad (2)$$

where *x* is the thickness of the sample in m, *k* is its thermal conductivity in W m^{-1} K^{-1} and *A* its cross-section area in m^2.

From equation (1) we see that if R_e is *doubled*, ϕ is *halved*, i.e. $\phi \propto 1/R_e$, and from equation (2) it follows that if *A* is *doubled*, R_e is *halved*, i.e. $R_e \propto 1/A$.

The **thermal resistance coefficient** *X* of a material is the thermal resistance of unit area and is related to the thermal resistance R_e by the equation

$$X = R_e A \qquad (3)$$

It is measured in m^2 K W^{-1} and is a constant for a given sample (since doubling *A* halves R_e). From equations (2) and (3) it follows that the relation between X and the thermal conductivity k is

$$X = R_eA = \frac{x}{kA} \cdot A = \frac{x}{k} \qquad (4)$$

If you use X in thermal conductivity calculations rather than U-values, the above four equations will be required.

VENTILATION

While it is important to reduce the energy loss from a building, ventilation is also necessary for the comfort of its occupants. A typical recommendation is that the air should be changed at a rate of $0.015\ \mathrm{m^3\,s^{-1}}$ per person in the room.

For ten people the rate would be $0.15\ \mathrm{m^3\,s^{-1}}$. If they were in a room measuring $10\ \mathrm{m} \times 6\ \mathrm{m} \times 3\ \mathrm{m}$, i.e. of volume $180\ \mathrm{m^3}$, a complete change of air would be required in $180\ \mathrm{m^3}/0.15\ \mathrm{m^3\,s^{-1}} = 1200\ \mathrm{s} = 20$ minutes.

In winter, the heating system for the room would have to supply energy to raise the temperature of the air drawn in to a comfortable temperature. Suppose a rise of 10 K was necessary, then, since the energy required to raise the temperature of $1\ \mathrm{m^3}$ of air by 1 K is 1.3 kJ, an air change every 1200 s would need a heater of power P for this rate of ventilation, where

$$P = \frac{1.3\ \mathrm{kJ\,m^{-3}\,K^{-1}} \times 180\ \mathrm{m^3} \times 10\ \mathrm{K}}{1200\ \mathrm{s}}$$

$$= \left(\frac{1.3 \times 180 \times 10}{1200}\right) \frac{\mathrm{kJ\,m^{-3}\,K^{-1}\,m^3\,K}}{\mathrm{s}}$$

$$\approx 2.0\ \mathrm{kW\ (kJ\,s^{-1})}$$

If the energy produced by each person's 'body heat' is roughly equivalent to that from a 100 W heater (see question 5a), P is halved ($10 \times 100\ \mathrm{W} = 1\ \mathrm{kW}$).

The relation between X and U-value can be derived from the thermal conductivity equation (p. 73)

$$\frac{Q}{t} = \frac{kA(\theta_2 - \theta_1)}{x}$$

We have

$$U\text{-value} = \frac{\text{rate of loss of energy}}{\text{surface area} \times \text{temperature difference}}$$

$$= \frac{Q}{tA(\theta_2 - \theta_1)} = \frac{k}{x}$$

But from equation (4) above left,

$$X = \frac{x}{k}$$

$$\therefore\ X = \frac{1}{U\text{-value}}$$

QUESTIONS

1. **a)** If energy is conserved, why is there an energy 'crisis'?
   b) Explain the terms 'high-grade' and 'low-grade' energy and give examples.
   c) What is meant by the statement 'the entropy of the universe is increasing'?
2. How much energy in J is supplied by the following?
   a) A 3 kW immersion heater in 2 hours.
   b) A car which uses 5 litres of petrol while travelling 80 km if 1 litre of petrol produces 10 kWh of energy when burnt.
   c) A 200 MW power station in a day.
   d) Stopping a 1 tonne car travelling at
   i) $40\ \mathrm{km\,h^{-1}}$
   ii) $80\ \mathrm{km\,h^{-1}}$.
   (1 tonne = 1000 kg)
   e) The world's annual energy consumption of about 10 000 Mtce (megatonne of coal equivalent) if 1 Mtce $\approx 30 \times 10^{15}$ J.
   f) One therm if 1 therm = 100 000 Btu (British thermal unit) and 1 Btu is the energy needed to raise the temperature of 1 lb (0.45 kg) of water by 1 °F ($\frac{5}{9}$ K). (Specific heat capacity of water = $4.2 \times 10^3\ \mathrm{J\,kg^{-1}\,K^{-1}}$)
3. If the world's recoverable reserves of coal, oil and gas and their annual consumption are as in the table below,

Fuel	Recoverable reserves/J	Annual consumption /J per year
Coal	20×10^{21}	84×10^{18}
Oil	7.0×10^{21}	120×10^{18}
Gas	3.0×10^{21}	55×10^{18}

work out the approximate lifetime of each resource.
4. A certain health diet claims that you can eat as much as you like and not gain weight if you drink enough cold water.
   Assume the body uses stored food energy to raise the temperature of ice-cold water to the body temperature of 37 °C.
   If 4200 J are needed to raise the temperature of 1 litre of water by 1 °C, use the data given below to calculate how much ice-cold water must be drunk to nullify a steak (100 g) and potato (200 g) meal.

Food	Steak	Potato
Energy value /J kg^{-1}	8.0×10^6	6.0×10^6

5. **a)** Assuming the human body is a machine that transfers to internal energy all the chemical energy it receives from food (about 10 MJ per day in the UK), what is its power as a heater?
   b) The human body is about 25% efficient at transforming chemical energy into mechanical energy (most of the rest is used to heat it). If someone develops a power output of 500 W when running up a flight of stairs, what is the power input to the body?
6. Draw an energy flow diagram for a thermal power station. State its approximate efficiency and say how it is affected by the temperatures at which the high-pressure turbine and the condenser operate. What are the advantages of a 'combined heat and power' (CHP) system?

7. Draw an energy flow diagram for a hydroelectric power station. Why does such a station have a much greater efficiency than a thermal power station? What is meant by 'pumped storage'?

8. a) Water flows in a waterfall at a rate of 1000 kg s^{-1} and takes 2 s to reach the stream vertically below. Calculate the power of the falling water just as it hits the stream.
   b) The reservoir for a hydroelectric power station is 40 m above the turbines. If the overall efficiency of the station is 50%, what mass of water must flow through the turbines per second to generate 1 MW of electrical power?

9. For each of the following tasks, indicate how each may be powered by a *different* energy resource and state whether the resource is finite or renewable. An example is given in the first row.

Grinding a sack of corn	Direct power from water wheel	Renewable
Heating a kettle of water		
Heating a factory		
Lighting an isolated hut		
Launching a satellite		

(*L*, A PH3, June 1998)

10. Taking the solar constant as 1.4 kW m^{-2} and assuming that on average this is reduced due to absorption in the atmosphere by 50%, calculate the total solar energy arriving at the earth's surface in an hour. Radius of earth ≈ 6400 km. (*Note.* Account must be taken of day and night and the fact that not all the earth's surface is normal to the radiation.)
    How does your answer compare with the world's total annual consumption of energy of about 3.7×10^{20} J at the present time?

11. Describe how energy is transferred from the Sun to the Earth.
    Explain the term **solar constant**.
    In Britain, on average, 200 W fall on each square metre of the Earth's surface during daylight hours. Estimate the total energy available from the Sun per square metre of land in Britain each year.
    Miscanthus (elephant grass) can be grown as a biofuel. It shows annual yields as high as 12 tonnes per hectare (10 000 m^2). Given that 1 tonne of miscanthus yields 1.7×10^{10} J when burnt, calculate the maximum annual energy yield per square metre of a miscanthus crop.
    Give one reason for the difference between the total energy available from the Sun per square metre of land in Britain each year and the maximum energy yield per square metre of miscanthus crop.

(*L*, A PH3, Jan 1998)

12. In the solar panel shown in Fig. 10.9, explain
    a) why the collector plates and tubes are made of copper and why they are thin-walled,
    b) why the collector plate and the tubes have a blackened surface, and
    c) how the narrow air space between the glass cover reduces energy loss by convection and radiation.

13. List the advantages and disadvantages of two primary energy sources which are
    a) finite,
    b) renewable.

14. If ρ is the density of air, A is the area swept out by the blades of a wind turbine with a horizontal shaft and v is the wind speed, prove that the maximum power P that can be developed by the turbine is given by

$$P = \tfrac{1}{2}\rho A v^3$$

Why cannot all this power be harnessed?

15. What are the main environmental implications of using as an energy source
    a) fossil fuels,
    b) nuclear fuels,
    c) active solar devices,
    d) wind energy,
    e) hydroelectric power,
    f) tidal energy?

16. This question is about a hydroelectric power scheme using tidal energy.
    Figure 10.22 shows a hydroelectric scheme constructed in the ocean near the shore. Built into the dam wall is a system of pipes and adjustable valves (not shown), to allow water to flow one way or the other, and a turbine connected to an electric generator.
    a) Explain in some detail how such a system might work to provide electrical energy from the tides.
    b) Tidal systems work only on a small scale and in certain places. Suggest **two** factors which make it impractical for such systems to provide electrical energy on a large scale or a widespread scale.
    c) The tides are the immediate source of energy for this hydroelectric system.
       i) Where does the energy of the tides come from?
       ii) Discuss briefly whether tidal energy systems give us something for nothing and whether the source of tidal energy can eventually be used up.

(*IB*, Subsidiary/Standard Paper 3, Nov 1998)

17. Calculate the rate of loss of energy through a window measuring 2.0 m by 2.5 m when the inside and outside temperatures are 18 °C and −2.0 °C respectively.
    a) If the window is single-glazed, 6.0 mm thick, and
    b) if the window is double-glazed with a 20 mm air gap.
    Use the *U*-values given in Table 10.7.

18. Figure 10.23 shows a cross-section of a double glazing unit and a graph of the temperatures at different points within the unit. The temperature difference across the unit is 13 K.
    The double glazing unit is in a steady state but it is not in thermal equilibrium.
    What is meant by the term **steady state**? What is meant by the term **thermal equilibrium**?
    A window has an area of 1.3 m^2 and the rate of heat flow through it is 61 W when the temperature difference across it is 13 K. What is the *U*-value for the double glazing unit used for the window?

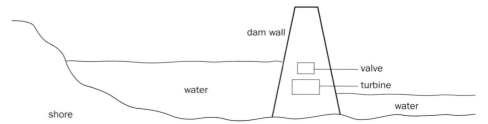

Fig. 10.22

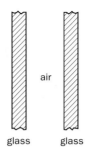

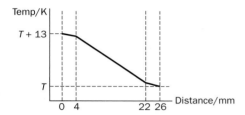

Fig. 10.23

If the two glass sheets in the double glazing unit were mounted further apart, the unit would be a poorer insulator. Suggest a reason for this.

(*L*, A PH3, June 1998)

19. a) Calculate the thermal resistance of a brick wall measuring 5.0 m by 4.0 m if the thickness of the brick is 0.20 m and k_{brick} is 0.60 W m^{-1} K^{-1}.

b) If the temperatures on either side of the brick wall in part a) are 25 °C and 5 °C respectively, what is the power loss through it?

20. A single-glazed window 6.0 mm thick measures 2.0 m by 1.0 m and has thermal conductivity 1.0 W m^{-1} K^{-1}. Calculate

a) its thermal resistance coefficient,

b) its thermal resistance, and

c) the power loss through it when the inside and outside temperatures are 18 °C and −3.0 °C respectively.

21. Use the data given in Tables 10.6*a* and *b* (p. 172) to construct pie charts showing how energy is lost from a building with and without insulation. A computer spreadsheet could be used.

22. a) Define **specific heat capacity**.

b) An open-air swimming pool of surface area 60 m² has a uniform depth of 1.5 m and is heated by the Sun. The rate at which energy arrives from the Sun per unit area of water surface is 800 W m^{-2}. Of this energy, 20% is reflected and the rest is absorbed by the water. Water has specific heat capacity 4200 J kg^{-1} K^{-1} and density 1000 kg m^{-3}. Calculate

i) the rate at which energy is absorbed by the pool,

ii) the mass of water to be heated,

iii) the mean rate of rise of temperature of the water,

iv) the time taken for the temperature of the water to rise by 3.0 K.

c) Suggest three reasons why, in cool climates, it is difficult to maintain a high water temperature in a swimming pool if it is only heated directly from the Sun.

(*OCR*, Basic 1, June 1999)

23. The passage below is taken from the marketing material supplied by a manufacturer of electrically heated showers.

Most electric showers draw cold water direct from the main supply and heat it as it is used — day or night. Not only are they particularly useful for those who do not have a stored hot water supply, but they are versatile because every home can have one.

Describe the energy transfer which takes place in an electric shower. Write an equation for the energy transfer you have described.

The technical data supplied by one manufacturer states that their most powerful shower system is fitted with a 10.8 kW heating element and can deliver up to 16 litres of water per minute.

Show that the showering temperature is about 25 °C if the temperature of the mains water is 15 °C and the shower is used at its maximum settings. (Specific heat capacity of water = 4200 J kg^{-1} K^{-1}.

Mass of 1 litre of water = 1.0 kg.)

The marketing material includes the statement:

Please remember that during the colder months, flow rates may need to be reduced to allow for the cooler temperature of incoming cold water.

Calculate the approximate flow rate required for an output temperature of 25 °C when the incoming water temperature is 5 °C.

The maximum steady current drawn by the unit is about 45 A. However, when the shower is first turned on the current is much higher for a short time. Suggest a possible explanation.

(*L*, 6551/P.01 PSA, June 1999)

11

Fluids at rest

INTRODUCTION

Liquids and gases can flow and are called **fluids**. Before considering their behaviour at rest, certain basic terms will be defined, by way of revision.

The **density** ρ of a sample of a substance of mass m and volume V is defined by the equation

$$\rho = \frac{m}{V}$$

In words, **density is the mass per unit volume**. The density of water (at 4 °C) is 1.00 g cm^{-3} or, in SI units, 1.00×10^3 kg m^{-3}; the density of mercury (at room temperature) is about 13.6 g cm^{-3} or 13.6×10^3 kg m^{-3}.

The term **relative density** is sometimes used and is given by

$$\text{relative density} = \frac{\text{density of material}}{\text{density of water}}$$

It is a ratio and has no unit. The relative density of mercury is thus 13.6.

If a force acts on a surface (like the weight of a brick on the ground) it is often more useful to consider the **pressure** exerted rather than the force. The pressure p caused by a force F acting normally on a surface of area A is defined by

$$p = \frac{F}{A}$$

Pressure is therefore **force per unit area**. The SI unit is the **pascal (Pa)** which equals a pressure of 1 newton per square metre.

PRESSURE IN A LIQUID

In designing a dam like that shown opposite in Fig. 11.2, the engineer has to know, among other things, about the size and point of action of the resultant force exerted on the dam by the water behind it. This involves making calculations based on the expression for liquid pressure.

(a) Expression for liquid pressure

The pressure exerted by a liquid is experienced by any surface in contact with it. It increases with depth because the liquid has weight and we define the **pressure at a point in a liquid as the force per unit area on a very small area round the point**. Thus if the force is δF and the small area δA then the pressure p at the point is given by

$$p = \frac{\text{force}}{\text{area}} = \frac{\delta F}{\delta A}$$

More exactly, in calculus terms, the defining equation is

$$p = \lim_{\delta A \to 0} \frac{\delta F}{\delta A}$$

An expression for the pressure p at a depth h in a liquid of density ρ can be found by considering an extremely small horizontal area δA, Fig. 11.1.

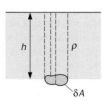

Fig. 11.1

The force δF acting vertically downwards on δA equals the weight of the liquid column of height h and uniform cross-section area δA above it. We can say

volume of liquid column $= h\,\delta A$

mass of liquid column $\quad= h\,\delta A\,\rho$

weight of liquid column $= h\,\delta A\,\rho g$

where g is the acceleration of free fall (or the strength of the earth's gravitational field). Hence

$$\delta F = h\,\delta A\,\rho g$$

and $\quad p = \dfrac{\delta F}{\delta A} = \dfrac{h\,\delta A\,\rho g}{\delta A}$

$$p = h\rho g$$

Fig. 11.2 The Hoover Dam on the Colorado river, Boulder, USA

Thus the pressure at a point in a liquid depends only on the depth and the density of the liquid. If h is in m, ρ in kg m^{-3} and g in m s^{-2} (or N kg^{-1}) then p is in Pa.

Notes. (**i**) In the above derivation, δA was considered to be horizontal but it can be shown that the same result is obtained for any other orientation of δA, i.e. **the pressure at a point in a liquid acts equally in all directions** — as experiment confirms.

(**ii**) The force exerted on a surface in contact with a liquid at rest is *perpendicular* to the surface at all points. Otherwise, the equal and opposite force exerted by the containing surface on the liquid would have a component parallel to the surface which would cause the liquid to flow.

(b) Transmission of pressure: the hydraulic principle

A liquid can transmit any external pressure applied to it to all its parts. Use is made of this property in a hydraulic system to produce a large

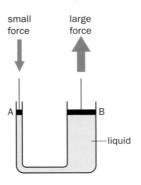

Fig. 11.3 The hydraulic principle

force from a small one. In its simplest form, Fig. 11.3, it consists of a narrow cylinder connected to a wide cylinder, both containing liquid (usually oil) and fitted with pistons A and B.

If A has a cross-section area of 1×10^{-4} m^2 (i.e. 1 cm^2) and a downward force of 1 N is applied to it, a pressure of 1×10^{-4} Pa (i.e. 1 N cm^{-2}) in excess of atmospheric pressure is transmitted through the liquid. If B has a cross-section area of 100×10^{-4} m^2 (i.e. 100 cm^2), it experiences an upward force of 100 N.

If piston B acts against a fixed plate it can be used, for example, to forge metal.

The same principle operates in hydraulic jacks for lifting cars, Fig. 11.4, and also in the hydraulic braking system of a car. In the latter, the force applied to the brake pedal causes a piston to produce an increase of pressure in an oil-filled cylinder and this is transmitted through oil-filled pipes to four other pistons, which apply the brake-shoes or discs to the car wheels. This results in the same pressure being applied to all wheels and minimizes the risk of the car pulling to one side or skidding.

Fig. 11.4 A hydraulic jack in use

(c) High-pressure water-jet cutting

This is a cutting technique which is entirely dust-free. It can be used with a wide range of materials such as slate, stone, brake-lining material, Formica, rubber and foams, and is especially advantageous for materials like asbestos, the dust from which can cause respiratory disease. The jet pressure might be about 3.5×10^8 Pa (3500 times normal atmospheric pressure).

LIQUID COLUMNS

(a) U-tube manometer

An open U-tube containing a suitable liquid can be used to measure pressures, for example the pressure of the gas supply, and is called a U-tube manometer. It uses the fact that the pressure in a column of liquid is directly proportional to the height of the column. In the U-tube manometer of Fig. 11.5*a* the pressure p to be measured acts on the surface of the liquid at A and balances the pressure of the liquid column BC of height h, *plus* atmospheric pressure P, acting on B. Therefore

$$p = P + h\rho g$$

where ρ is the density of the liquid in the manometer. For small pressures, water or a light oil is used; for medium pressures mercury is suitable. The amount by which p exceeds atmospheric pressure, i.e. $(p - P)$, equals the pressure due to the column of liquid BC, i.e. $h\rho g$. Consequently it is often convenient to state a pressure as a number of mm of water or mercury rather than in Pa.

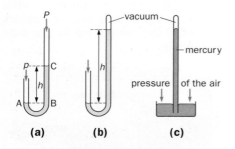

(a)   **(b)**   **(c)**

Fig. 11.5

When the absolute pressure is required rather than the excess over atmospheric, the limb of the tube open to the atmosphere is replaced by a closed, evacuated one, Fig. 11.5*b*. The height h of the liquid column then gives the absolute pressure directly. The principle is used in the measurement of atmospheric pressure by a mercury barometer, the short limb being replaced by a reservoir of mercury, Fig. 11.5*c*. The column of mercury is supported by the pressure of the air on the surface of the mercury in the reservoir and any change in this causes the length of the column to vary.

(b) Balancing liquid columns

If a U-tube contains two immiscible liquids of different densities the surfaces of the liquids are not level. In Fig. 11.6*a* the column of water AB is balanced by the column of paraffin CD.

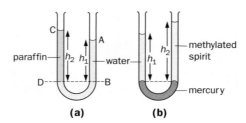

(a)   **(b)**

Fig. 11.6

So

pressure at B = pressure at D

$$\therefore \quad h_1\rho_1 g = h_2\rho_2 g$$

where h_1, ρ_1 and h_2, ρ_2 are the heights and densities of the water and paraffin respectively. It follows that

$$\frac{\rho_2}{\rho_1} = \frac{h_1}{h_2}$$

Measurement of h_1 and h_2 thus gives a simple way of finding the relative density of paraffin. The limbs of the U-tube need not have the same diameter (if not too small). Why not?

Miscible liquids can be separated by mercury as in Fig. 11.6*b*. In this case the heights of the columns are adjusted by adding liquid until the mercury surfaces are exactly level. Measurements of h_1 and h_2 are then made from this level.

ARCHIMEDES' PRINCIPLE

When a body is immersed in a liquid it is buoyed up and appears to lose weight. The upward force is called the **upthrust** of the liquid on the body and is due to the pressure exerted by the liquid on the lower surface of the body being greater than that on the top surface, since pressure increases with depth. The law summarizing such phenomena was discovered over 2000 years ago by the Greek mathematician Archimedes. It also applies to bodies in gases and is stated as follows.

When a body is completely or partly immersed in a fluid it experiences an upthrust, or apparent loss in weight, which is equal to the weight of fluid displaced.

A numerical case is illustrated in Fig. 11.7*a* below; more briefly we can say that **upthrust = weight of fluid displaced**.

The principle can be verified experimentally, or deduced theoretically, by considering the pressures exerted by a liquid on the top and bottom surfaces of a rectangular solid, Fig. 11.7*b* — as you can confirm for yourself.

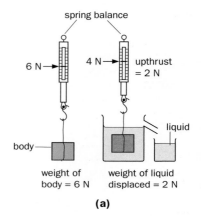

(a)

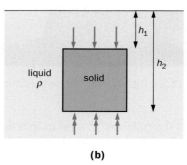

(b)

Fig. 11.7

(a) Floating bodies

If a body floats partly immersed in a liquid (e.g. a ship), or completely immersed in a liquid (e.g. a submarine) or a gas (e.g. a balloon), it appears to have zero weight and we can say

upthrust on body
$$= \text{weight of floating body}$$

By Archimedes' principle

upthrust on body
$$= \text{weight of fluid displaced}$$

Therefore

weight of floating body
= weight of fluid displaced

This result, sometimes called the 'principle of flotation', is a special case of Archimedes' principle and can be stated:

> A floating body displaces its own weight of fluid.

If the body cannot do this, even when completely immersed, it sinks.

The stability of a floating body, such as a ship, when it heels over depends on the relative positions of the ship's centre of gravity, G, through which its weight W acts, and the centre of gravity of the displaced water, called the **centre of buoyancy**, B, through which the upthrust U acts. In Fig. 11.8a the ship is on an even keel and B and G are on the same vertical line. If it heels over the shape of the displaced water changes, causing B to move and thereby setting up a couple which tends either to return the ship to its original position or to make it heel over more. The point of intersection of the vertical line from B with the central line of the ship is called the **metacentre**, M. If M is above G as in Fig. 11.8b, the couple has an anti-clockwise moment which acts to decrease the ship's heel and the equilibrium is stable. If M is below G as in Fig. 11.8c, equilibrium is unstable since the couple has a clockwise moment which causes further listing. For maximum stability G should be low (hence ships carry ballast) and M high.

(b) Hydrometer

This is an instrument that uses the principle of flotation to give a rapid measurement of the relative density of a liquid, e.g. the acid in a lead accumulator. It consists of a narrow glass stem, a large buoyancy bulb and a smaller bulb loaded with lead shot to keep it upright when floating, Fig. 11.9. The relative density is found by floating the hydrometer in the liquid and taking the reading on the scale inside the stem at the level of the liquid surface. The instrument shown is for use in the range 1.30 to 1.00; the numbers increase downwards (why?) and the scale is uneven.

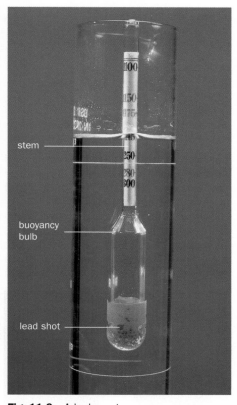

Fig. 11.9 A hydrometer

ATMOSPHERIC PRESSURE

The pressure due to a gas arises from the bombardment of the walls of the containing vessel by its molecules. In a small volume of gas the pressure is uniform throughout but in a large volume, such as the atmosphere, gravity causes the density of the gas and therefore its pressure to be greater in the lower regions than in the upper regions. In fact atmospheric pressure at a height of about 6 km is half its sea-level value even though the atmosphere extends to a height of 150 km or so.

The statement that atmospheric pressure is '760 mm of mercury' means that it is the same as the pressure at the bottom of a column of mercury 760 mm high. Sometimes pressure is expressed in torrs (after Torricelli, who made the first mercury barometer); 1 torr = 1 mmHg. To find the value of atmospheric pressure in SI units we use $p = h\rho g$ where $h = 0.760$ m, $\rho = 13.6 \times 10^3$ kg m^{-3} and $g = 9.81$ N kg^{-1}. Thus

$$p = 0.760 \text{ m} \times 13.6 \times 10^3 \text{ kg m}^{-3}$$
$$\times 9.81 \text{ N kg}^{-1}$$
$$= 1.01 \times 10^5 \text{ N m}^{-2}$$
$$= 1.01 \times 10^5 \text{ Pa}$$

Standard pressure or 1 atmosphere is defined as the pressure at the foot of a column of mercury 760 mm high of specified density and subject to a particular value of g; it equals 1.01325×10^5 Pa.

Another unit of pressure, used in weather forecasting, is the millibar (mb), 1 mb = 100 Pa.

PRESSURE GAUGES AND VACUUM PUMPS

The U-tube manometer can be used to measure moderate pressures, namely those in the region of atmospheric. High and low pressures require special gauges.

(a) Bourdon gauge

This measures pressures up to about 2000 atmospheres. It consists of a curved metal tube, sealed at one end, to which the pressure to be measured is applied, Fig. 11.10. As the pressure

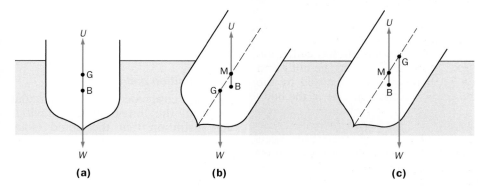

(a)   **(b)**   **(c)**

Fig. 11.8

increases the tube uncurls and causes a rack and pinion to move a pointer over a scale.

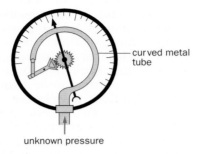

Fig. 11.10   A Bourdon gauge

(b) Rotary vacuum pump

One rotary pump can produce pressures of about 10^{-2} mmHg; with two in series 10^{-4} mmHg is attainable. The pump consists of an eccentrically mounted cylindrical rotor inside and in contact at one point with a cylindrical stator, Fig. 11.11. Two spring-loaded vanes attached to the rotor press against the wall of the stator. The whole is immersed in a special low vapour-pressure oil, which both seals and lubricates the pump. As the rotor revolves, driven by an electric motor, each vane in turn draws gas into the increasing volume of space A on the intake side and then compresses it in space B where it is ejected from the outlet valve. Lower pressures, down to 10^{-12} mmHg, require the use of a **diffusion pump**.

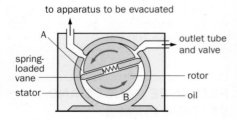

Fig. 11.11   A rotary vacuum pump

SURFACE TENSION

(a) Some effects

Various effects suggest that the surface of a liquid behaves like a stretched elastic skin, i.e. it is in a state of tension. For example, a steel needle will float if it is placed *gently* on the surface of a bowl of water, despite its greater density. (What else helps to

Fig. 11.12   Surface tension supports the weight of this pond-skater

support its weight?) The effect, called **surface tension**, enables certain insects to run over the surface of a pond without getting wet, Fig. 11.12.

Small liquid drops are nearly spherical, as can be seen when water drips from a tap, Fig. 11.13; a sphere has the minimum surface area for a given volume. (What distorts larger drops?)

Fig. 11.13   Surface tension causes small liquid drops to be almost spherical

This tendency of a liquid surface to shrink and have a minimum area can also be shown by the arrangement of Fig. 11.14*a*. When the soap film *inside* the loop of thread is punctured, the thread is pulled into the shape of a circle, Fig. 11.14*b*. Since a circle has the maximum area for a given perimeter, the area of the film *outside* the thread is a minimum.

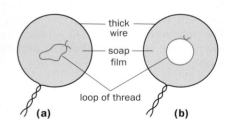

Fig. 11.14

(b) Definition and unit

We can conclude from the circular shape of the thread in the previous demonstration that the liquid (soap solution) is pulling on the thread at right angles all along its circumference, Fig. 11.15. This suggests that we might define the surface tension of a liquid in the following way.

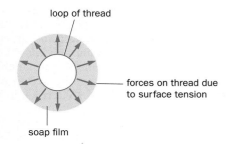

Fig. 11.15

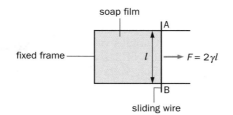

Fig. 11.17

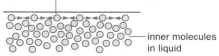

Fig. 11.18

Imagine a straight line of length l in the surface of a liquid. If the force at right angles to this line and in the surface is F, Fig. 11.16, then the surface tension γ of the liquid is defined by

$$\gamma = \frac{F}{l}$$

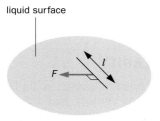

Fig. 11.16

In words, γ is **the force per unit length acting in the surface perpendicular to one side of a line in the surface**. The unit of γ is newton metre^{-1} (N m^{-1}). Its value depends on, among other things, the temperature of the liquid. At 20 °C, for water $\gamma = 72.6 \times 10^{-3}$ N m^{-1} and for mercury $\gamma = 465 \times 10^{-3}$ N m^{-1}.

It must be emphasized that normally surface tension acts equally on both sides of *any* line in the surface of a liquid; it creates a state of tension in the surface. The *effects* of surface tension are evident only when liquid is absent from one side of the line. For example, in Fig. 11.17, to keep wire AB at rest an external force F has to be applied to the right to counteract the unbalanced surface tension forces acting to the left. A film has two surfaces and so for a frame of width l, the surface tension force is $2\gamma l$ where γ is the surface tension of the liquid.

Or, when a drop of methylated spirit or soap solution is dropped into the centre of a dish of water whose surface has been sprinkled with lycopodium powder, the powder rushes out to the sides leaving a clear patch. The effect is due to the surface tension of water being greater than that of meths or soap solution, so that there is imbalance between the surface tension forces round the boundary of the two liquids. The powder is thus carried away from the centre by the water.

(c) Molecular explanation

Molecules in the surface of a liquid are farther apart than those in the body of the liquid, i.e. the surface layer has a lower density than the liquid in bulk. This follows because the increased separation of molecules that accompanies a change from liquid to vapour is not a sudden transition. The density of the liquid must therefore decrease *through* the surface.

The intermolecular forces in a liquid, like those in a solid, are both attractive and repelling and these balance when the spacing between molecules has its equilibrium value. However, from the intermolecular force–separation curve (see Fig. 5.11*a*, p. 71) we see that when the separation is greater than the equilibrium value (r_0), the attractive force between molecules exceeds the repelling force. This is the situation with the more widely spaced surface layer molecules of a liquid. They experience attractive forces on either side due to their neighbours, which puts them in a state of tension, Fig. 11.18. The liquid surface consequently behaves like a stretched elastic skin. If the tension-creating bonds between molecules are severed on one side by parting the liquid surface, then there is a resultant attractive force on the molecules due to the molecules on their other side. The effect of surface tension is then apparent.

The value of γ for a liquid does not increase when its surface area increases because more molecules enter the surface layer, thereby keeping the molecular separation constant. Otherwise, any increase of separation would increase the attractive force between molecules and so also the surface tension.

LIQUID SURFACES

(a) Shape of liquid surfaces

The surface of a liquid must be at right angles to the resultant force acting on it, otherwise there would be a component of this force parallel to the surface which would cause motion. Normally a liquid surface is horizontal, i.e. at right angles to the force of gravity, but where it is in contact with a solid it is usually curved.

To explain the shape of the surface in Fig. 11.19*a* consider the liquid at B adjoining a vertical solid wall. It experiences an attractive force BC due to neighbouring liquid molecules; this is the **cohesive** force which binds liquid molecules together and makes them behave as a liquid. An attractive force BA is also exerted by neighbouring molecules of the solid; this is called the **adhesive** force, and if it is greater than the cohesive force then the resultant force BR on the liquid at B will act to the left of the wall in the direction shown. The liquid surface at B has to be at right angles to this direction and therefore curves upwards, forming a concave meniscus. Since there is then equilibrium the resultant force must be balanced by appropriate intermolecular repulsive forces. At points on the liquid surface farther from the wall, the adhesive forces are smaller, the resultant force more nearly vertical and therefore the surface more nearly horizontal.

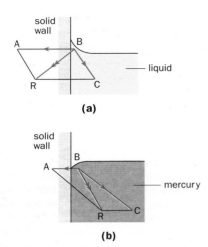

Fig. 11.19

By contrast, when the cohesive force between molecules of the liquid is greater than the adhesive force between molecules of the liquid and molecules of the solid, the resultant force BR acts as in Fig. 11.19*b* and the surface curves downwards at the wall, forming a convex meniscus. This is the case with mercury against glass.

The **angle of contact** θ is defined as the angle between the solid surface and the tangent plane to the liquid surface, measured *through the liquid*. The liquid in Fig. 11.20*a* has an acute angle of contact with this particular solid (θ < 90°), while that in Fig. 11.20*b* has an obtuse angle of contact (θ > 90°). Water, like many organic liquids, has zero angle of contact with a *clean* glass surface, i.e. the adhesive force is so much greater than the cohesive force that the water surface is parallel to the glass where it meets it, Fig. 11.20*c*. On a horizontal clean glass surface water tends to spread indefinitely and form a very thin film. Contamination of a surface affects the angle of contact appreciably; the value for water on greasy glass may be 10° and causes it to form drops rather than spread. Mercury in contact with clean glass has an angle of contact of about 140° and so tends to form drops instead of spreading over glass.

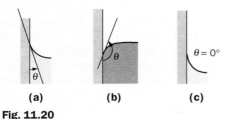

(a)   (b)   (c)

Fig. 11.20

Liquids with acute angles of contact are said to 'wet' the surface, those with obtuse angles of contact do not 'wet' it. Figure 11.21*a* shows a drop of a liquid which wets the surface and *b* shows a drop on a surface which it does not wet. The photographs, Figs 11.22*a* and *b*, show these situations in practice.

(a)   (b)

Fig. 11.21

(a) Water wets a smooth, clean surface

(b) Water forms droplets on a waxy surface

Fig. 11.22

(b) Practical applications of spreading

The behaviour of liquids in contact with solids is important practically. In soldering, a good joint is formed only if the molten solder (a tin–lead alloy) wets and spreads over the metal involved. Spreading will occur most readily if the liquid solder has a small surface tension. The use of a flux (e.g. resin) with the solder cleans the metal surface and acts as a 'wetting agent' which assists spreading. Metals like aluminium have an almost permanent oxide skin which resists the action of a flux and makes good soldered joints by normal methods very difficult.

Wetting agents play a key role in painting and spraying where the paint must not form drops but remain in a layer once spread out. The use of spreading agents (e.g. stearic acid) also assists lubricating oils to adhere to axles, bearings, etc.

If detergents are to remove the dirt particles that are held to fabrics usually by grease, they must be able to spread over the fabric before they can dislodge the grease. Detergent solutions should therefore, on this account, have low surface tensions and small angles of contact. By contrast, fabrics are weatherproofed by treatment with a silicone preparation which causes water to collect in drops.

CAPILLARITY

Surface tension causes a liquid with an angle of contact less than 90° to rise in a fine bore (capillary) tube above the level outside. The narrower the tube the greater the elevation, Fig. 11.23. The effect is called **capillarity** and is of practical importance.

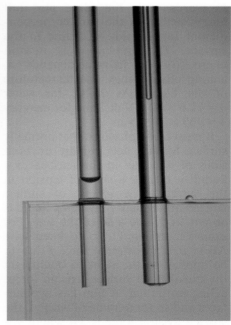

Fig. 11.23   Water in capillary tubes

Why does the rise occur? In Fig. 11.24a, round the boundary where the liquid surface meets the tube, surface tension forces exert a *downward* pull on the tube since they are not balanced by any other surface tension forces. The tube therefore exerts an equal but *upward* force on the liquid, Fig. 11.24b, and causes it to rise (by Newton's third law of motion). The liquid stops rising when the weight of the raised column acting vertically downwards equals the *vertical component* of the upward forces exerted by the tube on the liquid, Fig. 11.24c.

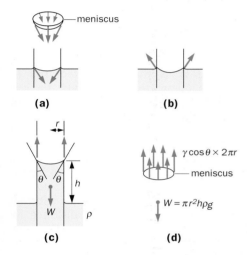

Fig. 11.24

If the liquid has density ρ, surface tension γ and angle of contact θ and if the rise is h in a tube of radius r then, neglecting the small amount of liquid in the meniscus,

weight W of liquid column
$$= \pi r^2 h \rho g$$

vertical component of supporting forces
$$= \gamma \cos \theta \times 2\pi r$$

since these forces act round a circumference of $2\pi r$, Fig. 11.24d. So

$$\pi r^2 h \rho g = 2\pi r \, \gamma \cos \theta$$

$$\therefore \quad h = \frac{2\gamma \cos \theta}{r\rho g}$$

This equation only holds for very fine bore tubes in which the curvature of the meniscus is everywhere spherical.

An estimate of the rise h can be obtained by substituting values for γ, r, ρ and g in the expression derived above. For example, for water in a very clean glass tube $\theta = 0°$, i.e. the water surface meets the tube vertically, and therefore $\cos \theta = 1$. Also $\rho = 1.0 \, \text{g cm}^{-3} = 1.0 \times 10^3 \, \text{kg m}^{-3}$ and $\gamma = 7.3 \times 10^{-2} \, \text{N m}^{-1}$. If the capillary tube has radius 0.50×10^{-3} m, then

$$h = \frac{2 \times 7.3 \times 10^{-2} \times 1}{0.50 \times 10^{-3} \times 1.0 \times 10^3 \times 9.8}$$
$$\times \frac{\text{N m}^{-1}}{\text{m kg m}^{-3} \, \text{N kg}^{-1}}$$
$$= 3.0 \times 10^{-2} \, \text{m}$$
$$= 30 \, \text{mm}$$

If θ is greater than 90°, the meniscus is convex upwards, $\cos \theta$ is negative and the expression shows that h will also be negative. This means the liquid falls in the capillary tube below the level of the surrounding liquid. Mercury in a glass capillary tube usually behaves in this way, Fig. 11.25. What happens when $\theta = 90°$?

Fig. 11.25 Mercury in capillary tubes

The drying action of a kitchen paper towel is due to fluid rising up the pores of the paper by capillarity. It also helps in soldering by causing the molten solder to penetrate any cracks. However, for this to happen the above expression for h indicates that the solder should have a high surface tension, a property which does not encourage spreading (see pp. 181–2). Compromise is clearly necessary here, as in the dyeing of fabrics where success depends largely on the dye penetrating into the fabric by capillarity.

BUBBLES AND DROPS

A study of bubbles and drops helps with the understanding of surface tension of liquids, and also has some practical importance. The formation of gas bubbles plays an important part in the manufacture of expanded plastics such as polystyrene. In oil-fired boilers pressure burners depend on droplet formation for fast and efficient burning of the vapour. In steam heating systems the efficiency of heat transfer from the steam would be higher if, instead of condensing as a film, which it does, it condensed in drops, and attempts are at present being made to achieve drop condensation.

A soap bubble blown on the end of a tube and then left open to the atmosphere gradually gets smaller, showing that the air is being forced out. Surface tension tries to make the film contract and causes the pressure inside the bubble to exceed that outside. An expression can be obtained for the *excess pressure* inside a spherical soap bubble, as follows.

Consider a bubble of radius r, blown from a soap solution of surface tension γ. Let atmospheric pressure be P and suppose the pressure inside the bubble exceeds P by p, i.e. is $(P + p)$, Fig. 11.26a. Consider the equilibrium of *one half* of the bubble; there are two sets of opposing forces.

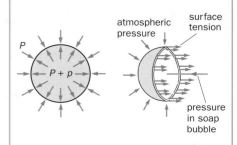

Fig. 11.26

(i) Atmospheric pressure acts in different directions over the surface of the hemisphere but the resultant force in Fig. 11.26b acts *horizontally to the right* since the vertical components cancel (if any vertical variation of atmospheric pressure is neglected). The projection of the hemisphere in a horizontal direction is a circle and the horizontal force due to atmospheric pressure equals the product of the pressure and the area of projection, that is, $P \times \pi r^2$. Also, surface tension forces are exerted by the right-hand hemisphere (not illustrated in Fig. 11.26b) on the circular rim of the left-hand hemisphere along *both* its inside and outside surfaces. This force equals $2\gamma \times 2\pi r$ and so the total horizontal force to the right is $P\pi r^2 + 4\gamma\pi r$.

(ii) The pressure $(P + p)$ acts on the curved inside surface of the left-hand hemisphere and produces a *horizontal force to the left* equal to $(P + p)\pi r^2$.

If the horizontal forces balance, we have

$$P\pi r^2 + 4\gamma\pi r = (P + p)\pi r^2$$

$$\therefore \quad 4\gamma\pi r = p\pi r^2$$

The excess pressure p inside the bubble is then

$$p = \frac{4\gamma}{r}$$

Taking γ for a soap solution as $2.5 \times 10^{-2}\,\text{N m}^{-1}$, the excess pressure inside a bubble of radius 1.0 cm or 1.0×10^{-2} m is

$$p = \frac{4 \times 2.5 \times 10^{-2}}{1.0 \times 10^{-2}} \frac{\text{N m}^{-1}}{\text{m}}$$

$$= 10\,\text{Pa}$$

If two soap bubbles of different radii are blown separately using the apparatus of Fig. 11.27, and then connected by opening taps T_1 and T_2 (T_3 being closed), the smaller bubble A gradually collapses while the larger bubble B expands. Why? Equilibrium is attained when A has become a small curved film of radius equal to that of bubble B.

A spherical drop of liquid in air or a bubble of gas in a liquid has only one

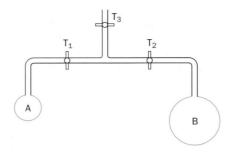

Fig. 11.27

surface and the excess pressure inside it is $2\gamma/r$, the derivation being similar to that given above for a soap bubble.

MEASUREMENT OF γ

The expression $\gamma = hr\rho g/(2\cos\theta)$ is used, where h is the capillary rise in a tube of radius r (see p. 183), and so θ must be known. In the case of a liquid for which $\theta = 0°$ the expression becomes

$$\gamma = \frac{hr\rho g}{2}$$

Knowing the density ρ of the liquid and g, only h and r remain to be determined.

The apparatus is illustrated in Fig. 11.28; the capillary tube is first cleaned thoroughly by immersing in caustic soda, dilute nitric acid and distilled water in turn. A travelling microscope is focused first on the bottom of the meniscus in the tube and then, with the beaker removed, on the tip of the pin which previously just touched the surface of the liquid in the beaker. Hence h is obtained.

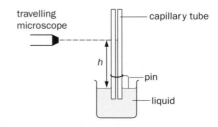

Fig. 11.28 Determination of γ

To find r the tube is broken at the meniscus level and the average reading of two diameters at right angles taken with the travelling microscope.

Surface tension decreases rapidly with temperature and so the temperature of the liquid should be stated.

SURFACE ENERGY

Molecules in the surface of a liquid are farther apart than those inside the liquid (see p. 181). The p.e.–separation curve for two molecules (Fig. 2.7b, p. 18) shows that if this is so, then the mutual intermolecular p.e. of surface molecules is greater, i.e. less negative, than that of molecules in the interior.

When a new surface is formed, energy must therefore be supplied to increase the separation of the new surface molecules. This energy is called the (free) **surface energy** of the liquid. It is denoted by σ and defined by the equation

$$\sigma = \frac{W}{A}$$

where W is the energy required to create a new area A of surface, i.e. σ is **the energy needed to create unit area of new surface**. Its unit is J m^{-2}.

In Fig. 11.29 a film of liquid, of surface tension γ, is shown stretched across a horizontal frame PQRS. The force on the sliding wire PQ of length l is $\gamma \times 2l$, since the film has two surfaces. If PQ is moved by an external force to P'Q' through a distance δx against the surface tension, the new surface area $A = 2l \times \delta x$ (the film has two sides), and

$$\begin{array}{l}\text{work done } W \\ \text{to enlarge surface}\end{array} = 2\gamma l \times \delta x$$

This equals the increase in the surface energy σ, and we then have

$$\sigma = \frac{W}{A} = \frac{2\gamma l \times \delta x}{2l \times \delta x}$$

$$\sigma = \gamma$$

i.e. σ **and** γ **are numerically equal**.

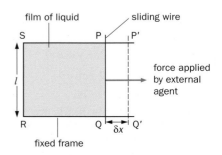

Fig. 11.29

QUESTIONS

Pressure; Archimedes

1. Define **pressure at a point** in a fluid. In what unit is it measured?

   State an expression for the pressure at a point at depth h in a liquid of density ρ. Does it also hold for a gas?

   What force is exerted on the bottom of a tank of uniform cross-section area 2.0 m^2 by water which fills it to a depth of 0.50 m? (Density of water = 1.0×10^3 kg m^{-3}; $g = 10$ N kg^{-1}.)

   Find the extra force on the bottom of the tank when a block of wood of volume 1.0×10^{-1} m^3 and relative density 0.50 floats on the surface.

2. **a)** State Archimedes' principle.
   b) A string supports a solid copper block of mass 1 kg (density 9×10^3 kg m^{-3}) which is completely immersed in water (density 1×10^3 kg m^{-3}). Calculate the tension in the string.

3. **a)** Define **density**.
   b) Copper has density 8930 kg m^{-3} and zinc has density 7140 kg m^{-3}. Brass is an alloy consisting of 70% copper and 30% zinc by volume. Assume that the volume of the alloy is equal to the sum of the volumes of the copper and zinc used. Consider one cubic metre of brass. Copy and complete the table below in order to find the density of the brass.

Metal	Volume /m^3	Mass /kg	Density /kg m^{-3}
copper	0.70		8930
zinc	0.30		7140
brass	1.00		

   c) Measurement shows that the density of the brass in b) is 8500 kg m^{-3}. Use your knowledge of the structure of crystalline solids to suggest why, in practice, it is possible that the volume of brass might be different from the volume of copper plus the volume of zinc.

   (*UCLES*, Basic 1, June 1998)

4. A simple hydrometer, consisting of a loaded glass bulb fixed at the bottom of a glass stem of uniform section, sinks in water of density 1.0 g cm^{-3} so that a certain mark X on the stem is 4.0 cm below the surface. It sinks in a liquid of density 0.90 g cm^{-3} until X is 6.0 cm below the surface. It is then placed in a liquid of density 1.1 g cm^{-3}. How far below the surface will X be? (Neglect surface tension effects.)

5. A physicist inflates a balloon with air to a volume of 1.5 litres and seals it. The density of the surrounding air at the time of the experiment was 1.30 g per litre. Calculate the upthrust on the balloon after the inflation process.

   Sketch and label a free-body force diagram to show the principal forces which act on the inflated balloon at rest on a balance. Given that the mass of air in the balloon is 2.15 g and the mass of the balloon fabric is 4.10 g, show the magnitudes of the forces.

   Calculate the density of the air in the balloon.

   Why is the density of the air in the balloon greater than the density of the air outside the balloon?

   (*L*, A PH3, Jan 1998)

Surface tension

6. Figure 11.30 shows a sealed container filled with air and attached to a manometer. The liquid in the manometer is an oil of density 860 kg m^{-3}. The vertical distance between the two oil surfaces is 11.2 cm. The atmospheric pressure is 101.2 kPa.

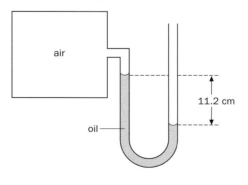

air

oil

11.2 cm

Fig. 11.30

   Calculate the pressure in the sealed container.

   The oil in the left hand arm has more potential energy than the oil in the right hand arm. Identify the source of this potential energy. Explain your answer.

   With reference to kinetic theory, explain why the pressure of the air in the container rises if the temperature rises.

   (*L*, A PH3, June 1998)

7. Explain, using a simple molecular theory, why the surface of a liquid behaves in a different manner from the bulk of the liquid.

   Giving the necessary theory, explain how the rise of water in a capillary tube may be used to determine the surface tension of water.

   A microscope slide measures 6.0 cm × 1.5 cm × 0.20 cm. It is suspended with its face vertical and with its longest side horizontal and is lowered into water until it is half immersed. Its apparent weight is then found to be the same as its weight in air. Calculate the surface tension of water, assuming the angle of contact to be zero.

8. A clean glass capillary tube of internal diameter 0.60 mm is held vertically with its lower end in water and with 80 mm of the tube above the surface. How high does the water rise in the tube?

   If the tube is now lowered until only 30 mm of its length is above the surface, what happens? (Surface tension of water is 7.2×10^{-2} N m^{-1}.)

9. Describe and explain *two* experiments of a different nature to illustrate the phenomenon of surface tension.

   Give a quantitative definition of **surface tension** and explain what is meant by **angle of contact**.

   The internal diameter of the tube of a mercury barometer is 3.00 mm. Find the corrected reading of the barometer after allowing for the error due to surface tension, if the observed reading is 76.56 cm. (Surface tension of mercury = 4.80×10^{-1} N m^{-1}; angle of contact of mercury with glass = 140°; density of mercury = 13.6×10^3 kg m^{-3}.)

10. Two spherical soap bubbles of radii 30 mm and 10 mm coalesce so that they have a common surface. If they are made from the same solution and if the radii of the bubbles stay the same after they join together, calculate the radius of curvature of their common surface.

12

Fluids in motion

- ■ Introduction
- ■ Viscosity
- ■ Coefficient of viscosity
- ■ Steady and turbulent flow
- ■ Motion in a fluid
- ■ Bernoulli's equation
- ■ Applications of Bernoulli
- ■ Flowmeters
- ■ Fluid flow calculations

INTRODUCTION

The study of moving fluids is important in engineering. A large quantity of liquid may have to flow rapidly through a pipe from one location to another, or air entering the inlet of a machine, such as a jet engine, may have to be transported to the outlet, undergoing changes of pressure, temperature and speed as it passes. In all such cases of **mass transport** a knowledge of the conditions existing at various points in the system is essential for efficient design.

VISCOSITY

If adjacent layers of a material are displaced laterally over each other as in Fig. 12.1*a*, the deformation of the material is called a **shear**. The simplest type of fluid flow basically involves shear.

All liquids and gases (except *very* low density gases) stick to a solid surface, so that when they flow the velocity must gradually decrease to zero as the wall of the pipe or containing vessel is approached. A fluid is therefore sheared when it flows past a solid surface and the **opposition set up by the fluid to shear** is called its **viscosity**. Liquids such as syrup and engine oil which pour slowly are more **viscous** than water.

Viscosity is a kind of internal friction exhibited to some degree by all fluids. It arises in liquids because the forced movement of a molecule relative to its neighbours is opposed by the intermolecular forces between them.

When the particles of fluid passing successively through a fluid follow the same path, the flow is said to be **steady**. 'Streamlines' can be drawn to show the direction of motion of the particles and are shown in Fig. 12.1*b* for steady flow of the water at various depths near the centre of a wide river. The layer of water in contact with the

bottom of the river must be at rest (or the river bed would be rapidly eroded) and the velocities of higher layers increase towards the surface. The length of the streamlines represents the magnitude of the velocities. The water suffers shear, a cube becoming a rhombus, Fig. 12.1*c*, as if acted on by tangential forces at its upper and lower faces. Steady flow thus involves parallel layers of fluid sliding over each other with different velocities. This creates viscous forces acting tangentially (as shear forces do) between the layers and impeding their motion.

COEFFICIENT OF VISCOSITY

To obtain a definition of viscosity we consider two plane parallel layers of liquid separated by a very small distance δy and having velocities $v + \delta v$ and v, Fig. 12.2. The **velocity gradient** (i.e. change of velocity/distance) in a direction perpendicular to the velocities is $\delta v/\delta y$. The slower, lower layer exerts a tangential retarding force F on the faster upper layer and experiences itself an equal and opposite tangential force F due to the upper layer (by Newton's third law). The **tangential stress** between the layers is therefore F/A where A is their area of contact. The **coefficient of viscosity** η is defined by the equation

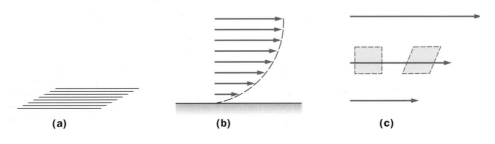

(a)   (b)   (c)

Fig. 12.1

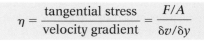

$$\eta = \frac{\text{tangential stress}}{\text{velocity gradient}} = \frac{F/A}{\delta v/\delta y}$$

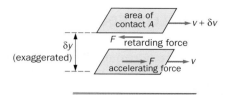

Fig. 12.2

If $\delta v/\delta y$ is small and F/A large, η is large and the fluid very viscous.

For many pure liquids (e.g. water) and gases η is independent of the velocity gradient at a particular temperature, i.e. η is constant and so the tangential stress is directly proportional to the velocity gradient. Fluids for which this is true are called **Newtonian fluids** since Newton first suggested this relationship might hold. For some liquids such as paints, glues and liquid cements, η decreases as the tangential stress increases and these are said to be **thixotropic**.

The equation defining η shows that it can be measured in newton second metre^{-2} (N s m^{-2}), i.e. in pascal seconds (Pa s). At 20 °C, η is 1.0×10^{-3} Pa s for water and 8.3×10^{-1} Pa s for glycerine. Experiment shows that the coefficient of viscosity of a liquid usually decreases rapidly with temperature rise.

Viscosity is an essential property of a lubricating oil if it is to keep apart two solid surfaces in relative motion, though too high a viscosity results in unnecessary resistance to motion. 'Viscostatic' oils have about the same value of η whether cold or hot.

Viscous forces act as soon as fluid flow starts. If the external forces causing the flow are constant, the rate of flow becomes constant and a steady state is attained with the resisting viscous forces equal to the applied force. The viscous forces stop the flow when the applied force is removed.

STEADY AND TURBULENT FLOW

The streamlines for steady flow in a circular pipe are shown in Fig. 12.3. Everywhere they are parallel to the axis of the pipe and represent velocities varying from zero at the wall of the pipe to a maximum at its axis. The surfaces of equal velocity are the surfaces of concentric cylinders.

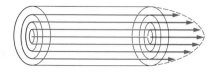

Fig. 12.3   Steady flow

The volume of liquid passing per second, V, through a pipe when the flow is steady depends on (*i*) the coefficient of viscosity η of the liquid, (*ii*) the radius r of the pipe and (*iii*) the pressure gradient p/l causing the flow, where p is the pressure difference between the ends of the pipe and l is its length, Fig. 12.4.

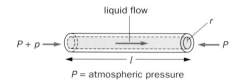

Fig. 12.4

The relationship is

$$V = \frac{\pi p r^4}{8 \eta l}$$

This is known as **Poiseuille's formula** since Poiseuille made the first thorough experimental investigation of the steady flow of liquid through a pipe, in 1844.

When the velocity of flow exceeds a particular critical value the motion becomes **turbulent**, the liquid is churned up and the streamlines are no longer parallel and straight. The change from steady to turbulent flow can be studied with the apparatus of Fig. 12.5. The flow of water along the tube T is controlled by the clip C.

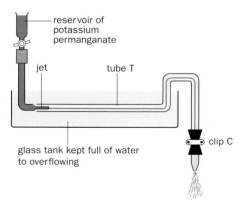

Fig. 12.5   Viewing turbulent flow

Potassium permanganate solution from the reservoir is fed into the water flowing through T by a fine jet. At low flow velocities a fine coloured stream is observed along the centre of T, but as the rate of flow increases it starts to break up and the colour rapidly spreads throughout T indicating the onset of turbulence.

Reynold's number (Re) is useful in the study of the stability of fluid flow. It is defined by the equation

$$Re = \frac{v l \rho}{\eta}$$

where η and ρ are the viscosity and density respectively of the fluid, v is the speed of the bulk of the fluid and l is a characteristic dimension of the solid body concerned. For a cylindrical pipe l is usually the diameter ($2r$) of the pipe. Experiment shows that for cylindrical pipes, when

$Re < 2200$	flow is steady
$Re \approx 2200$	flow is unstable (critical velocity v_c)
$Re > 2200$	flow is usually turbulent

Hence large η and small v, r and ρ promote steady flow. Poiseuille's formula for steady flow holds for velocities of flow less than v_c.

MOTION IN A FLUID

The streamlines for a fluid flowing *slowly* past a stationary solid sphere are shown in Fig. 12.6. When the sphere moves slowly, rather than the fluid, the pattern is similar but the streamlines then show the apparent motion of the fluid particles as seen by someone on the moving sphere. The layer of fluid in contact with the moving sphere moves with it, creating a velocity gradient between this layer and other layers of the fluid. Viscous forces are thereby brought into play and constitute the resistance experienced by the moving sphere.

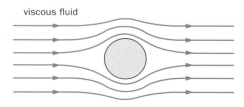

Fig. 12.6

The viscous retarding force F depends on (*i*) the viscosity η of the fluid and (*ii*) the velocity v and radius r of the sphere. An analysis, first done by Stokes, gives

$$F = 6\pi\eta v r$$

This expression, called **Stokes' law**, only holds for **steady motion** in a fluid of **infinite extent** (otherwise the walls and bottom of the vessel affect the resisting force).

Now consider the sphere falling vertically under gravity in a viscous fluid. Three forces act on it (Fig. 12.7):

- its weight, W, acting downwards;
- the upthrust, U, due to the weight of fluid displaced, acting upwards; and
- the viscous drag, F, acting upwards.

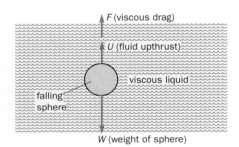

Fig. 12.7

The resultant downward force equals $(W - U - F)$ and causes the sphere to accelerate until its velocity and the viscous drag reach values such that

$$W - U - F = 0$$

The sphere then continues to fall with a constant velocity, known as its **terminal velocity**, of say v_t. Now

$$W = \tfrac{4}{3}\pi r^3 \sigma g$$

where σ is the density of the sphere,

and

$$U = \tfrac{4}{3}\pi r^3 \rho g$$

where ρ is the density of the fluid. Also, *if steady conditions still hold* when velocity v_t is reached then, by Stokes' law,

$$F = 6\pi\eta r v_t$$

Hence

$$\tfrac{4}{3}\pi r^3 \sigma g - \tfrac{4}{3}\pi r^3 \rho g - 6\pi\eta r v_t = 0$$

$$\therefore \quad v_t = \frac{2r^2(\sigma - \rho)g}{9\eta}$$

If the velocity of the sphere reaches a critical velocity v_c, however, the flow breaks up, eddies are formed as in Fig. 12.8*a* and the motion becomes turbulent. At velocities greater than v_c the resistance to motion, called the **drag**, increases sharply and is roughly proportional to the square of the velocity. The relation is given by

$$\text{drag force} = \frac{C\rho A v^2}{2}$$

where A is the cross-section area of the body perpendicular to its velocity v, ρ is the fluid density, and the drag coefficient C has a numerical value between 0 and 1.

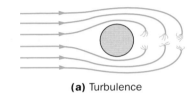

(a) Turbulence

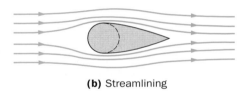

(b) Streamlining

Fig. 12.8 Turbulence can be eliminated by streamlining a body

By comparison, below v_c Stokes' law indicates that the resistance is proportional to the velocity.

Note that for turbulent flow the resistance is dependent on density, not on viscosity; this is the ordinary case of air resistance to vehicles.

By modifying the shape of a body the critical velocity can be raised. The drag is then reduced at a particular speed if steady flow replaces turbulent flow. This is called **streamlining** the body and Fig. 12.8*b* shows how it is done for a sphere. The pointed tail can be regarded as filling the region where eddies occur in turbulent motion, thus ensuring that the streamlines merge again behind the sphere. Streamlining is particularly important in the design of high-speed aircraft.

BERNOULLI'S EQUATION

The pressure is the same at all points on the same horizontal level in a fluid at rest; this is not so when the fluid is in motion. The pressure at different points in a liquid flowing through (*a*) a uniform tube and (*b*) a tube with a narrow part, is shown by the height of liquid in the vertical manometers in Figs 12.9*a* and *b*. In (*a*) the pressure drop along the tube is steady and maintains the flow against the viscosity of the liquid. In (*b*) the pressure falls in the narrow part B but rises again in the wider part C. If the liquid can be assumed to be incompressible, the same volume of liquid passes through B in a given time as enters A and so the velocity of the liquid must be greater in B than in A or C. Therefore a decrease of pressure accompanies an increase of velocity. This may be shown by blowing into a 'tunnel' made from a sheet of paper, Fig. 12.10. The faster you blow, the more the tunnel collapses.

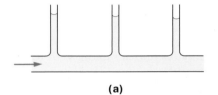

(a)

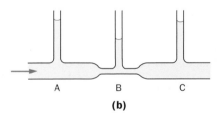

(b)

Fig. 12.9

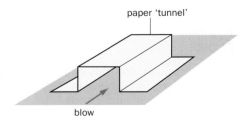

paper 'tunnel'

blow

Fig. 12.10

A useful relation can be obtained between the pressure and the velocity at different points of a fluid in motion.

Suppose a fluid flows through a non-uniform tube from X to Y, Fig. 12.11, and its velocity changes from v_1 at X where the cross-section area is A_1 to v_2 at Y where the cross-section is A_2. The flow of fluid between X and Y is caused by the forces acting on its ends which arise from the pressure exerted on it by the fluid on either side of it. At X, if the fluid pressure is p_1, there is a force p_1A_1 acting in the direction of flow, and at Y, if the fluid pressure is p_2, a force p_2A_2 opposes the flow.

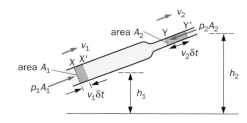

Fig. 12.11

Consider a small time interval δt in which the fluid at X has moved to X' and that at Y to Y'.

At X, work done during δt *on* the fluid XY by p_1A_1 pushing it into the tube

$$= \text{force} \times \text{distance moved}$$
$$= \text{force} \times \text{velocity} \times \text{time}$$
$$= p_1A_1 \times v_1 \times \delta t$$

At Y, work done during δt *by* the fluid XY emerging from the tube against p_2A_2

$$= p_2A_2 \times v_2 \times \delta t$$

Therefore

net work W done *on* the fluid $= (p_1A_1v_1 - p_2A_2v_2)\delta t$

Since the rate at which mass flows into the tube must equal the rate at which mass flows out, we have the **continuity equation**

$$\rho_1A_1v_1 = \rho_2A_2v_2 \quad \text{(mass = density} \times \text{volume)}$$

where ρ_1 and ρ_2 are the densities of the fluid at X and Y. If the fluid is incompressible then the density is the same at X and Y, so $\rho_1 = \rho_2 = \rho$ and

$$A_1v_1 = A_2v_2$$

Therefore we can write

$$W = (p_1 - p_2)A_1v_1\delta t$$

As a result of work done on it, the fluid gains p.e. and k.e. when XY moves to X'Y'.

gain of p.e. = p.e. of YY' − p.e. of XX'
$$= (A_2v_2\delta t\,\rho)gh_2 - (A_1v_1\delta t\,\rho)gh_1$$
$$\text{(since p.e.} = mgh)$$
$$= A_1v_1\delta t\,\rho g(h_2 - h_1) \quad \text{(since } A_1v_1\delta t = A_2v_2\delta t)$$

where h_1 and h_2 are the heights of XX' and YY' above an arbitrary horizontal reference level and ρ is the density of the fluid. Similarly,

gain of k.e. = k.e. of YY' − k.e. of XX'
$$= \tfrac{1}{2}(A_2v_2\delta t\,\rho)v_2^2 - \tfrac{1}{2}(A_1v_1\delta t\,\rho)v_1^2$$
$$\text{(since k.e.} = \tfrac{1}{2}mv^2)$$
$$= \tfrac{1}{2}A_1v_1\delta t\,\rho(v_2^2 - v_1^2)$$

If the fluid is non-viscous (i.e. inviscid) no work is done against viscous forces to maintain the flow, no change of internal energy of the fluid occurs and by the principle of conservation of energy we have

net work done *on* fluid = gain of p.e. + gain of k.e.

$$\therefore \; (p_1 - p_2)A_1v_1\delta t = A_1v_1\delta t\,\rho g(h_2 - h_1) + \tfrac{1}{2}A_1v_1\delta t\,\rho(v_2^2 - v_1^2)$$
$$p_1 - p_2 = \rho g(h_2 - h_1) + \tfrac{1}{2}\rho(v_2^2 - v_1^2)$$

or

$$p_1 + h_1\rho g + \tfrac{1}{2}\rho v_1^2 = p_2 + h_2\rho g + \tfrac{1}{2}\rho v_2^2$$

This is **Bernoulli's equation** and it is usually stated by saying that, **along a streamline in an incompressible inviscid fluid**

$$p + h\rho g + \tfrac{1}{2}\rho v^2 = \text{constant}$$

In deriving the equation we have in effect assumed that the pressure and velocity are uniform over any cross-section of the tube. This is not so for a real (viscous) fluid and so it only applies strictly to a single streamline in the fluid. In addition, actual fluids, especially gases, are compressible. The equation has therefore to be applied with care or the results will be misleading.

APPLICATIONS OF BERNOULLI

(a) Jets and nozzles

Bernoulli's equation suggests that for fluid flow where the potential energy change $h\rho g$ is very small or zero, as in a horizontal pipe, **the pressure falls when the velocity rises**. The velocity increases at a constriction — a slow stream of water from a tap can be converted into a fast jet by narrowing the exit with a finger — and the greater the change in cross-section area, the greater is the increase of velocity and so the greater is the pressure drop. Several devices with jets and nozzles use this effect, Fig. 12.12a–c.

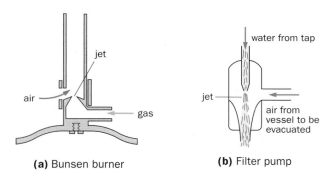

(a) Bunsen burner   **(b)** Filter pump

Fig. 12.12

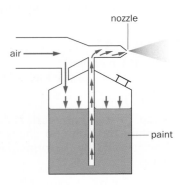

Fig. 12.12c   Paint spray

(b) Spinning ball

If a tennis ball is 'cut' or a golf ball 'sliced', it spins as it travels through the air and experiences a sideways force which causes it to curve in flight. This is due to air being dragged round by the spinning ball, thereby increasing the air flow on one side and decreasing it on the other. A pressure difference is thus created, Fig. 12.13. The 'swing' of a spinning cricket ball is complicated by its raised seam.

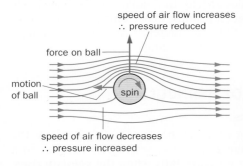

Fig. 12.13   Motion of a spinning ball

(c) Aerofoil

This is a device which is shaped so that the relative motion between it and a fluid produces a force perpendicular to the flow. Examples of aerofoils are aircraft wings, turbine blades and propellers.

The shape of the aerofoil section in Fig. 12.14*a* is such that fluid flows faster over the top surface than over the bottom, i.e. the streamlines are closer above than below the aerofoil. By Bernoulli, it follows that the pressure underneath is increased and that above reduced. A resultant upward force is thus created, normal to the flow, and it is this force which pro-

vides most of the 'lift' for an aeroplane, Fig. 12.14*b*. Its value increases with the angle between the wing and the airflow, called the 'angle of attack'. But at a certain critical angle the flow separates from the upper surface, the flow downstream becomes very turbulent, drag increases sharply, lift is lost almost completely and the aeroplane stalls.

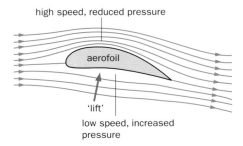

Fig. 12.14a   Principle of an aerofoil

Fig. 12.14b   Aeroplane wings act as aerofoils

The sail of a yacht 'tacking' into the wind is another example of an aerofoil, Fig. 12.15*a*. The air flow over the sail produces a pressure increase on the windward side and a decrease on the leeward side. The resultant force is roughly normal to the sail and can be resolved into a component *F* producing forward motion and a greater component *S* acting sideways, as in Fig. 12.15*b*. The keel produces a lateral force to balance *S*.

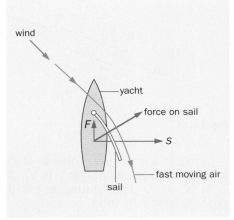

Fig. 12.15b   Force on the sail, which acts as an aerofoil

Fig. 12.15a   A yacht 'tacking' into the wind

FLOWMETERS

Flowmeters measure the rate of flow of a fluid through a pipe. Two types will be considered.

(a) Venturi meter

This consists of a horizontal tube with a constriction and replaces part of the piping of a system, Fig. 12.16. The two vertical tubes record the pressures (above atmospheric) in the fluid flowing in the normal part of the tube and in the constriction.

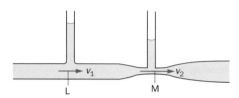

Fig. 12.16 Principle of the Venturi meter

If p_1 and p_2 are the pressures and v_1 and v_2 the velocities of the fluid (density ρ) at L and M on the same horizontal level, then assuming Bernoulli's equation holds

$$p_1 + \tfrac{1}{2}\rho v_1^2 = p_2 + \tfrac{1}{2}\rho v_2^2$$
$$\text{(since } h_1 = h_2)$$
$$\therefore \quad p_1 - p_2 = \tfrac{1}{2}\rho(v_2^2 - v_1^2)$$

If A_1 and A_2 are the cross-section areas at L and M and if the fluid is incompressible, the same volume passes each section of the tube per second.

$$\therefore \quad A_1 v_1 = A_2 v_2$$

Hence

$$p_1 - p_2 = \tfrac{1}{2}\rho v_1^2 \left(\frac{A_1^2}{A_2^2} - 1 \right)$$

Knowing A_1, A_2, ρ and $(p_1 - p_2)$, v_1 can be found and so also the rate of flow $A_1 v_1$.

(b) Pitot tube

The pressure exerted by a moving fluid, called the **total pressure**, can be regarded as having two components: the **static component**, which the fluid would have if it were at rest, and the **dynamic component**, which is the pressure equivalent of its velocity. A Pitot tube measures total pressure and in essence is a manometer with one limb parallel to the flow and open to the oncoming fluid, A in Fig. 12.17. The fluid at the open end is at rest and a 'stagnant' region exists there. The total pressure is also called the **stagnation pressure**. The static component is measured by a manometer connected at right angles to the pipe or surface over which the fluid is passing, B.

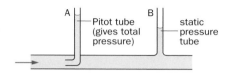

Fig. 12.17 A Pitot tube

In Bernoulli's equation

$$p + h\rho g + \tfrac{1}{2}\rho v^2 = \text{constant}$$

the static component is given by $p + h\rho g$, or p if the flow is horizontal, the dynamic component by $\tfrac{1}{2}\rho v^2$ and the total pressure by $p + \tfrac{1}{2}\rho v^2$. Hence

total pressure − static component =

$$p + \tfrac{1}{2}\rho v^2 - p = \tfrac{1}{2}\rho v^2$$

$$\therefore \quad v = \sqrt{\left\{ \frac{2}{\rho} \begin{pmatrix} \text{total pressure} \\ - \text{static component} \end{pmatrix} \right\}}$$

The expression derived above enables a value for the velocity of flow v of an incompressible inviscid fluid to be calculated from the readings of Pitot tubes. In real cases v varies across the diameter of the pipe carrying the fluid (because of its viscosity) but it can be shown that if the open end of the Pitot tube is offset from the axis of the pipe by 0.7 × radius of the pipe, then v is the *average* flow velocity.

FLUID FLOW CALCULATIONS

Example 1. A garden sprinkler has 150 small holes, each 2.0 mm² in area, Fig. 12.18. If water is supplied at the rate of 3.0×10^{-3} m³ s⁻¹, what is the average velocity of the spray?

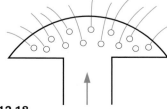

Fig. 12.18

Volume of water per second from sprinkler

= volume supplied per second
= 3×10^{-3} m³ s⁻¹
= total area of sprinkler holes
  × average velocity of spray
= 300×10^{-6} m²
  × average velocity of spray

Therefore average velocity of spray

$$= \frac{3 \times 10^{-3}}{300 \times 10^{-6}} \frac{\text{m}^3 \text{ s}^{-1}}{\text{m}^2}$$

$$= 10 \text{ m s}^{-1}$$

Example 2. Obtain an estimate for the velocity of emergence of a liquid from a hole in the side of a wide vessel 10 cm below the liquid surface.

Consider the general case in which the hole is at depth h below the surface of the liquid of density ρ, Fig. 12.19. If the liquid is incompressible and inviscid and the motion steady we can apply Bernoulli's equation to points A and B on the streamline AB.

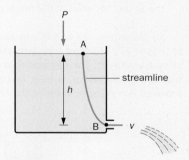

Fig. 12.19

At A

p_1 = atmospheric pressure = P

$$h_1 = h$$

$$v_1 = 0$$

(assuming the rate of fall of the surface can be neglected compared with the speed of emergence since the vessel is wide).

At B

p_2 = pressure of air into which the liquid emerges = P

$$h_2 = 0$$

$$v_2 = v$$

Substituting in Bernoulli's equation,

$$P + h\rho g + 0 = P + 0 + \tfrac{1}{2}\rho v^2$$

$$\therefore \quad h\rho g = \tfrac{1}{2}\rho v^2$$

from which we see that the potential energy lost by unit volume of liquid (mass ρ) in falling from the surface to depth h is changed to kinetic energy. The velocity of emergence is given by

$$v^2 = 2gh$$

and is the same as the vertical velocity that would be acquired in free fall — a statement known as **Torricelli's theorem**. In fact, v is always less than $\sqrt{(2gh)}$ owing to the viscosity of the liquid.

If $h = 10$ cm $= 0.1$ m and taking $g = 9.8$ m s^{-2} then

$$v = \sqrt{(2 \times 9.8 \text{ m s}^{-2} \times 0.1 \text{ m})}$$

$$= 1.4 \text{ m s}^{-1}$$

QUESTIONS

1. The volume flow rate of a fluid through a pipe of cross-sectional area A and length l is given by

$$\frac{V}{t} = \frac{cA^2 \Delta p}{\eta l}$$

where c is a constant having no unit.

a) i) What quantity does the term η represent?
ii) The SI unit of pressure in base units is kg m^{-1} s^{-2}. State the SI unit for each of the following quantities:

V $\quad t \quad$ A $\quad l$

iii) Determine the unit of η in terms of SI base units.

b) Figure 12.20 shows a system used to provide a patient with a blood transfusion. The required fluid pressure to enable blood to flow through the hollow needle, and into the patient, can be achieved by raising or lowering the container.

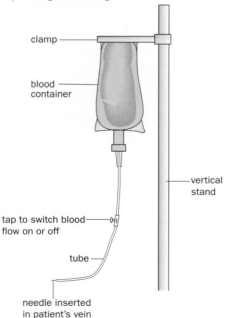

clamp

blood container

vertical stand

tap to switch blood flow on or off

tube

needle inserted in patient's vein

Fig. 12.20

In one case a needle, 0.035 m long and of internal cross-sectional area 3.0×10^{-7} m^2, is used. The required volume flow rate into the patient is 2.5×10^{-7} m^3 s^{-1}.

The magnitude of η in SI units $= 4.0 \times 10^{-3}$

The value of $c = 0.040$

i) Use the equation given to determine the pressure difference between the ends of the needle that will produce the required blood flow rate.
ii) Use your value from i) to determine the vertical distance between the top surface of the blood in the container and the end of the needle that will produce the required blood flow rate when the needle is inserted horizontally into a vein.

The density of blood $= 1050$ kg m^{-3}

The acceleration of free fall, $g = 9.8$ m s^{-2}

Average blood pressure of a patient above atmospheric pressure $= 140$ Pa

(AEB, 0635/2, Summer 1998)

2. Define **coefficient of viscosity** η.

Stokes' law for the viscous force F acting on a sphere of radius a falling with velocity v through a large expanse of fluid of coefficient of viscosity η is expressed by the equation

$$F = 6\pi a\eta v$$

State why this equation is true only for sufficiently low velocities.

Explain why a sphere released in a fluid will fall with diminishing acceleration until it attains a constant terminal velocity.

Calculate this velocity for an oil drop of radius 3.0×10^{-6} m falling through air of coefficient of viscosity 1.8×10^{-5} Pa s, given that the density of the oil is 8.0×10^2 kg m^{-3} and that the density of air may be neglected.

3. This question is about the drag force exerted on a car.

The drag force on a car is caused by the air flow around the car and is related to the speed, v, of the car by the following equation:

$$F_D = KA \rho v^2$$

where F_D is the drag force, A is the front cross-sectional area of the car, ρ is the density of air and K is a constant. The equation assumes that there is no wind blowing.

The graph in Fig. 12.21 shows how the drag force varies with the square of the speed for a certain car whose cross sectional area A is 2.0 m^2. The density of air $= 1.2$ kg m^{-3}.

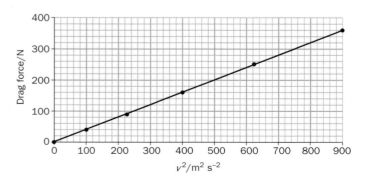

Fig. 12.21

a) From the graph determine
  i) the drag force on the car when its speed is 20 m s^{-1};
  ii) the value of K.

b) Calculate the engine power that is required to maintain a steady speed of 20 m s^{-1}.

c) By what factor would the engine power increase if the car were travelling at 20 m s^{-1} into a head wind of speed 20 m s^{-1}?

(*IB, Subsidiary/Standard Paper 2, Nov 1998*)

4. a) The Bernoulli equation can be written in the form

$$p_1 + \tfrac{1}{2}\rho v_1^2 = p_2 + \tfrac{1}{2}\rho v_2^2$$

  i) Explain what the symbols in this equation mean.
  ii) State two conditions which must apply for this equation to be true.

b) Explain how the rotating blades of a helicopter cause a lift force to be exerted.

(*UCLES, Physics of Transport, June 1998*)

5. a) State what is meant by
  i) equilibrium,
  ii) upthrust,
  iii) principle of flotation,
  iv) centre of buoyancy,
  v) metacentre of a floating body.

b) An ideal, incompressible fluid of density ρ flows through a horizontal tube which narrows between X, where the area of cross-section is A_X, and Y, where the area of cross-section is A_Y, as illustrated in Fig. 12.22.

The liquid has speed v_X at X. Deduce an expression for its speed v_Y at Y.

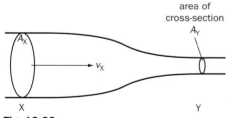

Fig. 12.22

c) Consider a mass m of liquid moving from X to Y in b).
  i) Deduce an expression for the increase in the kinetic energy of this mass.
  ii) Apply the principle of conservation of energy to this flow in order to show that the pressure in the fluid decreases as it passes from X to Y.
  iii) Hence show that the drop in pressure is given by

$$\Delta p = \tfrac{1}{2}\rho(v_Y^2 - v_X^2)$$

(*OCR, 9244/3, Nov 1999*)

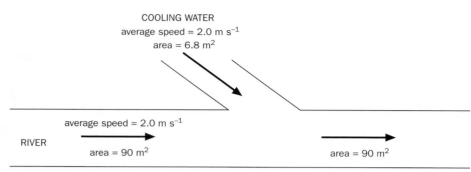

Fig. 12.23

6. Figure 12.23 shows the cooling water from a power station joining a river. The areas given on the diagram are the cross-sectional areas of the respective flow regions.

a) i) Calculate the volume flow rate of the water in the river before the cooling water joins it.
  ii) Calculate the river velocity after the cooling water has joined it.

b) One end of a tube is lowered into the water. The other end of the tube is connected to a pressure gauge as shown in Fig. 12.24.

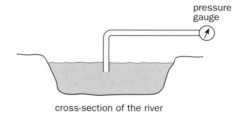

cross-section of the river

Fig. 12.24

State what will happen to the reading on the pressure gauge when the velocity of the water in the river increases. Give a reason for your answer.

c) Suggest a suitable type of thermometer for use when monitoring the temperature of the cooling water, giving a reason for your choice.

(*AEB, 0635/2, Summer 1997*)

7. When oil is pumped round a car engine it is essential that it flows at the correct rate. It must coat the moving parts and still move quickly through narrow passages. To ensure that this happens the viscosity of the oil must be measured.

One way of measuring the coefficient of viscosity is to use a falling ball viscometer. In this instrument a sphere is dropped into the liquid and its subsequent motion is recorded. This motion can be analysed in terms of the *viscous drag*, the upthrust and the weight of the sphere. Stokes' law can be used if the flow is *laminar*. When *terminal velocity* has been reached measurements can be made to determine the coefficient of viscosity.

Explain the meaning of the terms in italics.

Draw a labelled diagram showing the forces acting on the sphere when it is falling at constant velocity.

The sphere has a radius r and it is made from a material of density ρ. Write down an expression for the weight W of the sphere.

The sphere experiences an upthrust force $U = \tfrac{4}{3}\pi r^3 \sigma g$ where σ is the density of the oil. Explain briefly the cause of the upthrust force on the sphere.

Explain why, when the sphere is falling with terminal velocity, $F = W - U$ where F is the Stokes' law force.

Hence show that the viscosity η can be calculated from the expression

$$\eta = \frac{2r^2 g(\rho - \sigma)}{9v}$$

where v is the velocity of the sphere.

Explain why it is important that the experiment is performed using hot oil.

(*L, 6552/P.01 PSA2, June 1999*)

Objective-type revision questions 2

The first figure of a question number gives the relevant chapter, e.g. **7.3** is the third question for chapter 7.

MULTIPLE CHOICE

Select the response that you think is correct.

7.1 A pendulum bob suspended by a string from the point P, Fig. R11, is in equilibrium under the action of three forces: W, the weight of the bob; T, the tension in the string; and F, a horizontally applied force. Which one of the following statements is *untrue*?
A $F^2 + W^2 = T^2$
B F and W are the components of T
C $W = T \cos \theta$
D $F = W \tan \theta$

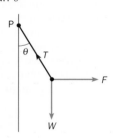

Fig. R11

7.2 Forces of 3 N, 4 N and 12 N act at a point in mutually perpendicular directions. The magnitude of the resultant force in newtons is
A 5 **B** 11 **C** 13 **D** 19
E indeterminate from information given

7.3 Which graph in Fig. R12 best represents the variation of velocity with time of a ball which bounces vertically on a hard surface, from the moment when it rebounds from the surface?

7.4 A ball is projected horizontally at 15 m s^{-1} from a point 20 m above a horizontal surface ($g = 10$ m s^{-2}). The magnitude of its velocity in m s^{-1} when it hits the surface is
A 10 **B** 15 **C** 20 **D** 25 **E** 35

7.5 A trolley of mass 60 kg moves on a frictionless horizontal surface and has kinetic energy 120 J. A mass of 40 kg is lowered vertically on to the trolley. The total kinetic energy of the system is now
A 60 J **B** 72 J **C** 120 J **D** 144 J
E another answer

7.6 The acceleration of free fall on the moon is 1.6 m s^{-2}. How long does a ball thrown vertically upwards from the moon's surface take to reach its maximum height of 80 m?
A 1.6 s **B** 10 s **C** 16 s **D** 20 s

7.7 A car moving with a velocity of 4 m s^{-1} on an icy road collides head-on with a stationary minibus whose mass is 4 times that of the car. They move off locked together. Assuming that momentum is conserved in the collision, what fraction of the initial kinetic energy is the final kinetic energy of the vehicles?
A $\frac{1}{10}$ **B** $\frac{1}{5}$ **C** $\frac{1}{4}$ **D** $\frac{1}{2}$

7.8 A uniform beam of weight mg and length l is attached to a pivot at one end. If an upward force of $2mg$ is applied to the free end of the beam, what is the torque on the beam?
A mgl **B** $1.5mgl$ **C** $2.0mgl$ **D** $2.5mgl$

8.1 A mass of 2.0 kg describes a circle of radius 1.0 m on a smooth horizontal table at a uniform speed. It is joined to the centre of the circle by a string which can just withstand 32 N. The greatest number of revolutions per minute the mass can make is
A 38 **B** 4 **C** 76 **D** 240 **E** 16

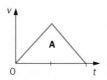 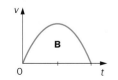 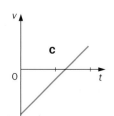 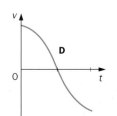 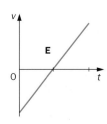

Fig. R12

8.2 In order to turn in a horizontal circle an aircraft banks so that
A there is a resultant force on the wings from the centre of the circle
B the weight of the aircraft has a component towards the centre of the circle
C the drag on the plane is reduced
D the lifting force on the wings has a component towards the centre of the circle.

8.3 If a small body of mass m is moving with angular velocity ω in a circle of radius r, what is its kinetic energy?
A $m\omega r/2$ **B** $m\omega^2 r/2$ **C** $m\omega r^2/2$ **D** $m\omega^2 r^2/2$

8.4 Planet X is twice the radius of planet Y and is of material of the same density. The ratio of the acceleration of free fall at the surface of X to that at the surface of Y is
A 1:4 **B** 1:2 **C** 2:1 **D** 4:1 **E** 8:1

8.5 A ball of mass m is suspended from a light string of length R. If the ball is released from a horizontal position P, as shown in Fig. R13, what is the tension in the string when the ball reaches Q?
A 0 **B** mg **C** $2mg$ **D** $3mg$

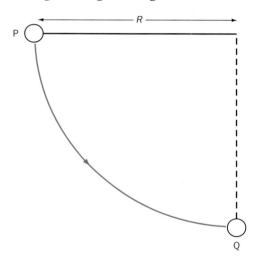

Fig. R13

8.6 A satellite S moves around a planet P in an elliptical orbit, Fig. R14. The ratio of the speed of the satellite at point a to that at point b is
A 1:9 **B** 1:3 **C** 1:1 **D** 3:1 **E** 9:1

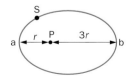

Fig. R14

8.7 A satellite moving round the Earth in a circular orbit of radius R has a period T. What would the period be if the orbit were of radius $R/4$?
A $T/8$ **B** $T/4$ **C** $T/2$ **D** $2T$ **E** $4T$

9.1 The frequency of oscillation of a mass m suspended at the end of a vertical spring having a spring constant k is directly proportional to
A mk **B** m/k **C** m^2k **D** $1/(mk)^{1/2}$ **E** $(k/m)^{1/2}$

9.2 The graph of Fig. R15 shows how the displacement of a particle describing s.h.m. varies with time. Which one of the following statements is, from the graph, false?
A The restoring force is zero at time $T/4$.
B The velocity is a maximum at time $T/2$.
C The acceleration is a maximum at time T.
D The displacement is a maximum at time T.
E The kinetic energy is zero at time $T/2$.

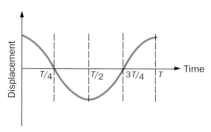

Fig. R15

9.3 A block of mass m is attached to two identical springs S_1 and S_2 as shown in Fig. R16. The force constant of the springs is k. If the block is made to execute simple harmonic motion, the period will be

A $2\pi\sqrt{\dfrac{m}{4k}}$ **B** $2\pi\sqrt{\dfrac{m}{2k}}$ **C** $2\pi\sqrt{\dfrac{m}{k}}$

D $2\pi\sqrt{\dfrac{2m}{k}}$ **E** $2\pi\sqrt{\dfrac{4m}{k}}$

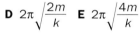

Fig. R16

9.4 An object moves vertically with simple harmonic motion just behind a wall. From the other side of the wall the object is visible in each cycle for 2.0 s and hidden behind the wall for 6.0 s. The maximum height reached by the object relative to the top of the wall is 0.30 m. The amplitude of the motion is
A 0.18 m **B** 0.51 m **C** 0.60 m **D** 1.02 m
E 1.20 m

11.1 When a capillary tube of uniform bore is dipped in water the water level in the tube rises 10 cm higher than in the vessel. If the tube is lowered until its open end is 5.0 cm above the level in the vessel, does the water in the tube appear as in (Fig. R17) **A**, **B**, **C**, **D** or **E**?

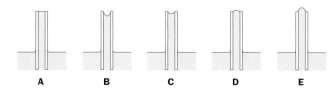

Fig. R17

12.1 Spheres X and Y of the same material fall at their terminal velocities through a liquid without causing turbulence. If Y has twice the radius of X, the ratio of the terminal velocity of Y to that of X is

A $1:4$ **B** $1:2$ **C** $1:1$ **D** $2:1$ **E** $4:1$

12.2 A piston of radius r and length l is pushed through a cylinder covered with a thin layer of oil of viscosity η. The effective thickness of the oil layer is d. When the speed of the piston is u, the resistive force it experiences is

A $2\pi rl\eta u/d$ **B** $2\pi rl\eta ud$ **C** $\pi r^2 l\eta u/d$
D $\pi r^2 l\eta ud$ **E** $2\pi rl\eta u$

MULTIPLE SELECTION

In each question one or more of the responses may be correct. Choose one letter from the answer code given.

Answer **A** *if i), ii) and iii) are correct.*
Answer **B** *if only i) and ii) are correct.*
Answer **C** *if only ii) and iii) are correct.*
Answer **D** *if i) only is correct.*
Answer **E** *if iii) only is correct.*

7.9 A ball rolls down an inclined plane, Fig. R18. The ball is first released from rest from P and then later from Q. Which of the following statements is/are correct?

i) The ball takes twice as much time to roll from Q to O as it does to roll from P to O.
ii) The acceleration of the ball at Q is twice as large as the acceleration at P.
iii) The ball has twice as much k.e. at O when rolling from Q as it does when rolling from P.

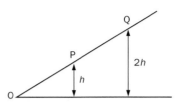

Fig. R18

7.10 In the equation $Ft = mv - mu$
i) the dimensions of F are MLT^2
ii) the dimensions of mv are MLT^{-1}
iii) the dimensions of all three terms are the same.

9.5 The period of a simple pendulum oscillating in a vacuum depends on
i) the mass of the pendulum bob
ii) the length of the pendulum
iii) the acceleration of free fall.

Fields

Electrical storm

13

Electric fields

- Simple electrostatics
- Electrostatics today
- Coulomb's law
- Electric field strength
- Field lines
- Electric potential

- Equipotentials
- Potential due to a point charge
- Potential due to a conducting sphere
- Potential difference
- Relation between *E* and *V*

- Gravitational analogy
- Field treatment of space travel
- Potential and field strength calculations

SIMPLE ELECTROSTATICS

(a) Electric charges

In general, when any two different materials are rubbed together they exert forces on each other and each is said to have acquired an 'electric charge'. **Electrostatics** is the study of electric charges at rest. Experiments show that there are two kinds of charge and that **like charges repel, unlike charges attract**. The two kinds cancel one another out and in this respect are opposite. One type is taken to be positive and the other negative.

The allocation of signs to charges was made quite arbitrarily many years ago and, with regard to the materials used today, the choice makes **polythene** rubbed with wool **negatively** charged and **cellulose acetate** (and also Perspex) rubbed with wool **positively** charged. The forces between charged strips can be investigated as in Fig. 13.1.

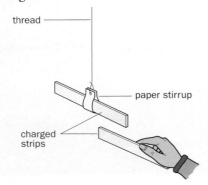

Fig. 13.1 Investigating electrostatic forces

The sign convention adopted led to the electron having a negative charge and the proton a positive one. A single atom normally contains equal numbers of electrons and protons, making it electrically neutral. Electrification by rubbing may be explained by supposing that electrons are transferred from one material to the other. For example, when cellulose acetate is rubbed with wool, electrons go from the surface of the acetate to the wool, thus leaving the acetate deficient of electrons, i.e. positively charged, and making the wool negatively charged. Equal amounts of opposite charges should therefore be produced (p. 224).

(b) Insulators and conductors

According to the electron theory, all the electrons in the atoms of electrical insulators (such as polythene, cellulose acetate, Perspex and glass) are firmly bound to their nuclei and the removal or addition of electrons at one place does not cause the flow of electrons elsewhere. This means that the charge is confined to the region where it was produced (e.g. by rubbing) or placed. Electrical conductors (e.g. metals) have electrons that are quite free from individual atoms (although fairly strongly bound within the material as a whole) and if such materials gain electrons, these can move about in them. Loss of electrons by a conductor causes a redistribution of those left. A charge on a conductor therefore spreads over the entire surface.

The human body and the earth are comparatively good conductors, so if we try to charge a metal rod, it must be well insulated and not held in the hand. Otherwise any charge produced is conducted away through the body to earth. Water also conducts and its presence on the surface of many materials (e.g. glass) that are otherwise insulators accounts for the charge leakage which often occurs. Many plastics (e.g. polythene, Perspex, cellulose acetate) are water-repellent.

(c) Electrostatic induction

A negatively charged polythene strip held close to an insulated, uncharged conductor, such as a small aluminized expanded-polystyrene ball, attracts it. This may be explained by saying that electrons are repelled to the far side of the ball leaving the near side positively charged, Fig. 13.2*a*. The attraction between the negative charge on the polythene strip and the induced positive charge is greater than the repulsion between the strip and the more distant negative charge. The effect is called **electrostatic induction**. It accounts for the attraction of scraps of paper by a plastic comb, charged by being drawn through the hair. It is also this effect that enables a party balloon, which has been rubbed on a woollen jumper, to stick to the ceiling.

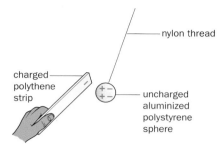

(a) Electrostatic induction

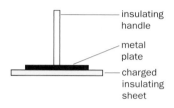

(b) Electrophorus

Fig. 13.2

(d) Electrophorus

This is a device for producing charges by electrostatic induction, Fig. 13.2b. It consists of a circular metal plate with an insulating handle, placed on an insulating sheet (e.g. of polythene) previously charged by rubbing. When the plate is earthed by touching it with the finger and then removed, it has a large charge of opposite sign to that on the insulating sheet. This charge can be transferred to another conductor and the electrophorus plate recharged as before, again and again.

It may seem strange that the metal plate does not become charged by contact with the insulating sheet and have the same sign of charge as it. However, it appears that contact between even plane surfaces occurs at only a few points so that, in fact, the metal plate is charged by induction. This would account for it becoming oppositely charged.

(e) Static and current electricity

Static charges produced by rubbing insulators (or insulated conductors) give the same effects when they move as do electric currents due to a battery. However, in electrostatics we are usually dealing with quite a small charge (a few microcoulombs) but a large p.d. (thousands of volts); in current electricity the opposite is usually true.

ELECTROSTATICS TODAY

Electrostatics was the first branch of electricity to be investigated and for a long time was regarded as a subject of no practical value. In recent years that has changed and it now has important industrial applications.

The electrostatic precipitation of flue-ash that would otherwise be discharged into the atmosphere from modern coal-fired power stations is a vital factor in the reduction of pollution. An average power station produces about 30 000 kg (30 tonnes) of flue-ash per hour. Power station precipitators are shown in Fig. 13.3 between the chimney and the main building, which contains the coal bunkers, boiler house and turbine hall. The precipitators are built to remove 99% of the ash from the flue-gases before they reach the power station chimney. A precipitator is made up of a number of wires and plates. Its wires are negatively charged and give a similar charge to the particles of ash which are then attracted to the positive plates. These are mechanically shaken to remove the ash which is collected and used as a by-product.

Electrostatic precipitation is also important in the steel, cement and chemical industries where flue-gas outputs are high.

Electrostatic spraying of plastics, paints and powders is employed and lends itself to automation.

In nuclear physics, electrostatic generators of the van de Graaff type (p. 226) are employed to produce p.ds of up to 14 million volts for accelerating sub-atomic particles, which can then be used for the production of radioisotopes used in medicine and industry, or for research in particle physics.

A knowledge of electrostatics is important in electrical prospecting for minerals and in surveying sites for large structures. Electrostatic loudspeakers and microphones are in widespread use, as are electrostatic copying machines.

Electric charges can build up because of friction on aircraft in flight and on plastic sheeting in industry, creating a potential explosion hazard unless preventive steps are taken. In the case of aircraft the rubber tyres are made slightly conducting so that the charge leaks away harmlessly at touch-down. The crackling that occurs when a

Fig. 13.3   Flue-ash precipitators are used at this power station

nylon garment is removed from the body or sometimes when stepping from a car is also due to static charges causing the insulation of the surrounding air to break down. A flash of lightning is nature's most spectacular electrostatic event.

COULOMB'S LAW

(a) Statement

A knowledge of the forces that exist between charged particles is necessary for an understanding of the structure of the atom and of matter. The magnitude of the forces between charged spheres was first investigated quantitatively in 1785 by Coulomb, a French scientist. The law he discovered may be stated as follows:

> The force between two point charges is directly proportional to the product of the charges divided by the square of their distance apart.

The law applies to point charges. Sub-atomic particles such as electrons and protons may be regarded as approximating to point charges. Later we shall see that a *uniformly* charged conducting sphere behaves — so far as external effects are concerned — as if the charges were concentrated at its centre. It is therefore sometimes considered to be a point charge, but there must not be any charges nearby to disturb the uniform distribution of charge on it, i.e. it must be an *isolated* charged spherical conductor. In practice two small spheres will only approximate to point charges when they are far apart. A point charge, like a point mass, is a convenient theoretical simplification.

Coulomb's law may be stated in mathematical terms as

$$F \propto \frac{Q_1 Q_2}{r^2}$$

where F is the electric (or Coulomb) force between two point charges Q_1 and Q_2, distance r apart.

(b) Experimental test

Two small metallized spheres X and Y can be used to test the law. X is glued to the bottom of a 'V' of a metre

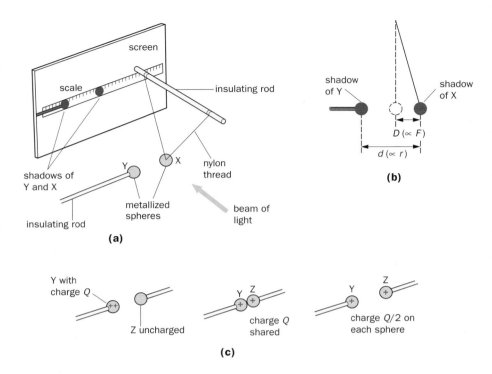

Fig. 13.4 Testing Coulomb's law

length of fine nylon thread so that it can only swing at right angles to the plane of suspension, Fig. 13.4a. The position of the centre (or edge) of the shadow of X on the scale on the screen is noted. Sphere Y is glued to the end of an insulating rod.

Both spheres are given the same charge by touching each in turn with, say, the plate of an electrophorus. When Y is brought up to X, repulsion occurs, Fig. 13.4b. The positions of the centres (or edges) of the shadows of X and Y are noted. Distance d is proportional to the separation (r) of the spheres and it can be shown that the deflection D (the distance between the centres — or edges — of the first and second positions of the shadow of X) is proportional to the force between X and Y. If several readings of d and D are taken and $D \times d^2$ found to be constant, then $F \propto 1/r^2$.

To test if $F \propto Q_1 Q_2$, Y is touched by another exactly similar but uncharged sphere Z, Fig. 13.4c, so that its charge is halved (Z taking the other half). Using one of the previous separations (d) between X and Y, the force should therefore become half what it was before. It will be one quarter of its original value if the charge on X is also halved by sharing its charge with the uncharged Z.

For reasonable success this experiment requires the insulators to be thoroughly dry (if necessary by using a hair dryer) to prevent loss of charge by leakage. The readings should therefore be taken in quick succession. The most convincing evidence for Coulomb's law, however, is provided not directly but indirectly by experimental verification of deductions from the law.

(c) Permittivity

The force between two charges also depends on what separates them; its value is always reduced when an insulating material replaces a vacuum. To take this into account a medium is said to have **permittivity**, denoted by ε (epsilon), and ε is included in the denominator of the expression for Coulomb's law. A material with high permittivity is one which reduces appreciably the force between two charges compared with the vacuum value.

SI units are said to be 'rationalized'. This means that the values of certain constants are adjusted so that π does not occur in formulae in frequent use, but it does occur in others. The formulae conveniently simplified in this way usually refer to situations in which there is plane symmetry (later we shall see that uniform electric and magnetic fields are in this category) and where we would not logically expect π to occur (in an unrationalized system it does). In a rationalized system 4π or π appears if there is spherical or cylindrical symmetry, respectively.

Spherical symmetry is associated with point charges (p. 204) and so the equation for Coulomb's law is rationalized by including 4π in the denominator. It becomes

$$F = \frac{1}{4\pi\varepsilon} \cdot \frac{Q_1 Q_2}{r^2}$$

If F is in newtons (N), r in metres, Q_1 and Q_2 in **coulombs** (C) — 1 coulomb being the charge flowing per second through a conductor in which there is a steady current of 1 ampere — then the unit of ε is $C^2 N^{-1} m^{-2}$, since $\varepsilon = Q_1 Q_2/(4\pi F r^2)$. A more widely used unit for permittivity is the **farad per metre** ($F m^{-1}$), as explained later (see p. 213).

The permittivity of a vacuum is denoted by ε_0 ('epsilon nought') and is called the **permittivity of free space**. The numerical value of ε_0 is found experimentally by an indirect method (p. 216) which does not involve the difficult task of measuring the force between known 'point' charges at a given separation in a vacuum. The result is

$$\varepsilon_0 = 8.85 \times 10^{-12} \, C^2 \, N^{-1} \, m^{-2}$$

We can also write

$$1/(4\pi\varepsilon_0) = 8.98 \times 10^9 \, N \, m^2 \, C^{-2}$$

and so $F \approx 9 \times 10^9 \, Q_1 Q_2/r^2$

The permittivity of air at s.t.p. is $1.0005\varepsilon_0$ and so we can usually take ε_0 as the value for air.

ELECTRIC FIELD STRENGTH

(a) Definition

A resultant force changes motion. Many everyday forces are pushes or pulls between bodies in contact. In other cases forces arise between bodies that are separated from one another. Electric, magnetic and gravitational effects involve such action-at-a-distance forces and to deal with them physicists find the idea of a **field of force**, or simply a **field**, useful.

Fields of these three types have common features as well as important differences.

An electric field is a region where an electric charge experiences a force. If a very small, positive point charge Q is placed at any point in an electric field and it experiences a force F, then the **field strength** E (also called the E-field) at that point is defined by the equation

$$E = \frac{F}{Q}$$

In words, the magnitude of E is the force per unit charge and its direction is that of F (i.e. of the force which acts on a positive charge). Field strength E is therefore a vector. Note that we refer to E as the force *per* unit charge and not as the force *on* unit charge. A finite charge (such as a unit charge) might affect the field by inducing charges on neighbouring bodies and so we must *imagine* a very small test charge $+Q$ to be placed at the point since we require to know E before $+Q$ was introduced into the field.

If F is in newtons (N) and Q is in coulombs (C) then the unit of E is the newton per coulomb ($N \, C^{-1}$). An equivalent but commoner unit is the **volt per metre** ($V \, m^{-1}$).

(b) E due to a point charge

The magnitude of E due to an isolated positive point charge $+Q$, at a point P distance r away, in a medium of permittivity ε, can be calculated by imagining a very small charge $+Q_0$ to be placed at P, Fig. 13.5. By Coulomb's law, the force F on Q_0 is

$$F = \frac{1}{4\pi\varepsilon} \cdot \frac{Q Q_0}{r^2}$$

But E is the force per unit charge, that is

$$E = \frac{F}{Q_0}$$

So

$$E = \frac{1}{4\pi\varepsilon} \cdot \frac{Q}{r^2}$$

E is directed away from $+Q$, as shown. If a point charge $-Q$ replaced $+Q$, E would be directed towards $-Q$ since unlike charges attract.

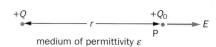

Fig. 13.5

The above expression shows that E decreases with distance from the point charge according to an inverse square law. The field due to an isolated point charge is therefore non-uniform but it has the same value at equal distances from the charge and so has spherical symmetry. In Fig. 13.6 if the magnitude of the field strength due to point charge $+Q$ is E at A, what is it at (*i*) B, (*ii*) C?

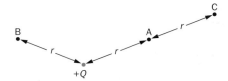

Fig. 13.6

(c) Field strength and charge density

So far as external effects are concerned, an isolated spherical conductor having a charge Q uniformly distributed over its surface behaves like a point charge Q at its centre (see p. 204). If r is the radius of the sphere, the field strength E at its surface is therefore given by

$$E = \frac{1}{4\pi\varepsilon} \cdot \frac{Q}{r^2}$$

The charge per unit area of the surface of the conductor is called the **charge density** σ (sigma) and since a sphere has surface area $4\pi r^2$ we have $\sigma = Q/(4\pi r^2)$. Therefore $Q = 4\pi r^2 \sigma$ and so

$$E = \frac{\sigma}{\varepsilon}$$

This expression will be used later (p. 215). It has been derived by considering a sphere but it gives E at the surface of *any* charged conductor. (It can be seen to apply to a plane surface if the radius of the sphere, which does not appear in the expression, is allowed to tend to infinity.)

FIELD LINES

An electric field can be represented and so visualized by electric field lines. These are drawn so that (*i*) the field line at a point (or the tangent to it if it is curved) gives the direction of *E* at that point, i.e. the direction in which a positive charge would accelerate, and (*ii*) the number of lines per unit cross-section area is proportional to *E*. The field line is imaginary but the field it represents is real.

Electric field patterns similar to the magnetic field patterns given by iron filings can be obtained using tiny 'needles' of an insulating substance such as semolina powder or grass seeds, suspended by previous stirring in fresh castor oil in a glass dish. An electric field is created by applying a high p.d. from a van de Graaff generator to metal electrodes dipping in the oil, Fig. 13.7. The powder or seed orientates itself to form different patterns according to the shape of the electrodes.

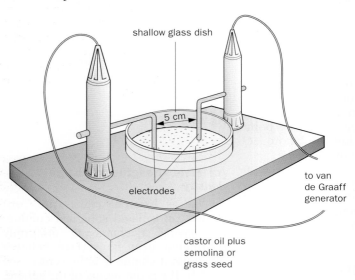

Fig. 13.7   Viewing electric field patterns

Some electric field patterns formed in this way are shown in Figs 13.8*a* and *b*. A uniform field is one in which *E* has the same magnitude and direction at all points, there is plane symmetry and the field lines are parallel and evenly spaced. In Fig. 13.8*a* the field is uniform between the plates away from the edges. When *E* varies in magnitude and direction with position, the field is non-uniform. In Fig. 13.8*b* the field is non-uniform but radial and there is spherical symmetry.

Electric field patterns are useful in designing electronic devices such as cathode ray tubes. The engineer is often able to sketch intuitively the pattern for a given electrode arrangement and so predict the probable behaviour of the device.

Field lines are also referred to as 'lines of force' and the term is appropriate when considering electric and gravitational fields because the field lines do indicate the direction in which a charge or mass experiences a force. This is not so in the magnetic case, as will be seen later, and therefore in general it is preferable to talk about field lines rather than lines of force.

(a)   Uniform field between two parallel plates

(b)   Radial field

Fig. 13.8   Electric field patterns

ELECTRIC POTENTIAL

Information about the field may be given by stating the field strength at any point; alternatively the **potential** can be quoted. Before discussing this idea some basic mechanics will be revised briefly.

(a) Work and energy

Work is done when the point of application of a force (or a component of it) undergoes a displacement in its own direction. The product of the force (or its component) *F* and the displacement *s* is taken as a measure of the work done *W*, i.e. $W = F \times s$. When *F* is in newtons and *s* in metres, *W* is in newton-metres, or joules.

If a body A exerts a force on body B and work is done, a transfer of energy occurs *which is measured by the work done*. So, if we raise a mass *m* through a vertical height *h*, the work done W by the force we apply (i.e. by mg) is $W = mgh$ (assuming the earth's gravitational field strength g is constant). The energy transfer is mgh and we consider that the system gains and stores that amount of gravitational potential energy in its gravitational field. This energy is obtained from the transfer of chemical energy by our muscular activity. When the mass falls the system loses gravitational potential energy and, neglecting air resistance, there is a transfer of kinetic energy to the mass equal to the work done by gravity.

(b) Meaning of potential

A charge in an electric field experiences a force and if it moves, work will, in general, be done. If a positive charge is moved from A to B in a direction opposite to that of the field E, Fig. 13.9a, an external agent has to do work against the forces of the field and energy has to be supplied. As a result, the system (of the charge in the field) gains an amount of electrical potential energy equal to the work done. This is analogous to a mass being raised in the earth's gravitational field g, Fig. 13.9b. When the charge is allowed to return from B to A, work is done by the forces of the field and the electrical potential energy previously gained by the system is lost. If, for example, the motion is in a vacuum, an equivalent amount of kinetic energy is transferred to the charge.

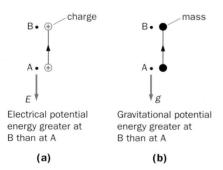

Electrical potential energy greater at B than at A

Gravitational potential energy greater at B than at A

(a)   **(b)**

Fig. 13.9 Analogy between electrical and gravitational potential energy

In general, the potential energy associated with a charge at a point in an electric field depends on the location of the point and the magnitude of the charge (since the force acting depends on the charge, i.e. $F = QE$).

Therefore if we state the magnitude of the charge we can describe an electric field in terms of the potential energies of that charge at different points. A unit positive charge is chosen and the change of potential energy which occurs when such a charge is moved from one point to another is called the change of **potential** of the field itself.

Hence the potential at point B in Fig. 13.9a exceeds that at A by the energy needed to take unit positive charge from A to B. To be strictly accurate, however, as we were when we defined E (p. 201), we should refer to the energy needed *per* unit charge when a *very small* charge moves from one point to the other since the introduction of a unit charge would in general modify the field.

If for theoretical purposes we select as the zero of potential the potential at an infinite distance from any electric charges, potential can be defined as follows.

> The potential at a point in a field is defined as the energy required to move unit positive charge from infinity to the point.

It is always assumed that the charge does not affect the field. The choice of the zero of potential is purely arbitrary and although infinity may be a few hundred metres in some cases, in atomic physics where distances of 10^{-10} m are involved it need only be a very small distance away from the charge responsible for the field.

Potential is a property of a *point* in a field and is a scalar since it deals with a quantity of work done or potential energy per unit charge. The symbol for potential is V and the unit is the joule per coulomb ($J\ C^{-1}$) or the **volt** (V).

Just as a mass moves from a point of higher gravitational potential to one of lower potential (i.e. it falls towards the earth's surface), so a positive charge is urged by an electric field to move from a point of higher electric potential to one of lower potential. Negative charges move in the opposite direction if free to do so.

(c) Potential and field strength compared

When describing a field, potential is usually a more useful quantity than field strength because, being a scalar, it can be added directly when more than one field is concerned. Field strength is a vector and addition (by the parallelogram law) is more complex. Also, it is often more important to know what energy changes occur (rather than what forces act) when charges move in a field and these are readily calculated if potentials are known (see p. 206).

EQUIPOTENTIALS

All points in a field that have the same potential can be imagined as lying on a surface — called an **equipotential** surface. When a charge moves on such a surface no energy transfer occurs and no work is done. The force due to the field must therefore act at right angles to the equipotential surface at any point and so equipotential surfaces and field lines always intersect at right angles.

A field can be represented pictorially by field lines and by equipotential surfaces (or lines in two-dimensional diagrams). Equipotential surfaces for a point charge are concentric spheres (or circles in two dimensions), Fig. 13.10a; there is spherical symme-

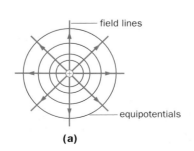

(a)

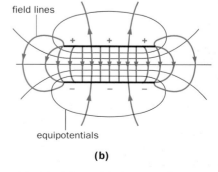

(b)

Fig. 13.10 Equipotential surfaces (in two dimensions) for (*a*) a point charge and (*b*) parallel plates

try. The plane symmetry of a uniform field is seen in Fig. 13.10*b*. If equipotentials are drawn so that the change of potential from one to the next is constant, then the spacing will be closer where the field is stronger. The closer the equipotentials, the shorter the distance that need be travelled to transfer a particular amount of energy.

The surface of a conductor in electrostatics (i.e. one in which no current is flowing) must be an equipotential surface since any difference of potential would cause a redistribution of charge in the conductor until no field existed in it.

Electrostatic mapping using conductive paper allows equipotentials around electrodes to be mapped in two dimensions. The electrode shapes are drawn on the paper with conducting silver paint and a p.d. applied between them. Points of equal potential are located using a probe attached to a multimeter. Equipotential lines can be drawn for simple electrode configurations such as a parallel plate capacitor (Fig. 13.10*b*).

POTENTIAL DUE TO A POINT CHARGE

The potential at a point A in the field of, and distance *r* from, an isolated point charge $+Q$ situated at O in a medium of permittivity ε, Fig. 13.11, can be calculated. Imagine that a very small point charge $+Q_0$ is moved by an external agent from C, distance *x* from A, through a very small distance δx to B without affecting the field due to $+Q$.

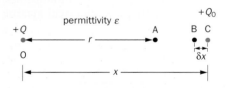

Fig. 13.11

Assuming the force *F* on Q_0 due to the field remains constant over δx, the work done δW by the external agent over δx against the force of the field is

$$\delta W = F(-\delta x)$$

The negative sign shows that the displacement δx is in the opposite direction to that in which *F* acts. By Coulomb's law

$$F = \frac{QQ_0}{4\pi\varepsilon} \cdot \frac{1}{x^2}$$

$$\therefore \quad \delta W = \frac{QQ_0}{4\pi\varepsilon} \cdot \frac{(-\delta x)}{x^2}$$

The total work done *W* in bringing Q_0 from infinity (∞) to A is

$$W = \frac{-QQ_0}{4\pi\varepsilon} \int_\infty^r \frac{dx}{x^2} = \frac{-QQ_0}{4\pi\varepsilon} \left[\frac{-1}{x}\right]_\infty^r$$

$$\therefore \quad W = \frac{QQ_0}{4\pi\varepsilon} \cdot \frac{1}{r}$$

The potential *V* at A is the energy needed to move unit positive charge from infinity to A.

$$V = \frac{W}{Q_0}$$

So

$$V = \frac{1}{4\pi\varepsilon} \cdot \frac{Q}{r}$$

What would be the analogous expression for the gravitational potential *V* at a distance *r* from a point mass *M*?

POTENTIAL DUE TO A CONDUCTING SPHERE

(a) An expression

A charge $+Q$ on an isolated conducting sphere is uniformly distributed over its surface (due to the repulsion of like charges) and has a radial electric field pattern, Fig. 13.12*a*. The field at any point outside the sphere is exactly the same as if the whole charge were concentrated as a point charge $+Q$ at the centre of the sphere, Fig. 13.12*b*. (This can be shown to follow from Coulomb's law; a spherical mass similarly behaves as if its whole mass were concentrated at its centre.)

From the expression already obtained for a point charge, we can say that the potential *V* at a point P, distance *r* from the sphere's centre, is

$$V = \frac{1}{4\pi\varepsilon} \cdot \frac{Q}{r}$$

If the radius of the sphere is *a*, the potential *V* at its surface is

$$V = \frac{1}{4\pi\varepsilon} \cdot \frac{Q}{a}$$

At all points inside the sphere the field strength is zero, otherwise field lines would link charges of opposite sign in the sphere and such a state of affairs is impossible under static conditions in a conductor. (This may also be shown to be a result of Coulomb's law.) It follows that no energy is transferred when a charge is moved between any two points inside the sphere. The potential is the same at all points throughout the sphere and equal to that at the surface. As well as talking about the potential at a point in a field we also consider an insulated conductor to have a potential.

The variation of *E* and *V* at points outside and inside a positively charged conducting sphere should therefore be as shown in the graphs of Fig. 13.13. Outside the sphere *E* varies as $1/r^2$ and, as we have just seen, theory predicts that *V* varies as $1/r$.

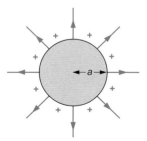

(a) Charged sphere with charge ($+Q$) on its surface

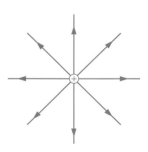

(b) Point charge ($+Q$)

Fig. 13.12

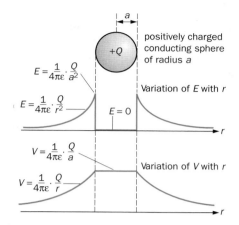

$E = \frac{1}{4\pi\varepsilon} \cdot \frac{Q}{a^2}$

$E = \frac{1}{4\pi\varepsilon} \cdot \frac{Q}{r^2}$

$E = 0$

$V = \frac{1}{4\pi\varepsilon} \cdot \frac{Q}{a}$

$V = \frac{1}{4\pi\varepsilon} \cdot \frac{Q}{r}$

Fig. 13.13 Electric field and potential due to a conducting sphere

(b) Flame probe investigation

The results for V outside may be investigated experimentally using the apparatus of Fig. 13.14. A probe,[1] in the form of a small gas flame at the point of a hypodermic needle, is connected to an electroscope calibrated to measure potentials (see p. 222). The potential measured is the value at the probe. The conducting sphere is

[1] The construction and action of the probe are described in Appendix 3.

charged to 1500 V from a high voltage power supply and the potential (V) noted from the electroscope when the probe is at different distances (r) from the centre of the sphere. (In particular, at a distance of twice the radius from the centre of the sphere the potential should be 750 V.) The sphere should be 'isolated' by being well away from walls, bench tops, the experimenter, etc., and the lead from the probe must also be clear of the bench. If the results confirm that $V \propto 1/r$ then this may be taken as indirect evidence of Coulomb's law since the $1/r$ law for potential is a consequence of a $1/r^2$ law for E.

(c) Effect of neighbouring bodies

The electric potential at a point in a field due to a charged body is not determined solely by the charge unless the body is 'isolated' as we assume in theory. In practice it is affected by the presence of other bodies, charged or uncharged, and the surrounding material. The potential at a point near a positively charged body increases when another positively charged body approaches and decreases when a negatively charged body is brought up.

In Fig. 13.15 the variation of potential with distance from a positively charged sphere A is shown before and after an uncharged conductor BC is brought near. Initially each point on BC is at the potential which previously existed at that point. Momentarily, therefore, B is at a higher potential than C and so, since BC is a conductor, electrons flow from C to B (i.e. from a lower to a higher potential) until the potential is the same all over BC — called the **potential of the conductor**. Electrostatic induction has occurred and the induced negative charge at B lowers the potential there, as well as at all points between A and B (including that of A). The induced positive charge at C raises the potential at C and at points beyond. The constant potential of BC lies between the original potentials at B and C.

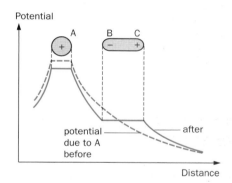

Fig. 13.15

POTENTIAL DIFFERENCE

The idea of potential applies not only to the electric fields produced in air or a vacuum by static charges (on an insulator or a conductor) but also to those in a wire having a battery or power supply across its ends and which cause charges to move as an electric current in the wire. The term **potential** is useful in both electrostatics and in current electricity.

However, there is an important difference between the two cases. In electrostatics when a charge moves in the direction of the field, the potential energy lost by the field–charge system can be regained if the charge is moved by an external agent in the opposite direction. In current electricity, energy lost by the electric field inside a con-

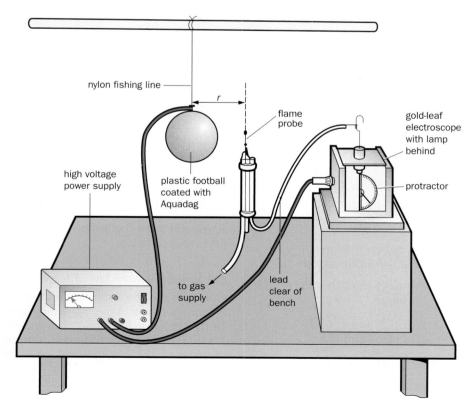

Fig. 13.14 Investigating the potential due to a charged conducting sphere

ductor is irrecoverable since the heat produced cannot be transferred back into other forms of energy by reversing the current.

In practice, especially in current electricity, we are usually concerned with the difference of potential or **p.d.** between two points in an electric field so that the zero of potential does not matter. The p.d. between two points is the **energy transformed when unit charge passes** from one point to the other. The same symbol is used for p.d. as for potential, i.e. V, and the same unit, i.e. volt (V) or joule per coulomb ($J C^{-1}$).

If the p.d. between two points in an electric field is 10 volts then in moving charge from one point to the other an energy transfer of 10 joules per coulomb occurs. In general if V is the p.d. (in V) between two points in an electric field, the energy transfer W occurring when a charge Q (in C) moves through the p.d. is given by

$$W = QV$$

Two examples follow to show how, using this expression, energy transfers in electric fields can be calculated.

First, suppose we wish to know the energy transferred when an electron 'falls' through a p.d. of 1 V in an electric field in a vacuum, i.e. it travels between two points whose p.ds differ by 1 V, Fig. 13.16. The electron is accelerated by the force acting on it due to the field and work is done. The energy transfer W is found from $W = QV$ where $Q = 1.6 \times 10^{-19}$ C (the electronic charge) and $V = 1.0$ V $= 1.0 J C^{-1}$.

$$\therefore \quad W = (1.6 \times 10^{-19} \text{ C})(1.0 \text{ J C}^{-1})$$
$$= 1.6 \times 10^{-19} \text{ J}$$

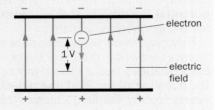

Fig. 13.16

This tiny amount of energy is called an **electronvolt** (eV) and is a unit of energy (not SI) much used in atomic and nuclear physics.

Second, an example from current electricity. If a p.d. of 12 V maintains a current of 3.0 A through a resistor, the electrical energy W (from the electric field in the resistor) transferred to heat per second is obtained from

$$W = QV = ItV \quad (\text{since } Q = It)$$

where $I = 3.0$ A $= 3.0$ C s^{-1}, $t = 1.0$ s and $V = 12$ V $= 12 J C^{-1}$.

$$\therefore \quad W = (3.0 \text{ C s}^{-1})(1.0 \text{ s})(12 \text{ J C}^{-1})$$
$$= 3.0 \times 1.0 \times 12 \text{ C s}^{-1} \times \text{s}$$
$$\times \text{ J C}^{-1}$$
$$= 36 \text{ J}$$

RELATION BETWEEN *E* AND *V*

Consider a charge $+Q$ at a point A in an electric field where the field strength is E, Fig. 13.17. The force F on Q is given by

$$F = EQ$$

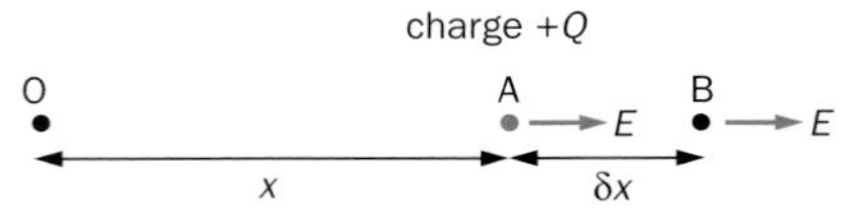

charge $+Q$

Fig. 13.17

If Q moves a very short distance δx from A to B in the direction of E, then (assuming E is constant over AB) the work done δW by the electric force on Q is

$$\delta W = \text{force} \times \text{distance}$$
$$= F \delta x$$
$$= EQ \delta x$$

If the p.d. between B and A is δV, we have by the definition of p.d.

$$\delta V = \text{energy transformed per unit charge}$$
$$= -\frac{\delta W}{Q} = -\frac{EQ \delta x}{Q}$$

That is, $\quad \delta V = -E \delta x$

The negative sign shows that if displacements in the direction of E are taken to be positive, then when δx is

positive, δV is negative, i.e. the potential decreases. On the other hand, if the charge is moved in a direction opposite to that of E, δx is negative and δV is positive, indicating an increase of potential as occurs in practice.

In the limit, as $\delta x \rightarrow 0$, E becomes the field strength at a point (A) and in calculus notation

$$E = -\frac{dV}{dx}$$

dV/dx is called the **potential gradient** in the x-direction and so the field strength at a point equals the negative of the potential gradient there. Potential gradient is a vector and is measured in volts per metre (V m^{-1}).

In a uniform field E is constant in magnitude and direction at all points, hence dV/dx is constant, i.e. the potential changes steadily with distance. The field near the centre of two parallel metal plates is uniform and if this is created by a p.d. V between plates of separation d, then

$$E = -\frac{V}{d}$$

where E is the field strength at any point in the **uniform** region (not at the edges).

For example, if $V = 2.0 \times 10^3$ V and $d = 1.0$ cm $= 1.0 \times 10^{-2}$ m,

$$E = V/d \text{ (numerically)}$$
$$= (2.0 \times 10^3 \text{ V})/(1.0 \times 10^{-2} \text{ m})$$
$$= 2.0 \times 10^5 \text{ V m}^{-1} \text{ (or N C}^{-1}\text{)}$$

GRAVITATIONAL ANALOGY

Analogies exist between electric and gravitational fields.

(a) Inverse square law of force

Coulomb's law is similar in form to Newton's law of universal gravitation. Both are inverse square laws, with $1/(4\pi\varepsilon)$ in the electric case corresponding to the gravitational constant G. The main difference is that electric forces can be attractive or repulsive but gravitational forces are always attractive. Two types of electric charge are known but there is only one type of gravitational mass. By comparison with electric forces, gravitational forces are extremely weak (p. 210).

Coulomb's law

$$F = \frac{1}{4\pi\varepsilon} \cdot \frac{Q_1 Q_2}{r^2}$$

Newton's law

$$F = G \cdot \frac{m_1 m_2}{r^2}$$

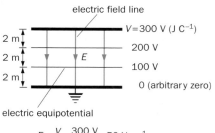

$$E = \frac{V}{d} = \frac{300 \text{ V}}{6 \text{ m}} = 50 \text{ V m}^{-1}$$

(a) Uniform electric field

$$g = \frac{V}{d} = \frac{60 \text{ J kg}^{-1}}{6 \text{ m}} = 10 \text{ N kg}^{-1}$$

(b) Uniform gravitational field

Fig. 13.18 Uniform fields

(b) Field strength

The field strength at a point in a gravitational field is defined as the force acting per unit mass placed at the point. Thus if a mass m in kilograms experiences a force F in newtons at a certain point in the earth's field, the strength of the field at that point will be F/m in newtons per kilogram. This is also the acceleration a the mass would have in metres per second squared if it fell freely under gravity at this point (since $F = ma$). The gravitational field strength and the acceleration of free fall at a point therefore have the same value (i.e. F/m), and the same symbol, g, is used for both. At the earth's surface $g = 9.8 \text{ N kg}^{-1} = 9.8 \text{ m s}^{-2}$ (vertically downwards).

Electric field strength
(at distance r from point charge Q)

$$E = \frac{1}{4\pi\varepsilon} \cdot \frac{Q}{r^2} \quad \text{(in N C}^{-1})$$

Gravitational field strength
(at distance r from point mass m)

$$g = G \cdot \frac{m}{r^2} \quad \text{(in N kg}^{-1})$$

We can also express the forces in terms of field strengths.

Electric force

$$F = QE$$

(on a charge Q)

Gravitational force

$$F = mg$$

(on a mass m)

(c) Field lines and equipotentials

These can also be drawn to represent gravitational fields but such fields are so weak, even near massive bodies, that there is no method of plotting field lines similar to those used for electric (and magnetic) fields. Field lines for the earth are directed towards its centre and the field is spherically symmetrical. Over a small part of the earth's surface the field can be considered uniform, the lines being vertical, parallel and evenly spaced. Figures 13.18a and b represent uniform electric and gravitational fields.

(d) Potential and p.d.

Electric potentials and p.ds are measured in joules per coulomb (J C^{-1}) or volts; gravitational potentials and p.ds are measured in joules per kilogram (J kg^{-1}). If the p.d. between two points in the earth's gravitational field is 20 J kg^{-1}, the work done and the change of potential energy will be 20 J when 1 kg moves from one point to the other. In general, if V is the p.d. between two points, the energy transfer W that occurs when a mass m moves from one point to the other is given by $W = mV$. From this expression the energy required to send a spacecraft from one point to another can be calculated (see p. 208).

Electric p.d. and energy transfer

$$W = QV$$

Gravitational p.d. and energy transfer

$$W = mV$$

As a mass moves away from the earth the potential energy of the earth–mass system increases, transfer of energy from some other source being necessary. If infinity is taken as the zero of gravitational potential (i.e. a point well out in space where no more energy is needed for the mass to move further away from the earth) then the potential energy of the system will have a negative value except when the mass is at infinity. At every point in the earth's field the potential is therefore negative (see expression below), a fact which is characteristic of fields that exert attractive forces. Fields giving rise to repulsive forces cause positive potentials.

Electric potential
(at distance r from point charge Q)

$$V = \frac{1}{4\pi\varepsilon} \cdot \frac{Q}{r}$$

Gravitational potential
(at distance r from point mass m)

$$V = -G \cdot \frac{m}{r}$$

Figure 13.19a shows the variation of gravitational potential V with distance r from the centre of a spherical mass or the variation of electric potential V with distance r from the centre of a negatively charged sphere, i.e. it applies to an attractive force field. Figure 13.19b shows the variation of electric potential V with distance r from the centre of a positively charged sphere, i.e. it refers to a repulsive force field.

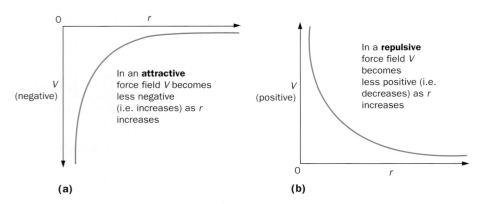

Fig. 13.19 Variation of potential in (a) attractive and (b) repulsive radial fields

(e) Shielding

While a body can be shielded from an electric field by surrounding it with a conductor, gravitational shielding is impossible.

FIELD TREATMENT OF SPACE TRAVEL

The motion of spacecraft and satellites in the earth's gravitational field has been considered in terms of forces (p. 144). As an alternative, a 'field' treatment will now be given which involves working out energy transfers from potential differences.

(a) Satellite orbits

The gravitational potential V at a distance r from the centre of the earth is $-GM/r$ where M is the earth's mass. Potential equals the potential energy per unit mass (or charge, in the electrical case, see p. 203) and so a satellite of mass m at a distance r from the centre of the earth has potential energy $E_p = -GMm/r$. If the satellite is moving with speed v in a circular orbit of radius r, Fig. 13.20, its kinetic energy $E_k = \frac{1}{2}mv^2$. Hence the total energy E of the satellite in orbit is given by

$$E = E_p + E_k = -GMm/r + \tfrac{1}{2}mv^2$$

From the example for centripetal force and the law of gravitation we have

$$\frac{mv^2}{r} = \frac{GMm}{r^2}$$

$$\therefore \quad mv^2 = \frac{GMm}{r}$$

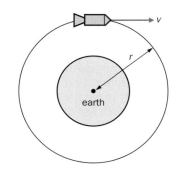

Fig. 13.20

Substituting for mv^2 in the expression for E, we can say

$$E = -\frac{GMm}{r} + \tfrac{1}{2}\cdot\frac{GMm}{r} = -\tfrac{1}{2}\cdot\frac{GMm}{r}$$

The potential energy E' of a mass m on the earth's surface is

$$E' = -GMm/R$$

where R is the radius of the earth. So to put a satellite into circular orbit of radius r it must be given the extra amount of energy $(E - E')$ where

$$E - E' = -\tfrac{1}{2}\cdot\frac{GMm}{r} - \frac{(-GMm)}{R}$$

$$= GMm\left(\frac{1}{R} - \frac{1}{2r}\right)$$

If the orbit is close to the earth's surface (e.g. at a height of 100–200 km), $r \approx R$ and

$$E - E' = \tfrac{1}{2}\cdot\frac{GMm}{R}$$

If this extra energy is all given in the form of kinetic energy ($\frac{1}{2}mu^2$) at lift-off, then

$$\tfrac{1}{2}mu^2 = \tfrac{1}{2}\frac{GMm}{R}$$

$$\therefore \quad u^2 = \frac{GM}{R}$$

Taking $G = 6.7 \times 10^{-11}$ N m^2 kg^{-2}, $M = 6.0 \times 10^{24}$ kg and $R = 6.4 \times 10^6$ m, and substituting in the expression for u^2, we get $u = 8.0$ km s^{-1}. The speed the satellite must be given at launching to go into a circular orbit close to the earth's surface will be greater than 8.0 km s^{-1} because air resistance has been neglected.

In practice circular orbits are seldom attained: most are elliptical.

(b) Speed of escape

Some values are given in Table 13.1 of the gravitational potential ($-GM/r$) at different distances r from the centre of the earth.

As r increases, the potential becomes less negative, i.e. it increases, and the potential difference between any two values of r equals the change of potential energy when a mass of 1 kg moves from one point to another. For example, if a spacecraft travels from the earth's surface ($r = 6.4 \times 10^6$ m) to the moon ($r = 400 \times 10^6$ m), the gain of potential energy per kg is $(63 - 1) \times 10^6 = 62 \times 10^6$ J kg^{-1}. The spacecraft loses this amount of kinetic energy and so the energy needed per kg to get the craft from the earth's surface to the distance of the moon is 62×10^6 J kg^{-1}.

In order to escape to 'infinity' 63×10^6 J kg^{-1} is required, and if this is given to a spacecraft of mass m at lift-off as kinetic energy, we can say

$$\tfrac{1}{2}mu^2 = 63 \times 10^6\,m$$

where u is the 'escape' speed from the surface of the earth. This gives

$$u \approx 11 \text{ km s}^{-1}$$

Table 13.1

r/m $\times 10^6$	6.4	6.6	20	40	400	∞
$(-GM/r)$/J kg^{-1} $\times 10^6$	−63	−61	−20	−10	−1.0	0.00

POTENTIAL AND FIELD STRENGTH CALCULATIONS

Example 1. Find (a) the potential and (b) the field strength at points A and B, Fig. 13.21, due to two small spheres X and Y, 1.0 m apart in air and carrying charges of $+2.0 \times 10^{-8}$ C and -2.0×10^{-8} C respectively. Assume the permittivity of air $= \varepsilon_0$ and $1/(4\pi\varepsilon_0) = 9.0 \times 10^9$ N m^2 C^{-2}.

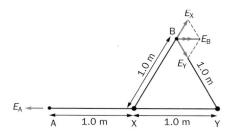

Fig. 13.21

(a) Potential at A due to X

$$= \frac{1}{4\pi\varepsilon_0} \cdot \frac{Q}{r}$$

$$= \frac{(9.0 \times 10^9 \text{ N m}^2 \text{ C}^{-2}) \times (+2.0 \times 10^{-8} \text{ C})}{1.0 \text{ m}}$$

$$= \frac{9.0 \times 10^9 \times 2.0 \times 10^{-8}}{1.0} \, \frac{\text{N m}^2 \text{ C}^{-2} \text{ C}}{\text{m}}$$

$$= +1.8 \times 10^2 \text{ N m C}^{-1} \quad (1 \text{ N m} = 1 \text{ J})$$

$$= +1.8 \times 10^2 \text{ V} \quad (1 \text{ J C}^{-1} = 1 \text{ V})$$

Similarly the potential at A due to Y

$$= \frac{9.0 \times 10^9 \times (-2.0 \times 10^{-8})}{2.0} \text{ V}$$

$$= -0.90 \times 10^2 \text{ V}$$

Potential is a scalar quantity and is added algebraically, therefore the potential V_A at A due to X and Y is

$$V_A = (1.8 - 0.90) \times 10^2 \text{ V}$$
$$= 0.90 \times 10^2 \text{ V} = 90 \text{ V}$$

Since B is equidistant from equal and opposite charges, the potential V_B at B due to X and Y is

$$V_B = (1.8 - 1.8) \times 10^2 \text{ V} = 0$$

(b) Field strength at A due to X

$$= \frac{1}{4\pi\varepsilon_0} \cdot \frac{Q}{r^2}$$

$$= \frac{9.0 \times 10^9 \times 2.0 \times 10^{-8}}{1.0^2} \, \frac{\text{N m}^2 \text{ C}^{-2} \text{ C}}{\text{m}^2}$$

$$= 1.8 \times 10^2 \text{ N C}^{-1} \text{ (or V m}^{-1}\text{)}$$
towards the left

Similarly the field strength at A due to Y

$$= 0.45 \times 10^2 \text{ N C}^{-1} \text{ (or V m}^{-1}\text{)}$$
towards the right

The resultant field strength E_A at A due to X and Y is

$$E_A = (1.8 - 0.45) \times 10^2 \text{ N C}^{-1} \quad \text{(or V m}^{-1}\text{)}$$
$$= 1.3(5) \times 10^2 \text{ N C}^{-1} \quad \text{(or V m}^{-1}\text{)}$$
towards the left

At B, the field strengths due to X and Y have the directions shown and the resultant field strength E_B is their vector sum. So,

$$E_B = E_X \cos 60° + E_Y \cos 60°$$
$$= E_X \text{ (since } E_X = E_Y \text{ and } \cos 60° = \tfrac{1}{2}\text{)}$$

$$= \frac{9.0 \times 10^9 \times 2.0 \times 10^{-8}}{1.0^2} \text{ N C}^{-1}$$

$$= 1.8 \times 10^2 \text{ N C}^{-1} \quad \text{(or V m}^{-1}\text{)}$$
towards the right

Example 2. Two large horizontal, parallel metal plates are 2.0 cm apart in a vacuum and the upper is maintained at a positive potential relative to the lower so that the field strength between them is 2.5×10^5 V m^{-1}. (a) What is the p.d. between the plates? (b) If an electron of charge 1.6×10^{-19} C and mass 9.1×10^{-31} kg is liberated from rest at the lower plate, what is its speed on reaching the upper plate?

(a) If E is the field strength (assumed uniform) and V is the p.d. between two plates distance d apart, we have from $E = V/d$ (p. 206) that

$$V = Ed$$
$$= (2.5 \times 10^5 \text{ V m}^{-1}) \times (2.0 \times 10^{-2} \text{ m})$$
$$= 5.0 \times 10^3 \text{ V}$$

(b) The energy transferred (i.e. work done) W when a charge Q moves through a p.d. of V in an electric field is given by $W = QV$. There is a transfer of electrical potential energy from the field to kinetic energy of the electron. We have

$$QV = \tfrac{1}{2}mv^2$$

where v is the required speed and m is the mass of the electron.
Therefore

$$v = \sqrt{\left(\frac{2QV}{m}\right)}$$

$$= \sqrt{\left\{\frac{(2 \times 1.6 \times 10^{-19} \text{ C})(5.0 \times 10^3 \text{ V})}{(9.1 \times 10^{-31} \text{ kg})}\right\}}$$

$$= \sqrt{\left\{\left(\frac{2 \times 1.6 \times 10^{-19} \times 5.0 \times 10^3}{9.1 \times 10^{-31}}\right) \times \left(\frac{\text{C J C}^{-1}}{\text{kg}}\right)\right\}}$$

$$= \sqrt{\left(\frac{16}{9.1} \times 10^{15} \text{ m}^2 \text{ s}^{-2}\right)}$$

$$(1 \text{ J} = 1 \text{ N m} = 1 \text{ kg m s}^{-2} \text{ m})$$

$$= 4.2 \times 10^7 \text{ m s}^{-1}$$

Example 3. Compare the electric and gravitational forces between a proton of charge $+e$ and mass M at a distance r from an electron of charge $-e$ and mass m (e.g. as in a hydrogen atom), given that

$$
\begin{aligned}
e &= 1.6 \times 10^{-19} \text{ C} \\
m &= 9.1 \times 10^{-31} \text{ kg} \\
M &= 1.7 \times 10^{-27} \text{ kg} \\
1/(4\pi\varepsilon_0) &= 9.0 \times 10^9 \text{ N m}^2 \text{ C}^{-2} \\
G &= 6.7 \times 10^{-11} \text{ N m}^2 \text{ kg}^{-2}
\end{aligned}
$$

Electric attraction $= \dfrac{1}{4\pi\varepsilon_0} \cdot \dfrac{Q_1 Q_2}{r^2} = \dfrac{1}{4\pi\varepsilon_0} \cdot \dfrac{e^2}{r^2}$

Gravitational attraction $= \dfrac{GmM}{r^2}$

So,

$$
\begin{aligned}
\frac{\text{electric attraction}}{\text{gravitational attraction}} &= \frac{1}{4\pi\varepsilon_0} \cdot \frac{e^2}{GmM} \\[2mm]
&= \frac{(9.0 \times 10^9 \text{ N m}^2 \text{ C}^{-2})}{(6.7 \times 10^{-11} \text{ N m}^2 \text{ kg}^{-2})} \\[2mm]
&\quad \times \frac{(1.6 \times 10^{-19} \text{ C})^2}{(9.1 \times 10^{-31} \text{ kg}) \times (1.7 \times 10^{-27} \text{ kg})} \\[2mm]
&\approx 10^{39} : 1
\end{aligned}
$$

The gravitational force is very small compared with the electric force.

QUESTIONS

1. a) Use a simple electron model to explain how an insulator, such as polythene, can be charged negatively by rubbing it with a duster.

b) Figure 13.22 shows a metal plate, initially uncharged, which has been placed on a polythene sheet. The polythene is negatively charged.

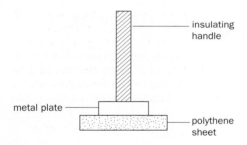

Fig. 13.22

i) Copy the diagram and show the distribution of charge on the metal plate.
ii) Explain the distribution of charge on the metal plate.
iii) A student touches the metal plate briefly with a finger. Explain why this leaves the plate positively charged.
iv) The insulating handle is then used to remove the plate from the sheet. Suggest why this raises the potential of the plate.
(*UCLES*, Further Physics, March 1998)

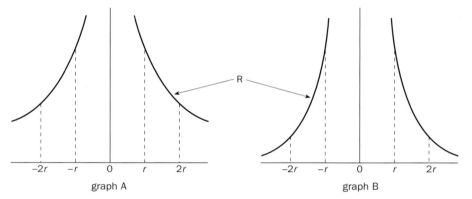

Fig. 13.23

2. Figure 13.23 shows graphs of the electric field strength and of the electric potential caused by a point charge. On each graph the vertical axis has a linear scale.
i) Which of the graphs, A or B, shows the variation of potential against distance?
ii) State why, in the regions marked R, the shapes of the graphs are different from each other.
iii) At a point where the distance, r, from the point charge is 60 mm, the electric field strength is 2.5×10^4 V m^{-1}. Calculate the potential at this point.
(*NEAB*, AS/A PHO3, Feb 1997)

3. Considering a hydrogen atom to consist of a proton and an electron at an average separation of 0.50×10^{-10} m, find
a) the electric potential at 0.50×10^{-10} m from the proton,
b) the potential energy of the electron,

c) the energy required to remove the electron from the atom assuming the electron is at rest.
(Charge on proton $= +1.6 \times 10^{-19}$ C; charge on electron $= -1.6 \times 10^{-19}$ C; $1/(4\pi\varepsilon_0) = 9.0 \times 10^9$ N m^2 C^{-2}.)
Note. In practice other factors as well as the potential energy have to be considered when calculating the energy needed to remove an electron from an atom. Suggest one.

4. a) Figure 13.24 shows two charged, parallel, conducting plates.

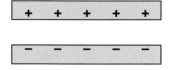

Fig. 13.24

Copy the diagram and add solid lines to show the electric field in the space between and *just beyond* the edges of the plates.

Add to your diagram dotted lines to show three equipotentials in the same regions.

b) Define **electric potential** at a point. Is electric potential a vector or a scalar quantity?

An isolated charged conducting sphere has a radius *a*. The graph in Fig. 13.25 shows the variation of electric potential *V* with distance *r* from the centre of the sphere.

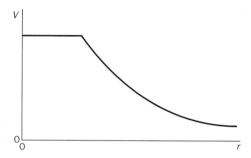

Fig. 13.25

Copy the graph and mark with an 'a' on the distance axis the point that represents the radius of the sphere.

Add to the graph a line showing how electric field strength *E* varies with distance for the same range of values of *r*.

(*L, A PH4, Jan 1996*)

5. a) i) Define **electric field strength**, and state whether it is a scalar quantity or a vector quantity.

ii) Copy and complete Fig. 13.26 to show the electric field lines in the region around two equal positive point charges. Mark with a letter N the position of any point where the field strength is zero.

Fig. 13.26

b) Point charges A, of +2.0 nC, and B, of −3.0 nC, are 200 mm apart in a vacuum, as shown by Fig. 13.27. The point P is 120 mm from A and 160 mm from B.

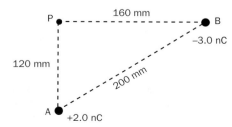

Fig. 13.27

i) Calculate the component of the electric field at P in the direction AP.

ii) Calculate the component of the electric field at P in the direction PB.

iii) Hence calculate the magnitude and direction of the resultant field at P.

c) i) Explain why there is a point X on the line AB in part b) at which the **electric potential** is zero.

ii) Calculate the distance of the point X from A.

(*AQA: NEAB, AS/A PHO3, March 1999*)

6. Many electrical insulators are ionic compounds consisting of arrays of dipoles, i.e. pairs of equal but oppositely charged ions a small distance apart. The properties of the dipoles account for the fact that such compounds have very strong internal electric fields but almost zero external fields.

In Fig. 13.28*a*, A and B form an ionic dipole with charges of $+1.6 \times 10^{-19}$ C and -1.6×10^{-19} C respectively and at a separation of 2.0×10^{-10} m. What is the field strength due to the dipole at

a) X, mid-way between A and B, and

b) Y, 50×10^{-10} m to the right of A?

Find the ratio of these two field strengths.

$(1/(4\pi\varepsilon_0) = 9 \times 10^9 \text{ N m}^2 \text{ C}^{-2})$

In Fig. 13.28*b*, A is a single charge of $+1.6 \times 10^{-19}$ C. What is the field at

c) X, and

d) Y? Again, find the ratio.

Compare the two ratios. Does the external field due to a dipole decrease more rapidly with distance than that of a single charge?

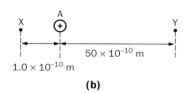

(a)

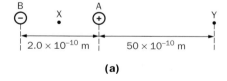

(b)

Fig. 13.28

7. Figure 13.29 shows a positively charged oil drop held at rest between two parallel conducting plates A and B.

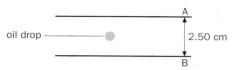

Fig. 13.29

The oil drop has a mass of 9.79×10^{-15} kg. The potential difference between the plates is 5000 V and plate B is at a potential of 0 V. Is plate A positive or negative?

Draw a labelled free-body force diagram which shows the forces acting on the oil drop. (You may ignore upthrust.)

Calculate the electric field strength between the plates.

Calculate the magnitude of the charge Q on the oil drop.

How many electrons would have to be removed from a neutral oil drop for it to acquire this charge?

(*L, A PH4, June 1998*)

8. An electric field is maintained between two parallel metal plates by a p.d. of 1000 V. A charge of 10^{-15} C is moved by the forces of the field from one plate to the other.

a) How much electrical energy is transferred from the field–charge system?

b) How much work is done by the electric force?

c) If the plates are 20 mm apart what is the value of the electric force assuming it is constant, i.e. the field between the plates is uniform?

d) What is the field strength in N C⁻¹?

e) What is the field strength in V m⁻¹?

9. What is the work done when a mass of 2.0 kg is taken from point A to point B, if A and B are at distances *r* and 3*r* respectively from the centre of the earth and if the gravitational potential at A is −12 MJ kg⁻¹?

10. a) Define **electric field strength** at a point in an electric field.

Figure 13.30 is a graph showing the relation of electric potential *V* with distance *x* from the centre of an isolated nucleus. X is a point 5.0×10^{-14} m from the centre of the nucleus.

b) i) Using the coordinates of point X and the permittivity of free space $(8.9 \times 10^{-12} \text{ F m}^{-1})$, determine the charge of the nucleus.

ii) Using *only* the coordinates of point X, calculate the electric field strength *E* at X.

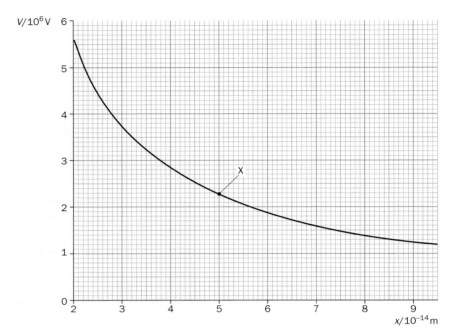

Fig. 13.30

iii) State a relation between the electric field strength E at X and the gradient G of the graph at X.

c) A proton, charge 1.6×10^{-19} C, is accelerated from rest in a particle-accelerator, and projected towards the nucleus. The proton travels in a vacuum and comes momentarily to rest at X. Assume that the nucleus remains at rest. Calculate:

i) the electric potential energy of the proton when at X;

ii) the kinetic energy given to the proton by the accelerator;

iii) the potential difference used to accelerate the proton.

(*OCR*, 6842, June 1999)

11. What is meant by the term **gravitational potential** V at a point? In what unit is it measured?

The gravitational potential at the surface of the Earth is given by the relationship

$$V = -\frac{GM}{r}$$

where G is the universal constant of gravitation

M is the mass of the Earth

r is the radius of the Earth

Explain why V at the surface of the Earth is negative.

Show that

$$g = \frac{GM}{r^2}$$

where g is the acceleration of free fall close to the Earth.

Use the relationships

$$V = -\frac{GM}{r} \quad \text{and} \quad g = \frac{GM}{r^2}$$

to help you to show that the escape speed for a projectile leaving the Earth is 11×10^3 m s^{-1}. Radius of Earth = 6400 km.

In calculating the escape speed for a space probe travelling outwards from the solar system what would the symbols M and r represent?

(*L*, A PH4, Jan 1998)

12. a) Define **gravitational field strength**.

b) Figure 13.31 shows a point X in the gravitational field of an isolated star S.

Fig. 13.31

The gravitational potential at the point X = -4.5×10^7 J kg^{-1}.

i) Explain why the gravitational potential at X is expressed as a negative number.

ii) On a copy of Fig. 13.31, label:

1. a point P where the gravitational potential is -9.0×10^7 J kg^{-1};

2. a point F where the gravitational field strength is twice that at X.

iii) Calculate the work done in moving a 750 kg mass from X to a region where the gravitational field of the star is negligible.

(*OCR*, 6842, March 1999)

13. Figure 13.32 illustrates a simple demonstration of the application of electrostatics in paint spraying. The paint is forced through the nozzle at high speed, producing a spray of charged droplets. The spray drifts towards the metal plate and sticks to it.

a) By reference to a simple electron model, explain briefly the difference in the properties of conductors and insulators.

b) The paint is an insulator, but the nozzle is a conductor. State and explain the type of charge acquired by the paint droplets produced at the nozzle.

c) Explain why the droplets are attracted to the earthed metal plate.

d) The demonstration is repeated without connecting the metal plate to earth. Suggest, with an explanation, what effect this has.

(*OCR*, Further Physics 4833, June 1999)

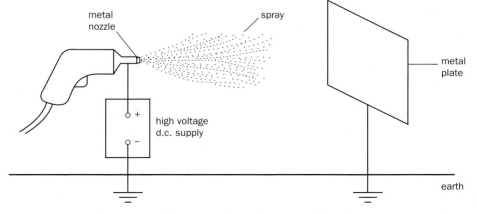

Fig. 13.32

14

Capacitors

CAPACITANCE

(a) Definition

Charge given to an isolated conductor may be thought of as being 'stored' on it. The amount it will take depends on the electric field created at the surface of the conductor. If this is too great there is breakdown in the insulation of the surrounding medium, resulting in sparking and discharge of the conductor.

In terms of potential we can say that the smaller the change of potential of a conductor when a certain charge is transferred to it, the more charge it can 'store' before breakdown occurs. The change in potential due to a given charge depends on the size of the conductor, the material surrounding it and the proximity of other conductors.

The idea that an insulated conductor in a particular situation has a certain **capacitance**, or charge-storing ability, is useful and is defined as follows: if the potential of an insulated conductor changes by V when given a charge Q, the capacitance C of the conductor is

$$C = \frac{Q}{V}$$

In words, **capacitance equals the charge required to cause unit change in the potential of a conductor**. If Q is in coulombs (C) and V in volts (V)

then the unit of C is coulomb per volt (C V^{-1}) or **farad** (F). This is a very large unit; the microfarad (1 μF = 10^{-6} F), the nanofarad (1 nF = 10^{-9} F) or the picofarad (1 pF = 10^{-12} F) are generally used.

Experiment (see p. 214) and theory both show that for an insulated conductor in a given situation, $V \propto Q$, i.e. C is a constant.

(b) Capacitance of an isolated conducting sphere

The potential V of such a sphere of radius a, in a medium of permittivity ε and having a charge Q, is given by

$$V = \frac{1}{4\pi\varepsilon} \cdot \frac{Q}{a} \qquad \text{(p. 204)}$$

Capacitance $C = \dfrac{Q}{V}$, therefore

$$C = 4\pi\varepsilon \cdot a$$

The capacitance of a sphere is therefore proportional to its radius. We also see from this expression that if C is in farads and a in metres then ε can be expressed in farads per metre (F m^{-1}), since $\varepsilon = C/(4\pi a)$.

(c) Practical zero of potential

The theoretical zero of potential is taken, as we have seen, as the potential of points at infinity. However,

when making actual measurements this is an impracticable zero and the potential of the earth (itself a conductor) is adopted as the practical zero.

When a charged conductor X of small capacitance and an uncharged conductor Y of large capacitance are connected, they end up with the same potential which is less than X's was originally, with X losing most of its charge to Y. There is charge flow until the potentials of X and Y are the same.

If a is the radius of the earth, its capacitance C is

$$
\begin{aligned}
C &= 4\pi\varepsilon \cdot a \\
&= (1/9 \times 10^{-9}\ \text{F m}^{-1}) \times (6 \times 10^6\ \text{m}) \\
&= 7 \times 10^{-4}\ \text{F} \\
&= 700\ \mu\text{F}
\end{aligned}
$$

This is a large capacitance compared with that of other conductors used in electrostatics. Consequently when a charged conductor is 'earthed' — say by touching it (the human body conducts) — it loses most of its charge to the earth; it is discharged and it acquires earth potential, i.e. zero potential. The earth has such a large capacitance that any change in its potential due to the loss or gain of charge, because of connection to another conductor, is negligible. As a result, it provides a satisfactory practical zero of potential.

CAPACITORS

A capacitor is designed to store electric charge and basically consists of two conductors, such as a pair of parallel metal plates, separated by an insulator. The symbol for a capacitor is ─┤├─ and the conductors are usually referred to as 'plates', whatever their form.

(a) Action of a capacitor

A capacitor is readily 'charged' by applying a p.d. across the plates from a battery or other power supply. Insight into the process can be obtained from the circuit of Fig. 14.1 using a 500 µF electrolytic capacitor (see p. 217).

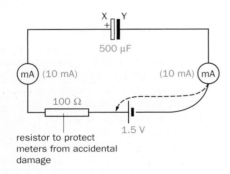

Fig. 14.1

When the circuit incorporating the supply is completed, equal-sized flicks occur on the meters indicating momentary but identical current flow through each. The current directions round the circuit are the same, which suggests that as much charge, in the form of a current pulse, leaves one plate of the capacitor as enters the other. So if charge $+Q$ has flowed on to X, an equal charge $+Q$ has flowed off Y.

In terms of electron flow we can say that when the supply is connected to the capacitor, electrons flow from the negative of the supply on to plate Y and from plate X towards the positive of the supply at the same rate. The positive and negative charges appearing on X and Y respectively oppose further electron flow and as these charges accumulate, the p.d. between X and Y increases until it equals the p.d. of the supply (1.5 V). Electron flow then stops. There is never a steady current and no permanent meter deflections.

We say the capacitor has charge Q, meaning that one plate has charge $+Q$ and the other charge $-Q$; as much charge flows off one plate as flows on to the other, Fig. 14.2.

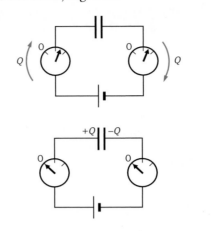

Fig. 14.2

Making the connection shown by the dashed line in Fig. 14.1 causes the meters to again give equal, momentary deflections but in the opposite direction to that indicated previously. Electrons flow back from Y to X until the positive charge on X is neutralized. This momentary pulse of current thus discharges the capacitor leaving zero charge on its plates.

(b) Factors affecting the charge stored

These may be investigated by the circuit of Fig. 14.3a using a parallel-plate capacitor in the form of two square metal plates each of side 25 cm, kept about 1.5 mm apart by four small polythene spacers (5 mm × 5 mm) at the corners, Fig. 14.3b. The capacitor is alternately charged, usually from a 12 V smooth d.c. supply (e.g. dry batteries), and discharged through a sensitive galvanometer 400 times a second by a **reed switch**.

A reed switch is shown in Fig. 14.3c. When *half-wave rectified* a.c. (see p. 283) passes through the coil surrounding it, the reed and magnetic contact become oppositely

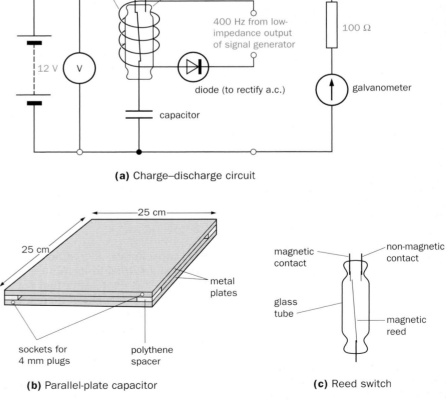

(a) Charge–discharge circuit

(b) Parallel-plate capacitor

(c) Reed switch

Fig. 14.3 Investigating the charge stored on a capacitor

magnetized and attract on the conducting half-cycle. On the non-conducting half-cycle the reed, no longer magnetized, springs back to the non-magnetic contact. The number of charge–discharge actions per second equals the frequency of the a.c. supply to the coil. If this is high enough current pulses follow one another so rapidly that the galvanometer deflection is steady and represents the average current I through it. When Q is the charge stored on the capacitor and released on discharge through the galvanometer and when f is the switching frequency (i.e. that of the a.c. supply) then

$$I = \text{charge passing per second}$$
$$= Qf$$

If f is constant then $I \propto Q$. The 100 kΩ protective resistor prevents excessive current pulses when the reed switch contacts close; the capacitor will not discharge completely if the resistor is too large. How could you check that discharge is complete?

If the charging p.d. V is varied, we find that the charge Q on the capacitor, i.e. the galvanometer reading I, is directly proportional to V.

When V is kept constant and the separation d of the plates varied (using small insulating spacers at the corners of the plates), it can be inferred that Q is inversely proportional to d if d is not too large. (A graph of Q against $1/d$ does not pass through the origin because of the capacitance between the upper plate and the bench or other nearby objects.)

If V and d are fixed and the area of overlap A of the plates is varied, we can conclude that Q is directly proportional to A. Finally, inserting a solid slab of insulator (e.g. Perspex) in the space between the plates causes Q to increase (if V, d and A are fixed).

This investigation may also be performed using a d.c. amplifier electrometer (p. 225).

Summarizing the results, we have

$$Q \propto V \qquad Q \propto 1/d \qquad Q \propto A$$

Therefore

$$Q \propto \frac{VA}{d}$$

The capacitance C of a capacitor is defined (as for an insulated conductor) as the charge stored per unit p.d. between its plates. That is,

$$C = \frac{Q}{V}$$

Therefore

$$C \propto \frac{A}{d} \quad \text{or} \quad C = \text{constant}\,.\,\frac{A}{d}$$

The capacitance of a parallel-plate capacitor is given by

$$C = \frac{\varepsilon A}{d} \qquad \text{(see below)}$$

from which we see that the constant of proportionality is ε, the permittivity of the insulating medium — also called the **dielectric** — between the plates.

MEASUREMENT OF CAPACITANCE

A reed switch is used as in the circuit of Fig. 14.3a but for capacitances of a few microfarads the galvanometer is replaced by a milliammeter (1 mA) and a protective resistor of 220 Ω is suitable. For smaller capacitances a microammeter (100 μA) with a 2 kΩ resistor is required. In both cases the reed switch can be operated from a 50 Hz supply.

The capacitance C is found from

$$C = \frac{Q}{V} = \frac{I}{fV} \qquad \text{(since } I = Qf\text{)}$$

where I and V are the two meter readings and f is the frequency of the supply.

The effect of connecting capacitors in series and in parallel may be investigated.

PARALLEL-PLATE CAPACITOR

An expression for the capacitance of a parallel-plate capacitor is required.

Consider a capacitor with plates of common area A, separated by a medium of thickness d and permittivity ε, Fig. 14.4. If one plate has charge $+Q$ and the other $-Q$, the charge density σ is Q/A. Assuming the field between the plates is uniform, the field strength E is the same at all points and is given by

$$E = \frac{\sigma}{\varepsilon} = \frac{Q}{A\varepsilon} \qquad \text{(see p. 201)}$$

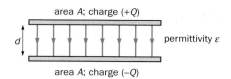

Fig. 14.4

(labels: area A; charge $(+Q)$; d; permittivity ε; area A; charge $(-Q)$)

Further, if V is the p.d. between the plates then

$$E = \frac{V}{d} \qquad \text{(see p. 206)}$$

Therefore

$$\frac{Q}{A\varepsilon} = \frac{V}{d}$$

$$\frac{Q}{V} = \frac{A\varepsilon}{d}$$

and since $Q = VC$

$$C = \frac{A\varepsilon}{d}$$

C will be in farads if A is in m^2, d in m and ε in F m^{-1}. It is worth noting that since the field between the plates is uniform, there is plane symmetry and π does not appear in the formula for C. It is an example of the simplification of a common formula by the use of rationalized units (p. 200). In practice, the expression is not strictly true owing to non-uniformity of the field at the edges of the plates.

Large values of capacitance are obtained by having plates (i) with a large overlapping area, (ii) very close together and (iii) separated by a dielectric with high permittivity.

PERMITTIVITY

Earlier we saw that the force between two charges depended on the intervening medium and the idea of permittivity ε was introduced to take this into account. We will now consider this idea further, especially in relation to capacitors.

(a) Relative permittivity or dielectric constant ε_r

Experiment shows that inserting an insulator or dielectric between the plates of a capacitor increases its capacitance. If C_0 is the capacitance of a capacitor when a vacuum separates its plates and C is the capacitance of the same capacitor with a dielectric filling the space between the plates, the **relative permittivity** ε_r of the dielectric is defined by

$$\varepsilon_r = \frac{C}{C_0}$$

Taking a parallel-plate capacitor as an example, we have

$$\varepsilon_r = \frac{C}{C_0} = \frac{\varepsilon A / d}{\varepsilon_0 A / d} = \frac{\varepsilon}{\varepsilon_0}$$

where ε is the permittivity of the dielectric and ε_0 is that of a vacuum (i.e. of free space). The expression for the capacitance of a parallel-plate capacitor with a dielectric of relative permittivity ε_r can therefore be written as

$$C = \frac{A \varepsilon_r \varepsilon_0}{d}$$

Relative permittivity has no units, unlike ε and ε_0 which have; it is a pure number without dimensions. For air at atmospheric pressure $\varepsilon_r = 1.0005$, which is near enough 1 and so for most purposes $\varepsilon_{air} = \varepsilon_0$. Table 14.1 gives some values of ε_r.

Table 14.1

Dielectric	Relative permittivity ε_r
Vacuum	1.0000
Air at s.t.p.	1.0005
Polythene	2.3
Perspex	2.6
Paper (waxed)	2.7
Mica	7
Water (pure)	80
Barium titanate	1200

The near impossibility of removing all the impurities dissolved in water makes it unsuitable in practice as a dielectric.

(b) Action of a dielectric

The molecules of a dielectric between the plates of a charged capacitor are in an electric field. The positive nuclei are urged in the direction of the field and the negative electrons in the opposite direction. As a result the molecules become distorted or **polarized** by the field, with one end having an excess of positive charge and the other of negative charge. Electric **dipoles** are thereby formed, Fig. 14.5a.

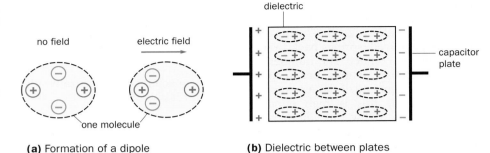

(a) Formation of a dipole

(b) Dielectric between plates of charged capacitor

Fig. 14.5 Action of a dielectric

Inside the dielectric the positive and negative ends of adjacent dipoles cancel each other's effects but at the surfaces of the dielectric unneutralized charges appear which have opposite signs to those on the plates, Fig. 14.5b. The positive potential of the positive plate is reduced by the negative charge at one surface of the dielectric and the negative potential of the negative plate is made less negative by the positive charge at the other surface of the dielectric. The p.d. between the plates is therefore less than it would otherwise be and so more charge is required before the p.d. between the plates equals the applied (i.e. charging) p.d. The capacitance is thus increased.

Certain molecules, called **polar** molecules, are polarized even in the absence of an electric field and consequently they increase the capacitance even more than does a dielectric with **non-polar** molecules. The high relative permittivity of water arises from the H_2O molecule being polar.

(c) Measurement of ε_0

The circuit is the same as that for investigating the factors affecting the charge stored by a capacitor (Fig. 14.3, p. 214). As before, a capacitor in the form of a large pair of parallel metal plates is alternately charged from a smooth d.c. supply (0–12 V) and discharged through a galvanometer (on its most sensitive range, e.g. ×1) about 400 times a second by a vibrating reed switch.

The capacitance C of a parallel-plate air capacitor having a plate overlap area A and plate separation d is given by

$$C = \frac{\varepsilon_{air} A}{d}$$

$$\therefore \quad \varepsilon_{air} = \frac{Cd}{A}$$

where ε_{air} is the permittivity of air which, to a good approximation, is ε_0. If Q is the charge stored on the capacitor when the applied p.d. is V (measured on the voltmeter) then since $Q = VC$ we have

$$\varepsilon_0 = \frac{Qd}{AV}$$

The steady current I recorded by the galvanometer is Qf, where f is the switching frequency, provided the capacitor is fully charged and discharged during each contact of the reed switch. So from $I = Qf$ we get

$$\varepsilon_0 = \frac{Id}{fVA}$$

Knowing the current sensitivity of the galvanometer (marked on the instrument or supplied by the manufacturer, and typically 20–25 mm μA^{-1}), I can be found in amperes. The switching frequency f is obtained from the dial reading on the signal generator and may be checked using a CRO and a low-voltage 50 Hz mains supply (see p. 415). The area A of one of the plates is readily measured, as is the separation d if the top capacitor plate rests on four small insulating spacers (each approximately 1.5 mm thick) at the corners of the bottom plate (Fig. 14.3b).

Very small charges are involved and, to prevent leakage, the connection from the top plate to the reed switch should not touch anything. The accepted value of ε_0 is 8.85 × 10^{-12} F m^{-1}.

The permittivity of different materials can also be found by this method if the experiment is done with the material completely filling the space between the plates. The permittivity ε

of the material is then $\varepsilon = I_1 d/(fVA)$ where I_1 is the galvanometer current. The ratio of the galvanometer currents with and without the material between the plates gives the relative permittivity ε_r. That is

$$\varepsilon_r = \frac{\varepsilon}{\varepsilon_0} = \frac{I_1 d/(fVA)}{Id/(fVA)} = \frac{I_1}{I}$$

Both ε_0 and ε_r may be determined in a similar manner using a d.c. amplifier electrometer (p. 225).

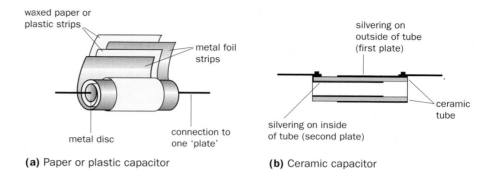

(a) Paper or plastic capacitor

(b) Ceramic capacitor

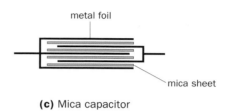

(c) Mica capacitor

Fig. 14.6 Capacitor construction

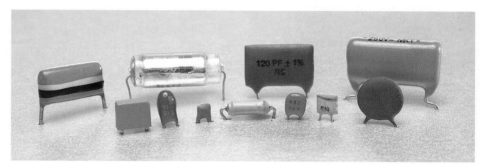

Fig. 14.7 Different types of small capacitor
Back row: polyester, polystyrene, silver mica, metallized polyester
Front row: metal film, tantalum, miniature ceramic, miniature polyester, polypropylene, ceramic, disc ceramic

TYPES OF CAPACITOR

Capacitors are used in electric circuits for various purposes, as we shall see later. Different types have different dielectrics. The choice of type depends on the value of capacitance and the stability (i.e. ability to keep the same value with age, temperature change, etc.) needed and on the frequency of the alternating current (a.c.) that will pass through the capacitor (this affects the power loss).

For every dielectric material there is a certain potential gradient at which it breaks down and a spark passes. The working p.d. of a capacitor is therefore determined by the thickness of the dielectric. Liquid and gaseous dielectrics recover when the applied p.d. is reduced below the breakdown value; only some solid dielectrics do.

(a) Paper, plastic, ceramic and mica capacitors

Waxed paper, plastics (e.g. polyester, polystyrene), ceramics (e.g. talc with barium titanate added) and mica (which occurs naturally and splits into very thin sheets of uniform thickness) are all used as dielectrics. Typical constructions are shown in Fig. 14.6 and actual capacitors in Fig. 14.7.

Polyesters are the most widely used for general use giving good stability and a wide range of values. Ceramics offer low cost and high capacitance with small size. Mica and polystyrene types are useful in high frequency circuits. Paper capacitors are now used mostly in mains supply applications.

Capacitance values for these four types rarely exceed a few microfarads.

(b) Electrolytic capacitors

These have capacitance values up to 100 000 µF and are quite compact because the dielectric can have a thickness as small as 10^{-4} mm and not suffer breakdown even for applied p.ds of a few hundred volts.

The dielectric is a film of aluminium oxide formed by passing a current through a strip of paper soaked with aluminium borate solution, separating two aluminium foil electrodes, Fig. 14.8a. The oxide forms on the anode, which acts as one plate of the capacitor. The borate solution, being an electrolyte, is the other; connection to it is made via the cathode (the other piece of aluminium foil).

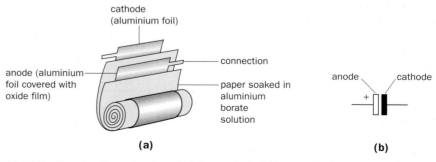

(a)

(b)

Fig. 14.8 Construction of and symbol for an electrolytic capacitor

In use, a 'leakage' current of the order of 1 mA should always pass through the capacitor, in the correct direction, to maintain the dielectric. The anode is marked with a + to show that it must be at a higher potential than the cathode when the capacitor is in circuit. Care must be taken to ensure that there is direct current (d.c.) in the circuit and that the capacitor is correctly connected.

Three electrolytic capacitors of different sizes are shown in Fig. 14.9; the case is often of aluminium and acts as the negative terminal. They are not used in a.c. circuits where the frequency exceeds about 10 kHz. Their stability is poor (10–20%) but in many cases this does not matter.

Fig. 14.9 Electrolytic capacitors

(c) Air capacitors

Air is used as the dielectric in variable capacitors. These consist of two sets of parallel metal plates: one set is fixed and the other moves on a spindle within — but not touching — the fixed set, Fig. 14.10. The interleaved area varies, thus changing the capacitance.

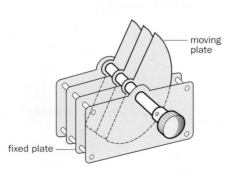

Fig. 14.10 Variable capacitor

Losses in an air dielectric are very small at all frequencies. However, relatively large thicknesses are needed because breakdown occurs at a potential gradient which is small compared with those of other dielectrics. Variable capacitors are used to tune radio receivers.

CAPACITOR NETWORKS

A network of capacitors has a combined or equivalent capacitance which can be calculated.

(a) Capacitors in parallel

In Fig. 14.11a the three capacitors of capacitance C_1, C_2 and C_3 are in parallel. *The applied p.d. V is the same across each but the charges are different* and are given by

$$Q_1 = VC_1 \qquad Q_2 = VC_2 \qquad Q_3 = VC_3$$

The total charge Q on the three capacitors is

$$Q = Q_1 + Q_2 + Q_3$$
$$= V(C_1 + C_2 + C_3)$$

If C is the capacitance of the single equivalent capacitor, it would have charge Q when the p.d. across it is V, Fig. 14.11b.

$$Q = VC$$

Hence

$$C = C_1 + C_2 + C_3$$

The combined capacitance of capacitors of 1 µF, 2 µF and 3 µF in parallel is 6 µF.

It should be noted that the charges on capacitors in parallel are in the ratio of their capacitances, i.e.

$$Q_1 : Q_2 : Q_3 = C_1 : C_2 : C_3$$

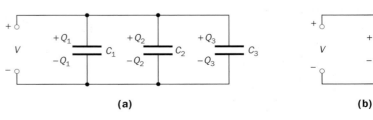

(a)

Fig. 14.11 Capacitors in parallel

The expression for capacitors in parallel is similar to that for resistors in series.

(b) Capacitors in series

The capacitors in Fig. 14.12a are in series and have capacitances C_1, C_2 and C_3. Suppose a p.d. V applied across the combination causes the motion of charge from plate Y to plate A so that a charge $+Q$ appears on A and an equal but opposite charge $-Q$ appears on Y. This charge $-Q$ will induce a charge $+Q$ on plate X if the plates are large and close together. The plates X and M and the connection between them form an insulated conductor whose net charge must be zero and so $+Q$ on X induces a charge $-Q$ on M. In turn this charge induces $+Q$ on L and so on.

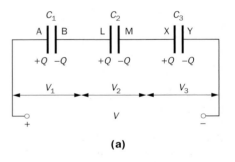

(a)

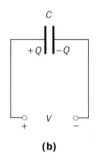

(b)

Fig. 14.12 Capacitors in series

Capacitors in series thus *all have the same charge* and the p.ds across each are given by

$$V_1 = \frac{Q}{C_1} \qquad V_2 = \frac{Q}{C_2} \qquad V_3 = \frac{Q}{C_3}$$

The total p.d. *V* across the network is

$$V = V_1 + V_2 + V_3$$

$$\therefore V = \frac{Q}{C_1} + \frac{Q}{C_2} + \frac{Q}{C_3}$$

$$= Q\left(\frac{1}{C_1} + \frac{1}{C_2} + \frac{1}{C_3}\right)$$

If *C* is the capacitance of the single equivalent capacitor, it would have charge *Q* when the p.d. across it is *V*, Fig. 14.12*b*.

$$V = \frac{Q}{C}$$

Therefore

$$\frac{Q}{C} = Q\left(\frac{1}{C_1} + \frac{1}{C_2} + \frac{1}{C_3}\right)$$

$$\frac{1}{C} = \frac{1}{C_1} + \frac{1}{C_2} + \frac{1}{C_3}$$

The combined capacitance of capacitors of 1 μF, 2 μF and 3 μF in series is $\frac{6}{11}$ μF, i.e. the equivalent capacitance is less than the smallest capacitance.

The expression for capacitors in series is similar to that for resistors in parallel.

Note that:

1. For capacitors in parallel the p.d. across each is the same.
2. For capacitors in series each has the same charge.

ENERGY OF A CHARGED CAPACITOR

Some basic facts can be established from experiments with a large electrolytic capacitor before the theory is developed.

(a) Discharge through a motor

If a 10 000 μF capacitor is charged using a 10 V supply and then discharged through a small electric motor, a light load can be raised, Fig. 14.13. Mechanical energy is obtained from energy stored in the capacitor. Is energy released in any other way?

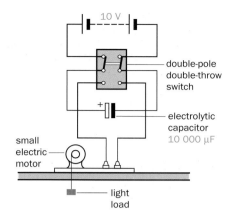

Fig. 14.13 Discharge through a motor

(b) Discharge through lamps

Using the circuit shown in Fig. 14.14*a* a 10 000 μF capacitor is charged from a 3 V supply then discharged through a 2.5 V, 0.3 A lamp, and the brightness is noted. If the procedure is repeated with two lamps in series (or in parallel) the brightness of each is much less than before. With a 6 V charging supply, two lamps in series flash brighter and longer than one did at 3 V. However, *four* lamps arranged as in Fig. 14.14*b* light up to about the same brightness and for about the

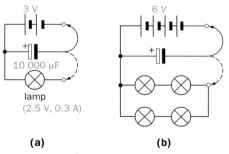

(a)   **(b)**

Fig. 14.14 Discharge through lamps

same time as one lamp did on 3 V. How many times more energy is stored in the capacitor at 6 V than at 3 V? What value of charging p.d. is needed to light *nine* lamps similarly to one lamp at 3 V?

(c) Discharge through a heating coil

The capacitor in Fig. 14.15 is charged and then discharged through the resistance coil made from 2 m of constantan wire (SWG 32). The resulting temperature rise in the coil is detected by a copper–constantan thermocouple and indicated by a sensitive galvanometer. One discharge at 20 V gives roughly the same deflection as four successive discharges at 10 V.

From this experiment and the previous one we can say that doubling the p.d. to which a capacitor is charged quadruples the energy stored, i.e. if the p.d. is *V* then the energy is proportional to V^2. Can you say why?

(d) Graphical treatment

The charge on a capacitor is directly proportional to the p.d. across it (i.e. $Q = VC$). Plotting p.d. against charge therefore yields a straight line through the origin, Fig. 14.16.

Suppose a capacitor has capacitance *C* and that when the p.d. is *V* the charge is *Q*. If the capacitor starts to discharge and initially a very small charge δQ passes from the negative to the positive plate, then by the definition of p.d. the resulting energy loss (or work done) is $V\delta Q$ — assuming δQ is so small that the decrease in *V* is negligible.

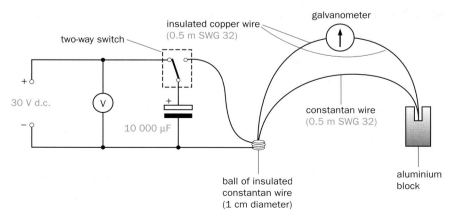

Fig. 14.15 Discharge through a heating coil

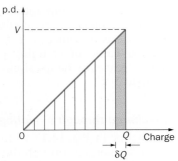

Fig. 14.16

So from the graph,

energy loss = area of shaded strip

If the capacitor discharges completely so that Q and V fall to zero, we have

total energy loss = area of all strips

= area of triangle below graph

= $\frac{1}{2}QV$

This is the energy stored in the capacitor. Since $Q = VC$ we can write

total energy = $\frac{1}{2}QV = \frac{1}{2}V^2C = \frac{1}{2}Q^2/C$

The energy stored is proportional to V^2, as suggested by the previous experiments. If Q is in coulombs, V in volts and C in farads, the energy is in joules. How much energy is stored in a 10 000 µF capacitor charged to 30 V? The energy of a charged capacitor is considered to be stored as electrical energy in the field in the medium between the plates.

The expression $\frac{1}{2}QV$ is analogous to $\frac{1}{2}Fe$ for the energy stored in a wire that obeys Hooke's law, when a tension F causes an extension e.

CAPACITOR CALCULATIONS

Example 1. In the circuit of Fig. 14.17, $C_1 = 2$ µF, $C_2 = C_3 = 0.5$ µF and E is a 6 V battery. For each capacitor calculate (*a*) the charge on it, and (*b*) the p.d. across it.

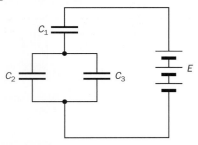

Fig. 14.17

The combined capacitance, C_4, of C_2 and C_3 in parallel, is given by

$C_4 = C_2 + C_3 = 0.5$ µF + 0.5 µF
= 1 µF

The circuit may now be redrawn as in Fig. 14.18 in which C_1 and C_4 are in series.

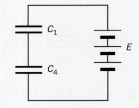

Fig. 14.18

Their charges Q_1 and Q_4 will be equal, hence

$Q_1 = Q_4 = V_1C_1 = V_4C_4$

where V_1 and V_4 are the p.ds across C_1 and C_4 respectively. Therefore

$$\frac{V_1}{V_4} = \frac{C_4}{C_1} = \frac{1}{2}$$

But $V_1 + V_4 = 6$ V

$\therefore$ $V_1 = 2$ V and $V_4 = 4$ V

Also

$Q_1 = V_1C_1 = (2\text{ V}) \times (2 \times 10^{-6}\text{ F})$

$= 4 \times 10^{-6}$ C

$\therefore$ $Q_4 = 4 \times 10^{-6}$ C = 4 µC

The p.d. across the combined capacitance C_4 equals that across each of C_2 and C_3. So

$V_2 = V_3 = V_4 = 4$ V

Now $Q_2 = V_2C_2$ and $Q_3 = V_3C_3$

$\therefore$ $\dfrac{Q_2}{Q_3} = \dfrac{C_2}{C_3} = \dfrac{0.5}{0.5} = 1$

(since $V_2 = V_3$)

But $Q_2 + Q_3 = Q_4 = 4 \times 10^{-6}$ C

$\therefore$ $Q_2 = Q_3 = 2 \times 10^{-6}$ C = 2 µC

Therefore

$Q_1 = 4$ µC $Q_2 = Q_3 = 2$ µC
$V_1 = 2$ V $V_2 = V_3 = 4$ V

Example 2. A 10 µF capacitor is charged from a 30 V supply and then connected across an uncharged 50 µF capacitor. Calculate (*a*) the final p.d. across the combination, and (*b*) the initial and final energies, Fig. 14.19*a* and *b*.

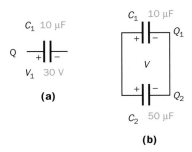

Fig. 14.19

(*a*) Initial charge Q on C_1

$= V_1C_1 = 30\text{ V} \times 10 \times 10^{-6}\text{ F}$
$= 3.0 \times 10^{-4}$ C = 300 µC

When C_1 and C_2 are connected (in parallel) the charge Q is shared in the ratio of their capacitances. That is,

$$\frac{Q_1}{Q_2} = \frac{C_1}{C_2} = \frac{10}{50} = \frac{1}{5}$$

But $Q = Q_1 + Q_2 = 300$ µC

$\therefore$ $Q_1 = 50$ µC and $Q_2 = 250$ µC

The final, common p.d. V is given by

$$V = \frac{Q_1}{C_1}\left(\text{or } \frac{Q_2}{C_2}\right) = \frac{50 \times 10^{-6}\text{ C}}{10 \times 10^{-6}\text{ F}}$$

$= 5.0$ V

(*b*) Initial energy

$= \frac{1}{2}QV_1 = \frac{1}{2} \times 3.0 \times 10^{-4}\text{ C} \times 30\text{ V}$

$= 4.5 \times 10^{-3}$ J

Final energy

$= \frac{1}{2}Q_1V + \frac{1}{2}Q_2V$

$= \frac{1}{2}V(Q_1 + Q_2) = \frac{1}{2}QV$

$= \frac{1}{2} \times 3.0 \times 10^{-4}\text{ C} \times 5.0\text{ V}$

$= 0.75 \times 10^{-3}$ J

Note. The apparent loss of energy is due to the production of heat when charge flows in the wires connecting the capacitors.

CHARGE AND DISCHARGE OF A CAPACITOR

Many electronic circuits involve capacitors charging or discharging through resistors. The way in which these processes occur is typical of other growth and decay effects in physics (e.g. radioactive decay), chemistry, biology and economics.

(a) Charge

In the circuit of Fig. 14.20, when the switch is in position 1, the capacitor of capacitance C charges through the resistor of resistance R from a 9 V battery. The microammeter records the charging current I and the voltmeter reads the p.d. V_C across the capacitor at different times t.

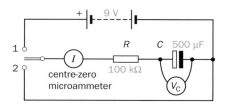

Fig. 14.20

If readings of I and V_C are taken at 10 second intervals and graphs of these quantities plotted against t, Fig. 14.21a and b, they show that:

1. I has its maximum value at the start when the capacitor begins to charge and then decreases more and more slowly until it becomes zero.
2. V_C rises rapidly from zero and slowly approaches its maximum value (9 V) which it reaches when the capacitor is fully charged and $I = 0$.

If the readings of the p.d. V_R across the resistor were also taken, a graph of V_R against t would have the same shape as that of I against t (since $V_R = IR$ at all times) and is shown as the dashed graph in Fig. 14.21b. All three graphs are **exponential** curves; those of I and V_R are decay curves and that of V_C is a growth curve.

A graph of the charge on the capacitor Q against t has the same shape as the V_C against t graph since $Q = V_C C$.

The characteristic of an exponential decay is that it represents a quantity which decreases by the *same fraction in successive equal time intervals*. So if I in Fig. 14.21a falls from its maximum value, say I_0, to $I_0/2$ in a time interval $t_{1/2}$, it will also fall from $I_0/2$ to $I_0/4$ in the next time interval $t_{1/2}$ and so on. Time interval $t_{1/2}$ is called the **half-life** of the decay process and for the graph shown in Fig. 14.21a is about 40 s. In general the time for I to fall to any fraction (not just half) is constant.

Also note that during charging the sum of the voltages across the resistor and capacitor equals the battery voltage V, that is

$$V = V_C + V_R$$

Initially $V_C = 0$, so $V = V_R$. Finally, when the capacitor is fully charged, $I = 0$, therefore $V_R = 0$ and $V_C = V$.

(b) Time constant

If the charging current I remained steady at its starting value, a capacitor would be fully charged after a time $T = C \times R$ seconds which, for the circuit of Fig. 14.20, equals $500 \times 10^{-6}\,\text{F} \times 100 \times 10^3\,\Omega = 50\,\text{s}$. In fact I decreases with time, as Fig. 14.21a shows, and the capacitor has only 0.63 of its full charge and p.d. after 50 s. Nevertheless, $T = CR$, called the **time constant**, is a useful measure of how long it takes a capacitor to charge through a resistor. The greater the values of C and R, the greater is T and the more slowly the p.d. across it rises.

As a check, the graph of V_C against t in Fig. 14.21b shows that after 50 s V_C is about 6 V, i.e. $\frac{2}{3} \times 9$ V. (For most purposes we can take $0.63 = \frac{2}{3}$.) After 100 s (i.e. after the second time constant), V_C rises by two-thirds of the p.d. remaining after 50 s, that is $\frac{2}{3} \times (9 - 6)\,\text{V} = \frac{2}{3} \times 3\,\text{V} = 2\,\text{V}$, thereby making $V_C = (6 + 2)\,\text{V} = 8\,\text{V}$. After approximately $5T$ (250 s) it is fully charged, i.e. $V_C = 9\,\text{V}$. In general, $T = CR$ is the time for V_C (and Q) to rise to two-thirds of the charging p.d. remaining at the start of the time.

(c) Discharge

In Fig. 14.20, when the switch is moved from position 1 to position 2, the capacitor discharges through the resistor. If the graphs of I, V_C and V_R are plotted as before, they are again exponential curves, as shown in Figs 14.22a and b. Note that:

1. The discharge current I, and so also V_R, are in the opposite direction to that during charge.
2. V_C and V_R are in opposition during discharge.

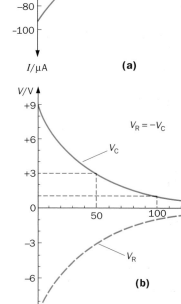

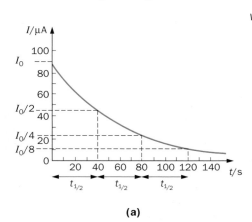

(a)

Fig. 14.21

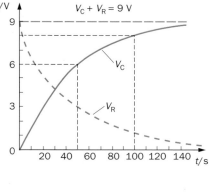

(b)

Fig. 14.22

In this case the time constant T is the time for V_C to fall *by* two-thirds (or fall *to* one-third) of its value at the start of the discharge. The V_C graph shows that in 50 s (T for the circuit), V_C falls from 9 V to 3 V and in the next 50 s it falls by $\frac{2}{3} \times 3$, i.e. by 2 V, making $V_C = 1$ V. After about $5T$ (250 s), $V_C = 0$.

Using calculus it can be shown that V_C decays according to the equation

$$V_C = V_0 e^{-t/CR}$$

where V_0 is the p.d. across the capacitor initially and e, the base of natural logarithms, equals 2.7. Substituting $t = T = CR$ we get

$$V_C = \frac{V_0}{e} = \frac{V_0}{2.7} = 0.37 \ V_0 \approx \tfrac{1}{3} V_0$$

Replacement of the voltmeter in Fig. 14.20 by a datalogger and a computer allows continuous monitoring of the voltage decay with time.

ELECTROSCOPES

The **gold-leaf electroscope**, shown in Fig. 14.23*a*, was used in most of the early work on electrostatics. The **Braun electroscope**, Fig. 14.23*b*, is a more robust instrument.

Basically an electroscope is a capacitor, the leaf (or pointer) and the rod to which it is attached forming one plate, and the case the other. When a p.d. is applied to the 'plates', a small charge flows on to the leaf and an equal but opposite charge is induced on the inside of the case. The forces of the resulting electric field between the

'plates' deflects the leaf, and the deflection is a measure of the p.d. between leaf, etc., and case. If the case is earthed, as it will be if it is made of wood and stands on a wooden bench, then the electroscope *records potential*. The deflection of the leaf is approximately proportional to the potential. This can be seen by connecting a calibrated high voltage supply across an electroscope having a protractor attached to it as a scale (see Fig. 13.14, p. 205).

THE D.C. AMPLIFIER AS AN ELECTROMETER

An electrometer is used to measure p.ds, including those produced electrostatically. An electroscope with a scale for reading deflections is an example of such an instrument.

The d.c. amplifier, an example of which is shown in Fig. 14.24, can also be used as an electrometer. It consists of an amplifier whose stages are *directly coupled* (hence 'd.c.') and are not joined by capacitors as in a normal amplifier. Essentially a d.c. amplifier acts as a very high resistance electronic voltmeter (about 10^{13} Ω) but it can be adapted to measure current and charge. In all cases, however, a p.d. of up to 1 V (for the instrument shown) has to be applied to the input terminals and this controls the output current which is recorded on a meter. The meter has to be calibrated to read the input p.d. directly, as described below.

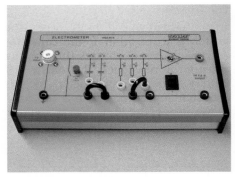

Fig. 14.24 A d.c. amplifier

To measure current, a resistor R is connected across the input as in Fig. 14.25*a* and the calibrated output meter then gives the p.d. V across R. The current I in R will be V/R. If $R = 10^{11}$ Ω, a meter reading corresponding to a 1 V input indicates that the current $I = V/R = 1/10^{11} = 10^{-11}$ A. (The much higher input resistance of the amplifier — about 10^{13} Ω — in parallel with R can be ignored.)

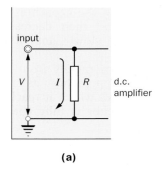

(a)

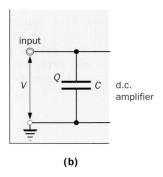

(b)

Fig. 14.25

The measurement of charge requires a known capacitor C across the input as in Fig. 14.25*b*. The p.d. V across C, i.e. the input p.d., is again shown by the calibrated output meter. The unknown charge Q on the capacitor (which is responsible for the p.d.) is obtained from $Q = VC$. For example, if $C = 10^{-8}$ F (0.01 μF) then when the

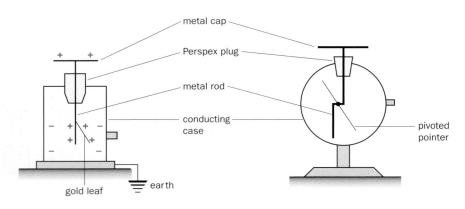

(a) Gold-leaf electroscope

(b) Braun electroscope

Fig. 14.23 Electroscopes

output meter gives a reading of 1 V, the input p.d. $V = 1$ V and therefore $Q = 10^{-8}$ C, the input capacitance of the amplifier being neglected. Almost complete transfer of charge to the amplifier capacitor C will occur if the capacitance of the object bringing the charge to the input terminal is small compared with that of the amplifier capacitor. Also, the input resistance of the amplifier is so high that leakage of charge from the amplifier capacitor through the instrument is negligible. There is, therefore, ample time to make a reading since the p.d. across the capacitor remains constant for a long time.

ELECTROSTATICS DEMONSTRATIONS WITH A D.C. AMPLIFIER

The instructions refer to the instrument shown in Fig. 14.24.

(a) Calibration

Connect a voltmeter to the *output sockets* and select the appropriate ×1 or ×10 switch setting. To obtain a record of the output connect a datalogger and computer across the output instead — these are particlarly useful for continuously monitoring the discharge of a capacitor (p. 225) but are not necessary for the other experiments.

Place both of the black connecting links in the earth line and switch on the electrometer; press the *push to zero* button and check the output reads zero. Connect the potential divider circuit of Fig. 14.26 so that 1 V is applied to the *input sockets* (as shown by the high-resistance voltmeter ⓥ). The output meter should

give a reading of 1 V; check the calibration is correct. (If using a datalogger and computer, choose settings which show the output conveniently on the computer display.) Remove the potential divider circuit. The instrument is ready for use.

(b) Positive and negative charges

Use a centre-zero meter for this experiment. Insert a brass rod (e.g. 25 mm long) in the input socket and set the electrometer on the 10^{-8} C range by making the link which connects the 10^{-8} F capacitor across the input.

Draw a cellulose acetate (or Perspex) strip, charged positively by rubbing, across the brass rod so that it charges the input capacitor; the meter deflects to the right.

Discharge the capacitor by pressing the 'push to zero' button; the meter should return to centre-zero. Repeat the procedure with a negatively charged polythene strip. The meter should deflect to the left.

(c) Spooning charge

This shows that charge is a *quantity* of something that can be measured and passed from place to place. Transfer a 'spoonful' of positive charge from the positive terminal of a high voltage supply (e.g. 1 kV) to the electrometer input using, say, a 30 mm diameter metal sphere on an insulating handle as the 'spoon', Fig. 14.27. Each additional 'spoonful' produces the same increase of deflection.

The effect of different sized spoons and different supply p.ds can be investigated.

(d) Electrostatic induction

Arrange two metal spheres (30 mm in diameter) on insulated stands, initially touching each other, Fig. 14.28. Bring a charged polythene strip near to one sphere, separate the spheres and then remove the strip. Show that the charges induced on the spheres are equal in magnitude but opposite in sign by touching each in turn on the brass rod in the input socket of the electrometer (meter still on centre-zero).

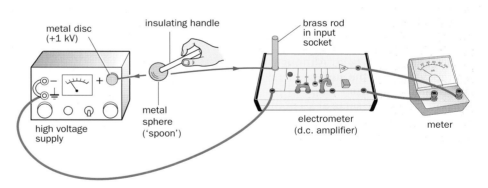

Fig. 14.27   Measuring a 'spoonful' of charge

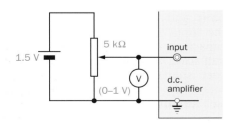

Fig. 14.26

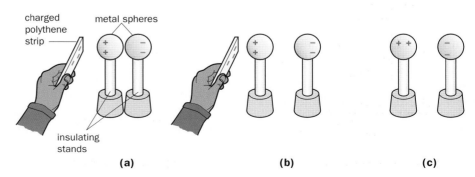

(a)   **(b)**   **(c)**

Fig. 14.28   Demonstrating electrostatic induction

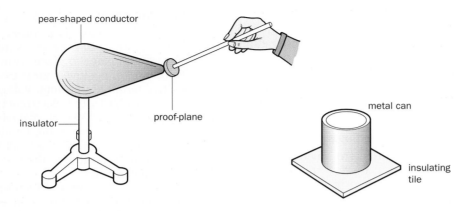

Fig. 14.29 Demonstrating variation of charge density on a conductor's surface

(e) Distribution of charge on conductors

The charge per unit area of the surface of a conductor, i.e. the charge density, is greatest on the most highly convex parts. Show this by transferring charge with a proof-plane (a small metal disc on an insulating handle) from different parts of various conductors on insulating stands, Fig. 14.29, to the electrometer. The conductors may be charged four or five times by an electrophorus.

The charge acquired by the proof-plane is proportional to the charge density on the surface touched since the proof-plane in effect becomes part of the charged surface. Although the charge density on the surface of a conductor may vary from place to place, the potential is the same at all points.

(f) Faraday's ice pail

Faraday investigated electrostatic induction using, by chance, a pail he had for storing ice. To repeat his experiment, plug a hollow metal sphere (75 mm diameter) into the input socket of the electrometer. Discharge a Perspex rod by quickly passing it through a flame. Plug a 30 mm sphere into the rod, charge the sphere from a polythene strip and lower it inside the hollow sphere without touching it, Fig. 14.30a. The meter deflects and does not change with the position of the charged sphere so long as it is well inside; nor does it alter when the charged body touches the inside of the hollow sphere and is then removed from it. Remove the sphere and show that it is completely discharged.

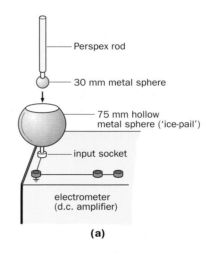

(a)

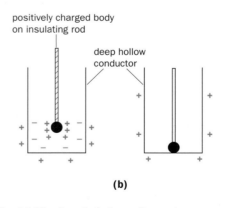

(b)

Fig. 14.30 Faraday's ice pail experiment

The conclusions to be drawn are:

- a charged body enclosed within a hollow conductor induces an equal charge of opposite sign on the inside of the conductor and an equal charge of the same sign as its own on the outside, Fig. 14.30b,

and
- inside a hollow conductor the net charge is zero; any excess charge resides on the outside surface.

The experiment also shows that all the charge on a conductor can be transferred to a hollow can if the conductor touches the can *well down inside*; only part of the charge would be transferred if the outside were touched.

(g) Conservation of charge

Use the 'ice pail' as above. First discharge strips of cellulose acetate and polythene in a flame, then rub them together inside the 'pail', Fig. 14.31. There should be little deflection until one or other of the strips is removed, when roughly equal and opposite deflections are obtained.

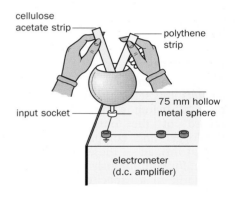

Fig. 14.31

(h) Charge sharing

Plug the 25 mm long brass rod into the input socket again and set the meter to 1 V full scale. Charge the internal 10^{-8} F capacitor by connecting 1 V to the input sockets; there should be a full-scale deflection on the meter. Remove the charging connections and connect externally across the input sockets a 10^{-8} F (0.01 μF) capacitor. The meter should fall to half-scale, showing the charge has been shared equally between capacitors of the same value. Try larger and smaller capacitors.

CAPACITOR EXPERIMENTS USING A D.C. AMPLIFIER

A calibrated d.c. amplifier can replace the reed switch in certain capacitor experiments considered earlier.

(a) Parallel-plate capacitor investigation *(also see p. 214)*

The arrangement is shown in Fig. 14.32, the capacitor consisting of two thick metal plates, each of side 25 cm, kept apart by four small polythene spacers (5 mm × 5 mm, about 1.5 mm thick) at the corners.

The capacitor is charged to a known p.d. using a flying lead from a protective resistance (10 MΩ) in series with the output from a high voltage power supply. The flying lead is removed and the charge on the capacitor measured by touching the upper plate with the tip of the screened extension cable from the *calibrated* electrometer set on the charge range. The input capacitor of the electrometer should be discharged (by pressing 'push to zero').

By varying the charging p.d. V in steps and noting the corresponding charges Q on the capacitor, it will be seen that $Q \propto V$.

If V is kept constant and the separation d of the plates changed by using more spacers at the corners, $Q \propto 1/d$ can be tested.

Finally, if V and d are fixed and the area of overlap A of the plates is varied, a test of $Q \propto A$ is possible.

(b) Measurement of ε_0 and ε_r *(also see p. 216)*

To find ε_0, the permittivity of free space, we use the expression for the capacitance of a parallel-plate capacitor, $C = A\varepsilon_0/d$ (p. 215) and experimental results from section (**a**) above. Since $Q = VC$ we have

$$\varepsilon_0 = \frac{Qd}{AV}$$

A mean value of Q/V can be obtained from a graph of Q against V; d and A are easily measured.

The relative permittivity ε_r of, say, polythene is found by charging the parallel-plate capacitor to the same p.d. — first with a sheet of polythene filling the space between the plates (giving capacitance C), then with the same thickness of vacuum or air between them (capacitance C_0) — and measuring with the electrometer the charges (Q and Q_0 respectively) stored in each case. The charges Q and Q_0 are in the ratio of the capacitances C and C_0, so

$$\varepsilon_r = \frac{C}{C_0} = \frac{Q}{Q_0}$$

(c) Decay curve for capacitor discharge *(also see p. 221)*

A capacitor–resistor combination is used, values of 10^{-8} F (0.01 μF) and 10^{10} Ω giving the convenient time constant of 100 s. The capacitor is first charged to 1 V to give a full-scale deflection on the meter, then the resistor is brought into circuit across it.

Meter readings are taken every 10 seconds while the capacitor discharges through the resistor. A decay curve is plotted from the results.

ACTION AT POINTS

The point of a pin is highly curved and if the charge density is sufficiently great, an intense electric field arises near the point. Background ionizing radiation produces electrons in the air which are accelerated by the intense field and cause ionization of the air by collision (p. 397). Ions having the same sign as the charge on the pin are strongly repelled from it to create an electric 'wind'. Ions with charges of opposite sign to that on the pin are attracted to the point and neutralize its charge. The net result of the 'action at points' is the apparent loss of charge from the pin to the surrounding air by what is termed a **point** or **corona discharge**.

The arrangement of Fig. 14.33*a* may be used to show a point discharge from a pin and collection of some of the charge by an electroscope in the path of the wind. The electric 'windmill' of Fig. 14.33*b* provides another demonstration of the electric wind. The windmill revolves rapidly when connected to a high potential due to the wind streaming away from each point. Why should apparatus working at high p.ds not have sharply curved parts?

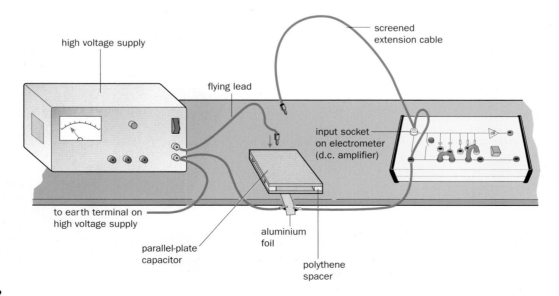

Fig. 14.32

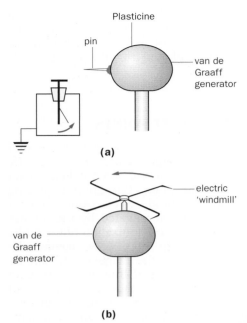

Fig. 14.33 Creating an electric 'wind'

The safe discharge of a thunder cloud by a lightning conductor on a tall building depends on action at points. It occurs at the tip of the metal spike (at the top of the building) which is connected by a thick copper conductor to a plate in the ground.

As well as losing charge, a sharp point can also collect it. When a negatively charged polythene strip is held close to the point of a pin taped to the cap of an electroscope, Fig. 14.34, a deflection is obtained which remains on removing the polythene. Electrostatic induction occurs as shown, then due to action at points a positive ion wind is created, neutralizing part of the charge on the strip. At the same time negative ions in the air are attracted to the point, neutralizing its induced positive charge. As a consequence the electroscope gains negative charge, apparently collected by the point of the pin.

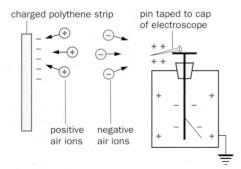

Fig. 14.34

VAN DE GRAAFF GENERATOR

(a) Action

Large van de Graaff generators are used to develop p.ds of up to 14 million volts for accelerating atomic particles in nuclear physics. In the 'tandem' van de Graaff of Fig. 14.35, two generators (brown and green) produce beams of ions and other charged particles, which are accelerated by a p.d. of up to 10 million volts. The large magnet in the foreground bends the beams and separates them.

A simplified version is shown in Fig. 14.36. Initially a positive charge is produced on the motor-driven Perspex roller by friction between it and the rubber belt. This charge induces a negative charge on the earthed comb of metal points P (by drawing electrons from earth) which is then sprayed off by action at points on to the outside of the belt and carried upwards. The comb of metal points Q is connected to the inside of an insulated metal sphere and when the negative charge on the upward-moving belt reaches Q, positive charge is induced on Q owing to the repulsion of negative charge to the surface of the sphere. By action at points Q has apparently drawn off negative charge from the belt on to the sphere. The belt becomes positively charged as it moves down because of (*i*) Q spraying charge on to it and (*ii*) friction between the belt and the polythene roller.

Fig. 14.35 Van de Graaff generator

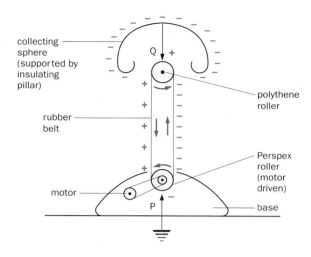

Fig. 14.36   Principle of the van de Graaff generator

A large force is set up as a result of repulsion between the negative charge on the collecting sphere and the negative charge on the upward-moving belt. The motor driving the roller therefore has to do work against this repulsion.

(b) Estimating the p.d.

Small generators can produce p.ds of up to a few hundred thousand volts, the insulation of the surrounding air being an important limiting factor.

A rough idea of the p.d. can be obtained by finding the greatest separation of the collecting sphere and a discharging sphere for a spark to pass between them. Under dry dust-free conditions at s.t.p., spark discharges of length 1 cm, 7 cm and 12 cm occur between two *rounded* electrodes when the p.ds are roughly 30 kV, 250 kV and 750 kV respectively.

Another method, not requiring the assumption of the above information, involves connecting a microammeter (100 μA) in series with the collecting sphere and base of the generator and measuring the steady current *I* which passes with the generator working at a certain speed, Fig. 14.37a. The meter is then removed and the discharging sphere brought up until regular sparking *just* occurs between it and the collecting sphere, Fig. 14.37b, with the generator running at the same speed as before. The number of sparks passing per second is then found with a stop-watch. The potential can be calculated as shown in the following example.

If Q is the charge that builds up on the collecting sphere and passes with each spark and t is the interval between sparks, then the discharge may be regarded as equivalent to a steady current of value Q/t. Assuming that this current is the same as the current I recorded by the microammeter, we can say $I = Q/t$. Suppose the discharge provides two sparks per second, then $t = 0.5$ s. If $I = 10\ \mu A = 1.0 \times 10^{-5}$ A, then

$$Q = It = (1.0 \times 10^{-5}\ \text{A}) \times (0.5\ \text{s})$$
$$= 5.0 \times 10^{-6}\ \text{C}$$

But $Q = VC$ where V is the potential of the collecting sphere and C is its capacitance. For an isolated sphere (which the collecting 'sphere' is not), $C = 4\pi\varepsilon_0 a$ where a is the radius of the sphere (p. 213). If $a = 13$ cm $= 13 \times 10^{-2}$ m, then $C = (4\pi \times 9.0 \times 10^{-12}\ \text{F m}^{-1}) \times (13 \times 10^{-2}\ \text{m})$ since $\varepsilon_0 \approx 9.0 \times 10^{-12}\ \text{F m}^{-1}$. Therefore $C \approx 1.5 \times 10^{-11}$ F, and

$$V = \frac{Q}{C} = \frac{5.0 \times 10^{-6}\ \text{C}}{1.5 \times 10^{-11}\ \text{F}}$$

$$\approx 3.3 \times 10^5\ \text{V}\quad (330\ \text{kV})$$

QUESTIONS

1. **a)** What is the capacitance of a small metal sphere of radius 1 cm? (Take $4\pi\varepsilon_0 = 10^{-10}$ F m^{-1}.)
   b) What charge is stored on it when it briefly touches the 1 kV terminal of a high voltage power supply?
   c) If the sphere now shares its charge with a 0.001 μF capacitor, what is the ratio of the charge on the capacitor to that on the sphere?
   d) What, in effect, will be the charge on
       i) the capacitor, and
       ii) the sphere?
   e) If the capacitor had a capacitance of 1 pF (10^{-12} F) how would the charge have been shared?

2. **a)** In the circuit of Fig. 14.38a what is the p.d. across each capacitor? What is the total charge stored? What is the capacitance of the single capacitor which would store the same charge as the two capacitors together?

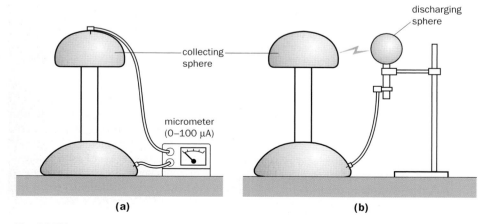

Fig. 14.37

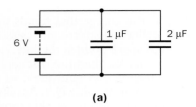

(a)

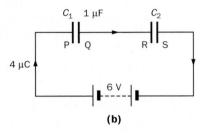

(b)

Fig. 14.38

b) In Fig. 14.38*b* if a charge of 4 µC flows from the 6 V battery to plate P of the 1 µF capacitor, what charge flows from
   i) Q to R, and
   ii) S to the battery?
What is the p.d. across
   iii) C_1, and
   iv) C_2?
What is the capacitance of
   v) C_2, and
   vi) the single capacitor which is equivalent to C_1 and C_2 in series, and what charge would it store?

3. a) i) On axes like those in Fig. 14.39, sketch a graph to show the variation with charge Q of the potential difference V across the plates of a capacitor.

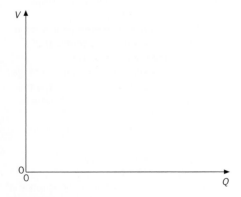

Fig. 14.39

   ii) Use your graph to show that the energy E stored in a capacitor of capacitance C with a p.d. V across its plates is given by

$$E = \tfrac{1}{2}CV^2$$

b) In Fig. 14.40, each capacitor has a capacitance of 2200 µF. The p.d. across each capacitor must not exceed 25 V.

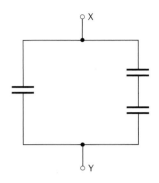

Fig. 14.40

For the arrangement of Fig. 14.40, calculate
   i) the total capacitance between X and Y,
   ii) the maximum energy that can be stored.
   (*UCLES*, Further Physics, March 1998)

4. a) A parallel-plate capacitor consists of two square plates each of side 25 cm, 3.0 mm apart. If a p.d. of 200 V is applied, calculate the charge on the plates with
   i) air, and
   ii) paper of relative permittivity 2.5, filling the space between them.
   ($\varepsilon_0 = 8.9 \times 10^{-12}\,\text{F m}^{-1}$)
b) A tubular 0.10 µF capacitor is to be made from a sheet of plastic 2.0 cm wide and 1.0×10^{-3} cm thick rolled between metal foil of the same width. What length of plastic is required if its relative permittivity is 2.4 and $\varepsilon_0 = 8.9 \times 10^{-12}\,\text{F m}^{-1}$?

5. The plates of an air-filled parallel plate capacitor are each of area $1.5 \times 10^{-2}\,\text{m}^2$ and are initially separated by 2.0 mm.
a) Calculate
   i) the capacitance of the capacitor,
   ii) the charge on the capacitor when connected across a 2500 V d.c. supply.
b) The plates of the capacitor are brought closer together while it is still connected to the 2500 V supply. State and explain what happens to the size of the charge on the plates.
c) The maximum electric field strength that the air can support before an electrical breakdown occurs is $3.0 \times 10^6\,\text{V m}^{-1}$. The plates of the capacitor are brought together, with the 2500 V supply connected, so that they

are as close as possible without causing breakdown.
   Calculate the separation of the plates when this condition is reached.
   (*NEAB*, AS/A PHO3, June 1998)

6. A 10 µF capacitor is connected across the terminals of a 100 V d.c. power supply and allowed to charge fully.
a) Calculate
   i) the charge on the capacitor,
   ii) the energy stored by the capacitor.
b) The fully-charged capacitor is disconnected from the power supply and connected via two wires across the terminals of an uncharged 10 µF capacitor as shown in Fig. 14.41.

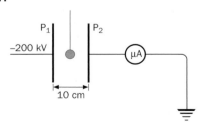

Fig. 14.41

The charge on the original 10 µF capacitor is shared equally between the capacitors in the parallel combination.
   i) Calculate the potential difference across the terminals of each capacitor.
   ii) Calculate the total energy stored by the two capacitors.
   iii) Account for the difference between the energy stored by the two capacitors in parallel and that stored by the original single 10 µF capacitor.
   (*AQA: NEAB*, AS/A PHO1, March 1999)

7.

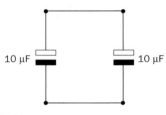

Fig. 14.42

Two parallel metal plates P_1 and P_2 are placed 10 cm apart in air, Fig. 14.42. P_1 is maintained at a potential of -200 kV whilst P_2 is connected to earth through a microammeter. Suspended between the plates by an insulating thread is a light plastic sphere, the surface of which is coated with conducting paint so that it will store charge.

a) Calculate the electric field strength between the plates.

b) The sphere, whose capacitance is 6.0×10^{-13} F, shuttles back and forth between the plates 360 times per minute, contacting each plate alternately. Calculate

　i) the magnitude of the charge it acquires every time it touches plate P_1, and state the sign of this charge,

　ii) the current flowing through the microammeter.

c) Explain why the ball shuttles between the plates.

(*NEAB*, AS/A PHO3, June 1997)

8. A 47 μF capacitor is used to power the flashgun of a camera. The average power output of the flashgun is 4.0 kW for the duration of the flash which is 2.0 ms.

i) Show that the energy output of the flashgun per flash is 8.0 J.

ii) Calculate the potential difference between the terminals of the capacitor immediately before a flash. Assume that all the energy stored by the capacitor is used to provide the output power of the flashgun.

iii) Calculate the maximum charge stored by the capacitor.

iv) Calculate the average current provided by the capacitor during a flash.

(*NEAB*, AS/A PHO1, March 1998)

9. Discuss the essential differences between current and static electricity.

Describe and explain the working of a device which can supply a current of the order of 10^{-6} A and maintain a potential difference of the order of 10^{6} V.

A high voltage generator continuously supplies charge to a spherical dome of radius 15 cm, which gives 10 sparks each minute to a nearby earthed sphere. While the generator is operating at the same rate, the dome is connected to earth through a galvanometer, which indicates a steady current of 2.0×10^{-6} A. Calculate

a) the charge passed in each spark,

b) the maximum potential difference between the dome and earth, and

c) the distance from the surface of the dome to the surface of the nearby earthed sphere.

Make the simplifying assumptions that an isolated spherical dome of 1 cm radius has a capacitance of 10^{-12} F, and that in this system the mean potential gradient needed for sparks to pass is 4×10^4 V cm^{-1}.

10. The circuit shown in Fig. 14.43 is used to charge a capacitor.

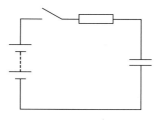

Fig. 14.43

The graph in Fig. 14.44 shows the charge stored on the capacitor whilst it is being charged.

Sketch as accurately as you can a graph of current against time. Label the current axis with an appropriate scale.

The power supply is 3 V. Calculate the resistance of the charging circuit.

(*L*, AS/A PH1, Jan 1997)

11. An uncharged capacitor of capacitance 2.2 μF is connected in series with a resistor of resistance 11 kΩ and a d.c. power supply of 12 V with negligible internal resistance, as shown in Fig. 14.45.

a) At one instant after the switch is closed, the charging current in the circuit is 0.40 mA. For this instant,

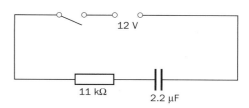

Fig. 14.45

calculate

　i) the potential difference across the resistor,

　ii) the potential difference across the capacitor,

　iii) the charge on the capacitor.

b) After some time, the capacitor becomes fully charged. Calculate the energy which is then stored in the capacitor.

(*UCLES*, Basic 1, March 1998)

12. A parallel-plate capacitor of capacitance C is charged by battery to a p.d. V and stores a charge Q and an amount of energy E. The battery is disconnected and the separation of the capacitor plates is reduced to one-quarter of its previous value. What will now be the value of

a) the capacitance in terms of C,

b) the p.d. in terms of V, and

c) the energy stored in terms of E?

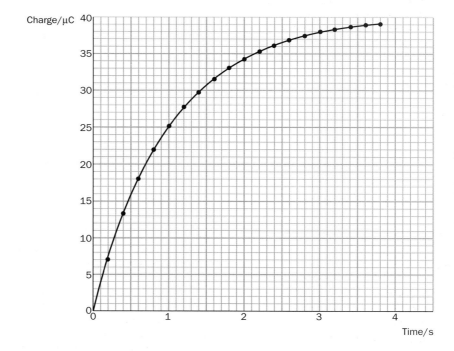

Fig. 14.44

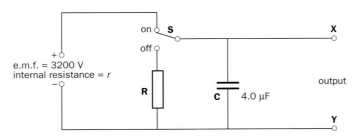

Fig. 14.46

13. Figure 14.46 shows the diagram of a
laboratory power supply which has an
e.m.f. of 3200 V and internal resistance
r. The capacitor **C** ensures a steady
output voltage between **X** and **Y** when
the switch **S** is in the **on** position. The
resistor **R** allows the capacitor to
discharge when **S** is switched to the **off**
position.

a) For safety reasons, the power
supply is designed with an internal
resistance, r, so that if **X** and **Y** are
short-circuited, the current that flows
does not exceed 2.0 mA.
   Calculate the value of r.

b) For safety reasons, when the supply
is switched off, the capacitor must
discharge from 3200 V to 50 V in 5.0
minutes.
   i) Calculate the time constant of the
   circuit when the capacitor
   discharges from 3200 V to 50 V in
   5.0 minutes.
   ii) Calculate the value of the
   discharge resistor **R**.
   iii) Explain briefly whether or not
   increasing the value of **R** would
   make the supply safer.

c) i) Calculate the charge on the
   capacitor when the potential
   difference across it is 3200 V.
   ii) Sketch a graph on axes like
   those in Fig. 14.47 to show the
   variation of the charge on the
   capacitor with time, as the capacitor
   discharges through the resistor.

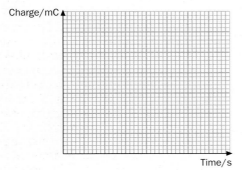

Fig. 14.47

Include appropriate scales on the
axes of your graph.
   iii) Sketch a graph on axes like
   those in Fig. 14.47 to show the
   variation of the charge on the
   capacitor with voltage across the
   capacitor as it discharges. Include
   appropriate scales on the axes of
   your graph.
   (AEB, 0635/6, Jan 1998)

14. **a)** By considering a relevant
   capacitance formula, show that the SI
   unit of permittivity is $F\,m^{-1}$.
   b) A demonstration-type parallel-plate
   capacitor has large flat plates with a
   fixed area of overlap and a fixed
   separation. Suggest how the
   capacitance of the capacitor can be
   varied.
   c) Figure 14.48 shows a fixed
   capacitor C_F of capacitance 100 pF and
   a variable capacitor C_V.

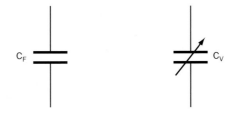

Fig. 14.48

The capacitance of C_V can be varied in
the range 10 pF to 50 pF. Draw
arrangements, both including C_F and C_V,
which will give a resultant capacitance
of
   i) 120 pF,
   ii) 20 pF.
   In each case state the capacitance of
C_V.
   (OCR, 6843, June 1999)

15. It is proposed to supply enough energy
from a charged capacitor so that, in the
event of an emergency, a hatch door
can be lifted a vertical distance of
0.31 m. The mass of the hatch door is
0.56 kg and the capacitor would be
automatically connected to a motor

which could lift the door in the
emergency.
   a) i) Calculate the gain in the potential
   energy of the hatch door when it is
   raised.
   ii) The maximum potential
   difference which can be applied
   across the capacitor is 30 V.
   Calculate the capacitance of the
   capacitor required to supply the
   energy in i).
   b) Give two reasons why, in practice, a
   capacitor of greater capacitance would
   be required.
   c) Suggest why the motor on such a
   system would initially run quickly but
   slow down during the lifting of the door.
   (OCR, Basic 1 4831, March 1999)

16. In the circuit in Fig. 14.49, switch A is
   initially closed and switch B is open.
   Calculate the energy stored in the 3 μF
   capacitor when it is fully charged.

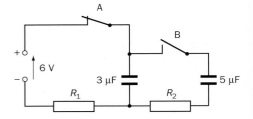

Fig. 14.49

Switch A is now opened and swtich B is
closed. Calculate the final value of the
total energy stored in the two capacitors
when the 5 μF capacitor is fully
charged.
   State briefly how you would account
for the decrease in stored energy.
   (L, AS/A PH1, June 1999)

15

Magnetic fields

- ■ Importance of electromagnetism
- ■ Fields due to magnets
- ■ Earth's magnetic field
- ■ Fields due to currents
- ■ Force on a current in a magnetic field
- ■ Magnetic flux density

- ■ The Biot-Savart law
- ■ Calculation of flux density
- ■ Magnetic field measurements
- ■ Force on a charge in a magnetic field
- ■ Hall effect
- ■ The ampere and current balances

- ■ Torque on a coil in a magnetic field
- ■ Moving-coil galvanometers
- ■ Moving-coil loudspeaker
- ■ Relay

IMPORTANCE OF ELECTROMAGNETISM

The discovery in 1819 by Professor Oersted at the University of Copenhagen that an electric current is accompanied by magnetic effects saw the birth of electromagnetism. Since then, the subject has been an extremely fruitful field of study for the imagination of scientists and the creative genius of engineers.

Faraday announced the discovery of electromagnetic induction in 1831 and developed the idea of a 'field' for dealing with such action-at-a-distance effects. Clerk Maxwell put Faraday's field ideas into mathematical form and predicted electromagnetic waves. At the beginning of the twentieth century, Einstein pondered over the mysteries of electromagnetism, especially the need for relative motion, and was led to the theory of relativity (see p. 504).

The dynamo, the motor and the transformer were the direct outcome of Faraday's work. Edison was responsible for opening the first power station in 1882, designed to supply electricity to domestic consumers in New York. Shortly afterwards Tesla invented the induction motor which does not require brushes (to supply current to its rotor) and which is now the indispensable servant of industry. In 1887 Hertz produced radio-type

electromagnetic waves and then Marconi, contrary to all theoretical predictions, transmitted them across the Atlantic. Remarkable developments in electrical engineering continued right through the twentieth century including, for example, the linear motor which, unlike conventional rotary motors, is flat and eliminates the energy losses occurring when rotary motion is converted to the often-required linear motion.

Modern civilization is heavily indebted to electromagnetism, the topic of this chapter and the next.

FIELDS DUE TO MAGNETS

The magnetic properties of a magnet appear to originate at certain regions in the magnet which we call the **poles**; in a bar magnet these are near the ends. Experiments show that (*i*) magnetic poles are of two kinds, (*ii*) like poles repel each other and unlike poles attract, (*iii*) poles always seem to occur in equal and opposite pairs, and (*iv*) when no other magnet is near, a freely suspended magnet sets so that the line joining its poles (i.e. its magnetic axis) is approximately parallel to the earth's north–south axis, i.e. to the magnetic meridian.

The last fact suggests that the earth itself behaves like a large permanent

magnet and it makes it appropriate to call the pole of a magnet that points (more or less) towards the earth's geographical North Pole, the **north pole** of the magnet and the other the **south pole**. What kind of magnetic pole must be near the earth's geographical North Pole?

The space surrounding a magnet where a magnetic force is experienced is called a **magnetic field**. The direction of a magnetic field at a point is taken as the direction of the force that acts on a *north* magnetic pole there. A magnetic field can be represented by magnetic **field lines** drawn so that (*i*) the line (or the tangent to it if it is curved) gives the direction of the field at that point, and (*ii*) the number of lines per unit cross-section area is an indication of the 'strength' of the field. Arrows on the lines show the direction of the field and, since a north pole is repelled by the north pole of a magnet and attracted by the south, the arrows point away from north poles and towards south poles.

Field lines can be obtained quickly with iron filings or accurately plotted using a small compass (i.e. a pivoted magnet). Some typical field patterns are shown in Fig. 15.1. The field round a bar magnet varies in strength and direction from point to point, i.e. is non-uniform. Locally the earth's magnetic field is uniform; the lines are parallel, equally spaced and point north.

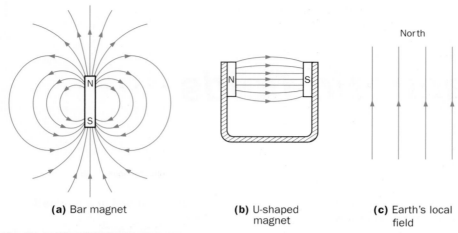

(a) Bar magnet

(b) U-shaped magnet

(c) Earth's local field

Fig. 15.1   Magnetic field patterns

EARTH'S MAGNETIC FIELD

The magnitude and direction of the earth's field varies with position over the earth's surface and it also seems to be changing gradually with time. The pattern of field lines is similar to that which would be given if there were a strong bar magnet at the centre of the earth, Fig. 15.2a. At present there is no generally accepted theory of the earth's magnetism but it may be caused by electric currents circulating in its core due to convection currents arising from radioactive heating inside the earth.

In Britain the earth's field is inclined downwards at an angle of about 70° to the horizontal — called the **angle of dip**, α. At the magnetic poles (which are near the geographical poles but

whose positions change slowly) the field is vertical, that is, $\alpha = 90°$. At the magnetic equator it is parallel to the earth's surface, i.e. horizontal, and $\alpha = 0°$.

It is convenient to resolve the earth's 'field strength' B_R into horizontal and vertical components, B_H and B_V respectively. We then have from Fig. 15.2b that

$$B_H = B_R \cos \alpha$$

and

$$B_V = B_R \sin \alpha$$

So

$$\tan \alpha = \frac{B_V}{B_H}$$

Compass needles, whose motion is confined to a horizontal plane, are affected by B_H only.

A **neutral point** is a place where two magnetic fields are equal and opposite and the resultant force is zero. Two such points are shown in Fig. 15.3 in the combined field due to the earth and a bar magnet with its N pole pointing north. Where will they be when the N pole of the magnet points south?

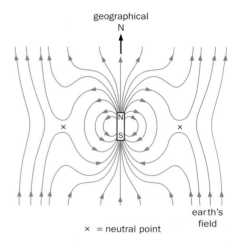

× = neutral point

Fig. 15.3   Location of neutral points

FIELDS DUE TO CURRENTS

A conductor carrying an electric current is surrounded by a magnetic field. The field patterns due to conductors of different shapes can be obtained as for a magnet using iron filings or a plotting compass.

The lines due to a straight wire are circles, concentric with the wire, Fig. 15.4a. The **right-hand screw rule** is a useful aid for predicting the direction of the field, knowing the direction of the current. It states:

If a right-handed screw moves forward in the direction of the current (conventional), then the direction of rotation of the screw gives the direction of the magnetic field lines.

Figures 15.4b and c summarize the rule; in b the current is flowing out of the paper and the dot in the centre of the wire is the point of an approaching arrow; in c the current is flowing into the paper and the cross is the tail of a receding arrow.

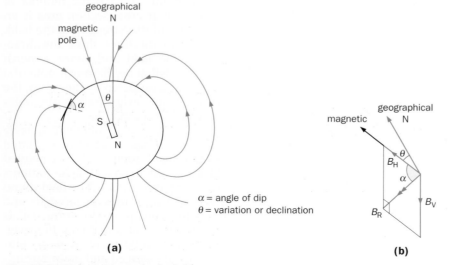

α = angle of dip
θ = variation or declination

(a)

(b)

Fig. 15.2   The earth's magnetic field

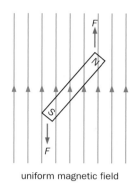

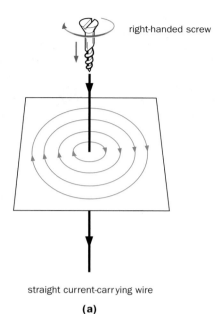

straight current-carrying wire

(a)

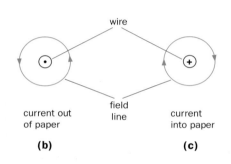

current out of paper   field line   current into paper

(b)   **(c)**

Fig. 15.4 The right-hand screw rule

The field pattern due to a current in a plane circular coil is shown in Fig. 15.5*a*, and one due to a current in a long cylindrical coil — called a **solenoid** — in Fig. 15.5*b*. In both cases the field direction is again given by the right-hand screw rule. A solenoid pro-

duces a field similar to that of a bar magnet; in Fig. 15.5*b* the left-hand end behaves like the north pole of a bar magnet and the right-hand end like the south pole. In general, the field produced by any magnet can also be produced by a current in a suitably shaped conductor. It is this fact that leads us to believe that *all* magnetic effects are due to electric currents, which in the case of permanent magnets arise from the motion of electrons within the atoms themselves.

In brief, we can say that a static electric charge gives rise to an electric field, and a moving electric charge creates a magnetic field as well as an electric field.

FORCE ON A CURRENT IN A MAGNETIC FIELD

A magnet in a uniform magnetic field experiences two equal but opposite parallel forces, i.e. a couple, which gives it an angular acceleration and — provided there is damping — it ultimately comes to rest with its axis parallel to the field, Fig. 15.6. In a non-uniform field, the poles of the magnet are acted on by unequal forces and these may cause translational as well as rotational acceleration. A current-carrying conductor behaves like a magnet and so it is not surprising to find that it too experiences a force in a magnetic field. In the next section we will see how this force can be used to define and measure the strength of a magnetic field. But first some basic experimental facts.

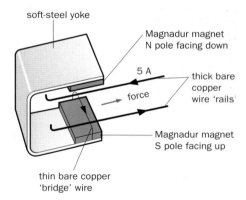

uniform magnetic field

Fig. 15.6

(a) Fleming's left-hand rule

Using the apparatus of Fig. 15.7 it can be observed that when current passes through the 'bridge' wire, the bridge shoots along the 'rails' if the magnetic field of the U-shaped magnet is perpendicular to it (as shown). There is no motion if the field is parallel to it. From such experiments we conclude that the force on a current-carrying conductor (*i*) is always perpendicular to the plane containing the conductor and the direction of the field in which it is placed, and (*ii*) is greatest when the conductor is at right angles to the field.

soft-steel yoke
Magnadur magnet N pole facing down
5 A
force
thick bare copper wire 'rails'
Magnadur magnet S pole facing up
thin bare copper 'bridge' wire

Fig. 15.7

The facts about the relative directions of current, field and force are summarized by **Fleming's left-hand (or motor) rule**. It states:

> If the thumb and first two fingers of the left hand are held each at right angles to the other, with the **F**irst finger pointing in the direction of the **F**ield and the se**C**ond finger in the direction of the **C**urrent, then the **Th**umb predicts the direction of the **Th**rust or force, Fig. 15.8.

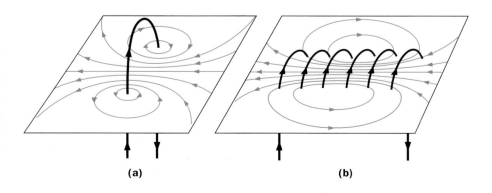

(a)   **(b)**

Fig. 15.5 Field patterns due to (*a*) a plane circular coil, (*b*) a solenoid

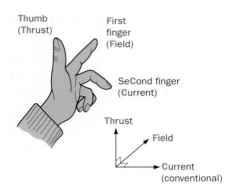

Fig. 15.8 Fleming's left-hand rule

(b) Factors affecting the force

The effect of the current, the length of the conductor and the strength and orientation of the magnetic field may be investigated by passing current through one arm PQ of a copper wire frame balanced on two razor blades through which the current enters and leaves, Fig. 15.9*a*. The ends of the wire in the arm RS of the frame are held together by an insulator and no current flows in RS. If the magnet is arranged so that PQ rises when current passes, the balance can be restored and the force measured by placing a suitable rider (a short length of wire) on PQ.

To see how the force depends on the *current* it is convenient to find the currents required to balance one, two and three identical riders on PQ. The effect on the force of doubling and trebling the *length* of conductor carrying the same current in the same magnetic field can be found by placing a second

and a third, equally strong, magnet beside the first, Fig. 15.9*b*. If the magnetic field is now produced by passing a current through a coil of many turns (for example 1100), Fig. 15.9*c*, increasing the coil current can reasonably be expected to increase the *field strength*. The response of the force on PQ to this may then be observed and also the effect of the *angle* between the direction of the field and PQ.

MAGNETIC FLUX DENSITY

(a) Definition of B

The experiments described in the previous section show that the force F on a wire lying at right angles to a magnetic field is directly proportional to the current I in the wire and to the length l of the wire in the field. It also depends on the magnetic field, a fact we can use to define the strength of the field.

Electric field strength E is defined as the force per unit charge; gravitational field strength g is force per unit mass. An analogous quantity for magnetic fields is the **flux density** or **magnetic induction** B (also called the B-field), defined as **the force acting per unit current length**, i.e. the force acting per unit length on a conductor which carries unit current *and is at right angles to the direction of the magnetic field*. In symbols, B is defined by the equation

$$B = \frac{F}{Il}$$

If $F = 1$ newton when $I = 1$ ampere and $l = 1$ metre then $B = 1$ newton per ampere metre, i.e. the SI unit of B is the newton per ampere metre ($\text{N A}^{-1} \text{m}^{-1}$), which is given the special name of 1 **tesla** (T).

B is a vector whose direction at any point is that of the field line at the point. Its magnitude may be represented pictorially by the number of field lines passing through unit area.

(b) Expression for force on a current

Rearranging the equation defining B we see that the force F on a conductor of length l, carrying a current I and lying at right angles to a magnetic field of flux density B, is given by

$$F = BIl$$

If the conductor and field are not at right angles, but make an angle θ with one another, Fig. 15.10, the expression becomes

$$F = BIl \sin \theta$$

When $\theta = 90°$, $\sin \theta = 1$ and $F = BIl$ as before. If $\theta = 0°$, the conductor and field are parallel and, as experiment confirms, $F = 0$.

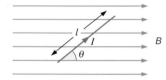

Fig. 15.10

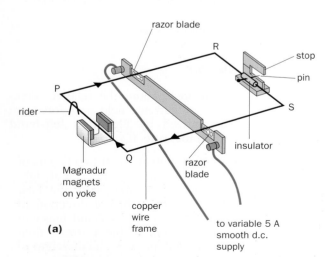

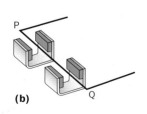

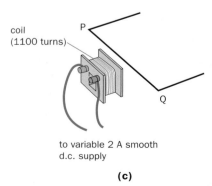

Fig. 15.9 Investigating the factors affecting the force

(c) Measuring B using a current balance

The copper wire frame used previously to investigate the factors affecting the force on a current in a magnetic field is a very simple 'current balance' (see p. 241) and enables us to measure the flux density of a field.

For example, suppose a rider of mass 0.084 g is required to counter-poise the frame when arm PQ, of length 25 cm and carrying a current of 1.2 A, is inside and in series with a flat, wide solenoid, Fig. 15.11.

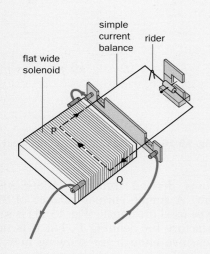

Fig. 15.11

We then have

F = force on arm PQ

= weight of rider

= mass of rider $\times$ g

= $(0.084 \times 10^{-3} \, \text{kg}) \times (10 \, \text{N kg}^{-1})$

= $0.084 \times 10^{-2} \, \text{N}$

I = 1.2 A

l = 0.25 m

Flux density B (assumed uniform) inside the solenoid is given by

$$B = \frac{F}{Il} = \frac{8.4 \times 10^{-4} \, \text{N}}{1.2 \, \text{A} \times 0.25 \, \text{m}}$$

$$= 2.8 \times 10^{-3} \, \text{T}$$

The field in the gap between two Magnadur magnets on a mild steel yoke (Fig. 15.9a) is about ten times stronger than this, while in the earth's magnetic field in Britain, $B_\text{R} \approx 5.3 \times 10^{-5} \, \text{T}$ and $B_\text{H} \approx 1.8 \times 10^{-5} \, \text{T}$. A flux density of 1 T would be produced by a fairly strong magnet.

THE BIOT-SAVART LAW

In the previous section we saw how B could be found experimentally; it can also be calculated utilizing the Biot-Savart law.

The calculation involves considering a conductor as consisting of a number of very short lengths, each of which contributes to the total field at any point. Biot and Savart *stated* that for a very short length δl of conductor, carrying a steady current I, the magnitude of the flux density δB at a point P distance r from δl is

$$\delta B \propto \frac{I \, \delta l \sin \theta}{r^2}$$

where θ is the angle between δl and the line joining it to P, Fig. 15.12. The direction of δB, given by the right-hand rule, is at right angles to the plane containing δl and P and, for the current direction shown, acts into the paper. The product $I\delta l$ is called a 'current element'. It is difficult to prove Biot and Savart's law directly for an infinitesimally small length of conductor, i.e. a current element, but experimentally verifiable deductions can be made from it for 'life-sized' conductors.

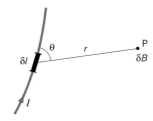

Fig. 15.12

Before the Biot-Savart law can be used in calculations it has to be expressed as an equation, and a constant of proportionality introduced. Now δB depends on one other factor besides I, δl, θ and r, and that is the medium through which the magnetic force acts. This factor can be incorporated in the Biot-Savart law if we regard the constant of proportionality as a property of the medium. The con-

stant is called the **permeability** of the medium and is denoted by μ (mu).

The permeability of a vacuum is denoted by μ_0 ('mu nought'), its value is *defined* (from the definition of the ampere, p. 241) to be $4\pi \times 10^{-7}$ and its unit is the **henry per metre** (H m^{-1}), as we shall see later. Air and most other materials except ferromagnetics have nearly the same permeability as a vacuum. (Note that while the value of ε_0 — the permittivity of free space — is found by experiment, that for μ_0 is defined.)

Rationalization (see p. 200) is accomplished by introducing 4π into the denominator. The Biot-Savart equation for a current element becomes

$$\delta B = \frac{\mu_0 I \delta l \sin \theta}{4\pi r^2}$$

It is an inverse square relation.

CALCULATION OF FLUX DENSITY

In most cases the calculation requires the use of calculus but for a circular coil the value of B at the centre can be found fairly simply.

(a) Circular coil

Suppose the coil is in air, has radius r, carries a steady current I and is considered to consist of current elements of length δl. Each element is at distance r from the centre O of the coil and is at right angles to the line joining it to O, i.e. $\theta = 90°$, Fig. 15.13.

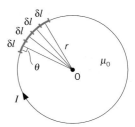

Fig. 15.13

At O the total flux density B is the sum of the flux densities δB due to all the elements. That is

$$B = \sum \frac{\mu_0 I \, \delta l \sin \theta}{4\pi r^2} = \frac{\mu_0 I \sin \theta}{4\pi r^2} \sum \delta l$$

But $\sum \delta l$ = total length of coil = $2\pi r$ and $\sin \theta = \sin 90° = 1$. Hence

$$B = \frac{\mu_0 I \, 2\pi r}{4\pi r^2} = \frac{\mu_0 I}{2r}$$

If the coil has N turns each of radius r

$$B = \frac{\mu_0 N I}{2r}$$

(b) Very long straight wire

It can be shown that the value of B at a perpendicular distance a from a very long straight wire carrying a current I, Fig. 15.14, is

$$B = \frac{\mu_0 I}{2\pi a}$$

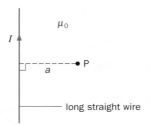

Fig. 15.14

In this case the field is non-uniform, there is cylindrical symmetry and 2π appears quite appropriately. If $I = 10$ A and $a = 1.0$ cm $= 0.010$ m, then $B = 2.0 \times 10^{-4}$ T.

(c) Very long solenoid

If the solenoid has N turns, length l and carries a current I, the flux density B at a point O on the axis near the centre of the solenoid, Fig. 15.15, is found to be given by

$$B = \frac{\mu_0 N I}{l}$$

or

$$B = \mu_0 n I$$

where $n = N/l$ = number of turns per unit length. B therefore equals μ_0 multiplied by the *ampere-turns per metre*. For a long solenoid the magnetic field is uniform over the cross-section for most of its length (see Fig. 15.5b, p. 233) and the flux density is given by the above expression. The absence of

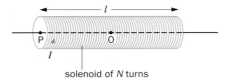

Fig. 15.15

π from it may be noted and is what we expect from the plane symmetry of a uniform field using a rationalized system.

For a solenoid of 400 turns, 25 cm long, carrying a current of 5.0 A, we have $n = 400/0.25 = 1600$ turns per metre, $I = 5.0$ A and therefore $B \approx 1.0 \times 10^{-2}$ T.

At P, a point at the end of a long solenoid,

$$B = \frac{\mu_0 n l}{2}$$

That is, B at a point at the end of the solenoid's axis is half the value at the centre.

These expressions are strictly true only for an infinitely long solenoid; in practice they are accurate enough for most purposes if the length of the solenoid is at least ten times its diameter.

(d) Helmholtz coils

Two flat coaxial coils having the same radius r, the same number of turns N and separated by a distance r, Fig. 15.16a, produce an almost uniform field over some distance near their common axis midway between them, Fig. 15.16b, when the same current I flows round each in the same direction. It is given by

$$B \approx 0.72 \frac{\mu_0 N I}{r}$$

Helmholtz coils are used in a school laboratory measurement of e/m for electrons (pp. 388–9).

MAGNETIC FIELD MEASUREMENTS

Measurement of B for a variety of current-carrying conductors can be made and compared with the values and variations predicted by expres-

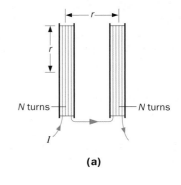

(a)

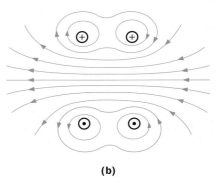

(b)

Fig. 15.16 Helmholtz coils and their field

sions derived from the Biot-Savart law. A simple current balance of the type used previously (Fig. 15.11) is satisfactory for defining B, but it is not sensitive enough for showing how B varies from one point to another in a field. Two other devices are convenient. One is the **search coil**, used with alternating fields produced by alternating currents, and the other is the **Hall probe**, for steady fields produced by direct currents.

(a) Search coil and a.c.

A typical search coil has 5000 turns of wire and an external diameter of no more than 1.5 cm so that it samples the field over a small area. If it is connected to the Y-plates of a CRO (as in Fig. 15.18) with its time base off, the length of the vertical trace is proportional to the maximum value of the alternating flux density, or any such component acting *along the axis of the search coil at right angles to its face*. (The technique uses an effect called **mutual induction** which will be considered in the next chapter. The CRO is in fact being used as a voltmeter to measure an e.m.f. induced in the search coil by the alternating magnetic field. The sensitivity of the method increases with the frequency of the a.c. in the conductor producing the field and 10 kHz is a convenient frequency to use.)

(b) Hall probe and d.c.

The Hall probe consists of a slice of semiconducting material on the end of a rod (Fig. 15.17), connected to a sensitive galvanometer via a circuit box. When a battery in the box is switched on, current is supplied to the slice. The 'balance' control is adjusted for zero deflection on the galvanometer and the probe is then inserted in the field to be measured (see Fig. 15.19*b*). The change in the galvanometer reading is proportional to the flux density (or its component) *at right angles to the slice*. The action of the probe depends on the Hall effect, to be discussed shortly (p. 239).

Fig. 15.17 Hall probe being used to measure the field strength between the poles of a magnet

(c) Investigations

If actual values of *B* are required, the search coil or Hall probe would have to be calibrated using known values of *B*. For studying comparisons and relationships this is not necessary.

A long straight wire may be investigated using the arrangement in Fig. 15.18. What is the direction of *B* near the wire? Is *B* the same all along the wire at a constant distance from it? Is *B* directly proportional to the current in the wire? Does *B* vary as $1/a$ (where *a* is the perpendicular distance from the wire), as theory suggests?

Using a Slinky spring as a solenoid, Fig. 15.19*a*, you can investigate whether the field is uniform across the

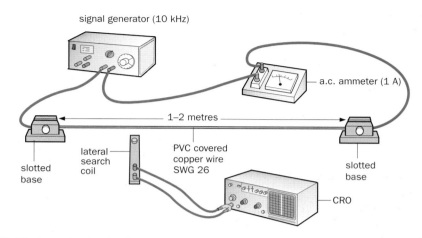

Fig. 15.18 Investigating the field due to a long straight wire

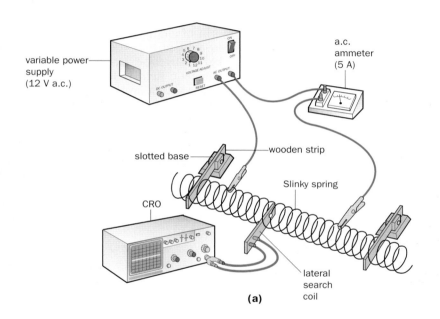

(a)

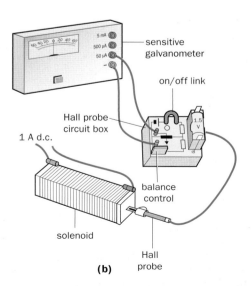

(b)

Fig. 15.19 Investigating the field due to a solenoid

width of the solenoid but zero outside. Is B constant along most of the length of the solenoid? Can we accept the theoretical prediction that for a long solenoid B at the end is half that at the centre? How does B at the centre alter if the solenoid is stretched, the current remaining constant? Answers to some of these questions may also be found with the apparatus of Fig. 15.19*b*.

A 'magnetic field board' like that in Fig. 15.20 enables circular coils to be studied. Do the direction and magnitude of B vary over the plane of the coil? When the current and radius are constant how does B at a given point vary with the number of turns? With a certain number of turns and current, is B at the centre inversely proportional to the radius of the coil? The field along the axis should vary with distance x along the axis from the centre as $r^2/(x^2 + r^2)^{3/2}$ where r is the coil radius. Does it?

For a compact coil of, say, 240 turns how do the values of B compare at P and Q, Fig. 15.21? Is there a relationship between B and x along the axis of the coil?

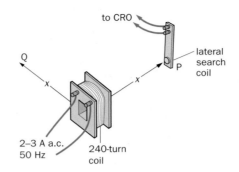

Fig. 15.21

FORCE ON A CHARGE IN A MAGNETIC FIELD

An electric current is regarded as a drift of charges and so it is reasonable to assume that the force experienced by a current-carrying conductor in a magnetic field is the resultant of the forces acting on the charges constituting the current. Accepting this, an expression for the force on a single charged particle moving in a magnetic field can be derived.

Consider a length of conductor containing n charged particles, each of charge Q and average drift velocity v, Fig. 15.22. Each particle will take the same average time t to travel distance l, and so

$$v = l/t$$

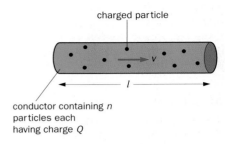

Fig. 15.22

The total charge passing through any cross-section of the conductor in time t is nQ. Therefore the current I is given by

I = total charge passing a section/time

$\quad = nQ/t$

$\quad = nQv/l \quad$ (since $t = l/v$)

If the conductor makes an angle θ with a uniform magnetic field of flux density B, then

force on conductor = $BIl \sin \theta$

$$\text{(p. 234)}$$

$$= B\left(\frac{nQv}{l}\right) l \sin \theta$$

$$= BnQv \sin \theta$$

Hence the force on one charged particle is $BQv \sin \theta$. *If the charge moves at right angles to the field*, $\theta = 90°$, $\sin \theta = 1$ and the force F is given by

$$F = BQv$$

If B is in teslas (T), Q in coulombs (C) and v in metres per second (m s^{-1}), then F is in newtons (N). The direction of the force is given by Fleming's left-hand rule (remembering that a negative charge moving one way is equivalent to a conventional current flowing in the opposite direction) and is at right angles both to the field and to the direction of motion.

This force is also responsible for the deflection of a beam of charged particles travelling through a magnetic field in a vacuum, e.g. electrons in a cathode ray tube. Since it is always at right angles to the path of the beam it only changes the direction of motion but not the speed. When the particles enter a uniform field at right angles, they are deflected into a circular path; for other angles (except 0°) they describe a helix, Figs 15.23*a* and *b*.

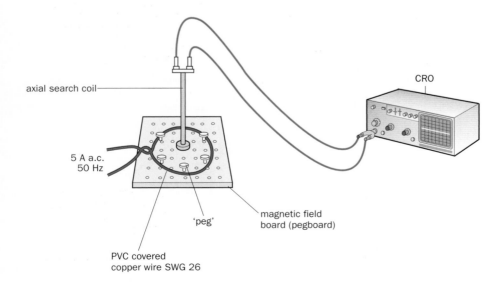

Fig. 15.20 Investigating the field due to a circular coil

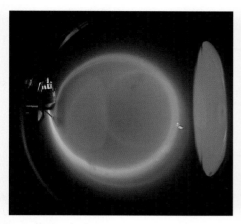

(a) Circular

(b) Helical

Fig. 15.23 Paths of electrons in a fine beam tube

Charged particles tend to become 'trapped' in magnetic fields and their existence in certain regions round the earth, known as the Van Allen radiation belts, is due to the earth's magnetic field. The behaviour of charged particles in magnetic fields is very important in atomic physics.

HALL EFFECT

A current-carrying conductor in a magnetic field has a small p.d. across its sides, in a direction at right angles to the field. In the slab of conductor shown in Fig. 15.24a, the p.d., called the **Hall p.d.**, appears across XY.

(a) Explanation

The effect was discovered in 1879 by Hall and can be attributed to the forces experienced by the charge carriers in the conductor. These act at right angles to the directions of the magnetic field and the current (as given by Fleming's left-hand rule) and cause the charge carriers to be pushed sideways, increasing their concentration towards one side of the conductor. As a result, a p.d. (and an electric field) is produced across the conductor.

With the magnetic field and conventional current having the directions shown in Fig. 15.24b and c, the forces act upwards and side X of the conductor develops the same sign as the charge carriers, i.e. positive in (b) and negative in (c). The Hall p.d. therefore reveals the sign of the charge carriers (more exactly, the majority charge carriers) in the conductor. In metals such as copper and aluminium they are negative; in semiconductors they may be positive or negative.

(b) Demonstration of Hall effect

Germanium, a semiconductor, is used because the effect is very much greater than with metals; p.ds of about 0.1 V are obtainable. Pure germanium itself has very few free electrons for current carrying but if it is 'doped' with a small proportion of atoms of another element (such as antimony) which provides extra electrons, n-type germanium is formed with enhanced conduction properties. In contrast, p-type germanium is doped with atoms (e.g. those of indium) which cause the germanium to behave as if it had positive charge carriers.

In the circuit of Fig. 15.25 a current of about 50 mA is passed through the Hall slice (a piece of doped germanium) via the current terminals P and Q. If the Hall p.d. connections X and Y on the slice are not directly opposite each other, a p.d. will exist between them even in the absence of a magnetic field and there will be a current through the microammeter. In this case the 'balance' control should be altered in order to zero the meter and bring X and Y to the same potential. If a Magnadur magnet is then placed face downwards over the slice, the microammeter should indicate a Hall p.d., the sign of which for p-type germanium will be opposite to that for n-type. Application of the left-hand rule enables the sign of the charge carriers to be determined.

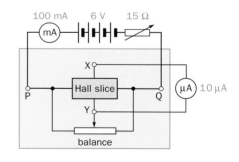

Fig. 15.25

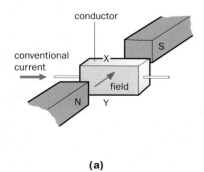

(a)

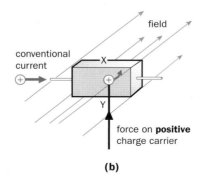

(b)

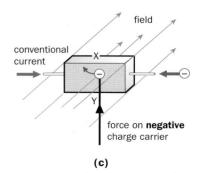

(c)

Fig. 15.24 The Hall effect

(c) Expression for Hall p.d.

The Hall p.d. and electric field build up until the repulsive electric force they cause on a charge carrier drifting through the conductor is equal and opposite to the deflecting magnetic force on it and no further charge displacement occurs.

If V_H is the Hall p.d. across the width d of the conductor and E is the final value of the electric field strength, then because electric field strength = potential gradient, we have

$$E = V_H/d$$

If the charge carriers each have charge Q and drift velocity v then

electric force on each drifting charge
$$= EQ = V_H Q/d$$

magnetic force on each drifting charge
$$= BQv$$

At equilibrium,

$$V_H Q/d = BQv$$
$$\therefore \quad V_H = Bvd$$

The case of positive charge carriers is illustrated in Fig. 15.26. Draw corresponding diagrams for negative charge carriers. It is possible to show (p. 41) that for a conductor of cross-section area A carrying current I and having n charge carriers per unit volume,

$$I = nAQv$$

Substituting for v we get

$$V_H = \frac{BId}{nAQ}$$

But $A = d \times t$ where t is the thickness of the conductor. Therefore

$$V_H = \frac{BI}{nQt}$$

The expression shows that V_H is greatest in materials for which n is small, i.e. semiconductors, since those have a smaller number of charge carriers per unit volume than metals.

(d) Number of charge carriers per atom

If V_H is measured for, say, a thin strip of aluminium foil carrying a large current in a strong magnetic field, n may be found as in the following example.

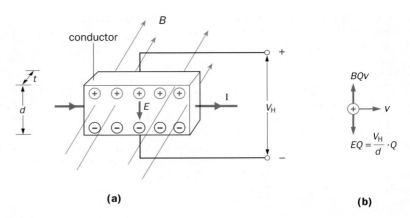

Fig. 15.26

If $B = 1.4$ T, $I = 10$ A, $t = 0.060$ mm $= 6.0 \times 10^{-5}$ m and $V_H = 10$ μV $= 10 \times 10^{-6}$ V, then taking Q = electronic charge $= 1.6 \times 10^{-19}$ C we get

$$n = \frac{BI}{V_H Q t}$$
$$= \frac{1.4 \times 10}{10 \times 10^{-6} \times 1.6 \times 10^{-19} \times 6.0 \times 10^{-5}}$$
$$= 1.4 \times 10^{29} \text{ electrons per m}^3$$

A calculation using the Avogadro constant (6.0×10^{23} atoms mol^{-1}), the relative atomic mass and the density of aluminium (27 and 2.7 g cm^{-3} respectively) shows that there are 6.0×10^{28} aluminium atoms per cubic metre. The number of free (conduction) electrons per atom must therefore be between 2 and 3, since $[1.4 \times 10^{29}]/[6.0 \times 10^{28}] = 2.3$. Chemical evidence suggests that aluminium is trivalent, i.e. has three 'outer' electrons (which participate in chemical reactions). Hall effect measurements do not conflict with this conclusion.

Check that the equation in the tinted example above gives the correct unit (m^{-3}) for n, remembering that 1 T $= 1$ N A^{-1} m^{-1}.

(e) Measuring B

From

$$B = \frac{nQtV_H}{I}$$

it can be seen that if the current in the Hall slice remains constant, B is proportional to the Hall p.d. and so a Hall probe may be used to measure magnetic flux density (see Fig. 15.19b, p. 237). The probe must first be calibrated in a known magnetic field.

THE AMPERE AND CURRENT BALANCES

(a) Force between currents

The ampere is the basic electrical unit of the SI system and has to be defined, like any other unit, so that it is accurately reproducible. This is achieved by basing the definition on the force between two long, straight, parallel, current-carrying conductors (i.e. by

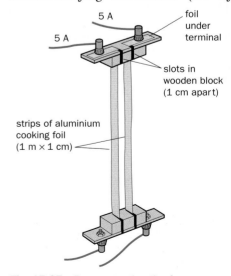

Fig. 15.27 Demonstrating the force between currents

using the magnetic effect), and making measurements with a current balance.

The forces between currents may be demonstrated with two long narrow strips of aluminium foil arranged as in Fig. 15.27. When the currents flow in opposite directions the strips repel and when in the same direction they attract. In brief, **unlike currents repel, like currents attract**.

To derive an expression for the force, consider two long, straight, parallel conductors, distance a apart in air, carrying currents I_1 and I_2 respectively, Fig. 15.28. The magnetic field at the right-hand conductor due to the current I_1 in the left-hand one is directed into the paper and its flux density B_1 is given by

$$B_1 = \frac{\mu_0 I_1}{2\pi a} \qquad \text{(p. 236)}$$

The force F acting on length l of the right-hand conductor (carrying current I_2) is

$$F = B_1 I_2 l$$

i.e.

$$F = \frac{\mu_0 I_1 I_2 l}{2\pi a}$$

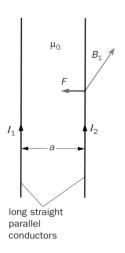

Fig. 15.28

The left-hand conductor experiences an equal and opposite force because of being in the field of the right-hand conductor.

(b) The ampere and μ_0

The definition of the ampere is based on the previous expression and may be stated as follows:

> The ampere is the constant current which, flowing in two infinitely long, straight, parallel conductors of negligible circular cross-section and placed in a vacuum 1 metre apart, produces between them a force equal to 2×10^{-7} newton per metre of their length.

Previously it was defined in terms of the chemical effect, and the choice of a force of 2×10^{-7} newton was made in the magnetic effect definition to keep the value of the new unit as near as possible to that of the old one.

Once the ampere has been defined, the value of μ_0 follows. We have from the definition

$$I_1 = I_2 = 1 \text{ A}$$
$$l = a = 1 \text{ m}$$
$$F = 2 \times 10^{-7} \text{ N}$$

Substituting into $F = \mu_0 I_1 I_2 l/(2\pi a)$ we get

$$2 \times 10^{-7} = \mu_0 \times 1 \times 1 \times 1/(2\pi \times 1)$$
$$\therefore \quad \mu_0 = 4\pi \times 10^{-7} \text{ H m}^{-1}$$

(c) Current balances and absolute measurement of current

A current balance enables a current to be measured by weighing and uses the fact that adjacent current-carrying conductors, whether straight wires (as considered in the definition of the ampere) or coils, exert forces on each other. The measurement process, when reduced to its essentials, consists of finding a mass, a length and a time and is said to be an *absolute* one; no electrical quantities as such have to be measured, only mechanical ones. Current balances are used primarily to calibrate more convenient types of current-measuring instruments.

A simple arrangement is illustrated in Fig. 15.29 in which the current balance used earlier (p. 235) measures the current I (about 5 A) in a long, straight wire, close and parallel to its

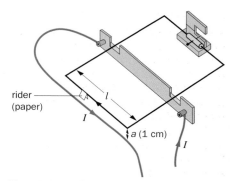

Fig. 15.29   Simple current balance

current-carrying arm of length l and in series with it. You should be able to show that $I^2 = mga/(2 \times 10^7 l)$, where m is the mass of the rider needed to counterpoise the balance. The accuracy of the result is poor because the force involved is very small. A larger force and greater accuracy are obtained if a wide, flat solenoid is used (see Fig. 15.11, p. 235) and I calculated from $I^2 = mg/(\mu_0 nl)$ where n is the number of turns per metre on the solenoid.

The principle of a practical current balance such as was formerly used in standardizing laboratories like the National Physical Laboratory (NPL) is shown in Fig. 15.30. The six coils carry the same current and are connected in series so that the forces on the two movable coils tip the balance to the same side. A rider of known mass is moved on the beam of the balance to restore equilibrium and enables the forces between the coils to be found. Knowing this and the value of B (from the dimensions of the coils), the current can be calculated to an accuracy of a few parts in a million. The latest NPL moving-coil balance is shown in Fig. 15.31.

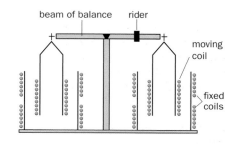

Fig. 15.30   Principle of a practical current balance

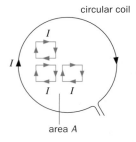

Fig. 15.31 The NPL moving-coil balance

TORQUE ON A COIL IN A MAGNETIC FIELD

Current-carrying coils within magnetic fields are essential components of electric motors and meters of various kinds.

(a) Expression for torque

Consider a rectangular coil PQRS of N turns pivoted so that it can rotate about a vertical axis YY′ which is at right angles to a *uniform* magnetic field of flux density B, Fig. 15.32*a*. Let the normal to the plane of the coil make an angle θ with the field, Fig. 15.32*b*.

When current I flows in the coil each side experiences a force (since all make some angle with B), acting perpendicularly to the plane containing the side and the direction of the field. The forces on the top and bottom (horizontal) sides are parallel to YY′ and for the current direction shown, they lengthen the coil. The forces on the vertical sides, each of length l, are

(a) Side view

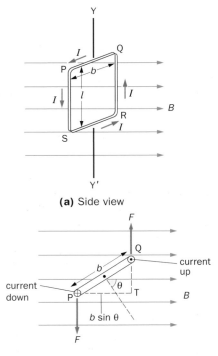

(b) Plan view

Fig. 15.32

equal and opposite and have value F where

$$F = BIlN$$

Whatever the position of the coil, its vertical sides are at right angles to B and so F remains constant. The forces constitute a couple whose torque C is given by

C = one force × perpendicular distance between lines of action of the forces

$= F \times \text{PT} = F \times b \sin \angle \text{PQT}$
     (b = breadth of coil)

$= Fb \sin \theta$

$= BIlN \times b \sin \theta$

i.e.

$$C = BIAN \sin \theta$$

where A = area of face of coil = $l \times b$. The torque causes angular acceleration of the coil which rotates until its plane is perpendicular to the field (i.e. $\theta = 0$) and then $C = 0$.

The expression for C can be shown to hold for a coil of *any* shape of area A. A circular coil, for example, carrying current I can be regarded as consisting of a large number of tiny rectangular coils, each with current I flowing in the same direction as in the circular coil, Fig. 15.33. The forces on all sides of the tiny coils cancel except where they lie on the circular coil itself.

Fig. 15.33

(b) Electromagnetic moment of a coil

From the above expression we see that C depends on, among other things, I, A and N. It is convenient to write

$$m = IAN$$

where m is a property of the coil and the current it carries, and is called the **electromagnetic moment** of the coil. We then have

$$C = Bm \sin \theta$$

or

$$m = \frac{C}{B \sin \theta}$$

If $B = 1\,\text{T}$ and $\theta = 90°$ then $m = C$ and we can define the electromagnetic moment of a coil as **the torque of the couple acting on it when it lies with its plane parallel to a magnetic field of unit flux density**, Fig. 15.34.

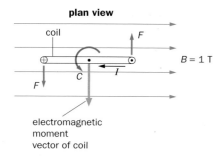

plan view

Fig. 15.34

Electromagnetic moment m is a vector whose direction is taken to be that of the flux density created along the axis of the coil by the current in it. We can therefore regard the couple that acts on a current-carrying coil as trying to align the electromagnetic moment of the coil with the direction of the flux density of the applied magnetic field.

MOVING-COIL GALVANOMETERS

A galvanometer detects (or measures, if its scale is calibrated) small currents passing through it or small p.ds across it; the addition of a shunt or multiplier converts it to an ammeter or a voltmeter.

(a) Construction

Basically a moving-coil galvanometer consists of a coil of fine, insulated copper wire which is able to rotate in a strong magnetic field. The field is produced in the narrow air gap

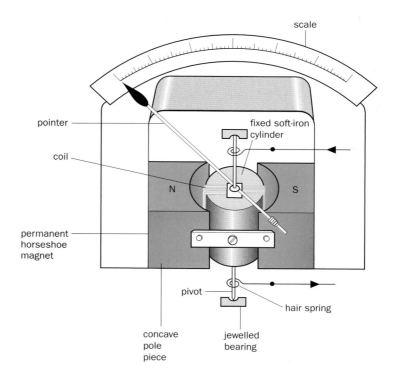

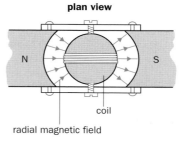

plan view

Fig. 15.35 Pointer-type moving-coil galvanometer

between the concave pole pieces of a permanent magnet and a fixed soft-iron cylinder and is *radial*, i.e. the field lines in the gap appear to radiate from the central axis of the cylinder and are always parallel to the plane of the coil as it rotates, Fig. 15.35.

In the *pointer-type* meter shown, the coil is pivoted on jewelled bearings and its rotation is resisted by hair springs both above and below it. The springs also lead the current in and out of the coil. In the more sensitive *light-beam* galvanometer, a beam of light is used as a weightless pointer.

(b) Theory

The magnetic field is *radial* and so the plane of the coil is parallel to it, whatever the deflection. The forces acting on the vertical sides are therefore always perpendicular to the sides and the deflecting couple consequently has a maximum value for all positions of the coil (since θ is fixed at $90°$ and so $\sin 90° = 1$). If the air gap is of constant width, the flux density B of the field is also nearly constant and the torque C of the deflecting couple due to current I in the coil is given by

$$C = BIAN$$

where A is the mean area of the coil and N is the number of turns on it.

The coil rotates until the resisting couple C' due to the suspension is equal and opposite to C. If the deflection then is α, and k is the torque of the couple needed to produce unit angular deflection (in newton metres per radian) of the suspension, we have

$$C' = k\alpha$$

(assuming Hooke's law holds for the suspension). Since at equilibrium $C = C'$,

$$BIAN = k\alpha$$

B, A, N and k are constants for a particular meter and so

$$\alpha \propto I$$

The use of a radial field and a uniform air gap results in the deflection α of the coil being directly proportional to the current I in it; the galvanometer scale is therefore linear.

(c) Sensitivity

The **current sensitivity** of a galvanometer is defined as the **deflection per unit current** and equals α/I. It follows from above that

$$\frac{\alpha}{I} = \frac{BAN}{k}$$

The **voltage sensitivity** is defined as α/V where α is the deflection when the p.d. across the galvanometer is V. If its resistance is R then as $I = V/R$, we have

$$\text{voltage sensitivity} = \frac{\alpha}{V} = \frac{BAN}{kR}$$

Sensitivities are usually expressed in mm per µA or mm per µV, this being the deflection produced on a mm scale by 1 µA or 1 µV.

MOVING-COIL LOUDSPEAKER

Most speakers are of this type. The construction is shown in Fig. 15.36a. Alternating current from the amplifier of a radio, record-player, etc., passes through a short cylindrical coil whose turns are at right angles to the magnetic field of a magnet with a central pole and a surrounding ring pole.

A force acts on the coil which, according to Fleming's left-hand rule, makes it move in and out. A paper cone attached to the former of the coil vibrates with it and sets up sound waves in the surrounding air with the same frequency as the alternating current.

Moving-coil microphones share a similar construction but they convert sound into alternating currents.

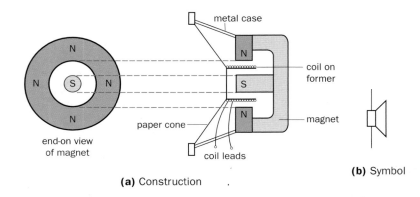

(a) Construction

(b) Symbol

Fig. 15.36 Moving-coil loudspeaker

RELAY

A relay is a switch worked by an electromagnet. It is useful if we want a small current in one circuit to control another circuit containing a device such as a lamp, electric bell or motor which requires a large current.

The structure of a relay is shown in Fig. 15.37. When the controlling current flows through the coil, the soft-iron core is magnetized and attracts the L-shaped soft-iron armature. This rocks on its pivot and closes the electrical contacts in the circuit being controlled.

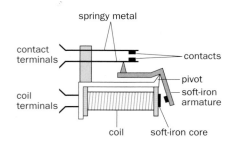

Fig. 15.37 Relay

1. In which of the four cases shown in Fig. 15.38 below will the arm XY of the current balance experience a tilting force and will it be upwards or downwards?

2. A current of 5 A flows in a straight wire in a uniform flux density of 2×10^{-3} T. Calculate the force per unit length on the wire if it is
   a) perpendicular to the field,
   b) inclined at 30° to it.

3. A child sleeps at an average distance of 30 cm from household wiring. The mains supply is 240 V r.m.s. Calculate the maximum possible magnetic flux density in the region of the child when the wire is transmitting 3.6 kW of power.
   Why might the magnetic field due to the current in the wire pose more of a health risk to the child than the Earth's magnetic field, given that they are of similar magnitudes?
   (*L*, A PH4, Jan 1997)

4. What current must be passed through a flat circular coil of 10 turns and radius 5.0 cm to produce a flux density of 2.0×10^{-4} T at its centre?
   ($\mu_0 = 4\pi \times 10^{-7}$ H m^{-1})

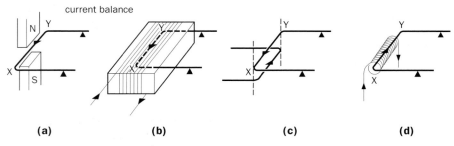

(a)   **(b)**   **(c)**   **(d)**

Fig. 15.38

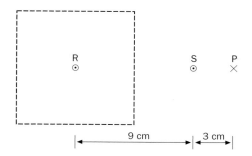

Fig. 15.39

5. Two long parallel wires R and S carry steady currents I_1 and I_2 respectively in the same direction. Figure 15.39 is a plan view of the arrangement. The directions of the currents are out of the page.

Copy the diagram and, in the region enclosed by the dotted lines, draw the magnetic field pattern due to the current in wire R alone.

The current I_1 is 4 A and I_2 is 2 A. Mark on your diagram a point N where the magnetic flux density due to the currents in the wires is zero.

Show on the diagram the direction of the magnetic field at P.

Calculate the magnitude of the magnetic flux density at P due to the currents in the wires.

(L, A PH4, June 1998)

6. a) Sketch the magnetic field pattern produced by a long, straight, current-carrying wire.

b) A U-shaped magnet sits on a top pan balance as shown in Fig. 15.40. A wire is fixed horizontally between the poles of the magnet such that a length of 4.20 cm lies in the field of the magnet. When there is no current in the wire, the balance reads 159.72 g and when a current of 2.00 A flows in the wire, the balance reads 159.46 g.

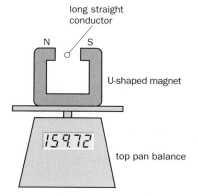

Fig. 15.40

i) Explain why the readings on the balance are different.
ii) Determine the direction of current flow.
iii) Determine the value of the magnetic field strength at the wire.
iv) Determine the reading on the balance if the direction of the current is reversed.
v) Suggest how a periodic force could be produced on the pan of the balance.
vi) Describe what would happen, with time, to both the wire and the balance reading when DC current is switched on, if the wire was not fixed in position.

(IB, Subsidiary/Standard Paper 2, Nov 1997)

7. Define the **ampere**.

Two long vertical wires, set in a plane at right angles to the magnetic meridian, carry equal currents flowing in opposite directions. Draw a diagram showing the pattern, in a horizontal plane, of the magnetic flux due to the currents alone — that is, neglecting for the moment the earth's magnetic field.

Next, taking into account the earth's magnetic field, discuss the various situations that can give rise to neutral points in the plane of the diagram.

Figure 15.41 shows a simple form of current balance. The 'long' solenoid S, which has 2000 turns per metre, is in series with the horizontal rectangular copper loop ABCDEF, where BC = 10 cm and CD = 3.0 cm. The loop, which is freely pivoted on the axis AF, goes well inside the solenoid, and CD is perpendicular to the axis of the solenoid. When the current is switched on, a rider of mass 0.20 g placed 5.0 cm from the axis is needed to restore equilibrium. Calculate the value of the current, I. ($\mu_0 = 4\pi \times 10^{-7}$ H m^{-1}; $g = 10$ N kg^{-1})

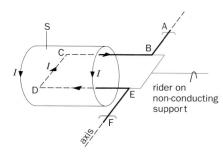

Fig. 15.41

8. Draw a diagram to show the magnetic field (magnetizing force) due to a long solenoid carrying a current. Write down

an equation for the flux density at its centre and at its ends, explaining your units. Describe an experiment you could perform to verify that the ratio of these flux densities is as predicted by your formulae.

A metal wire 10 m long lies east–west on a wooden table. What p.d. would have to be applied to the ends of the wire, and in what direction, in order to make the wire rise from the surface? Assume that the electrical connections to the wire cause no appreciable restraint.

(Density of the metal $= 1.0 \times 10^4$ kg m^{-3}, resistivity of the metal $= 2.0 \times 10^{-8}$ Ω m, horizontal component of earth's field $= 1.8 \times 10^{-5}$ T, $g = 9.8$ N kg^{-1}.)

9. An electron of charge e and mass m describes a circular path of radius r when it is projected with velocity v into a uniform magnetic field of flux density B. Derive an expression for the frequency of revolution.

How many orbits per second are made by an electron in a fine beam tube (Fig. 15.23*a*, p. 239) if the flux density in the field due to the Helmholtz coils is 1.0×10^{-3} T? ($e = 1.6 \times 10^{-19}$ C; $m = 9.1 \times 10^{-31}$ kg)

10. Figure 15.42 shows a beam of electrons which have been accelerated by a potential difference V, travelling in an evacuated tube. A magnetic field acts at right angles to their direction of motion in the shaded region and into the plane of the paper.

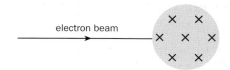

Fig. 15.42

Copy the diagram and draw the path of the electrons in the shaded region.

In Fig. 15.43 a pair of conducting plates, 2.5 cm apart, has been introduced into the shaded region. A potential difference is applied to the plates and is gradually increased until it reaches 400 V when the path of the electrons is a straight line.

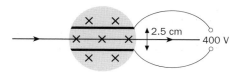

Fig. 15.43

Indicate the polarity of the plates.
Calculate

  i) the electric field strength in the region between the plates,

  ii) the force on an electron due to this field.

The magnetic flux density in the shaded region is 1×10^{-3} T. Show that the speed of the electrons must be 1.6×10^7 m s^{-1}.

Calculate the potential difference V required to accelerate electrons to this speed.

(*L*, A PH4, Jan 1997)

11. Figure 15.44 shows an arrangement in a vacuum to deflect protons into a detector using a magnetic field, which can be assumed to be uniform within the square shown and zero outside it. The motion of the protons is in the plane of the paper.

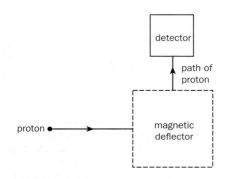

Fig. 15.44

a) Copy the diagram and sketch the path of a proton through the magnetic deflector. At any point on this path draw an arrow to represent the magnetic force on the proton. Label this arrow F.

b) State the direction of the uniform magnetic field causing this motion.

c) The speed of a proton as it enters the deflector is 5.0×10^6 m s^{-1}. If the flux density of the magnetic field is 0.50 T, calculate the magnitude of the magnetic force on the proton.

d) If the path were that of an electron with the same velocity, what **two** changes would need to be made to the magnetic field for the electron to enter the detector along the same path?

(*AQA: NEAB*, AS/A PHO3, March 1999)

12. a) In Fig. 15.45a electrons are shown drifting from right to left through a block of conductor. A flux density is applied as in Fig. 15.45b and the concentration of electrons increases near the front edge of the block. Why? If the current had consisted of positive charges

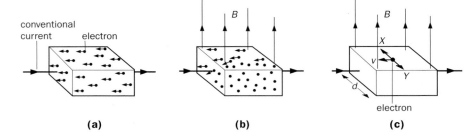

Fig. 15.45

drifting from right to left would they have been pushed towards the front?

b) An electric field and a p.d. (the Hall p.d.) are created across the block and soon attain maximum values. Why?

c) Figure 15.45c shows an electron (charge e, mass m) drifting with velocity v near the middle of the conductor (width d) when the electric field and Hall p.d. have their maximum values, say E and V_H respectively. It is undeflected because two equal and opposite forces X and Y act on it. What are they? Derive an equation relating them (let B be the magnetic flux density).

d) Hence calculate v if $V_H = 8$ μV, $B = 0.4$ T and $d = 2$ cm.

13. Figure 15.46 shows two horizontal wires, P and Q, 0.20 m apart, carrying currents of 1.50 A and 3.00 A respectively.

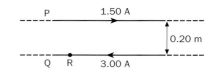

Fig. 15.46

i) On a copy of the outline diagram in Fig. 15.47, draw the **resultant** magnetic field produced by these two currents.

$\otimes$

$\odot$

$\otimes$ = wire P carrying current of 1.50 A into plane of paper

$\odot$ = wire Q carrying current of 3.00 A out of plane of paper

Fig. 15.47

ii) The flux density B of the magnetic field produced at a point by a current I flowing in a long straight wire is given by

$$B = \frac{\mu_0 I}{2\pi a}$$

where a is the perpendicular distance from the wire to the point. Calculate the flux density of the magnetic field at R due to the current in wire P.

iii) Hence calculate the force per metre on wire Q due to this field.

iv) State the direction of this force.

v) State the direction and magnitude of the force per metre on wire P due to the current in wire Q.

(*NEAB*, AS/A PHO3, March 1998)

14. Figure 15.48 shows a square coil of side length 0.25 m. The plane of the coil is parallel to a uniform magnetic field of strength 0.01 T. The coil carries a current of 2.0 A.

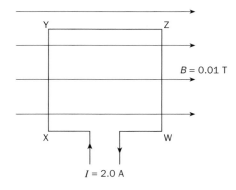

Fig. 15.48

a) What is the direction of the force on the side XY?

b) What is the magnitude of the force XY?

c) What is the magnitude of the force on the side marked YZ?

d) Describe the motion of the coil as a result of the magnetic forces acting upon it.

(*IB*, Subsidiary/Standard Paper 2, Nov 1998)

15. a) Describe a simple experiment which demonstrates that a force is experienced by a current-carrying conductor in a magnetic field. State the factors that determine
  i) the magnitude of the force, and
  ii) its direction.

b) A rectangular coil of wire carrying a steady current is pivoted on an axis which is at right angles to a radial magnetic field. Obtain an expression for the torque experienced by the coil, and explain the relevance of this result to the design of moving-coil galvanometers.

c) A moving-coil galvanometer has a resistance of 25 Ω and gives a full-scale deflection when carrying a current of 4.4×10^{-6} A. What current will give a full-scale deflection when the galvanometer is shunted by a 0.10 Ω resistance?

16. In Fig. 15.49, A and B are two long, straight, parallel conductors. Each carries the same current I into the plane of the page. Prove that at C in their combined magnetic field the magnetic flux density in the y-direction is zero.

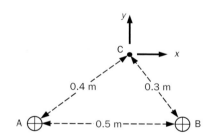

Fig. 15.49

17. a) Figure 15.50*a* shows the front view of a loudspeaker magnet. The dotted circle shows the position occupied by

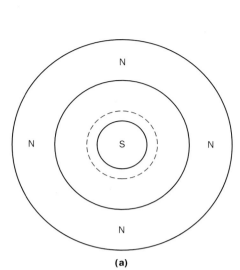

(a)

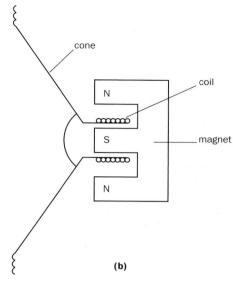

(b)

Fig. 15.50

the coil. Copy the diagram and add lines to show the shape and direction of the magnetic field between the poles of the magnet.

A cross-sectional view of the loudspeaker is shown in Fig. 15.50*b*. The coil, of mean diameter 0.030 m, consists of 200 turns of wire. At 0.015 m from the centre of the coil, the flux density of the magnetic field is 0.60 T.

b) i) Calculate the electromagnetic force acting on the coil when it carries a direct current of 2.0 mA.
  ii) Describe and explain the motion of the coil and the cone attached to it when the coil carries a sinusoidal alternating current.

(*OCR*, 6843, March 1999)

18. Figure 15.51 shows the cross-section of three long electric cables, X, Y and Z. The separation of X and Y, and of Y and Z, is *r*. X and Y carry currents I and $2I$ respectively, *away from the observer* at right angles to the plane of the paper. Initially, there is no current in Z.

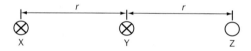

Fig. 15.51

a) Copy Fig. 15.51 and draw on it:
  i) the magnetic flux pattern created by the current in X. Include enough detail to show the varying strength and direction of the field;
  ii) arrows to show the direction of the electromagnetic forces experienced by X and Y.

b) The magnitude of the electromagnetic force acting on X is *F*. State, in terms of *F*, the magnitude of the electromagnetic force acting on Y.

c) State the magnitude and direction of a current in Z which would ensure that no resultant electromagnetic force acted on X.

(*OCR*, 6843, June 1999)

16

Electromagnetic induction

INDUCING E.M.Fs

An electric current creates a magnetic field; the reverse effect, of using magnetism to produce electricity, was discovered independently in 1831 by Faraday in England and Henry in America and is called **electromagnetic induction**. Induced e.m.fs can be generated in two ways.

(a) By relative movement (the generator effect)

If a bar magnet is moved in and out of a stationary coil of wire connected to a centre-zero galvanometer, Fig. 16.1, a small current is recorded *during the motion* but not at other times. Movement of the coil towards or away from the stationary magnet has the same results. It is relative motion between magnet and coil that is necessary.
Observation shows that the direction of the induced current depends on the direction of relative motion. Also the magnitude of the current increases with the speed of motion, the number of turns on the coil and the strength of the magnet.

Although it is current we detect in this demonstration, an e.m.f. (p. 51) must be induced in the coil to cause the current. The induced e.m.f. is the

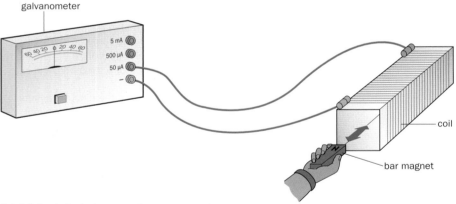

Fig. 16.1 Inducing an e.m.f.

more basic quantity and is always present even when the coil is not in a complete circuit. The value of the induced current depends on the resistance of the circuit as well as on the induced e.m.f. Replacing the galvanometer in Fig. 16.1 with a datalogger and computer, or a CRO, allows the induced e.m.f. to be recorded as the magnet is dropped through the coil (see question 6, p. 269).

We will be concerned here only with e.m.fs induced in conductors but they can be produced in any medium — even in a vacuum, where they play a basic role in the electromagnetic theory of radiation.

(b) By changing a magnetic field (the transformer effect)

In this case two coils are arranged one inside the other as in Fig. 16.2a. One coil called the **primary**, is in series with a 6 V d.c. supply, a tapping key and a rheostat. The other, called the **secondary**, is connected to a galvanometer. Switching the current on or off in the primary causes a pulse of e.m.f. and current to be induced in the secondary. Varying the primary current by quickly altering the value of the rheostat has the same effect. Electromagnetic induction therefore occurs when there is any *change* in the primary current and so also in the magnetic field it produces.

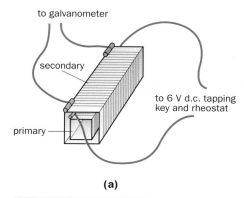

to galvanometer

secondary

to 6 V d.c. tapping key and rheostat

primary

(a)

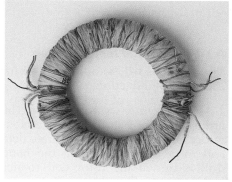

(b) Faraday's ring

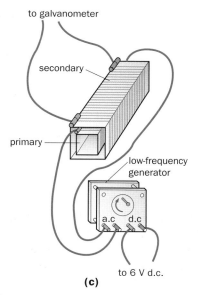

to galvanometer

secondary

primary

low-frequency generator

a.c   d.c

to 6 V d.c.

(c)

Fig. 16.2   Mutual induction coils

Cases of electromagnetic induction in which current changes in one circuit cause induced e.m.fs in a neighbouring circuit, not connected to the first, are examples of **mutual induction** — or the transformer principle. (Our use of search coils when investigating magnetic fields due to various current-carrying conductors in the previous chapter depended on it.)

The induced e.m.f. is increased by having a soft-iron rod in the coils or, better still, by using coils wound on a complete iron ring. The iron ring and coils used by Faraday in his original experiment are shown in Fig. 16.2*b*.

It is worth noting that the secondary current is in one direction when the primary current increases and in the opposite direction when it decreases. An alternating current is continually increasing and decreasing first in one direction and then in the opposite direction. The magnetic field accompanying it changes similarly and if a.c. is applied to the primary coil we would expect an induced e.m.f., also alternating, to be induced in the secondary. A simple arrangement for investigating the effect is shown in Fig. 16.2*c*; very low frequency a.c. is produced when the generator is hand-operated.

(c) Seismometer

This earthquake-detecting instrument, Fig. 16.3, is a good example of an application of electromagnetic induction. Any movement or vibration of the rock on which the seismometer rests (buried in a protective case) results in relative motion between the magnet (suspended by a spring from the frame) and the coil (also attached to the frame). The e.m.f. induced in the coil is directly proportional to the displacement associated with the earthquake and, after amplification, causes a trace on the paper of a pen recorder.

MAGNETIC FLUX

Electromagnetic induction is one of those action-at-a-distance phenomena whose mechanism is not revealed to our senses. Consequently we have to invent a conceptual model or theory which enables us to picture and to 'explain' (i.e. describe in terms of the model) what might be happening.

Faraday suggested a model based on magnetic field lines. He proposed that an e.m.f. is induced in a conductor *either* when there is a change in the number of lines 'linking' it (i.e. passing through it) *or* when it 'cuts' across field lines. The two statements often come to the same thing, as can be seen by considering Fig. 16.4 (and, later, the experiment described on p. 252). If the coil moves towards the magnet from X to Y, the number of lines linking or 'threading' it rises from three to five; alternatively we can say it cuts two lines in moving from X to Y.

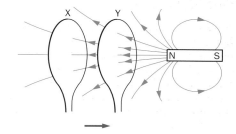

Fig. 16.4

Before expressing these ideas mathematically in the next section, we require some way of deciding how many lines link a coil or are cut by it. In the previous chapter (p. 231) it was

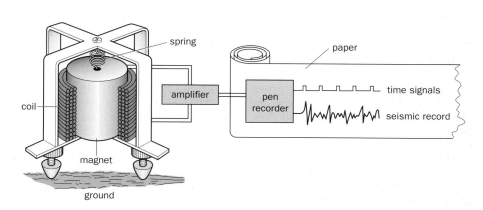

spring

paper

amplifier

pen recorder

time signals

seismic record

coil

magnet

ground

Fig. 16.3   Seismometer

suggested that magnetic field lines could be drawn so that the number per unit cross-section area represents the magnitude B of the flux density. It would then be reasonable to take the product, $B \times A$, as a measure of the number of lines linking a coil of cross-section area A.

This product is called the **magnetic flux** Φ. It is a more useful concept than B alone for quantifying our electromagnetic induction model and it is defined by the equation

$$\Phi = BA$$

where B is the flux density acting at right angles to and over an area A, Fig. 16.5a. In words, **the magnetic flux through a small plane surface is the product of the flux density normal to the surface and the area of the surface.** (Why a small surface?) If $B = 1$ tesla (T) and $A = 1$ square metre (m²) then Φ is defined to be 1 **weber** (Wb).

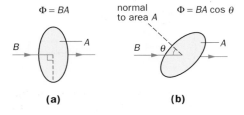

$\Phi = BA$ normal to area A $\Phi = BA \cos \theta$

(a)   **(b)**

Fig. 16.5

Since $B = \Phi/A$, the reason for calling B the flux density will now be evident. Another unit of B is therefore the weber per square metre (Wb m⁻²), i.e. $1\,\text{T} = 1\,\text{Wb m}^{-2}$.

In general, if the normal to the area A makes an angle θ with B, Fig. 16.5b, then the flux is

$$\Phi = BA \cos \theta$$

If Φ is the flux through the cross-section area A of a coil of N turns, the total flux through it, called the **flux-linkage**, is $N\Phi$ since the same flux Φ links each of the N turns.

Although the term *flux* suggests that something flows along the field lines, this is not so, but references to the flux 'entering', 'passing through' or 'leaving' a coil are in common use.

FARADAY'S LAW

(a) Statement

The law states that

> The induced e.m.f. is directly proportional to the rate of change of flux-linkage or rate of flux cutting.

In calculus notation it can be written

$$\mathscr{E} \propto \frac{\mathrm{d}}{\mathrm{d}t}(N\Phi)$$

or $\mathscr{E} = \text{constant} \times \dfrac{\mathrm{d}}{\mathrm{d}t}(N\Phi)$

where $\mathscr{E}$ is the induced e.m.f. and $\mathrm{d}(N\Phi)/\mathrm{d}t$ is the rate of change of flux-linkage or the rate of flux cutting. The law is found to be true for the generator and the transformer types of induction.

If instead of defining the weber in terms of the tesla we now redefine it as **the magnetic flux that induces in a one-turn coil an e.m.f. of 1 volt when the flux is reduced to zero in 1 second**, then the constant of proportionality in the above equation is 1. That is, if $\delta(N\Phi) = 1$ Wb when $\mathscr{E} = 1$ V and $\delta t = 1$ s, then we can write $1 = \text{constant} \times 1/1$. Then

$$\mathscr{E} = \frac{\mathrm{d}}{\mathrm{d}t}(N\Phi)$$

where $\mathscr{E}$ is in volts, $\mathrm{d}\Phi/\mathrm{d}t$ in webers per second and N is the number of turns if we are considering a coil.

(b) Numerical example

Suppose a single-turn coil of cross-section area 5.0 cm² is at right angles to a flux density of value 2.0×10^{-2} T, which is then reduced steadily to zero in 10 s. The flux-linkage change $\delta(N\Phi) = $ (number of turns) $\times$ (change in B) $\times$ (area of coil) $= (1) \times (2.0 \times 10^{-2}\,\text{T}) \times (5.0 \times 10^{-4}\,\text{m}^2) = 1.0 \times 10^{-5}$ Wb. The steady change occurs in time $\delta t = 10$ s, so the e.m.f. $\mathscr{E}$ induced in the coil is given by

$$\mathscr{E} = \frac{\mathrm{d}}{\mathrm{d}t}(N\Phi) = \frac{1.0 \times 10^{-5}}{10}\frac{\text{Wb}}{\text{s}}$$
$$= 1.0 \times 10^{-6}\,\text{V}$$

If the coil had 5000 turns, the flux-linkage would be 5000 times as great (namely, $N\,\delta\Phi = 5.0 \times 10^{-2}$ Wb) and

$$\mathscr{E} = 5000 \times 10^{-6} = 5.0 \times 10^{-3}\,\text{V}$$

If the normal to the plane of this coil made an angle of 60° (instead of 0°) with the field then $N\,\delta\Phi = 5.0 \times 10^{-2} \cos 60° = 2.5 \times 10^{-2}$ Wb and

$$\mathscr{E} = 2.5 \times 10^{-3}\,\text{V}$$

(c) Experimental test of the law

Two methods will be outlined, one for each type of induction.

1. Using a motor as a generator. The arrangement of Fig. 16.6a uses a laboratory motor (12 V) with separate field and rotor terminals. When the hand drill is turned the rotor coil rotates and cuts the flux due to the direct current (about 1 A) in the field coil.

To see if the e.m.f. induced in the rotor coil is proportional to the rate of flux cutting, i.e. to the speed of rotation of the rotor, the drill is turned steadily and the number of turns made in, say, 10 seconds is counted. The reading (steady) on the d.c. voltmeter across the rotor is noted. The procedure is repeated for different drill speeds.

The effect on the induced e.m.f. of increasing the field current (and so also B and Φ) can be found.

2. Using a.c., two coils and a CRO (or datalogger and computer). Alternating current from a signal generator's low-impedance output is passed through a solenoid; the peak-to-peak value of the e.m.f. induced (by mutual induction) in a ten-turn coil wound round the middle of the solenoid is measured on a CRO (used as a voltmeter and set on its most sensitive range, e.g. 0.1 V cm⁻¹) or by a datalogger and computer, for different frequencies between 1 kHz and 3 kHz, Fig. 16.6b. If the frequency of the a.c. in the solenoid is doubled, but the value of the current remains the same (as indicated by the a.c. ammeter), the rate of change of the flux linking the ten-turn coil doubles, as should the induced e.m.f.

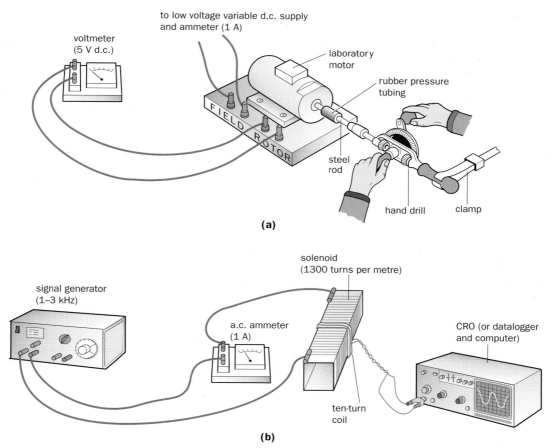

Fig. 16.6 Testing Faraday's law

(d) Further experimental investigations

These are concerned with properties of the coil itself which affect the e.m.f. induced in it. They help further with the understanding of Faraday's law.

(*i*) *Number of turns*. The effect of this can be investigated by comparing the vertical heights of the CRO traces due to ten-turn and five-turn coils round the middle of the solenoid, the same frequency of a.c. (say 2 kHz) being used in each case.

(*ii*) *Area*. In this case a.c. of frequency 2 kHz is passed through two solenoids in series, each having the same number of turns per metre, e.g. about 1300 (so that B [$= \mu_0 nI$] is the same inside both), but one with twice the cross-section area of the other, Fig. 16.7. The flux linking ten-turn coils round the middle of each will be in the ratio of the areas (i.e. 2 : 1), as should be the ratio of the e.m.fs induced in the coils and displayed on a CRO, preferably double-beam.

(*iii*) *Orientation*. If the plane of the coil is at an angle (other than 90°) to the magnetic field, the e.m.f. induced in it is less, as can be shown by the apparatus of Fig. 16.8. What angle will the handle of the search coil make with the plane of the coil when the induced e.m.f. has half its maximum value?

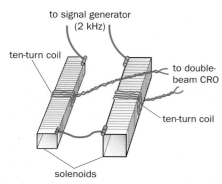

Fig. 16.7

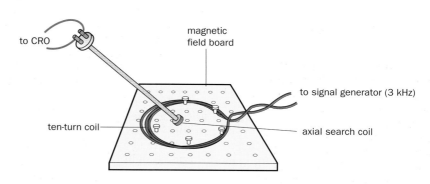

Fig. 16.8

(e) Equivalence of flux cutting and flux linking

This can be shown using the arrangement of Fig. 16.9. The outer solenoid is in series with a *smooth*, low-voltage variable d.c. supply and an ammeter, and it carries a steady current of 1–2 A. The inner solenoid is connected to a galvanometer and is pulled out slowly so that the galvanometer deflection remains steady. The time taken for the complete removal of the solenoid is noted.

The inner solenoid is re-inserted and the current reduced to zero in the outer solenoid by turning down the output control on the low-voltage supply. This is done at a rate which gives the same, steadiest possible, reading on the galvanometer as before. The time required is again measured.

In both cases the flux changes are the same and if the times are the same then the rate of flux cutting equals the rate of change of flux-linkage.

The actions are apparently different; the first involves relative motion and the second a changing magnetic field. Nevertheless, the same law describes both.

LENZ'S LAW

Faraday's law gives only the magnitude of the induced e.m.f.; its direction can be predicted by a law due to the Russian scientist Lenz. It may be stated as follows:

> The direction of the induced e.m.f. is such that it tends to oppose the flux change causing it, and does oppose it if induced current flows.

In Fig. 16.10*a* a bar magnet is depicted approaching the end of a coil, north pole first. If Lenz's law applies, the induced current should flow in a direction that makes the coil behave like a magnet with a north pole at the top. The downward motion of the magnet and the accompanying flux change will then be opposed. When the magnet is withdrawn, the top of the coil should behave like a south pole, Fig. 16.10*b*, and attract the north pole of the magnet, so hindering its removal and again opposing the flux change. The induced current is therefore in the opposite direction to that

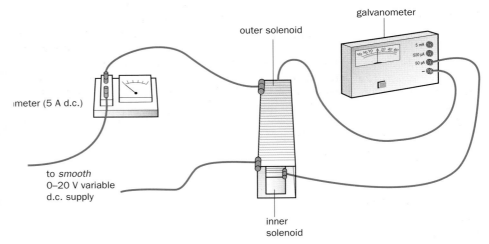

Fig. 16.9

when the magnet approaches. (Polarities may be checked using the right-hand screw rule if the direction of the windings on the coil are known and the current directions observed on the galvanometer.)

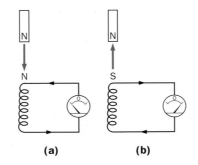

Fig. 16.10

Lenz's law is an example of the principle of conservation of energy; here, energy would be created from nothing if the e.m.fs and currents acted differently. For example, if a south pole were produced at the top of the coil in Fig. 16.10*a*, attraction would occur between coil and magnet and the latter would, if it was allowed to, accelerate towards the coil, gaining kinetic energy as well as generating electrical energy. In practice, work has to be done to overcome the forces that arise, i.e. energy transfer occurs which in this case is from the mechanical to the electrical form.

For straight conductors moving at right angles to a magnetic field a more useful version of Lenz's law is **Fleming's right-hand rule**.

If the thumb and first two fingers of the right hand are held so that each is at right angles to the other with the **F**irst finger pointing in the direction of the **F**ield and the thu**M**b in the direction of **M**otion of the conductor, then the se**C**ond finger indicates the direction (conventional) of the induced **C**urrent, Fig. 16.11.

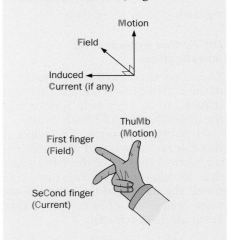

Fig. 16.11   Fleming's right-hand rule

Lenz's law is incorporated in the mathematical expression of Faraday's law by including a negative sign to show that current due to the induced e.m.f. produces an opposing flux change. So we write

$$\mathscr{E} = -\frac{\mathrm{d}}{\mathrm{d}t}(N\Phi)$$

EDDY CURRENTS

Any piece of metal moving in a magnetic field, or exposed to a changing one, has e.m.fs induced in it, as we might expect. These can cause currents, called **eddy currents**, to flow inside the metal and they may be quite large because of the low resistance of the paths they follow. Their magnetic and heating effects may be both helpful and troublesome.

(a) Magnetic effect

According to Lenz's law, eddy currents will circulate in directions such that the magnetic fields they create oppose the motion (or flux change) producing them. This acts as a brake on the moving body and may be shown simply with the arrangement of Fig. 16.12a. The solid copper cylinder quickly comes to rest if it is spun between the poles of the magnet. With the cylinder of coins (b) there is very little braking because dirt on the coins increases the electrical resistance of the cylinder as a whole, thereby reducing the eddy current flow.

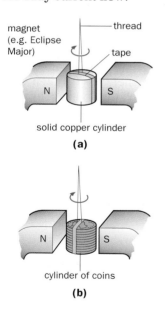

Fig. 16.12 Demonstrating the damping effect of eddy currents

Use is made of the effect for the electromagnetic damping of moving-coil meters so that the coil takes up its deflected position quickly, i.e. without overshooting and oscillating about its final reading. In most pointer instruments the coil is wound on a metal frame in which large eddy currents are induced and cause opposition to the motion of the coil as it cuts across the radial magnetic field of the permanent magnet. When oscillation is *just* prevented, the meter is said to be **critically damped** and its movement is 'dead-beat'. The curves in Fig. 16.13 show the effect of different degrees of damping.

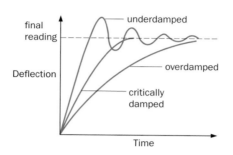

Fig. 16.13 Degrees of damping of a galvanometer

In light-beam galvanometers the coil is not usually wound on a frame but is glued together and near-critical damping is achieved by having appropriate internal shunts across the coil on all ranges (except 'direct'). Suitable eddy currents then flow round the coil and shunt, whatever the current to be measured. On the 'direct' setting, the electromagnetic damping is made a minimum (no internal shunts are connected) and the coil swings freely. Its first deflection can be shown to be proportional to the charge passing; the galvanometer is then said to be used **ballistically** (see p. 264). On 'short' the coil is short-circuited internally and the eddy currents induced in it bring it to rest quickly.

(b) Heating effect

In induction or eddy-current heating, a coil carrying high-frequency a.c. surrounds the material to be heated and the rapidly changing magnetic flux induces large eddy currents in the conducting parts of the material. For example, in the zone refining of metals and semiconductors a narrow crucible containing the material is passed very slowly through the heating coil, Fig. 16.14. The impurities tend to collect in the molten zone which moves to one end of the crucible. After cooling this end is removed, leaving a very pure, single crystal sample.

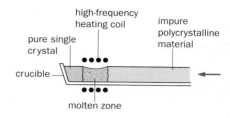

Fig. 16.14 Zone refining of metals and semiconductors by eddy-current heating

Electric motors, transformers and generators contain iron that experiences flux changes when the device is in use. To minimize energy loss through eddy currents the iron parts consist of sheets, called **laminations**, insulated from each other by thin paper, varnish or some other insulator. Eddy-current paths perpendicular to the laminations are then eliminated.

CALCULATION OF INDUCED E.M.Fs

(a) Straight conductor

Suppose a conducting rod XY, length l, moves sideways with steady velocity v through and at right angles to a uniform magnetic field of flux density B directed into the paper, Fig. 16.15a. The area swept out per second by XY is lv and therefore the flux cut per second is Blv. Assuming Faraday's law, we can say that the e.m.f. $\mathcal{E}$ induced in the rod is given numerically by

$$\mathcal{E} = \text{flux cut per second}$$

$$\mathcal{E} = Blv$$

The e.m.f. induced in a rod cutting magnetic flux can also be explained in terms of the forces acting on the charged particles in it. In a moving conductor both positive ions and electrons are carried along and both experience a force (magnetic) at right angles to the field and to the direction of motion of the conductor. However, only electrons are free to move inside the conductor and Fleming's left-hand (motor) rule indicates they will be forced to end X, making X negative and Y positive, Fig. 16.15b. (The conventional **C**urrent direction [se**C**ond finger] will be opposite to the motion of the conductor since we are dealing with negative charges.) As a result of the charge separation and electron accumulation, an electric field is created inside the conductor which causes a repulsive electric force to be exerted on other electrons being urged towards X by the magnetic force.

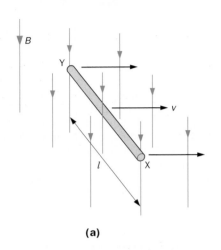

(a)

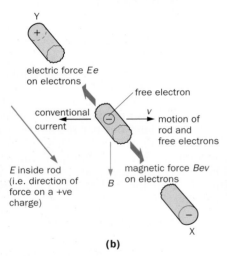

(b)

Fig. 16.15

These two forces act oppositely and when they become equal there is no further charge accumulation and we can say

$$Ee = Bev$$

where E is the equilibrium electric field strength, e the charge on an electron, B the magnetic flux density and v the velocity of the conductor. Therefore

$$E = Bv$$

If V is the p.d. developed between the ends X and Y of the conductor and l is its length, then E = potential gradient = V/l and so $V = Blv$. The rod is on open circuit, therefore V equals the induced e.m.f. $\mathscr{E}$. We therefore get, as before,

$$\mathscr{E} = Blv$$

(b) Spinning disc

The first generator was made by Faraday and consisted of a metal disc rotated in a magnetic field, Fig. 16.16. A modern version of his apparatus is shown in Fig. 16.17. When the disc is driven at a steady speed by the motor a steady deflection is obtained on the galvanometer, which is connected to two sliding contacts (e.g. 4 mm plugs), one held at the centre and the other at the edge of the disc between the poles of the magnet. The effect of changing the speed of rotation and the position of the contacts may be investigated.

Fig. 16.16   Faraday's disc generator

We can consider that an e.m.f. is induced in the circuit because the radius of the disc between the contacts at any instant is sweeping through the flux there, i.e. we regard the disc as a many-spoked wheel. If the disc makes f revolutions per second and has radius r, the area swept out per second by a radius is $\pi r^2 f$. The flux change per second = $B\pi r^2 f$ where B is the flux density (assumed uniform) between the contacts. Hence the induced e.m.f. $\mathscr{E}$ is, by Faraday's law,

$$\mathscr{E} = B\pi r^2 f$$

The 'homopolar generator' is a recent form of Faraday's disc and can deliver a very large direct current with very small e.m.f. for the production of powerful magnetic fields.

(c) Rotating coil

The coil in Fig. 16.18 has N turns each of area A and is being rotated about a horizontal axis in its own plane at right angles to a uniform magnetic field of flux density B. If the normal to the coil makes an angle θ with the field at time t (measured from the position where $\theta = 0°$) then the flux Φ linking each turn is given by

$$\Phi = BA \cos \theta$$

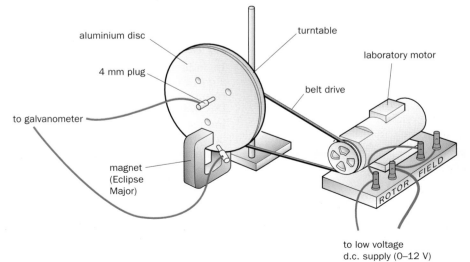

Fig. 16.17   A laboratory disc generator

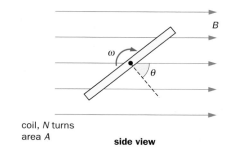

coil, N turns
area A

side view

Fig. 16.18

But $\theta = \omega t$ where ω is the steady angular velocity of the coil, therefore

$$\Phi = BA \cos \omega t$$

By Faraday's law, the e.m.f. $\mathscr{E}$ induced in N turns is

$$\mathscr{E} = -\frac{d}{dt}(N\Phi) = -N\frac{d}{dt}(BA \cos \omega t)$$

$$= -BAN\frac{d}{dt}(\cos \omega t)$$

$$= -BAN(-\omega \sin \omega t)$$

$$\mathscr{E} = BAN\omega \sin \omega t$$

The e.m.f. is an alternating one which varies sinusoidally with time and would cause a similar alternating current in an external circuit connected across the coil.

When the plane of the coil is parallel to B, $\theta = \omega t = 90°$, $\sin \omega t = 1$ and $\mathscr{E}$ has its maximum value $\mathscr{E}_0$ given by

$$\mathscr{E}_0 = BAN\omega$$

So we can write

$$\mathscr{E} = \mathscr{E}_0 \sin \omega t$$

When is $\mathscr{E}$ equal to (*i*) zero, and (*ii*) $\mathscr{E}_0/2$?

If a coil has area 1.0×10^{-2} m² (i.e. 100 cm²), 800 turns and makes 600 revolutions per minute in a magnetic field of flux density 5.0×10^{-2} T, then $\mathscr{E}_0$ is given by

$$\mathscr{E}_0 = BAN\omega$$

$$= (5.0 \times 10^{-2} \text{ T}) \times (1.0 \times 10^{-2} \text{ m}^2) \times (800) \times (2\pi \times 600/60 \text{ s}^{-1})$$

$$= 5.0 \times 10^{-2} \times 1.0 \times 10^{-2} \times 800 \times 20\pi \text{ Wb s}^{-1}$$

$$(1 \text{ T} = 1 \text{ Wb m}^{-2})$$

$$= 25 \text{ V}$$

GENERATORS (A.C. AND D.C.)

A generator produces electrical energy by electromagnetic induction. In principle it consists of a coil which is rotated between the poles of a magnet so that the flux-linkage is continuously changing, Fig. 16.19a. The flux Φ linking each turn of a coil of area A, having N turns, rotating with angular velocity ω in a uniform flux density B, is given at time t (measured from the vertical position) by

$$\Phi = BA \cos \omega t$$

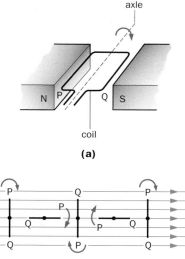

axle

coil

(a)

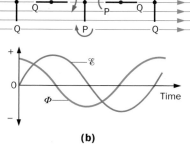

(b)

Fig. 16.19 Principle of a generator

By Faraday's law, the induced e.m.f. $\mathscr{E}$ is, as we found in the previous section,

$$\mathscr{E} = -\frac{d}{dt}(N\Phi) = BAN\omega \sin \omega t$$

Both Φ and $\mathscr{E}$ alternate sinusoidally and their variation with the position of the coil is shown in Fig. 16.19b.

We see that although Φ is a maximum when the coil is vertical (i.e. perpendicular to the field), $\mathscr{E}$ is zero because the *rate of change* of Φ is zero *at that instant*, i.e. the tangent to the Φ-graph is parallel to the time axis and so has zero gradient — in calculus terms $d\Phi/dt = 0$. Also, when $\Phi = 0$, its rate of change is a maximum, therefore $\mathscr{E}$ is a maximum.

The expression for $\mathscr{E}$ shows that its instantaneous values increase with B, A, N and the angular velocity of the coil. If the coil makes one complete revolution, one cycle of alternating e.m.f. is generated, i.e. for a simple, single-coil generator the frequency of the supply equals the number of revolutions per second of the coil.

In an **a.c. generator** (or **alternator**) the alternating e.m.f. is applied to the external circuit via two spring-loaded graphite blocks called 'brushes' which press against two copper *slip-rings*. These rotate with the axle, are insulated from one another and each is connected to one end of the coil, Fig. 16.20.

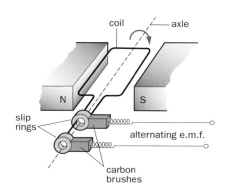

coil — axle

slip rings

alternating e.m.f.

carbon brushes

Fig. 16.20 A simple a.c. generator

In a **d.c. generator** (or **dynamo**) a *commutator* is used instead of slip-rings. This consists of a split-ring of copper, the two halves of which are insulated from each other and joined to the ends of the coil, Fig. 16.21a. The brushes are arranged so that the change-over of contact from one splitting ring to the other occurs when the coil is vertical. In this position the e.m.f. induced in the coil reverses and so one brush is always positive and the other always negative. The graphs of Fig. 16.21b show the e.m.f. in the coil and at the brushes; the latter, although varying, is unidirectional and supplies d.c. in an external circuit.

In actual a.c. and d.c. generators, several coils are wound in uniformly spaced slots in a soft-iron cylinder which is laminated to reduce eddy currents. The whole assembly is known as the **armature**. In the d.c. case the use of many coils and a correspondingly greater number of commutator segments gives a much steadier e.m.f. Also, in practical generators the magnetic field is produced by electro-

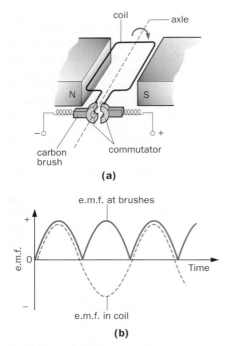

Fig. 16.21 A simple d.c. generator and its e.m.f.

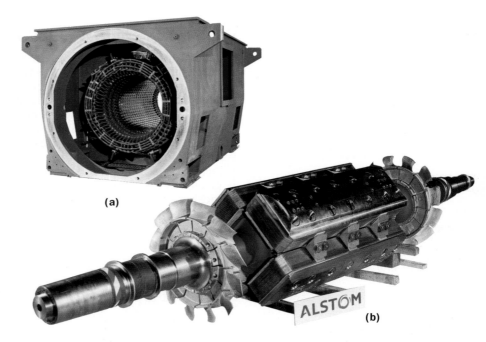

Fig. 16.23 The stator (*a*) and rotor (*b*) of a power-station alternator

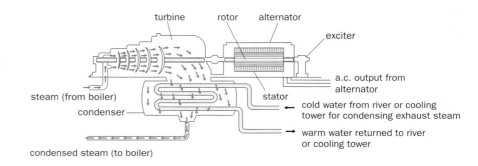

Fig. 16.24 Layout of a power station

magnets (except in a bicycle dynamo where a permanent magnet is used) and the coils that energize them are called **field coils**.

In power-station alternators the armature coils and their iron cores are stationary (and are called the **stator**) while the field coils and their cores (i.e. the electromagnets) rotate (and are called the **rotor**). Figure 16.22*a* shows a simplified alternator. The advantage of this is that only the relatively small direct current needed for the field coils is fed through the rotating slip-rings. The large p.ds and currents induced in the armature coils (25 kV and thousands of amperes in some modern alternators) are then led away through fixed connections. The rotor is driven by a steam, gas or water turbine which also powers a small

dynamo (called the **exciter**) for supplying current to the field coils.

The output from a power-station alternator is a *3-phase* a.c. supply, obtained by connecting the stator coils in three sets and having three rotor coils at 120° to each other. This gives three *live* wires (L_1, L_2, L_3) with a common return or *neutral* wire (N),

Fig. 16.22*b*, and makes current flow possible most of the time in all the live wires (each current is zero at a different time), Fig. 16.22*c*. A steadier power supply, more appropriate for powerful devices, is the result. Figure 16.23 shows a stator and a rotor of an alternator and Fig. 16.24 shows a simplified layout of a power station.

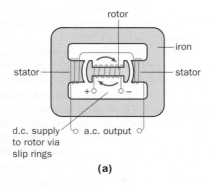

(a)

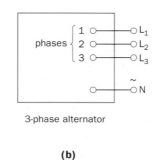

(b)

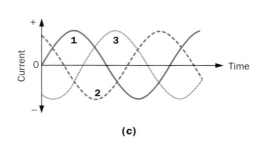

(c)

Fig. 16.22 Principle of a power-station alternator

ELECTRIC MOTORS (D.C. AND A.C.)

Electric motors form the heart of a whole host of devices ranging from domestic appliances such as vacuum cleaners and washing machines to electric trains and lifts. In a car the windscreen wipers are usually driven by one and the engine is started by another. There are many types, most of which work with a.c.

(a) d.c. motors

A d.c. motor consists of a coil on an axle, carrying a d.c. current in a magnetic field. The coil experiences a couple as in a moving-coil galvanometer (pp. 242 and 243) which makes it rotate. When its plane is perpendicular to the field, a split-ring commutator reverses the current in the coil and, as Fig. 16.25 shows, ensures that the couple continues to act in the same direction thereby maintaining the rotation.

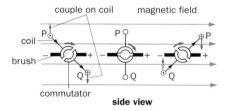

Fig. 16.25 Principle of a d.c. motor

In practice several coils are wound in equally spaced slots in a laminated soft-iron cylinder (the **armature** or **rotor**) and are connected to a commutator with many segments. Greater and steadier torque is thus obtained. Electromagnets with concave pole pieces frequently provide the magnetic field but the use of modern permanent magnets (e.g. Magnadur) for this is increasing, especially in small motors. When electromagnets are used the field coils may be in series with the armature (a series-wound motor), or in parallel (a shunt-wound motor), depending on what the motor is required to do. The construction of a d.c. motor is the same as that of a d.c. generator and in fact one can be used as the other (see p. 250).

When the armature coil in a motor rotates it cuts the magnetic flux of the field magnet and an e.m.f. $\mathscr{E}$, called the **back e.m.f.**, is induced in it (as in a generator); this e.m.f., by Lenz's law, opposes the applied p.d. V causing current I in the coil. If r is the armature coil resistance, then

$$V - \mathscr{E} = Ir$$

Multiplying by I we get

$$VI = \mathscr{E}I + I^2r$$

Now VI is the power supplied to the motor and I^2r is the power dissipated as heat in the armature coil. The difference, $\mathscr{E}I$, must be the mechanical power output of the motor; it is also the rate of working against the induced e.m.f.

The armature resistance r of a d.c. motor is small (e.g. 1 Ω or less) so as to make I^2r small and give high efficiency. However, when the motor is started, the armature is at rest and the back e.m.f. $\mathscr{E}$ is zero. The armature current I then equals V/r and would be so large as to burn out the armature coils. This is prevented by connecting a 'starting' resistance in series with the motor and gradually reducing it as the motor speeds up. The back e.m.f. then limits the current and will normally be only slightly less than the applied p.d. V. A motor driving a load takes a greater final steady current than an unloaded one since a greater torque ($BIAn$, p. 242) is required.

A d.c. motor transforms electrical energy into mechanical energy, the supply current and the induced (back) e.m.f. being in opposite directions. A d.c. generator, on the other hand, transforms mechanical energy into electrical energy, and in this case the induced current and e.m.f. act in the same direction. The 'direction' of energy transfer is thus determined by the sense of the current relative to the sense of the induced e.m.f.

(b) a.c. motors

A d.c. motor may be used on a.c. if the armature and field coils are in series. The current then reverses simultaneously in each and rotation in the same direction continues. (The torque developed in a shunt-wound motor on a.c. is very small owing to inductive effects causing the armature and field currents to reach their maxima at different times, p. 276.)

The **induction motor** is widely used in industry and is the commonest type of a.c. motor. Its action depends on the fact that a moving magnetic field can set a neighbouring conductor into motion. The converse is also true, that is, a moving conductor can cause a magnetic field to move, and this is readily demonstrated with the arrangement in Fig. 16.26 in which the bar magnet starts spinning when the copper disc is rotated rapidly. Whether it is the conductor or the field that moves, eddy currents are induced in the conductor and these tend to reduce the relative motion between conductor and field. If the conductor is stationary it starts moving in the same direction as the field and tries to catch it up in an (unsuccessful) attempt to eliminate the relative motion between them.

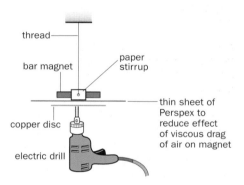

Fig. 16.26

Moving magnetic fields are produced in various ways in actual induction motors. In large rotary machines three pairs of stationary electromagnets (the stator) are arranged at equal angles round a conductor (the rotor) and each pair is connected to one phase of a 3-phase a.c. supply, Fig. 16.27*a*; the graph shows how the phases reach their maximum values one after the other. The rotor is generally of the 'squirrel-cage' pattern, comprising a number of copper rods in an iron cylinder, Fig. 16.27*b*, and it 'interprets' the alternations of the magnetic field as a field sweeping round it, i.e. a rotating field. The eddy currents induced in the copper rods set it into rotation as explained above. Linear induction motors operate on the same principle but the field travels in a straight line.

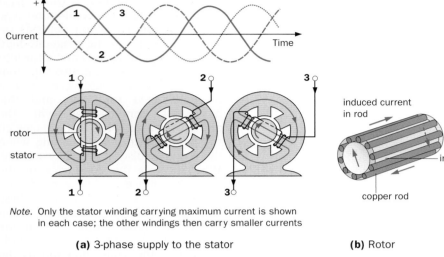

Note. Only the stator winding carrying maximum current is shown in each case; the other windings then carry smaller currents

(a) 3-phase supply to the stator   **(b)** Rotor

Fig. 16.27   The induction motor

The **shaded-pole** induction motor produces a 'rotating' magnetic field by covering *part* of the pole of an electromagnet carrying a.c. with a thick conducting plate. The alternating field of the electromagnet induces eddy currents in the plate and these create another field, adjacent to the main electromagnet field. There is a phase difference between the two fields and a nearby metal disc 'regards' this as a moving field and responds by rotating. A model shaded-pole motor is shown in Fig. 16.28.

Stepper motors are driven by a series of electrical pulses and rotate by a small exact amount for each pulse. They are used in robots and in computer disc drives.

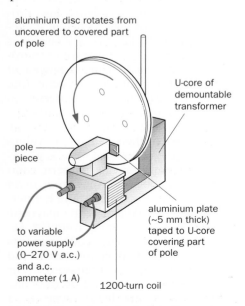

Fig. 16.28   A shaded-pole induction motor

TRANSFORMERS

(a) Action

A transformer changes, i.e. transforms, an alternating p.d. from one value to another of greater or smaller value using the principle of mutual induction (p. 249).

Two coils, called the **primary** and **secondary** windings, which are not connected to one another in any way, are wound on a complete soft-iron core, either one on top of the other as in Fig. 16.29*a* or on separate limbs of the core as in Fig. 16.29*b*. When an alternating p.d. is applied to the primary, the resulting current produces a large alternating magnetic flux which links the secondary and induces an e.m.f. in it. The value of this e.m.f. depends on the number of

turns on the secondary and we will show shortly that under certain conditions it is approximately true to say

$$\frac{\text{e.m.f. induced in secondary}}{\text{p.d. applied to primary}} = \frac{\text{secondary turns}}{\text{primary turns}}$$

A 'step-up' transformer has more turns on the secondary than the primary and the e.m.f. induced in the secondary is greater than the p.d. applied to the primary, e.g. if the turns are stepped up in the ratio 1 : 2, the secondary e.m.f. will be about twice the primary p.d. In a 'step-down' transformer the secondary e.m.f. is less than the primary p.d.

Three demonstrations to show the working of a transformer are illustrated in Figs 16.30*a*, *b* and *c*. In the first the lamp lights up to full brightness when a sufficient number of secondary turns have been wound on. The effect of placing an iron yoke across the U-core can be investigated. In the second demonstration the relation between the turns and p.d. ratios may be studied, a CRO being used as a voltmeter to measure the secondary e.m.f. The third demonstration shows the way in which current in the primary depends on that in the secondary, which is increased by connecting lamps across it.

If the p.d. is stepped up by a transformer, the current is stepped down, roughly in the same ratio. This follows from the conservation of energy, because, taking the transformer to be

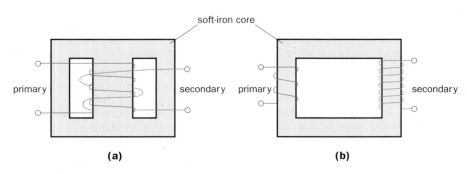

Fig. 16.29   Primary and secondary windings

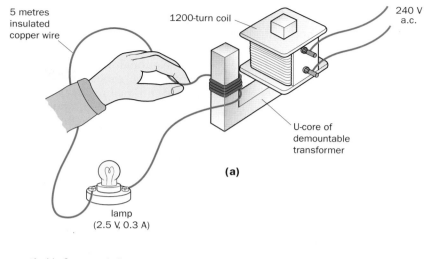

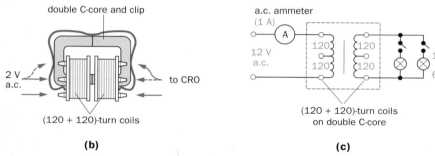

Fig. 16.30 Investigating the action of transformers

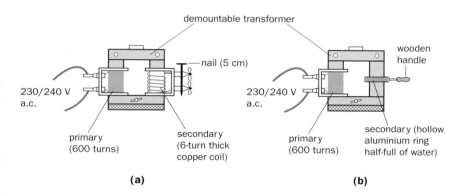

Fig. 16.31 Demonstrating the stepping up of current

100% efficient (many approach this), if all the electrical energy supplied to the primary appears in the secondary then

$$\frac{\text{power in}}{\text{primary}} = \frac{\text{power in}}{\text{secondary}}$$

that is,

primary p.d. × primary current = secondary e.m.f. × secondary current

or,

$$\frac{\text{secondary current}}{\text{primary current}} = \frac{\text{primary p.d.}}{\text{secondary e.m.f}}$$

The stepping up of current can be demonstrated effectively using the apparatus of Figs 16.31*a* and *b*. In the first the iron nail melts spectacularly and in the second the water boils very quickly.

A transformer can have more than one secondary and may step up and down simultaneously. A transformer (with two secondaries) for a mains-operated high voltage power supply unit is shown diagrammatically in Fig. 16.32.

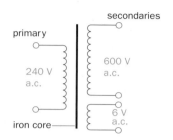

Fig. 16.32 Transformer in a mains-operated high voltage power supply

(b) Energy losses

Although transformers are very efficient devices, small energy losses do occur in them owing to four main causes:

(*i*) *Resistance of the windings.* The copper wire used for the windings has resistance and so ordinary (I^2R) heat losses occur. In high-current, low-p.d. windings these are minimized by using thick wire.

(*ii*) *Eddy currents.* The alternating magnetic flux induces eddy currents in the iron core and causes heating. The effect is reduced by having a laminated core (see p. 253).

(*iii*) *Hysteresis.* The magnetization of the core is repeatedly reversed by the alternating magnetic field. The resulting expenditure of energy in the core appears as heat and is kept to a minimum by using a magnetic material (such as Mumetal) which has a low hysteresis loss (p. 267).

(*iv*) *Flux leakage.* The flux due to the primary may not all link the secondary if the core is badly designed or has air gaps in it.

Very large transformers (like those in Fig. 16.35*a* on p. 261) have to be oil-cooled to prevent overheating.

(c) Theory

Consider initially a mains transformer whose primary winding has a resistance of 10 Ω; on a 240 V supply the primary current should therefore be 240/10 = 24 A. An a.c. ammeter connected in the primary records 0.10 A.

The difference is very large and is due to the fact that the alternating flux (arising from the a.c. in the primary), which induces an e.m.f. in the secondary, also induces an e.m.f. called a **back e.m.f.** in the primary. This e.m.f. opposes the applied p.d. and is nearly but not quite equal to it at every instant. The *net* e.m.f. in the primary is therefore quite small.

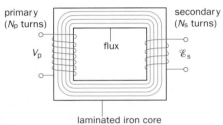

Fig. 16.33

Now consider an ideal transformer in which the primary has negligible resistance and all the flux in the core links both primary and secondary windings, Fig. 16.33. If Φ is the flux in the core at time t due to the current in the primary when a p.d. V_p is applied to it, then the back e.m.f. $\mathcal{E}_p$ induced in the primary of N_p turns is given by

$$\mathcal{E}_p = \frac{d}{dt}(N_p \Phi) = N_p \cdot \frac{d\Phi}{dt}$$

But, $\mathcal{E}_p = V_p$

If this were not so, the primary current would be infinite since the primary has zero resistance. Therefore

$$V_p = N_p \cdot \frac{d\Phi}{dt} \qquad (1)$$

The e.m.f. $\mathcal{E}_s$ induced in the secondary (N_s turns) by the same flux in the core is

$$\mathcal{E}_s = \frac{d}{dt}(N_s \Phi) = N_s \cdot \frac{d\Phi}{dt}$$

If the secondary is on open circuit or the current taken from it is small then, to a good approximation,

$$\mathcal{E}_s = V_s$$

where V_s is the p.d. across the secondary. Therefore

$$V_s = N_s \cdot \frac{d\Phi}{dt} \qquad (2)$$

From (1) and (2),

$$\frac{V_s}{V_p} = \frac{N_s}{N_p}$$

This expression is roughly true for an actual transformer if:

- the primary resistance and current are small;
- very little flux escapes from the core; and
- the secondary current is small.

When a smaller resistance is connected to the secondary, e.g. when another lamp is connected across the secondary in Fig. 16.30c, the secondary current increases and this acts to reduce the flux in the core (since the secondary current opposes the change producing it). The back e.m.f. in the primary therefore falls and so the primary current increases. Eventually the flux is restored to its previous value and as a result the back e.m.f. rises and becomes nearly equal to the applied p.d. The net effect is an increase of primary current, i.e. more energy is drawn from the source connected to the primary. The smaller resistance therefore acts as a greater load.

TRANSMISSION OF ELECTRICAL POWER

(a) Grid system

The National Grid in the UK is a network of cables, most of it supported on pylons, which links over 100 power stations, situated at convenient places throughout the country and carrying electrical energy from them to consumers.

Figure 16.34 shows the distribution system. In the largest modern power stations electricity is generated at about 25 kV (50 Hz) and stepped up by transformers to about 275 kV or 400 kV, Fig. 16.35a, for transmission over long distances. The p.d. is subsequently reduced in substations by step-down transformers for distribution to local users at suitable p.ds — 33 kV for heavy industry, 11 kV for light industry and 230 V for homes, schools, shops, farms, etc., Fig. 16.35b. For rail electrification systems working at 25 kV, there are special substations alongside the track taking their supply from the Grid. This is fed to step-down transformers in the electric locomotive and then rectified (see p. 282) for driving d.c. traction motors operating at about 900 V and 650 A.

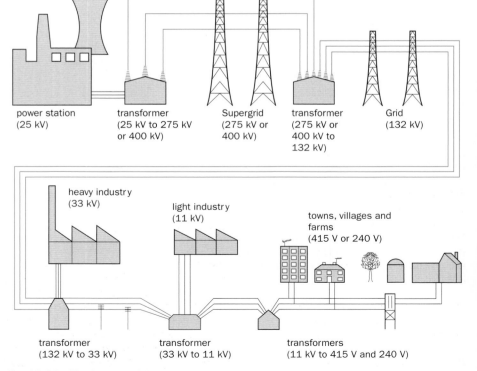

Fig. 16.34   The National Grid

Fig. 16.35a Step-up transformers at a power station

Fig. 16.35b Step-down transformer near a housing estate

To supervise the operation of the distribution system, engineers at the National Control Centre (Fig. 16.36) assess the demand, direct the flow and reroute it when breakdowns occur. In this way the electricity supply is made more reliable, requires less reserve plant to cover maintenance, etc., and costs are cut by taking output from smaller or less efficient power stations only at peak periods.

(b) Why high p.ds are used

Suppose electrical power P has to be delivered at a p.d. V by supply lines of total resistance R, Fig. 16.37. The current $I = P/V$ (since $P = IV$) and the power loss in the lines $= I^2R = (P/V)^2R$. Clearly, the greater V the smaller the loss — in fact, doubling V quarters the loss. Electrical power is thus transmitted more economically at 'high' p.ds but this creates insulation problems and raises installation costs. In the 400 kV Supergrid, currents of 2500 A are typical and the power loss is roughly 200 kW per kilometre of cable, i.e. a 0.02% loss per kilometre.

The ease and efficiency with which alternating p.ds are stepped up and down in a transformer and the fact that alternators produce much higher p.ds than d.c. generators (25 kV com-

Fig. 16.36 National Control Centre

pared with a few thousand volts) are the main considerations influencing the use of high alternating, rather than direct, p.ds in most situations. An exception to this is the cross-channel link between England and France where the underground cables favour a d.c. supply because of the high dielectric losses of a.c. in such cables.

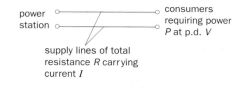

Fig. 16.37

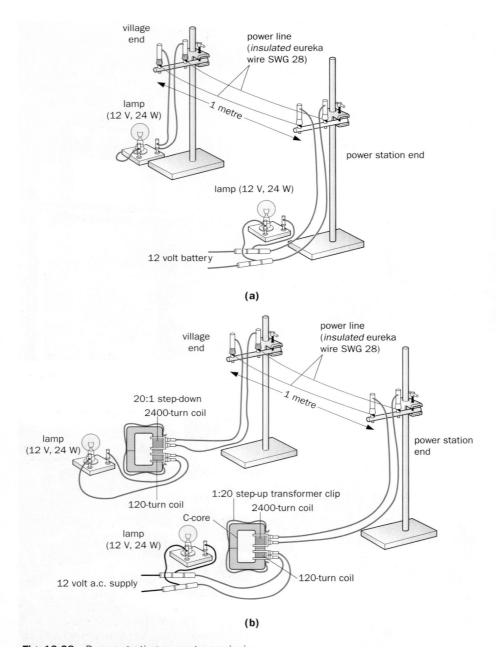

Fig. 16.38 Demonstrating power transmission

The advantages of high alternating p.d. power transmission may be illustrated with the model power line arrangements of Fig. 16.38*a* and *b*, *so long as the power line is well insulated* (using insulated Eureka wire) or the demonstration is located in an enclosure of, for example, clear polycarbonate, so that parts at high voltages cannot be touched.

SELF-INDUCTANCE

The flux due to the current in a coil links that coil and if the current changes the resulting flux change induces an e.m.f. in the coil itself. This changing-magnetic-field type of electromagnetic induction is called **self-induction**, and the coil is said to have **self-inductance**, or simply **inductance**, *L*. The coil is called an **inductor**, symbol ⎓⏦⎓ if air-cored and ⎓⏦⎓ if it has a core of magnetic material, when it may be called a

'choke'. The induced e.m.f. obeys Faraday's law like other induced e.m.fs.

(a) Some demonstrations

From Lenz's law we would expect the induced e.m.f. to oppose the current change causing it. That it does so in both d.c. and a.c. circuits may be demonstrated.

In Fig. 16.39*a*, *L* is an iron-cored inductor and *R* is a variable resistor adjusted to have the same resistance as *L*. When the current (d.c.) is switched on, the lamp in series with *L* lights up a second or two after that in series with *R*. This can be attributed to the induced e.m.f. in *L* opposing the change and tending to drive a current against the increasing current due to the battery. The growth of the current to its steady value is thereby delayed.

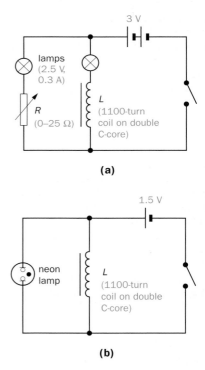

Fig. 16.39

The effect of the induced e.m.f. when the current is switched off is more striking and is shown with the circuit of Fig. 16.39*b*. Opening the switch causes the current to fall very rapidly to zero and the rate of change of flux is large. The induced e.m.f. is therefore large and in a direction tending to maintain the current in its original direction. It is sufficiently great in this case (more than 100 V) to produce a brief flash of the neon lamp across L. When circuits carrying large currents within large inductors are switched off, the induced e.m.f. can cause sparking between the switch contacts and may even fuse them together. In the Grid network special circuit breakers are used to prevent this.

Inductors are important components of a.c. circuits where the induced e.m.f. opposes the applied p.d. continuously. In Fig. 16.39a if the 3 V d.c. supply is replaced by a 3 V a.c. supply the lamp in series with L does not light — unless the iron core is removed.

(b) Definition and unit

It would seem reasonable to say that a coil (or circuit) has a large inductance if a small rate of change of current in it induces a large back e.m.f. This is the basis of the following definition of inductance. If the e.m.f. induced in a coil is $\mathscr{E}$ when the rate of change of current in it is dI/dt, the inductance L of the coil is defined by the equation

$$L = -\frac{\mathscr{E}}{dI/dt}$$

The negative sign is inserted to make L a positive quantity since $\mathscr{E}$ and dI/dt act in opposite directions and are given opposite signs.

The unit of inductance is the **henry** (H), defined as **the inductance of a coil (or circuit) in which an e.m.f. of 1 volt is induced when the current changes at the rate of 1 ampere per second.** That is, $1 \text{ H} = 1 \text{ V s A}^{-1}$.

(c) Inductance of a solenoid

Calculation of inductance is possible in certain cases, as it was of capacitance. Consider a long, air-cored solenoid of length l, cross-section area A, having N turns and carrying current I. The flux density B is almost constant over A and, neglecting the ends, is given by

$$B = \frac{\mu_0 N}{l} I \quad \text{(see p. 236)}$$

The flux Φ through each turn of the solenoid is BA and so for the flux-linkage we have $N\Phi = BAN$.

Therefore

$$N\Phi = \left(\mu_0 \frac{N}{l} I\right) AN$$
$$= \frac{\mu_0 A N^2}{l} \cdot I$$

If the current changes by δI in time δt causing a flux-linkage change $\delta(N\Phi)$ then by Faraday's law the induced e.m.f. $\mathscr{E}$ is

$$\mathscr{E} = -\frac{d}{dt}(N\Phi)$$
$$= -\frac{\mu_0 A N^2}{l} \cdot \frac{\delta I}{\delta t}$$
$$= -\frac{\mu_0 A N^2}{l} \cdot \frac{dI}{dt} \quad \text{as } \delta t \to 0$$

If L is the inductance of the solenoid, then from the defining equation we get

$$\mathscr{E} = -L\frac{dI}{dt}$$

Comparing these two expressions it follows that

$$L = \frac{\mu_0 A N^2}{l}$$

L depends only upon the geometry of the solenoid. If $N = 400$ turns, $l = 25 \text{ cm} = 25 \times 10^{-2} \text{ m}, A = 50 \text{ cm}^2 = 50 \times 10^{-4} \text{ m}^2$ and $\mu_0 = 4\pi \times 10^{-7} \text{ H m}^{-1}$, then $L = 4.0 \times 10^{-3} \text{ H} = 4.0 \text{ mH}$.

A solenoid having a core of a magnetic material would have a much greater inductance but the value would vary depending on the current in the solenoid.

(d) Energy stored by an inductor

A current-carrying inductor stores energy in the magnetic field associated with it and it can be shown that for current I and an inductance L this equals $\frac{1}{2}LI^2$. Compare this with the analogous case of $\frac{1}{2}Q^2/C$ for the energy stored in the electric field of a capacitor.

Since every current produces a field, every circuit must have some self-inductance. On switching on any circuit some time is necessary to provide the energy in the magnetic field and so no current can be brought instantaneously to a non-zero value. Similarly, on switching off any circuit, the energy of the magnetic field must be dissipated somehow, hence the spark. A capacitor across the switch can 'suppress' sparking.

UNIT OF μ_0

Reference was made to this on p. 235 and it is convenient to consider it now. The permeability of free space μ_0 was defined by the Biot-Savart law

$$\delta B = \frac{\mu_0 I \, \delta l \sin\theta}{4\pi r^2}$$

The unit of μ_0 from this equation is

$$\frac{(\text{Wb m}^{-2}) \times (\text{m}^2)}{(\text{A}) \times (\text{m})} \text{ or Wb A}^{-1}\text{m}^{-1} \quad (1)$$

From Faraday's law, $\mathscr{E} = \dfrac{d}{dt}(N\Phi)$, we can say that

$$1 \text{ Wb} = 1 \text{ V s} \quad (2)$$

In addition, from the inductance-defining equation $L = \mathscr{E}/(dI/dt)$ we have

$$1 \text{ H} = 1 \text{ V s A}^{-1} \quad (3)$$

From (2) and (3)

$$1 \text{ H} = 1 \text{ Wb A}^{-1}$$

Hence from (1), μ_0 can be expressed in

$$\text{H m}^{-1} \text{ (henry per metre)}$$

This is the SI unit of μ_0 (and μ); it may be compared to F m^{-1} (farad per metre) for the unit of ε_0, the permittivity of free space (and ε).

MUTUAL INDUCTANCE

(a) Definition and unit

In mutual induction, current changing in one coil or circuit (the primary) can induce an e.m.f. in a neighbouring coil or circuit (the secondary), as we saw earlier. The **mutual inductance** M of two coils, Fig. 16.40, is defined by the equation

$$M = -\frac{\mathscr{E}}{dI_p/dt}$$

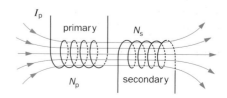

Fig. 16.40

where $\mathscr{E}$ is the e.m.f. induced in the secondary when the rate of change of current in the primary is dI_p/dt.

It follows from the definition that M has the same unit as L, i.e. henry (H). **Two coils are said to have a mutual inductance of 1 henry if an e.m.f. of 1 volt is induced in the secondary when the primary current changes at the rate of 1 ampere per second.**

It can be shown that the mutual inductance of two coils is the same if current flows in the secondary and flux links the primary, causing an induced e.m.f. when a change in flux-linkage occurs.

(b) Mutual inductance of a solenoid and a coil

In Fig. 16.41 the long air-cored solenoid (the primary) with N_p turns, length l_p and cross-section area A_p carries current I_p. The flux density B_p in the centre of the solenoid is nearly constant over A_p and is given by

$$B_p = \mu_0 \frac{N_p}{l_p} I_p \qquad \text{(see p. 236)}$$

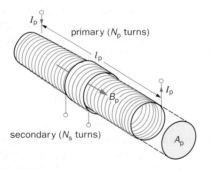

Fig. 16.41

The flux Φ_s linking each of the N_s turns of the short coil (the secondary) round the middle of the solenoid is $B_p A_p$. The flux-linkage of the short coil is therefore

$$N_s \Phi_s = B_p A_p N_s$$
$$= \left(\mu_0 \frac{N_p}{l_p} I_p\right) A_p N_s$$

If the current in the primary changes by dI_p in time dt causing a flux-linkage change $d(N_s \Phi_s)$ in the secondary, the e.m.f. $\mathscr{E}_s$ induced in the secondary is

$$\mathscr{E}_s = -\frac{d}{dt}(N_s \Phi_s)$$
$$= -\frac{\mu_0 A_p N_p N_s}{l_p} \cdot \frac{dI_p}{dt}$$

If M is the mutual inductance, we can also say

$$\mathscr{E}_s = -M \frac{dI_p}{dt}$$

Therefore

$$M = \frac{\mu_0 A_p N_p N_s}{l_p}$$

When the coils have a ferromagnetic core, the value of M is very much greater (especially if the core is complete as in a transformer), but it varies with the current.

INDUCED CHARGE AND FLUX CHANGE

When the flux linking a *complete* circuit changes, an e.m.f. is induced in it and current flows. A simple connection exists between the flux change and the total charge circulation that constitutes the current.

Consider a coil of N turns in a circuit of total resistance R in which the flux linking each turn is changing and has value Φ at time t. The induced e.m.f. at t will be (in magnitude)

$$\mathscr{E} = \frac{d}{dt}(N\Phi)$$

Also, the induced current I at time t will be

$$I = \frac{\mathscr{E}}{R} = \frac{1}{R} \cdot \frac{d}{dt}(N\Phi)$$

Since I = rate of flow of charge = dQ/dt,

$$\frac{dQ}{dt} = \frac{1}{R}\frac{d}{dt}(N\Phi) = \frac{N}{R}\frac{d\Phi}{dt}$$

If the flux changes from say Φ_1 to Φ_2 the total charge Q that passes is

$$Q = \int_0^Q dQ = \frac{N}{R}\int_{\Phi_1}^{\Phi_2} d\Phi$$

$$\therefore \quad Q = \frac{N(\Phi_2 - \Phi_1)}{R}$$

i.e.

$$Q = \frac{\text{flux-linkage change}}{R}$$

We see that Q does not depend on the time taken for the flux change.

Consider a numerical example. A search coil of average cross-section area 3.0 cm² has 400 turns and is in a circuit of total resistance 200 Ω. It is inserted into a magnetic field of flux density 2.5×10^{-3} T so as to produce the maximum flux change. We have $N = 400$ turns, $A = 3.0 \times 10^{-4}$ m², $R = 200$ Ω and $B = 2.5 \times 10^{-3}$ T. Hence $\Phi_2 = BA = 2.5 \times 10^{-3} \times 3.0 \times 10^{-4}$ Wb and $\Phi_1 = 0$. The induced charge Q is given by

$$Q = \frac{N(\Phi_2 - \Phi_1)}{R}$$
$$= \frac{400 \times 2.5 \times 10^{-3} \times 3.0 \times 10^{-4}}{200} \text{ C}$$
$$= 1.5 \times 10^{-6} \text{ C} = 1.5 \text{ μC}$$

MEASURING *B* BY BALLISTIC GALVANOMETER AND SEARCH COIL

(a) Ballistic galvanometer

A moving-coil galvanometer measures charge provided (*i*) the period of oscillation of the movement is large (e.g. 2 seconds) so that all the charge is able to pass through the coil before it moves appreciably, and (*ii*) the damping is very small. It is then called a ballistic galvanometer (since it is set into motion by an impulse, as is a projectile whose motion is under study in ballistics) and theory shows that the first deflection or 'throw' θ is proportional to the total charge Q that has passed:

$$\theta \propto Q$$

or

$$\theta = bQ$$

where b is a constant called the **charge sensitivity** of the galvanometer; it is expressed in mm per μC. Generally only light-beam galvanometers (see p. 243) are suitable for ballistic use.

(b) Measuring B

A search coil in series with a ballistic galvanometer and an appropriate high resistance (to reduce damping and adjust the sensitivity of the galvanometer) is placed in the magnetic field to be measured so that the flux links it normally, Fig. 16.42. It is then *quickly* removed (why?) from the field and the first 'throw' θ produced by the flux change is noted. The charge Q driven through the coil (e.g. 1–2 µC) is proportional to θ.

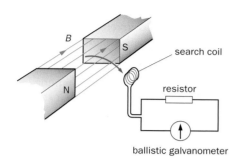

Fig. 16.42 Measuring B by ballistic galvanometer

If B is the flux density of the field and A is the cross-section area of the coil which has N turns, then

$$\text{flux-linkage change} = NAB$$

$$\therefore \quad Q = \frac{NAB}{R}$$

where R is the *total* resistance of the circuit. Hence

$$B = \frac{RQ}{NA}$$

R, N and A can be measured or are given; Q can be found from the charge sensitivity of the galvanometer.

ABSOLUTE MEASUREMENT OF RESISTANCE

In an absolute method an electrical quantity is measured in terms of the basic mechanical quantities, i.e. mass, length and time, and no electrical measurements are necessary. The absolute measurement of current was considered earlier (p. 241); the principle of a method for resistance, due to Lorenz and based on Faraday's disc dynamo (p. 254), is illustrated in Fig. 16.43.

A metal disc is rotated with its plane at right angles to the uniform flux density B at the centre of a long current-carrying solenoid having n turns per metre. The e.m.f. $\mathscr{E}$ induced between the centre and rim of the disc, for the radius joining the sliding contacts, is balanced against the p.d. across a low resistance R (a copper rod) which carries the same current I

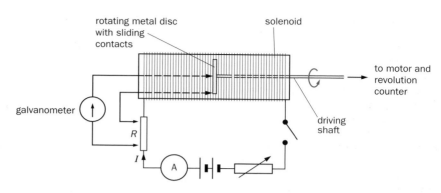

Fig. 16.43 Absolute measurement of resistance

as flows through the solenoid. R is to be measured.

If the disc makes f revolutions per second at balance and has radius r, the area swept out per second by a radius is $\pi r^2 f$. The flux change per second is $B\pi r^2 f$, hence when there is no deflection on the galvanometer,

$$\mathscr{E} = IR = B\pi r^2 f$$

$$\therefore \quad R = \frac{B\pi r^2 f}{I}$$

But at the centre of a long solenoid

$$B = \mu_0 nI \qquad \text{(p. 236)}$$

$$\therefore \quad R = \mu_0 n\pi r^2 f$$

R can be calculated if n, r and f are measured.

This method is used to measure the resistance of coils kept as standards in laboratories such as the NPL, but various modifications and precautions are necessary. Thermoelectric e.m.fs, comparable with the small induced e.m.f., arise at the sliding contacts owing to frictional heating and they must be allowed for. The earth's magnetic field has to be taken into account and allowance also made for the field inside the solenoid not being perfectly uniform over the disc.

FERROMAGNETIC MATERIALS

Iron, cobalt and nickel, and substances containing them, are strongly magnetic and are called **ferromagnetic** materials. Many other materials exhibit large magnetic effects at very low temperatures.

(a) Relative permeability

The flux density in a coil increases many times when it has a ferromagnetic core because the core becomes magnetized and contributes flux. The **relative permeability** μ_r of a material is defined by the equation

$$\mu_r = \frac{B}{B_0}$$

where B_0 is the flux density in a current-carrying **toroid** (an endless solenoid) containing air (more strictly, a vacuum) and B is the flux density when the same toroid is filled with the material, Fig. 16.44a. Since B and B_0 are both measured in teslas, μ_r has no units.

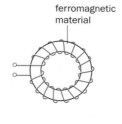

(a) Toroid winding

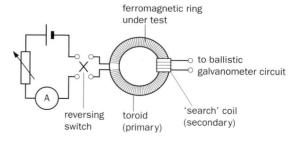

(b) Measuring μ_r

Fig. 16.44

A toroid rather than a solenoid is specified so that the magnetization of the material is nearly uniform, if the difference between the external and internal radii is relatively small. A rod-shaped specimen in a solenoid would have poles at its ends which tend to demagnetize the rod near the ends (hence explaining the use of keepers to store magnets and reduce self-demagnetization). Uniform magnetization, such as can be achieved all along the material under test when it forms a closed magnetic loop in a toroid, is impossible in a solenoid.

In a measurement of μ_r, B is found using a search coil, connected to a calibrated ballistic galvanometer and wound round part of the toroid, Fig. 16.44b. B_0 is calculated from $B_0 = \mu_0 nI$ (as for the middle of a long solenoid) where n is the number of turns per metre on the toroid and I is the current in it.

(b) Magnetization curve

If B is measured for a ferromagnetic material as the magnetizing current is increased from zero, a magnetization curve of B against B_0 (which is proportional to the current) can be obtained and has the form of Fig. 16.45a.

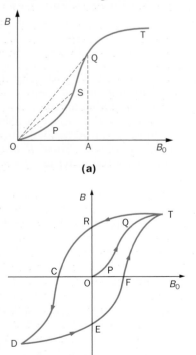

(a)

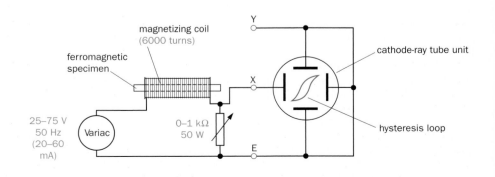

(b)

Fig. 16.45 Magnetization curve and hysteresis loop for a ferromagnetic material

Along OP the magnetization is small and reversible, that is, it returns to zero when the magnetizing field B_0 is removed. Between P and Q the magnetization increases rapidly as B_0 increases and is irreversible, i.e. the specimen remains magnetized when the field is reduced to zero. For values of B_0 beyond Q very little increase of B occurs and the specimen is said to be approaching full magnetization or 'saturation' along QT.

The relative permeability μ_r (that is, B/B_0) at a point such as S is the gradient of the line joining O to S. Its value varies along the graph and is a maximum at Q. The value of B_0 at Q (i.e. OA) therefore gives the most efficient flux production and would be achieved by a correct choice of n and I for the magnetizing toroid. When values of μ_r are quoted (and they can be as high as 10^5) they usually refer to point Q on the magnetization curve.

(c) Hysteresis loop

When a specimen of a ferromagnetic material has reached saturation and the magnetizing field is reduced to zero, it remains quite strongly magnetized. The flux density B_r it retains is called the **remanence** or **retentivity** of the material, OR in Fig. 16.45b. A reverse magnetizing field is required to demagnetize it completely and the value of B_0 which makes B zero is called the **coercivity** of the material, OC in Fig. 16.45b. If the reverse field is increased more, the specimen becomes saturated in the reverse direction (D). Decreasing the field and again reversing to saturation in the first direction gives the rest of the loop, DEFT.

The curve in Fig. 16.45b is called a **hysteresis loop**. It shows that the magnetization of a material (i.e. B) lags behind the magnetizing field (i.e. B_0) when it is taken through a complete magnetization cycle, an effect called **magnetic hysteresis**; the term *hysteresis* is derived from a Greek word meaning 'lagging behind'.

The shape of a hysteresis loop provides useful information to the designer of electrical equipment. For example, it can be shown that the area of the loop is proportional to the energy required to take unit volume of the material round one cycle of magnetization. This energy increases the internal energy of the specimen. It is called the **hysteresis loss** and is important when materials are subject to alternating fields which take them through many cycles of magnetization per second.

A hysteresis loop can be displayed on a CRO with an easily accessible tube from which any magnetic screen must be removed. The specimen (e.g. a strip of soft-iron tinplate or a length of steel clock-spring) is inserted in a magnetizing coil set at right angles to the oscilloscope tube and close to the deflecting plates, Fig. 16.46. When current flows in the coil the specimen is magnetized and deflects the electron beam in the Y-direction (Fleming's left-hand rule). The Y-deflection is thus a measure of the flux density B of the specimen. The magnetizing coil current also passes through a variable resistor, and the p.d. across this is applied to the X-plates. Therefore the X-deflection is proportional to the magnetizing current and also to B_0. With an a.c. input the specimen is taken through complete magnetization

Fig. 16.46 Observing a hysteresis loop on a CRO

cycles and the spot produces a hysteresis loop.

(d) Demagnetization

A simple but effective way of demagnetizing a magnetic material is to insert it in a multi-turn coil carrying a.c. and then either reduce the current to zero or withdraw the specimen from the coil. In both cases the material is taken through a series of ever-diminishing hysteresis loops.

MAGNETIC CIRCUIT: RELUCTANCE

Consider a coil of N turns wound round a toroid of ferromagnetic material of length l, cross-section area A and permeability μ where $\mu = \mu_r \times \mu_0$. If the current in the coil is I, the flux Φ through A is given by

$$\Phi = BA = \frac{\mu N I A}{l}$$

since for a toroid $B = \mu N I / l$.

Hence $$\Phi = \frac{NI}{l/\mu A}$$

Treating the ferromagnetic ring as a magnetic circuit carrying flux Φ (just as an electric circuit carries current I) and comparing the above equation with

$$I = \frac{\text{e.m.f.}}{\text{resistance}}$$

we see that NI is analogous to e.m.f. and is called the **magnetomotive force (m.m.f.)** of the magnetic circuit. The quantity $l/\mu A$ is analogous to resistance and is known as the **reluctance** $\mathcal{R}$ of the circuit; the unit of $\mathcal{R}$ is H^{-1}. Comparing the two equations

$$R = \frac{\rho l}{A} \quad \text{and} \quad \mathcal{R} = \frac{l}{\mu A}$$

we see that μ is analogous to $1/\rho$, i.e. the electrical conductivity. Hence μ can be regarded as a measure of the 'conductivity' of a ferromagnetic material for magnetic flux.

The reluctance of a magnetic circuit is greatly increased by an air gap: for air $\mu = \mu_0 = 4\pi \times 10^{-7} \text{ H m}^{-1}$ while for a ferromagnetic material $\mu = \mu_r \times \mu_0$. So, if μ_r is 1000, μA is large and therefore $\mathcal{R}_{\text{fe}}$ is small while $\mathcal{R}_{\text{air}}$ is large. The reluctance of the complete magnetic circuit, namely $\mathcal{R}_{\text{fe}} + \mathcal{R}_{\text{air}}$, is then large.

PROPERTIES AND USES OF MAGNETIC MATERIALS

Ferromagnetic materials can be classified into two broad groups — 'soft' and 'hard'. Soft magnetic materials are easily magnetized and demagnetized; hard materials require large magnetizing fields and retain their magnetization. Originally, the characteristics of the two groups were displayed by soft iron and hard steel but modern magnetic materials surpass these in performance. Typical hysteresis loops for soft and hard materials are shown in Fig. 16.47.

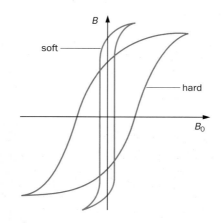

Fig. 16.47 Hysteresis loops for soft and hard ferromagnetic materials

(a) Permanent magnets

These are made of hard magnetic materials with high remanence to give them 'strength' and high coercivity so that they are not easily demagnetized by stray magnetic fields or mechanical ill-treatment. Such materials are either (*i*) *alloys* containing, for example, iron, aluminium, nickel, copper and cobalt, and having trade names like 'Ticonal', 'Alnico' and 'Alcomax', or (*ii*) *ceramics*, made by heat and pressure treatment from powders of iron oxide and barium oxide ($BaFe_{12}O_{19}$). The latter belong to a group of materials called **ferrites** and have hexagonal crystal structures; one is called 'Magnadur'.

Ceramic magnets are brittle, much like china, but the powder can be bonded with plastics and rubber to give flexible magnets of any shape. Very fine powder is used to coat tapes for audio and video cassette recorders.

(b) Electromagnets

These require a core of soft magnetic material which will give a strong but temporary magnet. Small coercivity is essential.

(c) Transformer cores

These are subject to many cycles of magnetization and must be 'soft' with a narrow hysteresis loop to minimize heating from hysteresis loss. They should also have high resistivity to reduce eddy current loss (p. 253) and must never be saturated when in normal use or the flux will not follow the changes in the primary current. Silicon iron (for example 'Stalloy') and Mumetal are used at mains frequencies, and ferrite materials with cubic crystal structures (e.g. 'Ferroxcube', general formula MFe_2O_4, where M is a divalent atom of copper, zinc, magnesium, manganese or nickel) and very high resistivities are appropriate for high-frequency applications in, for example, radio work.

DOMAIN THEORY OF FERROMAGNETISM

(a) Electrons, atoms and domains

The magnetic field produced by a magnet can, in general, also be produced by a current within a suitably shaped conductor. This suggests that possibly all magnetic effects, including permanent magnetism, may be due to electric currents. In fact the magnetic properties of materials are attributed to the motion of electrons inside atoms; each electron may be regarded as a tiny 'current-carrying coil' having a magnetic field.

In the atoms of some materials the magnetic effects of different electrons cancel; in others they do not and each atom has a resultant magnetic field. With most of the latter materials, the vibratory motion of the atoms (due to their internal energy) causes their magnetic axes to have random orientations and no appreciable magnetization is shown by the material as a whole, even when we attempt to align them all in the same direction by a strong applied field.

However, in ferromagnetic materials each atom has a resultant field and there is a force (explicable in terms of quantum mechanics) which causes *neighbouring* atoms to react on one another so that all their magnetic axes are lined up in the same direction even when there is no external magnetizing field. They do this in groups of about 10^{10} atoms to form regions called **domains** which behave like very small but very strong permanent magnets, each roughly 10^{-3} mm wide. The directions of alignment of the magnetic axes vary from one domain to another and in an unmagnetized specimen they form closed magnetic loops, Fig. 16.48*a*, with the 'closure' domains acting like the keepers on a pair of bar magnets. The magnetic fields of the domains thus neutralize one another and no detectable external magnetic effect is produced.

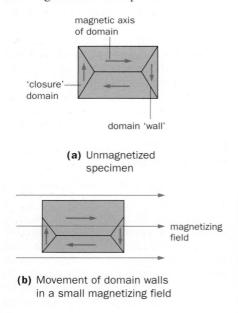

Fig. 16.48 Domain theory

(b) Explanation of magnetization and hysteresis

The domain theory offers the following explanation. When a small field is applied to an unmagnetized specimen, those domains whose magnetic axes are most in line with the field grow at the expense of others and a movement of domain 'walls' results, Fig. 16.48*b*. This first stage of magnetization, OP in the magnetization curve of Fig. 16.45*a* (p. 266), is almost wholly reversible and if the applied field is removed the walls return to their previous positions and the magnetization is again zero.

Larger magnetizing fields cause the magnetic axes of entire domains to 'jump' round quite suddenly in succession into alignment with the field and the magnetization increases sharply, PQ in Fig. 16.45*a*. This stage is largely irreversible and the specimen retains its magnetization if the field is reduced to zero. When the field is great enough, more or less all domains are in line with the field and saturation occurs; QT on the curve. If a sufficiently strong reverse field is applied the domains can be re-aligned and saturation obtained in the opposite direction.

Hysteresis is considered to result from domain walls being unable to move across grain boundaries and other defects in the polycrystalline specimen until a reverse field of sufficient strength is applied. The magnetization therefore lags behind the magnetizing field.

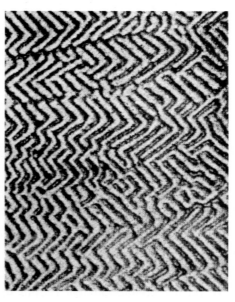

Fig. 16.49 Bitter pattern

(c) Evidence for domains

Various effects support their existence. Two will be considered briefly.

(*i*) *Bitter patterns*. These were first obtained by Bitter in 1931 when he allowed very fine iron powder in colloidal suspension to settle on a single ferromagnetic crystal with a smooth surface. At the domain walls there is slight leakage of magnetic flux; the powder collects there and a 'maze' pattern like that in Fig. 16.49 is seen through a microscope, whether the specimen is magnetized or not.

(*ii*) *Barkhausen effect*. This may be easily demonstrated using the apparatus of Fig. 16.50. When, say, the north pole of the magnet is drawn across and a little above the end of the bundle of ferromagnetic wires, a rushing noise is produced in the loudspeaker due to induced e.m.fs in the coil arising from the succession of 'jumps' made by the magnetic axes of domains during magnetization. There is no repetition of the effect on subsequent transits of the magnet unless the south pole of the magnet is nearest the wires. Copper wires give no effect.

(d) Curie temperature

On heating to a certain temperature, called the **Curie point**, a ferromagnetic material loses its ferromagnetic properties. This is attributed to the internal energy and vibration of the atoms becoming so vigorous as to destroy the domain structure. The Curie point of iron is 770 °C. In contrast, the ferromagnetic alloy JAE metal (70% Ni, 30% Cu) has a Curie point of about 70 °C, and so it can be used to show the effect, Fig. 16.51. The JAE metal drops off the magnet at the Curie point but becomes magnetic again below it.

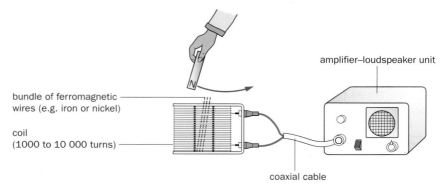

Fig. 16.50 Demonstrating the Barkhausen effect

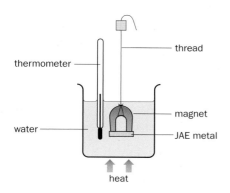

Fig. 16.51 Observing the Curie point

QUESTIONS

Induced e.m.f.; generators

1. Under what circumstances is an e.m.f. induced in a conductor? What factors govern the magnitude and direction of the induced e.m.f.? Describe a quantitative experiment which demonstrates how the magnitude of the induced e.m.f. depends on one of these factors.

   By considering any simple case, show that if the induced e.m.f. acted in the opposite direction to that in which it does act the law of conservation of energy would be contravened.

2. Write down an expression for the e.m.f. induced between the ends of a rod of length l moving with velocity v so as to cut a flux density B normally.

   A straight wire of length 50 cm and resistance 10 Ω moves sideways with a velocity of 15 m s^{-1} at right angles to a uniform magnetic field of flux density 2.0×10^{-3} T. What current would flow if its ends were connected by leads of negligible resistance?

3. What is meant by the term **electromagnetic induction**?

   Describe an experiment you could perform in a school laboratory to demonstrate Faraday's law of electromagnetic induction.

   An aircraft has a wing span of 54 m. It is flying horizontally at 860 km h^{-1} in a region where the vertical component of the Earth's magnetic field is 6.0×10^{-5} T. Calculate the potential difference induced between one wing tip and the other.

What extra information is necessary to establish which wing is positive and which negative?

(*L*, A PH4, June 1998)

4. A light aluminium washer rests on the end of a solenoid as shown in Fig. 16.52.

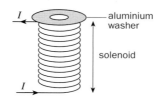

Fig. 16.52

A large direct current is switched on in the solenoid. Explain why the washer jumps and immediately falls back.

(*L*, A PH4, Jan 1998)

5. State Lenz's law of electromagnetic induction and describe an experiment by which it may be demonstrated.

   A circular disc of copper and a horseshoe magnet are mounted as shown in Fig. 16.53. The disc is free to rotate and the magnet can be rotated on the axle as shown. Describe and explain what happens when the magnet is set in rotation.

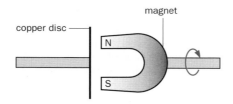

Fig. 16.53

6. **a)** State Lenz's law of electromagnetic induction

   b) Figure 16.54 shows a magnet being dropped through the centre of a narrow coil. The e.m.f. across the coil is monitored on an oscilloscope.

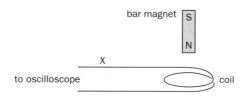

Fig. 16.54

   i) What is the induced polarity on the top side of the coil as the magnet falls towards the coil?

ii) Indicate the direction of current flow at position X as the magnet falls towards the coil.

   c) Figure 16.55 shows the variation of e.m.f. with time as the magnet falls.

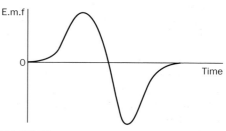

Fig. 16.55

State *two* ways in which the graph would change if the magnet were dropped from a greater height than that used to produce Fig. 16.55.

(*NEAB*, AS/A PHO3, March 1998)

7. A copper disc of radius 10 cm is situated in a uniform field of magnetic flux density 1.0×10^{-2} T with its plane perpendicular to the field.

   The disc is rotated about an axis through its centre parallel to the field at 3.0×10^{3} rev min^{-1}. Calculate the e.m.f. between the rim and centre of the disc.

   Draw a circle to illustrate the disc. Show the direction of rotation as clockwise and consider the field directed into the plane of the diagram.

   Explaining how you obtain your result, state the direction of the current flowing in a stationary wire whose ends touch the rim and centre of the disc.

8. **a)** State an expression for the e.m.f. induced in a conductor moving in a magnetic field and show in a diagram the directional relations involved.

   A rectangular coil 30.0 cm long and 20.0 cm wide has 25 turns. It rotates at the uniform rate of 3000 rev min^{-1} about an axis parallel to its long side and at right angles to a uniform magnetic field of flux density 5.00×10^{-2} T.

   b) Find
   i) the frequency, and
   ii) the peak value of the induced e.m.f. in the coil.

   c) Describe with the aid of diagrams how you would arrange for the rotating coil to supply to an external circuit
   i) direct current, and
   ii) alternating current.

9. State the laws of electromagnetic induction and describe briefly experiments (one in each case) by which they may be demonstrated.

An electromagnet is in series with a 12 V battery and a switch across which is connected a 230 V neon lamp. Explain why, when the switch is closed, the neon remains unlit but, when the switch is opened, it flashes momentarily. Explain the importance of this observation in connection with large power switching.

A flat circular coil of 100 turns of mean radius 5.0 cm is lying on a horizontal surface and is turned over in 0.20 s. Calculate the mean e.m.f. induced if the vertical component of the earth's magnetic flux density is 4.0×10^{-5} T.

10. The e.m.f. generated by a simple single-coil a.c. generator may be represented by the equation $E = E' \sin \omega t$.
   a) State the meanings of, and give the units for, the symbols employed.
   b) Draw diagrams showing the relative position of the coil and the magnetic field
      i) when $t = 0$, and
      ii) when $E = E'$.
   c) Discuss the factors which, in practice, determine the maximum current which may be generated by such a generator.
   d) Deduce a formula for the torque on the coil at the moment when the maximum current is flowing. Assume that the coil is rectangular, and state the units of any new symbols you employ.

11. **a)** The long solenoid shown in Fig. 16.56 has 1200 turns per metre and carries a current of 2.00 A.
      i) Copy the diagram and draw an arrow to show the direction of the magnetic field at X. Label your arrow **B**.
      ii) Describe the magnetic field inside the solenoid.
      iii) Arrange the letters W, X, Y and Z in the order of size of magnetic field strengths (weakest to strongest), at the points which correspond to these letters in Fig. 16.56.

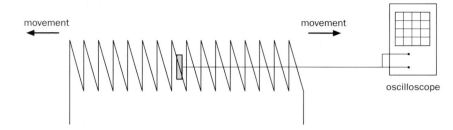

Fig. 16.57

   iv) The magnetic flux density at a point which is well inside and on the axes of a long solenoid is given by

$$B = \frac{\mu_0 N I}{l}$$

where N is the number of turns on the solenoid, l is the length of the solenoid, and μ_0 and I have their usual meaning. Show that the flux density of the magnetic field at X is 3.02×10^{-3} T.
   b) A narrow coil with a square cross-section of 2.00 cm by 2.00 cm and 1100 turns is placed inside the solenoid, close to X. The angle between the plane of this coil and the axis of the solenoid can be altered. State the angle θ between the plane of the coil and the axis of the solenoid, and calculate the magnetic flux through each turn of the coil, in the following cases.
      i) The flux is a *minimum*.
      ii) The flux is a *maximum*.
   c) The two ends of the narrow coil are connected to the terminals of an oscilloscope, as shown in Fig. 16.57 above. The solenoid, which was originally 40 cm long, is rapidly extended to become 140 cm long in 50 ms. Assume that the solenoid is pulled out at a uniform speed at both ends and that the coils remain uniformly distributed along its length. Calculate the average e.m.f. measured by the oscilloscope as the solenoid is extended.
      (NEAB, AS/A PHO3, June 1997)

12. Magnetic flux density B varies with distance beyond one end of a large bar magnet as shown on the graph in Fig. 16.58.

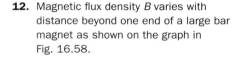

Fig. 16.58

A circular loop of wire of cross-sectional area 16 cm^2 is placed a few centimetres beyond the end of the bar magnet. The axis of the loop is aligned with the axis of the magnet.

   Calculate the *total* magnetic flux through the loop when it is 30 mm from the end of the magnet.

   Calculate the total magnetic flux through the loop when it is 10 mm from the end of the magnet.

   The loop of wire is moved towards the magnet from the 30 mm position to the 10 mm position so that a steady e.m.f. of 15 μV is induced in it. Calculate the average speed of movement of the loop.

   In what way would the speed of the loop have to be changed while moving towards the magnet between these two positions in order to maintain a steady e.m.f.?
      (L, A PH4, June 1999)

13. **a)** Figure 16.59 shows a long current-carrying solenoid with a 500 turn coil, C, wound round its central region. The flux density of the magnetic field along the axis of the solenoid is 4.5×10^{-3} T. The area of cross-section of the solenoid is 1.2×10^{-3} m^2.

W

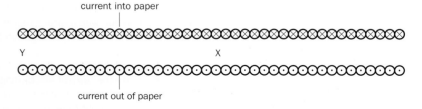

current into paper

current out of paper

Z   Y   X

Fig. 16.56

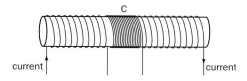

Fig. 16.59

i) Calculate the magnetic flux through a single turn of C. State your answer with the appropriate SI unit.

ii) Calculate the voltage induced in C if the current in the solenoid is reduced at a uniform rate to zero in 0.50 s.

b) Figure 16.60 shows the graph of current, I, against time, t, when a signal generator producing a voltage of triangular wave-form is connected to the solenoid in Fig. 16.59. Copy Fig. 16.60 and on the same axes, with the scale of the time axis unchanged, sketch the corresponding graph of the voltage induced in C. Assume that, at time $t = 0$, this voltage has a positive value.

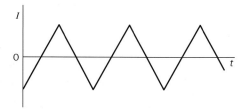

Fig. 16.60

(*OCR*, 6843, March 1999)

14. Figure 16.61 shows the principle of one type of residual current device (R.C.D.) that utilizes mutual induction. The function of the device is to switch off the mains supply when, owing to a fault, current flows from the live wire to earth.

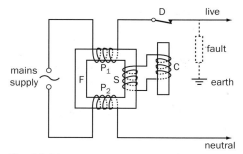

Fig. 16.61

Three coils, P_1, P_2 and S, are wound on a laminated iron former, F. Coils P_1 and P_2 are connected into the live and

neutral wires respectively of the supply. Coil S is in series with a relay coil C.

The currents in P_1 and P_2 flow in the same sense when viewed from the left of the diagram. Relay switch D is opened if C carries a current.

a) What is meant by **mutual induction**?

b) Explain:

i) why, in normal use, D experiences no electromagnetic force;

ii) why D opens when a fault occurs.

c) State and explain what would occur if a solid iron former were used instead of a laminated iron former.

(*OCR*, 6843, June 1999)

Motors and transformers

15. Draw a simple diagram to illustrate the essential electrical connections of a shunt-wound d.c. motor.

The armature resistance of such a motor is 0.75 Ω and it runs from a 240 V d.c. supply. When the motor is running freely, i.e. under no applied load, the current in the armature is 4.0 A and the motor makes 400 revolutions per minute. What is the value of the back e.m.f. produced in the motor and what is the rate of working?

When a load is placed on the motor the armature current increases to 60 A. What is now the back e.m.f., the rate of working, and the speed of rotation? It may be assumed that the field current remains constant. (*Hint.* For a shunt-wound motor, torque $\propto$ armature current since field current and so flux-linkage is constant.)

16. a) When an electric motor speeds up, what happens to

i) the current through it, and

ii) the torque (couple) acting on the rotating armature?

b) When a d.c. motor connected to a 12 V battery reaches a steady speed the current through it is 2.0 A and the back e.m.f. is 9.0 V. What is

i) the power output of the motor, and

ii) its efficiency?

17. a) Describe the construction of a simple form of alternating current transformer.

b) If the secondary coil is on open circuit explain, without calculation, the effect on the current flowing in the primary of

i) a fall in the supply frequency, and

ii) a reduction in the number of primary turns.

c) Calculate the current which flows in a resistance of 3 Ω connected to a secondary coil of 60 turns if the primary has 1200 turns and is connected to a 240 V a.c. supply, assuming that all the magnetic flux in the primary passes through the secondary and that there are no other losses.

18. A battery is connected, via a switch, to a coil P wound on a soft-iron core. Also wound on the core is a second coil Q with a large number of turns. Coil Q is connected to a voltmeter. The arrangement is shown in Fig. 16.62.

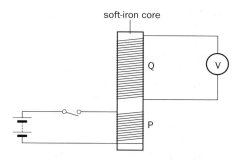

Fig. 16.62

a) The switch is operated so that the current in coil P changes with time in the way shown in Fig. 16.63*a*. The magnetic flux in the core is proportional to the current in coil P. Copy and complete Figs 16.63*b* and *c* to show the variations with time of the magnetic flux in the core and the corresponding voltmeter reading.

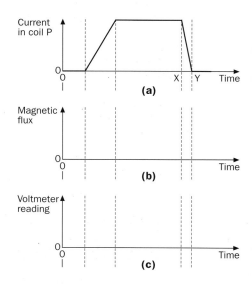

Fig. 16.63

b) How is the voltage across coil Q likely to compare in size with the battery voltage during the very short time interval XY?

c) Suggest, with a reason, a precaution which would need to be taken when demonstrating this experiment.

(*UCLES*, Basic 1, March 1998)

Self-inductance; ballistic galvanometers

19.

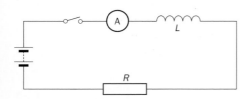

Fig. 16.64

In the series circuit shown in Fig. 16.64, the battery is of e.m.f. 10 V and internal resistance 1.0 Ω, *L* is a pure inductor of inductance 5.0 H, and the resistance of *R* is 11.0 Ω.

i) Immediately after the switch is closed, the ammeter reading slowly increases and eventually reaches a steady value. Explain why this happens.

ii) At a particular moment the current flowing is 50 mA. Calculate the energy stored in the inductor at this time.

iii) Calculate the current which flows when it has reached its eventual steady value.

(*NEAB*, AS/A PHO3, June 1998, part)

20. a) The switch in the circuit in Fig. 16.65*a* is closed at time *t* = 0. Figure 16.65*b* shows how the current *I* in the circuit subsequently varies with time *t*.

Explain, in terms of Faraday's and Lenz's laws, why the current does not rise immediately to its maximum value.

b) Use information from Figs 16.65*a* and *b* to determine the self-inductance of the coil.

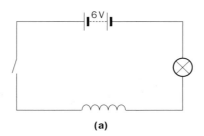

(a)

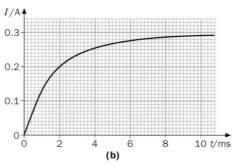

(b)

Fig. 16.65

c) Assuming that the battery and the coil have negligible resistance, determine the final resistance of the lamp.

d) The air-cored *plane circular coil* used as the inductor in the circuit in Fig. 16.65*a* has 600 turns and radius 0.050 m. The permeability of free space (air) is 1.26×10^{-6} H m^{-1}. Calculate the final magnetic flux density through the coil.

(*AEB*, 0635/2, Summer 1998)

21. The terminals of a moving-coil ballistic galvanometer are short-circuited after a charge has been passed through it. Explain why the oscillations of the coil are damped out and the coil returns slowly to its zero position.

A ballistic galvanometer is connected to a flat coil having 40 turns of mean area 3.0 cm^2 to form a circuit of total resistance 80 Ω. The coil, held between the poles of an electromagnet with its plane perpendicular to the field, is suddenly withdrawn from the field,

producing a throw of 30 scale divisions. Find the magnetic induction (flux density) of the field at the place where the coil was held, assuming that the sensitivity of the galvanometer under the conditions of the experiment is 0.40 division per microcoulomb.

Ferromagnetic materials

22. Explain what is meant by **cycle of magnetization**, and **hysteresis**.

Sketch on a single diagram the hysteresis loops for soft iron and hardened steel, indicating on the axes the physical quantities that have been plotted.

What information of practical importance can be obtained from a hysteresis loop? State, with reasons, which of the above metals would be suitable for

a) a permanent magnet, and

b) the core of a transformer.

Describe, with a circuit diagram, an effective electrical method of demagnetizing a steel bar magnet and explain what is happening during the process of demagnetization.

23. Give a general account of the magnetization of iron. Show how the processes of magnetization and demagnetization, the phenomenon of hysteresis, and the existence of a Curie temperature can be explained in terms of elementary magnets and a domain structure.

What magnetic properties are desirable for the material of

a) the core of a transformer,

b) the core of a relay electromagnet, and

c) the tape of a cassette recorder?

17

Alternating current

- ■ Introduction
- ■ Root-mean-square (r.m.s.) values
- ■ Meters for a.c.
- ■ Capacitance in a.c. circuits
- ■ Inductance in a.c. circuits

- ■ Phasor diagrams
- ■ Series circuits
- ■ Electrical resonance
- ■ Worked examples
- ■ Power in a.c. circuits

- ■ Electrical oscillations
- ■ Rectification of a.c.
- ■ Smoothing circuits

INTRODUCTION

(a) a.c. and d.c.

In a direct current (d.c.) the drift velocity superimposed on the random motion of the charge carriers (e.g. electrons) is in one direction only. In an alternating current (a.c.) the direction of the drift velocity reverses, usually many times a second.

The effects of a.c. are essentially the same as those of d.c. Both are satisfactory for heating and lighting purposes. The magnetic field due to a.c. fluctuates with time and although, for example, a torque is exerted on a current-carrying coil, the inertia of the coil may prevent it responding, unless the frequency of the a.c. is very low. Therefore a moving-coil meter gives a reading with d.c. but generally not with a.c. Chemical effects are observed with a.c. only in some cases. The electrolysis of acidified water by mains frequency a.c. produces a hydrogen–oxygen mixture at both platinum electrodes. There is no resultant effect when a.c. passes through copper sulphate solution using copper electrodes.

We have seen (p. 261) that a.c. is more easily generated and distributed than d.c. and for this reason the mains supply is a.c. However, processes such as electroplating and battery charging require d.c., and so does electronic equipment such as computers, televisions and radios. When necessary a.c. can be **rectified** to give d.c.

(b) Terms

An alternating current or e.m.f. varies periodically with time in magnitude and direction. One complete alternation is called a **cycle** and the number of cycles occurring in one second is termed the **frequency** (*f*) of the alternating quantity. The unit of frequency is the **hertz** (Hz), which is equivalent to cycle per second. The frequency of the electricity supply in the UK is 50 Hz which means that the duration of one cycle, known as the **period** (*T*), is $1/50 = 0.02$ s. In general $f = 1/T$.

The simplest and most important alternating e.m.f. can be represented by a sine curve and is said to have a **sinusoidal waveform**, Fig. 17.1.

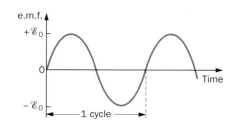

Fig. 17.1   Sinusoidally varying e.m.f.

It can be expressed by the equation

$$\mathcal{E} = \mathcal{E}_0 \sin \omega t$$

where $\mathcal{E}$ is the e.m.f. at time *t*, $\mathcal{E}_0$ is the peak or maximum e.m.f. and ω is a constant which equals $2\pi f$ where *f* is the frequency of the e.m.f.

Similarly, for a sinusoidal alternating current we may write

$$I = I_0 \sin \omega t$$

In the previous chapter we saw that a sinusoidal e.m.f. is induced in a coil rotating with *constant* speed in a *uniform* magnetic field. In that case ω was the angular velocity of the coil (in rad s^{-1}) and *f* equalled the number of complete revolutions of the coil per second. The mains supply is very nearly sinusoidal.

Alternating e.m.fs and currents of many different waveforms can be produced and have their applications, Fig. 17.2. All, however irregular, can be shown to be combinations of sinusoidal e.m.fs or currents.

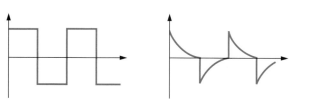

Fig. 17.2   Other waveforms

ROOT-MEAN-SQUARE (R.M.S.) VALUES

The value of an alternating current (and e.m.f.) varies from one instant to the next and the problem arises of what value we should take to measure it. The average value over a complete cycle is zero; the peak value is a possibility. However, the **root-mean-square (r.m.s.)** value is chosen because by using it many calculations can be done as they would be for direct currents.

> The **r.m.s.** value of an alternating current (also called the **effective** value) is the steady direct current which dissipates energy in a given resistance at the same rate as the a.c.

If the lamp in the circuit of Fig. 17.3 is lit first from a.c. and the brightness noted, then if 0.3 A d.c. produces the same brightness, the r.m.s. value of the a.c. is 0.3 A. A lamp designed to be fully lit by a current of 0.3 A d.c. will be fully lit by an a.c. of r.m.s. value 0.3 A. Although the value (I) of the a.c. is varying, the *average* rate at which it supplies energy to the lamp equals the *steady* rate of supply by the d.c. ($I_{d.c.}$) and in practice it is this aspect which is often important.

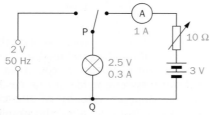

Fig. 17.3

In general, considering energy supplied to a resistance R, we can say

$$I_{d.c.}^2 R = \text{(mean value of } I^2) \times R$$

$$\therefore \quad I_{d.c.} = \sqrt{\text{(mean value of } I^2)}$$

$$= \text{square } root \text{ of the } mean \text{ value of the } square \text{ of the current}$$

$$= I_{r.m.s.}$$

If the a.c. is sinusoidal then

$$I = I_0 \sin \omega t$$

$$\therefore \quad I_{r.m.s.} = \sqrt{\text{(mean value of } I_0^2 \sin^2 \omega t)}$$

$$= I_0 \sqrt{\text{(mean value of } \sin^2 \omega t)}$$

From the graphs of $\sin \omega t$ and $\sin^2 \omega t$ in Fig. 17.4 it can be seen that $\sin^2 \omega t$ is always positive and varies between 0 and 1. The shaded areas above and below the dashed line are equal and there is symmetry; the mean value for $\sin^2 \omega t$ is therefore $\frac{1}{2}$. Hence

$$I_{r.m.s.} = I_0 \sqrt{\tfrac{1}{2}}$$

i.e.

$$I_{r.m.s.} = \frac{I_0}{\sqrt{2}} = 0.707 \, I_0$$

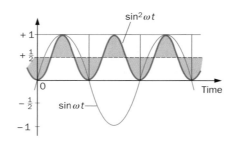

Fig. 17.4

The r.m.s. current is 0.707 times the peak current. Similar relationships also hold for e.m.fs and p.ds, and the r.m.s. value is the one usually quoted. In the UK 230 V is the r.m.s. value of the electricity supply, so the peak value $\mathscr{E}_0 = \sqrt{2}\,\mathscr{E}_{r.m.s.} = \sqrt{2} \times 230\,\text{V} = 325\,\text{V}$.

The circuit of Fig. 17.3 may be used to check roughly that the peak value of an alternating p.d. is 1.4 times its r.m.s. value. The lamp is adjusted to the same brightness on d.c. as on a.c. The CRO is then connected across PQ and is used as a voltmeter. It measures the r.m.s. value of the p.d. across the lamp when it is lit by d.c. and *twice* the peak value using the a.c. supply.

Most voltmeters and ammeters for a.c. use are calibrated to read r.m.s. values and give correct readings only if the waveform is sinusoidal.

METERS FOR A.C.

These may be analogue or digital. The deflection of an analogue a.c. meter must not depend on the direction of the current; this can be achieved by using a semiconductor rectifier. A rectifier is a device with a low resistance to current flow in one direction and a high resistance in the reverse direction. When connected to an a.c. supply it allows pulses of varying but direct current to pass. In an analogue meter the average value of these pulses is measured by a moving-coil meter, Figs 17.5a and b.

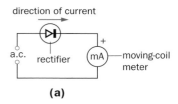

(a)

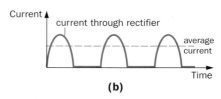

(b)

Fig. 17.5

The meter is calibrated to read r.m.s. values of currents and p.ds with sinusoidal waveforms.

Multimeters such as those shown in Fig. 4.21 (analogue) and Fig. 23.95 (digital) can measure alternating or direct currents and p.ds.

CAPACITANCE IN A.C. CIRCUITS

(a) Flow of a.c. 'through' a capacitor

If a 1000 μF capacitor is connected in series with a 2.5 V, 0.3 A lamp and a 2 V d.c. supply, Fig. 17.6a, the lamp, as expected, does not light. (Is there *any* current flow?) However, with a 2 V r.m.s., 50 Hz supply, Fig. 17.6b, it is nearly fully lit; the capacitor is being charged, discharged, then charged in the opposite direction and discharged again, 50 times per second, and the charging and discharging currents flowing through the lamp light it. No current actually passes through the capacitor (since its plates are separated by an insulator) but it appears to do so. A current would certainly be recorded by an a.c. milliammeter.

When the 1000 μF capacitor is replaced by one of 100 μF, the charging and discharging currents are too small to light the lamp. Larger capacitances give greater currents. Increasing the frequency of the a.c. (at constant p.d. and capacitance) also increases the current, since the same charge has to flow on and off the plates in a shorter time.

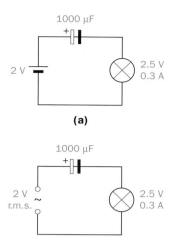

Fig. 17.6

(b) Phase relationships

When a.c. flows through a resistor (having no capacitance or inductance) the current and p.d. reach their peak values at the same instant, i.e. they are in phase. This is not so for a capacitor.

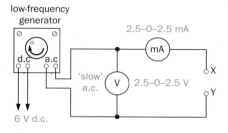

Fig. 17.7

The circuit shown in Fig. 17.7 enables phase relationships to be studied using 'slow a.c.' of frequency less than 1 Hz. With a 2000 Ω resistor between X and Y the current (shown on the milliammeter) and the p.d. (shown on the voltmeter) rise and fall together. With a 250 μF capacitor replacing the resistor, the current through the capacitor is seen to lead the p.d. across it by one-quarter of a cycle, i.e. the current reaches its maximum value one-quarter of a cycle before the p.d. reaches its peak value, as shown by the cosine and sine curves in Fig. 17.8.

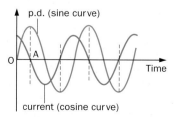

Fig. 17.8 Phase difference between current and p.d. for a capacitor

The circuit for an alternative demonstration at 50 Hz using a double-beam CRO (or datalogger and computer) is given in Fig. 17.9. The Y_2 trace is the p.d. across R and this gives the waveform of the current 'through' C since C and R are in series and the current in a resistor is in phase with the p.d. across it. The Y_1 trace is the p.d. across C and R in series and not just C, which accounts for the phase difference being less than a quarter of a cycle.

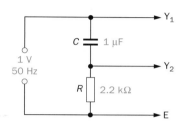

Fig. 17.9

Current and applied p.d. are out of step because current flow is a maximum immediately an uncharged capacitor is connected to a supply, whether it is d.c. (p. 214) or a.c. There is as yet no charge on the capacitor to oppose the arrival of charge. So at O (Fig. 17.8) the applied p.d., though momentarily zero, is increasing at its maximum rate (the slope of the tangent at O to the p.d. graph being a maximum) and so the rate of flow of charge — the current — is also a maximum. Between O and A the p.d. is increasing but at a decreasing rate, the charge on the capacitor is increasing but less quickly, which means that the charging current is less. At A the applied p.d. is a maximum and for a brief moment is constant. The charge on the capacitor will also be a maximum and constant; the rate of flow of charge is zero. The phase difference between V and I can be explained in this way.

(c) Capacitive reactance

Let a p.d. V be applied across a capacitance C and let its value at time t be given by

$$V = V_0 \sin \omega t$$

where V_0 is its peak value and $\omega = 2\pi f$ where f is the frequency of the supply.

The charge Q on the capacitance at time t is

$$Q = VC$$

For the current I flowing 'through' the capacitor we can write

$$
\begin{aligned}
I &= \text{rate of change of charge} \\
&= \frac{dQ}{dt} \\
&= \frac{d}{dt}(VC) = C\frac{dV}{dt} \\
&= C\frac{d}{dt}(V_0 \sin \omega t) \\
&= CV_0\frac{d}{dt}(\sin \omega t)
\end{aligned}
$$

$$\therefore \quad I = \omega C V_0 \cos \omega t$$

The current 'through' C (a cosine function) leads the applied p.d. (a sine function) by one-quarter of a cycle or, as is often stated, by $\pi/2$ radians or 90° (one cycle being regarded as 2π radians or 360°). The mathematics here has confirmed the results of the demonstrations and the 'physical' explanation outlined above. We can also write

$$I = I_0 \cos \omega t$$

where I_0 is the peak current given by $I_0 = \omega C V_0$.

$$\therefore \quad \frac{V_0}{I_0} = \frac{1}{\omega C}$$

But

$$\frac{V_{r.m.s.}}{I_{r.m.s.}} = \frac{V_0}{I_0}$$

$$\therefore \quad \frac{V_{r.m.s.}}{I_{r.m.s.}} = \frac{1}{\omega C} = \frac{1}{2\pi f C}$$

This expression resembles $V/I = R$ which defines resistance, with $1/(2\pi f C)$ replacing R. The quantity $1/(2\pi f C)$ is called the **capacitive reactance** X_C.

$$X_C = \frac{V_{r.m.s.}}{I_{r.m.s.}} = \frac{1}{2\pi f C}$$

The ohm is the unit of X_C if the unit of f is s^{-1} (hertz) and that of C is $C\,V^{-1}$ (farad). The term $1/(fC)$ then has units $V/(C\,s^{-1}) = V\,A^{-1} = \Omega$. It is clear that X_C decreases as f or C increases. The frequency dependence is shown in Fig. 17.10. A 10 μF capacitor has a reactance of 320 Ω at 50 Hz. Check this. What will it be at 1 kHz?

Reactance is not to be confused with resistance; electrical power is dissipated in a resistance but it is not in a reactance.

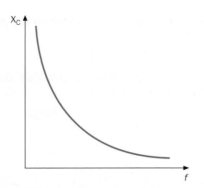

Fig. 17.10 Variation of capacitive reactance with a.c. frequency

INDUCTANCE IN A.C. CIRCUITS

(a) Phase relationships

An inductor in an a.c. circuit also causes a phase difference between the applied p.d. and the current. In this case, the current lags behind the p.d. by one-quarter of a cycle (i.e. 90°).

The effect may be observed using slow a.c. and the circuit of Fig. 17.7, p. 275, with a 12 000-turn coil on a complete iron core (from a demountable transformer) as the inductor (500 H) connected to XY. Alternatively, a double-beam CRO may be used with 50 Hz a.c. as in Fig. 17.11. In both cases the phase difference is less than 90° because of the resistance of the inductor, and in the CRO demonstration also because the Y_1 trace gives the p.d. across L and R in series; the graphs in Fig. 17.12 are for a pure inductor.

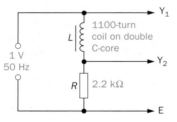

Fig. 17.11

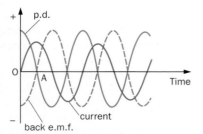

Fig. 17.12 Phase difference between p.d. and current for an inductor

The effects can be explained as follows. At O the current is zero but its rate of increase is a maximum (as given by the slope of the tangent to the current graph at O) which means, for an inductor of constant inductance, that the rate of change of flux is also a maximum. Therefore, by Faraday's law, the back e.m.f. is a maximum but, by Lenz's law, of negative sign since it acts to oppose the current change. At A the current and flux are momentarily a maximum and constant. Their rate of change is zero (slope of tangent to current graph is zero at A) and so the back e.m.f. is zero. If the inductor has negligible resistance, then at every instant the applied p.d. must be nearly equal and opposite to the back e.m.f. The applied p.d. curve is therefore as illustrated. The p.d. acts on the coil while the back e.m.f. acts back on the source, just like two forces acting on different bodies.

(b) Inductive reactance

Consider an inductance, L, through which current I flows at time t where

$$I = I_0 \sin \omega t$$

I_0 is the peak current and $\omega = 2\pi f$ where f is the frequency of the a.c. The back e.m.f. $\mathscr{E}$ in the inductor due to the changing current is

$$\mathscr{E} = -L \frac{dI}{dt} \quad \text{(p. 263)}$$

$$= -L \frac{d}{dt} (I_0 \sin \omega t)$$

$$= -\omega L I_0 \cos \omega t$$

Assuming the inductor has zero resistance, then for current to flow the applied p.d. V must be equal and opposite to the back e.m.f. So

$$V = -\mathscr{E} = \omega L I_0 \cos \omega t$$

The applied p.d. leads the current (a sine function) by 90°. We can also write

$$V = V_0 \cos \omega t$$

where V_0 is the peak value of the applied p.d. given by $V_0 = \omega L I_0$.

$$\therefore \quad \frac{V_{r.m.s.}}{I_{r.m.s.}} = \frac{V_0}{I_0} = \omega L = 2\pi f L$$

The quantity $2\pi f L$ is called the **inductive reactance** X_L of the inductor and like X_C it has the unit ohm.

$$X_L = \frac{V_{r.m.s.}}{I_{r.m.s.}} = 2\pi f L$$

X_L increases with f and with L. If $L = 10$ H for an inductor, X_L is 3.1 kΩ at 50 Hz and 63 MΩ at 1 MHz.

PHASOR DIAGRAMS

A sinusoidal alternating quantity can be represented by a rotating vector called a **phasor**. Suppose the graph in Fig. 17.13 represents an alternating quantity $y = Y_0 \sin \omega t$ where Y_0 is the peak value of the quantity and its frequency $f = \omega/2\pi$. If the line OP has length Y_0 and rotates in an anticlockwise direction about O with uniform angular velocity ω, the projection O'P' of OP on O'y at time t (measured from the time when OP passes through OO') is $Y_0 \sin \omega t$. If OP is directed as shown by the arrow on it, then we can say that the projection on O'y of the phasor OP gives the value at any instant of the sinusoidal quantity y. (Simple harmonic motion, being sinusoidal, can be derived similarly from uniform motion in a circle.)

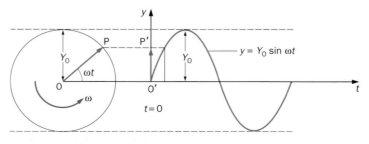

Fig. 17.13

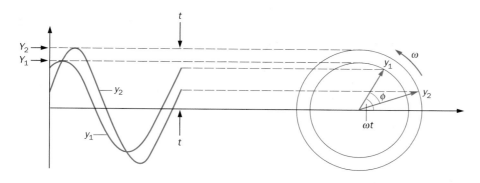

Fig. 17.14

The method is very useful for representing two sinusoidal quantities that have the same frequency but are not in phase. In Fig. 17.14 the waveforms of two such quantities y_1 and y_2 and the corresponding phasor diagram are shown for time t. The phase difference between them is ϕ, with y_2 lagging, and this phase angle is maintained between them as the phasors rotate. If the two phasors represent similar quantities, such as p.ds, they can be added by the parallelogram law for vectors.

Algebraically y_1 and y_2 are expressed by the equations

$$y_1 = Y_1 \sin \omega t$$

and

$$y_2 = Y_2 \sin (\omega t - \phi)$$

The phasor diagram for a pure capacitance (that is, infinite dielectric resistance) in an a.c. circuit is shown in Fig. 17.15*a*; the current I leads the applied p.d. V by 90°. That for a pure inductance (i.e. zero resistance) is given in Fig. 17.15*b*; in this case the current I lags behind the applied p.d. V by 90°.

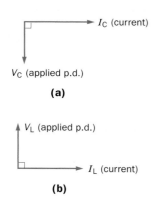

(a)

(b)

Fig. 17.15 Phasor diagrams for (*a*) pure capacitance and (*b*) pure inductance

Phasor diagrams are used to solve many problems involving capacitance, inductance and resistance in a.c. circuits.

SERIES CIRCUITS

When drawing phasor diagrams, a phasor representing a quantity that is the same for all the circuit components should be drawn first. For a series circuit it would be the current phasor. What would it be for a parallel circuit? This reference phasor is drawn horizontal, directed to the right, and the other phasors are then drawn so that their phases with respect to it are correct. The r.m.s. values of currents and p.ds are used.

(a) Resistance and capacitance

Suppose an alternating p.d. V is applied across a resistance R and a capacitance C in series, Fig. 17.16*a*. The same current I flows through each component and therefore the reference phasor will be that representing I. The p.d. V_R across R is in phase with I, and V_C (that across C) lags behind I by 90°. The phasor diagram is as shown in Fig. 17.16*b*, V_C and V_R being drawn to scale.

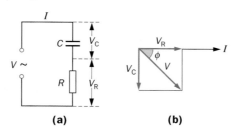

(a) **(b)**

Fig. 17.16 Resistance and capacitance in series

The vector sum of V_R and V_C equals the applied p.d. V, so

$$V^2 = V_R^2 + V_C^2$$

But $V_R = IR$ and $V_C = IX_C$ where X_C is the reactance of C and equals $1/\omega C$, so

$$V^2 = I^2(R^2 + X_C^2)$$
$$\therefore \quad V = I\sqrt{(R^2 + X_C^2)}$$

The quantity $\sqrt{(R^2 + X_C^2)}$ is called the **impedance** Z of the circuit. It has resistive and reactive components and like both it has the unit ohm.

$$Z = \frac{V}{I} = \sqrt{(R^2 + X_C^2)}$$

Also, from the phasor diagram we see that the current I leads V by a phase angle ϕ which is less than 90° and is given by

$$\tan \phi = \frac{V_C}{V_R} = \frac{X_C}{R}$$

(b) Resistance and inductance

The analysis is similar but in this case the p.d. V_L across L leads the current I and the p.d. V_R across R is again in phase with I, Fig. 17.17*a*. As before, the applied p.d. V equals the vector sum of V_L and V_R, Fig. 17.17*b*, and so

$$V^2 = V_R^2 + V_L^2$$

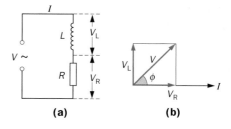

(a) **(b)**

Fig. 17.17 Resistance and inductance in series

But $V_R = IR$ and $V_L = IX_L$ where X_L is the reactance of L and equals ωL, so

$$V^2 = I^2(R^2 + X_L^2)$$
$$\therefore \quad V = I\sqrt{(R^2 + X_L^2)}$$

Here the impedance Z is given by

$$Z = \frac{V}{I} = \sqrt{(R^2 + X_L^2)}$$

The phase angle ϕ by which I lags behind V is given by

$$\tan \phi = \frac{V_L}{V_R} = \frac{X_L}{R}$$

(c) Resistance, capacitance and inductance

An RCL series circuit is shown in Fig. 17.18a; in practice R may be the resistance of the inductor. V_L leads the current (reference) phasor I by 90°, V_C lags behind it by 90°, and V_R is in phase with it. V_L and V_C are therefore 180° (half a cycle) out of phase, i.e. in antiphase. If V_L is greater than V_C, their resultant $(V_L - V_C)$ is in the direction of V_L, Fig. 17.18b. The vector sum of $(V_L - V_C)$ and V_R equals the applied p.d. V, therefore

$$V^2 = V_R{}^2 + (V_L - V_C)^2$$

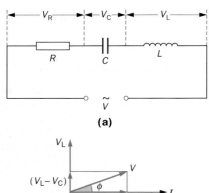

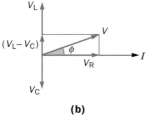

(b)

Fig. 17.18 *RCL* series circuit and phasor diagram

But $V_R = IR$, $V_L = IX_L$ and $V_C = IX_C$, so

$$V^2 = I^2[R^2 + (X_L - X_C)^2]$$
$$\therefore \quad V = I\sqrt{[R^2 + (X_L - X_C)^2]}$$

The impedance Z is given by

$$Z = \frac{V}{I} = \sqrt{[R^2 + (X_L - X_C)^2]}$$

The phase angle ϕ by which I lags behind V is given by

$$\tan \phi = \frac{V_L - V_C}{V_R} = \frac{X_L - X_C}{R}$$

ELECTRICAL RESONANCE

(a) Series resonance

The expression just derived for the impedance Z of an RCL series circuit shows that Z varies with the frequency f of the applied p.d. since X_L and X_C both depend on f. $X_L = 2\pi fL$ and increases with f; $X_C = 1/(2\pi fC)$ and decreases with f; R is assumed to be independent of f (but it can vary). Figure 17.19a shows how X_L, X_C, R and Z vary with f.

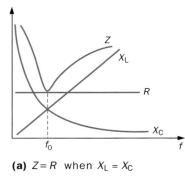

(a) $Z = R$ when $X_L = X_C$

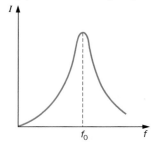

(b) Current resonance

Fig. 17.19

At a certain frequency f_0, called the **resonant frequency**, $X_L = X_C$ and Z has its minimum value, being equal to R. The circuit behaves as a pure resistance and the current I has a maximum value (given by $I = V/R$), Fig. 17.19b. The phase angle ϕ (given by $\tan \phi = (X_L - X_C)/R$) is zero, the applied p.d. V and the current I are in phase and there is said to be **resonance**. A series resonant circuit is called an **acceptor** circuit.

Series or current resonance may be shown using the arrangement of Fig. 17.20. As the frequency increases the milliammeter reading rises to a maximum and then falls.

An expression for f_0 is obtained from $X_L = X_C$, that is

$$2\pi f_0 L = \frac{1}{2\pi f_0 C}$$

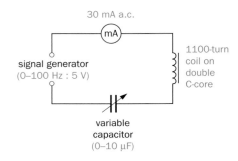

Fig. 17.20

or $\qquad 4\pi^2 f_0{}^2 LC = 1$

Therefore

$$f_0 = \frac{1}{2\pi\sqrt{(LC)}}$$

If L is in henrys and C in farads, f_0 will be in hertz.

At resonance V_L and V_C can both be very much greater than the total p.d. V applied across the whole circuit. Thus, if I is the resonance current, we have

$$I = \frac{V}{R}$$

$$\therefore \quad V_L = IX_L = \frac{V}{R}X_L$$

$$\therefore \quad \frac{V_L}{V} = \frac{X_L}{R}$$

R (which is mostly due in practice to the resistance of the inductor L) is usually very small compared with X_L, so V_L (and V_C) can be large compared with V. The magnification or **Q-factor** (Q for quality) of the circuit at resonance is defined by

$$Q = \frac{V_L}{V} = \frac{X_L}{R}$$

In actual circuits Q-factors of over 200 are realised.

It may seem strange for a small applied p.d., V, to give rise to two large p.ds, V_L and V_C. The explanation is that V_L and V_C are in antiphase (i.e. 180° out of phase) and so their vector sum can be quite small — zero, in fact — if they are equal in magnitude, as they are at resonance, Fig. 17.21. The circuit of Fig. 17.22a may be used to demonstrate on a double-beam CRO that at resonance V_L is much greater than V and leads it by nearly 90°, Fig. 17.22b. If C and L

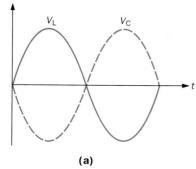

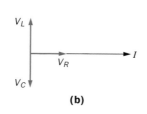

Fig. 17.21

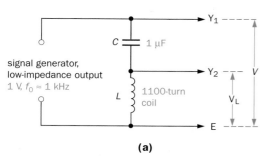

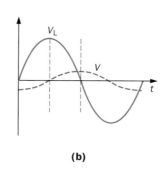

(a)

Fig. 17.22

e.m.fs of various frequencies in the aerial which cause currents to flow in the aerial coil. These induce currents of the same frequencies in coil L by mutual induction. If the capacitance C is adjusted ('tuned') so that the resonant frequency of circuit LC equals the frequency of the wanted station, a large p.d. at that frequency (and no other) is developed across C. This p.d. is then applied to the next stage of the receiver.

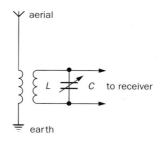

Fig. 17.25   Aerial circuit

(d) Mechanical analogy

The behaviour of a resonant circuit is similar to that of a vibrating system in mechanics. At one particular frequency the circuit responds and stores a large amount of energy which passes to and fro between the electric field of the capacitor and the magnetic field of the inductor. Kinetic and potential energy behave similarly in an oscillating mass–spring system. Energy has to be supplied only to compensate for that dissipated as heat in doing work against resistance — electrical or air.

are interchanged, V_C is seen to be equal to V_L and to lag behind V by almost 90°. V_L and V_C are approximately in antiphase.

(b) Parallel resonance

In a parallel *RCL* circuit resonance can also occur but in this case the impedance of the circuit becomes a maximum. The resonant frequency f_0 is again given to a good approximation by $f_0 = 1/2\pi\sqrt{(LC)}$. Large *currents* I_C and I_L, which are equal in magnitude but 180° out of phase, circulate to and fro *within* the circuit; the *supply* current I is thus small. The p.d. across the circuit at resonance is large.

In the circuit shown in Fig. 17.23 the brightness of the lamps indicates the flow of currents in different parts of the circuit. Their behaviour as the frequency of the p.d. applied from the signal generator increases (from about 0.5 kHz to 2 kHz) is interesting to study.

In Fig. 17.24 the p.ds across the 3.9 Ω resistors give the waveforms of the currents in C and L and show that they have equal amplitude but are in antiphase at resonance.

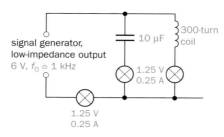

Fig. 17.23

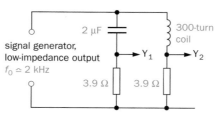

Fig. 17.24

(c) Tuned circuits

The ability of a resonant circuit to select and amplify a p.d. of one particular frequency (strictly, a very narrow band of frequencies) is used in radio and television reception. For example, in the aerial circuit of a radio receiver, Fig. 17.25, radio signals from different transmitting stations induce

WORKED EXAMPLES

Example 1. A 1000 µF capacitor is joined in series with a 2.5 V, 0.30 A lamp and a 50 Hz supply. Calculate (*a*) the p.d. (r.m.s.) of the supply to light the lamp to its normal brightness, and (*b*) the p.ds across the capacitor and the resistor respectively.

(*a*) Reactance X_C of capacitor

$= 1/(2\pi f C)$

$= 1/(2\pi \times 50 \text{ s}^{-1} \times 10^3 \times 10^{-6} \text{ F})$

$= (10/\pi) \ \Omega$

Resistance R of lamp

$$= 2.5/0.30 = 8.3 \; \Omega$$

Impedance Z of circuit

$$= \sqrt{(R^2 + X_C^2)}$$
$$= \sqrt{\{(8.3)^2 + (10/\pi)^2\}} \; \Omega$$
$$= \sqrt{(69 + 10)} \; \Omega$$
$$= 8.9 \; \Omega$$

The applied r.m.s. p.d. V to cause an r.m.s. current of 0.30 A to flow round the circuit is given by

$$V = IZ$$
$$= 0.30 \; A \times 8.9 \; \Omega$$
$$= 2.7 \; V$$

(*b*) p.d. V_C across the capacitor

$$= IX_C$$
$$= 0.30 \; A \times (10/\pi) \; \Omega = 0.96 \; V$$

p.d. V_R across the resistor

$$= IR = 2.5 \; V$$

Note. $V_C + V_R = 0.96 \; V + 2.5 \; V \approx 3.5 \; V$, which is greater than the applied p.d. V of 2.7 V. This is because V_C and V_R are not in phase. In fact $V^2 = V_C^2 + V_R^2$ — as you can check from the above figures.

Example 2. A 2.0 H inductor of resistance 80 Ω is connected in series with a 420 Ω resistor and a 240 V, 50 Hz supply. Find (*a*) the current in the circuit, and (*b*) the phase angle between the applied p.d. and the current.

(*a*) Reactance X_L of inductor

$$= 2\pi f L$$
$$= 2\pi \times 50 \; s^{-1} \times 2 \; H$$
$$= 200\pi \; \Omega$$

Total resistance R of circuit

$$= 80 \; \Omega + 420 \; \Omega$$
$$= 500 \; \Omega$$

Impedance Z of circuit

$$= \sqrt{(R^2 + X_L^2)}$$
$$= \sqrt{\{(500)^2 + (200\pi)^2\}} \; \Omega$$
$$= 800 \; \Omega$$

Current I in circuit

$$= V/Z$$
$$= 240 \; V/800 \; \Omega$$
$$= 0.30 \; A$$

(*b*) Phase angle ϕ between V and I is given by

$$\tan \phi = \frac{X_L}{R}$$
$$= \frac{200\pi}{500}$$
$$\therefore \quad \phi = 52°$$

V leads I by 52°.

Example 3. A circuit consists of a 200 µH inductor of resistance 10 Ω in series with a variable capacitor and a 0.10 V (r.m.s.), 1.0 MHz supply. Calculate (*a*) the capacitance to give resonance, (*b*) the p.ds across the inductor and the capacitor at resonance, and (*c*) the Q-factor of the circuit at resonance, Fig. 17.26.

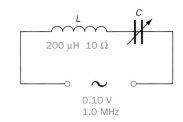

Fig. 17.26

(*a*) We have $L = 200 \times 10^{-6} \; H$, $R = 10 \; \Omega$, $f_0 = 10^6 \; Hz$. Also,

$$f_0 = \frac{1}{2\pi\sqrt{(LC)}} \quad \text{or} \quad 4\pi^2 f_0^2 LC = 1$$

$$\therefore \quad C = \frac{1}{4\pi^2 f_0^2 L}$$

$$= \frac{1}{4\pi^2 \times (10^6 \; s^{-1})^2 \times 200 \times 10^{-6} \; H}$$

$$= 0.000 \; 13 \; \mu F$$

(*b*) At resonance the impedance $Z = R$ and if V is the applied p.d. (0.10 V), the current I is given by

$$I = \frac{V}{R} = \frac{0.10 \; V}{10 \; \Omega} = 1.0 \times 10^{-2} \; A$$

If the inductor has reactance X_L, the p.d. V_L across it is

$$V_L = IX_L = I \times 2\pi f_0 L$$
$$= 1.0 \times 10^{-2} \; A \times 2\pi$$
$$\quad \times 10^6 \; s^{-1} \times 200 \times 10^{-6} \; H$$
$$= 4\pi \; V$$
$$\approx 13 \; V$$

Since $V_C = V_L$, we have

$$V_C \approx 13 \; V$$

(*c*) The Q-factor at resonance is given by

$$Q = \frac{V_L}{V} = \frac{13}{0.10} = 130$$

POWER IN A.C. CIRCUITS

(a) Resistance

The general expression for the power absorbed by a device at any instant is IV where I and V are the instantaneous values of the current through it and the p.d. across it respectively.

In a resistor I and V are in phase and we can write $I = I_0 \sin \omega t$ and $V = V_0 \sin \omega t$, therefore the *instantaneous* power absorbed at time t is $I_0 V_0 \sin^2 \omega t$. The *mean* power P will be given by

$$P = \text{mean value of } I_0 V_0 \sin^2 \omega t$$

We saw previously (p. 274, Fig. 17.4) that the mean value of $\sin^2 \omega t$ is $\frac{1}{2}$, so

$$P = \frac{I_0 V_0}{2}$$
$$= \frac{I_0}{\sqrt{2}} \cdot \frac{V_0}{\sqrt{2}}$$

For sinusoidal quantities

$$I_{\text{r.m.s.}} = \frac{I_0}{\sqrt{2}} \quad \text{and} \quad V_{\text{r.m.s.}} = \frac{V_0}{\sqrt{2}}$$

$$\therefore \quad P = I_{\text{r.m.s.}} \times V_{\text{r.m.s.}}$$

For a resistance

$$V_{\text{r.m.s.}} = I_{\text{r.m.s.}} \times R$$

$$\therefore \quad P = I_{\text{r.m.s.}}^2 \times R = \frac{V_{\text{r.m.s.}}^2}{R}$$

The power varies at twice the frequency of either V or I, as shown in Fig. 17.27.

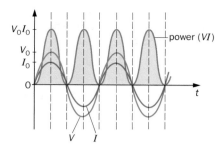

Fig. 17.27 Power absorbed by a resistor

(b) Inductance

In a pure inductor V leads I by 90° (or $\pi/2$ rad) and if $I = I_0 \sin \omega t$ then

$$V = V_0 \sin(\omega t + \pi/2) = V_0 \cos \omega t$$

Hence the *instantaneous* power absorbed at time t is $I_0 V_0 \sin \omega t \cos \omega t$. But $\sin 2\omega t = 2 \sin \omega t \cos \omega t$, therefore

$$\begin{aligned}\text{instantaneous}\\ \text{power absorbed}\end{aligned} = \tfrac{1}{2}I_0 V_0 \sin 2\omega t$$
$$= I_{\text{r.m.s.}} V_{\text{r.m.s.}} \sin 2\omega t$$

This represents a sinusoidal variation with mean value zero and frequency twice that of I and V, Fig. 17.28. It follows that the **power absorbed by an inductor in a cycle is zero**.

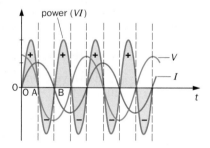

Fig. 17.28 Power absorbed by an inductor

To explain this we consider that during the first quarter-cycle of current, OA (Fig. 17.28), power is drawn from the source and energy is stored in the magnetic field of the inductor. In the second quarter-cycle AB, the current and magnetic field decrease and the e.m.f. induced in the inductor causes it to act as a generator, returning the energy stored in its magnetic field to the source. The power so restored is represented by the shaded area in Fig. 17.28. Although the mean power taken over a cycle is zero, large amounts of energy flow in and out of the inductor every quarter-cycle. In practice an inductor has resistance and so some energy drawn from the source is not returned.

(c) Capacitance

Similar reasoning shows that the power taken by a pure capacitor over a cycle is also zero — since V and I are 90° out of phase. In this case energy taken from the source is stored in the electric field due to the p.d. between the plates of the charged capacitor. During the next quarter-cycle the capacitor discharges and the energy is returned to the source.

(d) Formula for a.c. power

Power is only absorbed by the *resistive* part of a circuit, i.e. in the expression IV for power, V is that part of the applied p.d. across the total resistance in the circuit; it is in phase with the current. Thus if $V_{\text{r.m.s.}}$ is the p.d. applied to an a.c. circuit and it leads the current $I_{\text{r.m.s.}}$ by angle ϕ, Fig. 17.29, the component of p.d. in phase with $I_{\text{r.m.s.}}$ is $V_{\text{r.m.s.}} \cos \phi$. Hence the total power P expended in the circuit is given by

$$P = I_{\text{r.m.s.}} V_{\text{r.m.s.}} \cos \phi$$

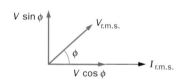

Fig. 17.29

The component $V_{\text{r.m.s.}} \sin \phi$ is that part of the applied p.d. across the total reactance of the circuit and is often called the 'wattless' component of the p.d.

The *apparent power* absorbed in an a.c. circuit is, by comparison with the d.c. case, $I_{\text{r.m.s.}} V_{\text{r.m.s.}}$. The *real power* is $I_{\text{r.m.s.}} V_{\text{r.m.s.}} \cos \phi$. The **power factor** of an a.c. circuit is defined by the equation

$$\text{power factor} = \frac{\text{real power}}{\text{apparent power}}$$
$$= \frac{I_{\text{r.m.s.}} V_{\text{r.m.s.}} \cos \phi}{I_{\text{r.m.s.}} V_{\text{r.m.s.}}}$$
$$= \cos \phi$$

In a purely resistive circuit $\cos \phi$ has its maximum value of 1.

ELECTRICAL OSCILLATIONS

Electrical oscillators are utilized in radio and television transmitters and receivers, in signal generators, oscilloscopes and computers; they produce a.c. with waveforms which may be sinusoidal, square, sawtooth, etc., and with frequencies from a few hertz up to millions of hertz.

(a) Oscillatory circuit

When a capacitor discharges through an inductor in a circuit of low resistance, a.c. flows. The circuit oscillates at its **natural frequency**, which we will show equals $1/[2\pi\sqrt{(LC)}]$, i.e. its resonant frequency f_0. Electrical resonance occurs when the applied frequency equals the natural frequency — as it does in a mechanical system.

In Fig. 17.30*a* a charged capacitor C is shown connected across a coil L. C immediately starts to discharge, current flows and a magnetic field is created which induces an e.m.f. in L. This e.m.f. opposes the current. C cannot therefore discharge instantaneously and the greater the inductance of L the longer the discharge takes. When C is discharged completely the energy originally stored in the electric field between its plates has been transferred to the magnetic field around L, Fig. 17.30*b*.

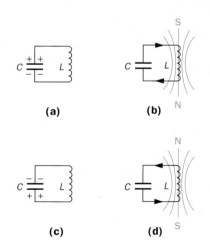

Fig. 17.30 An oscillatory *LC* circuit

At this instant the magnetic field begins to collapse and an e.m.f. is induced in L which tries to maintain the field. Current therefore flows in the same direction as before and charges C so that the lower plate is positive. By the time the magnetic field has collapsed, the energy is again stored in C, Fig. 17.30c. Once more C starts to discharge but current now flows in the opposite direction, creating a magnetic field of opposite polarity, Fig. 17.30d. When this field has decayed, C is again charged with its upper plate positive and the same cycle is repeated.

In the absence of resistance in any part of the circuit, an undamped, sinusoidal a.c. would be obtained. In practice, energy is gradually dissipated by the resistance as heat and a damped oscillation is produced, Fig. 17.31.

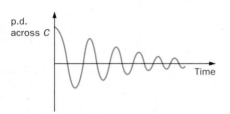

Fig. 17.31 Damped oscillation

(b) Demonstrations

Very slow damped electrical oscillations may be shown using the circuit of Fig. 17.32a. Oscillations are started by charging C from a 20 V d.c. supply. Several cycles of 'slow' a.c. are indicated on the moving-coil milliammeter. If either L or C is reduced, the frequency of oscillation increases.

Oscillations of slightly higher frequency (about 2 Hz) are obtained using the arrangement of Fig. 17.32b and a CRO on its slowest time-base speed.

(c) Frequency of the oscillations

In Fig. 17.33 the capacitor has capacitance C and is in series with an inductor of inductance L and negligible resistance. If Q is the charge on the capacitor at time t and I is the current flowing, then

$$\text{p.d. across inductor} = V_L = -L\frac{dI}{dt}$$

$$\text{p.d. across capacitor} = V_C = \frac{Q}{C}$$

The net p.d. in the circuit equals zero (since $R = 0$ and so there is no IR term), therefore $V_C = V_L$, that is,

$$\frac{Q}{C} = -L\frac{dI}{dt}$$

Also, current is rate of flow of charge, i.e. $I = dQ/dt$ and so

$$\frac{dI}{dt} = \frac{d}{dt}\left(\frac{dQ}{dt}\right) = \frac{d^2Q}{dt^2}$$

$$\therefore \quad \frac{Q}{C} = -L\frac{d^2Q}{dt^2}$$

$$\therefore \quad \frac{d^2Q}{dt^2} = -\frac{1}{LC}\cdot Q$$

Fig. 17.33

This equation is of the same form as that which represents s.h.m. (namely $d^2x/dt^2 = -\omega^2x$, p. 151) and indicates that the charge Q on the capacitor varies sinusoidally with time, having a period T given by $T = 2\pi/\omega = 2\pi\sqrt{(LC)}$. If f is the frequency of the oscillations, $f = 1/T$ and

$$f = \frac{1}{2\pi\sqrt{(LC)}}$$

As the resistance of an LC circuit increases, the oscillations decay more quickly, and when it is too large the capacitor discharge is unidirectional and no oscillations occur. To obtain undamped oscillations, energy has to be fed into the LC circuit in phase with its natural oscillations to compensate for the energy dissipated in the resistance of the circuit. This can be done in actual oscillators with the help of a transistor.

RECTIFICATION OF A.C.

Rectification is the conversion of a.c. to d.c. by a rectifier.

(a) Rectifiers

Rectifiers have a low resistance to current flow in one direction, known as the *forward* direction, and a high resistance in the opposite or *reverse* direction. When connection is made to a supply so that a rectifier conducts it is said to be **forward biased**; in the non-conducting state it is **reverse biased**, Fig. 17.34. The arrowhead on the symbol for a rectifier indicates the forward direction of conventional current flow.

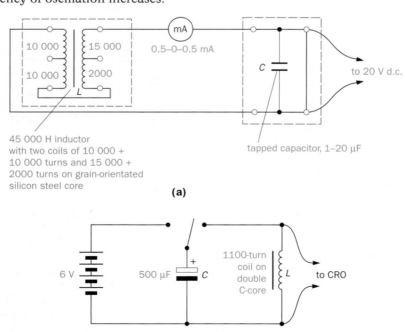

(a)

(b)

Fig. 17.32 Demonstrating electrical oscillations

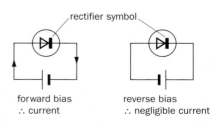

Fig. 17.34

The **semiconductor diode** was developed as a result of research aimed at producing rectifiers that could supply large currents, withstand high p.ds and be quite small but reliable and robust. The silicon diode rectifier is able to operate at a current density of 100 A cm^{-2} of rectifier surface (compared with 100 mA cm^{-2} for metal rectifiers), can withstand reverse p.ds of several hundred volts and has a working temperature of roughly 150 °C. Figure 17.35 shows a typical silicon diode rectifier. The construction and action of semiconductor diodes will be considered later (p. 416).

Fig. 17.35 Silicon diode rectifier

(b) Half-wave rectification

The rectifying circuit of Fig. 17.36 consists of a rectifier in series with the a.c. input to be rectified and the 'load' requiring the d.c. output. For simplicity the load is represented by a resistor R but it might be some piece of electronic equipment.

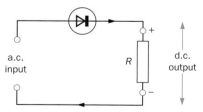

Fig. 17.36

In Fig. 17.37 the alternating input p.d. applied to the rectifier and load is shown. If the first half-cycle acts in the forward direction of the rectifier, a

pulse of current flows round the circuit, creating a p.d. across R which will have almost the same value as the applied p.d. if the forward resistance of the rectifier is small compared with R. The second half-cycle reverse-biases the rectifier, little or no current flows and the p.d. across R is zero. This is repeated for each cycle of a.c. input. The current pulses are unidirectional and so the p.d. across R is direct, for although it fluctuates it never changes direction.

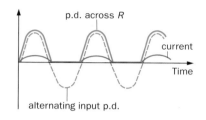

Fig. 17.37 Half-wave rectification

(c) Full-wave rectification

In this process both halves of every cycle of input p.d. produce current pulses and the p.d. developed across a load is like that shown in Fig. 17.38. There are two types of circuit.

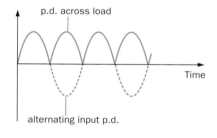

Fig. 17.38 Full-wave rectification

(*i*) *Centre-tap full-wave rectifier.* Two rectifiers and a transformer with a centre-tapped secondary are used, Fig. 17.39. The centre tap O has a potential half-way between that of A and F and it is convenient to take it as a reference point having zero potential. If the first half-cycle of input makes A positive, rectifier B conducts, giving a current pulse in the circuit ABC, R, OA. During this half-cycle the other rectifier E is non-conducting since the p.d. across FO reverse-biases it. On the other half of the same cycle F becomes positive with respect to O, and A negative. Rectifier E conducts to give current in the circuit FEC, R, OF; rectifier B is now reverse biased.

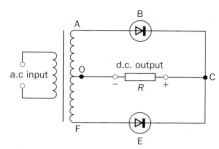

Fig. 17.39 Centre-tap full-wave rectifier

In effect the circuit consists of two half-wave rectifiers working into the same load on alternate half-cycles of the applied p.d. The current through R is in the same direction during *both* half-cycles and a fluctuating direct p.d. is created across R like that in Fig. 17.38.

(*ii*) *Bridge full-wave rectifier.* Four rectifiers are arranged in a bridge network as in Fig. 17.40a. If A is positive during the first half-cycle, rectifiers 1 and 2 conduct and current takes the path ABC, R, DEF. On the next half-cycle when F is positive, rectifiers 3 and 4 are forward biased and current follows the path FEC, R, DBA. Once again current flow through R is unidirectional during both half-cycles of input p.d. and a d.c. output is obtained.

Figure 17.40b shows a bridge rectifier.

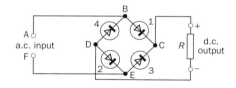

Fig. 17.40a Bridge full-wave rectifier circuit

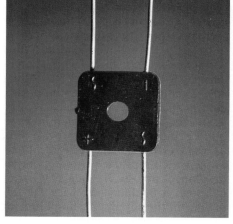

Fig. 17.40b Bridge rectifier

SMOOTHING CIRCUITS

To produce steady d.c. from the varying but unidirectional output from a half- or full-wave rectifier, 'smoothing' is necessary.

(a) Reservoir capacitor

The simplest smoothing circuit consists of a large capacitor — 16 µF or more — called a **reservoir** capacitor, placed in parallel with the load R. In Fig. 17.41*a*, C_1 is the reservoir capacitor and its action can be followed from Fig. 17.41*b* where V_1 represents the p.d. developed across C_1 and I is the rectifier (full-wave) current.

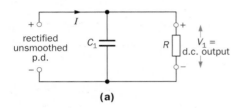

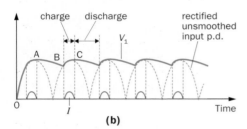

Fig. 17.41 Smoothing by a reservoir capacitor

The rectified input p.d. causes current to flow through R and at the same time C_1 becomes charged almost to the peak value of the input as shown by OA. At A, the input p.d. falls below V_1 and C_1 starts to discharge. It cannot do so through the rectifier since the polarity is wrong, but it does through the load and thus maintains current flow by its charge-storing (or reservoir) action. Along AB, V_1 falls. At B when the input p.d. equals the value to which V_1 has fallen, rectifier current I again flows to quickly recharge C_1 to the peak p.d., as shown by BC. The cycle of operations is then repeated. The d.c. output developed across R equals V_1 and, although it fluctuates at twice the frequency of the supply, the amplitude of the fluctuations is much less than when C_1 is absent.

The smoothing action of C_1 arises from its large capacitance, making the time constant C_1R large so that the p.d. across it cannot follow the variations of input p.d. A very large value of C_1 would give better smoothing but initially the uncharged reservoir capacity would act almost as a short-circuit and the resulting surge of current could damage the rectifier.

(b) Capacitor-input filter

A reservoir capacitor has a useful smoothing effect but it is usually supplemented by a filter circuit consisting of a choke L (i.e. an iron-cored inductor) having an inductance of about 15 H and a large capacitor C_2 arranged as in Fig. 17.42. The reservoir capacitor C_1 and the filter capacitor C_2 may be electrolytics enclosed in the same can.

C_1 behaves as explained above and the p.d. across it is similar to V_1 in Fig. 17.41*b*. The action of the filter circuit L–C_2 can be understood if V_1 is resolved into a steady direct p.d. (the d.c. component) and an alternating p.d. (the a.c. component). This procedure is often used when dealing with a varying d.c. and is illustrated in Fig. 17.43.

By redrawing the smoothing circuit as in Fig. 17.44 we can see that V_1 is applied across L and C_2 in series. L offers a much greater impedance to the a.c. component than C_2 and most of the unwanted 'ripple' p.d. appears across L. For the d.c. component, C_2 has infinite resistance and the whole of this component is developed across C_2 except for the small drop due to the resistance of the choke. The filter is acting as a potential divider, separating d.c. from a.c. and giving a steady d.c. output across C_2.

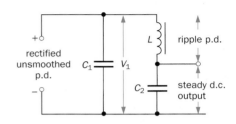

Fig. 17.44 Effect of the filter circuit

(c) Laboratory power supply unit ('power pack')

A typical circuit for a laboratory power pack is shown in Fig. 17.45; it uses a silicon-diode bridge rectifier fed by a step-up mains transformer and has a capacitor-input smoothing filter. A steady d.c. output of 0–400 V at 100 mA is produced as well as a 6.3 V a.c. heater supply for thermionic devices such as cathode ray tubes. The 25 kΩ variable resistor acts as a potential divider giving a continuously variable output and also allowing the capacitors to discharge when the pack

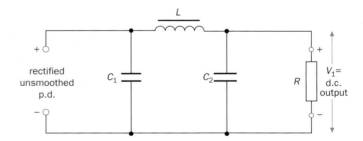

Fig. 17.42 Addition of a filter circuit

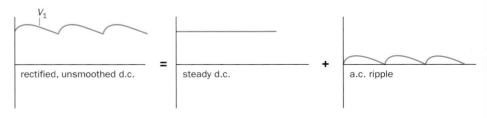

Fig. 17.43

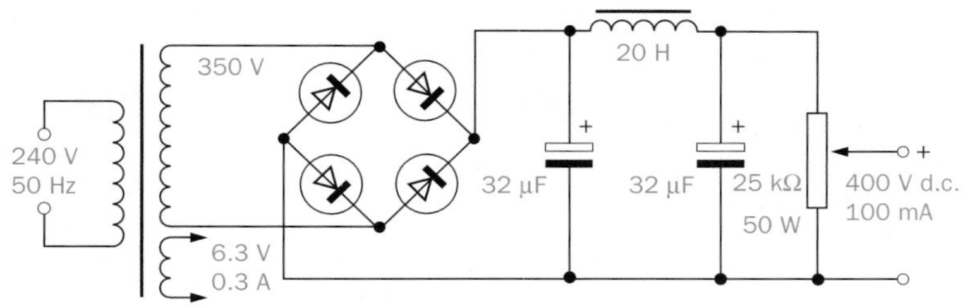

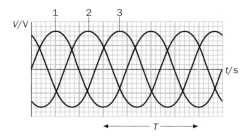

Fig. 17.45 Circuit of a laboratory power pack

is switched off — this eliminates risk of shock should the output terminals be touched subsequently.

It should be noted that r.m.s. p.ds are usually quoted for the secondary of a transformer but, depending on the current taken by the load, rectification and smoothing can result in a d.c. output that approaches the peak value, i.e. 1.4 times the r.m.s. value. In the circuit shown the nominal 350 V secondary gives a d.c. output of about 500 V at no load current and 400 V when 100 mA is supplied.

(d) Demonstrations

The circuits shown in Fig. 17.46 can be used to demonstrate rectification and smoothing using slow a.c. An a.c. of frequency about $\frac{1}{2}$ Hz is obtained by applying 6 V d.c. to the input of a low-frequency generator which is rotated by hand. The action of the various stages of a rectifying (half-wave) and smoothing circuit can be followed from the voltmeter reading using each arrangement in turn. The voltmeter acts as the load. A resistor is used here instead of a choke but in practice a choke is preferred in power units where currents of more than a few milliamperes have to be supplied.

QUESTIONS

1. When a certain a.c. supply is connected to a lamp it lights with the same brightness as it does with a 12 V battery.
   a) What is the r.m.s. value of the a.c. supply?
   b) What is the peak p.d. of the a.c. supply?
   c) The 12 V battery is connected to the Y-plates of a CRO and the gain adjusted so that it deflects the spot by 1.0 cm. What will be the total length of the trace on the CRO screen when the a.c. supply replaces the battery?

2. Electrical energy is supplied to a distant consumer through a power line which has a total resistance of 5 Ω. Explain why an input potential difference of 100 kV r.m.s. would be more satisfactory than one of 10 kV r.m.s., and calculate in each case
   a) the output voltage, and
   b) the output power when the power input is 20 MW.
   If the figure of 100 kV refers to the r.m.s. value of a sinusoidal a.c., what will be the maximum voltage for which the line must be insulated?

3. High voltage power lines and distribution cables may consist of four conductors. Three of these conductors supply the power and the fourth

conductor is a common return conductor called the neutral. The potential differences between each of the three conductors and Earth vary as shown in graphs 1, 2, 3 in Fig. 17.47.

The peak voltage of each supply conductor is 325 kV. What is the r.m.s. value of the potential difference between any one supply conductor and Earth?

What is the delay in terms of the time period *T* between the p.d. in conductor 1 and the p.d. in conductor 2?

Fig. 17.47

Estimate the maximum p.d. between any two supply conductors. Show on a copy of the graph how you obtained this value.

The graph can be used to show that the vector sum of the voltages in the three conductors at any instant is zero. Explain why the vector sum of the three currents in the return conductor might *not* be zero.

(*L*, A PH4, Jan 1998)

4. A sinusoidal p.d. of r.m.s. value 10 V is applied across a 50 μF capacitor.
   a) What is the peak charge on the capacitor?
   b) Draw a graph of charge *Q* on the capacitor against time.
   c) When is the current flowing into the capacitor
       i) a maximum, and
       ii) a minimum?
   d) Draw a graph of current against time.

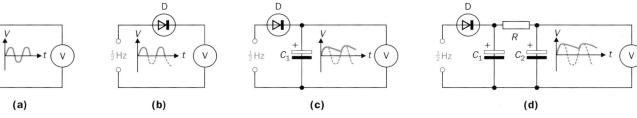

(a)   **(b)**   **(c)**   **(d)**

Ⓥ = 2.5–0–2.5 V moving-coil voltmeter

D = semiconductor diode (e.g. 1N4001)

$C_1 = C_2 = 500$ μF
$R = 1$ kΩ

Fig. 17.46

e) If the a.c. supply has a frequency of 50 Hz, calculate the r.m.s. current through the capacitor.

5. In Fig. 17.48 V_1 and V_2 are identical high-resistance a.c. voltmeters. Explain why the sum of the r.m.s. p.ds measured across XY and YZ by V_1 and V_2 exceeds the applied r.m.s. p.d. of 15.0 V. Calculate the value of C.

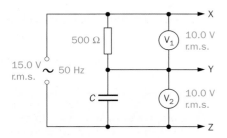

Fig. 17.48

6. As soon as the switch is closed in Fig. 17.49 what is the initial value of
  a) the p.d. across the capacitor,
  b) the current through the resistor,
  c) the rate at which the p.d. across the capacitor rises?

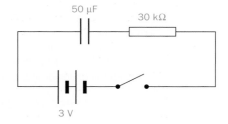

Fig. 17.49

7. a) State *two* variables which affect the reactance of inductors.
  b) i) Calculate the reactance of a coil of inductance 300 mH when carrying sinusoidal a.c. at a frequency of 50 Hz.
  ii) Calculate the r.m.s. current which would flow through the coil in part i) when the supply is 230 V r.m.s.
  iii) Explain why, in practice, the r.m.s. current in the coil would be less than that calculated in part ii).
  (*NEAB*, AS/A PHO3, March 1998)

8. What is the purpose of a **choke** in a circuit? Describe an experiment to demonstrate the action of a choke, explaining what happens.

  When an impedance, consisting of an inductance L and a resistance R in series, is connected across a 12 V, 50 Hz supply a current of 0.050 A flows which differs in phase from that of the applied potential difference by 60°. Find the values of R and L.

  Find the capacitance of the capacitor which, connected in series in the above circuit, has the effect of bringing the current into phase with the applied potential difference.

9. Figure 17.50a shows how the reactance X_C of a capacitor C varies with the frequency f of an applied alternating voltage.

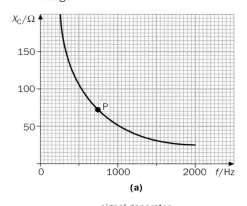

(a)

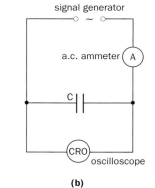

(b)

Fig. 17.50

  Figure 17.50b shows the circuit which was used to obtain data for the graph.

  Figure 17.51 shows the trace on the oscilloscope when the measurements for point P on Fig. 17.50a were being made. The Y-amplification setting on the oscilloscope was 5 V cm^{-1}.

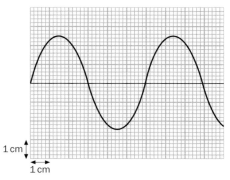

Fig. 17.51

a) i) Determine the root mean square (r.m.s.) voltage across C when the measurements for point P were being obtained.
  ii) What was the reading on the a.c. ammeter when the measurements for point P were being made?
  b) Copy Fig. 17.51 and draw on it a graph to show the corresponding variation of current with time in the circuit. It is not necessary to give a value for the maximum current.
  c) Use information from Fig. 17.50a to calculate the capacitance of the capacitor C in Fig. 17.50b.
  (*AEB*, 0635/2, Summer 1998)

10. Explain what is meant by **resonance** in an a.c. circuit containing inductance, resistance and capacitance in series. Give *one* practical application.

  A variable capacitor is connected in series with a coil and a sinusoidal alternating supply of 20 V (r.m.s.) at a frequency of 50 Hz. When the capacitor has a value of 1.0 µF, the current in the circuit reaches a maximum value of 0.50 A (r.m.s.). Find
  a) the resistance of the circuit,
  b) the self-inductance of the coil, and
  c) the potential difference across the capacitor.

11. a) State what is meant by the **frequency** of an alternating current.
  b) A component is connected across the terminals of a signal generator which supplies an a.c. output at constant voltage. If the frequency of this a.c. output is doubled, state the effect on the current flowing through the component if it is
    i) a pure inductor,
    ii) a capacitor.
  c) A pure inductor of inductance L and a capacitor of capacitance C are connected in series with the signal generator. The frequency is adjusted so that the reactance of the capacitor is equal to the reactance of the inductor. If L = 15 mH and C = 12 µF, calculate the frequency at which this will occur.
  (*NEAB*, AS/A PHO3, June 1997)

12. A loop of wire of negligible resistance is rotated in a magnetic field. A 4 Ω resistor is connected across its ends. A cathode ray oscilloscope measures the varying induced potential difference across the resistor as shown in Fig. 17.52.
  a) Copy the graph in Fig. 17.52. If the coil is rotated at twice the speed, show on the same axes how potential difference would vary with time.

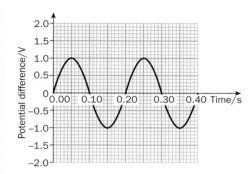

Fig. 17.52

b) What is the RMS value of the induced potential difference, V_{rms}, at the *original* speed of rotation?

c) On the axes like those in Fig. 17.53, draw a graph showing how the power dissipated in the resistor varies with time, at the *original* speed of rotation.

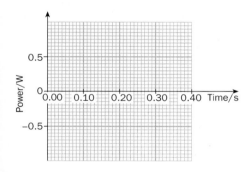

Fig. 17.53

(*IB*, Higher Paper 2, Nov 1998)

13. Figure 17.54 below shows an experimental arrangement used by a student investigating transformers. The signal generator provides a sinusoidal a.c. output of 7.0 V r.m.s.

a) Initially the paper is removed and the two C-shaped laminated soft iron cores are pushed firmly together. The switch is then closed. Stating any assumptions which you make, calculate the approximate **peak-to-peak** potential difference measured by the oscilloscope.

b) As part of the investigation the student places sheets of paper, to act as spacers, between the C-shaped cores and connects a 100 Ω resistor across the terminals of the oscilloscope. After each sheet is added, the cores are squeezed together and the peak-to-peak potential difference is measured from the oscilloscope. Two of the results are recorded in the table below.

Number of sheets of paper between cores	Peak-to-peak p.d./V	Power delivered/W
1	18	
3	12	

Copy the table and complete the third column to show the power delivered to the resistor. Give your working.

c) Explain why the core of a transformer
  i) is made from iron,
  ii) consists of laminated sheets.
    (*NEAB*, AS/A PHO3, Feb 1997)

14. The a.c. mains supply of root-mean-square (r.m.s.) voltage 230 V is to be converted into a much lower steady d.c. voltage for use with a high resistance load.

a) Draw a diagram of a suitable circuit. Your circuit should include a transformer, a full-wave rectifier using diodes, and a capacitor. Add arrows to your diagram to show the route of the current through the rectifier during one half-cycle.

b) If the output of the transformer in a) is 12 V r.m.s. explain why the steady d.c. output of the circuit you have drawn in a) is not 12 V. The diodes in the circuit can be considered ideal.
    (*OCR*, 6843, June 1999)

15. Draw a circuit and explain the action of components to provide full-wave rectification of an alternating supply. Explain the action of a capacitor which could be used to smooth the output.

16. a) Define
  i) inductive reactance,
  ii) capacitive reactance.

b) Consider the three circuits in Fig. 17.55. The source provides a potential difference $V = V_0 \cos \omega t$.
  i) Derive an expression for the current in each of the circuits.
  ii) Use the expression $\sin 2A = 2 \sin A \cos A$ to derive an expression for the power in each reactive circuit.
  iii) Sketch a graph of the power as a function of time for the reactive circuits, and show on the graph when energy is stored in the reactive element.

c) An inductor of inductance L and a charged capacitor of capacitance C are connected together.
  i) Derive an expression for the resonant frequency of the circuit.
  ii) Describe the energy transformations that occur at the resonant frequency.
  iii) Calculate the resonant frequency for $C = 10.0\ \mu F$ and $L = 1.0\ mH$.
    (*IB*, Higher Paper 2, Nov 1997)

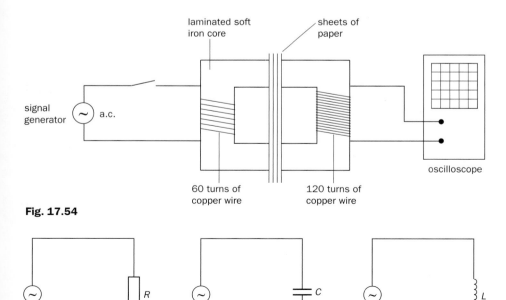

Fig. 17.54

Fig. 17.55

17. a) In the *LCR* circuit of Fig. 17.56*a* the applied alternating voltage of 13 V r.m.s. produces p.ds of 5 V r.m.s. and 4 V r.m.s. across *R* and *L* respectively. What is the p.d. across *C*?

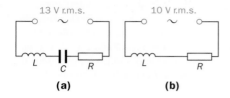

(a)   **(b)**

Fig. 17.56

b) Calculate the power dissipated in the circuit of Fig. 17.56*b* if *R* has zero reactance but resistance 3 Ω, *L* has zero resistance but reactance 4 Ω and the applied voltage is 10 V r.m.s.

18. In the half-wave rectifier smoothing circuit of Fig. 17.57 what is the effect on the ripple on the current passing through the load *R* of
a) decreasing *R*,
b) increasing *C*, and
c) decreasing the frequency of the a.c. input?

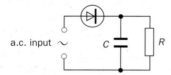

Fig. 17.57

19. The primary coil of an ideal transformer has 480 turns and is connected to a 120 V r.m.s., 60 Hz sinusoidal supply. The secondary coil has 40 turns and is connected to a resistor of resistance 2.0 Ω, as shown in Fig. 17.58.

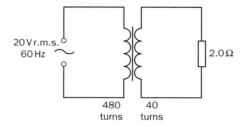

Fig. 17.58

a) Calculate
i) the r.m.s. voltage across the resistor,
ii) the peak power dissipated by the resistor.

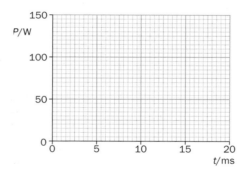

Fig. 17.59

b) On axes like those in Fig. 17.59, sketch a graph to show the variation with time *t* of the power *P* dissipated by the resistor. Show any calculations.
(*OCR*, Further Physics 4833, March 1999)

20. a) A circuit including a light-dependent resistor can produce a potential difference which increases with increasing light intensity.
    i) Describe the relevant property of the light-dependent resistor and, with the aid of a diagram, a suitable circuit.
    ii) Use the laws of d.c. circuits to explain the operation of the circuit.

b) A single diode may be used to produce direct current in a resistor connected to a source of alternating voltage.
    i) With the help of circuit diagrams and sketch graphs, explain how this can be done.
    ii) Describe and explain how a suitable capacitor may be used to smooth the current in the resistor. Include a discussion of the relationship of the amount of smoothing to the capacitance and the resistance.
(*OCR*, Further Physics 4833, March 1999)

Objective-type revision questions 3

The first figure of a question number gives the relevant chapter, e.g. **15.2** is the second question for chapter 15.

MULTIPLE CHOICE

Select the response that you think is correct.

13.1 A sphere carrying a charge of Q coulombs and having a weight of W newtons falls under gravity between a pair of vertical plates distance d metres apart. When a potential difference of V volts is applied between the plates the path of the sphere changes as shown in Fig. R19, becoming linear along CD. The value of Q is

A $\dfrac{W}{V}$ **B** $\dfrac{W}{2V}$ **C** $\dfrac{Wd}{V}$ **D** $\dfrac{2Wd}{V}$ **E** $\dfrac{Wd}{2V}$

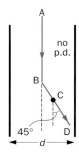

Fig. R19

13.2 Three charges $+Q$, $-2Q$ and $+Q$ are placed at the corners of a square of side a as shown in Fig. R20. The resultant electric field strength at P has magnitude

A $\dfrac{Q}{4\pi\epsilon a^2}$ **B** $\dfrac{\sqrt{2}Q}{4\pi\epsilon a^2}$ **C** $\dfrac{(\sqrt{2}-1)Q}{4\pi\epsilon a^2}$ **D** $\dfrac{(\sqrt{2}+1)Q}{4\pi\epsilon a^2}$

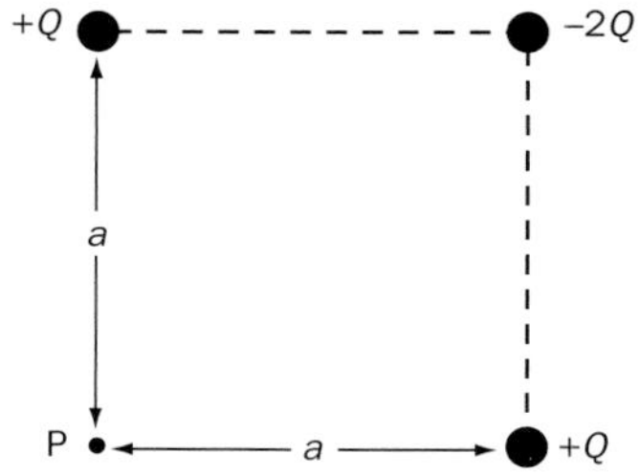

Fig. R20

14.1 A parallel-plate capacitor is to be made by inserting one of the sheets of dielectric material included below, between and in contact with two plates of copper.

	Relative permittivity	Thickness (mm)
Teflon	2	0.4
Quartz	3	0.8
Glass	4	1.0
Mica	5	1.2
Porcelain	6	1.3

The maximum capacitance will be obtained by using the sheet of
A Teflon **B** quartz **C** glass **D** mica **E** porcelain

14.2 Figure R21 represents a parallel-plate capacitor whose plate separation of 10 mm is very small compared with the size of the plates. If a potential difference of 5.0 kV is maintained across the plates, the intensity of the electric field in V m^{-1} at
i) X, and
ii) Y,
is
A 50 **B** 2.5×10^2 **C** 5.0×10^2 **D** 2.5×10^5
E 5.0×10^5

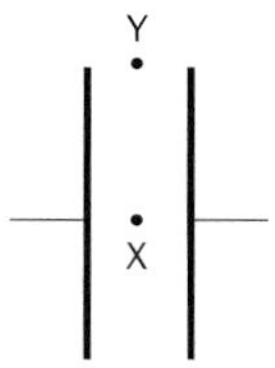

Fig. R21

14.3 A 1 MΩ resistor is connected across a 30 μF capacitor which has been charged to 240 V. Approximately how long does it take for the voltage across the capacitor to fall to 80 V?
A 120 s **B** 60 s **C** 30 s **D** 3 s

15.1 Two long solenoids, P and Q, have n_1 and n_2 turns per unit length respectively and carry currents I_1 and I_2 respectively. The magnetic flux density on the axis of P at a point near the middle is four times that at a corresponding point in Q. The value of I_1/I_2 is

A $\dfrac{2n_1}{n_2}$ **B** $\dfrac{2n_2}{n_1}$ **C** $\dfrac{4n_1}{n_2}$ **D** $\dfrac{4n_2}{n_1}$ **E** $\dfrac{16n_1}{n_2}$

15.2 The wire X in Fig. R22 is at right angles to the plane of the paper and carries a current into the paper. The magnetic flux density due to this current will be in the same direction as that of the horizontal component of the earth's field at a point on the diagram
  A labelled 1 **B** labelled 2
  C labelled 3 **D** labelled 4
  E whose position is not deducible from the data without a knowledge of the value of the current.

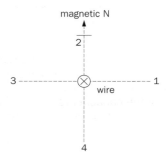

Fig. R22

15.3 An experimental arrangement for the Hall effect is shown in Fig. R23. The Hall voltage
  A is parallel to the magnetic field
  B decreases if the current increases
  C decreases if the width XY of the specimen decreases
  D increases if the magnetic flux density increases.

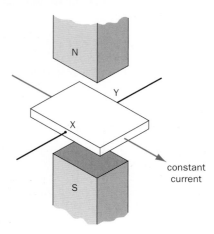

Fig. R23

16.1 The e.m.f. induced in a coil of wire, which is rotating in a magnetic field, does not depend on
  A the angular speed of rotation
  B the area of the coil

  C the number of turns on the coil
  D the resistance of the coil
  E the magnetic flux density

16.2 When the speed of an electric motor is increased due to a decreasing load, the current flowing through it decreases. Which of the following is the best explanation of this?
  A The resistance of the coil changes.
  B Frictional forces increase as the speed increases.
  C Frictional forces decrease as the speed increases.
  D The induced back e.m.f. increases.
  E At high speeds it is more difficult to feed current into the motor.

16.3 When an ammeter is well damped,
  A large currents can be measured
  B small currents can be measured
  C readings can be taken quickly
  D it is very accurate
  E it is very robust.

16.4 Which feature of a moving-coil ammeter is of most importance in making it well damped?
  A The strength of the hairsprings.
  B The number of turns on the coil.
  C The former on which the coil is wound.
  D The size of the iron cylinder.
  E The material used for the coil.

16.5 Alternating current is preferable to direct current for the transmission of power because
  A it can be rectified
  B it is easier to generate
  C thinner conductors can be used
  D no question of polarity arises with equipment
  E it is safer.

16.6 A short bar magnet is dropped through a coil of wire of a similar length, Fig. R24. Which of the graphs in Fig. R25 shows best how the current through the coil varies with time?

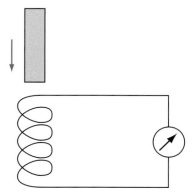

Fig. R24

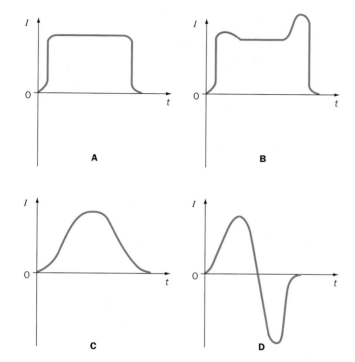

Fig. R25

17.1 The direct current which would give the same heating effect in an equal constant resistance as the current shown in Fig. R26, i.e. the r.m.s. current, is
A zero **B** $\sqrt{2}$ A **C** 2 A **D** $2\sqrt{2}$ A **E** 4 A

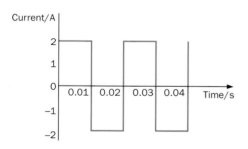

Fig. R26

17.2 What value of L, in henrys, will make the circuit in Fig. R27 resonate at 100 Hz?

A $\dfrac{1}{4\pi^2}$ **B** $\dfrac{1}{2\pi}$ **C** 1 **D** $4\pi^2$

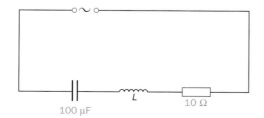

Fig. R27

17.3 The diagram in Fig. R28 shows a half-wave rectifier with reservoir capacitor and load resistance. The time constant of C and R should be
A small compared with the time of one cycle
B independent of the time of one cycle
C large compared with the time of one cycle
D the same as the time of one cycle.

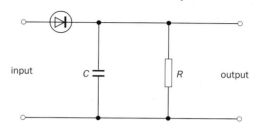

Fig. R28

17.4 An alternating voltage which can be represented by the equation $V = V_0 \sin \omega t$ is connected across a capacitor as shown in Fig. R29. If ω is increased to 2ω, the reactance of the capacitor is
A unchanged **B** doubled **C** halved
D $\sqrt{2}$ of original value

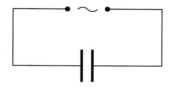

Fig. R29

MULTIPLE SELECTION

In each question one or more of the responses may be correct. Choose one letter from the answer code given.

> *Answer **A** if i), ii) and iii) are correct.*
> *Answer **B** if only i) and ii) are correct.*
> *Answer **C** if only ii) and iii) are correct.*
> *Answer **D** if i) only is correct.*
> *Answer **E** if iii) only is correct.*

13.3 When a positive charge is given to an isolated hollow conducting sphere
i) a positively charged object anywhere inside the sphere experiences an electric force acting towards the centre of the sphere
ii) the potential to which it is raised is proportional to its radius
iii) the potential gradient outside the sphere is independent of its radius.

13.4 The intensity of an electric field
i) is a vector quantity
ii) can be measured in N C^{-1}
iii) can be measured in V m^{-1}.

14.4 A parallel-plate capacitor C is charged by connection to a battery using the switch S as shown in Fig. R30. S is opened and the plate separation is then increased. Increases occur in the following:
i)   the charge stored
ii) the potential difference between the plates
iii) the energy stored.

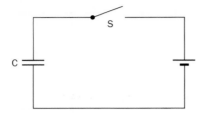

Fig. R30

15.4 The magnetic flux density well inside a long uniformly wound solenoid depends on
i)   the number of turns per unit length
ii) the area of cross-section
iii) the distance from the axis of the solenoid.

16.7 The circuit of Fig. R31 shows a cell of negligible internal resistance in series with a pure inductor and a pure resistor. When the switch is closed
i)   the current rises initially at the rate of $6.0\,A\,s^{-1}$
ii) the final value of the current is 1.5 A
iii) the final energy stored in the space in and around the coil is 2.3 (2.25) J.

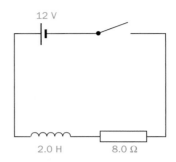

Fig. R31

Waves

Interference in a layer of bubbles of washing-up liquid

18

Wave motion

- ■ Progressive waves
- ■ Describing waves
- ■ Huygens' construction
- ■ Principle of superposition

- ■ Interference
- ■ Diffraction
- ■ Stationary waves
- ■ Polarization

- ■ Speed of mechanical waves
- ■ Reflection and phase changes
- ■ Using reflected waves
- ■ Wave equations

PROGRESSIVE WAVES

The concept of a **wave** is useful for dealing with a wide range of phenomena and is one of the basic concepts of physics. A knowledge of wave behaviour is also important to engineers.

A **progressive** or travelling wave consists of a disturbance moving from a source to surrounding places as a result of which **energy is transferred from one point to another**.

There are two types of progressive wave. In the **transverse** type the direction associated with the disturbance is at right angles to the direction of travel of the wave, Fig. 18.1*a*. In the **longitudinal** type the disturbance is in the same direction as that of the wave, Fig. 18.1*b*. Transverse and longitudinal pulses (that is, waves of short duration) can be sent along a Slinky spring, Figs 18.2*a* and *b*. In both cases the disturbance generated by the hand is passed on from one coil of the spring to the next which performs the same motion but at a slightly later time. This pulse travels along the spring, with the coils propagating the disturbance merely by vibrating to and fro (transversely or longitudinally) about their undisturbed positions. A succession of disturbances creates a continuous train of waves, i.e. a wavetrain. The wave demonstrators shown in Figs 18.3*a* and *b* are useful for modelling transverse and longitudinal waves.

Waves may be classified as **mechanical** or **electromagnetic**. Mechanical waves are produced by a disturbance (e.g. a vibrating body) in a material medium and are transmitted by the particles of the medium oscillating to and fro. Such waves can be seen or felt and include waves on a spring, water waves, waves on stretched strings (e.g. in musical instruments) and sound waves in air and in other materials. Many of the properties of mechanical waves can be shown using water waves in a ripple tank. Such *surface water waves* are less complex than sea waves and are transmitted by the surface layer. However, the displacement of the water is not a simple up-and-down, transverse motion (although the effects we shall describe will be due to this transverse component of the motion); the water particles move in elliptical or circular paths in the direction of the wave at a crest and in the opposite direction in a trough.

Electromagnetic waves (see p. 349) consist of a disturbance in the form of varying electric and magnetic fields. No medium is necessary and they travel more easily in a vacuum than in matter. Radio signals, light and X-rays are examples of this type. In this chapter microwaves will be used to investigate those properties of electromagnetic radiation which can be explained in terms of waves. Microwaves are radar-type waves and are generated by an oscillator of very high frequency a.c. (10 000 MHz). In the transmitter on the left of Fig. 18.4 the oscillator feeds a small aerial in a rectangular metal tube called a **wave guide** which opens into a horn at one end and is closed at the other. The radiation emitted has a wavelength of about 3 cm. The horn receiver (right) has a wave guide and horn like those of the transmitter, and contains a silicon diode mounted to 'detect' (see p. 456) the signal. The probe receiver (bottom, left) is less sensitive and is also non-directional; it simply consists of a diode. The signal from the

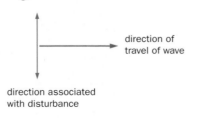

(a) Transverse wave

Fig. 18.1

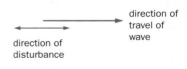

(b) Longitudinal wave

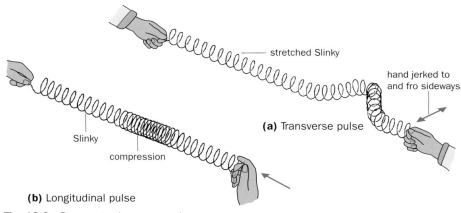

Fig. 18.2 Demonstrating progressive waves

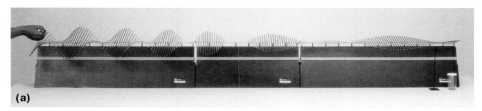

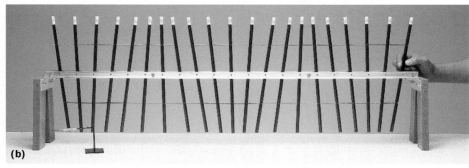

Fig. 18.3 Wave machines showing (*a*) transverse and (*b*) longitudinal waves

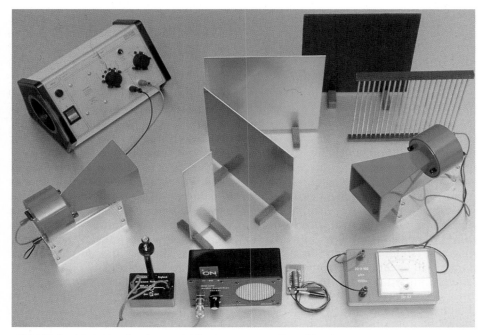

Fig. 18.4 Apparatus for demonstrating wave properties with microwaves

transmitter can be modulated so that when picked up by either of the receivers, a note is heard in an amplifier–loudspeaker unit. Alternatively it can produce current in a microammeter connected to the receiver.

The effects of mechanical and electromagnetic waves are explicable by the same general principles.

DESCRIBING WAVES

(a) Graphical representation

Two kinds of graph may be drawn. A **displacement–distance** graph for a transverse mechanical wave shows the displacements y of the vibrating particles of the transmitting medium at different distances x from the source *at a certain instant*. The red dots in Fig. 18.5*a*, p. 296, show the positions of the particles at a particular time and the corresponding displacement–distance graph is given in Fig. 18.5*b*.

A longitudinal wave can also be represented by a graph of transverse displacement–distance, the displacements in this case, however, being those of the vibrating particles in the line of travel of the wave. Thus y and x are in the same direction in the wave, Fig. 18.5*c*, but at right angles to each other in the graph, i.e. the graph represents a longitudinal displacement as if it were transverse. Regions of high particle density are called **compressions** and regions of low particle density are called **rarefactions**. Both move in the direction of travel of the wave while the particles of the medium vibrate to and fro about their undisturbed positions.

The maximum displacement of each particle from its undisturbed position is the **amplitude** of the wave; in Fig. 18.5*b* this is OA or OB. The **wavelength** λ of a wave is the distance between two consecutive points on it which are in step, i.e. have the same phase. For a transverse wave it is generally taken to be the distance between two successive crests or two successive troughs; for a longitudinal wave it is the separation of consecutive compressions or consecutive rarefactions. In Fig. 18.5*b*, $\lambda = \text{OS} = \text{PT} = \text{RV}$.

Electromagnetic waves behave as transverse waves (see p. 349) and may be represented by the same kind of displacement–distance graph, y being

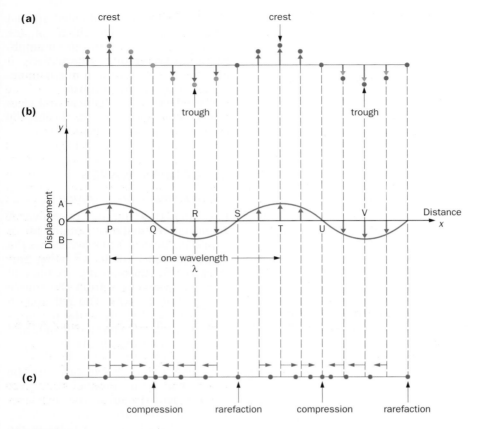

Fig. 18.5

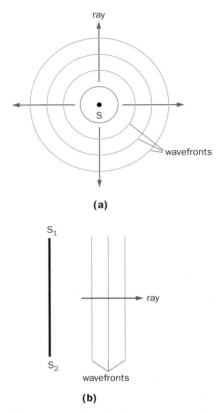

Fig. 18.6 Wavefronts and rays for (a) a point source and (b) a line source

the instantaneous value of the electric field strength E or the magnetic flux density B at different distances from the source.

A **displacement–time** graph may also be drawn for a wave motion, showing how the displacement of one particle (or the value of E or B) at a particular distance from the source varies with time. If this is a simple harmonic variation the graph is a sine curve.

(b) Wavelength, frequency, speed

If the source of a wave makes f vibrations per second, so too will the particles of the transmitting medium in the case of a mechanical wave and E or B in an electromagnetic wave. That is, the frequency of the wave equals the frequency of the source.

When the source makes one complete vibration, one wave is generated and the disturbance spreads out a distance λ from the source. If the source continues to vibrate with constant frequency f, then f waves will be produced per second and the wave advances a distance $f\lambda$ in one second. If v is the wave speed then

$$v = f\lambda$$

This relationship holds for all wave motions.

(c) Wavefronts and rays

A wavefront is a line or surface on which the disturbance has the same phase at all points. If the source is periodic it produces a succession of wavefronts, all of the same shape. A point source, S, generates circular wavefronts in two dimensions, as in Fig. 18.6a, and spherical wavefronts in three dimensions. A line source S_1S_2 (e.g. the straight vibrator in a ripple tank) will create wavefronts that are straight in two dimensions, Fig. 18.6b, and cylindrical in three dimensions. Plane wavefronts are produced by a plane source or by *any* source at a distant point.

A line at right angles to a wavefront which shows its direction of travel is called a **ray**.

HUYGENS' CONSTRUCTION

The construction proposed by Huygens (a contemporary of Newton) enables the new position of a wavefront to be found, knowing its position at some previous instant. It can be used to explain the reflection, refraction and dispersion of waves as we shall see shortly.

According to Huygens, **every point on a wavefront may be regarded as a source of secondary spherical wavelets which spread out with the wave speed. The new wavefront is the envelope of these secondary wavelets**, that is, the surface which touches all the wavelets.

As a simple example, suppose AB is a straight wavefront travelling from left to right, Fig. 18.7. The wavefront at time t later is the common tangent CD of the spherical wavelets of radius

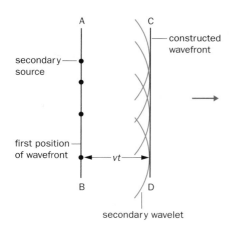

Fig. 18.7 Huygens' construction for a straight wavefront

vt, drawn with centres at every point on AB, where v is the speed of the waves in the medium.

The construction does suggest that waves should also travel backwards to the left. This does not occur in practice and the assumption (which can be justified) has to be made that the amplitude of the secondary wavelets varies from a maximum in the direction of travel of the wave to zero in the opposite direction. This is indicated in Fig. 18.7 by drawing only arcs of the wavelets in the forward direction.

(a) Reflection

In Fig. 18.8 the wavefront AB is incident obliquely on the reflecting surface and A has just reached it. To find the new position of the wavefront when B is about to be reflected at B', we draw a secondary wavelet with centre A and radius BB'. The tangent B'A' from B' to this wavelet is the required reflected wavefront.

In triangles AA'B' and ABB'

angle AA'B' = angle ABB' = 90°

AB' is common

AA' = BB' (by construction)

Therefore the triangles are congruent (rt : h : s), and

angle BAB' = angle A'B'A

These are the angles made by the incident and reflected wavefronts, AB and A'B' respectively, with the reflecting surface. The incident rays and reflected rays, e.g. CA and B'D, are at right angles to the wavefronts, as are the normals to the surface (the dotted lines at A and B'), and so the angle of incidence i equals the angle of reflection r. This is the law of reflection. It may be easily shown to hold for the reflection of straight water waves in a ripple tank at a straight barrier, and it may be tested for the reflection of microwaves from a metal plate with the apparatus of Fig. 18.4 arranged as in Fig. 18.9.

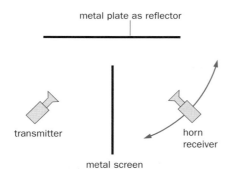

Fig. 18.9 Demonstrating reflection of microwaves

(b) Refraction

Figure 18.10, p. 298, shows the end A of a plane wavefront AB about to cross the boundary between media 1 and 2 in which its speeds are v_1 and v_2 respectively. The new position of the wavefront at time t later when B has travelled a distance BB' = v_1t and reached B' is found by drawing a secondary wavelet with centre A and radius AA' = v_2t. The tangent B'A' from B' to this wavelet is the required wavefront and the ray AA', which is normal to the wavefront, is a refracted ray. If v_2 is less than v_1 the refracted ray is bent towards the normal in medium 2, as indicated. When v_2 is greater than v_1 refraction away from the normal occurs.

In triangles BAB' and A'AB'

$$\frac{\sin i_1}{\sin i_2} = \frac{\sin \angle CAN}{\sin \angle A'AN'} = \frac{\sin \angle BAB'}{\sin \angle A'B'A}$$

$$= \frac{BB'/AB'}{AA'/AB'} = \frac{BB'}{AA'} = \frac{v_1t}{v_2t} = \frac{v_1}{v_2}$$

But v_1 and v_2 are constants for given media and a particular wavelength, so

$$\frac{\sin i_1}{\sin i_2} = \text{a constant}$$

This is Snell's law of refraction and it holds for electromagnetic, sound and water waves. For waves passing from medium 1 to medium 2, $n_1 \sin i_1 = n_2 \sin i_2$ where n_1 and n_2 are the refractive indices of media 1 and 2 (see p. 82), so

$$\frac{v_1}{v_2} = \frac{\sin i_1}{\sin i_2} = \frac{n_2}{n_1}$$

Huygens' construction thus explains the refraction of, for example, light, on the assumption that the speed of light decreases when it passes into a medium in which the refracted ray is bent towards the normal, i.e. when it enters an optically denser medium. This assumption is verified by experiment.

Refraction occurs whenever a wave passes from one medium into another in which it has a different speed. The speed of surface waves depends primarily on depth in shallow liquids, i.e. when the depth is not large compared with the wavelength (in deep liquids it depends on wavelength, not on depth). Water ripples travel more slowly in shallow than in deeper water

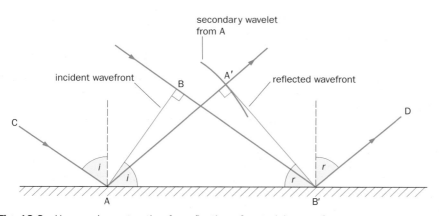

Fig. 18.8 Huygens' construction for reflection of a straight wavefront

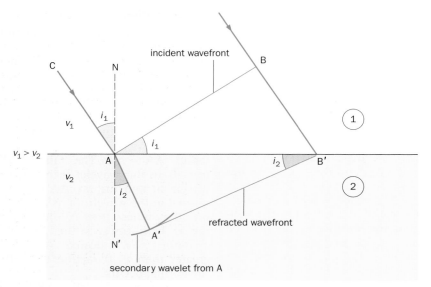

Fig. 18.10 Huygens' construction for refraction of a straight wavefront

and are refracted when they pass from one to the other. In Fig. 18.11 a glass plate on the bottom of the lower half of a ripple tank reduces the depth. As well as the change of direction of the refracted wavefront in the shallower water, note the shorter wavelength (and hence smaller speed since the frequency is unchanged — how would you check that it is so?). The refraction of microwaves may be demonstrated as in Fig. 18.12.

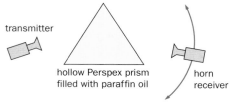

Fig. 18.12 Demonstrating refraction of microwaves

(c) Dispersion

Huygens' construction predicts that if the speed of waves in a given medium depends on the frequency (and therefore on the wavelength) of the waves, then when refraction occurs its extent will also depend on the frequency. In Fig. 18.13a the waves of frequencies f_A and f_B travel with the same speed in medium 1 but in medium 2, v_A (the

speed of wave A with frequency f_A) is greater than v_B and so wave B suffers greater deviation than wave A. Dispersion occurs; medium 1 is called a **non-dispersive** medium and medium 2 a **dispersive** medium. Red light travels faster than blue light in glass and if white light travels from air to glass, the blue wavefront is refracted more than the red one, that is, **dispersion** is observed, Fig. 18.13b.

PRINCIPLE OF SUPERPOSITION

What happens when waves meet? Is their motion changed, as when solid objects collide? The answers to these questions may be obtained by producing pulses on a long narrow spring.[1]

The drawings in Fig. 18.14a show two transverse pulses (of slightly different shapes) approaching each other on a spring. Where they cross (diagram 3) the resultant displacement of the spring is evidently the sum of the displacements which each would have caused at that point, i.e. the pulses superpose. After crossing, each pulse travels along the spring as if nothing had happened and each retains its original shape. The result is always the same, whatever shapes the pulses have. Figure 18.14b shows the superposition of two equal and opposite pulses. In this case, when the pulses meet (diagram 3) they cancel out and the net displacement of the spring is zero.

[1] A PSSC-type spring is better than a Slinky for this.

Fig. 18.11 Refraction of water waves at a depth change in a ripple tank

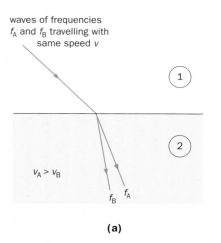

waves of frequencies f_A and f_B travelling with same speed v

$v_A > v_B$

f_B f_A

(a)

Fig. 18.13 Dispersion

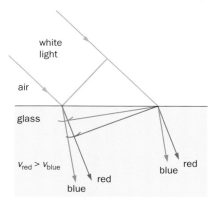

white light

air

glass

$v_{red} > v_{blue}$

blue red

blue

red

(b)

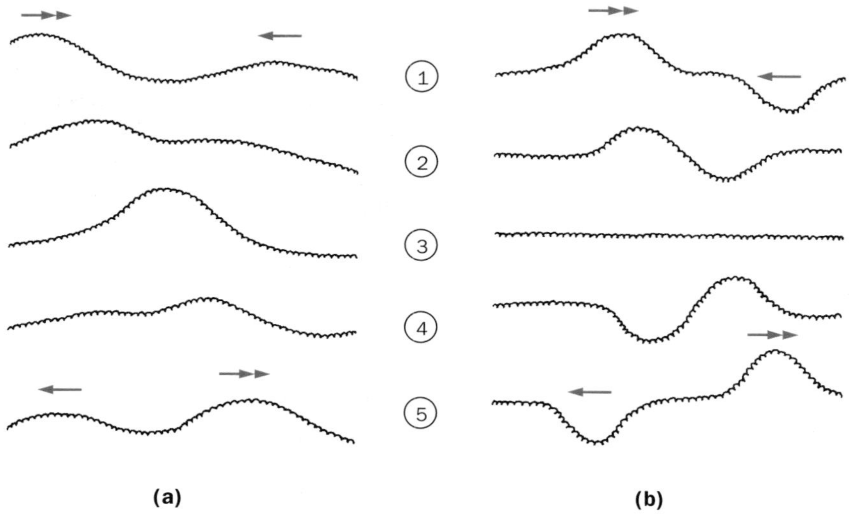

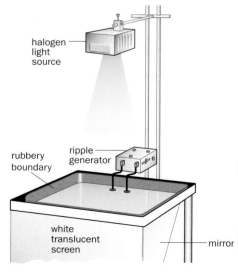

(a) **(b)**

Fig. 18.14 Superposition of oppositely directed pulses

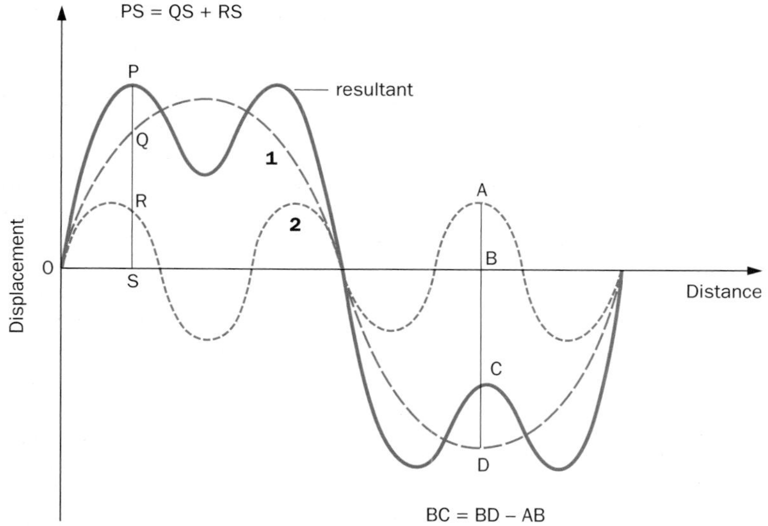

$PS = QS + RS$

resultant

P

Q

1

R

2

S

O

Distance

A

B

C

D

$BC = BD - AB$

Displacement

Fig. 18.15 Superposition of waves

halogen
light
source

rubbery
boundary

ripple
generator

white
translucent
screen

mirror

Fig. 18.16 Demonstrating interference of circular water waves in a ripple tank

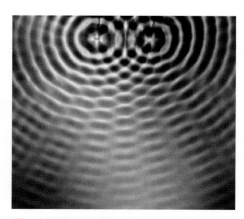

Fig. 18.17 Interference pattern due to two coherent sources of circular waves

In general we can conclude that:

> Pulses and waves (unlike particles) pass through each other unaffected and, **where they cross, the total displacement is the vector sum of the individual displacements due to each pulse at that point**.

This is called the **principle of superposition** and can be used to explain many wave effects. It is applied in Fig. 18.15, where the resultant of two waves, 1 and 2, of different amplitudes and frequencies, is obtained by adding the displacements. Wave 2 has half the amplitude of 1 and three times its frequency. The shape of the resultant is often quite different from those of its components, as here.

INTERFERENCE

In a region where wave-trains from *coherent* sources cross, superposition occurs, giving reinforcement of the waves at some points and cancellation at others. The resulting effect is called an **interference pattern** or a **system of fringes**.

Coherent sources have a **constant phase difference**, which means they must have the same frequency, and for complete cancellation to occur the amplitudes of the superposing waves they produce must be about equal. If their phase difference is not constant, at a certain point there may be reinforcement at one instant and cancellation at the next. If these variations follow one another rapidly, the effect at the point is to produce uniformity, i.e. the interference pattern changes so quickly with the continuously changing phase difference between the wave sources that the detector, for example, the eye, cannot follow the alterations and records an average effect.

In practice coherent sources are derived from a single source.

(a) Mechanical waves

Interference is readily shown by the transverse component of surface water waves in a ripple tank, using two small spheres attached to the same vibrating bar, Fig. 18.16. Circular waves are produced and give a pattern like that of Fig. 18.17.

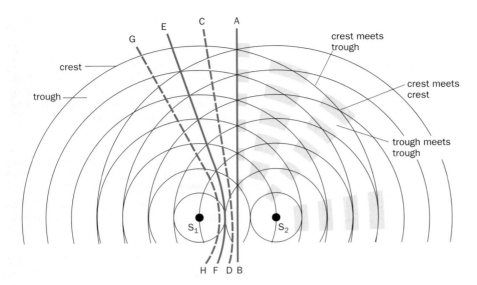

Fig. 18.18

To explain this interference pattern, consider Fig. 18.18. All points on AB are equidistant from the sources S_1 and S_2 and since the vibrations of these are in phase, crests (or troughs) from S_1 arrive at the same time as crests (or troughs) from S_2. Along AB reinforcement occurs by superposition and a wave of double amplitude is formed, Fig. 18.19a. Points on CD are half a wavelength (a wavelength being the distance between the centres of successive bright (or dark) bands in Fig. 18.17) nearer to S_1 than to S_2, i.e. there is a path difference of half a wavelength. As a result, crests (or troughs) from S_1 arrive simultaneously with troughs (or crests) from S_2 and the waves cancel, Fig. 18.19b.

(a) Reinforcement   **(b)** Cancellation

Fig. 18.19

Along EF in Fig. 18.18 the difference of the distances from S_1 and S_2 to any point is one wavelength. EF is a line of reinforcement, i.e. **constructive interference** occurs. GH is a line of cancellation where **destructive interference** occurs since at every point on GH the distance to S_1 is one and a half wavelengths less than to S_2. Lines such as CD and GH are called **nodal**

lines; AB and EF are examples of **antinodal** lines. Both sets are hyperbolic curves.

It is possible to show that the separation of nodal (and antinodal) lines increases:

- as the distance from sources S_1 and S_2 increases;
- the smaller the separation of S_1 and S_2; and
- as the wavelength increases (i.e. as the frequency decreases).

(b) Electromagnetic waves

Two arrangements for showing the interference of microwaves are given in Figs 18.20a and b, using the apparatus of Fig. 18.4. In (a) interference occurs between the two wave-trains emerging from the 3 cm wide slits which act as two coherent sources. The receiver detects the maxima and

minima of the fringe pattern as it is moved round; if it is a minimum when the waves from one slit are cut off by a metal plate, the signal *increases* — clear evidence of destructive interference.

In (b), the glass plate partially reflects and partially transmits the microwaves. The transmitted wave-train traverses the air 'film' to the metal plate where it is reflected back across the film and again transmitted (partially) by the glass. Two wave-trains (derived from the same source) thus reach the receiver and the thickness of the air 'film' decides the path difference between them (i.e. how much farther one has travelled than the other). If 'crests' of both arrive simultaneously at the receiver, there is reinforcement but if a 'crest' arrives with a 'trough' there is cancellation. Changing the path difference, e.g. by moving the metal plate towards or away from the glass, produces maxima and minima at the receiver. What is the minimum change in the thickness of the air 'film' that will cause a maximum to be replaced by another maximum?

Navigation systems on aircraft and ships are based on the interference of radio waves (see p. 306).

Summing up, effects in which disturbances when added produce no disturbance can be regarded as being due to some kind of wave motion. Therefore radio signals and light are treated as wave motions because radio signals sometimes 'fade' and light plus light can result in darkness. However, no energy disappears: it is redistributed (p. 313). In the case of invisible waves where the nature of the disturbance is not revealed directly to our

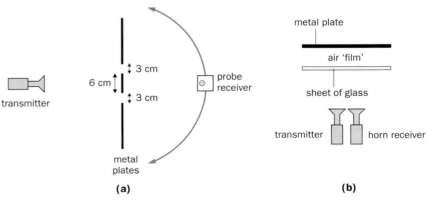

Fig. 18.20   Demonstrating interference of microwaves

senses we are really using a wave model, based on our experience of visible waves, to enable us to explain their behaviour.

(c) Rectilinear propagation

Interference shows the mechanism by which Huygens' secondary wavelets superpose to form the new wavefront. Rectilinear propagation for an infinite plane wave is then explained by the destructive interference of out-of-phase wavelets in all directions except forwards.

DIFFRACTION

The spreading of waves — when they pass through an opening or round an obstacle — into regions where we would not expect them is called **diffraction**.

(a) Mechanical waves

The diffraction of water ripples at an opening in a barrier may be studied in a ripple tank, a vibrating bar being used to produce straight ripples.

In Fig. 18.21a, the opening is wide and the incident waves emerge almost unchanged, i.e. straight. A fairly sharp 'shadow' of the opening is obtained and diffraction is not marked. Most of the energy propagated through the opening is in the same direction as the incident waves. In Fig. 18.21c, the opening is narrow and the emerging waves are circular, i.e. the opening behaves like a point source. In this case diffraction is appreciable and the energy of the diffracted waves is more or less equally distributed over 180°.

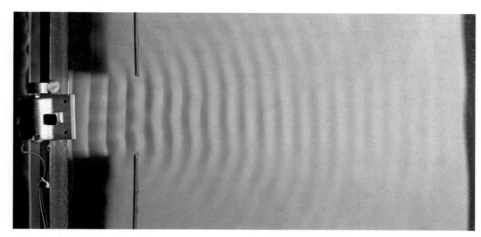

Fig. 18.22a Diffraction of straight waves at a slit

Fig. 18.22b Interference due to a line of equally spaced point sources

In Fig. 18.21b some diffraction is evident at the 'medium-sized' gap. Diffraction effects are thus greatest when the width of the opening is comparable with the wavelength of the waves.

Huygens' procedure for predicting the future position of a wavefront by replacing it by point sources is able to account for diffraction, as well as for reflection and refraction. The patterns obtained can be considered to arise from the interference (superposition) of the secondary wavelets produced by the point sources imagined to exist at the unrestricted part of the wavefront which falls on the opening. The two ripple tank photographs above, Figs 18.22a and b, support this view. The diffraction pattern of straight waves passing through a slit is shown

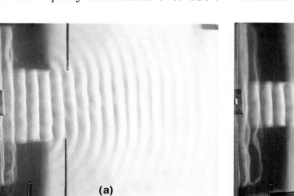

(a)

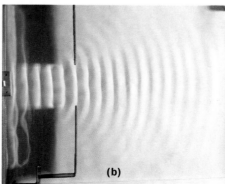

(b)

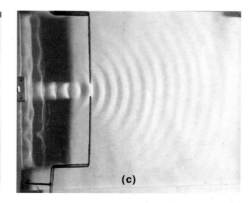

(c)

Fig. 18.21 Diffraction of straight water waves at an opening in a barrier

in (*a*); (*b*) is the interference pattern of a line of equally spaced point sources occupying the same position and width as the slit and vibrating together with the same frequency as that of the waves in (*a*). The two patterns are very similar except near the sources in (*b*).

(b) Electromagnetic waves

Diffraction of microwaves at a slit may be shown with the apparatus of Fig. 18.4, arranged as in Fig. 18.23*a*. The spreading into the regions of geometrical shadow behind the slit is greatest when the slit width is comparable with the wavelength of the microwaves, i.e. 3 cm. (It should now be apparent why it is possible to demonstrate the interference of microwaves using the double-slit arrangement of Fig. 18.20*a*.)

It is also instructive to position a single metal plate as in Fig. 18.23*b* so that it just cuts off the signal to the receiver, placed in the shadow of the plate. When a second plate is slid up (along the dotted line) to make a slit 3–6 cm wide with the first plate, the received signal increases, showing that a *smaller* wavefront (less energy) can produce a *stronger* signal. This is due to destructive interference between different parts of the larger wavefront and by removing some of them (with the second plate) the resultant signal increases in this particular direction.

If a circular metal disc (about 15 cm diameter) is used as an 'obstacle', a signal is received along the axis in the centre of the shadow, indicating appreciable diffraction round the edges of the disc.

STATIONARY WAVES

(a) Mechanical waves

If one end of a long, narrow, stretched spring (or a Slinky) is fixed and the other end is moved continuously from side to side, a progressive transverse wave is generated. At the fixed end the wave is reflected, travels back to the vibrating end and repeated reflection occurs. Two progressive trains of waves travel along the spring in opposite directions. If the shaking frequency is slowly increased, at certain frequencies one or more vibrating loops of large amplitude are formed in the spring. A **stationary** or **standing wave** is said to have been produced since the waveform does not seem to be travelling along the spring in either direction.

A more effective demonstration, called Melde's experiment, is shown in Fig. 18.24*a* using a $\frac{1}{2}$ m length of rubber cord (3 mm square section) stretched to 1 m. As the frequency of the a.c. from the low-impedance output of the signal generator is increased from 10 Hz to about 100 Hz, stationary waves with one, two, three or four loops are obtained. If the cord is illuminated in the dark by a lamp stroboscope and the flashing rate adjusted to be nearly but not quite equal to that of the vibrator, the cord can be seen moving up and down slowly, the crests of the stationary wave becoming troughs, then crests again and so on, Fig. 18.24*b*. Note the following.

(*i*) There are points on the cord such as N, called **nodes** of the stationary wave, where the displacement is always zero.

(*ii*) Within a single loop all particles oscillate in phase but with different amplitudes (how does this compare with the behaviour of a progressive wave?) and so all points except nodes have their maximum displacement simultaneously; points such as A with the greatest amplitude are called **antinodes**.

(*iii*) The oscillations in one loop are in antiphase with those in an adjacent loop.

(*iv*) The frequency of the particle vibration in both the standing wave and the progressive waves is the same, and the wavelength of the stationary wave is *twice* the distance between successive nodes or successive antinodes and equals the wavelength of either of the progressive waves. This offers a solution to the otherwise difficult problem of measuring the wavelength of a progressive wave. We simply produce a stationary wave from the progressive wave and measure the wavelength directly.

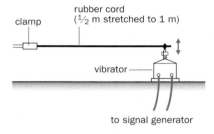

(a) Melde's experiment

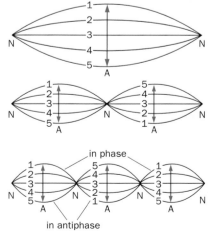

(b) Movement seen with a stroboscope

Fig. 18.24 Demonstrating stationary waves on a cord

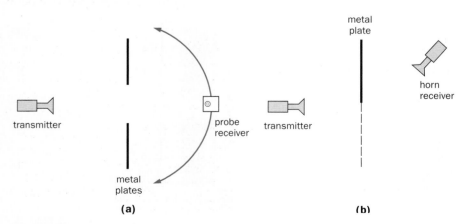

(a)

(b)

Fig. 18.23 Demonstrating diffraction of microwaves

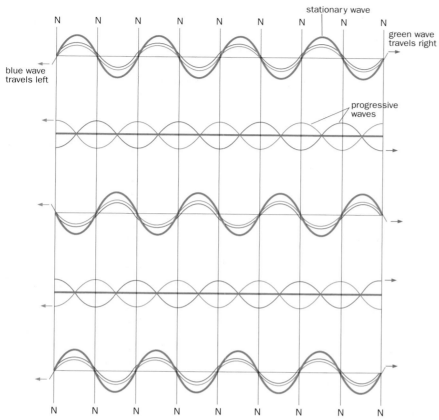

N N N N N N N N N

stationary wave

green wave travels right

blue wave travels left

progressive waves

N N N N N N N N N

Fig. 18.25 Formation of stationary waves

(b) Explanation of stationary waves

Stationary waves arise from the super-position of two trains of progressive waves of equal amplitude and frequency travelling with the same speed in opposite directions. In Fig. 18.25 the blue and green curves are the displacement–distance graphs of two such waves at successive equal time intervals. The red curve in each case is their resultant at these instants formed by superposition. The formation of the stationary wave loops of Fig. 18.24b can now be understood.

In the two previous demonstrations one wave-train is obtained by a reflection of the other and if, in the time for a wave to travel along the spring or cord and back again, the vibrator is about to send off the second wave, the latter will reinforce the first. This will also be true if the vibrator produces exactly two, three or any integral number of waves in the time for a wave to make a return journey on the spring or cord. The amplitude of the waves builds up since they are returned in phase to the source, i.e. there is resonance. The coupling is thus improved between the source and the medium, producing 'stimulated emission'. If the stationary wave has one loop at frequency f then a vibrator frequency of $2f$ (called the **second harmonic**) will give two loops and so on.

We can regard the stretched spring or cord as having a number of natural frequencies of vibration and when the applied frequency from the vibrator is near one of these, there is a large-amplitude standing-wave vibration, that is, resonance occurs. A stretched spring has a number of natural frequencies. By contrast, a mass hanging from the end of a spring, for example, has only one natural frequency. (We will see later, p. 304, that a system can only vibrate if it has mass and elasticity. The difference between the two systems arises from the fact that in the mass–spring system, the mass is concentrated in one part of the system and the elasticity property in another part; in the stretched spring or cord system all parts have both properties. The first is said to have *lumped* elements and the second *distributed* elements.)

Standing-wave resonance can occur only in systems with 'boundaries' (e.g. the ends of a spring or cord) that restrict wave travel. Progressive waves reflected from the boundaries interfere with progressive waves travelling to the boundaries and the resulting standing waves have to 'fit' into the system, e.g. have nodes at the fixed ends of the cord in Fig. 18.24a, otherwise the pattern is not fixed in form. Large amounts of energy are stored locally in standing waves; **there is no energy transmission** as with progressive waves. This is an important difference between stationary and progressive waves. In 'unbounded' systems the waves are not confined and so travel on, unrestricted.

Some other demonstrations of stationary waves in mechanical systems are illustrated in Figs 18.26a, b, c and d. The first two should be viewed in the dark in stroboscopic light (see pp. 324–5).

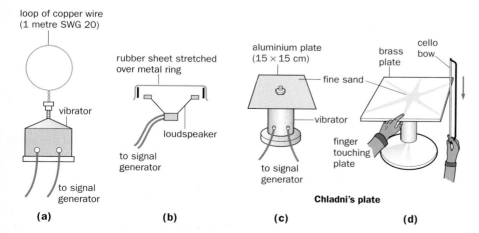

loop of copper wire (1 metre SWG 20)

vibrator

to signal generator

(a)

rubber sheet stretched over metal ring

loudspeaker

to signal generator

(b)

aluminium plate (15 × 15 cm)

fine sand

vibrator

to signal generator

Chladni's plate

(c)

brass plate

cello bow

finger touching plate

(d)

Fig. 18.26 Viewing stationary wave patterns

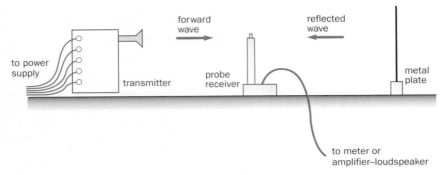

Fig. 18.27 Demonstrating stationary waves with microwaves

(c) Electromagnetic waves

Stationary microwave patterns can be obtained with the arrangement of Fig. 18.27. Waves from the transmitter are superposed on those reflected from the metal plate. If the plate is moved slowly towards or away from the transmitter, the signal at the receiver varies and the distance moved by the reflector between two consecutive minima (nodes) equals half the wavelength of the microwaves. Why is the signal not quite zero at a node?

(d) Importance of stationary waves

Stationary waves are a feature of many physical phenomena. The mechanical type is produced on strings and in air columns of musical instruments when sounding a note (pp. 318 and 316). Electromagnetic examples occur in radio and television aerial systems and, as we shall see later (p. 407), the idea that an electron in an atom behaves like a standing wave is used to account for the fact that atoms have definite energy levels.

Engineers are often confronted with standing-wave problems when dealing with systems such as turbines, propellers, aircraft wings, car bodies, etc., which vibrate and have edges.

POLARIZATION

This effect occurs only with transverse waves.

(a) Mechanical waves

In Fig. 18.28 the rope PQRS is fixed at end S and passes through two vertical slits at Q and R. If end P is moved to and fro in all directions (as shown by the short arrowed lines), vibrations of the rope occur in every plane and transverse waves travel towards Q. At Q only waves due to vibrations in a vertical plane can emerge from the slit and the wave between Q and R is said to be **plane polarized** (in the vertical plane containing the slit at Q), in contrast to that between P and Q which is unpolarized. If the slit at R is vertical (as shown) the wave travels on, but if it is horizontal, i.e. the slits are 'crossed', the wave is stopped.

(b) Electromagnetic waves

Electromagnetic radiation from radio, television and radar aerials is plane polarized. This may be demonstrated for 3 cm microwaves using the microwave transmitter (10 000 MHz = 10 GHz) and receiver. When the receivers are rotated through 90° in a vertical plane

from the maximum signal position, the signal decreases to a minimum. Polarization provides the evidence for the transverse nature of electromagnetic waves.

Consideration of the way in which a transmitting aerial produces radio and other types of waves shows that the electric field comes off parallel to the aerial and the magnetic field at right angles to it. If the transmitting aerial is vertical, the signal in the receiving aerial is a maximum when it too is vertical. In this case the radio waves are said to be **vertically polarized**, the direction of polarization being given by the direction of the *electric* field. This follows the practice adopted with light waves, where experiments show that the coupling of light with matter is more often through the electric field. The polarization of light will be considered later.

SPEED OF MECHANICAL WAVES

(a) Factors affecting speed

A mechanical wave is transmitted by the vibration of particles of the propagating medium. A system can vibrate only if it has **mass** and **elasticity**.

Consider the case of a load hung from the lower end of a spring. When the load is pulled down and released, the spring — due to its elasticity — pulls the load up again. However, owing to the load having mass it acquires kinetic energy and overshoots its original position. The opposition of the spring to compression gradually brings it to rest and it starts to fall. It again overshoots its original position and stretches the spring. The cycle of events is then repeated. Proper vibrations of the spring would not occur if it was very light (i.e. had no mass on the end) or was inelastic.

Mechanical waves can consequently only be transmitted by a medium if it has mass and elasticity, and it is these two factors which determine the speed of a wave. This last point may be demonstrated using a wave model in which trolleys represent particles of the transmitting medium and springs (slightly stretched) represent the bonds between the particles. A line of twelve trolleys are linked together and the

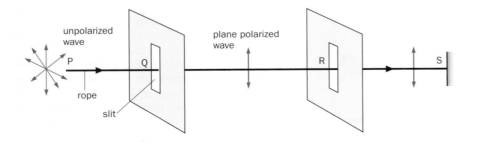

Fig. 18.28 Demonstrating polarization of a mechanical wave

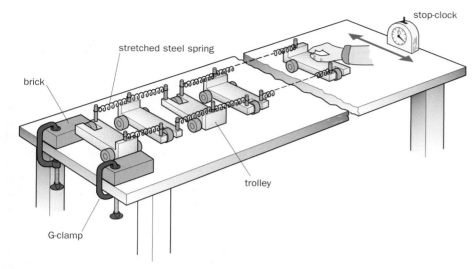

Fig. 18.29   Model of a mechanical wave

REFLECTION AND PHASE CHANGES

(a) Mechanical waves

The behaviour of a wave at a boundary can be studied by sending pulses along a long narrow spring.

In Fig. 18.30a, the left-hand end of the spring is fixed and a transverse 'upward' pulse travelling towards it is reflected as a trough. A phase change of 180° or π rad has occurred and there is a phase difference of half a wavelength (λ/2) between the incident and reflected pulses. In Fig. 18.30b the left-hand end of the spring is attached to a heavier spring and at the boundary the pulse is partly transmitted and partly reflected, the reflected pulse being inverted.

In Fig. 18.30c a pulse passes from a heavy spring on the left to a light spring. Partial reflection and transmission again occur but the reflected pulse is not turned upside down. In Fig. 18.30d the left-hand end of the long narrow spring is fastened to a length of thin string and is in effect 'free'. Here almost the whole of the incident pulse is reflected the right way up, i.e. a crest is reflected as a crest and no phase change occurs.

These results may be summarized by saying that when a transverse wave on a spring is reflected at a 'denser' medium (e.g. a fixed end or a heavier spring) there is a phase change of 180° (or π rad or λ/2). The phase change occurs in the case of the spring with one end fixed, for example, because there can be no displacement of the

time is found for a transverse pulse to travel along to the fixed end and back again, Fig. 18.29.

Doubling the mass of each trolley (by adding a load — typically 1 kg) *decreases* the wave speed since it takes longer for the larger mass to acquire a certain speed. Doubling the tension (by connecting another spring in parallel with each one in the model) *increases* the wave speed because the larger force causes the next trolley to respond more quickly. If mass and tension are both doubled, the wave speed has its original value since both quantities change the speed by the same factor.

Note how the energy of the vibrating 'particles' (trolleys) changes from k.e. to p.e. to k.e. and so on as the wave progresses.

Longitudinal waves along masses (e.g. trolleys) linked by springs:

$$v = x \sqrt{\left(\frac{k}{m}\right)}$$

x = spacing between mass centres
k = spring constant
m = one mass

Short wavelength ripples on surface of deep water:

$$v = \sqrt{\left(\frac{2\pi\gamma}{\lambda\rho}\right)}$$

γ = surface tension
λ = wavelength
ρ = density

(b) Expressions for wave speed

Expressions may be derived for different types of wave using basic mechanical ideas. Some are quoted below.

Note that in each case the numerator contains an 'elasticity' term and the denominator a 'mass' term.

Transverse waves on a taut string or spring:

$$v = \sqrt{\left(\frac{T}{\mu}\right)}$$

T = tension
μ = mass per unit length

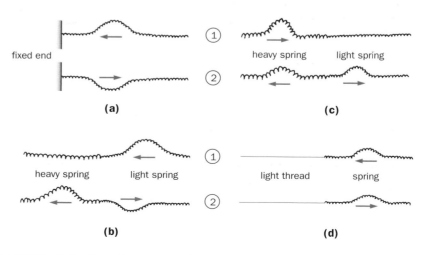

Fig. 18.30   Reflection of transverse pulses

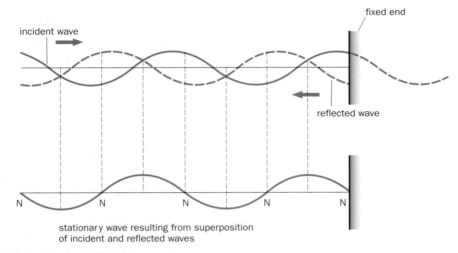

Fig. 18.31 Reflection of a transverse wave at a denser medium

fixed end — it must be a node. The incident and reflected waves therefore cause equal and opposite displacements at the fixed end so that they superpose to give resultant zero displacement, as shown in Fig. 18.31.

Phase changes also occur whenever longitudinal waves are reflected. This can be shown by sending pulses along a Slinky spring to 'denser' and 'less dense' boundaries, i.e. to fixed and free ends. At a fixed end a compression is reflected as a compression; at a free end it is reflected as a rarefaction. Similar effects are obtained when sound waves are reflected in pipes with closed and open ends: a compression is reflected as a compression at a closed end and as a rarefaction at an open end. In the latter case the air in the compression is able to expand outwards suddenly at the end of the pipe and a rarefaction travels back along the pipe.

(b) Electromagnetic waves

The phase change in microwaves (half a wavelength ($\lambda/2$)) when they are reflected at a metal plate can be demonstrated with the arrangement in Fig. 18.32. Interference occurs between the waves reaching the probe receiver directly and those arriving at it by reflection from the metal plate. Consequently, maxima and minima are detected when the receiver is moved along PQ. At Q the geometrical path difference between the direct and reflected waves is zero and the two

sets of waves should reinforce to give a maximum. In fact a *minimum* is obtained owing to the $\lambda/2$ phase change of the reflected wave.

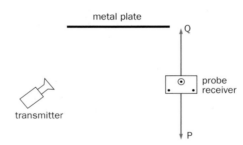

Fig. 18.32 Demonstrating phase change on reflection with microwaves

USING REFLECTED WAVES

A number of techniques have been developed for locating the positions of 'objects' by reflecting from them waves of known speed.

In **radar** (acronym for **ra**dio **d**etection **a**nd **r**anging), radio waves (e.g. 3 cm microwaves) are emitted in short pulses by a transmitter and picked up after reflection from the object, Fig. 18.33. The received pulse is displayed on a CRO with a calibrated time base which is triggered to start by the transmitted pulse. The time for the waves to travel twice the distance from the transmitting station to the object is thus found (the interval between transmitted pulses must be greater than this to avoid confusion). Objects might include military aircraft and missiles. Radar is also used to help control aeroplanes waiting to land. In ships equipped with radar to assist navigation in fog and at night, a narrow microwave beam is swept continuously through 360° by a rotating aerial. The pulses reflected from land, other ships and buoys are shown on a CRO, called a 'plan position indicator' (PPI), which has the time base origin in the centre of the screen and represents the ship. Figure 18.34 shows a ship's navigation screens, including a PPI.

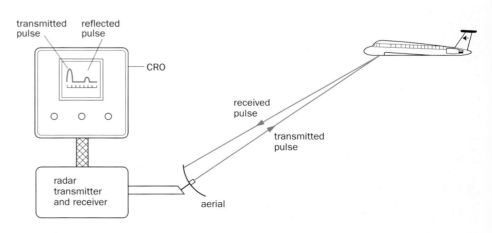

Fig. 18.33 Principle of radar

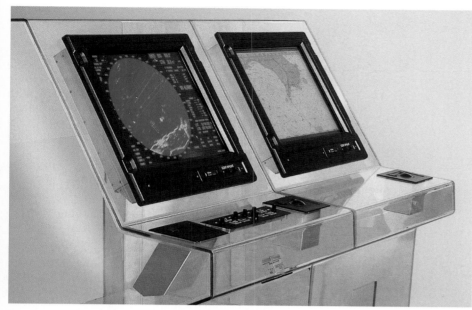

Fig. 18.34 Racal-Decca 'Bridgemaster' navigation system — a plan position indicator (PPI) and a video plotter

In **radar astronomy** pulses of radio waves (1–10 cm wavelength) are reflected from distant objects such as the moon and planets. The distance from the earth can be calculated from the time of travel of the reflected pulses. For large distances the radar beam may span the entire surface of the planet and the radius R is then estimated from the time spread Δt of reflected pulses; reflections from the pole will travel a distance $2R$ further than reflections from the equator so $\Delta t = 2R/c$ where c is the speed of the light. If spacecraft or satellites are used, large-scale topographical features of the planet can be mapped by scanning the radar beam across smaller distances of the surface. Figure 18.35 shows a radar map of the surface of Venus, constructed from data collected by the Magellan space probe. A high mountain range (bright spot) can be seen, and a plateau to its left. Faults are also visible, above the centre. The surface cannot be seen with optical wavelengths because of the thick clouds covering the planet.

Sonar, or **s**ound **n**avigation **a**nd **r**anging, is similar to radar but employs ultrasonic waves, i.e. waves with frequencies above the maximum audible frequency of about 20 kHz. It is used to measure the depth of the sea (i.e. in echo sounding) and to detect shoals of fish.

In the non-destructive testing of materials ultrasonic waves can detect flaws. If three pulses are obtained on the display CRO they correspond to the transmitted pulse A, the pulse B reflected from the flaw and the pulse C reflected from the boundary of the specimen, Fig. 18.36.

Essentially the same technique is used in **medical ultrasound imaging** to image soft tissue not readily seen with X-rays. Ultrasonic pulses are partially reflected from the boundary between different media, such as tissue, fluid or bone, and are used to image internal organs, a fetus, a tumour or a pocket of fluid in the body. The portion of the pulse that is reflected is determined by the values of the **acoustic impedance** Z of the media either side of the boundary; Z depends on both the density of a material and the speed of sound in it. When the reflected ultrasonic pulses are displayed on a CRO as a function of time, an 'A-scan' like that shown in Fig. 18.37a is produced, p. 308; the position of each reflection indicates the depth inside the body at which reflection occurs. To form an image, the signal is displayed as a 'B-scan', Fig. 18.37b, where each reflection is represented by a point, the brightness of which corresponds to the strength of the reflection. If the ultrasonic transducer (which acts as both transmitter and receiver) is scanned across the body, Fig. 18.37c, a detailed image of internal features can be built up as shown in Fig. 18.37d. In practice a gel is used between the transducer and the skin to reduce the strong reflection and resulting loss of signal that occurs at the first air/body boundary. The speed of ultrasonic waves in body tissue is similar to that for water (1500 m s^{-1}) and since the maximum frequency utilized is typically 10^6 Hz (higher frequencies are strongly absorbed) the minimum wavelength, $\lambda = v/f = 1500/10^6 = 1.5 \times 10^{-3}$ m = 1.5 mm, sets the lower limit to the object size that can be successfully imaged. Figure 18.38 shows an image of a fetus produced in this way.

Ultrasonic waves reflected from a moving object undergo a change in frequency due to the Doppler effect (see p. 322) and this can be used to monitor heart beats and the motion of blood in the veins.

Fig. 18.35 Radar map of the surface of Venus

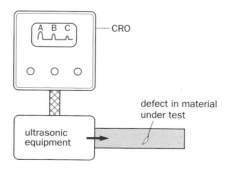

Fig. 18.36 Using ultrasonic waves to test materials

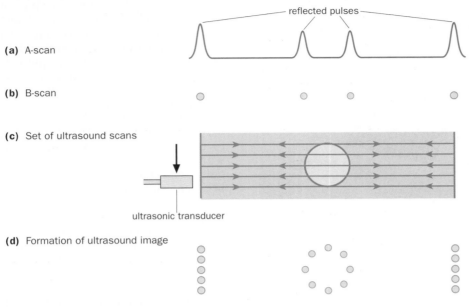

(a) A-scan

(b) B-scan

(c) Set of ultrasound scans

ultrasonic transducer

(d) Formation of ultrasound image

reflected pulses

Fig. 18.37 Medical ultrasound image formation

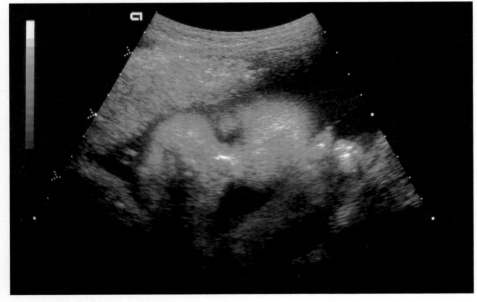

Fig. 18.38 Ultrasound scan image of the face of a nine-month old fetus

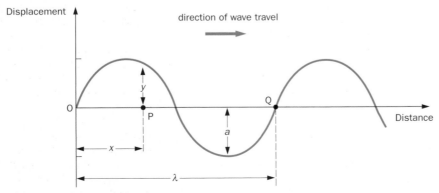

Fig. 18.39

WAVE EQUATIONS

(a) Progressive wave

Suppose the oscillation of the particle at O in Fig. 18.39 is simple harmonic of frequency f. Its displacement with time t will be given by $a \sin \omega t$ where a is the amplitude of the oscillation and $\omega = 2\pi f$. If the wave generated travels from left to right a particle at P, distance x from O, will lag behind the particle at O, say by phase angle ϕ. For the displacement y of the particle at P we can write

$$y = a \sin (\omega t - \phi) \quad \text{(p. 277)}$$

But $\phi/2\pi = x/\lambda$ since at Q, distance λ from O, the phase difference between the motions of particles at O and Q is 2π (rad). Substituting for ϕ in the above equation we get

$$y = a \sin (\omega t - 2\pi x/\lambda)$$
$$= a \sin (\omega t - kx)$$

where $k = 2\pi/\lambda$.

If the wave travelled in the opposite direction, the vibration at P would lead on that at O by ϕ and the displacement y at P would be written

$$y = a \sin (\omega t + kx)$$

This is the equation for a wave travelling from *right to left*.

(b) Stationary wave

Let the two progressive waves of equal amplitude a and frequency f ($\omega = 2\pi f$) travelling in opposite directions be represented by

$$y_1 = a \sin (\omega t + kx) \quad \text{(to the left)}$$

and

$$y_2 = a \sin (\omega t - kx) \quad \text{(to the right)}$$

By the principle of superposition, the resultant displacement y is given by

$$y = y_1 + y_2$$
$$= a \sin (\omega t + kx) + a \sin (\omega t - kx)$$

Using the trigonometrical transformation for converting the sum of two sines to a product

$$\left(\text{i.e.} \quad \sin \alpha + \sin \beta \right.$$
$$\left. = 2 \sin \frac{\alpha + \beta}{2} \cos \frac{\alpha - \beta}{2} \right)$$

we get

$$y = 2a \sin \omega t \cos kx$$

This is the equation of the stationary wave. It can be written

$$y = A \sin \omega t$$

where $A = 2a \cos kx = 2a \cos (2\pi x/\lambda)$ is the amplitude of oscillation of the various particles. We see that when $x = 0, \lambda/2, \lambda, 3\lambda/2$, etc., A is a maximum and equal to $2a$. These are the antinodes. Nodes occur midway between antinodes since $A = 0$ when $x = \lambda/4, 3\lambda/4, 5\lambda/4$, etc.

QUESTIONS

1. State Snell's law of refraction and define refractive index.

   Show how refraction of light at a plane interface can be explained on the basis of the wave theory of light.

   Light travelling through a pool of water in a parallel beam is incident on the horizontal surface. Its speed in water is 2.2×10^{10} cm s^{-1}.

   Calculate the maximum angle which the beam can make with the vertical if light is to escape into the air where its speed is 3.0×10^{10} cm s^{-1}.

   At this angle in water, how will the path of the beam be affected if a thick layer of oil, of refractive index 1.5, is floated on to the surface of the water?

2. S_1 and S_2 in Fig. 18.40 are two sources of circular water waves of the same frequency, phase and amplitude. There is a maximum disturbance at P, a minimum at Q and a maximum at R.
   a) Write *two* expressions for the wavelength λ of the ripples.
   b) How does the pattern change if
   i) S_1 and S_2 are moved farther apart, and
   ii) the frequency of the waves decreases?

Fig. 18.40

3. **a)** What is meant by
   i) diffraction, and
   ii) superposition of waves?

Describe *one* phenomenon to illustrate each in the case of sound waves.
   b) The floats of two men fishing in a lake from boats are 21 m apart. A disturbance at a point in line with the floats sends out a train of waves along the surface of the water, so that the floats bob up and down 20 times per minute. A man in a third boat observes that when the float of one of his colleagues is on the crest of a wave that of the other is in a trough, and that there is then one crest between them. What is the velocity of the waves?

4. **a)** Explain what is meant by the principle of superposition of waves.
   b) i) State what is meant by **node** and **antinode** of a stationary wave.
   ii) State the necessary conditions for the establishment of a stationary wave.
   (*NEAB*, AS/A PHO2, Feb 1997)

5. Explain the differences between an undamped progressive transverse wave and a stationary transverse wave, in terms of
   i) amplitude,
   ii) phase, and
   iii) energy transfer.
   (*AQA: NEAB*, AS/A PHO2SP, March 1999)

6. The range of sound frequencies which can be detected by a certain person is 17.0 Hz to 20.0 kHz. What is the corresponding range of wavelengths if the speed of sound in air is 340 m s^{-1}?

7. **a)** BBC Radio 4 broadcasts on a wavelength of 1.50 km. What is its frequency?
   b) What are the wavelengths of a television station which transmits vision on 500 MHz and sound on 505 MHz?
   c) What is the frequency of yellow light of wavelength 0.6 μm?
   (Speed of electromagnetic waves in free space = 3×10^8 m s^{-1};
   1 μm (micrometre) = 10^{-6} m.)

8. The equation $y = a \sin (\omega t - kx)$ represents a plane wave travelling in a medium along the x-direction, y being the displacement at the point x at time t. Deduce whether the wave is travelling in the positive x-direction or in the negative x-direction.

   If $a = 1.0 \times 10^{-7}$ m, $\omega = 6.6 \times 10^3$ s^{-1} and $k = 20$ m^{-1}, calculate
   a) the speed of the wave, and
   b) the maximum speed of a particle of the medium due to the wave.

9. In Fig. 18.41 a pulse is shown travelling to the right along a rope fixed at the end to a wall. Which of the graphs of vertical displacement s against time t in

Fig. 18.42 represents the behaviour of point X on the rope?

Fig. 18.41

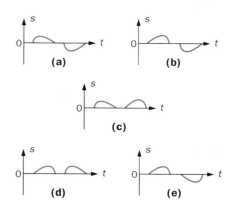

Fig. 18.42

10. A pulse of ultrasound is sent into a person. Figure 18.43 illustrates part of the trace, on a display, of the received signal at the transducer.

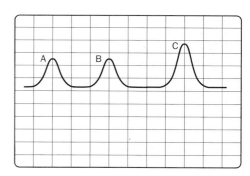

Fig. 18.43

   a) Explain the process in the body which gives rise to a peak on the trace.
   b) Peaks A and B are produced on the screen as a result of the passage of ultrasound through a section of soft tissue. Each horizontal division on the display corresponds to a time interval of 20 μs. Ultrasound travels at a speed of 1500 m s^{-1} in the soft tissue. Calculate the thickness of this section of soft tissue.
   c) i) Copy the trace, shown as a dotted line in Fig. 18.44, then sketch on it the shape of the trace if the ultrasound pulse is about four times longer than that which produced the original trace.

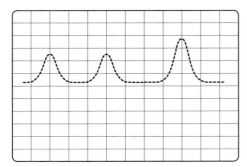

Fig. 18.44

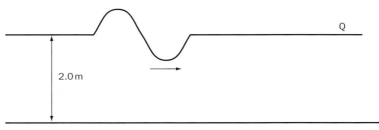

2.0 m

Q

Fig. 18.45

ii) With reference to your sketch, explain the need for short ultrasound pulses for the purposes of imaging.
(*OCR*, Health Physics 4835, June 1999)

11. A water wave of amplitude 0.50 m is travelling in water which is 2.0 m deep, as illustrated in Fig. 18.45.

Water waves travel with a speed v which is dependent on the depth of water h and is given by the equation

$$v = \sqrt{gh}$$

where g is the acceleration of free fall. As there is a greater depth of water beneath the crest of a water wave than beneath the trough, wave crests will travel faster than wave troughs.

a) Determine the depth of water beneath the crest of the wave.

b) For the wave illustrated in Fig. 18.45, calculate the speed of travel of
  i) the crest,
  ii) the trough.

c) Copy Fig. 18.45 and draw a suggested shape of the wave a little later as it passes Q.
(*OCR*, 9244/2, June 1999)

12. Figure 18.46 shows wavefronts spreading out from two identical sources, S_1 and S_2.

Describe how such a pattern could be produced and observed using a ripple tank.

Trace the diagram and draw the following:
  i) a line labelled A joining points where the waves from S_1 and S_2 have travelled equal distances,
  ii) a line labelled B joining points where the waves from S_1 have travelled one wavelength further than the waves from S_2,
  iii) a line labelled C joining points where the waves from S_2 have travelled half a wavelength further than the waves from S_1.

Copy and complete each of the sentences below by selecting an appropriate term from the following:

increase   decrease   stay the same

If only the separation of the sources were increased, the angle between lines A and B would . . .

If only the wavelength of the waves were increased, the angle between lines A and B would . . .

If only the depth of the water in the ripple tank were increased, the angle between lines A and B would . . .
(*L*, AS/A PH2, June 1999)

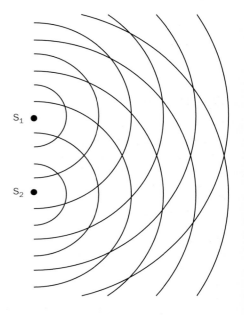

S_1 ●

S_2 ●

Fig. 18.46

19

Sound

SOUND WAVES

Sound exhibits all the properties of waves discussed in the previous chapter, except polarization. This suggests it is a longitudinal wave motion and, since it cannot travel in a vacuum but requires a material medium, it must be of the mechanical type.

A sound wave is produced by a vibrating object which superimposes, on any movement the particles of the transmitting medium have, an oscillatory to-and-fro motion along the direction of travel of the wave. The frequency of these oscillations — i.e. of the sound wave — equals that of the vibrating source and to be audible to human beings must be roughly in the range 20 Hz to 20 000 Hz. The speed of sound in air is approximately 330 m s^{-1} and the corresponding wavelength limits are therefore (using $v = f\lambda$) 17 m and 17 mm respectively.

A progressive sound wave carries energy from the source to the receiver where, in the case of the ear, the succession of compressions and rarefactions causes the ear-drum to vibrate. The sensation of what we also call 'sound' is then produced by impulses sent to the brain.

It can be informative to connect a loudspeaker to a signal generator, Fig. 19.1, varying the frequency from a very low to a very high value (e.g. 10 Hz to 30 kHz) and observing the effect on (*i*) the cone of the loudspeaker, (*ii*) the ear and (*iii*) a CRO with a microphone feeding its Y-input. At the lowest frequencies the vibrations of the cone can be seen and felt with the finger, but not heard (except for a weak thudding). As the frequency increases, the pitch of the note rises from that of a hum to a whistle and then to a hiss before it becomes inaudible. Above a few hundred hertz the movement of the speaker cone cannot be felt. The CRO responds throughout.

Sound waves which have frequencies greater than 20 kHz or so are called **ultrasonic** waves. A correctly cut quartz crystal generates ultrasonic waves when an alternating p.d. (of ultrasonic frequency) is applied across its faces. The uses of ultrasonic waves in the non-destructive testing of materials, in navigation and ranging, and in medical imaging have already been mentioned (p. 307). Ultrasonic vibrators can be used to remove grease and dirt from materials, and tartar from teeth. Bats emit and detect ultrasonic waves to navigate at night.

GRAPHICAL REPRESENTATION

In Fig. 19.2*a* the circled figures ①, ②, ③, etc., represent parts of the air in which a progressive longitudinal wave is generated if the blade P is plucked. It can be seen that as P and the air oscillate to and fro about their undisturbed positions, compressions (C) and rarefactions (R) travel to the right. The following conclusions may be drawn.

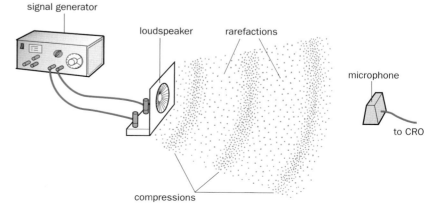

Fig. 19.1 Demonstrating sound waves

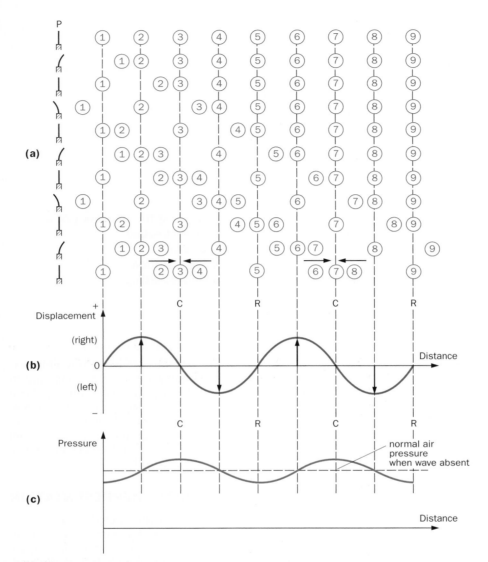

Fig. 19.2 A progressive sound wave

(a) Displacement

Air at the centres of compressions and rarefactions has zero displacement. A displacement–distance graph for the instant of time corresponding to the last line of Fig. 19.2*a* can be drawn as in Fig. 19.2*b*, where the longitudinal displacements are represented as if they were transverse and displacements to the right are taken as positive. The amplitude of vibration of the air is much exaggerated in the graph; even for a very loud sound it is only of the order of 10^{-2} mm. The velocities of the central parts are a maximum but they are directed away from the source in a compression and towards it in a rarefaction, i.e. the air moves forwards in a compression and backwards in a rarefaction. The variation with time of

the displacement of the air is thus 90° ($\pi/2$ rad) out of phase with its variation of velocity with time. Draw two graphs to show this for a sinusoidal wave.

(b) Pressure excess

The crowding together of the air in a compression causes the density and pressure of the air to be greater there than normal; both are a maximum at the centre of the compression. Conversely, density and pressure have minimum values at the centre of a rarefaction. Figure 19.2*c* shows how the pressure (or density) varies with *distance* in a progressive longitudinal wave. The variations of the graph are exaggerated: actual variations are

small, 30 Pa being typical for a very loud sound (normal atmospheric pressure is about 10^5 Pa). The ear responds to pressure changes rather than displacements and is able to detect variations as small as 2×10^{-5} Pa.

REFLECTION AND REFRACTION OF SOUND

(a) Reflection

Sound waves obey the laws of reflection. Regular reflection occurs at a surface if it treats all parts of the incident wavefront similarly. To do this it must be 'flat' to within a fraction of the wavelength of the waves falling on it. Speech sound waves have wavelengths of several metres and so it is not surprising that surfaces as 'rough' as cliffs, large buildings, etc., reflect sound regularly to give **echoes**, i.e. sound 'images'. (Light requires a highly polished surface for regular reflection. Why?)

In a hall or a large room, sound reaches a listener by many paths — some much longer than others — as a result of reflection at walls, ceilings, etc. **Reverberation** occurs when the sound produced at one instant by the source persists at the listener for some time afterwards. Too short a reverberation time makes a room sound 'dead', but if it is too long, 'confusion' results. The best value depends on the function of the building; for speech about half a second is acceptable, whereas for music between one and two seconds is required.

In a modern concert hall the reverberation time is made the same irrespective of the size and the audience by upholstering the seats with material having similar absorption to clothing. In addition, to distribute the sound as equally as possible, walls tend to be built up from many small, flat surfaces at various inclinations. Curved surfaces, which might focus the sound strongly in one place, are avoided.

(b) Refraction

If the speed of sound waves changes at a boundary between two media, refraction occurs, i.e. the direction of travel changes as does the wavelength (since the frequency is fixed by the source). Snell's law (p. 82) is obeyed and so for waves of a given frequency, having speed v_1 in medium 1 and v_2 in medium 2,

$$\frac{v_1}{v_2} = \frac{n_2}{n_1}$$

where n_1 and n_2 are the refractive indices of media 1 and 2. Distant sounds are more audible at night than during the day because the speed of sound in warm air exceeds that in cold air (see p. 320) and refraction occurs. At night the air is usually colder near the ground than it is higher up and refraction towards the earth occurs, Fig. 19.3*a*. During the day the reverse is usually true, Fig. 19.3*b*.

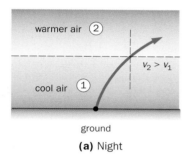

(a) Night

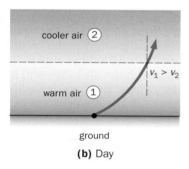

(b) Day

Fig. 19.3   Refraction of sound

INTERFERENCE AND DIFFRACTION OF SOUND

(a) Interference

Interference between sound waves may be demonstrated with two loud-speakers connected in parallel to the low-impedance output of a signal generator set at 4 kHz, Fig. 19.4. The fringes are detected by moving a microphone feeding a CRO, preferably via a pre-amplifier, along XY. (The effect of unwanted reflections is minimized by working between two shields (e.g. benches) or, better still, out of doors.)

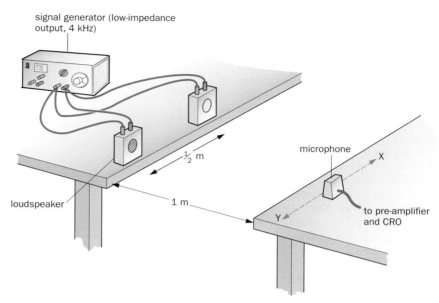

Fig. 19.4   Demonstrating interference of sound

When the microphone is at the central maximum of the interference pattern (i.e. along AB in Fig. 18.18, p. 300) and one of the loudspeakers is covered up, the output p.d. from the microphone (as shown on the CRO) falls to *half* of the value it had with both speakers uncovered. If the CRO were replaced by a resistor, the current through it would also be halved on blocking off the sound from one speaker and so the rate of energy conversion in it would be a quarter of its value with two speakers (since power = p.d. × current). Conversely, the energy at a central maximum will be *four* times that emitted in a given time by one of the speakers. We must conclude that since there is zero energy at minima, the energy from them goes to the maxima when two waves superpose to produce interference fringes. That is, energy is redistributed.

A tuning fork, Fig. 19.5*a*, may also be used to show interference of sound. If it is struck and held close to the ear with the prongs vertical while being slowly rotated about a vertical axis, at four times in each revolution no sound is heard. When the prongs PP are moving outwards, a rarefaction is produced in the air between them and travels along OY and OY' as in Fig. 19.5*b*; simultaneously, compressions are generated outside the prongs and move along OX and OX'. The compressions and rarefactions spread out as spherical waves and destructive interference occurs along AOB and COD where compressions and rarefactions superpose.

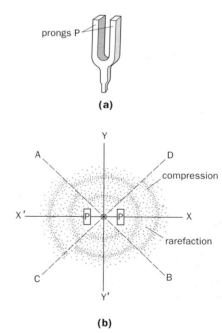

Fig. 19.5

(b) Diffraction

The spreading of sound waves round the corners of an open doorway is an everyday event and suggests that the width of such an aperture (say one metre) is of the same order as the wavelength of sound. A note of frequency 300 Hz has a wavelength in air of about one metre.

BEATS

When two notes of slightly different frequencies but similar amplitudes are sounded together, the loudness increases and decreases periodically and **beats** are said to be heard. The effect may be demonstrated with two signal generators, each feeding a loudspeaker. As an alternative, two tuning forks of the same frequency can be used, one having a small piece of Plasticine stuck on to one prong to lower its frequency slightly. They are struck simultaneously and the stems pressed against the bench top. If the difference between the frequencies of the two sources is increased, the beat frequency also rises until it is too high to be counted.

The production of beats is a wave effect explained by the principle of superposition. The displacement–time graphs for the wave-trains from two sources of nearly equal frequency are shown in Fig. 19.6a for a certain observer. At an instant such as A the waves from the sources arrive in phase and reinforce to produce a loud sound. The phase difference then increases until a compression (or rarefaction) from one source arrives at the same time as a rarefaction (or compression) from the other. The observer hears little or nothing, point B. Later the waves are in phase again (point C) and a loud note is heard. Figure 19.6b gives the resultant variation of amplitude. Beats are therefore due to interference but because the sources are not coherent (frequencies different) there is sometimes reinforcement at a given point and at other times cancellation.

We will now show that the **beat frequency equals the difference of the two almost equal frequencies**. Suppose the beat period (i.e. the time between two successive maxima) is T and that one wave-train of frequency f_1 makes one cycle more than the other of frequency f_2. Then

$$\text{number of cycles of frequency } f_1 \text{ in time } T = f_1 T$$

and

$$\text{number of cycles of frequency } f_2 \text{ in time } T = f_2 T$$

$$\therefore \quad f_1 T - f_2 T = 1$$

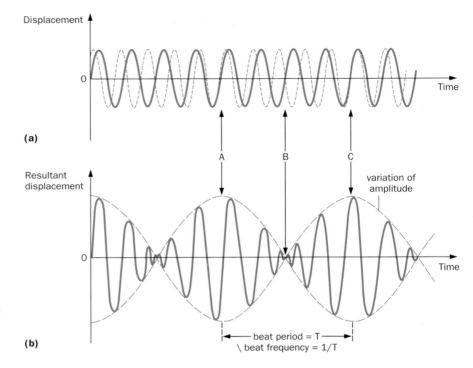

Fig. 19.6 Formation of beats

$$\therefore \quad f_1 - f_2 = \frac{1}{T}$$

But one beat has occurred in time T and so $1/T$ is the beat frequency, so

$$\text{beat frequency} = f_1 - f_2$$

STATIONARY SOUND WAVES

(a) Production

The production of stationary or standing waves by the superposition of two trains of progressive waves of equal amplitude and frequency, travelling with the same speed in opposite directions, was considered in the last chapter for transverse waves. We saw that nodes and antinodes were formed. At a node the displacement of the vibrating particle was always zero but at an antinode it varied periodically from zero to maxima in opposite directions (see Fig. 18.24b, p. 302). Also, the wavelength of the stationary wave equalled that of either progressive wave and was twice the distance between two consecutive nodes or antinodes.

Stationary sound waves can be obtained in air using apparatus called Kundt's tube, Fig. 19.7. The speaker, driven by a signal generator, produces progressive longitudinal waves which travel through the air to the end of the cylinder where they are reflected to interfere with the incident waves. The frequency of the sound is varied and resonance occurs when it equals one of the natural frequencies of vibration of the air column. Stationary waves are then formed and are revealed by the lycopodium powder which swirls away from the antinodes (where the air is vibrating strongly) and after a short time forms into heaps at the nodes.

(b) Displacement and pressure graphs

Earlier we discussed the pressure variations in a progressive sound wave (see p. 312); now we will consider how it varies in a stationary sound wave. In Fig. 19.8a, curves 1 and 2 are displacement–distance graphs for a stationary wave for times when the displacements are greatest. As before (Fig. 19.2b) displacements to the right are taken to be positive. Consider

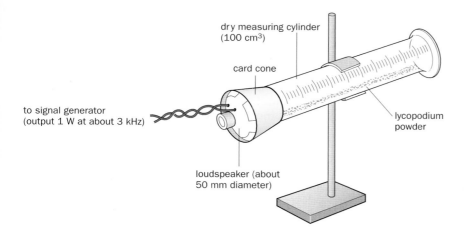

Fig. 19.7 Kundt's tube

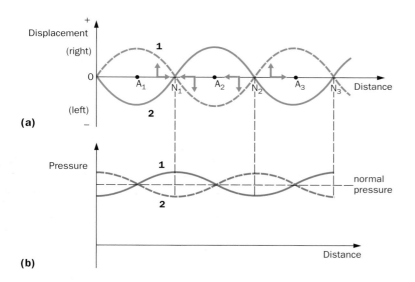

Fig. 19.8 Pressure variation for a stationary sound wave

MUSICAL NOTES

A musical note is produced by vibrations that are regular and repeating, i.e. by periodic motion. Non-periodic motion results in noise which is not pleasant to the ear.

A musical note has three characteristics.

(a) Loudness

Loudness is a subjective sensation (i.e. dependent on the listener) which is determined by the intensity of the sound and by the sensitivity of the listener's ear. **Intensity**, unlike loudness, is a physical quantity and is defined as the **rate of flow of energy through unit area perpendicular to the direction of travel of the sound** at the place in question. It is measured in watts per square metre ($W\,m^{-2}$). If P is the total energy emitted per second equally in all directions by a point source, then neglecting absorption in the surrounding medium, the intensity I at a distance r from the source is given by

$$I = \frac{P}{4\pi r^2}$$

This is an inverse square relationship and it holds for the intensity at a point due to any source of spherical waves of any type.

The energy of a particle describing simple harmonic motion (s.h.m.) is directly proportional to the square of the amplitude of vibration if other factors are held constant (p. 155). Any periodic motion which is not simple harmonic can be resolved into a number of such motions and can conversely be thought of as formed from these separate s.h.ms. Sound energy is carried by periodic vibrations of the air and so the **intensity (and loudness) of a sound must also be directly proportional to the square of the amplitude of vibration of the air** and, in turn, **of the source**. This is true for the intensity of any type of wave motion.

Amplitudes are alternately positive and negative and their average value is zero. The squares of amplitudes are always positive and their average is not zero, which further confirms their use to measure the rate at which a wave transports energy, i.e. its intensity. The demonstration described on

graph 1. Air to the left of node N_1 is displaced towards N_1 since this displacement is positive; air to the right has a negative displacement but is also displaced towards N_1. At N_1 the air is compressed and the pressure greater than normal. At node N_2 the air is again displaced in opposite directions but away from N_2. In this case the air is less dense and the pressure less than normal. On either side of a node the air is vibrating in antiphase. Approach causes pressure increase and separation makes it decrease. Parts of the air on either side of antinode A_2 that are equidistant from it have the same displacement to the left, i.e. they vibrate in phase, and the air pressure is nor-

mal. The pressure–distance graph corresponding to displacement–distance graph 1 is therefore given by curve 1 in Fig. 19.8b. By treating displacement graph 2 in the same way we obtain pressure–distance graph 2.

From Fig. 19.8b we can conclude that in a stationary longitudinal wave the pressure *variation* is a maximum at a displacement node and is always zero at a displacement antinode. It is sometimes stated that displacement nodes and 'pressure antinodes' coincide, as do displacement antinodes and 'pressure nodes'. Adjacent node and antinode are a quarter of a wavelength apart.

p. 313 using two loudspeakers, a microphone and a CRO to show interference with sound waves offers some experimental evidence for this point. An analogous situation occurs with alternating currents, where the rate of delivery of energy is proportional to the square of the peak current.

(b) Pitch

This is also a sensation experienced by a listener. It depends mainly on the frequency of vibration of the air, which of course is the same as that of the source of sound. Frequency is a physical quantity and is measured in hertz (Hz), one vibration per second being one hertz. A high frequency produces a high-pitched note and a low frequency gives a low-pitched note. A note of frequency 500 Hz sounded in front of a microphone connected to a CRO gives twice as many waves (for the same time-base setting) as one of frequency 250 Hz. The notes are said to differ in pitch by one **octave**. Pitch of sound is analogous to colour of light.

(c) Quality or timbre

The same note played on two different instruments does not sound the same and different waveforms are obtained on a CRO. The notes are said to have different **quality** or **timbre**. This is due to the fact that with one exception (a tuning fork), sounds are never 'pure notes', i.e. of one frequency; their waveforms are not sinusoidal. They consist of a main or **fundamental** note

— which usually predominates — and others, generally with smaller intensities, called **overtones**. The fundamental is the component of lowest frequency and the overtones have frequencies that are multiples of the fundamental frequency. **The number and intensities of the overtones determine the quality of a sound** and depend mostly on the instrument producing the sound. The fundamental is also called the **first harmonic** and if it has frequency f, the overtones with frequencies $2f$, $3f$, etc., are the second, third, etc., harmonics. With some instruments certain harmonics are not present as overtones in a note (p. 317).

Waveforms for the same note of fundamental frequency 440 Hz played on a violin and a piano are shown in Fig. 19.9a. They are very different and provide visual evidence that the notes are of different 'quality'. Using a mathematical technique known as Fourier analysis, the frequencies and amplitudes of the harmonics present as overtones can be worked out in each case and sound spectra obtained, Fig. 19.9b. Loud higher harmonics (notably the second and fifth) are present in the violin spectrum. The analysis depends on the fact that a periodic waveform, however complex, can be resolved into a number of sine waveforms with frequencies that are multiples of the fundamental. For example, a waveform like that in Fig. 18.15 (p. 299) has two components — the fundamental **1** and an overtone **2** (the third harmonic) with three times the frequency and half the amplitude of the fundamental.

VIBRATING AIR COLUMNS

(a) Wind instruments

Stationary longitudinal waves in a column of air in a pipe are the source of sound in wind instruments. To set the air into vibration a disturbance is created at one end of the pipe; the opposite end can be open or closed (stopped). In the pipe of an organ, Fig. 19.10, a jet of air strikes a sharp edge; in a recorder the player blows across a hole, while in an oboe a reed vibrates when blown. Progressive sound waves travel from the source to the end of the pipe where they are reflected and interfere with the incident waves. The wavelength of some waves will be such that the length of the pipe produces resonance, i.e. a compression in the reflected wave arrives back at the source just as it is producing another compression. The amplitude of these waves will be large and a stationary longitudinal wave is formed.

The possible modes of vibration of the air in closed and open pipes will now be considered.

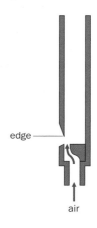

Fig. 19.10 An organ pipe

(b) Closed pipe

In the simplest stationary wave vibration possible, there will be a displacement node N at the closed end of the pipe, since the air there must be at rest, and a displacement antinode A at the open end where the air can vibrate freely, Fig. 19.11a. (Where will the pressure node and antinode be formed?) Figure 19.11b shows the displacement–distance graph for this vibration and from Fig. 19.11c we see that the length l of the pipe equals the

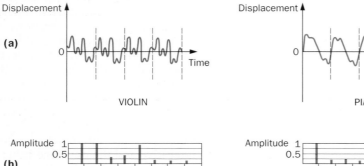

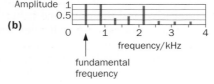

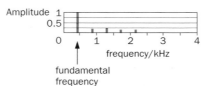

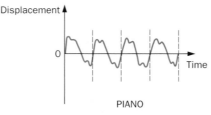

Fig. 19.9 Harmonics in musical notes

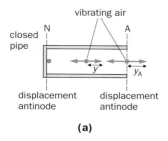

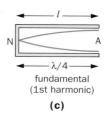

Fig. 19.11 Fundamental vibration in a closed pipe

distance between a node and the next antinode, which is $\lambda/4$ (λ being the wavelength of the stationary wave):

$$l = \frac{\lambda}{4}$$

$$\therefore \quad \lambda = 4l$$

The frequency f of the note is given by $v = f\lambda$ where v is the speed of sound in air. Therefore

$$f = \frac{v}{4l}$$

This is the lowest frequency produced by the pipe, i.e. the **fundamental frequency** f or **first harmonic**.

The next two simplest modes by which the air may vibrate in the same pipe are shown in Figs 19.12*a* and *b*. In both cases there are displacement nodes at the closed ends and displacement antinodes at the open ends. In (*a*), if the wavelength is λ_1, we have

$$l = \frac{3\lambda_1}{4}$$

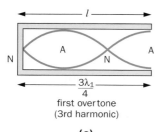

(a)

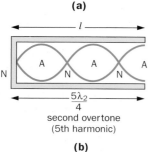

(b)

Fig. 19.12 Overtones in a closed pipe

The frequency f_1 is therefore given by

$$f_1 = \frac{3v}{4l}$$

This is the **first overtone** and since $f_1 = 3f$ it is the *third* harmonic. Similarly from (*b*) we find that the frequency f_2 of the second overtone is $5f$, i.e. it is the fifth harmonic, and in general **a closed pipe gives only odd-numbered harmonics**.

(c) Open pipe

Here, both ends of the pipe are open and both are displacement antinodes A. The simplest stationary wave is shown in Fig. 19.13*a*. There is a displacement node N midway between the two antinodes, so

$$l = \frac{\lambda}{2}$$

The fundamental frequency f is given by $f = v/\lambda$, i.e.

$$f = \frac{v}{2l}$$

The next simplest mode of vibration, Fig. 19.13*b*, gives the first overtone and we have

$$l = \lambda_1$$

$$f_1 = \frac{v}{l}$$

This is the second harmonic since $f_1 = 2f$. The second overtone is obtained from Fig. 19.13*c* and is the third harmonic. In an open pipe all harmonics are possible as overtones.

(d) Further points

(*i*) The fundamental frequency of an open pipe is *twice* that of a closed pipe of the same length.

(*ii*) The note from an open pipe is richer than that from a closed pipe owing to the presence of extra overtones.

(*iii*) Higher overtones are encouraged by blowing harder.

(*iv*) The actual vibration of the air in a pipe is the resultant, by superposition, of the various modes that occur.

(*v*) In practice the air just outside the open end of a pipe is set into vibration and the displacement antinode of a stationary wave occurs a distance c — called the **end correction** — beyond the end. The effective length of the air column is therefore greater than the length of the pipe. For an expression of the fundamental of a closed pipe, Fig. 19.14*a*, we should write $\lambda/4 = l + c$ and for an open pipe, Fig. 19.14*b*, we have $\lambda/2 = l + 2c$. The value of c is generally taken to be $0.6r$ where r is the radius of the pipe.

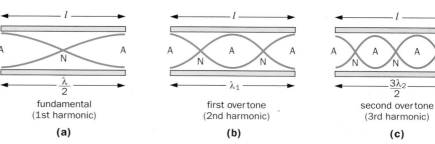

Fig. 19.13 Vibrations in an open pipe

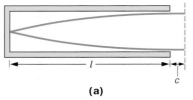

Fig. 19.14 The 'end correction'

Fig. 19.15 Organ pipes at the Westerkerk, Amsterdam

The impressive array of organ pipes at the Westerkerk, Amsterdam, is shown in Fig. 19.15.

VIBRATING STRINGS

(a) String instruments

The 'string' is a tightly stretched wire or length of gut. When it is struck, bowed or plucked, progressive transverse waves travel to both ends, which are fixed, where they are reflected to meet the incident waves. A stationary wave pattern is formed for waves whose wavelengths 'fit' into the length of the string, i.e. resonance occurs. A progressive sound wave (i.e. a longitudinal wave) is produced in the surrounding air with a frequency equal to that of the stationary transverse wave on the string.

The weak sound produced by the vibrating strings is transmitted to a sounding board in a piano or to the hollow body of instruments such as the violin and the guitar. A larger mass of air is thereby set vibrating and the loudness of the sound increased.

(b) Modes of vibration

The fixed ends must be displacement nodes N. If the string is plucked in the middle the simplest mode of vibration occurs, as in Fig. 19.16a, A being a displacement antinode. This vibration creates the fundamental note of frequency f, and

$$l = \frac{\lambda}{2}$$

where l is the length of the string. The frequency $f = v/\lambda$ is therefore

$$f = \frac{v}{2l}$$

where v is the speed of the transverse wave along the string.

By plucking the string at a point a quarter of its length from one end, it can vibrate in two segments, Fig. 19.16b. Then, if λ_1 is the wavelength of the resulting stationary wave,

$$l = \lambda_1$$

$$\therefore \quad f_1 = \frac{v}{\lambda_1} = \frac{v}{l} = 2f$$

This is the first overtone or second harmonic. If the string vibrates in three segments, Fig. 19.16c, the second overtone or third harmonic is obtained.

A string can vibrate in several modes simultaneously, depending on where it is plucked, etc., and the frequencies and relative intensities of the overtones produced determine the quality of the note emitted.

Melde's experiment (p. 302) shows the various stationary wave patterns of a vibrating string (or rubber cord). It will be seen that, rather unexpectedly, the driven point of the string is nearly a node; certainly the amplitude of vibration is much less than at an antinode. The motion there is no more than is required for the vibrator to make good the energy dissipated in the system. Also, the frequency of the wave is the same as that of the vibrator.

(c) Laws of vibration of stretched strings

It was stated earlier (p. 305) that the speed v of a transverse wave travelling along a stretched string is given by

$$v = \sqrt{\frac{T}{\mu}}$$

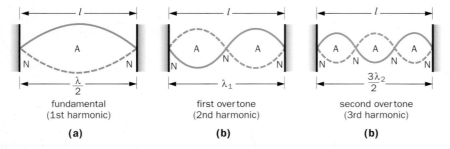

fundamental
(1st harmonic)

(a)

first overtone
(2nd harmonic)

(b)

second overtone
(3rd harmonic)

(b)

Fig. 19.16 Vibrations on a string

where T is the tension in the string and μ is its mass per unit length. If the string has length l, the frequency f of the fundamental note emitted by it has just been shown to be

$$f = \frac{v}{2l}$$

Therefore

$$f = \frac{1}{2l}\sqrt{\frac{T}{\mu}}$$

It follows that:

- $f \propto 1/l$ if T and μ are constant;
- $f \propto \sqrt{T}$ if l and μ are constant;
- $f \propto 1/\sqrt{\mu}$ if l and T are constant.

These three statements are known as the **laws of vibration of stretched strings**. They indicate that short, thin wires under high tension emit high notes. The laws may be verified experimentally using a **sonometer**. This consists of a wire under tension, arranged on a hollow wooden box as in Fig. 19.17. The vibrations of the wire are passed by the movable bridges to the box and then to the air inside it.

In order to investigate the relationship between f and l, the bridges are moved so that different lengths between them emit their fundamental frequencies when plucked at the centre. The frequencies may be found by one or other of the methods outlined on p. 324 or by taking lengths of wire which vibrate in unison (as judged by the ear) with tuning forks of known frequencies. The product $f \times l$ should be constant within the limits of experimental error.

To check that $f \propto \sqrt{T}$, the same length of wire is subjected to different tensions by changing the hanging mass.

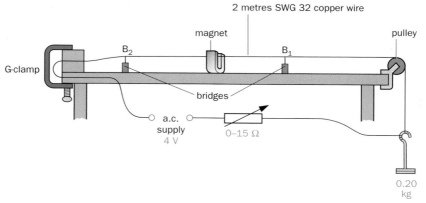

Fig. 19.18 Measuring the frequency of a.c.

The relation between f and $1/\sqrt{\mu}$ requires the use of wires of different diameters and materials but of the same vibrating length and under the same tension.

(d) Measuring the frequency of an a.c. supply

The wire carrying the a.c., Fig. 19.18, experiences an alternating transverse force when the magnet is placed so that the wire passes between its poles. Bridge B_1 is arranged near B_2 at the fixed end of the wire and then slowly separated from it, keeping the magnet approximately midway between them, until the wire vibrates in one loop with maximum amplitude, the current being reduced to a low value. The frequency f of the a.c is given by

$$f = \frac{1}{2l}\sqrt{\frac{T}{\mu}}$$

where l = distance B_1B_2, T = (0.20 kg) $\times$ (10 N kg^{-1}) = 2.0 N and μ is found by weighing a two-metre length of the SWG 32 copper wire. If T is kept constant and the magnet left in position, doubling the distance between the bridges should produce two loops and a second value of f may be calculated (in this case $l = B_1B_2/2$).

SPEED OF SOUND

It was shown on pp. 304–5 that the speed of a mechanical wave in a medium depends upon 'elasticity' and 'mass' factors and formulae for the speeds of different waves were quoted. An expression for the speed of sound in *any* medium was derived by Newton and may be written

speed =

$$\sqrt{\left(\frac{\text{elastic modulus of medium}}{\text{density of medium}}\right)}$$

The elastic modulus (i.e. stress/strain) concerned depends on the type of strain produced by the wave in the medium.

(a) Solids

In a solid rod of small diameter compared with the sound wavelength, the compressions and rarefactions of the sound wave cause changes of length and the Young modulus E is appropriate in the above expression. So for a solid rod of density ρ in which the speed of sound is v, we have

$$v = \sqrt{\frac{E}{\rho}}$$

When E is in N m^{-2} (i.e. Pa) and ρ in kg m^{-3}, v is in m s^{-1}. For steel $E = 2.0 \times 10^{11}$ Pa and $\rho = 7.8 \times 10^3$ kg m^{-3} giving $v = 5.1 \times 10^3$ m s^{-1}.

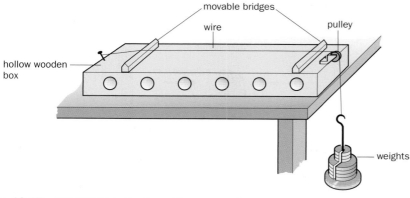

Fig. 19.17 Investigating vibration with a sonometer

(b) Gases

Volume changes occur in the case of gases (and liquids) and the bulk modulus K of the gas is the relevant elastic modulus. Under the conditions in which a sound wave travels in a gas, $K = \gamma p$ where p is the pressure of the gas and γ is the ratio of its principal heat capacities, i.e. C_p/C_V (see p. 375). We then have for the speed of sound v in a gas of density ρ

$$v = \sqrt{\left(\frac{\gamma p}{\rho}\right)}$$

For air at s.t.p., $\rho = 1.3 \text{ kg m}^{-3}$, $\gamma = 1.4$ and $p = 1.0 \times 10^5 \text{ Pa}$, so $v = 3.3 \times 10^2 \text{ m s}^{-1}$, a value which agrees well with the experimental result.

(c) Effect of pressure, temperature and humidity

For one mole of an ideal gas having volume V and pressure p at temperature T, we can write (p. 365)

$$pV = RT$$

where R is a constant. If M is the molar mass of the gas

$$\rho = \frac{M}{V} = \frac{Mp}{RT}$$

From above $v = \sqrt{\frac{\gamma p}{\rho}}$

Therefore

$$v = \sqrt{\left(\frac{\gamma RT}{M}\right)}$$

The final expression does not contain p and so **the speed of sound in a gas is independent of pressure**. This is not unexpected because changes in density are proportional to changes of pressure, i.e. p/ρ is a constant.

Also, since R has the same value for all gases and γ and M are constants for a particular gas

$$v \propto \sqrt{T}$$

This is, **the speed of sound in a gas is directly proportional to the square root of the absolute temperature of the gas** (provided γ is independent of temperature).

For moist air ρ is less than that for dry air and γ is slightly greater. The net result is that **the speed of sound increases with humidity**.

(d) Effect of wind

If the air carrying sound waves is itself moving, i.e. there is a wind, the wind velocity has to be added vectorially to that of the sound to obtain the velocity of the latter with respect to ground. Since wind velocities usually increase with height, wavefronts become distorted as in Fig. 19.19, resulting in greater audibility downwind than upwind.

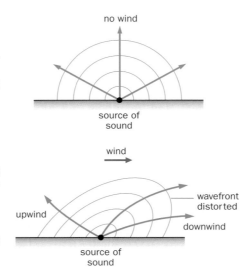

Fig. 19.19 The effect of wind on sound

MEASURING THE SPEED OF SOUND

(a) In air by progressive waves using a CRO

The CRO must have an output terminal from its time base and the latter must be calibrated. The time base will run at the speed indicated by the time/div control. This is adjusted to 100 ms/div. The volts/div knob is set at a Y-amplifier gain of 0.1 V. The CRO is then switched on and the brightness control adjusted. The X-shift needs to be adjusted so that the start of the trace at the left of the screen is visible.

The sweep output and earth (E) terminals on the CRO are connected to the input of a loudspeaker. The method uses the fact that as the time base initiates the trace at the left of the screen, a pulse comes from the sweep output and produces a noise in the loudspeaker. That this is so should be checked.

To make a measurement the time-base speed is increased to 1 ms/div and a microphone is connected to the input and E terminals on the CRO via an amplifier. The signal received by the microphone from the loudspeaker causes a 'wavy' trace on the CRO. (The volume control on the amplifier may have to be turned up and the position of the trace adjusted by the X-shift.) When the microphone is moved away from the speaker the wavy trace moves to the right on the screen. The distance that the microphone has to be moved to cause the trace to move one screen division to the right is found. The sound must have travelled that distance in one millisecond, and so its speed in air can be calculated.

(b) In air by stationary waves

The apparatus is arranged as shown in Fig. 19.20. The signal generator delivers a note of known frequency f (2 kHz is suitable) to a loudspeaker which is directed towards a metal sheet. Interference occurs between

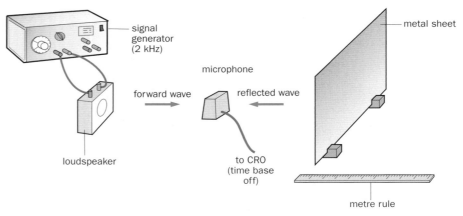

Fig. 19.20 Using a CRO to measure the speed of sound by stationary waves

the forward and reflected progressive sound waves and a stationary wave pattern with nodes and antinodes is established. If the reflector is moved slowly towards or away from the microphone the vertical trace on the CRO varies from a maximum to a minimum. The distance moved by the reflector between two consecutive maxima or minima equals $\lambda/2$ where λ is the wavelength of the sound wave (progressive or stationary). The speed v is obtained from $v = f\lambda$. In practice the reflector is moved through several maxima and minima so that a greater distance is measured.

The stationary wave pattern produced in Kundt's tube (p. 315) — indicated by the lycopodium powder — may also be used to find the speed of sound in air or in different gases.

(c) In air using a resonance tube

The arrangement is illustrated in Fig. 19.21a. The tuning fork is struck and held over the top of the tube whose position in the water is raised or lowered until the note is at its loudest. The fundamental frequency of the air column then equals the frequency of the fork, i.e. there is resonance. A stationary wave now exists in the tube, with a displacement node N at the closed end and a displacement antinode A near the open end. We then have

$$l_1 + c = \frac{\lambda}{4} \quad (1)$$

where l_1 is the distance from the water level to the top of the tube and c is the end correction.

A second, weaker resonance can be obtained with the same fork by slowly raising the tube out of the water. The air column of greater length l_2 is then producing its first overtone (which equals the frequency of the fork) and from the stationary wave pattern of Fig. 19.21b we have

$$l_2 + c = \frac{3\lambda}{4} \quad (2)$$

Subtracting (1) from (2) eliminates c and we get

$$l_2 - l_1 = \frac{\lambda}{2}$$

Knowing the frequency f of the tuning fork, v can be calculated from $v = f\lambda$.

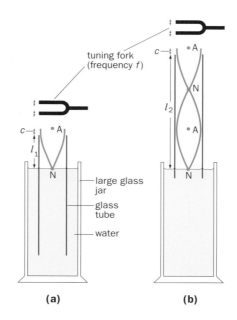

(a)   **(b)**

Fig. 19.21 Measuring the speed of sound using a resonance tube

Alternatively, using several forks of different known frequencies f, the fundamental resonance lengths l can be found for each and we have

$$l + c = \frac{\lambda}{4} = \frac{v}{4f}$$

$$\therefore \quad l = \frac{v}{4f} - c$$

A graph of l against $1/f$ should be a straight line of slope $v/4$ (and intercept $-c$ on the $1/f$ axis).

(d) In a metal rod

This method provides an approximate value and allows a rough check to be made on the expression $v = \sqrt{(E/\rho)}$ (p. 319) for, say, steel. In Fig. 19.22 the output from the signal generator is applied to the input of the CRO when the near end of the suspended rod is hit with the hammer. A compression pulse travels to the far end and is reflected as an expansion (rarefaction) pulse to the near end where *it breaks the contact between the hammer and the rod*. The time of contact is calculated from the length of the trace on the CRO (using its calibrated time scale of $100\ \mu s\ cm^{-1}$) and is the time for sound to travel *twice* the length of the rod. (*Note.* The stability control on the CRO should be turned only as far anticlockwise as is needed to give a trace when contact is made.)

A demonstration in slow motion with the apparatus of Fig. 19.23, p. 322, may help to reinforce the idea that the reflected expansion pulse breaks the contact between the hammer and the rod in Fig. 19.22. The four trolleys linked by springs are *all pushed together* to the left. The front trolley stops on striking the 'wall' and a compression pulse travels along the row as each trolley stops in turn. The rear trolley then moves to the right, starting an expansion pulse which travels left to the front trolley; on reaching it, contact with the wall is broken. The disturbance thus travels twice along the row of trolleys during

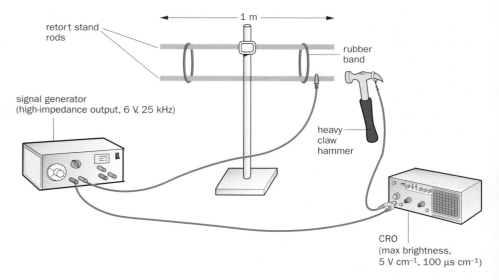

Fig. 19.22 Measuring the speed of sound in a metal rod

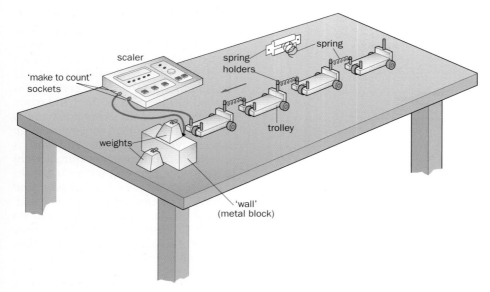

Fig. 19.23 A slow-motion model

the time the front trolley is in contact with the wall. The scaler records the time of contact and if the speed of the pulses is required the distance from the front of the first trolley to the front of the rear one has to be measured.

DOPPLER EFFECT

The pitch of the note from the siren of a fast-travelling ambulance or police car appears to a stationary observer to drop suddenly as it passes. This *apparent* change in the frequency of a wave motion when there is relative motion between the source and the observer is called the **Doppler effect**. It occurs with electromagnetic waves as well as with sound. The expressions derived below apply to sound, and to electromagnetic waves — although only if the relative speed of source and observer is small compared with the speed of light (otherwise relativistic effects have to be considered). The microwave Doppler effect is used in police radar speed checks and in tracking satellites.

(a) Source moving

In Fig. 19.24*a* S is the source of waves of frequency f and velocity v, and O is the stationary observer. If S were at rest, the f waves emitted per second would occupy a distance v and the wavelength would be v/f. When S is moving towards O with velocity u_s, f

waves are now compressed into the smaller distance $(v - u_s)$ because S moves a distance u_s towards O per second, Fig. 19.24*b*. To O the effect therefore appears to be a *decrease of wavelength* to a value λ_o given by

$$\lambda_o = \frac{(v - u_s)}{f}$$

So if f_o is the apparent frequency we have

$$f_o = \frac{\text{velocity of waves}}{\text{apparent wavelength}}$$
$$= \frac{v}{(v - u_s)/f}$$

Therefore

$$f_o = \left(\frac{v}{v - u_s}\right)f$$

The apparent frequency is therefore greater than the true frequency since $(v - u_s) < v$.

If S is moving away from the stationary observer O, the apparent wavelength $\lambda_o = (v + u_s)/f$, Fig. 19.24*c*, and the apparent frequency f_o is

$$f_o = \left(\frac{v}{v + u_s}\right)f$$

In this case $f_o < f$ since $(v + u_s) > v$.

Note that if, for example, a source of sound is approaching an observer with *constant velocity*, the apparent pitch of the note heard does not increase as the source gets nearer; it is constant but higher than the true pitch. Similarly, if the source recedes with constant velocity the observer hears a note of constant but lower pitch. The apparent change heard by the observer occurs *suddenly as the source passes*. What expression gives the value of the apparent change?

(b) Observer moving

In this case the wavelength is unchanged and is given by v/f since the f waves sent out per second by the stationary source S occupy a distance v, Fig. 19.25. If the observer O has

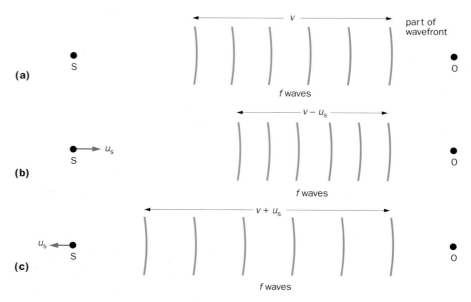

Fig. 19.24 Doppler effect: moving source

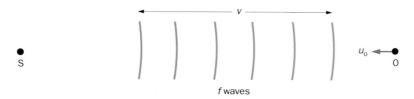

Fig. 19.25 Doppler effect: moving observer

velocity u_o towards S, the velocity of the waves relative to O is $(v + u_o)$. The apparent frequency f_o is given by

$$f_o = \frac{\text{relative velocity of waves}}{\text{wavelength}}$$

$$= \frac{v + u_o}{v/f}$$

Therefore

$$f_o = \left(\frac{v + u_o}{v}\right)f$$

In this case $f_o > f$ since $(v + u_o) > v$.

If O is moving away from the stationary source S, the velocity of the waves relative to O is $(v - u_o)$ and the apparent frequency f_o is

$$f_o = \left(\frac{v - u_o}{v}\right)f$$

Here $f_o < f$.

(c) Source and observer moving

Motion of the source affects the *apparent wavelength* and the motion of the observer affects the *velocity of the waves* received. If the source and observer are approaching each other with velocities u_s and u_o respectively then, as before, we can say

$$\frac{\text{velocity of waves}}{\text{relative to O}} = v_o = v + u_o$$

apparent wavelength $\lambda_o = (v - u_s)/f$

So for the apparent frequency f_o we have

$$f_o = \frac{v_o}{\lambda_o} = \frac{v + u_o}{(v - u_s)/f}$$

$$\therefore \quad f_o = \left(\frac{v + u_o}{v - u_s}\right)f$$

When source and observer are moving away from one another

$$\lambda_o = \frac{v + u_s}{f} \quad \text{and} \quad v_o = v - u_o$$

Therefore

$$f_o = \frac{v_o}{\lambda_o} = \frac{v - u_o}{(v + u_s)/f}$$

$$\therefore \quad f_o = \left(\frac{v - u_o}{v + u_s}\right)f$$

In general

$$f_o = \left(\frac{v \pm u_o}{v \mp u_s}\right)f$$

where the upper signs correspond to approach and the lower signs to separation.

In the context of medical ultrasound scanning where waves are reflected from an object moving with velocity u, such as blood flowing in a vein, the equation becomes

$$f_o = \left(\frac{v \pm u}{v \mp u}\right)f$$

For $u \ll v$, the change in observed frequency

$$\Delta f = (f_o - f) \approx \pm 2\frac{uf}{v}$$

so

$$u = \pm \frac{v \Delta f}{2f}$$

This equation also holds for the reflection of radar pulses in police radar speed checks and in radar astronomy where the relative velocity u between source and receiver is much less than the speed of light.

(d) Doppler effect in light

Because of the very great speed of light the effect is negligible for most terrestrial sources but it can be used to measure the speeds of stars relative to the earth. The positions of certain wavelengths (i.e. spectral lines) due to an identifiable element in the star's spectrum are compared with their positions in a spectrum of the element produced in a laboratory. **Red shift** indicates recession of the star from the earth and from its size the speed may be calculated. For electromagnetic waves travelling at speed c, if u_o is interpreted as the relative velocity v between source and observer and $v \ll c$, then

$$f_o = f\left(1 \pm \frac{v}{c}\right)$$

and

$$(f_o - f) = \pm \frac{fv}{c}$$

The magnitude of the frequency shift

$$\Delta f = \frac{fv}{c}$$

or, in terms of the shift in wavelength $\Delta\lambda$,

$$\Delta\lambda = \frac{\lambda v}{c}$$

The recession speed is then

$$v = \frac{c \Delta f}{f} = \frac{c \Delta\lambda}{\lambda}$$

Measurements of Doppler-effect red shift on the light from galaxies (each comprising many millions of stars) provide evidence that the universe is expanding, with each galaxy retreating from every other at speeds of up to one-third that of light.

The speed of rotation of the sun has also been found from the apparent difference in wavelength between the Fraunhöfer lines (p. 344) seen in the spectra of the western and eastern edges of its disc (which are respectively moving towards the earth and receding from it). The value obtained agrees with that deduced from observations of sunspots.

One cause of the broadening of spectral lines is due to the Doppler effect. Molecules of a gas or vapour which is emitting light are moving at different angles towards and away from the observer with various (very high) speeds. As a result the wavelength of a particular spectral line has a range of apparent values. How will the molecules that are responsible for the edges of the line be moving?

SOUND CALCULATIONS

Example 1. What are the first two successive resonance lengths of a closed pipe containing air at 27 °C for a tuning fork of frequency 341 Hz? Take the speed of sound in air at 0 °C to be 330 m s^{-1}.

The speed of sound in a gas is directly proportional to the square root of the absolute temperature,

$$\therefore \quad \frac{v_1}{v_0} = \sqrt{\frac{(273 + 27)\text{K}}{273 \text{ K}}} = \sqrt{\frac{300}{273}}$$

v_1 and v_0 being the speeds of sound at 27 °C and 0 °C respectively.

$$\therefore \quad v_1 = 330 \sqrt{\frac{300}{273}} = 346 \text{ m s}^{-1}$$

For the first resonance length l_1, the closed pipe emits its fundamental frequency f_1 where $f_1 = 341$ Hz. The wavelength of the fundamental note is given by

$$\lambda_1 = \frac{v_1}{f_1} = \frac{346 \text{ m s}^{-1}}{341 \text{ s}^{-1}} = 1.02 \text{ m}$$

The stationary wave pattern in the tube is then as in Fig. 19.26a, so

$$l_1 = \frac{\lambda_1}{4} = \frac{1.02 \text{ m}}{4} = 0.255 \text{ m}$$

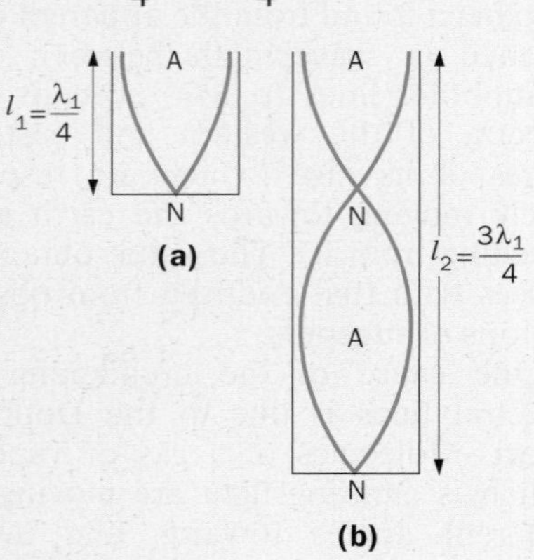

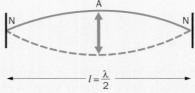

(a)

(b)

Fig. 19.26

At the second resonance length l_2 the stationary wave is as shown in Fig. 19.26b. The resonating note has frequency f_1 and wavelength λ_1 as before, but it is now the first overtone for length l_2. Therefore

$$l_2 = \tfrac{3}{4}\lambda_1 = \frac{3 \times 1.02 \text{ m}}{4} = 0.765 \text{ m}$$

Example 2. A sonometer wire, length 0.50 m and mass per unit length 1.0×10^{-3} kg m^{-1}, is stretched by a load of 4.0 kg. If it is plucked at its mid-point, what will be (a) the wavelength and (b) the frequency, of the note emitted? Take $g = 10$ N kg^{-1}.

(a) The wire vibrates as shown in Fig. 19.27 and emits its fundamental frequency f of wavelength λ. If l is the length of the wire then

$$l = \frac{\lambda}{2}$$

$$\therefore \quad \lambda = 2l = 2 \times 0.50 \text{ m} = 1.0 \text{ m}$$

Fig. 19.27

(b) The fundamental frequency f emitted by a wire of length l, mass per unit length μ and under tension T is given by

$$f = \frac{1}{2l}\sqrt{\frac{T}{\mu}}$$

$T = 4.0$ kg $\times 10$ N kg^{-1} = 40 N, so

$$f = \frac{1}{2 \times 0.50}\sqrt{\left(\frac{40}{1.0 \times 10^{-3}}\right)} \text{ Hz}$$

$$= 2.0 \times 10^2 \text{ Hz}$$

Example 3. Explain why beats are heard by a stationary observer when a source of sound of frequency 100 Hz moves directly away from her with a speed of 10.0 m s^{-1} towards a vertical wall, and calculate the beat frequency. (Speed of sound in air = 340 m s^{-1}.)

The observer hears beats because of interference between the sound coming to her directly from the source and that reflected from the wall. The apparent frequency of the direct sound is less than 100 Hz since the source is moving away from her, while the apparent frequency of the reflected sound is greater than 100 Hz. The motion of the source towards the wall causes the waves incident on it to have a shorter wavelength than they would if the source were at rest and so, to the observer, the reflected waves appear to come from a source moving towards her.

Apparent frequency of sound direct from source

$$= \left(\frac{340}{340 + 10}\right) 100 \text{ Hz}$$

$$= 97.1 \text{ Hz}$$

Apparent frequency of reflected sound

$$= \left(\frac{340}{340 - 10}\right) 100 \text{ Hz}$$

$$= 103 \text{ Hz}$$

Therefore number of beats per second

$$= 5.90$$

MEASURING THE FREQUENCY OF A SOURCE OF SOUND

Three methods will be outlined using instruments that are assumed to be correctly calibrated.

(a) Using a signal generator and a loudspeaker

The loudspeaker is fed from a signal generator whose frequency is altered until the notes from the speaker and the source of sound are judged to have the same pitch.

(b) Using a microphone and a CRO

The sound is received by the microphone and fed to the CRO (if need be via a pre-amplifier) which is set on a suitable, known time-base range. By noting the number of cycles of a.c. on a certain length of the time scale, the frequency can be worked out. A data-logger and computer can be used instead of the CRO.

(c) Using a stroboscope

A stroboscope makes an object that is vibrating appear to be at rest. A lamp stroboscope, Fig. 19.28, consists of a lamp containing xenon gas, which emits brief but intense flashes of light from about 2 to 250 times a second according to the setting of the speed control. If, for example, 200 is the *highest* flashing speed to make a sonometer wire appear to be at rest and in one position, then 200 Hz is the frequency of the stationary wave on the wire and also of the progressive sound wave emitted by it into the surrounding air. What would be seen if the flashing speed were (*i*) 100 Hz, and (*ii*) 400 Hz?

Stroboscopes should not be used in the presence of anyone susceptible to epileptic fits.

(a) Lamp

(b) Speed setting

Fig. 19.28   Xenon stroboscope

THE DECIBEL

To the human ear the change in loudness when the power of a sound increases from 0.1 W to 1 W is the same as when it increases from 1 W to 10 W. The ear responds to the *ratio* of the powers and not to their difference. Taking logs (to base 10) of the ratios above:

$$\log_{10}\left(\frac{1}{0.1}\right) = \log_{10} 10 = 1$$

and

$$\log_{10}\left(\frac{10}{1}\right) = \log_{10} 10 = 1$$

The log (to base 10) of the ratio of the powers is the same for each change and this accounts for the way the unit of the **change of power**, called the **bel** (B), is defined. If a power changes from P_1 to P_2, then we define:

$$\text{number of bels change} = \log_{10}\frac{P_2}{P_1}$$

In practice the bel is too large and the **decibel** (dB) is used. It is one-tenth of a bel, so

number of decibels change
$$= 10 \log_{10}\frac{P_2}{P_1}$$

For example, if the power output from an amplifier increases from 100 mW to 200 mW,

gain, in dB $= 10 \log_{10}(200/100)$

$= 10 \log_{10} 2$

$= 10 \times 0.30 = 3$ dB

Figure 19.29 shows how a very wide range of power ratios is reduced to the more compact range of dBs. Note that a power ratio of 10 (whatever the two powers involved) is a 10 dB change, while a power ratio of 100 is a 20 dB change.

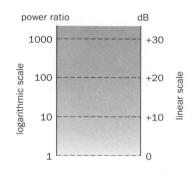

Fig. 19.29   The decibel scale

Various changes of sound level are given in Table 19.1, taking as zero dB the threshold of hearing, i.e. the sound the average human ear can just detect. Measurements are made through use of a sound-level meter.

Table 19.1

Sound	dB level
Threshold of hearing	0
Whispering	30
Normal conversation	60
Busy street	70
Noisy factory	90
Jet plane overhead	100
Loud thunderclap	110
Threshold of pain	120

NOISE POLLUTION

Unwanted sounds are called noises. High-pitched noises are usually more irritating than low-pitched ones. Noise can damage the ears, cause tiredness and loss of concentration and, if it is very loud, result in sickness and temporary deafness. Some of the main noise 'polluters' are aircraft, motor vehicles, very loud music and many types of machinery including domestic appliances.

Ways of reducing unwanted noise include designing quieter engines and better exhaust systems. For example, rotating shafts in machinery can be balanced better so that they do not cause vibration. Car engines are often mounted on metal brackets via rubber blocks which absorb vibrations and do not pass them on to the car body.

The use in the home of sound-insulating materials such as carpets, curtains and double-glazed windows also helps. The farther away the noise originates the weaker it is, so distance is a natural barrier; another example of a natural barrier is the planting of trees between houses and a noisy road. Tractor drivers, factory workers, pneumatic drill operators and others exposed regularly to loud noise should wear ear protectors.

SOUND RECORDING

(a) Magnetic tape

Sound can be stored by magnetizing a coating on plastic tape, consisting of finely powdered iron oxide or chromium oxide. The **recording head** is an electromagnet having a tiny gap filled with a strip (a 'shim') of non-magnetic material between its poles, Fig. 19.30. When a.c. from the recording microphone passes through the windings, it causes an alternating magnetic field in and just outside the gap. This produces in the tape a chain of permanent magnets, in a magnetic pattern which represents the original sound.

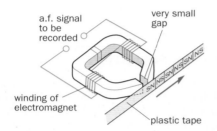

Fig. 19.30 Recording on magnetic tape

The strength of the magnet produced at any part of the tape depends on the value of the a.c. when that part passes the gap and this in turn depends on the loudness of sound being recorded. The length of a magnet depends on the frequency of the a.c. (which equals that of the sound) and on the speed of the tape. The effect of high- and low-frequency a.c. is shown in Fig. 19.31.

High frequencies and low speeds create shorter magnets; very short ones tend to lose their magnetism immediately. Also, each part of the tape being magnetized must have moved on past the gap in the recording head before the magnetic field reverses. Good quality recording, particularly of high frequencies, therefore requires high tape speeds. Chromium oxide tapes record a wider range of frequencies and their output is greater on playback.

The principle of two-track tape is shown in Fig. 19.32*a*. Half the tape width is used for a recording, and the other half when the tape is turned over. With stereo tape, the recording head has two electromagnets, each using one-quarter of the tape width, Fig. 19.32*b*.

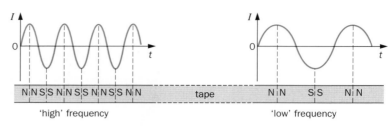

Fig. 19.31 The higher the frequency, the shorter the magnets produced on the tape

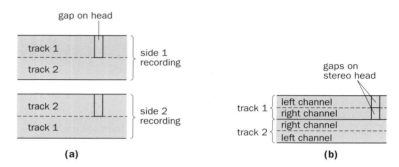

Fig. 19.32 Two-track tape

Fig. 19.33 Checking the position of a CD master, about to be 'written on' by a laser. The rapid laser pulses create a pattern of tiny hard areas in a 'photo-resist' plastic layer; unhardened photo-resist is then etched off

On playback, the tape runs past another head — the **playback head** — which is similar to the recording head (sometimes the same head records and plays back). The varying magnetization of the tape induces a small audio-frequency voltage in the windings of the playback head. After amplification, this voltage is used to reproduce the original sound in a loudspeaker.

(b) Compact disc (CD)

The sound is recorded as a series of tiny pits of different length and spacing on a plastic disc with a reflecting surface, Fig. 19.33. The 'pick-up' on the disc player sends a very narrow intense beam of light from a laser through a partly reflecting mirror and a lens system which focuses it on the disc, Fig. 19.34. If the beam falls on a pit, it is scattered, otherwise it is reflected back to a photodiode which produces an electronic signal.

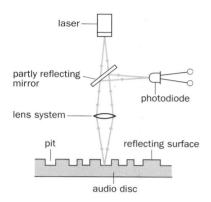

Fig. 19.34 Laser CD player

QUESTIONS

1. Figure 19.35 shows the variation with time *t* of the displacement *x* of the cones of two identical loudspeakers A and B placed in air.

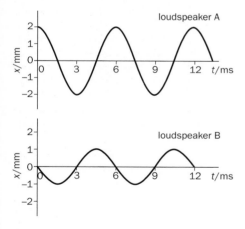

Fig. 19.35

a) Calculate the frequency of vibration of the loudspeaker cones.
b) Deduce the phase angle by which the vibrations of cone B lead those of cone A.
c) State the type of wave produced in the air in front of each loudspeaker.
d) Suggest, with a reason, which loudspeaker is likely to be producing the louder sound.
e) The speed of the sound waves in air is 340 m s^{-1}. Use your answer in a) to calculate the wavelength of these waves.

(*UCLES*, Physics Foundation, March 1998)

2. Figure 19.36 shows two small loudspeakers positioned a few metres apart with a small microphone, M, connected to an oscilloscope, positioned exactly midway between them. The loudspeakers are connected in parallel to a signal generator producing oscillations at a frequency of 3300 Hz.

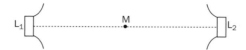

Fig. 19.36

a) Calculate the wavelength of the sound waves produced. Speed of sound waves in air = 330 m s^{-1}.
b) The microphone is now moved slowly to the right.
   i) Describe and explain how the trace height on the oscilloscope will change as the position of the microphone is changed.
   ii) Explain why this occurs.
(*NEAB*, AS/A PHO2, June 1998)

3. Figure 19.37 shows a loudspeaker which sends a note of constant frequency towards a vertical metal sheet. As the microphone is moved between the loudspeaker and the metal sheet the amplitude of the vertical trace on the oscilloscope continually changes several times between maximum and minimum values. This shows that a stationary wave has been set up in the space between the loudspeaker and the metal sheet.

How has the stationary wave been produced?

State how the stationary wave pattern changes when the frequency of the signal generator is doubled. Explain your answer.

What measurements would you take, and how would you use them, to calculate the speed of sound in air?

Suggest why the minima detected near the sheet are much smaller than those detected near the loudspeaker.
(*L*, AS/A PH2, Jan 1996)

4. Distinguish between **free vibrations** and **forced vibrations**, and explain the term **resonance**. Discuss the meanings of these terms by considering a specific vibrating system chosen from any branch of physics other than sound.

A small loudspeaker, actuated by a variable frequency oscillator, is sounded continuously over the open end of a vertical tube 40 cm long and closed at its lower end. At what frequencies will resonance occur as the frequency of the note emitted by the loudspeaker is increased from 200 Hz to 1200 Hz, given that the velocity of sound in air is 3.44 × 10^4 cm s^{-1}? Neglect the end effect of the tube.

5. Two open organ pipes of lengths 50 cm and 51 cm give beats of frequency 6.0 Hz when sounding their fundamental notes together. Neglecting end corrections, what value does this give for the velocity of sound in air?

6. The apparatus in Fig. 19.38 is used to set up a stationary wave on a stretched string. When the frequency of the vibrator is 60 Hz, resonance occurs and the stationary wave shown is produced.

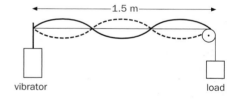

Fig. 19.38

For each of the following statements indicate whether the statement is true or false.

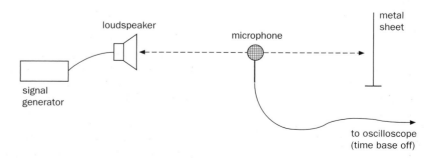

Fig. 19.37

Statement

When the frequency of the vibrator is 160 Hz there will be eight loops on the string.

The speed of the wave is 30 m s^{-1}.

Resonance will also occur when the frequency is 40 Hz.

If the load is doubled and the frequency is kept constant, there will be twice as many loops on the string.

(L, AS/A PH2, June 1996)

7. What is meant by a **wave motion**? Define the terms **wavelength** and **frequency** and derive the relationship between them.

   Given that the velocity v of transverse waves along a stretched string is related to the tension T and the mass μ per unit length by the equation

   $$v = \sqrt{\frac{T}{\mu}}$$

   derive an expression for the natural frequencies of a string length l when fixed at both ends.

   Explain how the vibration of a string in a musical instrument produces sound and how this sound reaches the ear. Discuss the factors which determine the quality of the sound heard by the listener.

8. Explain why the note emitted by a stretched string can easily be distinguished from that of a tuning fork with which it is in unison. How would you justify your answer by experiment?

   Describe how, using a set of standard forks, you would verify experimentally the relationship between the frequency of the note emitted by a string of fixed length and tension, and the mass per unit length of the wire.

   A sonometer wire of length 1.0 m emits the same fundamental frequency as a given tuning fork. The wire is shortened by 0.05 m, tension remaining unaltered, and 10 beats per second are heard when the wire and fork are sounded together. What is the frequency of the fork? If the mass per unit length of the wire is 1.4 × 10^{-3} kg m^{-1}, what is the tension?

9. Describe the motion of the particles of a string under constant tension and fixed at both ends when the string executes transverse vibrations of

a) its fundamental frequency, and
b) the first overtone (second harmonic). Illustrate your answer with suitable diagrams.

   A horizontal sonometer wire of fixed length 0.50 m and mass 4.5 × 10^{-3} kg is under a fixed tension of 1.2 × 10^2 N. The poles of a horse-shoe magnet are arranged to produce a horizontal transverse magnetic field at the midpoint of the wire, and an alternating sinusoidal current passes through the wire. State and explain what happens when the frequency of the current is progressively increased from 100 to 200 Hz.

   Indicate how you would use such an apparatus to measure the fixed frequency of an alternating current.

10. a) A radio source of frequency 95 MHz is set up in front of a metal plate. The distance from the plate is adjusted until a standing wave is produced in the space between them. The distance between any node and an adjacent antinode is found to be 0.8 m.
    Calculate the wavelength of the wave.
    Calculate the speed of the radio wave.
    What does this suggest about the nature of radio waves?
    b) The minimum intensity that can be detected by a given radio receiver is 2.2 × 10^{-5} W m^{-2}. Calculate the maximum distance that the receiver can be from a 10 kW transmitter so that it is *just* able to detect the signal.

    (L, AS/A PH2, Jan 1997)

11. Show that when a source emitting sound waves of frequency f moves towards a stationary observer with velocity u, the observer hears a note of frequency $fv/(v - u)$, where v is the velocity of sound.

    Describe the effect of a steady wind blowing with velocity w directly from source to observer
    a) in the case above, and
    b) if the source and the observer are both at rest.

    A model aircraft on a control line travels in a horizontal circle of 10 m radius, making 1 revolution in 3.0 s. It emits a note of frequency 300 Hz. Calculate the maximum and minimum frequencies of the note heard at a point 20 m from the centre of the circle and in the plane of the path, and find the time interval between a maximum and the minimum that next succeeds it. (Take the velocity of sound in air to be 330 m s^{-1}.)

12. Show that two identical progressive wave-trains travelling along the same straight line in opposite directions in a given medium set up a system of stationary waves. Compare the properties of a stationary wave system in air with those of a progressive wave-train in respect of
    a) amplitude,
    b) phase, and
    c) pressure variation.

    An observer moving between two identical sources of sound along the straight line joining them, hears beats at the rate of 4.0 s^{-1}. At what velocity is he moving if the frequency of each source is 500 Hz and the velocity of sound when he makes the observations is 3.40 × 10^4 cm s^{-1}?

13. A car travelling normally towards a cliff at a speed of 30 m s^{-1} sounds its horn which emits a note of frequency 100 Hz. What is the apparent frequency of the echo as heard by the driver? (Speed of sound in air = 330 m s^{-1})

14. A large asteroid recently passed the Earth at a distance similar to that of the distance between the Moon and Earth. A radar system was used to track its path. The frequency of the radar signal was 1.1 × 10^{10} Hz.
    a) At a certain point in the path of the asteroid, the time between the transmission of a radio pulse and the return of its echo was 2.9 s. Calculate the distance of the asteroid from Earth.
    b) The increase in frequency between the outgoing and returning radio signal was 6.6 × 10^5 Hz.
    i) Why did the return signal differ in frequency from the transmitted signal?
    ii) Calculate the speed of the asteroid relative to the Earth at this point.
    iii) Deduce whether the asteroid was moving away from or towards the Earth.
    c) In an attempt to estimate the radius of the asteroid, individual radar echoes were studied. The time between receiving the first part and last part of the echo of a particular pulse was 1.5 µs.
    i) Assuming that the asteroid was spherical and that the duration of the outgoing pulse was much shorter than 1.5 µs, explain how this information leads to a value for the radius of the asteroid.
    ii) Calculate the radius of the asteroid.

    (NEAB, AS/A PHO4, March 1998)

15. Compare the mode of propagation of sound with that of light.

How is the speed of sound in a gas affected by

a) an increase in the temperature of the gas, and

b) a decrease in the pressure of the gas?

A generator of ultrasonic waves of frequency 1.00 MHz is set up in a rectangular tank of paraffin facing a detector, which is connected to an amplifier and oscilloscope. It is found that the amplitude of the oscilloscope trace varies periodically as the detector is moved away from the generator. A series of consecutive maxima occurs at 2.99 mm, 3.65 mm, 4.33 mm, 5.00 mm and 5.66 mm from the generator. Explain the variation of the oscilloscope trace, and calculate the speed of sound in paraffin.

16. Explain the formation of 'beats' and show that if two notes of frequencies f_1 and f_2 are sounded together the frequency of the beats is given by $f_1 - f_2$. How would you determine, other than by ear, which frequency is the higher?

Explain with the aid of diagrams how sound waves may be refracted by

a) a wind gradient, and

b) a temperature gradient.

What is the effect of these refractions on the audibility of a sound?

A man standing close to an iron railing consisting of evenly-spaced uprights makes a sharp sound and hears a note of frequency 640 Hz. Calculate the spacing between the uprights. (Speed of sound in air = 330 m s^{-1} at 0 °C. Ambient temperature = 17.0 °C.)

17. A vibrating tuning fork, observed by a light which flashes at regular intervals, has the same appearance as it has when at rest. A circular disc with 70 equally-spaced radii drawn on it is rotated from rest with gradually increasing speed. When viewed in the same light the disc first appears at rest when it rotates at one revolution per second. What are the possible values of the frequency of the fork?

18. Write an essay on the topic of waves and wave motion.

a) Describe the motions of particles in transverse and longitudinal waves, and explain why polarisation is a phenomenon associated with transverse waves.

b) Explain the terms **frequency**, **wavelength**, **speed**, **period**, **displacement** and **amplitude**, and show how these quantities may be obtained from graphical representations of both transverse and longitudinal waves.

c) Describe an experiment to determine the frequency of a sound wave in air.

(*UCLES*, Physics Foundation, June 1998)

19. How many times approximately will the power emitted by a source have increased if the sound level increases by

a) 3 dB,

b) 10 dB?

20. If the volume control on a radio receiver is turned down, so decreasing the power output from the loudspeaker from 500 mW to 100 mW, what is the fall in sound level in dB? (log 5 = 0.7)

21. a) What term is used to describe a region in a stationary wave at which a maximum amplitude of oscillation occurs?

Figure 19.39 shows an organ pipe. To operate the pipe, compressed air is pumped into its lower end. This air hits the edge E, producing vibrations of many frequencies. Some of these vibrations create resonances in the air column Y–Y, producing the characteristic notes of the pipe.

b) The air column Y–Y is open at the lower end and closed at the upper end. Copy Fig. 19.39 and draw arrows to represent the directions and amplitudes of the oscillations of the air molecules on the axis of the pipe when the pipe is sounding its lowest note.

c) The effective length of the air column is 2.5 m. The speed of sound in air is 340 m s^{-1}. Calculate the frequencies of:

i) the lowest note produced by the air column;

ii) the next higher note produced.

(*OCR*, 6841, June 1999)

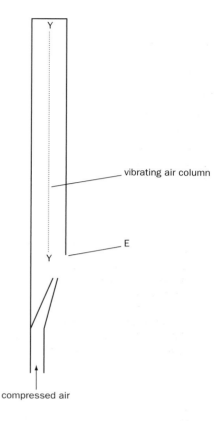

vibrating air column

E

compressed air

Fig. 19.39

22. a) Explain what is meant by:

i) threshold intensity of sound;

ii) loudness of sound.

b) At a given distance from an alarm clock, the intensity of sound from the clock is 10^{-4} W m^{-2} giving an intensity level of 80 dB. Use this information to show that the lowest intensity of sound which may be heard by the human ear is 10^{-12} W m^{-2}.

(*OCR*, 6846, June 1999)

20

Physical optics

- ■ Nature of light
- ■ Speed of light
- ■ Interference of light
- ■ Young's double slit
- ■ Fresnel's biprism; Lloyd's mirror
- ■ Wedge fringes
- ■ Newton's rings
- ■ Using interference
- ■ Everyday examples of interference

- ■ Optical path length
- ■ Interference calculations
- ■ Diffraction of light
- ■ Diffraction at a single slit
- ■ Diffraction at multiple slits
- ■ Diffraction grating
- ■ Optical spectra
- ■ Resolving power
- ■ Polarization of light
- ■ Producing polarized light

- ■ Using polarized light
- ■ Electromagnetic waves
- ■ Infrared radiation
- ■ Ultraviolet radiation
- ■ Thermal radiation
- ■ Radio waves
- ■ Holography

NATURE OF LIGHT

Two apparently contradictory theories of the nature of light were advanced in the seventeenth century.

The **corpuscular theory** regarded light as a stream of tiny particles or corpuscles travelling at high speed in straight lines. Newton supported this view which accounted for rectilinear propagation, reflection and refraction — the latter by assuming that on entering an optically denser medium the corpuscles are attracted, causing bending towards the normal. On the other hand, the **wave theory** proposed by Huygens around 1680 considered light to travel as waves. As we have seen (pp. 296–7), a wave model can also account satisfactorily for reflection and refraction.

Opponents of the wave theory argued that waves require a transmitting medium and there did not appear to be one for light, which was able to travel in a vacuum. Subsequently a medium, called the 'ether', was invented but it defied all attempts at detection.

An apparently crucial difference was that whereas the corpuscular theory required light to have a greater speed in a material than in air, the wave theory predicted a lower speed. It was not until 1862 that the wave theory prediction was confirmed when Foucault found the speed of light in water to be less than that in air. In the meantime, interference and diffraction effects had been discovered which were more readily explicable in terms of waves than in terms of corpuscles.

The need for the ether disappeared when Maxwell suggested in 1864 that light was an electromagnetic 'wave' consisting of a fluctuating electric field coupled with a fluctuating magnetic field (p. 349). By the end of the nineteenth century Maxwell's electromagnetic wave theory of radiation had established itself as one of the great intellectual pillars of physics, a unifying principle linking electricity, magnetism and light. However, as we shall see later, it did not on its own give a completely satisfactory explanation of all the properties of electromagnetic radiation.

In this chapter we will deal with 'physical optics' — the subject concerned with those effects that make sense if we regard light as having a wave-like nature: such effects are interference, diffraction and polarization.

SPEED OF LIGHT

A knowledge of the speed of light is important for several reasons. First, if the experimental value agrees with the theoretical value predicted by Maxwell for electromagnetic waves (see p. 349) then it is reasonable to assume that light is an electromagnetic wave and that the electromagnetic theory of radiation is valid. Second, it occurs in certain basic expressions of atomic and nuclear physics, such as Einstein's mass–energy equation $E = mc^2$ (see p. 480). Third, the measurement of distance by radar techniques become possible (p. 306).

The rotating mirror method of measuring the speed of light is based on one due to Foucault (1862). In a simplified method, an attempt is made to estimate, in effect, the time taken by light to travel a distance of about four metres. This is of the order of 10^{-8} second; the result obtained is consequently subject to a large error. The principle of the method is shown by Fig. 20.1. Light from a source S illuminates a cross-wire C in the viewing box and travels on through the glass plate P to the rotating mirror R where it is reflected to the large fixed con-

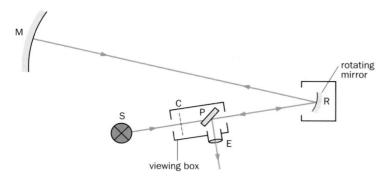

Fig. 20.1   Measuring the speed of light by Foucault's rotating mirror method

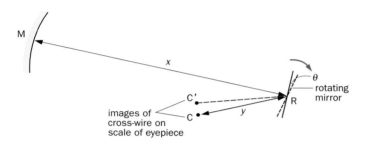

Fig. 20.2

cave mirror M, placed so that R is at its centre of curvature. M reflects the light back to R and into the viewing box where it is partially reflected by P into the eyepiece E. An image of C is formed on the eyepiece scale. When R is rotated at high speed (by a small electric motor), the image of C is displaced by a very small distance.

To obtain an expression for the speed of light c, consider the simplified diagram in Fig. 20.2. Let MR = x, CR = y, and when the rotating mirror makes n revolutions per second, suppose the displacement of the image of C is CC' = d. Then

distance travelled by light between reflections at R = $2x$

time for rotating mirror to make one revolution = $1/n$

time for mirror to rotate angle θ (in rad) = $\theta/(2\pi n)$

and

$$c = \frac{\text{distance travelled by light}}{\text{time taken}}$$

$$= \frac{2x}{\theta/(2\pi n)} = \frac{4\pi n x}{\theta}$$

Additionally, since the reflected ray turns through twice the angle of rotation θ of the mirror,

angular rotation of image of C (i.e. angle CRC') = $2\theta = d/y$

Substituting for θ,

$$c = \frac{8\pi n x y}{d}$$

In practice d is extremely small (about 0.05 mm) and the error in measuring it is reduced by noting the displacement $2d$ of the cross-wire when the rotation of the mirror is reversed.

Measurements of the speed of other members of the electromagnetic family confirm that they all — whatever their wavelength — travel in a vacuum with the *same* speed of 3.00×10^8 m s^{-1}. In other media the speed varies with the wavelength; for example, red light travels faster in glass than blue light.

INTERFERENCE OF LIGHT

Interference occurs whenever waves from two coherent sources cross. Such sources produce waves having the same frequency, equal or comparable amplitudes and a phase difference that does not alter with time.

The wavelength of light must be very small, or else diffraction effects would be more evident than they are.

It therefore follows from the experiments with water waves in a ripple tank (p. 299) that in order to obtain nodal and antinodal lines (called **interference fringes**) sufficiently far apart to be seen, we must have

- the sources very close together;
- the screen (or eyepiece) as far as possible from the sources.

Also, if the sources are to be coherent they must be derived, in practice, from the same source. If an attempt is made to produce interference with two separate light sources, uniform illumination results instead of regions of light and dark. This is due to the fact that in a light source the phase is constantly changing because light is emitted in short bursts that last about 10^{-9} s (when electrons in individual atoms suffer energy changes that occur very quickly and randomly). Phase changes happen abruptly as different atoms come into action and the eye is unable to follow the rapidly changing interference pattern.

This is also true for light coming from different parts of the same source (with one exception). What is necessary then is that two wave-trains arriving at a given point, say by different paths, should have come from the *same point* of the *same source*. Any phase change occurring in the source then occurs in both wave-trains and stationary interference effects result. To obtain two coherent wave-trains from a point of a single source one of two methods is adopted: **division of wavefront**, as is done in Young's double slit (p. 332), Fresnel's biprism and Lloyd's mirror (pp. 333–4); or **division of amplitude**, usually by partial reflection and transmission at some boundary, as occurs in wedge fringes and Newton's rings (pp. 334, 335).

The exception mentioned above is the **laser** which does produce coherent light because the atoms are made to act in unison and all undergo energy changes simultaneously (see chapter 24). If a screen with two small holes is placed in a laser beam so that different parts of the source are used, an interference pattern is obtained.

Interference accounts for the colours of soap bubbles and of thin films of oil on a wet road; it also has practical applications. See pp. 336–7.

YOUNG'S DOUBLE SLIT

One of the first to demonstrate interference of light was Thomas Young in 1801.

(a) Principle

The principle of his method is shown in Fig. 20.3a. Monochromatic light (i.e. of one colour) from a narrow vertical slit S falls on two other narrow slits S_1 and S_2 which are very close together and parallel to S. S_1 and S_2 act as two coherent sources (both being derived from S) and if they (as well as S) are narrow enough, diffraction causes the emerging beams to spread into the region beyond the slits. Superposition occurs inside the shaded area of Fig. 20.3a where the diffracted beams overlap. Alternate bright and dark equally spaced vertical bands (interference fringes) can be observed on a screen or at the cross-wires of an eyepiece, Fig. 20.3b. If either S_1 or S_2 is covered the bands disappear.

(b) Theory

An expression for the separation of two bright (or dark) fringes can be obtained from Fig. 20.4a.

The path difference between waves reaching O from S_1 and S_2 is zero, i.e. $S_1O = S_2O$; they therefore arrive in phase and so there is a bright fringe at O, in the centre of the pattern. At P, distance x_1 from O, there will be a bright fringe if the path difference is a whole number of wavelengths, i.e. if

$$S_2P - S_1P = n\lambda$$

where n is an integer (or zero) and λ is the wavelength of the light. We say the nth bright fringe is formed at P.

The distance from the screen or cross-wire to the double slit is d and the slit separation is a (Fig. 20.4a). In practice a is very small (e.g. 0.5 mm) compared with d (e.g. 1 m) and if P is near O then S_1P and S_2P are approximately parallel to CP, so

$$S_2P - S_1P \approx a \sin \theta$$

For small θ, $\sin \theta \approx \tan \theta = x_1/d$

$$\therefore \quad S_2P - S_1P = ax_1/d$$

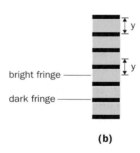

(a)

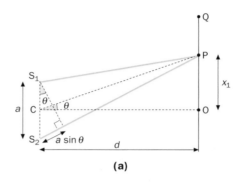

(b)

Fig. 20.4

For the nth bright fringe at P we have

$$n\lambda = ax_1/d \qquad (1)$$

If the next bright fringe, namely the $(n + 1)$th, is formed at Q where OQ = x_2 then

$$S_2Q - S_1Q = (n + 1)\lambda$$
$$(n + 1)\lambda = ax_2/d \qquad (2)$$

Subtracting (1) from (2) we get

$$\lambda = a(x_2 - x_1)/d$$

If y is the distance between two adjacent bright (or dark) fringes, called the **fringe spacing**, Fig. 20.4b, then $y = x_2 - x_1$ and so

$$\lambda = \frac{ay}{d}$$

We see that:

- $y \propto 1/a$ if λ and d are constant (therefore the slit separation should be small);
- $y \propto d$ if λ and a are constant (therefore the fringes should be viewed from a distance); and
- $y \propto \lambda$ if a and d are constant.

If a dark fringe were formed at P then $(S_2P - S_1P)$ would equal an odd number of half-wavelengths.

(c) Measurement of λ

One arrangement is illustrated in Fig. 20.5, in which the lamp filament acts as the single slit. The fringes are viewed in a blacked-out room at the cross-wire of a travelling eyepiece (e.g. a travelling microscope with the objective removed). Filters are used to obtain coloured light from the white light source.

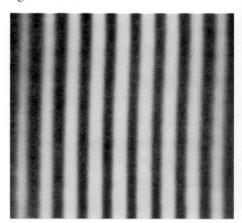

(b) Fringes

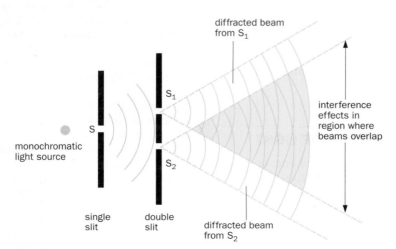

(a) Principle

Fig. 20.3 Young's double-slit interference

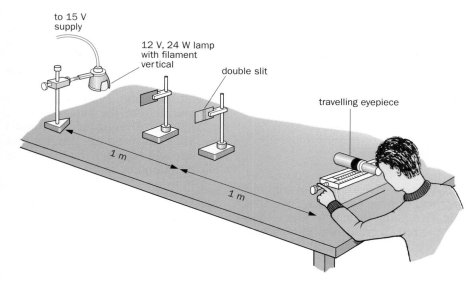

Fig. 20.5 Measuring wavelength by interference fringes

(*vi*) The fringe spacing for red light is greater than for blue light, Fig. 20.6. Red light must therefore have a greater wavelength than blue light since $y \propto \lambda$ (if a and d are constant).

(*vii*) The number of fringes obtained depends on the amount of diffraction occurring at the slits and this in turn depends on their width. The narrower the slits, the greater the number of fringes because of increased diffraction; however, they will also be fainter, because less light gets through. In practice, to give easily seen fringes, the slits have to be many wavelengths wide; the case is similar to that shown for water waves in Fig. 18.21*b* (p. 301).

The average fringe spacing y is found by measuring across as many fringes as possible with the travelling eyepiece. A metre rule is used to measure d and the slit separation a is measured directly with a travelling microscope. The value of λ obtained is approximate: for violet light it is about 4×10^{-7} m (0.4 µm) and for red light about 7×10^{-7} m (0.7 µm).

(d) Further points

(*i*) The overlapping beams which interfere are produced by diffraction at S_1 and S_2. Since the fringes can be observed anywhere in the overlapping region they are called *non-localized* fringes.

(*ii*) Point sources would give the same fringe system but slits (i.e. a line of point sources) give brighter fringes by reinforcing the pattern.

(*iii*) The fringes are really the intersections with a vertical plane of hyperboloids of revolution having S_1 and S_2 as foci. (These correspond in the three-dimensional case to the two-dimensional hyperbolic nodal and antinodal lines; see Fig. 18.18, p. 300.)

(*iv*) The interference is incomplete because for all fringes — except the central bright one — the amplitudes of the two wave-trains are not exactly equal. Why?

(*v*) Using white light fewer fringes are seen and each colour produces its own set of fringes which overlap. (The pattern is similar to that produced by a diffraction grating, Fig. 20.33*b*, but

much less bright.) Only the central fringe is white, its position being the only one where the path difference is zero for all colours. The first coloured fringe is bluish near the central fringe and red at its far side. Why?

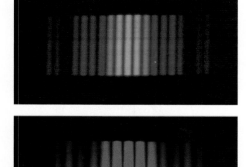

Fig. 20.6 The fringe spacing is proportional to the wavelength

FRESNEL'S BIPRISM; LLOYD'S MIRROR

We will now outline two other ways of producing interference fringes from two coherent sources that are derived by division of the wavefront from a single source.

(a) Fresnel's biprism

Monochromatic light from a narrow slit S falls on a double glass prism arranged as in Fig. 20.7. Two virtual images S_1 and S_2 are formed of S, by refraction at each half of the prism, and these act as coherent sources which are close together because of the small refracting angles (about 0.5°) of the prism. An interference pattern, similar to that given by the double slit but brighter, is obtained in the shaded region where the two refracted beams overlap. The fringes can be observed as before, using a travelling eyepiece focused on the fringes at its cross-wire.

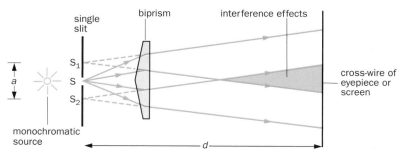

Fig. 20.7 Fresnel's biprism

The theory and the expression for the fringe spacing y are the same as for Young's method, i.e. $y = d\lambda/a$. To obtain a, a convex lens is moved between the biprism and the eyepiece until a real magnified image of the virtual 'objects' S_1 and S_2 is in focus. The distance b between the images of S_1 and S_2 (on the cross-wire of the eyepiece) is measured; then knowing the object and image distances u and v we can say that the magnification m is given by $m = v/u = b/a$. The wavelength λ can then be calculated if y and d are also determined, although this is not one of the most accurate ways of measuring the wavelength of light.

(b) Lloyd's mirror

A plane glass plate (acting as a mirror) is illuminated at almost grazing incidence by a light from a slit S_1, parallel to the plate, Fig. 20.8. A virtual image S_2 of S_1 is formed close to S_1 by reflection and these two act as coherent sources. In the shaded area direct waves from S_1 cross reflected waves which appear to come from S_2 and interference fringes can be seen.

The expression giving the fringe spacing is the same as for the double slit and the biprism experiments, but the fringe system differs in one important respect. In Lloyd's mirror, if the point P, for example, is such that the path difference $(S_1A + AP) - S_1P$ (or $S_2P - S_1P$) is a whole number of wavelengths, the fringe at P is *dark*, not bright. This is because of the 180° phase change that occurs when light is reflected at a rare–dense boundary. This is equivalent to adding an extra half-wavelength to the path of the reflected wave and is similar to the phase change a pulse on a spring

undergoes at a fixed end, or to that when microwaves are reflected at a metal plate (p. 300). At grazing incidence a fringe is formed at C, where the geometrical path difference between the direct and reflected waves is zero and it follows that it will be dark rather than bright.

WEDGE FRINGES

Interference fringes are produced by a thin wedge-shaped film of air, the thickness of which gradually increases from zero along its length. The wedge can be formed from two microscope slides clamped at one end and separated by a thin piece of paper at the other so that the wedge angle is very small. In Fig. 20.9a monochromatic light from an extended source (e.g. a sodium lamp or flame) is partially reflected vertically downwards by the glass plate G. When the microscope is focused on the wedge, light and dark equally spaced fringes are seen, parallel to the edge of contact of the wedge, Fig. 20.9b.

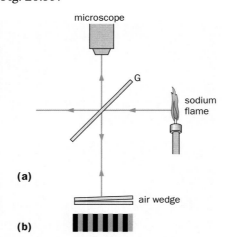

(a)

(b)

Fig. 20.9 Wedge interference

Some of the light falling on to the wedge is reflected upwards from the bottom surface of the top slide and some of the rest, which is transmitted through the air wedge, is reflected upwards from the top surface of the bottom slide. Both wave-trains have arisen from the same point P, Fig. 20.10, by **division of the amplitude**. They are therefore coherent and when brought together (by the eye or a microscope) they can interfere.

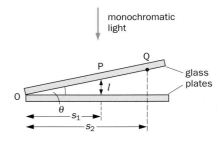

Fig. 20.10

If l is the thickness of the air wedge at P then, since the incidence is nearly normal, the path difference between the rays at P is $2l$. At O, where the path difference is zero, we would expect a bright band — but a *dark* band is observed. This is due to the 180° phase change which occurs when the wave-train *in* the air wedge is reflected at the top surface of the bottom slide, i.e. at a denser medium. The phase change, in effect, adds an extra path of half a wavelength (i.e. a crest is reflected as a trough). The path difference between the two wave-trains at P is thus $(2l + \lambda/2)$ where λ is the wavelength of the light. A bright fringe is formed at P if

$$2l + \lambda/2 = n\lambda$$

or $\qquad 2l = (n - \tfrac{1}{2})\lambda$

where $n = 1$ gives rise to the first bright fringe, $n = 2$ gives the second bright fringe, etc. ($n = 0$ is impossible). A dark fringe is formed at P if

$$2l = n\lambda$$

where $n = 0$ gives the first dark fringe, $n = 1$ gives the second dark fringe, $n = 2$ gives the third dark fringe, etc.

The microscope — or the unaided eye — has, for normal incidence, to be focused on the top surface of the air wedge. This ensures that superposition of the two interfering wave-trains then occurs at a particular point in the retina. The fringes in this case are

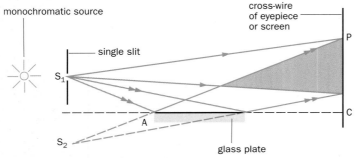

monochromatic source

single slit

cross-wire of eyepiece or screen

Light from S_1 should fall at grazing incidence on the glass plate; the incidence is exaggerated here

Fig. 20.8 Lloyd's mirror

called **localized** fringes. Each fringe is the locus of points of equal air-wedge thickness (i.e. same path difference) and they are often referred to as 'fringes of equal thickness'. There are also reflections from the other surfaces; these are usually out of focus and since they involve large path differences they do not spoil the fringes.

The angle θ of the wedge can be found if the reading on a travelling microscope is taken when the cross-wire is on a dark fringe at P, say, the nth from O. We then have from Fig. 20.10 that

$$2l = n\lambda$$

But $l = s_1\theta$ (if θ is in radians), so

$$2s_1\theta = n\lambda \qquad (1)$$

If the microscope is now moved until it is on the $(n + k)$th dark fringe, say at Q, and the reading again taken, then

$$2s_2\theta = (n + k)\lambda \qquad (2)$$

Subtracting (1) and (2),

$$2\theta(s_2 - s_1) = k\lambda$$

Therefore

$$\theta = \frac{k\lambda}{2(s_2 - s_1)}$$

where $(s_2 - s_1)$ is the distance moved by the microscope and λ is the wavelength of the light.

NEWTON'S RINGS

This system of interference fringes, also produced by division of amplitude, was discovered by Newton but it was Young who first gave a satisfactory explanation of their formation in terms of waves.

(a) Principle

The arrangement is illustrated in Fig. 20.11. Monochromatic light (e.g. from a sodium lamp or flame) is reflected by the glass plate G so that it falls normally on the air film formed between the convex lens of long focal length (about one metre) and the flat glass plate. The thickness of the air film gradually increases outwards from zero at the point of contact B but is the same at all points on any circle with centre B.

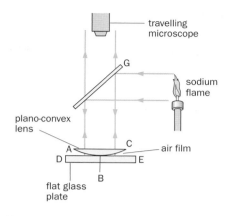

Fig. 20.11 Viewing Newton's rings

Interference occurs between light reflected from the lower surface ABC of the lens and the upper surface DBE of the plate. A series of bright and dark rings is seen through G when a travelling microscope is focused on the air film, Fig. 20.12. The rings are fringes of 'equal thickness', 'localized' in the air film (like those formed by a wedge) and as their radii increase, the separation decreases. At the centre B

of the fringe system, where the geometrical path difference between the two wave-trains is zero, there is a *dark* spot. As with wedge fringes, this is due to the 180° phase change which occurs when light is reflected at an optically denser medium. In effect, the path difference at B is not zero but half a wavelength.

(b) Radius of a ring

In Fig. 20.13, p. 336, the complete circular section of a sphere of radius R is shown, R being the radius of curvature of the lower surface ABC of the lens. Let r_n be the radius OP of the ring at P, where the thickness PQ of the film is l. If BO is produced to meet the circle at L then BOL is a diameter. By the theorem of intersecting chords we have

$$SO \cdot OP = LO \cdot OB$$

$$\therefore \quad r_n \cdot r_n = (2R - l)l$$

$$\text{(since OB = PQ = } l\text{)}$$

$$\therefore \quad r_n^2 = 2Rl - l^2$$

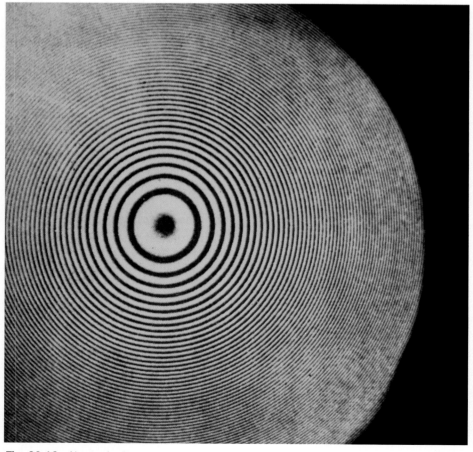

Fig. 20.12 Newton's rings

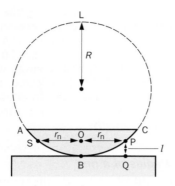

Fig. 20.13

But l^2 is very small compared with $2Rl$ since R is large, so

$$r_n^2 = 2Rl$$

The path difference between the two interfering wave-trains at P is $2l$ for light incident normally on the air film. If a dark ring is formed at P then

$$2l = n\lambda$$

where n is an integer and λ is the wavelength of the light. Even though the geometrical path difference is a whole number of wavelengths, the $\lambda/2$ phase change at Q results in destructive interference.

From the two previous expressions we can write for the radius r_n of a *dark* fringe:

$$r_n^2 = Rn\lambda$$

$n = 0$ gives the central dark spot, $n = 1$ gives the first dark ring, $n = 2$ gives the second dark ring and so on.

A *bright* ring would be formed at P if $2l$ equalled $\frac{1}{2}\lambda$, or $1\frac{1}{2}\lambda$, or $2\frac{1}{2}\lambda$, etc., because the phase change at the glass plate effectively increases the path length of the light reflected there by $\lambda/2$. If the nth bright ring is formed at P, then

$$2l = (n - \tfrac{1}{2})\lambda$$

The radius r_n of a *bright* ring is therefore given by

$$r_n^2 = R(n - \tfrac{1}{2})\lambda$$

where $n = 1$ gives the first bright ring, $n = 2$ gives the second bright ring and so on.

(c) Measurement of λ

It is better to measure the diameters of rings rather than their radii because of the uncertainty of the position of the centre of the system. If d_n is the diameter of the nth *dark* ring we have

$$r_n^2 = (d_n/2)^2 = Rn\lambda$$

$$\therefore \quad d_n^2 = 4Rn\lambda$$

A graph of d_n^2 against n is a straight line of slope $4R\lambda$. If it does not pass through the origin, n may have been miscounted each time *or* the contact between lens and plate is poor (and may give a bright spot in the centre). Neither effect, however, alters the slope of the graph. A travelling microscope is used to measure d_n, and R can be calculated if the focal length and refractive index of the lens are known (p. 90).

USING INTERFERENCE

(a) Testing of optical surfaces

Fringes of equal thickness are useful for testing optical components. For example, in the making of optical 'flats', the plate under test is made to form an air wedge with a standard plane glass surface. Any uneven parts of the surface which require more grinding will show up as irregularities in what should be a parallel, equally spaced, straight set of fringes.

The grinding of a lens surface may be checked if it is placed on an optical flat and Newton's rings then observed in monochromatic light. The rings should be exactly circular if the lens is spherical.

(b) Non-reflecting glass

Inside optical instruments containing lenses or prisms light is lost by reflection at each refracting surface; this results in reduced brightness of the final image. There will also be a loss of contrast against such a background of stray light.

The amount of light reflected at a surface can be appreciably reduced by coating it (by evaporation in a vacuum) with a film of transparent material, e.g. magnesium fluoride, to a thickness of one quarter of a wavelength of light in the film, Fig. 20.14. Light reflected from the top and bottom surfaces of the film (rays 1 and

2 respectively) then interfere destructively since ray 2 has to travel twice the thickness of the film, i.e. the path difference is $2l$. The refractive index of the film is less than that of glass and so for each reflection, occurring as it does at a rare–dense boundary, there is a 180° phase change. The net effect of the phase changes on the path difference is therefore zero. The refractive index of the film should be as nearly as possible the mean of the refractive indices for air and glass so that the amounts of light reflected at the two surfaces are almost equal.

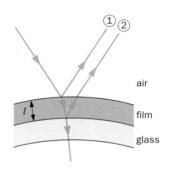

Fig. 20.14

For light of wavelength λ_a in air and λ_f in the film, the condition for destructive interference, at normal incidence, is

$$2l = \tfrac{1}{2}\lambda_f$$

$$\therefore \quad l = \tfrac{1}{4}\lambda_f$$

When light passes from one medium into another, its speed and wavelength change but not its frequency f. Therefore if v_a and v_f are the speeds of light in air and the film respectively, we have

$$v_a = f\lambda_a \quad \text{and} \quad v_f = f\lambda_f$$

Also, if n_f is the refractive index of the material of the film and n_a is the refractive index of air then

$$\frac{n_f}{n_a} = \frac{v_a}{v_f} \quad \text{(see p. 297)}$$

$$= \frac{\lambda_a}{\lambda_f}$$

So $$\lambda_f = \lambda_a \frac{n_a}{n_f}$$

The previous expression for l may then be written more conveniently as

$$l = \frac{\lambda_a n_a}{4 n_f}$$

The interference is completely destructive for one wavelength only, usually taken to be that at the centre of the visible range (i.e. yellow-green). For red and blue light the reflection is weakened but not eliminated, so a coated or 'bloomed' lens appears purple in white light. Energy which would have been wasted as reflected light increases the amount of transmitted light when destructive interference occurs.

If $\lambda_a = 550$ nm, $n_a = 1.0$ and $n_f = 1.38$, then $l = 100$ nm.

(c) Measurement of length

The SI unit of length, the metre, is defined as a certain number of wavelengths (1 650 763.73) in a vacuum of a particular line in the spectrum of an isotope of the inert gas krypton. The measurement required to set up this standard involves using a **Michelson interferometer** (Fig. 20.15) to split a light beam into two paths by means of a beam splitter M; some light is reflected up to mirror M_1 and some is transmitted through to M_2. After reflection at M_1 and M_2 the light is recombined at M and if the distance $2(L_1 - L_2) = \lambda$, the beams interfere constructively to form a bright image; if $2(L_1 - L_2) = \lambda/2$, destructive interference occurs and the image is dark; a movement of M_1 that changes L_1 by $\lambda/4$ causes a change from bright to dark. If M_1 is tilted slightly, bright and dark fringes are obtained which move across the field of view as L_1 is varied.

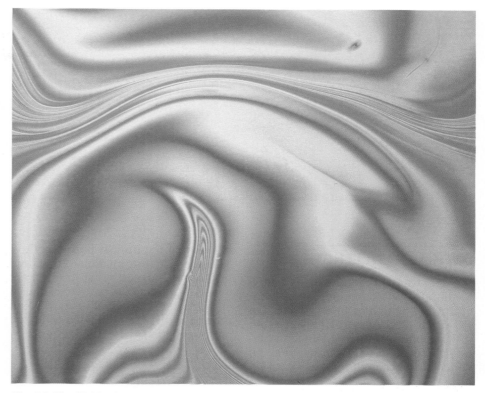

Fig. 20.16 Multi-coloured interference pattern in an oil film on water

The movement of one-tenth of a fringe can just be detected and so the accuracy attainable is about 1 part in 10^7.

A similar technique is used to measure the very small expansion of a crystal when it is heated, the motion of interference fringes again being observed.

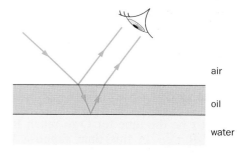

Fig. 20.17 Division of amplitude in an oil film

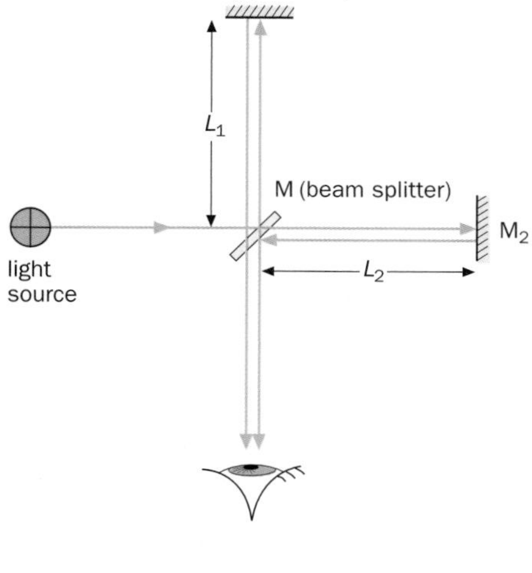

EVERYDAY EXAMPLES OF INTERFERENCE

(a) Colours of oil films on water

This beautiful natural effect, Fig. 20.16, is produced by interference occurring between two wave-trains — one reflected from the surface of the oil and the other from the oil–water interface, Fig. 20.17. When the path difference gives constructive interference for light of one wavelength, the corresponding colour is seen in the film. The path difference varies with the thickness of the film and the angle of viewing, both of which affect the colour produced at any one part. The colours in soap bubbles, p. 293, arise similarly.

(b) Pulsing of the picture on a television

This occurs when an aircraft passes low overhead. The signal travelling directly from the transmitting to the receiving aerial interferes with that reflected from the aircraft, Fig. 20.18. Usually the reflected wave is significantly weaker and so the interference is never completely destructive.

Fig. 20.15 Michelson interferometer

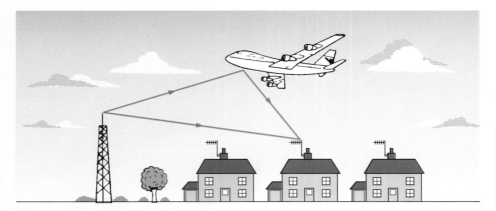

Fig. 20.18

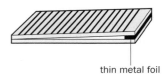

thin metal foil

Fig. 20.20

OPTICAL PATH LENGTH

We can show that a length l in a medium of refractive index n is equivalent to a length nl in a vacuum. Let PQ in Fig. 20.19 be a plane wavefront falling obliquely on the surface separating a vacuum and the medium (in which the speeds of light are c and v respectively).

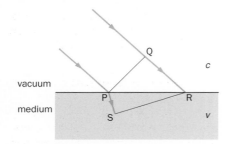

Fig. 20.19

Let t be the time for Q to reach R, i.e. QR $= ct$. In this time suppose P travels a distance PS in the medium, then PS $= vt$. If n is the absolute refractive index of the medium then

$$n = \frac{c}{v} \quad \text{and} \quad v = \frac{c}{n}$$

Therefore

$$\text{PS} = \frac{ct}{n} = \frac{\text{QR}}{n}$$

$$\therefore \quad \text{QR} = n \cdot \text{PS}$$

That is, if light travels a distance PS in the medium then in the same time it would travel a distance $n \times$ PS in a vacuum. In general:

> Length l in medium of refractive index n is optically equivalent to length nl in a vacuum.

nl is called the **optical path length** of distance l in the medium. It follows that if a thickness l of transparent material of refractive index n is placed in the path of a beam of light, the path difference between the new and the previous paths is $nl - l = (n - 1)l$.

INTERFERENCE CALCULATIONS

Example 1. In a Young's double-slit experiment the distance between the slits and the screen is 1.60 m; using light of wavelength 5.89×10^{-7} m the distance between the centre of the interference pattern and the fourth bright fringe on either side is 16.0 mm. What is the slit separation?

> From $\lambda = ay/d$ (p. 332) we have
>
> $$a = \frac{d\lambda}{y}$$
>
> where $\lambda = 5.89 \times 10^{-7}$ m, $d = 1.60$ m and $y =$ fringe spacing $= 16.0$ mm/4 $= 4.0$ mm $= 4.0 \times 10^{-3}$ m. So
>
> $$a = \frac{1.60 \times 5.89 \times 10^{-7}}{4.0 \times 10^{-3}} \frac{\text{m} \times \text{m}}{\text{m}}$$
>
> $$= 0.236 \times 10^{-3} \text{ m}$$
>
> $$= 0.236 \text{ mm}$$

Example 2. An air wedge is formed between two glass plates which are in contact at one end and separated by a piece of thin metal foil at the other end. Calculate the thickness of the foil if 30 dark fringes are observed between the ends when light of wavelength 6.0×10^{-7} m is incident normally on the wedge, Fig. 20.20.

The *first* dark fringe is formed at the end where the plates are in contact and the thickness l of the air wedge is zero.

The *second* dark fringe occurs where $l = \lambda/2$; the geometrical path difference is then $2l = \lambda$ and the $\lambda/2$ phase change at the bottom plate results in destructive interference.

The *third* dark fringe is formed where $l = 2(\lambda/2)$, the *fourth* where $l = 3(\lambda/2)$ and the *thirtieth* where $l = 29(\lambda/2)$.

$\therefore$ Thickness of foil

$$= 29 \times \frac{6.0 \times 10^{-7}}{2} \text{ m}$$

$$= 8.7 \times 10^{-3} \text{ mm}$$

Example 3. Calculate the radius of curvature of a plano-convex lens used to produce Newton's rings with a flat glass plate if the diameter of the tenth dark ring is 4.48 mm, viewed by normally reflected light of wavelength 5.00×10^{-7} m. What is the diameter of the twentieth bright ring?

> For the nth dark ring of radius r_n
>
> $$r_n^2 = Rn\lambda \quad \text{(see p. 336)}$$
>
> where R is the radius of curvature of the lens surface and λ is the wavelength of the light.
>
> $$\therefore \quad R = \frac{r_n^2}{n\lambda}$$
>
> $$= \frac{(4.48 \times 10^{-3}/2)^2}{10 \times 5.00 \times 10^{-7}} \frac{\text{m}^2}{\text{m}}$$
>
> (because $n = 10$ for the tenth dark ring)
>
> $$\therefore \quad R = 1.00 \text{ m}$$
>
> The radius r_n of the twentieth bright ring is given by
>
> $$r_n^2 = R(n - \tfrac{1}{2})\lambda$$

where $n = 20$ and $R = 1.00$ m. So

$$r_n{}^2 = 1(20 - \tfrac{1}{2})5.00 \times 10^{-7} \text{ m}^2$$

$$\therefore \quad r_n = \sqrt{(9.75)} \times 10^{-3} \text{ m}$$

$$= 3.12 \text{ mm}$$

$$\therefore \quad d_n = 6.24 \text{ mm}$$

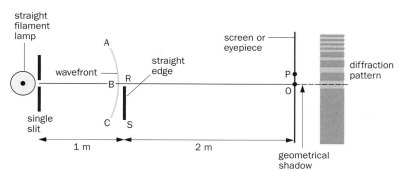

Fig. 20.22 Diffraction by a straight edge

DIFFRACTION OF LIGHT

Light can spread round obstacles into regions that would be in shadow if it travelled exactly in straight lines, i.e. it exhibits the typical wave-like property of diffraction. Therefore the edges of shadows are not sharp. Behind the obstacles or apertures at which diffraction occurs, a diffraction pattern of dark and bright fringes is formed which, under the correct conditions, may be seen on a screen or at the cross-wire of an eyepiece.

A striking example of a diffraction pattern is shown in Fig. 20.21. It was produced by placing a razor blade midway between an illuminated pinhole and a photographic film. In most cases, small sources (for example a pinhole or a slit) are needed to observe diffraction effects; these are more evident if obstacles and apertures have linear dimensions comparable with the wavelength of light. With a large source each point gives rise to its own diffraction pattern and there is uniform illumination when these overlap.

Diffraction is regarded as being due to superposition of secondary wavelets from coherent sources on the unrestricted part of a wavefront that has been obstructed by an obstacle or aperture (p. 301). So, whereas interference involves the superposition of waves on two *different* wavefronts, in diffraction there is superposition of waves from different parts of the *same* wavefront.

(a) Straight edge (e.g. a razor blade edge)

It can be shown that the fringes at points, such as P in Fig. 20.22, which are close to the region of geometrical shadow, are due to the superposition of secondary wavelets from point sources on the unrestricted wavefront (e.g. near R in the diagram).

(b) Circular obstacle (e.g. a ball-bearing)

In this case there is, rather surprisingly, a *bright* spot at the centre of the geometrical shadow, Fig. 20.23.

(c) Straight narrow obstacle (e.g. a pin)

Between P and Q in Fig. 20.24 the fringes arise largely from the superposition of secondary wavelets from points near R, i.e. the pattern is a diffraction one due to coherent sources on the same wavefront. Similarly, secondary wavelets from points near S create the pattern between L and M. The evenly spaced fringes inside the geometrical shadow LP are formed by the superposition of secondary

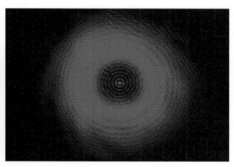

Fig. 20.23 Diffraction by a circular obstacle

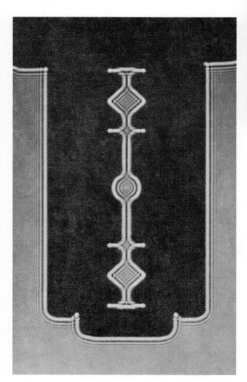

Fig. 20.21 Diffraction by a razor blade

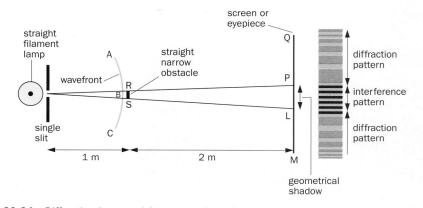

Fig. 20.24 Diffraction by a straight narrow obstacle

wavelets from points around R and S acting as *two* coherent sources; this is an interference pattern similar to that obtained with Young's double slit.

DIFFRACTION AT A SINGLE SLIT

(a) Experimental arrangement

The arrangement is shown in Fig. 20.25. The lamp has a straight filament, arranged to be vertical, and the slit, whose width can be adjusted by a screw, is mounted with its length parallel to the filament. Initially the slit is opened wide and the lens moved to give a sharp image of the filament on the translucent screen. The shield round the lamp and the large stop cut out stray light.

The diffraction pattern depends on the slit width and is observed from behind the screen, preferably through a magnifying glass. For widths of a few millimetres a rectangle of light is produced, the result of near-rectilinear propagation. As the slit is narrowed,

a pattern is obtained with a white central band having dark bands either side, fringed with colour. Inserting a red filter between lamp and screen gives red and black bands and the effect is as in Fig. 20.26a. In the diagrammatic depiction of Fig. 20.26b, bright red bands are formed at A, B, C, D and E. With blue light the bands are closer together. Just before the slit closes the pattern dims and the central bright band widens, causing illumination well into the geometrical shadow of the slit, i.e. the diffraction is very marked as the width of the slit approaches the wavelength of light and the slit behaves like a secondary source of spherical wavelets.

(b) Theory

To account for the maxima and minima of the diffraction pattern we use Huygens' construction and consider each point of the slit as a source of secondary wavelets. The slit is imagined to consist of strips of equal width, parallel to the length of the slit. The total effect in a particular dir-

ection is then found by adding the wavelets emitted in that direction by all the strips, using the superposition principle. In practice this operation presents mathematical difficulties too complex to be dealt with here, but a simplified two-dimensional treatment based on dividing the slit into strips is possible.

Suppose plane waves (e.g. from a distant light source) fall normally on a narrow rectangular slit of width a, Fig. 20.27. Consider the first dark band (i.e. the first minimum) where there is no light. It will be formed at an angle θ to the incident beam if the path difference for the secondary wavelets from the strip just below A and the strip just below C (the midpoint of the slit) is $\lambda/2$, where λ is the wavelength of the light. Destructive interference will thereby occur for wavelets from this pair of strips since a crest from one strip reaches the observer with a trough from the other. This happens for all pairs of *corresponding* strips in AC and CB because the same path difference of $\lambda/2$ exists. So there is no light in direction θ when

$$CD = \lambda/2 \quad (\text{or } BE = \lambda)$$

But $\sin \theta = \sin \angle CAD = CD/(a/2)$, so

$$CD = \frac{a \sin \theta}{2} \quad (\text{or } BE = a \sin \theta)$$

That is to say, the direction of the first minimum is given by

$$a\frac{\sin \theta}{2} = \frac{\lambda}{2}$$

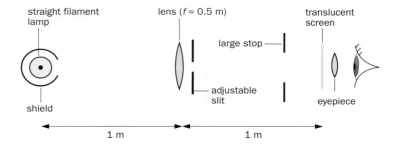

Fig. 20.25 Viewing diffraction due to a single slit

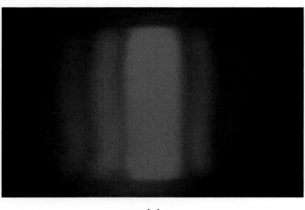

(a)

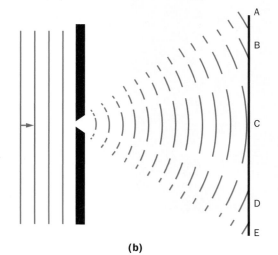

(b)

Fig. 20.26 Diffraction of monochromatic light by a single slit

or
$$\sin \theta = \frac{\lambda}{a}$$

It can be shown by similar 'pairing' processes that other minima occur when

$$\sin \theta = \frac{n\lambda}{a}$$

where $n = \pm1$, ±2, ±3, etc.; the $\pm$ signs indicate that, for example, there are two 'first-order' minima, one on each side of the original direction of the incident beam. If θ is small and in radians we can write $\theta \approx n\lambda/a$.

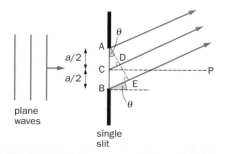

Fig. 20.27

At P in Fig. 20.27, where the central bright band is formed, wavelets from all the imaginary strips in the slit arrive in phase since they have the same path length and the intensity of light is greatest. Other maxima occur roughly half-way between the minima at angles such that $\sin \theta$ has values $\pm3\lambda/2a$ (i.e. $(\lambda/a + 2\lambda/a)/2$), $\pm5\lambda/2a$, etc. The first maximum is explained if the slit is divided into three equal parts and a direction considered in which the path differences between their ends are $\lambda/2$, Fig. 20.28. Wavelets from strips in two adjacent parts then cancel (as above), leaving only wavelets from one part, to give a much

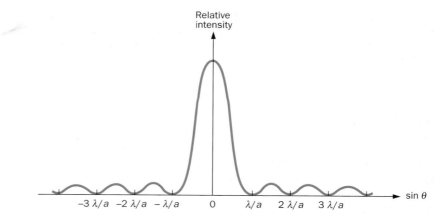

Fig. 20.29 Intensity distribution for diffraction at a single slit

less bright band. A graph of the relative intensity distribution for a single slit diffraction pattern is shown in Fig. 20.29.

(c) Further points

From $\sin \theta = \lambda/a$ it follows that if the slit is wide, λ is much less than a and so $\sin \theta$, and therefore θ, are very small. The directions of the first (and all other) minima are then extremely close to the middle of the central maximum. Most of the light emerging from the slit is in the direction of the incident light and there is little diffraction. Propagation is almost rectilinear and the laws of geometrical optics are applicable. On the other hand, if the slit is one wavelength wide, $\lambda = a$, $\sin \theta = \sin 90° = 1$ and the central bright band spreads completely into the geometrical shadow. In this case the behaviour of light requires a wave model.

The width of the central bright band can be seen from Fig. 20.29 to be twice that of any other bright band. (Also see Fig. 20.26a.)

DIFFRACTION AT MULTIPLE SLITS

Effects similar to those demonstrated by Young's double slit are obtained if more than two slits are used. They may be produced with the apparatus of Fig. 20.30, set up as for observing diffraction at a single slit. It can be seen that as the number of slits is increased, the bright bands become both brighter and sharper. With equally spaced slits the pattern is in effect the same as a two-slit one.

Diffraction occurs at the slits and these, being very narrow, behave as sources of secondary wavelets (semicircular in two dimensions) which superpose beyond the slits, Fig. 20.31. In certain directions the wavelets interfere constructively to form a new, straight wavefront (a bright band), while in others they interfere destructively (a dark band).

As well as interference occurring between secondary wavelets from *different* slits, it also occurs between secondary wavelets from the *same* slit.

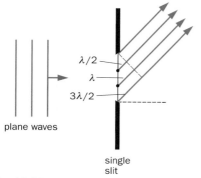

Fig. 20.28

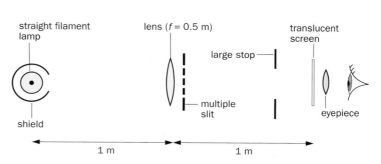

Fig. 20.30 Viewing diffraction due to multiple slits

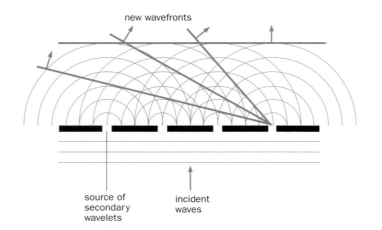

new wavefronts

source of
secondary
wavelets

incident
waves

Fig. 20.31

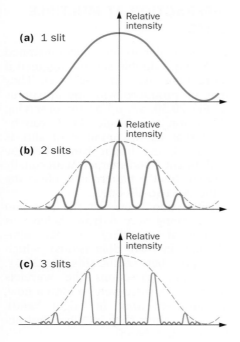

(a) 1 slit

Relative
intensity

(b) 2 slits

Relative
intensity

(c) 3 slits

Relative
intensity

Fig. 20.32 Intensity distributions for
multiple slits

Each slit therefore produces its own
diffraction pattern; these are similar,
in the same direction and will coincide exactly if focused by a lens. The
diffraction pattern due to a single slit
is superimposed on the interference
pattern and determines the variations
of intensity of the bright bands in the
latter, as shown by the dotted curves
in Figs 20.32*b* and *c* and by the
double-slit interference fringes in the
photograph of Fig. 20.3*b* (p. 332).

DIFFRACTION GRATING

A diffraction grating consists of a large
number of fine, equidistant, closely
spaced parallel lines of equal width,
ruled on glass or polished metal by a
diamond point. In **transmission gratings** glass is used; the lines scatter the
incident light and are more or less
opaque, while the spaces between
them transmit light and act as slits.
Such gratings are very expensive and
cheaper plastic replicas are made. In
reflection gratings the lines, ruled on
metal, are again opaque but the
unruled parts reflect regularly. This
type has the advantage that radiation
absorbed by transmission grating material can be studied; if it is ruled on a
concave spherical surface it focuses
the radiation as well as diffracting it
and no lenses are needed.

Diffraction gratings are used for
producing spectra and for measuring
wavelengths accurately. They have replaced the prism in modern spectroscopy. Their usefulness arises from
the fact that they give very sharp
spectra, most of the incident light
being concentrated in certain directions.

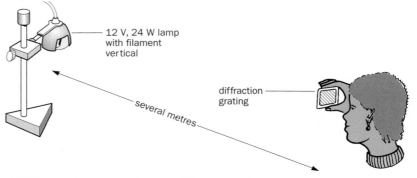

12 V, 24 W lamp
with filament
vertical

several metres

diffraction
grating

Fig. 20.33 Viewing spectra due to a diffraction grating

(a) 'Fine' and 'coarse' gratings

To observe spectra produced by a diffraction grating, a low-voltage lamp
with its filament vertical is viewed at a
distance of a few metres through a
transmission grating held near the eye
and with its lines parallel to the filament, Fig. 20.33. A 'fine' grating (e.g.
300 lines per mm) and a 'coarse'
grating (e.g. 100 lines per mm) should
be tried in turn.

A typical pattern for a 'fine' grating
is shown in Fig. 20.34. The central
bright band, called the **zero-order
image**, is white (W) but on either side
of it are brilliant bands of colour,
known as **first-** and **second-order
spectra**, like those given by a prism
but having red light (R) deviated more
than violet (V) and the dispersion
increasing with order. Each spectrum
consists of a series of adjacent images
of the filament formed by the constituent colours of white light. A
'coarse' grating forms many more

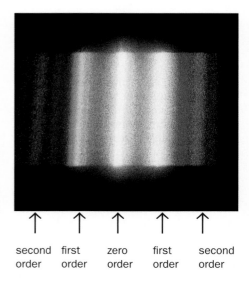

second   first   zero   first   second
order   order   order   order   order

Fig. 20.34 Spectra produced by a 'fine'
grating (300 lines/mm)

orders of spectra, closer together. The effect of allowing a narrower band of wavelengths to fall on the grating can be observed by placing red and green filters in turn in front of the lamp.

A 'very coarse' grating (e.g. 10 lines per mm) gives a pattern that resembles Young's fringes.

Grating spectra can be projected on to a screen by inserting a plate with a single narrow slit into a slide projector and using this arrangement to illuminate the grating, placed in front of a translucent screen.

(b) Theory

Suppose plane waves of monochromatic light of wavelength λ fall on a transmission grating in which the slit separation (called the **grating spacing**) is d, Fig. 20.35. Consider wavelets coming from corresponding points A and B on two successive slits and travelling at an angle θ to the direction of the incident beam. The path difference AC between the wavelets is $d \sin \theta$, as it is for all pairs of wavelets from other corresponding points in these two slits and in all pairs of slits in the grating.

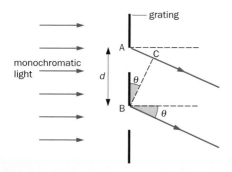

Fig. 20.35

So if

$$d \sin \theta = n\lambda$$

where n is an integer giving the order of the spectrum, then reinforcement of the diffracted wavelets occurs in direction θ and a maximum will be obtained when the wavelets are brought to a focus by a lens.

When $n = 0$, $\theta = 0$ and we observe in the direction of the incident light the central bright maximum, i.e. the

zero-order image, for which the path difference of diffracted wavelets equals zero. First-order, second-order, etc., spectra are given by $n = 1, 2$, etc., but these are much less bright than the $n = 0$ maximum.

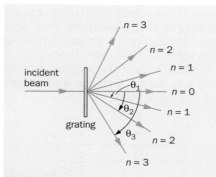

Fig. 20.36

As an example, consider a grating with 500 lines per mm on which yellow light, wavelength 6×10^{-7} m, falls normally. Since there are 500 lines and 500 spaces per mm of the grating, the grating spacing (i.e. 1 line + 1 space), $d = 1/500$ mm $= 10^{-3}/500$ m $= 2 \times 10^{-6}$ m (this is about three times the wavelength of yellow light). For the first order, $n = 1$,

$$\sin \theta_1 = \frac{\lambda}{d} = \frac{6 \times 10^{-7} \text{ m}}{2 \times 10^{-6} \text{ m}}$$

$$= 0.3$$

$$\therefore \quad \theta_1 = 17°$$

For the second order, $n = 2$,

$$\sin \theta_2 = \frac{2\lambda}{d} = 0.6$$

$$\therefore \quad \theta_2 = 37°$$

With this grating a third-order spectrum is obtained, Fig. 20.36, but not a fourth. Why?

(c) Measurement of wavelength

The wavelength of monochromatic light, for example from a sodium lamp or flame, can be measured to four-figure accuracy using a spectrometer (p. 101) and a transmission diffraction grating having 600 lines per mm (which is typical of many gratings).

After moving the eyepiece so that the cross-wires are seen clearly, the spectrometer is adjusted for parallel light by first focusing the telescope on

a distant object and then moving the position of the collimator slit so that a sharp image is obtained when it is viewed through the telescope.

The telescope is turned through 90° exactly, from the position T_1, in which it is directly opposite the illuminated collimator slit, to position T_2, Fig. 20.37a.

The grating is placed on the spectrometer table, at right angles to the line joining two of the levelling screws, say A and B, and the table turned until the image of the slit reflected from the grating is in the centre of the field of view. Adjustment of A or B may be necessary to achieve this. The plane of the grating is now parallel to the axis of rotation of the telescope.

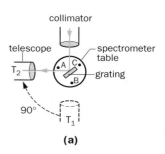

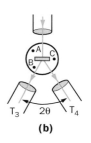

Fig. 20.37

The table (and grating) are next turned through 45° exactly, so that the incident light falls on the grating normally. The telescope is rotated, say to T_3, Fig. 20.37b, where the first-order image is seen. If the lines of the grating are parallel to the axis of rotation of the telescope the image will be central; otherwise levelling screw C will require altering. The first-order reading on the other side of the normal, at T_4, should also be taken. Half the angle between these two telescope settings gives θ, from which λ can be calculated using $\lambda = d \sin \theta$ where d is the grating spacing.

OPTICAL SPECTRA

Optical spectra fall into two basic groups, as do the spectra of all types of electromagnetic radiation when analysed by an appropriate spectrometer. The study of spectra, known as **spectroscopy**, provides information about the structure of atoms and molecules (p. 403).

(a) Emission spectra

These are obtained when light from a luminous source undergoes dispersion (formerly by a prism, nowadays often by a diffraction grating) and is observed directly. There are three types.

Line spectra consist of quite separate bright lines of definite wavelengths on a dark background and are given by luminous gases and vapours at low pressure. Sodium vapour emits two bright yellow lines which are very close together (wavelengths of 0.5896 and 0.5890 μm), Fig. 20.38a. Hydrogen emits red, blue-green and violet lines, Fig. 20.38b. Each line is an image of the slit (on the collimator) of the spectrometer on which the light falls. No two elements produce the same line spectrum and spectroscopic methods are so sensitive that they can reveal the presence of and identify the most minute quantities of materials. In general line spectra are due to the individual *atoms* concerned since in a gas (especially at low pressure) the atoms are far enough apart not to interact. They are also called **atomic** spectra.

Fig. 20.38 Line spectra

A convenient source for producing line spectra is the discharge tube. This comprises a glass tube containing a gas (e.g. hydrogen, helium or neon) or vapour (e.g. mercury or sodium) at low pressure and two metal electrodes, across which a p.d. of several thousand volts is applied. The gas or vapour conducts and a luminous discharge occurs — an effect used in neon advertising signs and certain types of street lighting.

Band spectra have several well-defined groups or bands of lines. The lines are close together at one side of each band, making this side sharper and brighter than the other, Fig. 20.39. Band spectra are complex and are obtained from molecules of glowing gases or vapours, heated or excited electrically at low pressure. They arise from interaction between atoms in each molecule. The blue inner cone of a Bunsen burner flame gives a band spectrum, as do nitrogen and oxygen in a discharge tube.

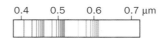

Fig. 20.39 Band spectrum

Continuous spectra are emitted by hot solids and liquids and also by hot gases at high pressures. The atoms are then so close that interaction is inevitable, and all wavelengths are emitted, Fig. 20.40. A continuous spectrum is not characteristic of the source and is conveniently produced by a white-hot tungsten lamp filament.

(b) Absorption spectra

These form the second basic group of spectra and are observed when part of the radiation emitted is absorbed by a material between the source and observer. Line, band and continuous spectra are again obtained.

A line absorption spectrum occurs when white light passes through a cool gas or vapour. Dark lines occur, against the continuous spectrum of white light, exactly at those wavelengths which are present in the line emission spectrum of the gas or vapour. The absorption spectrum of an element is therefore the same as its emission spectrum except that the latter consists of bright lines on a dark background and the former of dark lines on a bright background. The atoms of the cooler gas absorb light of the wavelengths which they can emit, and then re-radiate the same wavelengths almost immediately but in all directions. Consequently, the parts of the spectrum corresponding to those wavelengths appear dark by comparison with wavelengths which are not absorbed.

The production of the line absorption spectrum of iodine vapour is described on p. 000. The presence of a layer of relatively cooler gas round the sun causes the so-called **Fraunhöfer lines** in the solar spectrum, Fig. 20.41, which is therefore an example of a line absorption spectrum. The lines indicate the presence of hydrogen, helium, sodium, etc., in the sun's chromosphere.

Band absorption spectra are formed when a continuous emission spectrum is observed through a material which itself can emit a band spectrum. In a continuous absorption spectrum the brightness of one region of the spectrum may be reduced but not another.

Fig. 20.40 Continuous spectrum

Fig. 20.41 Fraunhöfer lines in the solar spectrum

RESOLVING POWER

This is the ability of an optical system to form separate images of objects that are very close together, and so reveal detail.

(a) Factors affecting resolving power

It can be shown that resolving power depends on the size of the slit or hole through which the objects are viewed and on the colour, i.e. the wavelength, of the light used. If the multiple light source of Fig. 20.42, with a green filter in front of it, is observed through an adjustable slit it will be found that if the slit is narrow or the viewing distance large the lamps are not seen separately but appear to be continuous.

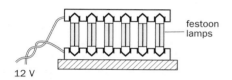

festoon lamps

12 V

Fig. 20.42

If the distance and the slit width are arranged so that the lamps can *just* be resolved, replacing the green filter by a blue one improves the resolution whilst a red one worsens it. Evidently the shorter the wavelength of the light the easier it is to see detail.

(b) Resolving power of the eye

The smallest angle θ which two points can subtend at the eye and still be seen as separate is taken as a measure of the resolving power of the eye. For example, if two well-lit black lines 2 mm apart on a card can *just* be distinguished separately by a certain observer at a distance of 5 m, then from Fig. 20.43 and using the fact that θ will be small, we have $\theta = a/d = 2 \times 10^{-3}$ m/5 m $= 4 \times 10^{-4}$ rad. That is, the resolving power of the observer's eye is 4×10^{-4} rad. The smaller θ is, the *greater* the resolving power.

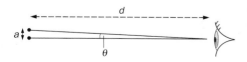

Fig. 20.43

(c) Resolving power and diffraction

A limit on the resolving power of an optical system is set by diffraction at the slit or hole through which observation occurs. A point object does not give a point image (even in an aberration-free system as we assume in geometrical optics); instead a diffraction pattern is obtained with its centre where the point image would be formed.

Rayleigh suggested that we should consider objects to be *just* resolved when the first minimum of the diffraction pattern of one falls on the central maximum of the other, Fig. 20.44. In the case of a slit of width a, the angular separation of these two fringes is λ/a (see p. 341) and in general we can take this as being *roughly* true for other diffraction apertures of width a, e.g. a circle of diameter a.

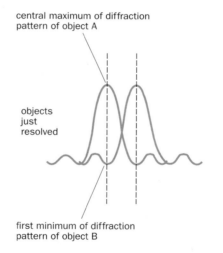

central maximum of diffraction pattern of object A

objects just resolved

first minimum of diffraction pattern of object B

Fig. 20.44   Rayleigh's criterion

Assuming the pupil of the eye has a diameter of two millimetres and that the average wavelength of light is 6×10^{-7} m, we might expect the eye to have a resolving power equal to 6×10^{-7} m/$(2 \times 10^{-3}$ m$) = 3 \times 10^{-4}$ rad. This agrees with experimental values — see section *(b)* above.

(d) Resolving power and magnifying power of optical instruments

When a lens forms an image of a small object, it acts as a circular aperture and the image is a diffraction pattern like that in Fig. 20.45. From *(c)* above it follows that the resolving power

(λ/a) of a lens is improved by increasing its diameter a or by decreasing the wavelength λ of the light used. By having a large diameter, a telescope objective gathers a large amount of light and forms a small diffraction pattern of a distant star, i.e. the image is bright, sharp and detailed.

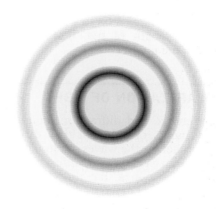

Fig. 20.45   Image of a small object through a circular lens

Note, however, that in some optical instruments, in particular many cameras, the resolving power is limited more by aberrations than by diffraction and so the lens system has to be 'stopped down' for finer resolution.

Some microscopes obtain greater resolution by using ultraviolet radiation, which has a smaller wavelength than light. In an electron microscope moving electrons behave as waves with wavelengths about 10^5 times less than those of light and so extremely small objects can be resolved.

The resolving power of an instrument sets a limit on its useful magnifying power. For example, the objective of the Mount Palomar telescope has a diameter of 5 m — an approximate resolving power of $6 \times 10^{-7}/5 = 10^{-7}$ rad. The resolving power of the eye is about 3×10^{-4} rad, so the magnifying power required to increase the angular separation from 10^{-7} to 3×10^{-4} rad is given by

$$\frac{\text{resolving power of eye}}{\text{resolving power of telescope}}$$

$$= \frac{3 \times 10^{-4}}{10^{-7}}$$

$$= 3 \times 10^3$$

Any magnification beyond this reveals no further detail; it is like stretching a rubber sheet on which there is a picture. The size of the picture increases but not the resolution.

In an electron microscope with a resolving power of 10^{-12} rad, a magnifying power of the order of 10^5 could be usefully employed. In any type of microscope the magnifying (and the resolving) power is limited basically by the wavelength of the 'illumination' used in the instrument.

POLARIZATION OF LIGHT

Interference and diffraction require a wave model of light but they do not show whether the waves are longitudinal or transverse. Polarization of light provides evidence that the waves are transverse in character.

If a lamp is viewed through a piece of Polaroid (used for example in sunglasses), apart from it appearing marginally less bright there is no effect when the Polaroid is rotated about an axis perpendicular to itself. However, using two pieces, one of which is kept at rest and the other rotated slowly, Fig. 20.46a, the light is cut off more or less completely in one position. The Polaroids are then said to be 'crossed'. Rotation through a further 90° allows maximum light transmission.

This experiment is analogous to that shown in Fig. 20.46b where transverse waves are sent along a spring to two slots B and C. The waves are generated by moving end A of the spring to and fro in *all* directions perpendicular to the direction of travel. Slot B passes only waves due to vibrations in a vertical plane. The transmitted wave is said to be **plane polarized** in a vertical plane and is unable to pass through the horizontal slot C. In this position slots B and C are 'crossed'. A longitudinal wave would emerge from both slots whatever their relative position, i.e. it cannot be polarized.

The optical experiment above can be explained if light is regarded as a transverse wave motion which the first Polaroid plane polarizes by transmitting only the light 'vibrations' in one particular plane and absorbing those in a plane at right angles. The second Polaroid transmits or absorbs the plane polarized light incident on it — depending on its orientation with

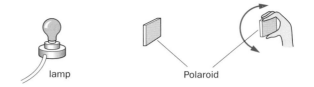

(a) Demonstrating plane polarization of light

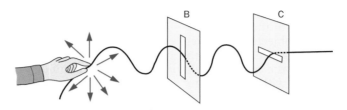

(b) A mechanical analogy

Fig. 20.46

respect to the first Polaroid. It thus acts as a detector of polarized light.

Light from the sun — and other sources — is unpolarized and consists of vibrations in every plane perpendicular to the direction of travel. Figure 20.47a is an end-on representation of unpolarized light; the arrowed lines show some of the directions in which vibrations may occur. At one point and time the vibration is in just one direction and is not the same for different points at that time. It also changes randomly and rapidly (about 10^9 times a second) at each point. Any vibration can be resolved into two perpendicular components; another end-on picture of unpolarized light is given in Fig. 20.47b. A side view of an unpolarized ray is shown in Fig. 20.47c, the dots representing vibrations at right angles to the paper and the arrowed lines vibrations in the plane of the paper.

Like other forms of electromagnetic radiation, light is also regarded as a varying electric field (E) coupled with a varying magnetic field (B), at right angles to each other and to the direction of travel. Figure 20.48 is an attempt to indicate diagrammatically an electromagnetic wave.

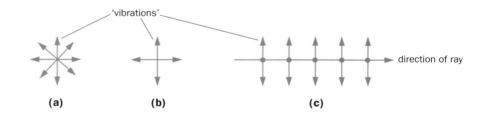

Fig. 20.47 Unpolarized light

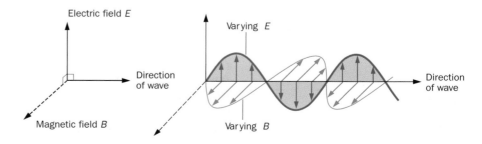

Fig. 20.48 An electromagnetic wave

Experiment shows that the coupling of light with matter is more often through the electric field, e.g. affecting a photographic film. For this reason light 'vibrations' are taken to be the variations of the electric field strength E. If light is vertically polarized we mean that the plane containing E and the direction of travel of the wave, called the **plane of vibration**, is vertical, as in Fig. 20.48.

PRODUCING POLARIZED LIGHT

(a) By Polaroid

Polaroid is made from tiny crystals of quinine iodosulphate all lined up in the same direction in a sheet of nitro-cellulose. Crystals such as these which transmit light vibrations (i.e. electric field variations) in one specific plane and absorb those in a mutually perpendicular plane are said to be **dichroic**. They produce polarization by **selective absorption**. The effect is analogous to the absorption of vertically polarized microwaves by a grid of vertical wires, Fig. 20.49a, and transmission when the wires are horizontal, Fig. 20.49b.

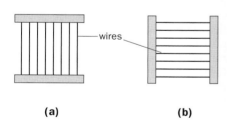

--- wires

(a)   **(b)**

Fig. 20.49

(b) By reflection

When unpolarized light falls on glass, water and some other materials the reflected light is, in general, partially plane polarized. But at one particular angle of incidence, called the **polarizing angle** i_p, the polarization is complete. At this angle, the reflected ray and the refracted ray in a transparent medium are found to be at right angles to each other. The vibrations in the reflected ray are parallel to the surface.

Applying Snell's law to Fig. 20.50 we have

$$n_1 \sin i_p = n_2 \sin r$$

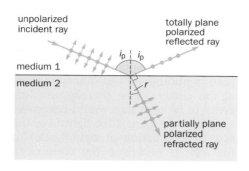

Fig. 20.50   Polarization by reflection

where n_1 and n_2 are the absolute refractive indices of media 1 and 2. Also

$$i_p + 90° + r = 180°$$
$$\therefore \quad r = 90° - i_p$$

So

$$n_1 \sin i_p = n_2 \sin (90° - i_p)$$
$$= n_2 \cos i_p$$
$$\therefore \quad \frac{n_2}{n_1} = \frac{\sin i_p}{\cos i_p} = \tan i_p$$

If medium 1 is air or a vacuum $n_1 = 1$ and so

$$n_2 = \tan i_p$$

This is **Brewster's law**. For glass of refractive index 1.5, $i_p = 57°$.

(c) By double refraction

When a crystal of calcite (a form of calcium carbonate also called 'Iceland spar') is placed over, say, a straight line, two images are seen, Fig. 20.51. Such crystals exhibit double refraction — an incident unpolarized ray is split into two rays, called the **ordinary** and **extraordinary** rays, which are plane polarized in directions at right angles to each other. The ordinary ray obeys Snell's law, but the extraordinary ray does not because it travels with different speeds in different directions in the crystal.

In a **Nicol prism**, Fig. 20.52, which is made by cutting a calcite crystal in a certain way, only the extraordinary ray (E) is transmitted. The prism may therefore be used to produce plane polarized light, as Polaroid does. If the

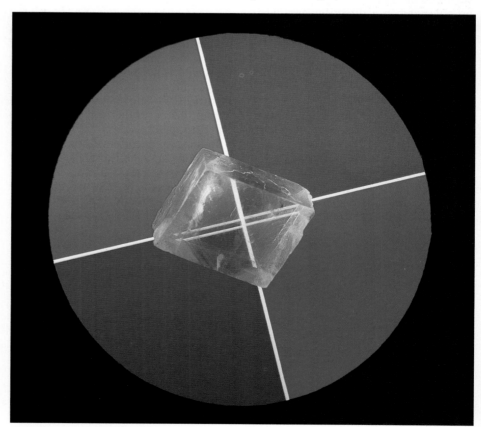

Fig. 20.51   Double refraction in calcite

angle of incidence exceeds the critical angle, total internal reflection of the ordinary ray (O) occurs at the transparent layer of Canada balsam cement; this is because it is optically less dense than the crystal for the ordinary ray but not for the extraordinary ray.

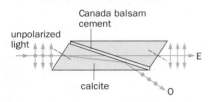

Fig. 20.52 Polarization by a Nicol prism

(d) By scattering

This may be shown by sending a narrow beam of unpolarized light through a tank of water in which scattering particles are obtained by adding one drop of milk. Polarization of the scattered light is detected by holding a piece of Polaroid in different positions, shown by the dotted lines in Fig. 20.53*a*.

The polarization arises from the transverse wave nature of light. That scattered along PA in Fig. 20.53*b* must be polarized in a direction parallel to PB (otherwise light would be longitudinal in character) and along PB the direction of polarization must be parallel to PA. What will happen if a piece of Polaroid is placed between the tank and the lamp?

USING POLARIZED LIGHT

(a) Reducing glare

Glare caused by light reflected from a smooth surface can be reduced by using polarizing materials since the reflected light is partially or completely polarized. Polaroid coatings on sunglasses reduce glare by absorbing a component of the polarized light. Polaroid discs, suitably orientated, are used in photography as 'filters' in front of the camera lens, thereby enabling detail to be seen that would otherwise be hidden by glare. Blue light from the sky, in a direction at right angles to the sun, is polarized (by scattering), so the brightness of the sky may be reduced in a colour photograph by a polarizing filter.

(b) Optical activity

Certain crystals (e.g. quartz) and liquids (e.g. sugar solutions) rotate the plane of vibration of polarized light passing through them and are said to be 'optically active'. For a solution the angle of rotation depends on its concentration and in a **polarimeter** this is used to measure concentration.

(c) Stress analysis

When glass, Perspex, polythene and some other plastics are under stress (e.g. by bending, twisting or uneven heating) they become doubly refracting and if viewed in white light between two crossed Polaroids, coloured fringes are seen round the regions of strain. The effect is called **photoelasticity** and is used to analyse stresses in plastic models of various structures. Figure 20.54 shows the fringes in a model beam under stress (see also Fig. 3.15, p. 32).

In a more recent development of the technique, actual structures are coated with a photoelastic plastic to which strains created in the structure are transmitted directly. In this case a reflection polariscope is used.

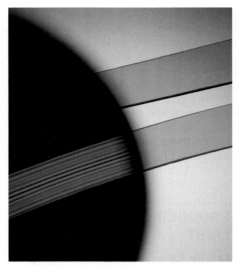

Fig. 20.54 Photoelastic stress analysis of a model beam — *top* unstressed, *bottom* stressed

(d) Liquid crystal displays (LCDs)

These are used as numerical indicators, for example in digital watches, where their very small current requirement makes them preferable to LED displays (p. 418). Normally they transmit light but when a voltage is applied the light is cut off and the crystal goes 'dark'.

Liquid crystals are organic compounds (such as cyanobiphenyls) that exhibit both solid and liquid properties. Their molecules, which are long, narrow, stiff and colourless, are arranged regularly as in a crystal structure (tending to align with each other) but the material in bulk flows like a liquid. Their use in displays depends on two effects. First, when a voltage is applied, the molecules line up with the resulting electric field (like the semolina powder in Fig. 13.8, p. 202). Second, they can rotate the plane of

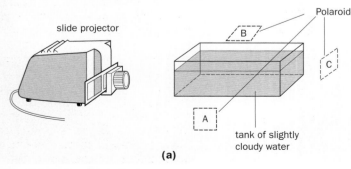

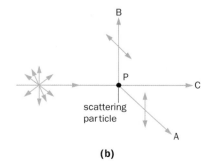

(a)   (b)

Fig. 20.53 Polarization by scattering

vibration of polarized light passing through them.

In a display, a thin liquid crystal cell is used in which the molecules have been arranged so that their alignment gradually changes through 90° from one face of the cell to the other. (One of several ways of achieving this is to gently rub a piece of glass so that fine scratches, invisible to the naked eye, are produced on it all in the same direction; the molecules of a liquid crystal poured on to the glass then tend to lie along the scratches — like matches in the grooves of corrugated paper. Two such pieces of glass with a very thin gap between them, having their scratch lines at right angles, produce a 90° twist in the alignment of the molecules.)

If the cell is set between two *crossed* polarizing films, light passes through because of the 90° twist produced by the liquid crystal, Fig. 20.55a. The glass faces of the cell each have a very thin, transparent, conducting coating and when a p.d. is applied across them, the direction of the resulting electric field is such that it alters the twist of the liquid crystal molecules which no longer rotate the polarized light, and the display goes from light to dark. When the p.d. (and field) is removed, the molecules return to their twisted state. The operation can be repeated almost indefinitely.

The pattern of the conductive coating in a 7-segment liquid crystal display for producing numbers 0 to 9 is shown in Fig. 20.55b: only the liquid crystal under those segments of the coating to which the voltage is applied

is untwisted by the electric field. The display has a silvered background which reflects back incident light; it is continuously visible (except at night when it has to be illuminated); by contrast an LED display is lit only when required, to save the battery.

ELECTROMAGNETIC WAVES

The concept that fields of force exist in the space which surrounds electric charges, current-carrying conductors and magnets is useful for 'explaining' electrical and magnetic effects. The flow of current in a conductor may also be considered to be due to the existence of an electric field in the conductor and since current can be induced by a changing magnetic field it follows that *a changing magnetic field creates an electric field* in the medium where the change occurs (whether it is free space or a material medium). This effect is electromagnetic induction and was discovered by Faraday in 1831 (p. 248).

The converse effect, i.e. *a changing electric field sets up a magnetic field*, was proposed by Maxwell in 1864 on the grounds that if this assumption is made, the mathematical equations expressing the laws of electricity and magnetism (in terms of Faraday's ideas of fields of force) become both simple and symmetrical. There is no easy way of demonstrating Maxwell's assumption directly on account of the difficulty of detecting a weak magnetic field, but one consequence is that any electric field produced by a changing

magnetic field must inevitably create a magnetic field and vice versa, i.e. one effect is coupled with the other.

On this basis Maxwell predicted that when a moving electric charge is oscillating it should radiate an electromagnetic wave consisting of a fluctuating electric field accompanied by a fluctuating magnetic field of the same frequency and phase, the fields being at right angles to each other and to the direction of travel of the wave. The intensities (E and B) of the fields vary periodically with time *like* the amplitude of a wave motion of the transverse type but, in fact, no medium seems to be involved in the transmission. A diagrammatic representation of an electromagnetic wave was given in Fig. 20.48 (p. 346).

Maxwell also showed that the speed c of all electromagnetic waves in free space, irrespective of their wavelength or origin, should be given by

$$c = \frac{1}{\sqrt{(\mu_0 \varepsilon_0)}}$$

where μ_0 and ε_0 are the permeability and permittivity respectively of free space.

Substituting numerical values in this expression gives $c = 3.0 \times 10^8 \text{ m s}^{-1}$, namely, the speed of light. This led Maxwell to suggest that light might be one form of electromagnetic wave motion.

The search for other forms of electromagnetic radiation was taken up and resulted in the discovery by Hertz in 1887 of waves having the same speed as light but with wavelengths of several metres compared with a very small fraction of a millimetre for light. These waves, now called **radio waves**, were generated by Hertz using a spark gap transmitter which produced bursts of high frequency a.c.

A whole family of electromagnetic radiations is now known, extending from gamma rays of very short wavelength to very long radio waves. Figure 20.56 shows the wavelength ranges of the various types but there is no rapid change of properties from one to the next. Although methods of production and detection differ from member to member, all exhibit reflection, refraction, interference, diffraction and polarization. When they fall on a body, partial reflection, transmission

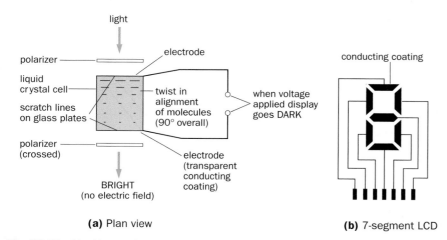

(a) Plan view

(b) 7-segment LCD

Fig. 20.55 Liquid crystal display

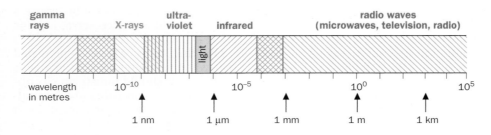

Fig. 20.56 The electromagnetic spectrum

or absorption occurs; the part absorbed becomes internal energy.

For any wave motion, $v = f\lambda$. For all electromagnetic waves in a vacuum $v = c = 3.0 \times 10^8 \text{ m s}^{-1}$ and so if f or λ for a particular radiation is known, the other can be calculated. For violet light of wavelength $\lambda = 0.40 \text{ μm} = 4.0 \times 10^{-7} \text{ m}$ we have $f = c/\lambda = (3.0 \times 10^8 \text{ m s}^{-1})/(4.0 \times 10^{-7} \text{ m}) = 7.5 \times 10^{14}$ Hz. By contrast, a 1 MHz radio signal has a wavelength of 300 m. Electromagnetic waves may be distinguished either by their wavelength or frequency but the latter is more fundamental since, unlike wavelength (and speed), it does not change when the wave travels from one medium to another. The production of various types of electromagnetic radiation will be considered later.

A wave model accounts for many of the properties of electromagnetic radiation but it cannot explain some other aspects of its behaviour (see p. 352).

INFRARED RADIATION

The wavelength of light varies from 0.4 μm for violet light to about 0.7 μm for red light. This defines the 'visible' region of the electromagnetic spectrum. Longer wavelength radiation, extending from 0.7 μm to about one millimetre, is called **infrared** and is not detected by the eye. It has all the properties of electromagnetic waves.

(a) Sources

The surfaces of all bodies emit infrared radiation in a continuous range of wavelengths. The relative amount of each wavelength depends mainly on the temperature but also on the nature of the surface of the body. At low temperatures, long-wavelength infrared is emitted; as the temperature of the body rises the emission is more copious, and shorter and shorter wavelengths are present. At about 500 °C, red light is emitted as well as long- and short-wave infrared, i.e. the body is red-hot. Increasing the temperature adds orange, yellow, green, blue and violet light in turn and at around 1000 °C the body becomes white-hot. Eventually ultraviolet radiation (p. 352) is emitted as well.

The presence of infrared in the radiation from a white-hot filament lamp may be shown with the arrangement of Fig. 20.57, in which a phototransistor is used as a detector. When this is moved slowly through the spectrum produced by the glass prism, the milliammeter gives a reading beyond the visible red region, i.e. in the infrared.

(b) Absorption

When absorbed by matter, infrared — like all other types of electromagnetic radiation — causes an increase of internal energy which usually results in a temperature rise. This is why infrared falling on the skin produces the sensation of warmth. Infrared lamps are used for the treatment of muscular complaints and to dry the paint on cars during manufacture. Remote-control keypads for televisions and video recorders use infrared signalling.

Except for wavelengths near to that of red light, infrared is absorbed by glass but transmitted by rock salt. The action of a greenhouse depends on the fact that sunlight passes through the glass, is absorbed by and warms up the plants, soil, etc., which re-radiate long-wavelength infrared (because of their comparatively low temperature). Most of this infrared cannot penetrate the glass and is trapped in the greenhouse.

Water vapour and carbon dioxide in the lower layers of the atmosphere exhibit the same 'selective absorption' effect and prevent infrared emitted by the earth from escaping. It has been estimated that if the average temperature of the earth rose by 3.5 °C (due, for example, to the combustion of fossil fuels causing an increase in carbon dioxide in the atmosphere), dramatic climatic and geographical changes could occur. Some scientists believe that 'thermal' pollution due to this greenhouse effect is as great a threat as any other form of pollution.

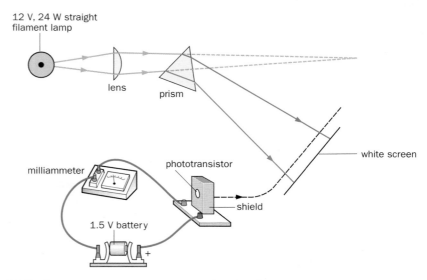

Fig. 20.57 Demonstrating the presence of infrared radiation

Fig. 20.58 Aerial infrared photograph showing rainforest clearance

(c) Detection

There are three types of detector — photographic, photoelectric (e.g. phototransistors) and thermal.

Special photographic films sensitive to infrared may be used. They enable pictures to be taken in the dark or in hazy conditions since infrared is scattered less than light by particles in the atmosphere (because of its longer wavelength). Figure 20.58 is an aerial infrared photograph of areas of a rainforest in Argentina cleared for sugar plantations; forest is brown and bare ground is grey.

Very sensitive photoelectric devices which can detect the infrared emitted by a person buried alive in a collapsed building are used to help locate and rescue victims of earthquakes. Such devices can also be used to detect the infrared emitted by a rocket and give early warning of its launching. Similar detectors are used in weather satellites to obtain cloud formation patterns over the whole of the earth's surface and are especially useful for detecting hurricanes.

Types of thermal detector include the thermometer, the thermopile and the bolometer.

A **thermopile** consists of many thermocouples in series. In a thermocouple an e.m.f. is produced between the junction of two different metals or semiconductors if one junction is at a higher temperature than the other, Fig. 20.59a. The hot junctions of a thermopile, Fig. 20.59b, are blackened to make them good absorbers of the incident radiation to be measured, while the cold junctions are shielded from it. The hot junction temperature rises until the rate of gain of heat equals the rate of loss of heat to the surroundings and then the e.m.f. is measured with a potentiometer. Alternatively it may be enough to observe the reading produced on a galvanometer connected across the thermopile.

A **bolometer** is a device such as a thermistor in which a small temperature change causes a large change of electrical resistance. In practice two thermistors are used in a Wheatstone bridge circuit, Fig. 20.60, with radiation being allowed to fall on the 'active' one but not on the 'compensating' one. The resulting temperature change creates an out-of-balance p.d. and the bridge has to be re-balanced. Greater sensitivity is achieved if the incident radiation is 'chopped', i.e. interrupted by a rotating disc with slots, so that the output from the bridge is an alternating p.d. which can be amplified.

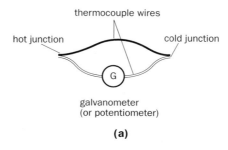

(a)

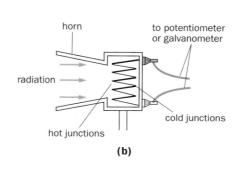

(b)

Fig. 20.59 Principle of the thermopile

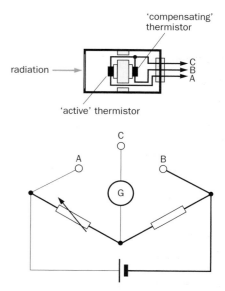

Fig. 20.60 Principle of the bolometer

ULTRAVIOLET RADIATION

Shortly after infrared was discovered in 1800, radiation beyond the violet end of the sun's visible spectrum was found. This radiation, which has typical electromagnetic wave properties, has a shorter wavelength than light and covers the range 0.4 μm to about 1 nm.

The existence of ultraviolet in the spectrum formed from an overrun filament lamp may be shown by allowing it to fall on strips of fluorescent paper and non-fluorescent white paper, one above the other, Fig. 20.61, within a darkened room. The fluorescent paper glows outside (as well as at) the violet of the visible spectrum, the latter being revealed by the non-fluorescent white paper. If an ultraviolet filter is inserted it cuts off most of the visible light.

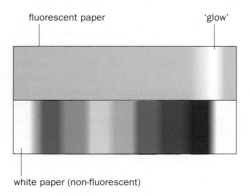

Fig. 20.61 Demonstrating the presence of ultraviolet radiation

The sun emits ultraviolet but much of it is absorbed by a layer of ozone in the earth's atmosphere. This is fortunate, for although ultraviolet produces vitamins in the skin and causes 'suntan', an overdose can be harmful and cause skin cancer, particularly if it is short-wavelength radiation. The eyes are especially vulnerable. Damage to the ozone layer by chlorofluorocarbons (CFCs) — utilized in aerosols and refrigerators — has been of recent concern.

Fluorescent materials absorb 'invisible' ultraviolet radiation and re-radiate 'visible' light. Fluorescent lights contain mercury vapour and their inner surfaces are coated with fluorescent powders which emit light of a characteristic colour when struck by ultraviolet.

A convenient source of ultraviolet radiation is an ultraviolet lamp, which contains mercury vapour and usually has a quartz bulb or window — not glass, which would absorb much of the ultraviolet.

An ultraviolet lamp should never be viewed directly.

Photoelectric devices and photographic films can detect ultraviolet.

THERMAL RADIATION

Thermal radiation is energy that moves in electromagnetic wave form, having been produced by a source because of its temperature. Its chief component is usually infrared radiation but light and ultraviolet may also be present. It is energy in the process of transfer and may be regarded as 'heat'. (See chapter 5.)

At the start of the twentieth century certain aspects of the emission of thermal radiation were responsible, along with the photoelectric effect to be discussed later (p. 390), for a revolutionary change of view about the nature of light and other types of electromagnetic radiation. This resulted in the emergence of the **quantum theory**. Here we shall deal mainly with the facts concerning the emission and absorption of thermal radiation and other related ideas.

(a) Prévost's theory of exchanges

In Fig. 20.62 the small body A, at temperature T_A, is suspended by a non-conducting thread inside the box B whose walls are at a constant, different temperature T_B. If B is then evacuated, energy exchange between A and B can occur only by radiation. If $T_A > T_B$, A's temperature falls until it is also T_B, but if $T_A < T_B$, it rises to T_B. In either case A acquires B's temperature and it might appear that energy exchange then ceases.

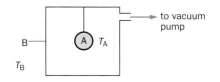

Fig. 20.62

Prévost suggested in 1792 that, on the contrary, **when a body is at the same temperature as its surroundings its rate of emission of radiation to the surroundings equals its rate of absorption of radiation from the surroundings**. That is, there is dynamic equilibrium and energy exchange continues, at a rate depending on the temperature. This view is now generally accepted.

It follows that a body which is a good absorber of radiation must also be a good emitter of radiation, otherwise its temperature would rise above that of its surroundings. Conversely, a good emitter must be a good absorber. Experiments confirm these conclusions and indicate that a dull, black surface is the best absorber and, as we might anticipate, also the best emitter.

(b) Black body radiation

The theoretical concept of a perfect absorber, called a **black body**, which **absorbs all the radiation of every wavelength falling on it**, is like the idea of an ideal gas in kinetic theory (p. 366), a useful standard for judging the performance of other bodies. At normal temperatures it must appear black since it does not reflect light — hence the term 'black body'. We would expect a black body to be the best possible emitter at any given temperature. The radiation emitted by it is called **black body radiation**, or 'full radiation' or 'temperature radiation'; this last term is used because the relative intensities of the various wavelengths present depend only on the temperature of the black body.

In practice an almost perfect black body consists of an enclosure, such as a cylinder, with dull black interior walls and a small hole, Fig. 20.63. Radiation entering the hole from outside has little chance of escaping since any energy not absorbed when the wall is first struck, will be absorbed subsequently. The *hole* in the enclosure thereby acts as a black body because it absorbs all the radiation falling on it.

When the enclosure is heated to a certain temperature, say by an electric heating coil wrapped round it, black body or 'cavity' radiation emerges from the hole (which may appear red or yellow or even white if the temperature is high enough). A spectrum of

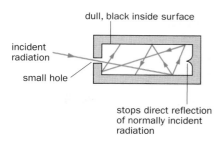

Fig. 20.63   A black body

the radiation can be formed if it falls on the slit of a suitable spectrometer and in general infrared, light and ultraviolet will be present. All three have a heating effect when absorbed and can therefore be detected by a thermopile or a bolometer. If a narrow slit is placed in front of one of these instruments and moved along the spectrum, the energy carried by the small band of wavelengths passing through the slit at different mean wavelengths can be obtained. Repeating this for different temperatures enables a family of curves to be drawn showing how the energy in the spectrum of a black body is distributed among the various wavelengths.

The curves have the form shown in Fig. 20.64; note the following points.

(*i*) As the temperature rises the energy emitted in each band of wavelengths increases, that is, the body becomes 'brighter'.

(*ii*) Even at 1000 K only a small fraction of the radiation is light. (The maximum does not lie in the visible region until about 4000 K.)

(*iii*) At each temperature T the energy radiated is a maximum for a certain wavelength λ_{max}, which decreases with rising temperature; in fact $\lambda_{max} \propto 1/T$, i.e.

$$\lambda_{max}T = \text{constant}$$

— a statement known as **Wien's displacement law**. This explains why a body appears successively red-hot, yellow-hot and white-hot, etc. Sirius (the Dog Star) looks blue, not white. Why?

It is interesting to compare the spectral emission curve of a tungsten filament lamp (a non-black body) with that of a black body at the same temperature (2000 K), Fig. 20.65.

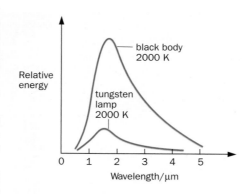

Fig. 20.65

(c) Stefan's law

This states that if E is the *total* energy radiated of all wavelengths per unit area per unit time by a black body at the thermodynamic temperature T (see p. 364), then

$$E \propto T^4$$

or

$$E = \sigma T^4$$

where σ is a constant, called **Stefan's constant**. If E is in $W\,m^{-2}$ (i.e. $J\,s^{-1}\,m^{-2}$) and T is in K (kelvins) then σ has units $W\,m^{-2}\,K^{-4}$. Its value is

$$\sigma = 5.7 \times 10^{-8}\,W\,m^{-2}\,K^{-4}$$

The area under each spectral emission curve, Fig. 20.64, represents the total energy radiated per unit area per unit time at temperature T and is found to be proportional to T^4 — agreeing with Stefan's law.

A black body at temperature T in an enclosure at a lower temperature T_0 loses energy by emission but also gains some from the enclosure. The net loss of energy E from the black body per unit area per unit time is thus given by

$$E = \sigma T^4 - \sigma T_0{}^4$$
$$= \sigma(T^4 - T_0{}^4)$$

If $T \gg T_0$ then $T_0{}^4$ can be neglected and we have

$$E = \sigma T^4$$

For a non-black body Stefan's law can be applied in the form

$$E = \varepsilon\sigma T^4$$

where ε is called the **total emissivity** of the body and has a value between 0 and 1.

(d) Optical pyrometer

A pyrometer is an instrument for measuring high temperatures and different types use all or part of the thermal radiation emitted by the hot source. The principle of the 'disappearing filament' optical pyrometer, which responds to light only, is shown in Fig. 20.66. Light from the source and the tungsten filament lamp passes through a red filter of a known wavelength range before reaching the eye; both appear red. The current through the filament is adjusted until it appears

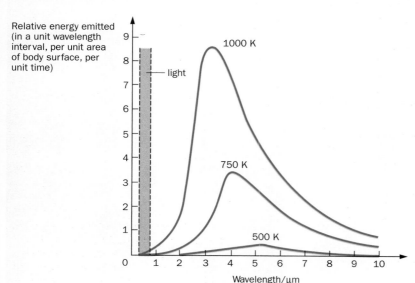

Fig. 20.64   Black body radiation

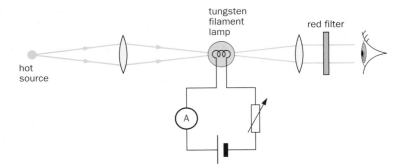

Fig. 20.66 'Disappearing filament' optical pyrometer

as bright as the background of light from the source. The temperature is then read off from the ammeter, previously calibrated in K.

Calibration may be done against a gas thermometer up to 1800 K. The range of a pyrometer can be extended up to about 3000 K by having a sectored disc rotating in front of the source which cuts down the radiation in a known proportion.

Pyrometers can be used to measure, from a safe distance, the temperature of the interiors of furnaces and the surfaces of molten metals.

(e) Modern radiation thermometers

These instruments use semiconductor infrared radiation detectors, for example silicon or germanium photodiodes, and can measure temperatures over the range −50 °C to 3500 °C to an accuracy within 1%. With electronic circuitry and LED or LCD digital displays, they are less cumbersome than their predecessors and their use is not confined just to high temperature measurements.

Th plastics and food industries use these 'non-contact' thermometers for low temperatures (0–300 °C); the steel and glass industries use them for high temperatures (300–3000 °C).

RADIO WAVES

The wavelengths of radio waves extend from 1 mm upwards (see Fig. 20.56, p. 350). Their role in astronomy is mentioned in chapters 6 and 26, and their role in telecommunications is considered in chapter 24.

Short-wavelength radio waves are utilized in radar (see p. 306) and in microwave ovens. In the latter the microwave radiation of frequency 2.45 GHz and wavelength 123 mm causes water molecules in the moisture of the food to vibrate vigorously at this particular frequency. Internal energy is thereby generated inside the food which allows it to cook itself instead of by heat from an outside source.

HOLOGRAPHY

A **hologram** is a three-dimensional image recorded on a special photographic plate by light from a laser. When developed and illuminated the image not only looks 'real' but it appears to float in space and to move when the viewer moves, as with a real object.

The term 'holography' comes from the Greek words 'holos' meaning 'whole' and 'graphis' meaning 'image'. Holography was invented in Britain in 1948 by Dr Dennis Gabor, who was later awarded a Nobel prize in recognition of the importance of his discovery. Only towards the end of the twentieth century was its full potential appreciated in the industrial, commercial and art worlds.

Figure 20.67 shows the impressive hologram of the 'Venus of Milo' statue, produced by the Optical

Fig. 20.67 Hologram of the 'Venus of Milo' statue

Laboratory at Besançon, France. At 1.5 m tall it is one of the largest holograms in the world.

(a) Making a hologram

A laser beam is first split into an object beam and a reference beam by a beam splitter, Fig. 20.68. The reference beam is reflected by a mirror on to a lens which spreads it out on to a holographic plate, set at a certain angle. The object beam is also reflected and spread out but falls on the subject, here a hand. The light waves reflected from the subject meet those of the reference beam at the surface of the plate where, being coherent (p. 299), they 'interfere' to produce the hologram.

The plate is developed in much the same way as for a photograph. When it is illuminated by light falling on it at the same angle as the reference beam, a three-dimensional image appears, floating in space and looking almost as real as the actual subject.

(b) Types of hologram

There are two main types: the transmission type, viewed by light passing through it towards the observer, and the reflection type, viewed by light reflected back from the plate to the observer. They are made by slightly different techniques.

Using a curved photographic plate, an image is obtained that can be viewed from all angles, while multiple exposures produce the impression of movement. It is also possible to magnify the image, turn it inside out or back to front. The colour of the image depends on the gas used in the laser; helium–neon gives red, argon gives green or blue, krypton gives gold-colour. With the development of more sensitive emulsions it is now possible to view holograms with ordinary white light, whereas previously laser light had to be used.

(c) Uses

Holography is being used increasingly in all walks of life. Some examples follow.

(*i*) A phonecard — used instead of cash to make a telephone call on a cardphone — contains an invisible hologram which keeps a check on the number of units used.

(*ii*) The scanners at a supermarket checkout are based on a holographic disc which reads and interprets the bar code on the merchandise.

(*iii*) Some aircraft engines and car tyres are tested for safety holographically.

(*iv*) The instrument panel in an aircraft may have a holographic display so that the pilot does not lose vital seconds in moving his or her head around.

(*v*) A credit card may contain a hologram which allows a shopkeeper to check it is not a counterfeit (since holograms cannot be forged).

(*vi*) In medicine a hologram of a living brain can reveal to surgeons the exact size and location of a tumour. Holography is also useful in the design of artificial joints and can replace 'pickled' specimens used in teaching.

(*vii*) A fuel rod in position in a working nuclear reactor can be holographed and the hologram projected and magnified for inspection to detect any distortion.

(*viii*) Holograms are used for the display and marketing of goods in the commercial world.

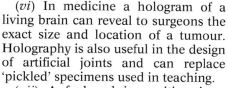

QUESTIONS

Speed of light; interference

1. A plane mirror rotating at 35 revolutions per second reflects a narrow beam of light to a stationary mirror 200 m away. The stationary mirror reflects the light normally so that it is again reflected from the rotating mirror. The light now makes an angle of 2.0 minutes of arc with the path it would travel if both mirrors were stationary. Calculate the velocity of light.

   Give *two* reasons why it is important that an accurate value of the velocity of light should be known.

2. A beam of monochromatic light of wavelength 6.0×10^{-7} m in air passes into glass of refractive index 1.5. If the speed of light in air is 3.0×10^8 m s^{-1}, calculate
   a) the speed of the light in glass,
   b) the frequency of the light, and
   c) the wavelength of the light in glass.

3. Monochromatic light illuminates a narrow slit which is 4.0 m away from a screen. Two very narrow parallel slits 0.50 mm apart are placed midway between the single slit and the screen so that interference fringes are obtained. If the spacing of five fringes is 10 mm, calculate the wavelength of the light.

   What will be the effect of
   a) halving the distance between the double slit and the screen,
   b) halving the slit separation,
   c) covering one of the double slits, and
   d) using white light?

   Discuss whether interference, diffraction or both are involved in the formation of the fringes.

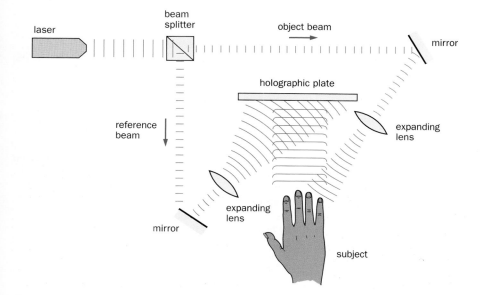

Fig. 20.68   Making a hologram

4. a) Describe, with the aid of a diagram, how you would set up the apparatus to produce and observe optical interference fringes using a double slit and any other essential apparatus.

b) A student set up a two-slit interference experiment in the laboratory to determine the wavelength of the yellow light from a sodium lamp. A travelling microscope was used to make measurements on the fringes produced and the readings obtained are shown below. The slit separation was 0.40 mm and the distance from the double slit to the plane in which the fringes were viewed was 1.80 m.

Fringe number	1	2	3	4	5	6
Scale reading/ mm	26.9	29.6	32.3	35.0	37.8	40.5

Use these values to calculate the wavelength of the yellow light.

c) For the apparatus used in b), describe and explain the effect on the appearance of the fringes of
   i) reducing the separation of the double slits but keeping the width of each slit constant,
   ii) making each of the double slits wider but keeping the slit separation constant.

d) You have been asked to demonstrate two-source interference with sound waves. Describe, with the aid of a diagram which shows the approximate distances involved, how you would do this.
(NEAB, AS/A PHO2, June 1997)

5. a) Describe a double-slit experiment to measure the wavelength of red light. Include in your answer a labelled diagram indicating appropriate dimensions, details of how the measurements are made, and an explanation of how the wavelength is calculated.

b) Explain the changes which would need to be made to the arrangement in a) to measure the wavelength of microwaves.

c) A student suggests that a double-slit experiment could be used to measure the wavelength of X-rays. Discuss why this suggestion is inappropriate.
(UCLES, Further Physics, March 1998)

6. Explain what is meant by the term **path difference** with reference to the interference of two wave-motions.

Why is it not possible to see interference where the light beams from the headlamps of a car overlap?

Interference fringes were produced by the Young's slits method, the wavelength of the light being 6.0×10^{-5} cm. When a film of material 3.6×10^{-3} cm thick was placed over one of the slits, the fringe pattern was displaced by a distance equal to 30 times that between two adjacent fringes. Calculate the refractive index of the material. To which side are the fringes displaced?

(When a layer of transparent material, whose refractive index is n and whose thickness is d, is placed in the path of a beam of light, it introduces a path difference equal to $(n - 1)d$.)

7. Give an account of the evidence that leads you to believe that light is propagated as a transverse wave motion.

In an interferometer, two coherent beams are made to follow separate paths, one traversing the length of a glass tube A, the other the length of a glass tube B, before being brought together to interfere. The tubes, each 20 cm long, are closed by plane glass plates, and initially contain air at atmospheric pressure. Fringes are obtained with light of wavelength 5.6×10^{-5} cm. Explain why the fringes move across the field of view while one of the tubes is being evacuated, and calculate the number which cross a fixed point in the field of view before a good vacuum is finally reached.

(Take the refractive index of air to be 1.000 28 at atmospheric pressure.)

8. State the conditions necessary for the effects of interference in optics to be observed.

Two microscope slides, each 7.5 cm in length are placed with two of their faces in contact. A cover glass is then inserted between them at one end to enclose a wedge-shaped layer of air. When illuminated normally by sodium light and the reflected light viewed, parallel interference bands are seen at a uniform distance apart of 0.11 mm. Explain the formation of these bands and calculate the thickness of the cover glass.

(Wavelength of sodium light may be taken as 5.9×10^{-5} cm.)

9. a) State the conditions necessary for overlapping light from two sources to exhibit interference.

b) Figure 20.69 shows light hitting the face of a glass lens at approximately normal incidence. The lens is coated with a thin layer of transparent magnesium fluoride in order to reduce the amount of light reflected by the surface.

Light reflected by the air/magnesium fluoride surface is marked **A**. Light reflected by the magnesium fluoride/glass surface is marked **B**.

When rays **A** and **B** interfere destructively, no energy is lost by reflection and so the final image formed by the lens is brighter. The magnesium fluoride layer is made of an appropriate thickness for this destructive interference to occur.

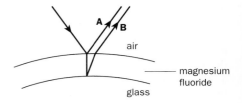

Fig. 20.69

The refractive index for light passing from air to magnesium fluoride = 1.4. The speed of light in air, $c = 3.0 \times 10^8$ m s^{-1}.
   i) Calculate the speed of light in magnesium fluoride.
   ii) Calculate the wavelength, in magnesium fluoride, of light of frequency 5.7×10^{14} Hz.
   iii) Suggest a value for the thickness of the magnesium fluoride layer such that light of frequency 5.7×10^{14} Hz will not be reflected from the lens. Briefly explain your answer.
(AEB, AS/A 0635/7, Jan 1998)

10. Describe how you would set up the necessary apparatus to observe and measure Newton's rings. Derive an expression for the difference in the squares of the radii of two consecutive bright rings in the pattern.

A set of Newton's rings was produced between one surface of a biconvex lens and a glass plate, using green light of wavelength 5.46×10^{-5} cm. The diameters of two particular bright rings, of orders of interference p and $p + 10$, were found to be 5.72 mm and 8.10 mm respectively. When the space between the lens surface and the plate was filled with liquid the corresponding values were 4.95 mm and 7.02 mm.

Determine the radius of curvature of the lens surface and the refractive index of the liquid.

Diffraction; spectra; polarization

11. Calculate the angular separation of the red and violet rays of the first-order spectrum when a parallel beam of white light is incident normally on a diffraction grating with 5000 lines per cm. Take the wavelengths of red and violet light to be 7×10^{-7} m and 4×10^{-7} m respectively.

12. When a source of monochromatic light of wavelength 6.0×10^{-7} m is viewed 2.0 m away through a diffraction grating, two similar lamps are 'seen' — one on each side of it and each 0.30 m away from it. Calculate the number of lines in the grating.

13. Explain the action of a plane diffraction grating. Giving the necessary theory and experimental detail, explain how it may be used to measure the wavelength of light from a monochromatic source. It may be assumed that the initial adjustments to the spectrometer have been made.

With the given grating used at normal incidence a yellow line, $\lambda = 6.0 \times 10^{-7}$ m, in one spectrum coincides with a blue line, $\lambda = 4.8 \times 10^{-7}$ m, in the next order spectrum. If the angle of diffraction for these two lines is 30° calculate the spacing between the grating lines.

14. Describe *two* experiments to illustrate the diffraction of light. How, in general, do the effects produced depend on the size of the obstacle or aperture relative to the wavelength of the light used?

A diffraction grating consists of a number of fine equidistant wires each of diameter 1.00×10^{-3} cm with spaces of width 0.75×10^{-3} cm between them. A parallel beam of infrared radiation of wavelength 3.00×10^{-4} cm falls normally on the grating, behind which is a converging lens made of material transparent to the infrared. Find the angular positions of the second order maxima in the diffraction pattern.

15. a) Explain why the apertures (or lines) of a diffraction grating for visible light should be
  i) narrow compared to the wavelength of visible light,
  ii) close together,
  iii) equally spaced.

b) Sodium light of wavelength 589 nm falls normally on a diffraction grating which has 600 lines per mm. Calculate the angle between the directions in which the first order and second order maxima, on the same side of the straight through position, are observed.
(NEAB, AS/A PHO2, Feb 1997)

16. a) Distinguish between emission spectra and absorption spectra. Describe the spectrum of the light emitted by
  i) the sun,
  ii) a car headlamp fitted with yellow glass, and
  iii) a sodium vapour street lamp.

b) What are the approximate wavelength limits of the visible spectrum? How would you demonstrate the existence of radiations whose wavelengths lie just outside these limits?

17. The space probe Magellan used radar astronomy to carry out a survey in order to produce a map of the surface of Venus. The receiving dish of the space probe had a diameter of 3.7 m and, at a height of 200 km above the surface of the planet, it could resolve features with a minimum diameter of 120 m.

a) Describe the technique of **radar astronomy**.

b) i) Calculate the angle subtended at the dish of the space probe by an object having a diameter of 120 m.
  ii) Hence calculate the approximate wavelength used by the space probe to obtain the data.

c) Suggest a reason why data about the surface of Venus could *not* be obtained by optical measurements.

d) Radar can also be used to determine the speed of rotation of a planet. Explain, giving physical principles, how this can be done.
(AQA: NEAB, AS/A PHO4, March 1999)

18. Copy and complete Fig. 20.70 to show the different regions of the electromagnetic spectrum.

State *four* differences between radio waves and sound waves.

Two radio stations broadcast at frequencies of 198 kHz and 95.8 MHz. Which station broadcasts at the longer wavelength?

Why do obstacles such as buildings and hills present less of a problem for the reception of the signal from the

station transmitting at the longer wavelength?
(L, AS/A PH2, June 1998)

19. Describe and explain the action of one device for producing a beam of plane polarized light.

How would you make an experimental determination of the refractive index of a polished sheet of black glass using this device?

Give two practical applications of polarized light.

20. a) Explain what is meant by
  i) unpolarized light,
  ii) plane polarized light.

b)

Fig. 20.71

A light source appears bright when viewed through two pieces of polaroid, as shown in Fig. 20.71. Describe what is seen when B is slowly rotated through 180° in its own plane.

c) i) Which of the following categories of waves can be polarized?
  radio
  ultrasonic
  microwaves
  ultra-violet
  ii) State your criterion for deciding which.
(NEAB, AS/A PHO2, June 1997)

21. a) The light from a new type of laser appears to be yellow. This yellow light is, in fact, a mixture of red and green light, each of a single wavelength.
  i) Describe how a diffraction grating may be used to make accurate determinations of the wavelengths of the red and green light.
  ii) Discuss quantitatively the relevant properties of the grating.
  iii) Explain why a double-slit method would not be suitable for making an accurate determination of these wavelengths.

Radio waves	

Fig. 20.70

b) Visible light is an electromagnetic wave. Compare the properties of light waves with those of infrared and ultraviolet waves.

(*OCR*, Further Physics 4833, March 1999)

22. a) Figure 20.72 shows a plane transmission grating and its effect on a beam of laser light.

What two optical processes are involved when a grating produces multiple beams such as those shown in Fig. 20.72?

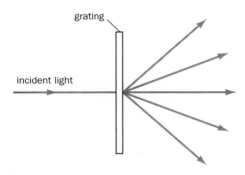

Fig. 20.72

b) Another grating has 500 lines per millimetre and is used with laser light of wavelength 633 nm at normal incidence. Calculate how many beams appear in the multiple pattern that emerges from the right-hand side of the grating.

(*OCR*, 6844, June 1999)

Thermal radiation

23. The cylindrical element of a 1.0 kW electric fire is 30.0 cm long and 1.0 cm in diameter. Taking the temperature of the surroundings to be 20 °C and Stefan's constant to be 5.7×10^{-8} W m^{-2} K^{-4}, estimate the working temperature of the element.

Give *two* reasons why the actual temperature is likely to differ from your estimate.

24. State **Stefan's law** and explain what is meant by **Prévost's theory of exchanges**.

What is meant by a **black body radiator**? Draw labelled curves which show how the energy in the spectrum of the radiation from such a body varies with wavelength at various temperatures.

A copper sphere, at a temperature of 727 °C, is suspended in an evacuated enclosure the walls of which are maintained at 27 °C. If the radius of the sphere is 5.0 cm, and if conditions are such that Stefan's law is obeyed, calculate the initial rate of fall of temperature of the sphere.

(Density of copper = 9.0×10^3 kg m^{-3}; specific heat capacity of copper = 400 J kg^{-1} K^{-1}; Stefan's constant = 5.8×10^{-8} W m^{-2} K^{-4}.)

25. Discuss the nature of the processes by which a hot body may lose heat to its surroundings.

A blackened platinum strip of area 0.20 cm^2 is placed at a distance of 200 cm from a white-hot iron sphere of diameter 1.0 cm, so that the radiation falls normally on the strip. The radiation causes the temperature, and hence the resistance, of the platinum to increase. It is found that the same increase in resistance can be produced under similar conditions, but in the absence of radiation, when a current of 3.0 mA is passed through the platinum strip (the potential difference between its ends being 24 mV). Estimate the temperature of the iron sphere.

(Stefan's constant = 5.7×10^{-8} W m^{-2} K^{-4})

26. a) Give the equation which represents **Stefan's law** for a black body source of radiation, defining the symbols used.

b) i) The space probe Voyager I is travelling out of the solar system and is, at present, approximately 8000×10^6 km from the Sun. The power source on board is a nuclear generator providing 400 W. If this power had to be obtained from solar panels, calculate the area of the solar panels which would be necessary.

power received on Earth from the Sun
= 1400 W m^{-2}
distance of the Earth from the Sun
= 150×10^6 km

ii) State *one* assumption you have made in the calculation.

(*AQA: NEAB*, AS/A PHO4, March 1999)

Objective-type revision questions 4

The first figure of a question number gives the relevant chapter, e.g. **18.2** is the second question for chapter 18.

MULTIPLE CHOICE

Select the response that you think is correct.

18.1 Which one of the following statements is *not* correct? For interference to occur between two sets of waves
A each set must have a constant wavelength
B the two sets must have the same wavelength
C the two sets must not be polarized or must have corresponding polarizations
D the waves must be transverse
E the waves must have similar amplitudes.

18.2 Which of the graphs in Fig. R32 best represents the variation of the frequency of waves with wavelength if their speed remains constant?

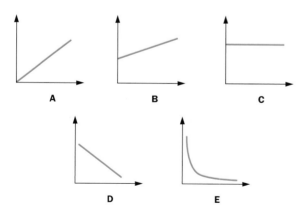

Fig. R32

18.3 Figure R33 shows two coherent sources of waves vibrating in phase and separated by a distance of 8.0 cm. P is the nearest point to the axis at which constructive interference occurs in a plane 24 cm from the line joining the sources. If S_1P is 26 cm, the wavelength of the waves in cm is
A 4.0 **B** 8.0 **C** 10 **D** 16 **E** 18

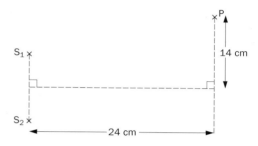

Fig. R33

18.4 Two waves each of amplitude 1.5 mm and frequency 10 Hz are travelling in opposite directions with velocity 20 mm s^{-1}. The distance in mm between adjacent nodes is
A 1.0 **B** 1.5 **C** 2.0 **D** 5.0 **E** 10

18.5 Figure R34 shows how the transverse displacement y of a slinky varies with the distance x from one end at a particular time. If the end of the slinky moves through two complete oscillations each second, the speed at which the wave travels along the slinky is
A 0.25 m s^{-1} **B** 0.5 m s^{-1} **C** 1.0 m s^{-1}
D 2.0 m s^{-1}

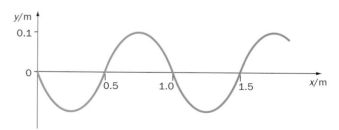

Fig. R34

19.1 A steel piano wire 0.5 m long has a total mass of 0.01 kg and is stretched with a tension of 800 N. The frequency when it vibrates in its fundamental mode is
A 2 Hz **B** 4 Hz **C** 100 Hz **D** 200 Hz
E 20 000 Hz

19.2 A closed pipe resonates at its fundamental frequency of 300 Hz. Which one of the following statements is *not* correct?

A If the pressure rises the fundamental frequency increases.

B If the temperature rises the fundamental frequency increases.

C The first overtone is of frequency 900 Hz.

D An open pipe with the same fundamental frequency has twice the length.

E If the pipe is filled with a gas of lower density the fundamental frequency increases.

19.3 If the threshold of hearing is 10^{-12} W m^{-2}, a sound level of one microwatt per m^2 is above threshold by

A 6 dB   **B** 11 dB   **C** 60 dB   **D** 110 dB   **E** 120 dB

20.1 For a monochromatic light wave passing from air to glass, which one of the following statements is true?

A Both frequency and wavelength decrease.

B Frequency increases and wavelength decreases in the same proportion.

C Frequency stays the same but wavelength increases.

D Frequency stays the same but wavelength decreases.

E Frequency and wavelength are unchanged.

20.2 In Young's double-slit interference experiment using green light the fringe width was observed to be 0.20 mm. If red light replaces green light the fringe width becomes

A 0.31 mm   **B** 0.25 mm   **C** 0.20 mm   **D** 0.16 mm
E 0.13 mm

(Wavelength of green light = 5.2×10^{-7} m; of red light = 6.5×10^{-7} m.)

20.3 Newton's rings are formed in the air gap between a plane glass surface and the convex surface of a lens which is in contact with it. If the first bright ring has a radius of 2.0×10^{-4} m, the radius of the second bright ring in m is

A 1.4×10^{-4} **B** 2.8×10^{-4}
C 3.5×10^{-4} **D** 4.0×10^{-4} **E** 6.0×10^{-4}

20.4 In the diffraction of light round an obstacle, the angle of diffraction is increased when

A the wavelength of the incident light wave is increased

B the wavelength of the incident light wave is decreased

C the amplitude of the incident light wave is increased

D the amplitude of the incident light wave is decreased

E the width of the obstacle is increased

MULTIPLE SELECTION

In each question one or more of the responses may be correct. Choose one letter from the answer code given.

> *Answer **A** if i), ii) and iii) are correct.*
> *Answer **B** if only i) and ii) are correct.*
> *Answer **C** if only ii) and iii) are correct.*
> *Answer **D** if i) only is correct.*
> *Answer **E** if iii) only is correct.*

20.5 When monochromatic light is incident normally on a wedge-shaped thin air film, Fig. R35, an interference pattern may be seen by reflection. Which of the following is/are correct?

i) Parallel fringes are observed.

ii) If water is introduced into the region between the plates, the fringe separation decreases.

iii) If the angle of the wedge is increased, the fringe separation decreases.

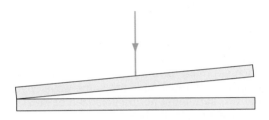

Fig. R35

20.6 When parallel light is incident at the Brewster angle in air on the surface of a glass block, which of the following statements is/are correct?

i) The refracted light is plane polarized.

ii) The reflected light is plane polarized.

iii) The reflected ray and the refracted ray are at right angles to each other.

Atoms

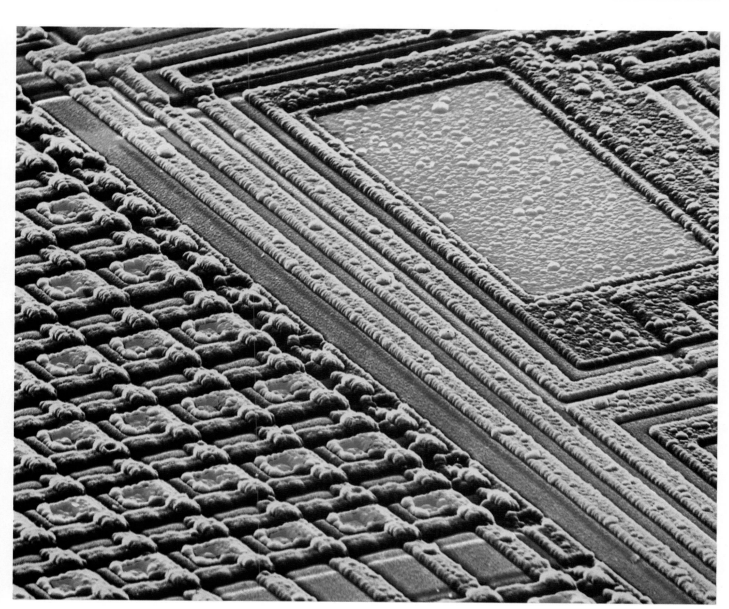

Integrated circuitry on a silicon chip

21

Kinetic theory and thermodynamics

- Introduction
- Gas laws
- Temperature scales
- Ideal gas equation
- Kinetic theory of an ideal gas
- Temperature and kinetic theory
- Deductions from kinetic theory

- Molecular magnitudes
- Properties of vapours
- Van der Waals' equation
- Thermodynamics, heat and work
- Laws of thermodynamics
- Work done by an expanding gas
- Principal heat capacities of a gas

- Isothermal, adiabatic and other processes
- Carnot's ideal heat engine
- Some calculations
- Thermodynamics and chance
- Entropy

INTRODUCTION

The behaviour of matter can be described in terms of quantities such as density, pressure and temperature. One of the aims of modern science is to relate these macroscopic (i.e. large-scale) properties of matter to the masses, speeds, energies, etc., of the constituent atoms and molecules. An attempt is made to explain the macroscopic in terms of the microscopic. If the atomic model of matter is valid this should be possible since the same thing is being considered from two different points of view. In this way we hope to obtain deeper understanding of the phenomenon under investigation.

Here we pursue this approach by studying **thermodynamics** and **kinetic theory** (a branch of statistical mechanics). In thermodynamics, thermal effects are considered using macroscopic quantities like pressure, temperature, volume and internal energy. Kinetic theory covers similar ground but assumes the existence of atoms and molecules and applies the laws of mechanics to large numbers of them, using simple averaging techniques.

We will be concerned mainly with gases. In these, conditions are simpler and the mathematical problems less difficult as a result.

GAS LAWS

Experiments have established three laws describing the thermal behaviour of gases.

(a) Boyle's law

This relates the pressure and volume of a gas and arose from work done by Boyle in about 1660.

The pressure of a fixed mass of gas is inversely proportional to its volume if the temperature is kept constant.

In symbols, if p is the pressure and V the volume then

$$p \propto \frac{1}{V} \quad \text{or} \quad p = \frac{\text{constant}}{V}$$

Therefore

$$pV = \text{constant}$$

The value of the constant depends on the mass of gas and the temperature. Graphically the law may be represented by plotting p against $1/V$, Fig. 21.1a, which is a straight line through the origin, or pV against p, Fig. 21.1b, giving a straight line parallel to the p-axis, or p against V, Fig. 21.1c, to obtain a rectangular hyperbola. If a fixed mass of gas is investigated at different temperatures, a series of graphs each called an **isothermal** (i.e. a curve of constant temperature), can be drawn as shown.

Measurements made since Boyle's time have indicated that the law is obeyed only when the density of the gas is low. Gases such as oxygen, nitrogen, hydrogen and helium (known as 'permanent' gases) follow the law well at normal temperatures and pressures but they deviate from it at high pressures (e.g. above several hundred atmospheres) when their density is high. Their p–V graphs become more

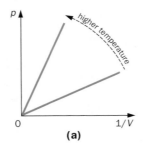

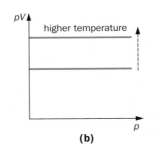

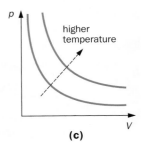

Fig. 21.1 Boyle's law

like those of liquids and the volume decrease is quite small for large pressure increases, as shown by the dashed curve in Fig. 21.2. At high pressures, carbon dioxide and some other gases and vapours are close enough to being liquid at ordinary temperatures to show marked deviations from the law.

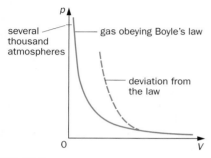

Fig. 21.2

(b) Charles' law

The connection between the volume change with temperature of a fixed mass of gas kept at constant pressure was published by Charles in 1787 (and independently by Gay-Lussac in 1802). It states that the volume V of the gas, of fixed mass and pressure, measured at temperature θ by, say, a mercury-in-glass thermometer is related approximately to its volume V_0 at the ice point (0 °C) by

$$V = V_0(1 + \alpha\theta)$$

where α is the **cubic expansivity** of the gas *at constant pressure*, which has roughly the same value of 0.00366 °C^{-1} (1/273) for all gases at low pressure. This is a linear relationship and a graph of V against θ is a straight line but does not pass through the volume origin (Fig. 21.3). How would you find α from such a graph?

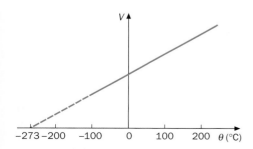

Fig. 21.3 Charles' law

(c) Pressure law

A similar relationship is found to exist between the pressure p of a fixed mass of a gas, kept at constant volume, and the temperature θ, again measured by the mercury-in-glass thermometer. It is

$$p = p_0(1 + \beta\theta)$$

where p_0 is the pressure at the ice point and β is a constant, known as the **pressure coefficient** of the gas. It has practically the same value as α.

TEMPERATURE SCALES

(a) Disagreement between thermometers

The early experimenters measured temperatures with mercury thermometers and often found that different instruments gave different readings for the same object. Later, when thermometers based on other temperature-dependent properties of matter were used (e.g. electrical resistance of a wire, pressure of a gas at constant volume) there was disagreement between the different kinds of thermometer as well (except at the fixed points where they had to agree by definition). (See p. 64.)

It became clear that one type of thermometer based on one scale of temperature would have to be chosen as a standard. Gas thermometers, either constant-volume, Fig. 21.4, or constant-pressure, showed the closest agreement over a wide range of temperatures, especially at low pressures.

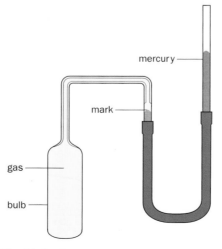

Fig. 21.4 Constant-volume gas thermometer

They were also very sensitive, accurate and gave highly reproducible readings. Although the readings depended on the nature of the gas used and on its pressure, all indicated the same reading as the pressure was lowered and approached zero.

(b) Ideal gas scale

Kelvin suggested that the standard scale of temperature should be based on an imaginary ideal gas with properties that were those of real gases at very low pressure, i.e. one that obeyed Boyle's law. He proposed that the product of the pressure and the volume of this ideal gas be used as the thermometric property. The simplest procedure is then to say that if p_1 and p_2 are the pressures and V_1 and V_2 are the corresponding volumes of a fixed mass of the gas at two temperatures T_1 and T_2 on this scale, then these temperatures are defined by

$$\frac{T_1}{T_2} = \frac{p_1 V_1}{p_2 V_2}$$

A linear relationship between pV and temperature is thus assumed and means that we have *chosen* to make equal changes of pV represent equal changes of temperature. That is, the ideal gas scale of temperature has been *defined* so that pV for a fixed mass of ideal gas is directly proportional to the temperature of the gas measured on this scale. The measurements can be imagined to be made either by a constant-volume or a constant-pressure gas thermometer or by an experiment in which neither pressure nor volume remains constant. In all cases the ratio T_1/T_2 will be the same.

If the volume of the gas is kept constant we have

$$\frac{p_1}{p_2} = \frac{T_1}{T_2}$$

which is the pressure law for an ideal gas and is a consequence of the way temperature is defined on the ideal gas scale. Similarly, at constant pressure we can say

$$\frac{V_1}{V_2} = \frac{T_1}{T_2}$$

(c) Fixed points

The unit of temperature on any scale has to be defined so that thermometers can be calibrated. This is done by choosing two fixed points and assigning numbers to them. Until the middle of the twentieth century the ice and steam points were used and, on the Celsius scale, given the values 0 °C and 100 °C respectively.

For practical reasons (one of which is the extreme sensitivity of the steam point to pressure change) the upper fixed point is now taken as the only temperature at which ice, liquid water and water vapour coexist in equilibrium and is called the **triple point** of water. It is given the value of 273.16 K (which is nearly 0.01 °C), K being the symbol for the standard (SI) unit of temperature, i.e. the kelvin. This number was chosen so that there are 100 kelvins between the ice and steam points. The temperatures of the ice, triple and steam points in °C and in K are given in Table 21.1.

Table 21.1

	°C	K
Steam point	100	373.15
Triple point	0.01	273.16
Ice point	0	273.15

The lower fixed point, to which a value of 0 K is assigned, (0 K = −273.15 °C), is called **absolute zero** and more will be said about this later.

If we use a constant-volume gas thermometer, and in the equation $T_1/T_2 = p_1/p_2$ we put

$T_1 = T$ (the unknown temperature)
$T_2 = 273.16$ (triple point)
$p_1 = p$ (pressure of the ideal gas in the constant-volume thermometer at temperature T) and
$p_2 = p_{tr}$ (pressure at the triple point)

we get

$$T = 273.16\left(\frac{p}{p_{tr}}\right)$$

(d) Measuring temperature on the ideal gas scale

To find an unknown temperature T using a constant-volume gas thermometer, the pressure p_{tr} at the triple point has to be found when the bulb is surrounded by water at 273.16 K in a triple-point cell, Fig. 21.5. The pressure p at the unknown temperature must also be determined (keeping the volume constant) and then the value of $273.16(p/p_{tr})$ calculated.

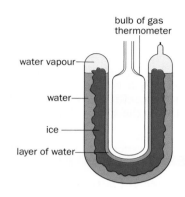

Fig. 21.5 Triple-point cell

Some of the gas in the thermometer is then removed, so reducing the pressure, and the new values for p_{tr} and p measured. The value of $273.16(p/p_{tr})$ is again obtained and is found to be different. This process is repeated for smaller and smaller amounts of gas in the bulb and a graph of the calculated value of $273.16(p/p_{tr})$ plotted against p_{tr}.

Graphs for different gases at the temperature of steam in equilibrium with water at standard pressure are shown in Fig. 21.6. When extrapolated to $p_{tr} = 0$, they all indicate the same value, which is the temperature T of the steam point on the ideal gas scale. A constant-volume thermometer containing a real gas, e.g. helium, hydrogen or nitrogen, can thus be used to measure temperatures on the ideal gas scale.

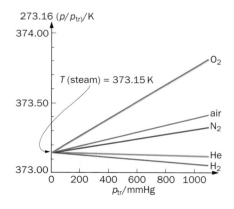

Fig. 21.6

(e) Absolute thermodynamic scale

The ideal gas scale is independent of the properties of any particular gas, but it does depend on the properties of gases in general. The absolute thermodynamic or Kelvin scale is quite independent of the properties of any substance — it is based on the efficiency of an ideal reversible heat engine (p. 376). It is a theoretical scale and the SI unit of temperature, i.e. the kelvin, is defined in terms of it, as is absolute zero (0 K).

The ideal gas scale and the Kelvin scale can be shown, in a more advanced treatment, to be theoretically identical at all temperatures. We are therefore justified in writing 'K' after an ideal gas scale temperature as we have done. The symbol for temperature in kelvins is T.

Theory also predicts that temperatures below absolute zero do not exist and certainly so far the lowest temperature reached is of the order of 10^{-6} K.

IDEAL GAS EQUATION

The equation defining temperature on the ideal gas scale is (p. 363)

$$\frac{T_1}{T_2} = \frac{p_1 V_1}{p_2 V_2}$$

or

$$\frac{p_1 V_1}{T_1} = \frac{p_2 V_2}{T_2}$$

Hence for an *ideal* gas we can say

$$\frac{pV}{T} = \text{constant}$$

where p and V represent any pair of simultaneous values of the pressure and volume of a given mass of ideal gas at the ideal gas scale temperature T.

This equation is *not* based on experiment but incorporates two definitions. The first is that an ideal gas obeys Boyle's law ($pV = $ constant) and the second is that temperature on the ideal gas scale is a quantity proportional to pV.

The value of the constant depends on the mass of gas, and experiments with real gases at low enough pressures show that it has the same value, R, for all gases if one mole is considered. R is called the **universal molar gas constant** and has a value 8.31 J mol^{-1} K^{-1}. (Check that the units are correct.) We can now write, *for one mole*,

$$pV = RT$$

This is called the **equation of state for an ideal gas**, the state being determined by the values of pressure and temperature. The ideal gas does not exist of course, but since real gases behave almost ideally at low pressure the concept is sufficiently close to actual fact to be useful.

Note the following.

(*i*)   The **mole** is the amount of substance that contains the same number of elementary entities as there are atoms in 12 grams of carbon-12. Experiment shows this to be 6.02×10^{23} — a value denoted by L and called the **Avogadro constant**, i.e. $L = 6.02 \times 10^{23}$ particles per mole, where the particles may be atoms, molecules, electrons, etc.

(*ii*)   The **relative molecular mass** M_r (known previously as the *molecular weight*) of a substance is defined by

$$M_r = \frac{\text{mass of a molecule of substance}}{\text{mass of carbon-12 atom}} \times 12$$

The values for hydrogen, carbon and oxygen are 2, 12 (by definition) and 32 respectively.

(*iii*)   The **molar mass** M_m of a substance is the mass of one mole and if expressed in kg mol^{-1} it is related (in SI units) to M_r by the equation

$$M_m = M_r \times 10^{-3}$$

M_m is then the mass of a substance (in kg) containing L molecules; its value given in SI units for oxygen is $32 \times 10^{-3}\ \text{kg mol}^{-1}$. ('Molar' means 'per unit amount of substance' and terms such as 'gram-molecule' are now obsolescent.)

(*iv*)   In $pV = RT$, V is the molar volume given by

$$V = \frac{\text{molar mass (in kg mol}^{-1}\text{)}}{\text{density (in kg m}^{-3}\text{)}}$$

$$= \frac{M_m}{\rho} \text{ (in m}^3\text{ mol}^{-1}\text{)}$$

We can also write

$$R = \frac{pM_m}{T\rho}$$

$$= \frac{pM_r}{T\rho} \times 10^{-3}$$

Fig. 21.7

Inside the hot air balloon shown in Fig. 21.7 the air stays at atmospheric pressure and so its density is inversely proportional to temperature. When the temperature of the air inside the balloon increases, the density (and hence the total weight of the balloon) falls as air is lost and so the balloon rises.

(*v*)   One mole of an ideal gas at s.t.p. has a volume of $22.4 \times 10^{-3}\ \text{m}^3$, as can be shown if we substitute $p = 1.01 \times 10^5\ \text{Pa}$, $R = 8.31\ \text{J mol}^{-1}\ \text{K}^{-1}$ and $T = 273\ \text{K}$ into $pV = RT$. $(1 \times 10^{-3}\ \text{m}^3 = 1\ \text{litre})$

(*vi*)   For n moles of an ideal gas occupying volume V,

$$pV = nRT$$

where

$$n = \frac{\text{mass of gas in kg}}{\text{molar mass (kg mol}^{-1}\text{)}}$$

KINETIC THEORY OF AN IDEAL GAS

So far, the ideal gas has been considered from the macroscopic point of view and treated in terms of quantities such as pressure, volume and temperature, which are properties of the gas as a whole. The kinetic theory, on the other hand, attempts to relate the macroscopic behaviour of an ideal gas with the microscopic properties (e.g. speed, mass) of its molecules.

(a) Assumptions

The main assumption is that the range of intermolecular forces (both attractive and repulsive) is small compared with the average distance between molecules. From this it follows that:

- the intermolecular forces are negligible except during a collision;
- the volume of the molecules themselves can be neglected compared with the volume occupied by the gas (see p. 370);
- the time occupied by a collision is negligible compared with the time spent by a molecule between collisions; and
- between collisions a molecule moves with uniform velocity.

We also assume that there is a large number of molecules even in a small volume and that a large number of collisions occurs in a small time.

(b) Calculation of pressure

An expression for the pressure exerted by an ideal gas due to molecular bombardment can be derived and the model made quantitative.

Suppose a gas in a cubic box of side l contains N molecules each of mass m. Consider one molecule moving with velocity c, as shown in Fig. 21.8. We can resolve c into components u, v and w in the three directions Ox, Oy and Oz respectively, parallel to the edges of the box. First consider motion along Ox. The molecule has momentum mu to the right and on striking the shaded wall X of the box, we can assume for mathematical simplicity that on average (since the system is in equilibrium) it rebounds with the same velocity reversed. Its momentum is now mu to the left, i.e.

$-mu$ and so, by providing an impulse (i.e. a force for a short time), the wall has caused a change of momentum of

$$mu - (-mu) = 2mu$$

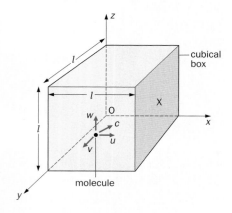

Fig. 21.8

The momentum components mv and mw have no effect on the momentum exchanges at wall X since they are directed at right angles to Ox.

If the molecule travels to the wall opposite X and rebounds back to X again without striking any other molecule on the way, it covers a distance $2l$ in time $2l/u$. This is the time interval between successive collisions of the molecule with wall X. So

rate of change of momentum at X

$$= \frac{\text{change of momentum}}{\text{time}}$$

$$= \frac{2mu}{2l/u} = \frac{mu^2}{l}$$

Newton's second law of motion states that force equals rate of change of momentum. Therefore the *total* force on wall X due to impacts by all N molecules is given by

$$\text{force on X} = \frac{m}{l}(u_1^2 + u_2^2 + \ldots + u_N^2)$$

where u_1, u_2, etc. are the Ox components of the velocities of molecules 1, 2, etc.

Since pressure is force per unit area, we can say that the pressure p on wall X of area l^2 is given by

$$p = \frac{m}{l^3}(u_1^2 + u_2^2 + \cdots + u_N^2)$$

If $\overline{u^2}$ represents the mean value of the squares of all the velocity components in the Ox direction, then

$$\overline{u^2} = \frac{u_1^2 + u_2^2 + \cdots + u_N^2}{N}$$

and $\quad N\overline{u^2} = u_1^2 + u_2^2 + \cdots + u_N^2$

$$\therefore \quad p = \frac{Nm\overline{u^2}}{l^3}$$

For any molecule, the application of Pythagoras' theorem twice to Fig. 21.9 shows that

$$c^2 = u^2 + v^2 + w^2$$

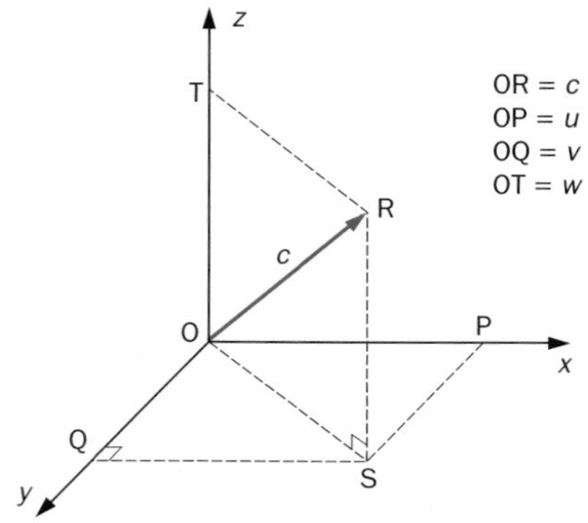

In rt. angled △ ORS: $OR^2 = OS^2 + RS^2$
In rt. angled △ OQS: $OS^2 = QS^2 + OQ^2$
∴ $OR^2 = QS^2 + OQ^2 + RS^2 = OP^2 + OQ^2 + OT^2$
∴ $c^2 = u^2 + v^2 + w^2$

Fig. 21.9

This will also hold for the mean square values, therefore

$$\overline{c^2} = \overline{u^2} + \overline{v^2} + \overline{w^2}$$

But since N is large and the molecules move randomly (i.e. they show no preference for moving parallel to any one edge of the cube, see section *(c)*, paragraph *(v)* below), it follows that the mean values of u^2, v^2 and w^2 are equal. Therefore

$$\overline{c^2} = 3\overline{u^2}$$

$$\overline{u^2} = \frac{\overline{c^2}}{3}$$

and so $\qquad p = \frac{Nm\overline{c^2}}{l^3}$

But $l^3 =$ volume of the cube $=$ volume of gas $= V$ and so

$$pV = \tfrac{1}{3}Nm\overline{c^2}$$

This important result links the macroscopic property pressure with the number, mass and speed of the molecules. We will see shortly what further information it yields.

Alternatively, if ρ is the density of the gas we can write

$$p = \tfrac{1}{3}\rho\overline{c^2}$$

since $\rho = Nm/V$, Nm being the total mass of gas.

(c) Further points

The following points about the above derivation should be noted.

(*i*)   A container of any shape could have been chosen but a cube simplifies matters.

(*ii*)   Intermolecular collisions were ignored but these would not affect the result because the average momentum of the molecules on striking the walls is unaltered by their collisions with each other.

(*iii*)   The mean-square speed $\overline{c^2}$ is not the same as the square of the mean speed. If five molecules have speeds of 1, 2, 3, 4, 5 units, their mean speed is $(1 + 2 + 3 + 4 + 5)/5 = 3$ and its square is 9. The mean-square speed is $(1^2 + 2^2 + 3^2 + 4^2 + 5^2)/5 = 55/5 = 11$

(*iv*)   The pressure on one wall only was found but it is the same on all sides (neglecting the weight of the gas).

(*v*)   Figure 21.10 shows a horizontal table with lots of small holes through which air is blown, forming a cushion on which lightweight pucks can move almost without friction. A vibrating wire around the 'air table' keeps the pucks moving randomly and compensates for the inelastic colli-

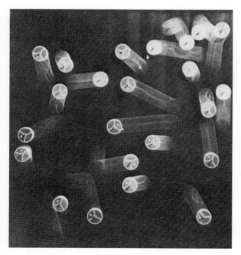

Fig. 21.10   Two-dimensional model of a gas

sions they have with each other and with the sides. The apparatus is a two-dimensional model of a gas. (The photograph was taken from above by opening a camera shutter about one second before the air supply was cut off and closing it one second afterwards. The pucks stop immediately the air supply stops and 'tracks' of their direction of motion are obtained.)

TEMPERATURE AND KINETIC THEORY

We have found from the kinetic theory that, for a mole of an ideal gas,

$$pV = \tfrac{1}{3}Lm\overline{c^2}$$

where L is the Avogadro constant. This may be rewritten as

$$pV = \tfrac{2}{3}L(\tfrac{1}{2}m\overline{c^2})$$

The ideal gas equation of state, also for a mole, is

$$pV = RT$$

where R is the universal molar gas constant. If we combine these last two equations we get

$$\tfrac{2}{3}L(\tfrac{1}{2}m\overline{c^2}) = RT$$

$$\therefore \quad \tfrac{1}{2}m\overline{c^2} = \tfrac{3}{2}\frac{R}{L}T$$

Now $\tfrac{1}{2}m\overline{c^2}$ is the average kinetic energy (k.e.) of translational motion[1] per molecule. From the above equation we see, since R and L are constants, that **for an ideal gas, the average translation k.e. is proportional to the thermodynamic temperature, T.**

The kinetic theory thus offers a microscopic basis for the macroscopic quantity of temperature. It suggests that an increase in the temperature of a gas is due to an increase of molecular speed and so of translational k.e.

The ratio R/L is called **Boltzmann's constant** and is denoted by k. It is the gas constant per molecule and has the value 1.38×10^{-23} J K^{-1}. Then we can say

$$\tfrac{1}{2}m\overline{c^2} = \tfrac{3}{2}kT$$

[1] This is motion in which every point of the body has the same instantaneous velocity and there is no rotational or vibrational motion.

The **total translational k.e. per mole of an ideal gas** is given by

$$\tfrac{1}{2}Lm\overline{c^2} = \tfrac{3}{2}RT$$

The above expression indicates that at absolute zero the molecular k.e. of an *ideal gas* is zero, but this is not so for real substances. The molecular k.e. of a real substance can be shown to have a *minimum* value, called the **zero-point energy**, which suggests that not all molecular motion stops at 0 K.

DEDUCTIONS FROM KINETIC THEORY

(a) Avogadro's law (or hypothesis)

Consider two ideal gases, 1 and 2. We can write

$$p_1 V_1 = \tfrac{1}{3}N_1 m_1 \overline{c_1^2}$$

$$p_2 V_2 = \tfrac{1}{3}N_2 m_2 \overline{c_2^2}$$

If their pressures, volumes and temperatures are the same, then

$$p_1 = p_2$$
$$V_1 = V_2$$
$$\tfrac{1}{2}m_1\overline{c_1^2} = \tfrac{1}{2}m_2\overline{c_2^2}$$
$$\therefore \quad N_1 = N_2$$

Equal volumes of different ideal gases existing under the same conditions of temperature and pressure contain equal numbers of molecules.

This statement is **Avogadro's law**. It cannot be proved directly for real gases but has to be assumed to explain the volume relationships which exist between combining gases.

(b) Dalton's law of partial pressures

The kinetic theory attributes gas pressure to bombardment of the walls of the containing vessel by molecules. In a mixture of ideal gases we might therefore expect the total pressure to be the sum of the partial pressures due to each gas. This statement is Dalton's law of partial pressures, which holds for real gases when they behave ideally. Figure 21.11, p. 368, illustrates the law.

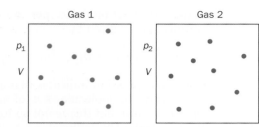

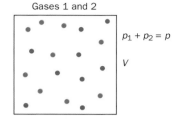

Fig. 21.11 Dalton's law

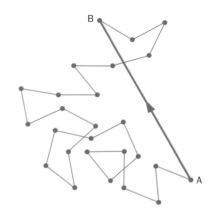

Fig. 21.13 A 'random walk'

MOLECULAR MAGNITUDES

Information about gas molecules can be obtained from the kinetic theory if use is also made of knowledge obtained from macroscopic measurements on gases.

(a) Speed of air molecules

For a gas of density ρ, exerting pressure p, we have

$$p = \tfrac{1}{3}\rho\overline{c^2}$$

where $\overline{c^2}$ is the mean-square speed of the molecules. Assuming we can apply this equation to air, for which $\rho = 1.29$ kg m^{-3} at s.t.p. (i.e. 273 K and 1.01×10^5 Pa), then

$$\overline{c^2} = \frac{3p}{\rho} = \frac{3 \times 1.01 \times 10^5}{1.29} \frac{\text{Pa}}{\text{kg m}^{-3}}$$
$$= 2.35 \times 10^5 \text{ m}^2 \text{ s}^{-2}$$

The square root of $\overline{c^2}$ is called the **root-mean-square (r.m.s.) speed**. It is denoted by $\sqrt{\overline{c^2}}$ or $c_{\text{r.m.s.}}$ (is greater than the mean speed, see p. 367) and for air has the value

$$\sqrt{\overline{c^2}} = \sqrt{(2.35 \times 10^5)} = 485 \text{ m s}^{-1}$$

This is a surprisingly high speed. Some molecules will have greater speeds, others smaller, and these will change due to energy exchanges in intermolecular collisions. Figure 21.12 shows a random distribution of molecular speeds in a fixed mass of an ideal gas at two temperatures; at the higher temperature more molecules have a higher velocity so the high temperature peak is shifted towards a higher r.m.s. speed, is broader and reduced in height compared with the low temperature peak.

The speeds of gas molecules have been measured directly and the results are in good agreement with calculated values. The fact that the speed of sound in a gas is of the same order as

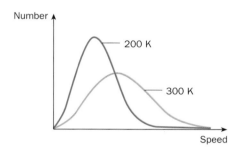

Fig. 21.12

the speed of its molecules offers some confirmation. We would not expect the speed with which the energy of a sound wave is passed on from one air molecule to the next to be greater than the speed of the air molecules themselves. In fact, it could well be less due to the random nature of molecular motion. (A 'message' carried by a 'runner' can only be transmitted as fast as he can 'run' and may travel even more slowly if there is 'traffic' congestion.) The speed of sound in air at 0 °C is about 330 m s^{-1}, which compares with the 485 m s^{-1} we calculated for the r.m.s. speed of air molecules at 0 °C.

(b) Mean free path

Despite their high average molecular speeds, gases diffuse very slowly into air, as we know from the fact that a 'smell' (even when greatly aided by convection) takes time to travel. This suggests that gas molecules, because of their finite size, suffer frequent collisions with other molecules. At each collision the direction of motion changes and the path followed is a succession of zig-zag steps — called a 'random walk'. The progress of a molecule is slow, although the sum of all the separate steps it has taken between collisions is a very large distance (Fig. 21.13).

The average length of a step between collisions is called the **mean free path**. The mean free path of an air molecule in air at atmospheric pressure is about 10^{-7} m = 100×10^{-9} m = 100 nm.

PROPERTIES OF VAPOURS

(a) Saturated and unsaturated vapours

If a very small quantity of a volatile liquid such as ether is introduced at the bottom of a mercury barometer, Fig. 21.14a, it rises to the top of the mercury column. Here it evaporates into the vacuum and the pressure exerted by the vapour causes the mercury level to fall, Fig. 21.14b. When more ether is injected, the level drops further but stops as soon as liquid ether appears on top of the column. The ether vapour is then said to be **saturated** and exerting its **saturation vapour pressure (s.v.p.)** at the temperature of the surroundings, Fig. 21.14c. Before liquid ether gathers above the mercury, the ether vapour is **unsaturated**.

> The s.v.p. of a substance is the pressure exerted by the vapour in equilibrium with the liquid.

If a vapour is in a closed vessel above the liquid, equilibrium will be reached and the vapour will then exert its s.v.p. If the space above the mercury in Fig. 21.14a contained some air initially, the s.v.p. of the ether would be the same but the evaporation would take place more slowly. At 15 °C the s.v.p. of water is 1.3 cmHg and of ether 35 cmHg.

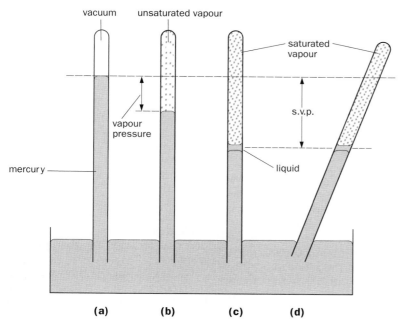

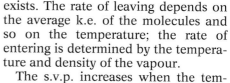

Fig. 21.14

(b) Vapours and the gas laws

Unsaturated vapours obey Boyle's law roughly up to near saturation point, i.e. along AB in the p–V graph of Fig. 21.15a. At B condensation of the vapour starts, liquid and saturated vapour exist together along BC and, since the mass of vapour is changing, Boyle's law is no longer relevant. The pressure along BC is the s.v.p. at the temperature concerned and we see that it does not change as the volume of saturated vapour decreases. This may be shown by tilting a barometer tube containing saturated vapour as in Fig. 21.14d; the vertical depth of the mercury surface below its level in Fig. 21.14a indicates the s.v.p. and remains the same as the volume of saturated vapour diminishes.

Unsaturated vapours approximately obey Charles' law and the pressure law; saturated vapours do not. The s.v.p. of a vapour does increase with temperature but much more rapidly than is required by Charles' law, as Fig. 21.15b shows for water.

A mixture of gas and unsaturated vapour obeys the gas laws fairly well, each of them exerting the same pressure as it would exert if it alone occupied the total volume (by Dalton's law of partial pressures, p. 367). A graph of p against $1/V$ for such a mixture is shown in Fig. 21.15c. What occurs at B? What does OD represent?

(c) Kinetic theory explanations

In an unsaturated vapour the rate at which molecules leave the liquid surface exceeds that at which they enter it from the vapour, i.e. evaporation occurs more rapidly than condensation. In a saturated vapour the rates are equal and dynamic equilibrium exists. The rate of leaving depends on the average k.e. of the molecules and so on the temperature; the rate of entering is determined by the temperature and density of the vapour.

The s.v.p. increases when the temperature of the liquid is raised because as the average k.e. of the molecules increases, more are able to escape from the surface. The rate of evaporation becomes greater, thereby increasing the density of the vapour and so also the rate of condensation. Eventually equilibrium and saturation are re-established at a greater s.v.p. than before.

The s.v.p. is not affected by changes of volume (at a constant temperature). If the volume available to the vapour decreases, its density momentarily increases and more molecules return to the liquid in a given time than previously. The rate at which molecules leave the surface remains steady, however, and so the rate of condensation now exceeds the rate of evaporation until the density of the vapour falls to its original value and dynamic equilibrium is restored once more, with the s.v.p. having its initial value. What happens if the volume of the vapour increases?

(d) Boiling

Whereas evaporation occurs from the surface of a liquid at all temperatures, boiling takes place at a temperature determined by the external pressure and consists in the formation of bubbles of vapour throughout the liquid. The pressure inside such bubbles must at least equal the pressure in the surrounding liquid. The pressure inside is the s.v.p. at the temperature of the boiling-point (since the vapour is in contact with liquid). The pressure outside may be taken as practically equal to atmospheric pressure for a liquid in an open vessel (if we neglect the hydrostatic pressure of the liquid itself and surface tension effects). Hence, from this qualitative treatment it would seem that **a liquid boils when its s.v.p. equals the external pressure.**

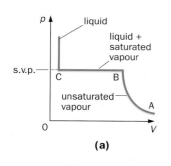

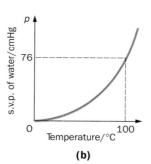

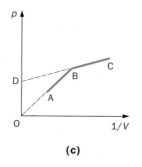

Fig. 21.15 Behaviour of vapours

VAN DER WAALS' EQUATION

The equation of state for an ideal gas $pV = RT$ holds fairly well for real gases (and unsaturated vapours) at low pressures but not at moderate to high pressures. An accurate knowledge of the deviations of real gases from ideal behaviour is important for two reasons. First, they are required to make the necessary corrections to constant-volume gas thermometer readings so that temperatures can be expressed on the thermodynamic scale. Second, they provide information about the nature of intermolecular forces and the structure of molecules.

The ideal gas equation is consistent with the kinetic theory. As a step towards obtaining an equation of state for real gases and explaining their deviations, we might consider if any of the assumptions of the theory are invalid for real gases. Van der Waals assumed that when real gases are *not* at low pressure, the range of intermolecular forces is not small compared with the average distance between the molecules (see *Assumptions*, p. 366). There are then two points to consider.

(i) *Effect of repulsive intermolecular forces.* These act at very short range, i.e. when molecules approach each other very closely, and in effect cause a reduction in the volume in which the molecules can move. This is often expressed by saying that the molecules themselves have a certain volume which is not negligible in relation to the volume V occupied by the gas. A factor called the 'co-volume' and denoted by b is introduced, making the 'free volume' $(V - b)$.

(ii) *Effect of attractive intermolecular forces.* These act over slightly greater distances and may cause two or more molecules to form loose associations or 'complexes', thereby reducing the number of particles in the gas. The observed pressure p is less than it would be if this did not happen. (The 'complexes' do not have greater momentum than single molecules because, being in thermal equilibrium with the latter they have the same average translation k.e.) It can be shown that to allow for this 'pressure defect', p, which still represents the pressure, should be replaced by $(p + a/V^2)$ where a, like b, is a 'con-

stant' that varies from gas to gas and is found by experiment.

Van der Waals' equation of state for a mole of a real gas is

$$\left(p + \frac{a}{V^2}\right)(V - b) = RT$$

The quantities a/V^2 and b become important at high pressures when the molecules are close together.

The isothermals (i.e. graphs of p against V at different, constant temperatures) predicted by van der Waals' equation for three temperatures are shown in Fig. 21.16a. Those in Fig. 21.16b were obtained experimentally by Andrews (in 1863) for carbon dioxide. The top curve is a rectangular hyperbola, the middle one (which represents a lower temperature) has a point of inflexion C and occurs at the **critical temperature**, i.e. the temperature above which a gas cannot be liquefied by increasing the pressure. The upper two van der Waals curves are very similar to the corresponding Andrews curves but the third one has a curved section PQRST whereas the lowest Andrews curve has a straight section PRT (which represents the change from liquid to vapour), although parts PQ and TS have been realised in practice.

The van der Waals' equation thus describes fairly well the behaviour of real gases above their critical temperature, but not below it. Evidently the modifications we have made to the kinetic theory are still over-simplified. Other equations of state have been tried but there is no simple one that applies to all gases under all conditions.

THERMODYNAMICS, HEAT AND WORK

Thermodynamics deals with processes that cause energy changes as a result of heat flow to or from a system and/or of work done on or by a system. Heat engines such as a petrol engine, a steam turbine and a jet engine all contain thermodynamic systems designed to transform heat into mechanical work. Heat pumps and refrigerators are thermodynamic devices for transferring energy from a cold body to a hotter one.

Heat and **work** are terms used to describe **energy in the process of transfer**.

Heat is energy that flows by conduction, convection or radiation from one body to another because of a temperature difference between them.

Note that the energy is only 'heat' as it flows. The notion that heat is something *in* a body is wrong. For example, an unlimited amount of heat can be obtained by rubbing two surfaces together and the system remains the same during the process. No definite meaning can therefore be given to the expression 'the heat in the system'.

Work is energy that is transferred from one system to another by a force moving its point of application in its own direction. The force may have a mechanical, electrical, magnetic, gravitational, etc., origin but no temperature difference is involved.

Again, experience shows that an indefinite amount of work can be put into a system without it changing and so the phrase 'work in a body' is meaningless.

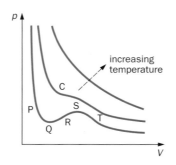

(a) Theoretical (van der Waals)

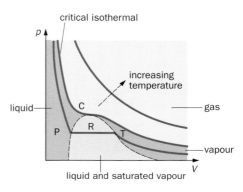

(b) Experimental (Andrews)

Fig. 21.16   Isothermals for a real gas

Energy may be transferred to or from a system by heat or by work. If the air in a bicycle pump is hot, it could either have been compressed by the piston or held near a body at a higher temperature.

The energy in a system, whether transferred to it by heat or by work, is called **internal energy** and will be considered shortly.

LAWS OF THERMODYNAMICS

(a) Zeroth law and thermal equilibrium

This basic law was proposed after the first law had been stated. It stems from the fact that all energy exchange appears to stop after a time when hot and cold bodies are brought into contact. The bodies are then said to be in 'thermal equilibrium' and we call the common property which they have **temperature**. The zeroth law is stated as follows.

> If bodies A and B are each separately in thermal equilibrium with body C, then A and B are in thermal equilibrium with each other.

For example, if C is a thermometer and reads the same when in contact with two bodies A and B, then A and B are at the same temperature. This apparently simple fact is not altogether obvious (in the realm of human relations two people who both know a third person may not know each other). In effect, the law says that there exists a useful quantity called 'temperature' and if the law were not true, taking thermometer readings would be a pointless exercise.

(b) First law and internal energy

Heat supplied to a gas (or a liquid or a solid) may

- raise its **internal energy**; and/or
- enable it to expand and do **external work** by pushing back the atmosphere — or, if it is in a cylinder, by moving a piston against a force.

In general, the internal energy of a gas consists of two components:

(i) *kinetic energy* due to the translational, rotational and vibrational[2]

[2]In a monatomic gas there is only translational motion.

motion of the molecules, all of which depend only on the temperature;

(ii) *potential energy* due to the intermolecular forces; this depends on the separation of the molecules, i.e. the volume of the gas.

In an ideal gas only the kinetic components is present (why?) and the kinetic theory shows that the translational part of it equals $3RT/2$ per mole. For real gases, both components are present, with the kinetic form predominating, and their internal energy depends on the temperature and the volume of the gas. (In a solid, k.e. and p.e. occur in roughly equal amounts.)

If δQ is the heat supplied to a mass of gas and if δW is the external work done by it then the increase of internal energy δU equals $(\delta Q - \delta W)$, since energy is conserved. So

$$\delta Q = \delta U + \delta W$$

This equation is taken as the first law of thermodynamics and is a particular case of the principle of conservation of energy. δQ is taken as positive if heat is supplied to the gas and negative if heat is transferred from it. δW is positive when external work is done by the gas (expansion) and negative if work is done on it (compression).

(c) Second law, heat engines and heat pumps

The thermodynamic processes that occur in a heat engine are represented schematically in Fig. 21.17. A 'reservoir' in thermodynamics is a source or sink of infinite heat capacity. Its temperature remains constant however much heat it supplies or receives.

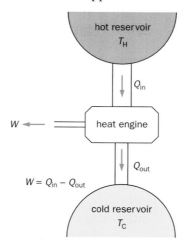

Fig. 21.17 Heat engine

In a petrol engine (p. 373) the hot reservoir (source) is provided by the combustion of fuel and the cold reservoir (sink) by the atmosphere at the exhaust. Heat Q_{in} flows from the source at temperature T_H to the engine, which transforms part of this energy into mechanical work, W, and transfers the remainder, Q_{out}, to the sink at the lower temperature, T_C. Conservation of energy requires that

$$Q_{in} = W + Q_{out}$$

or

$$W = Q_{in} - Q_{out}$$

The **efficiency** of the engine is defined as the ratio over a cycle of

$$\frac{W}{Q_{in}} = \frac{(Q_{in} - Q_{out})}{Q_{in}} = 1 - \frac{Q_{out}}{Q_{in}}$$

which increases as the temperature ratio T_H/T_C increases (see p. 376).

Use of materials that can withstand very high operating temperatures can lead to improved engine efficiency (Fig. 21.18).

Fig. 21.18 Ceramic engine rotors

In a power station, steam enters the turbines at about 830 K (560 °C) but, even so, the efficiency attained is only about 30% which means that about two-thirds of the heat absorbed is rejected to the atmosphere, usually via the cooling towers.

The heat-engine version of the second law may be stated as follows.

> No continually working heat engine can take heat from a source and use it all for work.

While all the work done on a system may become heat, the transformation of heat into mechanical energy can only occur to a limited extent, no matter how good the engine design may be.

The action of heat pumps and refrigerators is the reverse of that occurring in a heat engine. In these devices work W is done on the system which enables it to transfer heat Q_{in} from a low-temperature reservoir at T_C, to a high temperature reservoir at T_H, Fig. 21.19. Here

$$Q_{in} + W = Q_{out}$$

or

$$W = Q_{out} - Q_{in}$$

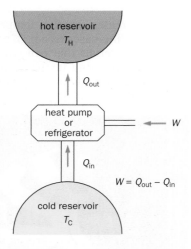

Fig. 21.19 Heat pump

The function of a heat pump is to supply as much heat as possible to the hot body, i.e. to make Q_{out} large; its **coefficient of performance** is measured by Q_{out}/W (which is greater than 1). A refrigerator, on the other hand, is designed to remove heat from the cold body, i.e. to make Q_{in} large; its coefficient of performance is given by Q_{in}/W.

Heat naturally flows from a higher temperature to a lower one and the heat-pump–refrigerator statement of the second law may be expressed in the following way.

Heat cannot be transferred continually from one body to another at a higher temperature unless external work is done.

The two statements of the second law are equivalent: they state the same thing in different ways. They cannot be proved directly but their consequences can. The law has been applied to a very wide range of phenomena.

(d) Degradation of energy

All energy ultimately becomes internal energy of the surroundings, e.g. through doing work against friction. This energy cannot be used unless we have something cooler to which to transfer it. Although energy is always conserved, it does become inaccessible and we say it is **degraded** as internal energy (see p. 160). The oceans of the world are an enormous energy reservoir but their use requires another reservoir at a lower temperature and this we unfortunately do not have.

WORK DONE BY AN EXPANDING GAS

(a) Expression for work done

Consider a mass of gas enclosed in a cylinder by a frictionless piston of cross-section area A which is in equilibrium under the action of an external force F acting to the left and a force due to the pressure p of the gas acting to the right, Fig. 21.20. We have

$$F = pA$$

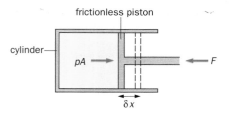

Fig. 21.20

Let the gas expand, moving the piston outwards through a distance δx which is so small that p remains practically constant during the expansion. The external work done δW by the gas against F will be

$$\delta W = F\,\delta x$$
$$= pA\,.\,\delta x$$
$$= p\,.\,\delta V$$

where $\delta V (= A\,\delta x)$ is the increase in volume of the gas. The work done by the gas in this small expansion equals its pressure multiplied by the increase of volume.

The total work done W by the gas in a finite expansion from V_1 to V_2 is given, in calculus notation, by

$$W = \int dW = \int_{V_1}^{V_2} p\,dV$$

To evaluate this integral and obtain a value for W requires the relation between p and V to be known. If p is constant during the expansion from V_1 to V_2, then

$$W = p(V_2 - V_1)$$

When a gas is compressed, work is done on it, also given by

$$\int_{V_1}^{V_2} p\,dV$$

(b) Reversible changes

To calculate the work done from the above expressions the pressure of the gas must have a determinable value. This will only be so if the changes of volume occur infinitely slowly. Suppose the external force F in Fig. 21.20 is reduced rapidly during an expansion, there is then not enough time for the gas pressure to be uniform at any stage of the expansion. It will have different values throughout the cylinder, being least near the piston and no proper value can be assigned to it.

If the quantities (p, V, T) defining the state of the system are to be expressed at every stage of the change, the system should always be in equilibrium. In other words, we have to regard it as passing through an infinite series of states of equilibrium. Such a change is called a **reversible** one because an infinitesimal change of the controlling conditions will reverse the direction of the change at every stage. In Fig. 21.20 this may be done by increasing F very slightly. The energy transformations must also be reversed exactly in a reversible change and so there must be no energy 'losses'. This means that as well as proceeding infinitely slowly, friction and heat losses must not occur.

In practice the perfect reversible change does not exist but the idea is a useful theoretical standard for judging real changes. A change that is nearly reversible is the very small alternating pressure changes in a sound wave, the departures from equilibrium being very small. The reactions in some electrical cells can also be approximately reversible. In other cases, while the initial and final states are equilibrium ones, which can be expressed in terms of p, V and T, the intermediate ones are not, especially if rapid changes occur.

(c) Indicator diagrams

An indicator diagram is a graph showing how the pressure p of a gas varies with its volume V during a change. The work done in any particular case can be derived from it.

In Fig. 21.21a below, if the pressure is $p = AB$ at the start of a very small expansion $\delta V = BC$, the work done, $p\delta V$, is represented to a good approximation by the area of the shaded strip ABCD. The total work done *by* the gas in a large expansion from V_1 to V_2 is therefore the sum of the areas of all such strips, that is, area $p_1 p_2 V_2 V_1$ and clearly depends on the shape of the p–V graph. If the graph had represented a compression of the gas from V_2 to V_1 the work done *on* it would be represented by the same area. When the gas suffers changes which eventually return it to its initial state, it is said to have undergone a **cycle** of operations and the indicator diagram is then a closed loop like that in Fig. 21.21b. The net work done *by* the gas in this case is represented by the shaded area.

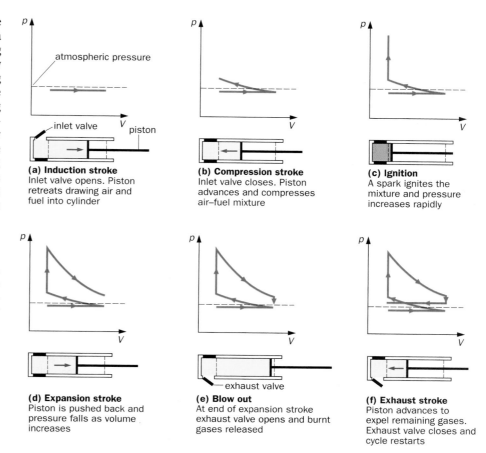

(a) Induction stroke
Inlet valve opens. Piston retreats drawing air and fuel into cylinder

(b) Compression stroke
Inlet valve closes. Piston advances and compresses air–fuel mixture

(c) Ignition
A spark ignites the mixture and pressure increases rapidly

(d) Expansion stroke
Piston is pushed back and pressure falls as volume increases

(e) Blow out
At end of expansion stroke exhaust valve opens and burnt gases released

(f) Exhaust stroke
Piston advances to expel remaining gases. Exhaust valve closes and cycle restarts

Fig. 21.22 The Otto cycle for a four-stroke petrol engine

Indicator diagrams play a very important part in the theory of heat engines. Those in Fig. 21.22 show the theoretical relations between p and V for different positions of the piston during one complete cycle of operations of a four-stroke petrol engine, known as an **Otto cycle** after the inventor of the internal combustion engine.

A practical p–V diagram for a complete cycle is similar to that in Fig. 21.22f but the sharp corners tend to be rounded, because neither the combustion of the fuel nor the opening and closing of the valves is instantaneous. Indicator diagrams are produced while the engine is working by an electronic or mechanical device called an **engine indicator**.

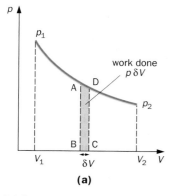

(a)

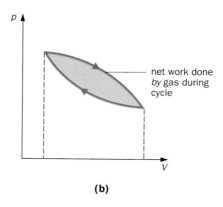

(b)

Fig. 21.21 Indicator diagrams

left-hand compartment once in every $2^6 = 64$ throws, so it would not be too common an occurrence even with only six objects.

With a very large number of objects the likelihood of them all ever being in the same compartment together becomes very remote. For example, if we have a gas jar of bromine vapour containing, say, 10^{22} molecules and we place another jar over the first, the chance of seeing all the bromine molecules back in the first jar at some future instant is so unlikely as to be impossible. Diffusion is evidently a one-way process, the result of chance when large numbers of particles are involved. By spreading, more arrangements become possible, which is what chance seems to favour.

Many mixing processes such as diffusion, the addition of milk to a cup of tea, etc. occur in one direction only; they appear ridiculous if seen in a film run backwards. Can you think of any *unmixing* processes that occur spontaneously?

(b) Thermal equilibrium

Thermal equilibrium is attained when there is no net flow of energy from one body to another. A picture, at the atomic level, of how it might occur in solids and of the difference between hot and cold bodies can be obtained using a model proposed by Einstein. We will develop the picture by means of a game that simulates the model.

In Einstein's solid the atoms vibrate more or less independently about their mean positions in the crystal lattice with internal energies (k.e. + p.e.) which, as we shall see later (p. 402), can only have certain values. If we assume the oscillation is simple harmonic, the values are simple integral multiples of a certain minimum called a **quantum** of energy. It is helpful to think of a ladder whose *equally spaced* rungs represent the definite energy values or levels, any one of which an atom can occupy. When energy is gained the atom 'jumps' from a lower to a higher level and vice versa when energy is lost. (A simple experiment supporting this idea is described on p. 405.)

If we assume quanta can move from atom to atom and do so randomly, the next step is to find out what numbers of atoms in the solid have any particular energy, i.e. what are the relative numbers of atoms in each energy level. An idea of whether there is any pattern in this *distribution* can be obtained from the following game.

Each of the 36 spaces in the grid of Fig. 21.29a represents an atom whose energy will be indicated by the number of counters in the space. An atom in its lowest energy level has no counters; one in its first level has one counter; in its second level two counters, and so on. We will start with each atom having one counter, as shown. Two dice are then thrown to give a random choice of the atom which is to lose a counter (quantum of energy). If the numbers obtained are 2 and 3 then the atom indicated may be taken as the second from the left and third up from the bottom. The counter on that space (atom) is then moved to the space (atom) given by the next throw of the two dice. This is repeated at least 100 times.

A distribution curve (in the form of a histogram) can be drawn but when chance is involved the results are most reliable if large numbers are concerned. Figure 21.29b shows the curve that would be obtained by programming a computer to make 10 000 moves with 900 spaces and 900 counters. A further 10 000 moves make little difference and the result is similar if we start with a different initial distribution or if 400 atoms and 400 quanta are used. The distribution depends on the ratio of the number of atoms to the number of quanta and not on their actual numbers.

We see from Fig. 21.29b that the number of atoms with no quanta is twice that with one quantum and the number with one quantum is twice that with two quanta and so on. The curve is an exponential one and decreases by a constant fraction as the number of quanta increases by one. It represents a **Boltzmann distribution**.

The curve for an Einstein-type solid with 300 quanta shared among 900 atoms is shown in Fig. 21.29c. The shape is still exponential but steeper, and the ratio of the numbers of atoms with a certain number of quanta to those with one more quantum is still constant, but greater (four compared with two). If we regard temperature as the quantity that determines the likelihood of energy flowing from one body to another, then the 900-atom solid with 900 quanta is hotter than the 900-atom solid with 300 quanta (it has more energy per atom), and so it is at a higher temperature.

Temperature can therefore be associated with the distribution of atoms among energy levels, i.e. with the steepness of the distribution curve. Theory shows that, in general, if q quanta are shared among N atoms, the distribution curve ratio is $(1 + N/q)$. (Check this for the cases considered.) The *larger* the value of $(1 + N/q)$ the greater the steepness and the *lower* the temperature of the solid, and vice versa. In other words, the temperature is high if the proportion of atoms in higher energy levels is high.

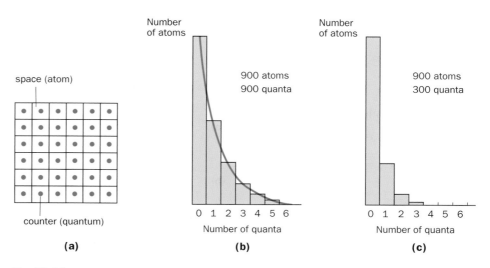

Fig. 21.29

With this interpretation it would be reasonable to suppose that heat flows from an Einstein-type solid with a small distribution curve slope to one with a large slope until both slopes are the same. When this has happened thermal equilibrium will have been reached, although quanta will still be exchanged. The computer, suitably programmed, does in fact show that the ratio of the resultant distribution curve is greater than that of the 'hot' solid but less than that of the 'cold' one.

Attaining thermal equilibrium is then a one-way process, like diffusion, and happens inevitably if left to chance. Heat flows due to the random sharing out of internal energy among the atoms of the bodies involved.

It must be remembered that Einstein's model is a simplification. In fact the atoms in a solid do not vibrate independently and consequently the energy levels of one atom affect those of others, so that to talk of 'the' energy levels of 'an' atom is not quite realistic. Neither are the atoms generally harmonic oscillators; but many of the results derived hold for more complex and 'life-like' models which require advanced mathematical treatment. The use of a model to represent a real system is a common technique in science.

ENTROPY

One-way processes such as the attainment of thermal equilibrium and diffusion can be described statistically using the difficult but fundamental concept of **entropy**. This is a measure of the 'disorder' in a system (see p. 160). It can be obtained by combining ideas about 'numbers of ways' and the 'distribution curve steepness ratio'.

(a) Effect of more quanta on W

When energy is added to a substance the number of ways W of arranging the quanta among the atoms increases. In Fig. 21.30 we see that there are 3 ways of sharing 1 quantum among 3 atoms and 6 ways of sharing 2 quanta between them. It can be shown for an Einstein solid that, in general, if there are W ways in which q quanta can be distributed among N atoms, the addition of one more

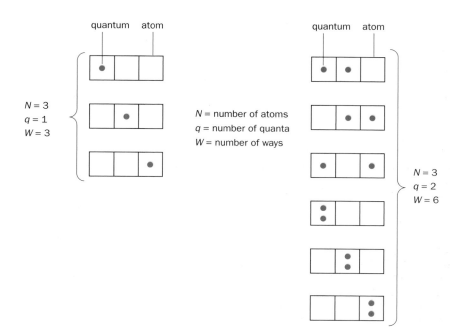

Fig. 21.30

N = number of atoms
q = number of quanta
W = number of ways

quantum increases the number of ways to W_1 where

$$\frac{W_1}{W} = \frac{N + q}{q + 1}$$

If N and q are both large (as they are in practice) then

$$\frac{W_1}{W} = \frac{N + q}{q} = 1 + \frac{N}{q}$$

The factor $(1 + N/q)$ was defined in the previous section as a measure of the steepness of the Boltzmann distribution curve, linked with temperature. We now see that it, and therefore temperature also, is related to the *change* in the number of ways of distributing quanta when energy is added (or removed).

Suppose that in a certain solid A, $q = N$, then $W_1/W = 2$ so that the loss (or gain) of one quantum would *halve* (or *double*) the number of ways of distributing the remaining quanta. On the other hand, in solid B having $q = N/3$, $W_1/W = 4$ and in this case the gain (or loss) of one quantum *quadruples* (or quarters) the number of ways of arranging the quanta.

The transfer of a quantum has a greater effect on W in solid B than in solid A. Chance favours those events which lead to an increase in W and so we would expect that bringing A and B together would lead to the net transfer of quanta from A (the hotter, since q/N is greater) to B. Thermal equilib-

rium would be established when exchange of quanta between A and B has the same effect on W for each and the process can be regarded as one which must happen inevitably since chance and large numbers are involved.

(b) Temperature and change in number of ways

From what has been said it follows that every quantum added to an Einstein solid increases W by the same factor, say a, where $a = (1 + N/q)$ and depends only on the temperature of the solid. Then if W_n is the number of ways after adding n quanta

$$\frac{W_n}{W} = a \times a \times a \times \dots (n \text{ times})$$

This is conveniently and usually expressed in natural logarithms (i.e. to base e) and if the extra quanta do not materially affect the ratio (i.e. the temperature) we have, writing $\log_e = \ln$,

$$\ln(W_n/W) = \ln W_n - \ln W$$

$$= \ln a + \ln a + \ln a + \dots (n \text{ times})$$

The number of terms on the right-hand side equals the number of quanta added and the right-hand side is therefore proportional to the change δU in the internal energy of the solid. Writing $\delta \ln W$ for $\ln W_n - \ln W$,

$$\delta \ln W \propto \delta U$$

the volume being constant and the temperature nearly so.

We associate a *high* temperature T with a *small* change in W and so if T is to be incorporated in the above expression we must write

$$\delta \ln W \propto \frac{\delta U}{T}$$

Introducing a constant k, the relation between W and T is, under the conditions given above,

$$\delta \ln W = \frac{\delta U}{kT}$$

$$\therefore \quad k \, \delta \ln W = \frac{\delta U}{T}$$

This expression, which is applicable to solids, liquids and gases as well as Einstein solids, defines temperature in terms of energy and number of ways and could, if desired, be used as the basis of a scale of temperature. However, as we have seen, T is defined on the Kelvin scale and then k is found to have the value 1.38×10^{-23} J K^{-1} — which is Boltzmann's constant (p. 367).

(c) Entropy and the direction of a process

The quantity $(k \, \delta \ln W)$ is called the **change of entropy** δS.

$$\delta S = k \, \delta \ln W = \frac{\delta U}{T}$$

It can be calculated by measuring δU and T and gives us information about the increase in the number of ways resulting from adding energy δU to matter (in any phase) at temperature T and constant volume.

Absolute zero is taken as the arbitrary zero of **entropy** S and we can write

$$S \doteq k \ln W$$

Knowing S we then have information about the direction processes will take, for we have seen that *chance favours those for which W tends to increase*. If W increases so does the entropy $(k \ln W)$ and an alternative but more basic statement of the second law of thermodynamics than those given earlier (pp. 371 and 372) is as follows.

The direction of a process is such as to increase the total entropy.

This means that any physical or chemical change only occurs if the number of ways W and the entropy S do not diminish as a result. To take a very simple example, ice and water at s.t.p. have entropies of 41 J K^{-1} mol^{-1} and 63 J K^{-1} mol^{-1}, respectively, and so when heat is supplied to melt ice, the entropy increases.

(d) Entropy and heat engines

The efficiency of a heat engine, previously discussed in terms of the Carnot cycle (p. 376), can also be viewed from the standpoint of entropy. Suppose a steam turbine receives Q_1 joules of energy per second from steam at temperature T_1. The entropy *decrease* δS_1 of the steam for the *removal* of this amount of energy is

$$\delta S_1 = \frac{\delta U}{T_1} = \frac{Q_1}{T_1}$$

The entropy in the *whole* process must increase and does so as a result of the smaller amount of energy, say Q_2 per second, which is *supplied* to the atmosphere as 'exhausted' steam at temperature T_2. The entropy *increase* δS_2 of the atmosphere for this part must be greater than δS_1 and is

$$\delta S_2 = \frac{Q_2}{T_2}$$

The *net* change of entropy δS is thus

$$\delta S = \frac{Q_2}{T_2} - \frac{Q_1}{T_1}$$

If δS is to be positive then

$$\frac{Q_2}{T_2} > \frac{Q_1}{T_1}$$

That is, for the turbine to operate without a net decrease in entropy, T_1 must be high and T_2 low so that even though $Q_1 > Q_2$, the term Q_1/T_1 is smaller than Q_2/T_2. The difference $(Q_1 - Q_2)$ is used by the turbine to do external work, e.g. to drive a dynamo and we are again led to conclude that an engine must necessarily have a maximum efficiency, i.e. it must be inefficient to some extent, since Q_2 cannot be zero.

QUESTIONS

Kinetic theory; gas laws

1. If a mole of oxygen molecules occupies 22.4×10^{-3} m^3 at s.t.p. (i.e. 273 K and 1.00×10^5 Pa), calculate the value of the molar gas constant R in J mol^{-1} K^{-1}.

2. a) Assuming the equation of state for an ideal gas, show that the number of molecules in a volume V of such a gas at pressure p and temperature T is $pVL/(RT)$ where L is the Avogadro constant and R is the molar gas constant.
b) Hence find the number of molecules in 1.00 m^3 of an ideal gas at s.t.p. (One standard atmosphere = 1.00×10^5 Pa.)

3. a)

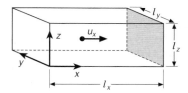

Fig. 21.31

A single gas molecule of mass m is moving in a rectangular box with a velocity of u_x in the positive x-direction as shown in Fig. 21.31. The molecule moves backwards and forwards in the box, striking the end faces normally and making elastic collisions.

i) Show that the time, t, between collisions with the shaded face is

$$t = \frac{2l_x}{u_x}$$

ii) If it is assumed that the box contains N identical molecules, each of mass m, all moving parallel to the x-direction with speed u_x and making elastic collisions at the ends, show that the average force, F, on the shaded face is given by

$$F = \frac{Nmu_x^2}{l_x}$$

iii) In a better model of molecular motion in gases, molecules of mean square speed $\overline{c^2}$ are assumed to move randomly in the box. By considering this random motion, show that a better expression for F is

$$F = \frac{Nm\overline{c^2}}{3l_x}$$

and hence derive the equation

$$pV = \tfrac{1}{3}Nm\overline{c^2}$$

b) For 1.0 mol of helium at a temperature of 27 °C, calculate
  i) the total kinetic energy of the molecules,
  ii) the root mean square speed of the molecules.

c) 1.0 mol of the gas neon is mixed with the 1.0 mol of helium in part b). Calculate the total kinetic energy of the molecules at 27 °C.

(*NEAB, AS/A PHO3, June 1998*)

4. a) With reference to the appropriate physical principles, explain the following in terms of the motion of the gas molecules.
  i) A gas in a container exerts a pressure on the container walls.
  ii) The pressure increases if the temperature of a gas is increased, keeping the mass and volume constant.

b) i) State what is meant by the root mean square speed of the molecules of a gas.
  ii) Calculate the r.m.s. speed of four molecules travelling at speeds of 400, 450, 500 and 550 m s^{-1}, respectively.

c) For a constant mass of gas, explain how the r.m.s. speed of the molecules changes, if at all, when
  i) the gas expands at constant temperature,
  ii) the gas expands by pushing back a piston so that work is done without heat entering or leaving the system.

d) The broadening of a line in the visible spectrum is proportional to the r.m.s. speed of the atoms emitting the light. Determine which source would have the greater broadening: a mercury source at 300 K or a krypton source at 77 K. Support your answer with a calculation.
molar mass of mercury = 0.200 kg mol^{-1}
molar mass of krypton = 0.083 kg mol^{-1}

(*NEAB, AS/A PHO3, June 1997*)

5. If the density of nitrogen at s.t.p. is 1.25 kg m^{-3}, calculate the root-mean-square speed of nitrogen molecules at 227 °C. (One standard atmosphere = 1.00 × 10^5 Pa.)

6. Two gas containers, A and B, have equal volumes and contain different gases at the same temperature and pressure. Use the ideal gas equation to show that there are equal numbers of molecules in the two containers.

The molecules in container A have mass four times as great as the molecules in container B. Sketch and

label graphs on the same axes (as in Fig. 21.32) to show how the molecular speeds are distributed in the two containers, given that the speed distribution curve for the molecules in container A peaks at 200 m s^{-1}.

Fig. 21.32

How does kinetic theory account for the equal pressures at the same temperature and volume?

(*L, A PH3, June 1997*)

7. a) One of the assumptions of the kinetic theory of gases is that gas molecules are in a state of random motion. State two features of random motion.

b) State three further assumptions which are made in the kinetic theory of gases.

c) The Earth's atmosphere is heated in the daytime by the Sun.
  i) State the mechanism by which energy is transferred from the Sun to the Earth.
  ii) Explain, in molecular terms, why the temperature of the atmosphere changes when sunlight is absorbed.

(*UCLES, Basic 2, June 1998*)

8. A vessel of volume 50 cm^3 contains hydrogen at a pressure of 1.0 Pa at a temperature of 27 °C. Estimate
a) the number of molecules in the vessel, and
b) their root-mean-square speed.
  (*R* = 8.3 J mol^{-1} K^{-1}; Avogadro constant = 6.0 × 10^{23} mol^{-1}; mass of 1 mole of hydrogen molecules = 2.0 × 10^{-3} kg mol^{-1}.)

9. a) i) State what is meant by **thermal equilibrium**.
  ii) Explain thermal equilibrium by reference to the behaviour of the molecules when a sample of hot gas is mixed with a sample of cooler gas and thermal equilibrium is reached.

b) A sealed container holds a mixture of nitrogen molecules and helium molecules at a temperature of 290 K. The total pressure exerted by the gas on the container is 120 kPa.

molar mass of helium
  = 4.00 × 10^{-3} kg mol^{-1}
molar gas constant *R*
  = 8.31 J K^{-1} mol^{-1}
the Avogadro constant N_A
  = 6.02 × 10^{23} mol^{-1}

  i) Calculate the root mean square speed of the helium molecules.
  ii) Calculate the average kinetic energy of a nitrogen molecule.
  iii) If there are twice as many helium molecules as nitrogen molecules in the container, calculate the pressure exerted on the container by the helium molecules.

(*AQA: NEAB, AS/A PHO3, June 1999*)

10. a) Write down the equation of state for an ideal gas.

b) The table below shows data recorded by a weather station early in the morning on a summer's day. At that time, there was negligible vertical movement of the air in the atmosphere.

Height/m	Atmospheric pressure/kPa	Atmospheric temperature/K
0 (sea level)	105.0	283
1000	89.0	278

  i) Assuming air to be an ideal gas, calculate, for these conditions, the volume of 1.00 mol of air
    1. at sea-level,
    2. at a height of 1000 m.
  ii) Hence suggest why there was negligible vertical movement of air in the atmosphere at this time.

c) Later in the day, the air at sea level increased in temperature and expanded at a constant external pressure of 105 kPa. For an initial air volume of 2000 m^3 and a temperature rise of 1.0 K, calculate
  i) the volume increase,
  ii) the work done against the external pressure.

(*OCR, Basic 2 (4832), Nov 1999*)

Thermodynamics

---

11. The working part of a freezer is a heat pump which pumps energy from the inside of a freezer to the outside. Figure 21.33 shows the energy flows for one day of operation.

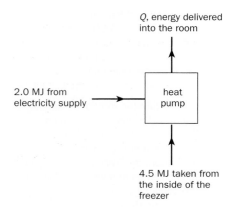

Fig. 21.33

a) What is the value of *Q*, the energy flow out of the freezer?

b) State the physical law you used to calculate your answer.

c) Suggest a reason why you need an energy source to pump energy from the inside of the freezer to the outside.

d) What is the power flow through the walls of the freezer?

e) The freezer has an internal temperature of −16 °C and is in a room whose temperature is 20 °C. The walls of the freezer are insulating foam, 5 cm thick, and have an area of 4.1 m². Calculate the thermal conductivity of the walls.

f) Four containers of liquid milk, each having mass of 2.3 kg and initially at 0 °C, are placed in the freezer. It takes 24 hours before the milk freezes. (The specific latent heat of fusion of milk is 334 kJ kg⁻¹.) What is the extra energy that the heat pump must pump out of the freezer as the milk freezes?

g) Inside the freezer there are no cooling fins at the bottom, but there are a large number of them towards the top. Explain how these fins cool the freezer and why there are none at the bottom.

(*L*, A PH3, Jan 1997)

12. Copy and label the diagram of a thermocouple thermometer shown in Fig. 21.34.

A thermocouple operates as a simple heat engine. Explain where the energy comes from, where it goes to and whether the energy transfers are heating or working.

Explain one way of increasing the efficiency of a thermocouple when considered as a heat engine.

(*L*, A PH3, Jan 1998)

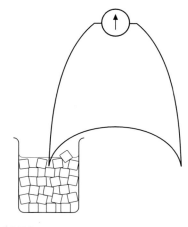

Fig. 21.34

13. The graph in Fig. 21.35 shows the three curves relating the pressure and the volume of a fixed mass of a perfect gas at three different temperatures. Curve 2 is for 0 °C.

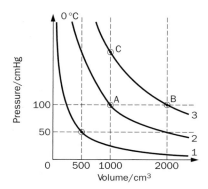

Fig. 21.35

a) Name and state the law that any one of these three curves represents.

b) What relationship exists between all the points on the three curves?

c) Name the law that connects all points on the line through A and B and express it mathematically.

d) Deduce the pressure at the point C, explaining your reasoning. (*Do not attempt to read it from the graph.*)

e) What temperatures do the curves 1 and 3 refer to?

f) How much external work (in joules) would have to be done to take the gas from the state represented at A to that at B?

g) What other energy would have to be supplied during this process?

h) Under what conditions does an ordinary gas behave as a perfect gas?

i) What properties must be postulated for the molecules of an ordinary gas to account for its deviations, under other conditions, from perfect gas behaviour? (Density of mercury = 13.6 × 10³ kg m⁻³; *g* = 10 N kg⁻¹.)

14. a) State what is meant by the **internal energy** of a system.

b) 200 kJ of thermal energy is required to vaporise 1.0 kg of liquid oxygen at atmospheric pressure (100 kPa). This vaporisation is accompanied by an increase in volume of 0.23 m³. For this vaporisation, calculate

   i) the work done on the atmosphere by the expanding oxygen,

   ii) the change in internal energy of the oxygen.

c) The volume of the gaseous oxygen is measured as its temperature increases, the pressure remaining constant. At 100 K the volume is 0.25 m³ and at 290 K the volume is 0.79 m³. Show, by means of a calculation, whether oxygen behaves as an ideal gas during this temperature change.

(*UCLES*, Basic 2, June 1998)

15. a) The equation of state of an ideal gas is

$$pV = nRT$$

For each of these symbols, state the physical quantity and the SI unit. Draw up a table like the one below.

Symbol	Physical quantity	Unit
p		
V		
n		
R		
T		

b) An ideal gas of volume 1.0 × 10⁻⁴ m³ is trapped by a piston in a cylinder as shown in Fig. 21.36. There is negligible friction between the piston and the cylinder. Initially, the temperature of the gas is 20 °C and the external atmospheric pressure acting on the piston is 100 kPa.

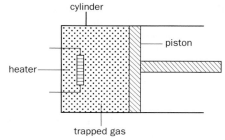

Fig. 21.36

The gas expands slowly when heat is supplied by an electric heater inside the cylinder.

   i) Calculate the work done by the gas when its volume slowly increases by 5.0×10^{-5} m³, at a constant pressure, while being heated.

   ii) What is the temperature of the gas, in °C, following its expansion?

   iii) Describe *two* changes that occur in the motion of a typical molecule of the gas during the expansion.

   (*AQA: NEAB*, AS/A PHO3, March 1999)

16. a) The equation below represents the first law of thermodynamics. State the meaning of each term.

$$\Delta U = Q + W$$

b) In an engine, 0.025 mol of air in a cylinder is compressed adiabatically by a piston. In this process the volume of the trapped air is reduced from 400 cm³ to 20 cm³ and the temperature rises from 300 K to 980 K.

   i) State which of the quantities in the first law is zero for an **adiabatic change**.

   ii) The initial pressure of the air is 1.01×10^5 Pa. Assuming air to behave as an ideal gas, calculate the pressure of the air after it has been compressed.

   iii) The work done in compressing the air is 350 J. Calculate the work which has to be done to raise the temperature of 1 mol of air by 1.0 K during the compression.

   iv) For gases, the molar heat capacity at constant volume (C_v) is different from the molar heat capacity at constant pressure (C_p). Explain briefly why these values are different.

   (*AEB*, 0635/2, Summer 1998)

17. 0.80 mole of an ideal monatomic gas is enclosed in a cylinder by a frictionless, perfectly-fitting piston. Conditions are such that the gas expands at constant pressure (A to B); it is then compressed at constant temperature (B to C) until its volume returns to the original value. These changes are represented, together with relevant numerical data, on the graph in Fig. 21.37.

a) i) Taking the molar gas constant to be 8.31 J K⁻¹ mol⁻¹, show that the temperature of the gas at A is 300 K.

   ii) Calculate the temperature of the gas at B.

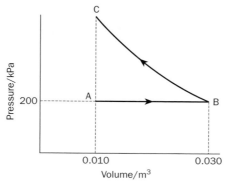

Fig. 21.37

   iii) Calculate the pressure of the gas at C.

b) Defining the terms used in your explanations, explain how the first law of thermodynamics applies to the changes

   i) from A to B,
   ii) from B to C.

c) Calculate the change in the internal energy of the gas when it is returned from C to A at constant volume. Explain whether this is an increase or a decrease.

   (*NEAB*, AS/A PHO3, Feb 1997)

18. Explain why, when quoting the specific heat capacity of a gas, it is necessary to specify the conditions under which the change of temperature occurs. What conditions are normally specified?

   A vessel of capacity 10 litres contains 1.3×10^2 grams of a gas at 20 °C and 10 atmospheres pressure. 8.0×10^3 joules of heat energy is suddenly released in the gas and raises the pressure to 14 atmospheres. Assuming no loss of heat to the vessel, and ideal gas behaviour, calculate the specific heat capacity of the gas under these conditions.

In a second experiment the same mass of gas, under the same initial conditions, is heated through the same rise in temperature while it is allowed to expand slowly so that the pressure remains constant. What fraction of the heat energy supplied in this case is used in doing external work? (Take 1 atmosphere = 1.0×10^5 pascals.)

19. Define **isothermal** and **adiabatic** changes and give the equation relating the pressure and volume of an ideal gas for each type of change.

   Why has it been concluded that the pressure and volume changes accompanying the passage of sound waves through a gas are adiabatic?

   Air occupying a volume of 10 litres at 0.0 °C and atmospheric pressure is compressed isothermally to a volume of 2.0 litres and is then allowed to expand adiabatically to a volume of 10 litres. Show the process on a *p–V* diagram and calculate the final pressure and temperature of the air. At what volume was the pressure momentarily atmospheric? Mark this point on your diagram.

   Assume $\gamma = 1.4$ for air (and that all changes are reversible).

20. a) The 1st law of thermodynamics suggests two ways in which the internal energy of a gas may be **increased**. State these *two* ways.

b) Figure 21.38 shows the four strokes in a cycle of a petrol engine. Figure 21.39 shows the variation of pressure *p* and volume *V* within the cylinder during the cycle.

   i) Sketch Fig. 21.39 and identify, with an arrow labelled **C**, the part of the graph that relates to the compression stroke of the cycle.

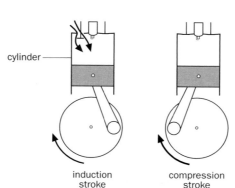
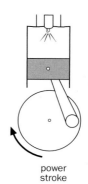
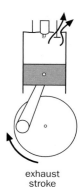

cylinder

induction stroke   compression stroke   power stroke   exhaust stroke

Fig. 21.38

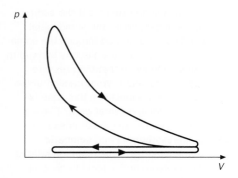

Fig. 21.39

ii) State and explain briefly what happens to the internal energy of the gases in the cylinder during the power stroke.

(AEB, 0635/2, Summer 1997)

21. Which of the following statements are correct?

a) In deriving the kinetic theory of gases it is assumed there is negligible change of momentum when molecules collide with the walls of the container.

b) The mean free path of a gas molecule equals the average distance between collisions of gas molecules.

c) The work done by an ideal gas when it expands isothermally equals the energy added to it.

d) When a fixed mass of an ideal gas in a cylinder is compressed by pushing in a piston *quickly*, the pressure, temperature and internal energy of the gas all increase.

22. a) Define:
i) an **isothermal** expansion;
ii) an **adiabatic** compression.

b) Figure 21.40 shows the variation of the pressure p with the volume V of a fixed mass of gas expanding isothermally from initial values $p_1 V_1$ to final values $p_2 V_2$.
i) Copy Fig. 21.40 and on it sketch the graph for an adiabatic expansion from V_1 to V_2 from the same initial pressure p_1. Label the final pressure p_3.
ii) By referring to the first law of thermodynamics, explain why the final pressure p_3 for the adiabatic expansion is different from the final isothermal pressure p_2.
iii) Suggest what could be done to restore the gas to the pressure p_2 at volume V_2.

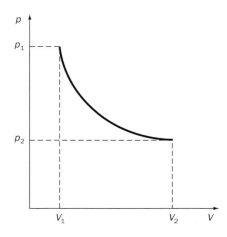

Fig. 21.40

c) State and explain an everyday example of an adiabatic change.

(OCR, 6842, March 1999)

22

Atomic physics

THERMIONIC EMISSION

In a metal each atom has a few loosely attached outer electrons which move randomly through the material as a whole. The atoms exist as positive ions in a 'sea' of free electrons. If one of these electrons near the surface of the metal tries to escape, it experiences an attractive inward force from the resultant positive charge left behind. The electron cannot escape unless an external source does work against the attractive force and thereby increases the kinetic energy of the electron. If this is done by heating the metal, the process is called **thermionic emission**.

> The **work function** Φ of a metal is the minimum energy that must be supplied to enable an electron to escape from its surface.

Φ is conveniently expressed in electronvolts (p. 206).

The smaller the work function of a metal the lower the temperature of thermionic emission; usually the temperature has to be too near the melting-point to be useful. Two materials used are (*i*) thoriated tungsten, having $\Phi = 2.6$ eV and giving good emission at about 2000 K, and, in most cases, (*ii*) a metal coated with barium oxide for which $\Phi = 1$ eV, copious emission occurring at 1200 K.

In many thermionic devices a plate, called the **anode**, is at a positive potential with respect to the heated metal and attracts electrons from it. The heated metal is then called a **hot cathode**. Hot cathodes are heated electrically either directly or indirectly. In direct heating current passes through the cathode (or filament) itself which is in the form of a wire, Fig. 22.1*a*. An indirectly heated cathode consists of a thin, hollow metal tube with a fine wire, called the **heater**, inside and separated from it by an electrical insulator, Fig. 22.1*b*. Indirect heating is most common since it allows a.c. to be used without the potential of the cathode continually varying. A typical heater supply for many thermionic devices is 6.3 V a.c., 0.3 A.

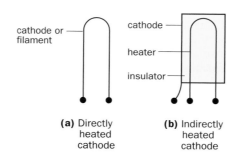

(a) Directly heated cathode

(b) Indirectly heated cathode

Fig. 22.1 Hot cathodes

CATHODE RAYS AND THE ELECTRON

(a) Properties

Streams of electrons moving at high speed are called **cathode rays**. They exhibit several important and useful effects, some of which can be demonstrated with the Maltese cross tube of Fig. 22.2, p. 386. It consists of a hot cathode and a hollow cylindrical anode enclosed in an evacuated glass bulb and with a coating of fluorescent material on the inside of the bulb. The anode is connected to the positive of a high voltage supply of 2–3 kV so that electrons from the cathode are accelerated along the tube in a divergent beam. Most bypass the anode and a dark shadow of the cross appears on the screen against a blue or green fluorescent background. This suggests that the rays are travelling in straight lines from the cathode and those not intercepted by the cross cause the screen to fluoresce.

When a magnet is brought near the side of the tube level with the anode, the beam is deflected vertically and the shadow can be made partially or wholly to disappear. Using Fleming's left-hand rule, the direction of the deflection shows that the rays behave like a flow of negative charge travelling from cathode to anode.

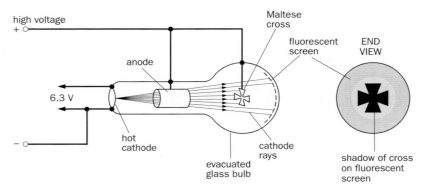

Fig. 22.2 Maltese cross tube

The beam also carries energy since the end of the tube struck by it becomes warm.

These and other properties of cathode rays may be summarized as follows.

- They travel from the cathode in straight lines.
- They cause certain substances to fluoresce.
- They possess kinetic energy.
- They can be deflected by a magnetic field.
- They can be deflected by an electric field (p. 387).
- They produce X-rays on striking matter (p. 394).

(b) Discovery of the electron

The first evidence to establish the existence of the electron is usually considered to be provided by J. J. Thomson's experiment in 1897 in which he measured the speeds and the charge-to-mass ratio (e/m), called the **specific charge**, of cathode rays. The view that cathode rays were not electromagnetic waves was based on the facts that (*i*) the speeds were typically one-tenth that of light, and (*ii*) they could be deflected by electric and magnetic fields.

The value of e/m obtained by Thomson was always the same, whatever the source or method of production of the cathode rays. This suggested that electrons are all alike and that they are universal constituents of matter. By assuming that the charge carried was equal to that on a monovalent ion in electrolysis, Thomson estimated the mass m of an electron, knowing e/m. Using modern values m is 9.11×10^{-31} kg, i.e. it is 1837 times smaller than the mass of

the hydrogen atom. Other interpretations of the value of e/m were possible; the electron might have had the same mass as a hydrogen ion but a much greater charge. However, there is now no doubt that an electron carries the fundamental unit of electric charge.

For most purposes the electron can be regarded as a sub-atomic particle with a negative charge of value e, the electronic charge, and a very small mass (since force is required to accelerate it). The value of the mass quoted above is known as the 'rest mass' — it cannot be measured directly. The term has arisen because it has been found, as predicted by Einstein in the theory of relativity (p. 506), that the mass of a particle accelerated to a speed approaching that of electromagnetic waves in a vacuum increases with speed.

Ordinary particles are characterized by size and shape but such properties cannot be stated precisely for sub-atomic particles. To some extent the size of the electron depends on the method of determination and while some measurements indicate that it is a sphere of diameter 10^{-15} m, it is also satisfactory on occasions to regard it as a dimensionless point. By contrast its charge and mass can be uniquely specified. More will be said about the nature of the electron at the end of this chapter.

ELECTRON DYNAMICS

If cathode rays are assumed to consist of particles (electrons) to which the laws of mechanics apply, we can obtain information about their speed and specific charge from their behaviour in electric and magnetic fields.

(a) Speed of electrons

Consider an electron of charge e and mass m which is emitted from a hot cathode and then accelerated by an electric field towards an anode. It experiences a force due to the field and work is done on it. The system (of field and electron) loses electrical potential energy and the electron gains kinetic energy. Let V be the p.d. between anode and cathode responsible for the field. If the electron starts from the cathode with zero speed and moves in a vacuum attaining speed v as it reaches the anode, the energy change W is given by

$$W = eV \qquad \text{(p. 206)}$$

But

$$W = \tfrac{1}{2}mv^2$$
$$\therefore \quad eV = \tfrac{1}{2}mv^2$$

From this 'energy equation' it follows that

$$v = \sqrt{\left(\frac{2eV}{m}\right)}$$

Substitution of numerical values for e/m and V shows that v is about 1.9×10^7 m s^{-1} (one-sixteenth of the speed of light) when $V = 1000$ V.

(b) Deflection of electrons by a magnetic field

It was shown earlier (p. 238) that the force F on charge Q moving with speed v at right angles to a uniform magnetic field of flux density B is

$$F = BQv$$

For an electron $Q = e$, so

$$F = Bev$$

The direction of the force is given by Fleming's left-hand rule (remembering that a negative charge moving one way is equivalent to conventional current in the opposite direction) and is always at right angles to the field and to the direction of motion. Therefore at X in Fig. 22.3 the speed of the electron remains unaltered but it is deflected from its original path to, say, Y. Here the force acting on it still has the same value, Bev, and since the direction of motion and the field continue to be mutually perpendicular, the force is perpendicular to the new direction. The force therefore only changes the direction of motion but not the speed and so the electron of mass m describes a *circular arc* of radius r. The constant radial force Bev is the centripetal force (p. 133) and so

$$Bev = \frac{mv^2}{r}$$

This equation describes the path of the electron in the magnetic field.

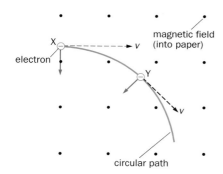

Fig. 22.3

(c) Deflection of electrons by an electric field

If electrons enter an electric field acting at right angles to their direction of motion, they are deflected from their original path. In Fig. 22.4 a p.d. applied between plates P and Q of length l creates a uniform electric field of strength E, non-uniformities at the edges of the plates being ignored.

Consider an electron of charge e, mass m and horizontal speed v on entering the field. If the upper plate is positive the electron experiences a force Ee and an acceleration Ee/m (by the second law of motion) both acting

vertically upwards. Since the field is uniform, the acceleration is uniform and combines with the initial horizontal speed v, which the electron retains during the whole of its journey in the vacuum between the plates, to give a path which we shall show is a *parabola*. The behaviour of the electron is similar to that of a projectile fired horizontally; its path (neglecting air resistance) is also a parabola, the resultant of a uniform horizontal velocity and the vertical acceleration of free fall.

The vertical displacement y of the electron at any time t is given by

$$y = \tfrac{1}{2}at^2$$

But $a = Ee/m$,

$$\therefore \quad y = \frac{1}{2} \cdot \frac{Ee}{m} \cdot t^2$$

The corresponding horizontal displacement x is given by

$$x = vt$$

Eliminating t,

$$y = \left(\frac{Ee}{2mv^2}\right) \cdot x^2$$

This equation is of the form $y = kx^2$ (where $k = Ee/(2mv^2) = $ a constant) and so the path of the electron between the plates is a parabola.

The deflection D of the electron (i.e. its displacement from the original direction) on a screen distance L from the centre of the plates, can be obtained using the fact that it continues in a straight line after leaving the field. From Fig. 22.4, $\tan \theta \approx D/L$, where $\tan \theta$ is the slope of the tangent at the end of the parabolic path. The slope equals, in calculus terms, the differential coefficient ds/dx when $x = l$. By differentiating $s = kx^2$ we obtain

$ds/dx = 2kx$; therefore the slope is $2kl$ and so $D/L = 2kl$, giving $D = 2klL$; and, substituting for k, we get $D = EelL/(mv^2)$. If V is the p.d. which has accelerated the electron to speed v then $eV = \tfrac{1}{2}mv^2$ and so

$$D = \frac{ElL}{2V}$$

This shows that D is proportional to E if V is constant, and inversely proportional to V if E is constant.

(d) Worked example

(i) An electron emitted from a hot cathode in an evacuated tube is accelerated by a p.d. of 1.0×10^3 V. Calculate the kinetic energy and speed acquired by the electron. ($e = 1.6 \times 10^{-19}$ C, $m = 9.1 \times 10^{-31}$ kg.)

We have
$$\tfrac{1}{2}mv^2 = eV$$
$$= (1.6 \times 10^{-19} \text{ C}) \times (1.0 \times 10^3 \text{ J C}^{-1})$$
$$= 1.6 \times 10^{-16} \text{ J}$$
$$\therefore \quad v = \sqrt{\left(\frac{2 \times 1.6 \times 10^{-16} \text{ J}}{9.1 \times 10^{-31} \text{ kg}}\right)}$$
$$= 1.8 \times 10^7 \text{ m s}^{-1}$$

(ii) The electron now enters at right angles a uniform magnetic field of flux density 1.0×10^{-3} T. Determine its path.

The magnetic force Bev on the electron makes it describe a circular path of radius r given by

$$Bev = \frac{mv^2}{r}$$

so
$$r = \frac{mv}{Be}$$

where
$B = 1.0 \times 10^{-3}$ T,
$m = 9.1 \times 10^{-31}$ kg,
$e = 1.6 \times 10^{-19}$ C,
and from (i),
$v = 1.8 \times 10^7$ m s^{-1}.
Therefore

$$r =$$
$$\frac{(9.1 \times 10^{-31} \text{ kg})(1.8 \times 10^7 \text{ m s}^{-1})}{(1.0 \times 10^{-3} \text{ T})(1.6 \times 10^{-19} \text{ C})}$$

$$= 0.10 \text{ m}$$

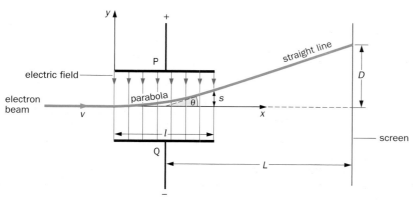

Fig. 22.4

Check the units to ensure that $\mathrm{kg\,m\,s^{-1}\,T^{-1}\,C^{-1}}$ reduces to m, using the fact that $1\,\mathrm{T} = 1\,\mathrm{N\,A^{-1}\,m^{-1}}$.

(*iii*) Find the intensity of the uniform electric field which, when applied perpendicular to the previous magnetic field so as to be co-terminous with it, compensates for the magnetic deflection. If the electric field plates are $2.0 \times 10^{-2}\,\mathrm{m}$ apart what is the p.d. between them?

> For the electric and magnetic forces to balance
>
> $$Ee = Bev$$
>
> So
>
> $$E = Bv$$
> $$= (1.0 \times 10^{-3}\,\mathrm{T})(1.8 \times 10^{7}\,\mathrm{m\,s^{-1}})$$
> $$= 1.8 \times 10^{4}\,\mathrm{V\,m^{-1}}$$
>
> (Check that $1\,\mathrm{V\,m^{-1}}$ is the same as $1\,\mathrm{T\,m\,s^{-1}}$, using the facts that $1\,\mathrm{V} = 1\,\mathrm{J\,C^{-1}}$ and $1\,\mathrm{C} = 1\,\mathrm{A\,s}$.)
>
> If V_1 is the p.d. between the plates and d their separation, then $E = V_1/d$ and so
>
> $$V_1 = Ed$$
> $$= (1.8 \times 10^{4}\,\mathrm{V\,m^{-1}})$$
> $$\times (2.0 \times 10^{-2}\,\mathrm{m})$$
> $$= 3.6 \times 10^{2}\,\mathrm{V}$$

SPECIFIC CHARGE OF THE ELECTRON

Two methods of determining the specific charge of the electron that can be performed in a school laboratory will be outlined. The value is

$$\frac{e}{m} = 1.76 \times 10^{11}\,\mathrm{C\,kg^{-1}}$$

(a) Cathode ray tube method using crossed fields

This method is similar in principle to that developed by J. J. Thomson in which an electron beam is subjected simultaneously to mutually perpendicular, i.e. crossed, electric and magnetic fields. Here a vacuum-type cathode ray tube is used, Fig. 22.5, having a hot cathode C and an anode A with a horizontal collimating slit from which the electrons emerge in a flat beam. The beam produces a narrow lumi-

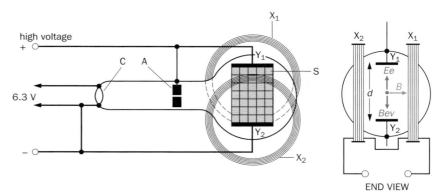

Fig. 22.5 Determining e/m with a cathode ray tube

nous trace when it hits a vertical fluorescent screen S, marked in squares. S is supported by two parallel deflecting plates Y_1, Y_2, across which a p.d. (about 3 kV) is applied, thereby creating an electric field between them.

Helmholtz coils X_1, X_2 (p. 236) mounted on opposite sides of the bulb produce a magnetic field between the plates, at right angles to both the direction of travel of the beam and the electric field. The coils are separated by a distance equal to their radius and when connected in series — so that the current has the same direction in each — they give an almost uniform field for a short distance along their common axis midway between them.

Consider an electron of charge e and mass m which emerges from the slit in the anode having been accelerated to speed v. Let E be the electric field strength between Y_1 and Y_2 and B the magnetic flux density along the axis of X_1 and X_2. When E and B are such that the electron suffers no deflection, the electric force Ee on it must be equal and opposite to the magnetic force Bev.

$$Ee = Bev \qquad (1)$$

If the electron is emitted from the cathode with zero speed and moves in a good vacuum, its kinetic energy $\frac{1}{2}mv^2$ is given by

$$\tfrac{1}{2}mv^2 = eV \qquad (2)$$

where V is the accelerating p.d. between anode and cathode. Eliminating v from (1) and (2),

$$\frac{e}{m} = \frac{E^2}{2B^2V}$$

If the p.d. between Y_1 and Y_2 creating the electric field equals the accelerating p.d., then $E = V/d$ (i.e. field

strength = potential gradient) where d is the separation of Y_1 and Y_2. Therefore

$$\frac{e}{m} = \frac{V}{2B^2d^2}$$

So e/m can be found if V, B and d are known. B may be determined experimentally by removing the tube and investigating the region between the coils with a current balance (p. 235) or it can be calculated from the expression $B \approx 0.72\,\mu_0 NI/r$ (p. 236) where $\mu_0 = 4\pi \times 10^{-7}\,\mathrm{H\,m^{-1}}$, N is the number of turns on one coil, I is the current and r the radius of the coil.

In the above simple treatment the fields are assumed to be uniform and co-terminous, i.e. to extend over the same length of the electron beam. In practice such conditions are not achieved and this partly accounts for only an approximate value of e/m being obtained.

(b) Fine beam tube method

The fine beam tube, Fig. 22.6, is a special type of cathode ray tube containing a small quantity of gas (often hydrogen) at a pressure of $\sim 10^{-2}$ mmHg. Electrons from a hot cathode emerge as a narrow beam from a small hole at the apex of a conical-shaped anode and collide with atoms of the gas in the tube. As a result, the gas atoms may lose electrons (i.e. inelastic collisions occur) and they then form positive gas ions.

The electrons created by ionization are easily scattered but the relatively heavy gas ions form a line of positive charge along the path of the beam which attracts the fast electrons from the anode, focusing them into a 'fine' beam. It will be seen later (pp. 402–3)

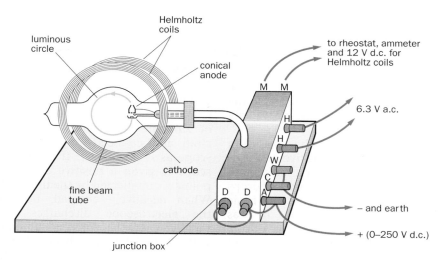

Fig. 22.6 Determining e/m with a fine beam tube

that a gas atom which has lost an electron can emit light when it re-captures an electron. The gas, therefore, not only focuses the beam but also reveals its path (see Fig. 15.23, p. 239).

Helmholtz coils are arranged one on each side of the tube and produce a fairly uniform magnetic field at right angles to the beam. If the field is sufficiently strong, the electrons are deflected into a circular orbit and a luminous circle of low intensity appears when the tube is viewed in the dark. The diameter of the circle may be measured by placing a large mirror with a scale behind the tube so that the circle, its image and the corresponding marks on the scale can all be seen in line. The circle diameter is altered by varying either the anode voltage or the current in the coils.

If B is the magnetic flux density between the coils in the region of the tube, r the radius of the luminous circle and e, m and v are the charge, mass and orbital speed of the electron then the circular motion equation gives

$$Bev = \frac{mv^2}{r} \qquad (3)$$

We will assume electrons are emitted from the cathode with zero speed and that their orbital speed v is constant and equal to that with which they emerge from the anode after being accelerated by the p.d. V between anode and cathode. The energy equation then gives

$$\tfrac{1}{2}mv^2 = eV \qquad (4)$$

Eliminating v from (3) and (4),

$$\frac{e}{m} = \frac{2V}{B^2 r^2}$$

As in the cathode ray tube method, B is obtained experimentally or by calculation.

ELECTRONIC CHARGE

(a) Electrolysis and the Faraday

During the second half of the nineteenth century it was suggested that electricity, like matter, was atomic and that a natural unit of electric charge existed. The basis for this belief was Faraday's work (1831–34) on electrolysis, which may be summarized by the statement that a **mole of monovalent ions of any substance is liberated by 9.65×10^4 coulombs**. This electric charge is called the **Faraday constant** and is denoted by F.

The number of atoms (ions) in a mole of any substance is 6.02×10^{23} and if we assume every atom is associated with the same charge during electrolysis, it follows that each monovalent ion carries a charge of $9.65 \times 10^4/(6.02 \times 10^{23}) = 1.60 \times 10^{-19}$ C. It appeared that the natural unit of charge, now called the **electronic charge**, had this value. For a monovalent ion we therefore have

$$F = Le$$

where L is the Avogadro constant $(6.02 \times 10^{23} \text{ mol}^{-1})$ and e is the electronic charge. Nowadays, the charge indicated by the Faraday is referred to as a 'mole of electrons'. A mole of divalent ions is liberated by two 'moles of electrons' (i.e. $2 \times 9.65 \times 10^4$ C) and of trivalent ions by three 'moles of electrons' (i.e. $3 \times 9.65 \times 10^4$ C).

The above expression provides one of the most accurate ways of measuring e. F is found by electrolysis and L from X-ray crystallography measurements.

(b) Millikan's oil-drop experiment

In 1909 Millikan started a series of experiments lasting many years which supplied evidence for the atomic nature of electricity and provided a value for the magnitude of the electronic charge. The principle of this method is to observe very small oil drops, charged either positively or negatively, falling in air under gravity and then either rising or being held stationary by an electric field.

The essential features of the apparatus are shown in Fig. 22.7. A spray of oil drops is formed above a tiny hole in the upper of two parallel metal plates and some find their way into the space between them. The drops are strongly illuminated and appear as bright specks on a dark background when viewed through a microscope.

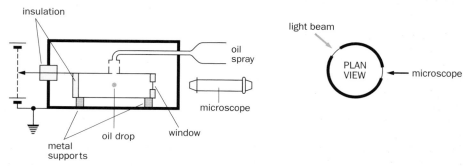

Fig. 22.7 Millikan's oil-drop experiment for determining e

With no electric field between the plates, one drop is selected and its velocity of fall found by timing it over a convenient number of divisions on a scale in the eyepiece of the microscope. (To find the actual distance fallen, the eyepiece scale is calibrated by viewing a millimetre scale through the microscope and comparing it with the eyepiece scale.) For a spherical drop of radius r, moving with uniform velocity v through a homogeneous medium having coefficient of viscosity η, Stokes' law (p. 188) states that the viscous force retarding its motion is $6\pi\eta v r$. In falling, the drop attains its terminal velocity almost at once because it is so small. The retarding force acting up then equals its weight, given by $\frac{4}{3}\pi r^3 \rho g$, where ρ is the density of the oil and g the acceleration of free fall. If v is the terminal velocity and the small upthrust of the air is neglected,

$$6\pi\eta v r = \tfrac{4}{3}\pi r^3 \rho g \qquad (1)$$

From this the radius r of the drop can be found.

Some of the drops become charged either by friction in the process of spraying or from ions in the air. Suppose the drop under observation has a negative charge Q. When a p.d. is applied to the plates so that the top one is positive, an electric field is created which exerts an upward force on the drop. If V is the p.d. and E is the intensity of the field required to keep the drop at rest, then the electric force experienced by it is EQ (since E is the force per unit charge). The electric force on the drop then equals its weight and so

$$EQ = \tfrac{4}{3}\pi r^3 \rho g \qquad (2)$$

$E = V/d$ where d is the distance between the plates, r is known from (1) and so Q can be calculated.

Certain measures were adopted by Millikan to improve accuracy.

(*i*) He used non-volatile oil to prevent evaporation altering the mass of the drop.

(*ii*) Convection currents between the plates and variation of the viscosity of air due to temperature change were eliminated by enclosing the apparatus in a constant-temperature oil bath.

(*iii*) Stokes' law assumes fall in a homogeneous medium. The air consists of molecules and, as Millikan put it, very small drops 'fall freely through the holes in the medium'. He investigated this effect and corrected the law to allow for it.

(*iv*) To simplify the theory we have considered the drop held at rest by the electric field but Millikan reversed the motion and found the upward velocity of the drop.

Millikan found that the charge on an oil drop, whether positive or negative, was always an integral multiple of a basic charge. He studied drops having charges many times the basic charge and by using X-rays he was able to change the charge on a drop. The same minimum charge, equal to that on a monovalent ion, was always involved. The value of the 'atom' of electric charge, i.e. the electronic charge e, is

$$e = 1.60 \times 10^{-19}\,\text{C}$$

PHOTOELECTRIC EMISSION

In photoelectric emission electrons are ejected from metal surfaces when electromagnetic radiation of high enough frequency falls on them. The effect is given by zinc when exposed to X-rays or ultraviolet light. Sodium gives emission with X-rays, ultraviolet and all colours of light except orange and red, while preparations containing caesium respond to infrared as well as to higher frequency radiation.

(a) Simple demonstration of the photoelectric effect

Ultraviolet from a mercury vapour lamp is allowed to fall on a small sheet of zinc, freshly cleaned with emery cloth and connected to an electroscope as in Fig. 22.8a. If the electroscope is given a positive charge the pointer is unaffected by the ultraviolet. When negatively charged, however, the electroscope discharges quite rapidly when the zinc is illuminated with ultraviolet. A sheet of glass between the lamp and the zinc plate halts the discharge.

The photoelectric effect was discovered in 1887 but was not explained until the electron had been 'discovered' by J. J. Thomson. When the zinc plate is positively charged the electrons ejected from it by the ultraviolet fail to escape, being attracted back to the plate, Fig. 22.8b. A negative charge on the zinc repels the emitted electrons and both the zinc and the electroscope lose negative charge, Fig. 22.8c. The insertion of the sheet of glass cuts off much of the ultraviolet but allows the passage of violet light, also emitted by the lamp, and shows that violet light does not produce the effect with zinc.

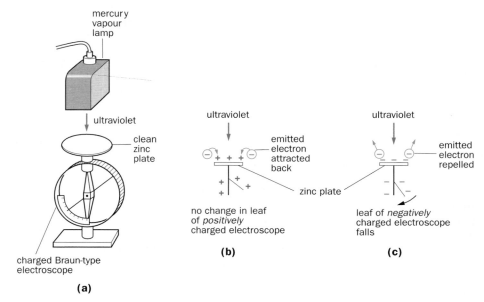

Fig. 22.8 Demonstrating the photoelectric effect

(b) Laws of photoelectric emission

An experimental study of the photoelectric effect yields some surprising results which may be summarized as follows.

Law 1. The *number* of photoelectrons emitted per second is proportional to the *intensity* of the incident radiation.

Law 2. The photoelectrons are emitted with a range of kinetic energies from zero up to a *maximum* which increases as the *frequency* of the radiation increases and is independent of the intensity of the radiation. (For example, a faint blue light produces electrons with a greater maximum kinetic energy than those produced by a bright red light, but the bright red light releases a greater number.)

Law 3. For a given metal there is a certain minimum frequency of radiation, called the **threshold frequency**, below which no emission occurs irrespective of the intensity of the radiation. For zinc, the threshold frequency is in the ultraviolet.

Most of these facts appear to be inexplicable on a wave theory of electromagnetic radiation. Law 1 can be vindicated in terms of waves because if the radiation has greater intensity, more energy is absorbed by the metal and it is possible for more electrons to escape. Also, it is reasonable to suppose that the range of emission speeds (and kinetic energies) from zero to a maximum is due to electrons having a range of possible kinetic energies inside the metal. Those with the highest kinetic energy are emitted with the maximum speeds. However, we would expect a certain number of photoelectrons to be ejected with greater speeds when the radiation intensity increases but this is not so according to Law 2.

The increase of maximum kinetic energy with frequency and the existence of the threshold frequency are even more enigmatic. Furthermore, according to the wave theory, radiation energy is spread over the wavefront and since the amount incident on any one electron would be extremely small, some time would elapse before an electron gathered enough energy to escape. No such time lag between the start of radiation and the start of emission is observed, even when the radiation is weak.

QUANTUM THEORY

By the end of the nineteenth century the wave theory, despite its earlier notable successes, was unable to account for most of the known facts concerning the interaction of electromagnetic radiation with matter. One of these, as we have just seen, was photoelectric emission and another was black body radiation (chapter 20).

(a) Planck's theory

In 1900 Planck tackled the problem of finding a theory that would fit the facts of black body radiation. Whereas others had considered the radiation to be emitted continuously, Planck supposed the emission to occur intermittently in integral multiples of an 'atom' or **quantum** of energy, the size of which depended on the frequency of the oscillator producing the radiation. A body could emit one, two, three, etc. quanta of energy but no fractional amounts.

According to Planck, the quantum E of energy for radiation of frequency f is given by

$$E = hf$$

where h is a constant, now called the **Planck constant**. For electromagnetic radiation of wavelength λ, $c = f\lambda$, where c is its speed in a vacuum and so we also have $E = hc/\lambda$. The energy of a quantum is therefore inversely proportional to the wavelength of the radiation but directly proportional to the frequency. It is convenient to express many quantum energies in electronvolts; the quantum for red light has energy of about 2 eV and for blue light about 4 eV.

Using the equation $E = hf$, Planck derived an expression for the variation of energy with wavelength for a black body which agreed with the experimental curves (Fig. 20.64, p. 353) at all wavelengths and temperatures. At the time the quantum theory was too revolutionary for most scientists and little attention was paid to it. Nevertheless the interpretation of black body radiation was the first of its many successes.

(b) Einstein's photoelectric equation

Einstein extended Planck's ideas in 1905 by deriving an equation that explained in a completely satisfactory way the laws of photoelectric emission. He assumed that not only were light and other forms of electromagnetic radiation emitted in whole numbers of quanta but that they were also absorbed as quanta, called **photons**. This implied that electromagnetic radiation could exhibit particle-like behaviour when being emitted and absorbed, and led to the idea that light has a dual nature; under some circumstances it behaves as waves and under others as particles. **Wave–particle duality** will be considered later in the chapter.

When dealing with thermionic emission it was explained that to liberate an electron from the surface of a metal a quantity of energy, called the **work function** Φ, which is characteristic of the metal, has to be supplied. In photoelectric emission, Einstein proposed that a photon of energy hf causes an electron to be emitted if $hf \geq \Phi$. The excess energy $(hf - \Phi)$ appears as kinetic energy of the emitted electron which escapes with a speed having any value up to a maximum v_{max}. The actual value depends on how much energy the electron has inside the metal. So

$$hf - \Phi = \tfrac{1}{2}mv_{max}^2$$

This is Einstein's photoelectric equation.

If the photon has only just enough energy to liberate an electron, the electron gains no more kinetic energy. It follows that since Φ is constant for a given metal, there is a minimum frequency, the threshold frequency f_0, below which no photoelectric emission is possible. It is given by

$$hf_0 = \Phi$$

Einstein's equation can also be written

$$h(f - f_0) = \tfrac{1}{2}mv_{max}^2$$

The increase of maximum emission speed with higher frequency radiation can now be seen to be due to the greater photon energy of the radiation. Note that it is assumed that a photon imparts all its energy to one electron and then no longer exists.

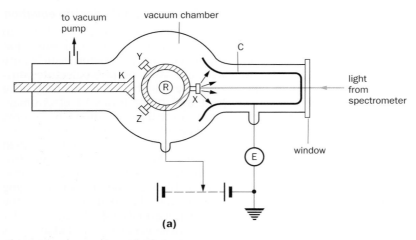

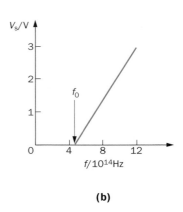

Fig. 22.9   Millikan's photoelectric experiment

(c) Millikan's experiment on photoelectric emission

In 1916 Millikan verified Einstein's photoelectric equation experimentally and provided irrefutable evidence for the photon model of light. He was the first to obtain photoelectrically a value for the Planck constant which agreed with values from other methods. His apparatus is shown in simplified form in Fig. 22.9a.

Monochromatic light from a spectrometer entered the window of a vacuum chamber and fell on a metal X mounted on a turntable R controlled from outside the chamber. The photoelectrons emitted were collected by an electrode C and detected by a sensitive current-measuring device E. The minimum positive potential, called the **stopping potential**, which had to be applied to X to prevent the most energetic photoelectrons reaching C and causing current flow, was found for different frequencies of the incident radiation. The procedure was repeated with Y and then Z opposite C. X, Y and Z were made from the alkali metals lithium, sodium and potassium, since these emit photoelectrons with light and each one can therefore be studied over a wide range of frequencies. Immediately before taking a set of readings, R was rotated and the knife K adjusted so that a fresh surface was cut on the metal, so eliminating the effects of surface oxidation.

In such experiments the p.d. between the electrodes, as measured by any form of voltmeter, is not the same as the p.d. in the space between them unless their work functions are equal. The error is called the **contact p.d.** In

Millikan's experiment this was a few volts, that is, the correction was as large as the effect measured. One feature of his experiment was an ingenious device (not shown in Fig. 22.9a) by which an accurate correction was made.

The relation between the stopping potential V_s and the maximum kinetic energy of the photoelectrons is given by the energy equation $eV_s = \frac{1}{2}mv_{max}^2$. Einstein's equation may then be written

$$\frac{1}{2}mv_{max}^2 = eV_s = hf - \Phi$$

$$\therefore \quad V_s = \frac{h}{e} \cdot f - \frac{\Phi}{e}$$

The graph of V_s against f should be a straight line, a fact that Millikan's results confirmed. Einstein's relation was verified. Furthermore, the stopping potential for a given frequency of light was independent of the intensity of the light. One of Millikan's graphs is shown in Fig. 22.9b. He found that whatever the metal all graphs had the same slope and from the above equation it is seen to be h/e. Knowing e, h can be calculated and the value obtained agrees with that found from black body radiation experiments. The threshold frequency f_0 and the work function Φ are characteristic for each metal and may also be deduced from the graph. The values of Φ are in good agreement with those found from thermionic emission.

The Planck constant h is a fundamental physical constant and occurs in many formulae in atomic physics. Its value to three figures is

$$h = 6.63 \times 10^{-34} \, \text{J s}$$

It is because of the smallness of the Planck constant that quantum effects are not normally apparent.

Because of the need for very clean surfaces, quantitative photoelectric experiments are difficult to perform with simple apparatus in a school laboratory but one which gives fair results is outlined in Appendix 4.

(d) Worked example

If a photoemissive surface has a threshold wavelength of 0.65 μm, calculate (i) its threshold frequency, (ii) its work function in electronvolts, and (iii) the maximum speed of the electrons emitted by violet light of wavelength 0.40 μm. (Speed of light $c = 3.0 \times 10^8 \, \text{m s}^{-1}$, $h = 6.6 \times 10^{-34} \, \text{J s}$, $e = 1.6 \times 10^{-19} \, \text{C}$ and mass of electron $m = 9.1 \times 10^{-31} \, \text{kg}$.)

(i) $\lambda_0 = 0.65 \, \mu\text{m} = 6.5 \times 10^{-7} \, \text{m}$

$$f_0 = c/\lambda_0$$
$$= \frac{3.0 \times 10^8 \, \text{m s}^{-1}}{6.5 \times 10^{-7} \, \text{m}}$$
$$= 4.6 \times 10^{14} \, \text{Hz}$$

(ii) We have

$\Phi = hf_0$
$= (6.6 \times 10^{-34} \, \text{J s})(4.6 \times 10^{14} \, \text{s}^{-1})$
$= 6.6 \times 10^{-34} \times 4.6 \times 10^{14} \, \text{J}$

But $1 \, \text{eV} = 1.6 \times 10^{-19} \, \text{J}$

$$\therefore \quad \Phi = \frac{6.6 \times 4.6 \times 10^{-20} \, \text{eV}}{1.6 \times 10^{-19}}$$

$$= 1.9 \, \text{eV}$$

(iii) For violet light

$$\lambda = 0.40\ \mu\text{m} = 4.0 \times 10^{-7}\ \text{m}$$

$$\begin{aligned}
f &= c/\lambda \\
&= \frac{3.0 \times 10^{8}\ \text{m s}^{-1}}{4.0 \times 10^{-7}\ \text{m}} \\
&= 7.5 \times 10^{14}\ \text{Hz}
\end{aligned}$$

From the photoelectric equation

$$\begin{aligned}
\tfrac{1}{2}mv^2 &= hf - \varPhi \\
&= (6.6 \times 10^{-34} \times 7.5 \times 10^{14} \\
&\quad - 1.9 \times 1.6 \times 10^{-19})\ \text{J} \\
&= 1.9 \times 10^{-19}\ \text{J} \\
\therefore\ v_{\text{max}} &= \sqrt{\left(\frac{2 \times 1.9 \times 10^{-19}}{9.1 \times 10^{-31}}\right)}\ \text{m s}^{-1} \\
&= 6.5 \times 10^{5}\ \text{m s}^{-1}
\end{aligned}$$

PHOTOCELLS AND THEIR USES

There are several kinds of photocell.

(a) Photoemissive cell

The construction of a typical cell and its symbol are shown in Fig. 22.10. Two electrodes are enclosed in a glass bulb which may be evacuated or contain an inert gas at low pressure. The cathode, often called the **photocathode**, is a curved metal plate having an emissive surface facing the anode, here shown as a single metal rod. When electromagnetic radiation falls on the cathode, photoelectrons are emitted and are attracted to the anode if it is at a suitable positive potential. A current of a few microamperes flows and increases with the intensity of the incident radiation. An inert gas in the cell gives greater current but causes a time lag in the response of the cell to very rapid changes of radiation which may make it unsuitable for some purposes.

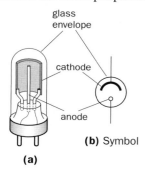

glass envelope

cathode

anode

(b) Symbol

(a)

Fig. 22.10 A photoemissive cell

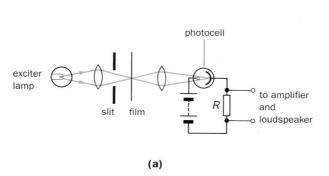

photocell

exciter lamp

slit film

R

to amplifier and loudspeaker

(a)

sound track

film

picture

picture

(b)

Fig. 22.11 Reproduction of sound using a photocell

The choice of material for the cathode surface depends on the frequency range over which the cell is to operate, and should be such that a good proportion of the incident photons yield electrons. Ideally, every photon should release one electron but in practice the yield is much lower. Pure metals are rarely used because of their high reflecting power; most photocathode surfaces are composite materials. Caesium on oxidized silver has a peak response near the red end of the spectrum and a threshold wavelength of 12 μm which makes it suitable for use with infrared.

In the reproduction of sound from a film, light from an exciter lamp, Fig. 22.11a, is focused on the 'sound track' at the side of the moving film and then falls on a photocell. The sound track, Fig. 22.11b, varies the intensity of the light passing through it so that the photocell creates a varying current which represents the digital signal obtained from the recording microphone at the time the film was made. The fluctuating p.d. developed across a load R is amplified, decoded and converted to sound by a loudspeaker.

(b) Photoconductive cell or light-dependent resistor (LDR)

The resistance of certain semiconductors such as cadmium sulphide decreases as the intensity of the light falling on them increases. The effect is due to light photons setting free electrons in the semiconductor, so increasing its conductivity, i.e. reducing its resistance.

A popular LDR is the ORP12 shown in Fig. 22.12. There is a 'window' over the grid-like metal structure to allow light to fall on a thin layer of cadmium sulphide. Its resistance varies from about 10 MΩ in the dark to 1 kΩ or so in daylight.

An alarm circuit using an LDR is described later (p. 426). LDRs are also used in photographic exposure meters.

(a)

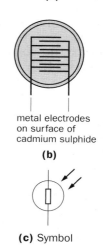

metal electrodes on surface of cadmium sulphide

(b)

(c) Symbol

Fig. 22.12 An LDR

X-RAYS

X-rays, so named because their nature was at first unknown, were discovered in 1895 by Röntgen. They are a type of electromagnetic radiation produced whenever cathode rays (high-speed electrons) are brought to rest by matter.

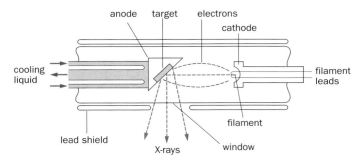

Fig. 22.13 X-ray tube

(a) Production

A modern X-ray tube is highly evacuated and contains an anode and a tungsten filament connected to a cathode, Fig. 22.13. Electrons are obtained from the filament by thermionic emission and are accelerated to the anode by a p.d., typically up to 100 kV. The anode is a copper block inclined to the electron stream and having a small target of tungsten, or another high-melting-point metal, on which electrons are focused by the concave cathode. The tube has a lead shield with a small window to allow the passage of the X-ray beam.

Less than $\frac{1}{2}$% of the kinetic energy of the electrons is converted into X-rays. The rest of the kinetic energy becomes internal energy of the anode, which has to be kept cool by circulating oil or water through channels in it or by the use of cooling fins.

The *intensity* of the X-ray beam increases when the number of electrons hitting the target increases and this is controlled by the filament current. The *quality* or penetrating power of the X-rays is determined by the speed attained by the electrons and increases with the p.d. across the tube. 'Soft' X-rays only penetrate such objects as flesh; 'hard' X-rays can penetrate much more solid matter.

The p.d. required to operate an X-ray tube may be obtained from a half-wave rectifying circuit containing a step-up transformer, in which the X-ray tube itself acts as the rectifier.

(b) Properties

These may be summarized as follows.
- They travel in straight lines.
- They readily penetrate matter; penetration is least in materials containing elements of high density and high atomic number. For example, while sheets of cardboard, wood and some metals fail to stop them, all but the most penetrating are absorbed by a sheet of lead 1 mm thick. Lead glass is a much better absorber than ordinary soda glass.
- They are not deflected by electric or magnetic fields.
- They eject electrons from matter by the photoelectric effect and other mechanisms. The ejected electrons are responsible for the next three effects.
- They ionize a gas, permitting it to conduct. An electrified body such as an electroscope charged positively or negatively is discharged when the surrounding air is irradiated by X-rays.
- They cause certain substances to fluoresce, e.g. barium platinocyanide.
- They affect a photographic emulsion in a similar manner to light.

(c) Nature: von Laue's experiment

X-rays cannot be charged particles since they are not deflected by electric and magnetic fields. The vital experiment which established their electromagnetic nature was initiated by von Laue in 1912. After unsuccessful attempts had been made to obtain X-ray diffraction patterns with apparatus similar to that used for light, it was realised that the failure might be due to X-rays having much smaller wavelengths.

Von Laue's suggestion was that if the regular spacing of atoms in a crystal is of the same order as the wavelength of the X-rays, the crystal should act as a three-dimensional diffraction grating. This idea was put to the test with success by Friedrich and Knipping, two of von Laue's colleagues. The arrangement of their apparatus is shown in Fig. 22.14a. A narrow beam of X-rays from two slits S_1 and S_2 fell on a thin crystal in front of a photographic plate. After a long exposure the plate (on developing) revealed that most of the radiation passed straight through to give a large central spot, but some of it produced a distinct pattern of fainter spots around the central one, Fig. 22.14b.

This demonstrated the wave behaviour of X-rays and analysis of diffraction patterns confirmed they had wavelengths of about 10^{-10} m (0.1 nm). They are considered to be electromagnetic waves because (i) their method of production involves

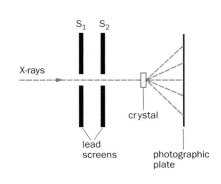

(a) Apparatus

Fig. 22.14 Von Laue's experiment

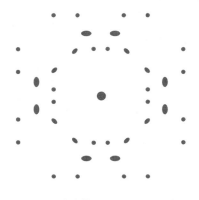

(b) Diffraction pattern

accelerated charges, (*ii*) they eject electrons from matter (implying strong electric fields), and (*iii*) they give line spectra similar in general character to the optical spectrum of hydrogen (p. 344).

(d) Uses

The usefulness of X-rays is largely due to their penetrating power.

(*i*) *Medicine*. X-ray photographs, or 'radiographs', are used for a variety of purposes. Since X-rays can damage healthy cells of the human body, great care is taken to avoid unnecessary exposure. In radiography the film is sandwiched between two screens which fluoresce when subjected to a small amount of X-radiation. The film is affected by the fluorescent light rather than by the X-rays; formerly X-rays were used to sensitize the film directly and much longer exposures were required.

Suspected bone fractures can be investigated since X-rays of a certain hardness can penetrate flesh but not bone. In the detection of lung tuberculosis by mass radiography, use is made of the fact that diseased tissue is denser than healthy lung tissue which consists of air sacs and so the former absorbs X-rays more strongly. When an internal organ is being X-rayed whose absorptive power is similar to that of the surrounding tissue, a 'contrast' medium such as a barium meal is given to the patient, orally or by injection. This is less easily penetrated by the X-rays and enables a shadow of the organ to be obtained on a radiograph.

Overlap of structures may also make them difficult to distinguish and to overcome this it is necessary to use the three-dimensional imaging provided by a **computerized axial tomography** (CAT) scan. In this technique a rotating source exposes the body to a narrow beam of X-rays from a wide range of directions. An array of detectors measure the intensity of X-rays that have passed through the body along many different paths and this allows the computer to build up 3-D density information, from which detailed images of internal organs can be constructed and displayed.

In the treatment of cancer by radiotherapy, very hard (i.e. short wavelength) X-rays are used to destroy the cancer cells whose rapid multiplication causes malignant growth.

(*ii*) *Industry*. Castings and welded joints can be inspected for internal imperfections using X-rays. A complete machine may also be examined from a radiograph without having to be dismantled.

(*iii*) *X-ray crystallography*. The study of crystal structure by X-rays is now a powerful method of scientific research. The first crystals to be analysed were of simple compounds such as sodium chloride. In recent years the structure of very complex organic molecules, including DNA, has been unravelled.

X-RAY DIFFRACTION

(a) Bragg's law

In von Laue's method of producing X-ray diffraction, the crystal is used as a transmission grating and interpretation of the patterns is not easy. Sir William Bragg and his son Sir Lawrence developed a simpler technique in which the crystal acts as a reflection diffraction grating.

When X-rays fall on a single plane of atoms in a crystal, each atom scatters a small fraction of the incident beam and may be regarded as the source of a weak secondary wavelet of X-rays. In most directions destructive interference of the wavelets occurs but in the direction for which the angle of incidence equals the angle of 'reflection', there is reinforcement and a weak reflected beam is obtained. The X-rays behave as if they were weakly 'reflected' by the layer of atoms, Fig. 22.15*a*.

Other planes of atoms, such as p, q and r in Fig. 22.15*b*, to which the X-rays penetrate, behave similarly.

The reflected beams from all the planes involved interfere and the resultant reflected beam is only strong if the path difference between successive planes is a whole number of wavelengths of the incident X-radiation. That is, reinforcement only occurs for planes p and q when

$$AB + BC = n\lambda$$

where *n* is an integer and λ is the wavelength of the X-rays. If *d* is the distance between planes of atoms and θ the angle between the X-ray beam and the crystal surface, called the **glancing angle**, then $AB + BC = 2d \sin \theta$ and the reflected beam has maximum intensity when

$$2d \sin \theta = n\lambda$$

This equation is **Bragg's law**; note that it refers to the glancing angle and not to the angle of incidence. Intensity maxima occur for several glancing angles. The smallest angle is given by $n = 1$ and is the first-order reflection; $n = 2$ gives the second-order, and so on.

This process is sometimes termed X-ray 'reflection' and although it may appear that reflection occurs it is in fact a diffraction effect, since interference takes place between X-rays from secondary sources.

The atoms in a crystal can be considered to be arranged in several different sets of parallel planes, from all of which strong reflections may be obtained to give a pattern of spots characteristic of the particular structure.

A microwave analogue of X-ray 'reflection' from the planes of 'atoms' in a polystyrene ball crystal can be demonstrated (see p. 23.)

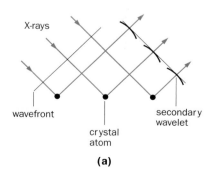

(a)

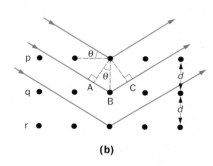

(b)

Fig. 22.15

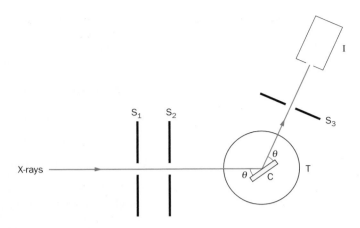

Fig. 22.16 Principle of the Bragg spectrometer

X-RAY SPECTRA AND THE QUANTUM THEORY

(a) Continuous and line spectra

The radiation from an X-ray tube can be analysed with a spectrometer and an intensity–wavelength graph obtained to show the spectral distribution. A typical X-ray spectrum is given in Fig. 22.18. It has two parts.

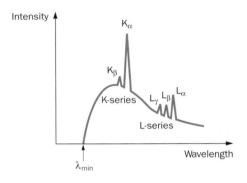

Fig. 22.18 X-ray spectrum

(b) X-ray spectrometer

The Bragg X-ray spectrometer was developed to measure (*i*) X-ray wavelengths, and (*ii*) the spacing of atoms in crystals. The principle of the instrument is shown in Fig. 22.16. X-rays from the target of an X-ray tube are collimated by two slits S_1 and S_2 (made in lead sheets) and the narrow beam so formed falls on a crystal C set on the table T of the spectrometer. The reflected beam passes through a third slit S_3 into an ionization chamber I (p. 466) where it creates an ionization current which is a measure of the intensity of the reflected radiation.

As the crystal and the ionization chamber are rotated, the angle of reflection always being kept equal to the angle of incidence, the ionization current is found. Strong reflection occurs for glancing angles satisfying Bragg's law $2d \sin \theta = n\lambda$. Knowing either *d* or λ, the other can be calculated.

X-ray wavelengths can now be measured directly using mechanically ruled diffraction gratings (similar to optical gratings) provided the X-rays strike the grating at a glancing angle much less than 1°.

(c) X-ray powder photography

If instead of a single crystal, a polycrystalline specimen or a crystalline powder is used, many planes are involved at once in X-ray 'reflection' and thousands of spots are produced from 'reflection' at all possible angles.

As a result, circles or circular arcs are obtained on a suitably placed X-ray film. Figure 22.17*a* shows the arrangement in the X-ray powder (or polycrystalline) technique using a strip of film and Fig. 22.17*b* indicates how the 'lines' are formed. Figures 22.17*c* and *d* are photographs for sodium chloride powder and a polycrystalline copper wire.

(*i*) A *continuous* spectrum which has a definite lower wavelength limit, increases to a maximum and then decreases gradually in the longer wavelengths. All targets emit this type of radiation.

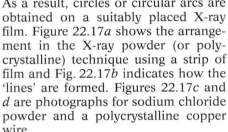

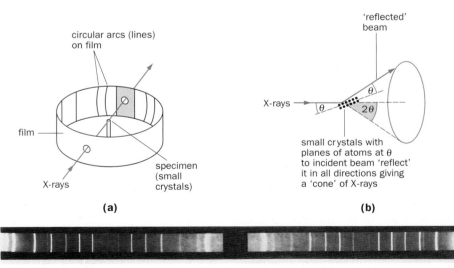

(c) Sodium chloride powder

(d) Polycrystalline copper

Fig. 22.17 X-ray powder photography

(*ii*) A *line* spectrum consisting of groups or series of two or three peaks of high-intensity radiation superimposed on the continuous spectrum. The series are denoted by the letters K, L, M, etc., in order of increasing wavelength, and the peaks by α, β, γ. The wavelengths of the peaks are characteristic of the target element; all the series are not normally given by one element.

(b) Quantum theory and the continuous spectrum

Certain features of the continuous spectrum are readily explained by the quantum theory. The existence of a definite *minimum* wavelength can be justified if we assume that this radiation consists of X-ray photons produced by electrons that have given up all their kinetic energy in a single encounter with a target atom. If such an electron has mass m and speed v on striking the target, the energy hf of the photon is given by

$$hf = \tfrac{1}{2}mv^2 \qquad (1)$$

h being the Planck constant and f the frequency of the radiation. With a p.d. V across the X-ray tube, an electron of charge e has work eV done on it by the electric field and so

$$eV = \tfrac{1}{2}mv^2 \qquad (2)$$

From (1) and (2)

$$hf = eV \qquad (3)$$

The value of f given by (3) is the maximum frequency of the X-rays emitted at p.d. V, since all the energy of the electron is converted to the photon. The corresponding wavelength will have a minimum value, and if this is λ_{min} then $c = f\lambda_{min}$ where c is the speed of travel of X-rays. It follows that

$$\lambda_{min} = \frac{hc}{eV}$$

As V increases we see that λ_{min} decreases, i.e. X-rays of higher frequency and greater penetrating power are emitted. The values of λ_{min} calculated from this equation agree with those found experimentally.

Most of the electrons responsible for the X-radiation usually have more than one encounter before losing all their energy. Several photons are produced with smaller frequencies than f and therefore with greater wavelengths than λ_{min}. Different electrons lose different amounts of energy and so a continuous spectrum covering a range of wavelengths is obtained. The great majority of electrons, however, lose their kinetic energy too gradually for X-rays to be emitted and merely increase the internal energy of the target.

(c) Explanation of line spectra

This will be considered later (p. 403).

(d) Worked example

An X-ray tube operates at 30 kV and the current through it is 2.0 mA. Calculate (*i*) the electrical power input, (*ii*) the number of electrons striking the target per second, (*iii*) the speed of the electrons when they hit the target, and (*iv*) the lower wavelength limit of the X-rays emitted.

(*i*) If V is the p.d. across the tube and I the tube current then

$$\text{power input} = VI$$
$$= (30 \times 10^3\,\text{V})(2.0 \times 10^{-3}\,\text{A})$$
$$= 60\,\text{W}$$

(*ii*) Current through the tube is given by $I = ne$, where n is the number of electrons striking the target per second and e is the electronic charge (i.e. 1.6×10^{-19} C).

$$\therefore\ n = \frac{I}{e} = \frac{2.0 \times 10^{-3}\,\text{A}}{1.6 \times 10^{-19}\,\text{C}}$$
$$= 1.3 \times 10^{16}\,\text{s}^{-1}$$

(*iii*) If m is the mass of an electron (i.e. 9.0×10^{-31} kg) and v its speed at the target, then from equation (2) $\tfrac{1}{2}mv^2 = eV$. Therefore

$$v = \sqrt{\left(\frac{2eV}{m}\right)}$$
$$= \sqrt{\left(\frac{2 \times 1.60 \times 10^{-19}\,\text{C} \times 30 \times 10^3\,\text{V}}{9.0 \times 10^{-31}\,\text{kg}}\right)}$$
$$= 1.0 \times 10^8\,\text{m s}^{-1}$$

(Check that C V reduces to m s^{-1}.)

(*iv*) The lowest X-ray wavelength emitted is given by

$$\lambda_{min} = \frac{hc}{eV}$$

where $h = 6.6 \times 10^{-34}$ J s and $c = 3.0 \times 10^8$ m s^{-1}. Therefore

$$\lambda_{min} = \frac{(6.6 \times 10^{-34}\,\text{J s})(3.0 \times 10^8\,\text{m s}^{-1})}{(1.6 \times 10^{-19}\,\text{C})(30 \times 10^3\,\text{V})}$$
$$= 0.41 \times 10^{-10}\,\text{m}$$

ELECTRICAL CONDUCTION IN GASES

(a) Ionized gases

Gases at s.t.p. are very good electrical insulators and about 30 kV is required to produce a spark discharge between two rounded electrodes 1 cm apart in air; the value for pointed electrodes 1 cm apart is 12 kV. To conduct, a gas must be ionized.

Ionization occurs when an electron is removed from an atom (or molecule) by supplying a certain amount of energy to overcome the attractive force securing the electron to the atom. The two resulting charged particles form an **ion-pair**, the electron being a negative ion and the atom, now deficient of an electron, a positive ion. When a p.d. is applied between two electrodes in the ionized gas, positive ions moves towards the cathode and electrons towards the anode. The ions act as charge carriers and current flows in the gas. If an atom gains an electron it becomes a heavy negative ion but these are generally few in number.

Various agents can ionize a gas and these include a flame, sufficiently energetic electrons, ultraviolet, X-rays and the radiation from radioactive substances, i.e. α-, β- and γ-rays. The few ions always present in the atmosphere are caused by cosmic rays from space and by radioactive minerals in the earth. They are responsible for a charged, insulated body (such as an electroscope) gradually discharging.

(b) Current–p.d. relationship

The gas between two parallel plates P and Q, Fig. 22.19a, is ionized by a beam of ionizing radiation. When a p.d. is applied the resulting ionization current is recorded by a sensitive current detector, e.g. a d.c. amplifier, as described on p. 467. If the p.d. is varied while the intensity of the radiation remains constant, the current variation is shown by the curve OABCD in Fig. 22.19b.

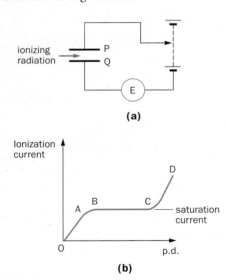

(a)

(b)

Fig. 22.19 Ionization of a gas

The shape of the curve is explained as follows. Small p.ds cause the electrons and positive ions to move slowly to the anode P and the cathode Q respectively. On the way some ions recombine to form neutral atoms. As the p.d. increases, the ions travel more quickly and there is less opportunity for recombination; the ionization current increases. At B all the ions

produced by the radiation reach the electrodes, no recombination occurs and further increase of p.d. between B and C does not affect the current. Along BC the current has its *saturation value* and is independent of the applied p.d.

Beyond C the current rises rapidly with p.d. and indicates that a new source of ion-pairs has become operative. The original ions due to the radiation are now accelerated sufficiently by the large p.d. to form new electrons and new positive ions by collisions with neutral gas molecules between the plates. Thus along CD each original ion-pair creates several other ion-pairs, the process being known as *ionization by collision*. It can be shown that electrons are mainly responsible for this.

Two points should be noted:

(*i*) Ohm's law is obeyed between O and A;

(*ii*) the saturation current is proportional to the rate of production of ions by the radiation and so measures the intensity of the radiation.

TERMS AND DEFINITIONS

(a) Atomic or proton number Z

After it had been established that the negatively charged electron was one of the basic constituents of all atoms, the search began for a positively charged counterpart.

Experiments on the bombardment of atoms by suitable high-speed particles revealed that in certain cases a particle with the properties of a hydrogen ion is ejected from the nucleus of an atom (pp. 477–8). This provided fairly conclusive evidence for the belief that such particles are one of the fundamental particles from which atoms are made. As they were the lightest and simplest positively charged particles then known, they were called **protons** (from the Greek *protos* meaning first).

The proton, denoted by the symbol p, has a charge equal in magnitude but opposite in sign to that of an electron and its rest mass is 1836 times the rest mass of the electron.

The number allotted to an element in the periodic table was called its **atomic number**. The significance of this term was not apparent until the nuclear theory of the atom had been

proposed and evidence obtained about the structure of the nucleus.

> The **atomic** or **proton number** Z of an atom is the number of protons in its nucleus.

An atom is normally electrically neutral and so the atomic number is also the number of electrons in a neutral atom of the element. Hydrogen with $Z = 1$ has 1 proton and 1 electron, while uranium with $Z = 92$ has 92 protons and 92 electrons.

(b) Mass or nucleon number A

Since the atomic number of an element is about half its relative atomic mass, it follows that if (as the nuclear theory supposes, pp. 400–401) the mass of an atom is due to its nucleus, then there must be other constituents besides protons in the nucleus. The possibility of the existence of an electrically neutral particle with a similar mass to that of the proton, was suggested in 1920. This particle, named the **neutron**, symbol n, remained undiscovered until 1932 (p. 478), partly on account of the difficulty of detecting a particle which, being uncharged, is not deflected by electric or magnetic fields and produces no appreciable ionization in its path.

Neutrons can also be expelled from certain nuclei by bombardment and there is now no doubt that they are basic constituents of matter. The rest mass of the neutron is 1839 times that of the electron. Protons and neutrons are collectively called **nucleons**.

> The **mass** or **nucleon number** A of an atom is the number of nucleons in the nucleus.

It follows that if N is the **neutron number** of a nucleus, i.e. the number of neutrons it contains, then

$$A = Z + N$$

The simplest nucleus is that of hydrogen which consists of 1 p; in symbolic notation it is written ^1_1H, where the superscript gives the nucleon number and the subscript gives the proton number. Helium has 2 p and 2 n, giving $A = 4$ and $Z = 2$ and symbol ^4_2He. Lithium ^7_3Li has $A = 7$ and $Z = 3$

and its atom has 3 p and 4 n. In general:

atom X is represented by $^A_Z X$

The neutron is written ^1_0n since it has $A = 1$ and zero charge, i.e. $Z = 0$; the proton can be written ^1_1p and the electron $^0_{-1}e$. Sometimes the proton number is omitted and the name or symbol of the element is given followed by the mass number, e.g. lithium-7 or Li-7.

(c) Isotopes

If two atoms have the same number of protons but different numbers of neutrons, their atomic (proton) numbers are equal but not their mass (nucleon) numbers. Each atom is said to be an **isotope** of the other. They are chemically indistinguishable (since they have the same number of electrons) and occupy the same place in the periodic table.

Few elements consist of identical atoms; most are isotopic mixtures. Hydrogen has three isotopes: ^1_1H with 1 p, **heavy hydrogen** or **deuterium** ^2_1H with 1 p and 1 n, and **tritium** ^3_1H with 1 p and 2 n. Ordinary hydrogen contains 99.99% ^1_1H atoms. Isotopes are not as a rule given separate names and symbols; an exception is made in the case of hydrogen because there is an appreciable difference in the physical properties of the three forms. Water made from deuterium is called **heavy water**; it has density 1.108 g cm^{-3}, a freezing-point of 3.82 °C and a boiling-point of 101.42 °C. The nucleus of the deuterium atom is called a **deuteron**.

Isotopes account for fractional atomic masses. For example, chlorine with atomic mass 35.5 has two forms: $^{35}_{17}Cl$ and $^{37}_{17}Cl$ and these are present in ordinary chlorine in the approximate ratio of three atoms of Cl-35 to one of Cl-37. Chemically they are identical but one is slightly denser than the other.

Isotopes were discovered among the radioactive elements in 1906 but their nature was not understood. Although there are only just over 100 elements (natural and artificial), each one has isotopes and the total number known at present is about 1500. Of these about 300 occur naturally and the rest are artificial; all artificial ones and some natural ones are radioactive.

The term **nuclide** is used to specify an atom with a particular proton–neutron combination. ^6_3Li and ^7_3Li are nuclides and also isotopes; ^9_4Be and $^{10}_5B$ are nuclides but not isotopes.

ATOMIC MASS; MASS SPECTROGRAPH

(a) Atomic mass

Atomic mass or, more correctly, **relative atomic mass**, is denoted by A_r. It is defined as the ratio:

$$\frac{\text{mass of an atom}}{\frac{1}{12}\text{mass of }^{12}_6C\text{ atom}}$$

Initially atomic masses were quoted relative to the hydrogen atom as 1, and later to the oxygen isotope $^{16}_8O$ as 16, but in 1960 physicists and chemists agreed to adopt the carbon-12 scale. This was chosen because in measuring atomic masses by mass spectroscopy (to be explained shortly), carbon-12 is a convenient standard for comparison since it forms many compounds.

Atomic masses were first measured to a high degree of accuracy by Aston who established, between 1919 and 1927, that (i) most elements exhibit isotopy, and (ii) most isotopic masses are very nearly but not quite whole numbers.

In Aston's apparatus the deflecting electric and magnetic fields were arranged so that all particles of the same mass, irrespective of their speed, were brought to a line focus. When ions of different masses were present, a series of lines, i.e. a mass spectrum, Fig. 22.20, was obtained on a photographic film. The relative intensities of the lines enabled an estimate to be made of the relative amounts of isotopes. Aston called his instrument a **mass spectrograph**.

Fig. 22.20   A mass spectrum

(b) Mass spectrograph

Many types have been constructed for the accurate determination of atomic mass; the essential features of that due to Bainbridge are shown in Fig. 22.21.

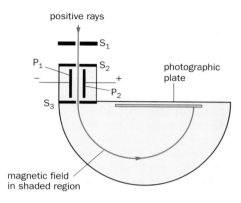

Fig. 22.21   Principle of Bainbridge's mass spectrograph

A stream of positive ions, called positive rays, is produced by passing electrons from a hot cathode into the gas or vapour to be investigated. The positive ions so formed are directed through slits S_1 and S_2 and emerge as a narrow beam with a range of speeds and specific charges. In the region between S_2 and S_3 crossed uniform electric and magnetic fields are applied. The electric field between P_1 and P_2 exerts a force acting to the left on the positive ions. The magnetic field, which acts normal to and into the plane of the diagram, tends to deflect the ions to the right. If Q is the charge of an ion and E the electric field strength, the electric force is EQ. The magnetic force is BQv where v is the velocity of the ion and B the flux density. When the forces are equal

$$BQv = EQ$$
$$\therefore \quad v = \frac{E}{B} \qquad (1)$$

The ion will be undeflected and will emerge from the selector system, as S_2–S_3 is called, if its velocity equals the ratio E/B. All ions leaving S_3 therefore have the same velocity v whatever their specific charge and **velocity selection** is said to have occurred.

Beyond S_3 only the magnetic field acts, the ions describe circular arcs and strike the photographic plate. For particles of mass M, the radius r of the

path is given by

$$BQv = \frac{Mv^2}{r}$$

$$\therefore \quad r = \frac{Mv}{BQ} \qquad (2)$$

From (1) and (2)

$$r = \frac{M}{Q} \cdot \frac{E}{B^2}$$

If B and E are constant, r is directly proportional to M (assuming Q is the same for all ions). When ions with different masses are present each set produces a definite line and from their positions the masses can be found. In a mass spectrograph the masses of individual atoms are being measured; by contrast chemical methods give the average atomic mass for a large number of atoms.

The mass spectrograph, which uses photographic detection, is employed primarily for precise mass determinations and can measure the very small differences of mass occurring in nuclear reactions when one nuclide changes to another. The **mass spectrometer** uses an electrometer as a detector and gives the exact relative abundances, as in gas analysis. Such an instrument is shown in Fig. 22.22. The man is operating the ion source. The beam of ions is deflected by electric and magnetic fields (top right) and a computer analyses the results, displaying them graphically.

NUCLEAR MODEL OF THE ATOM

(a) Rutherford's nuclear atom

The nuclear atom is the basis of the modern theory of atomic structure; it was proposed by Rutherford in 1911. He had observed that the passage of alpha particles through a very thin metal foil was accompanied by some scattering of the particles from their original direction. Two of his assistants, Geiger and Marsden, made a more detailed study of this effect. They directed a narrow beam of alpha particles on to gold foil about 1 μm thick and found that while most of the particles passed straight through, some were scattered appreciably and a very few — about 1 in 8000 — suffered deflections of more than 90 °. In effect they were reflected back towards the radioactive source.

To account for this very surprising result Rutherford suggested that **all the positive charge and nearly all the mass were concentrated in a very small volume or nucleus at the centre of the atom**. The large-angle scattering of alpha particles would then be explained by the strong electrostatic repulsion to which the alpha particles (also positively charged) are subjected on approaching closely enough to the tiny nucleus; the closer the approach the greater the scattering. We now know that protons are responsible for the positive charge on the nucleus,

and protons and neutrons together form the nuclear mass.

Rutherford considered the electrons to be outside the nucleus and at relatively large distances from it so that their negative charge did not act as a shield to the positive nuclear charge when an alpha particle penetrated the atom. The electrons were supposed to move in circular orbits round the nucleus (like planets round the sun), the electrostatic attraction between the two opposite charges being the required centripetal force for such motion.

In this planetary model it would be reasonable to expect that many physical and chemical properties of atoms could be explained in terms of the number and arrangement of the electrons, on account of their greater accessibility.

(b) Geiger–Marsden scattering experiment

To test his theory Rutherford derived an expression for the number of alpha particles deflected through various angles. The derivation was complex, and involved, among other factors, the charge on the nucleus of the scattering atom, the thickness of the foil, and the charge, mass and speed of the bombarding alpha particles, and was based on the assumption that the repulsive force between the two positive charges obeys an inverse square law. The path predicted for the scattered alpha particle was a hyperbola, Fig. 22.23.

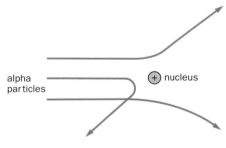

alpha particles (+) nucleus

Fig. 22.23

The test was performed by Geiger and Marsden using the apparatus shown in Fig. 22.24. A fine beam of alpha particles from a radioactive source fell on a thin foil of gold, platinum or other metal in an evacuated box. The angular deflection of the particles was measured by using a microscope to observe the scintillations

Fig. 22.22 A mass spectrometer

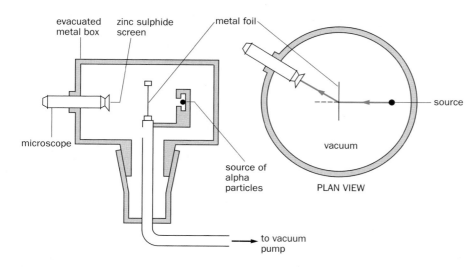

Fig. 22.24 Geiger–Marsden experiment

(flashes of light) on a glass screen coated with zinc sulphide. The screen and microscope could be rotated together relative to the foil and source, which were fixed.

Geiger and Marsden spent many hours in a darkened room counting the scintillations for a wide range of angles. Their results completely confirmed Rutherford's deductions and vindicated use of the inverse square law. (We will see later, however, p. 478, that in certain cases where the alpha particle approaches extremely close to the nucleus the inverse square law no longer holds.)

This experiment represents one of the great landmarks in physics. As well as putting the nuclear model on a sound footing, it inaugurated the technique of using high-speed particles as atomic probes. The subsequent exploitation of the technique was responsible for profound discoveries in nuclear physics.

(c) Nuclear size

The maximum angle of scattering will occur when the distance between the centres of an alpha particle and the atomic nucleus involved in the encounter is a minimum. This distance gives an upper limit for the sum of the radii of an alpha particle and the nucleus. It may be calculated by considering the energy changes occurring as an alpha particle approaches a nucleus. The potential at a distance r from a nucleus carrying charge Ze will be $V = Ze/4\pi\varepsilon r$ (p. 204). The energy change when charge Q moves through potential V is given (p. 206) by $W = QV = 2Ze^2/4\pi\varepsilon r$ (for an alpha particle of charge $2e$). Equating W to the original kinetic energy of the alpha particle then allows the distance of closest approach to be calculated. It is about 10^{-14} m; therefore the radius of the nucleus must be of the order of 10^{-15} m.

Taking into account the reservation mentioned earlier about the size of atomic particles (p. 386), if the radius of the nucleus is compared with that of the atom (about 10^{-10} m) it is evident, since electrons are similar in size to nuclei, that most of the atom is empty. The volume of the nucleus and the electrons in an atom is roughly 10^{-12} of the total volume of the atom. The penetration of thin foil, with negligible deflection, by most alpha particles is not surprising; close approaches to the nucleus are rare.

(d) Moseley and X-ray spectra

The determination of the charge on the nucleus was a vital problem in the development of models of the atom. It was thought that the position of an element in the periodic table, i.e. its atomic number Z, was equal to the number of protons in the nucleus and so also to the number of extra-nuclear electrons. The measurements made by Geiger and Marsden, later improved upon by Chadwick in 1920, showed that this was approximately true. However, in 1913 Moseley, a young physicist who was also working with Rutherford, obtained convincing evidence using a different technique.

He carried out a detailed study of X-ray spectra using a Bragg X-ray spectrograph (i.e. a spectrometer with a photographic plate instead of an ionization chamber to detect the X-rays). It has already been explained (pp. 396–7) that in the production of X-rays a characteristic line spectrum is superimposed on a continuous spectrum. Moseley measured the frequencies of the lines in the line spectra of nearly 40 elements and found that the frequency of any particular line, such as the K_α of Fig. 22.18, increased progressively from one element to the next in the periodic table. If f is the frequency of this line for a certain element, it is related to a number Z which always works out to be the atomic number, by the expression

$$\sqrt{f} = a(Z - b)$$

where a and b are constants for this particular line.

Moseley wrote, 'We have here a proof that there is in the atom a fundamental quantity which increases by regular steps from one element to the next.' He supported his belief by theoretical arguments that this fundamental quantity was the positive charge on the nucleus.

THE BOHR ATOM

Rutherford's model of the atom, although strongly supported by evidence for the nucleus, is inconsistent with classical physics. An electron moving in a circular orbit round a nucleus is accelerating and according to electromagnetic theory it should emit radiation continuously and so lose energy. If this happened the radius of the orbit would decrease and the electron would spiral into the nucleus. Evidently either this model of the atom or the classical theory of radiation requires modification.

In 1913, in an effort to overcome this paradox, Bohr, drawing inspiration from the success of the quantum theory in solving other problems involving radiation and atoms, made two revolutionary suggestions.

(i) Electrons can revolve round the nucleus only in certain 'allowed orbits' and while they are in these orbits they do not emit radiation. An electron in an orbit has a definite amount of energy. It possesses kinetic energy because of its motion and potential energy on account of the attraction of the nucleus. Each allowed orbit is therefore associated with a certain quantity of energy, called the 'energy of the orbit', which equals the total energy of an electron in it.

(*ii*) An electron can 'jump' from one orbit of energy E_2 to another of lower energy E_1 and the energy difference is emitted as one quantum of radiation of frequency f given by Planck's equation $E_2 - E_1 = hf$.

By choosing the allowed orbits correctly Bohr was able to explain quantitatively why particular wavelengths appeared in the line spectrum of atomic hydrogen and this provided evidence for his ideas concerning how electromagnetic radiation originates in an atom.

Despite its considerable achievements the Bohr atom had certain shortcomings. First, it could not interpret the details of the optical spectra of atoms containing more than one electron. Second, the very arbitrary method of selecting allowed orbits had no theoretical basis. Third, it involved quantities, such as the radius of an orbit, which could not be checked experimentally. Nevertheless, great credit is due to Bohr for linking spectroscopy and atomic structure and for introducing quantum ideas into atomic theory.

Bohr's model of the atom has been superseded by a theory based on **wave mechanics**, in which there is no need to make arbitrary assumptions to give correct results. But whereas the Bohr atom was easily visualized and involved fairly simple mathematics, its successor is abstract and the mathematics more difficult. Wave mechanics preserves the general idea of a hollow, nuclear atom but it discards the Bohr picture of electrons moving in allowed orbits. However, the essential characteristic of Bohr's orbits, i.e. definite energy values, is retained.

ENERGY LEVELS IN ATOMS

Wave mechanics permits the electrons in an atom to have only certain energy values. These values are called the **energy levels** of the atom. They are not something we can observe in the usual sense but, as we will see shortly, there is fairly direct experimental evidence to support our belief in their existence. Any theory of atomic structure must be able to explain how they arise; the wave mechanical justification will be given later.

The levels can be represented by horizontal lines, arranged one above the other to form an energy level diagram (or a ladder, of unequally spaced rungs), each line indicating by its position a particular energy value. Every *atom* has a characteristic set of energy levels whose values can be found experimentally or calculated using wave mechanics. While an electron is permitted to pass from one level to another by gaining or losing energy, it is not allowed to have an amount of energy that would put it between two levels. An atom can thus only accept 'parcels' of energy of certain definite sizes, i.e. its energy is **quantized**.

All levels have negative energy level values because the energy of an electron at rest outside an atom is taken as zero and when the electron 'falls' into the atom energy is lost as electromagnetic radiation (compare the loss of p.e. when a body falls in the earth's attractive gravitational field). In effect the electron is passing from a higher to a lower energy level and the lower

the level the larger the negative energy value. The most stable or **ground state** of the atom is the condition in which every electron is in the lowest energy state available. An atom is in an **excited state** when an electron is in a state of energy above that of a state which is unoccupied.

Energy level diagrams can be drawn for every atom — that for atomic hydrogen is shown in Fig. 22.25a. In this case the lowest level has energy −13.6 eV and is the one normally occupied by the single electron of a hydrogen atom. Above this state are the excited states to which the hydrogen atom may be raised by absorbing the correct amount of energy. If the energy absorbed is sufficient to allow the electron to escape from the atom, the atom becomes **ionized**; for hydrogen the ionization energy is 13.6 eV (21.8×10^{-19} J).

Electrons in the lower energy levels are held strongly by the nucleus. Those in higher levels need less energy to escape and are more loosely held in the atom. We may think of them as being near the outside of the electron cloud which surrounds the nucleus and partly screened from the attraction of the positive nuclear charge by inner, lower energy electrons.

Detailed study of the periodic variation of the chemical properties and the ionization energies (Fig. 22.28, p. 404) of the elements and of their optical and X-ray spectra suggests that the electrons in an atom fall into groups, called **shells**. All the energy states which the electrons in one group can occupy have *about* the

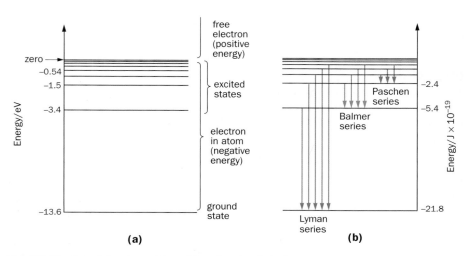

Fig. 22.25 Energy levels and transitions for atomic hydrogen

same value. Each shell is given a **principal quantum number** n and they are labelled in ascending order of energy by the letters K, L, M, etc.

Wave mechanics indicates that the number of electrons that can be accommodated in a shell is $2(n)^2$. For example, the K-shell ($n = 1$) can take at most $2(1)^2 = 2$ electrons, the L-shell ($n = 2$) can take a maximum of $2(2)^2 = 8$ electrons and so on. For the first five shells the numbers are 2, 8, 18, 32, 50. Normally the electrons occupy the lowest energy shells first, i.e. starting with K, but this is not always so. Argon with 18 electrons has its full complement of 2 and 8 electrons in the K- and L-shells respectively and 8 electrons in the incomplete M-shell, but in potassium (atomic number 19) the electronic configuration is 2–8–8–1, i.e. the extra electron is in the N-shell although the M-shell can take 10 more electrons.

EVIDENCE OF ENERGY LEVELS

(a) Optical line spectra

A line in an optical emission spectrum indicates the presence of a particular frequency (and wavelength) of light and is considered to arise from the loss of energy that occurs in an excited atom when an electron 'jumps', directly or in stages, from a higher to a lower level. The frequency f of the quantum of electromagnetic radiation emitted in a transition between levels of energies E_2 and E_1 ($E_2 > E_1$) is given by

$$E_2 - E_1 = hf$$

where h is the Planck constant. The transitions for some of the lines in the visible spectrum of atomic hydrogen, called the **Balmer series**, are shown in Fig. 22.25b. An electronic transition from the -2.4×10^{-19} J level to the -5.4×10^{-19} J level represents an energy loss to the atom of

$$[-2.4 - (-5.4)] \times 10^{-19} = 3.0 \times 10^{-19} \text{ J}$$

The wavelength λ of the radiation emitted is given by

$$3.0 \times 10^{-19} \text{ J} = hf = \frac{hc}{\lambda}$$

where c is the speed of light $= 3.0 \times 10^8$ m s^{-1}. So

$$\lambda = \frac{hc}{3.0 \times 10^{-19} \text{ J}}$$

$$= \frac{(6.6 \times 10^{-34} \text{ J s}) \times (3.0 \times 10^8 \text{ m s}^{-1})}{3.0 \times 10^{-19} \text{ J}}$$

$$= 6.6 \times 10^{-7} \text{ m}$$

$$= 0.66 \text{ μm}$$

The spectrum of atomic hydrogen contains a line of this wavelength (Fig. 20.38, p. 344). The lines of the **Lyman series** are in the ultraviolet and those of the **Paschen series** in the infrared. Using energy levels for hydrogen calculated from wave mechanics, the optical spectrum of hydrogen can be explained.

(b) X-ray line spectra

The energy of a light photon is a few electronvolts; that of an X-ray photon is several kiloelectronvolts. This suggests that whereas optical line spectra are due to transitions of loosely held electrons in higher energy levels, X-ray line spectra originate from electronic transitions in the lowest energy levels which have their full complement of electrons. They are only produced when materials are bombarded by high-energy electron beams (of the order of thousands of electronvolts) which are able to penetrate deep into atoms and displace electrons from very 'deep' energy levels. The subsequent fall of an electron from a higher level into one of these gaps in an otherwise complete, but appreciably lower energy level, causes the emission of a high-energy X-ray photon.

The K-series of X-ray lines (see Fig. 22.18) is produced when an electron is knocked out of the lowest or K-shell. The return of the same or another electron to the gap in the K-shell causes the emission of a line of the K series. If the electron falls from the L-shell the K$_\alpha$ line occurs; if it falls from the M-shell we have the K$_\beta$ line and so on, Fig. 22.26. Similarly, the L-series arises when electrons return to vacancies in the L-shell. This series is excited by smaller energies than the K-series because the L-electrons are less strongly held; the X-rays emitted have lower frequencies and longer wavelengths and are less penetrating.

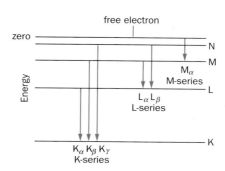

Fig. 22.26 Transitions giving rise to X-ray line spectra

The progressive increase of frequency with increasing atomic number, in say the K$_\alpha$ line, is due to small energy differences in the K- and L-shells of different atoms. This arises from the increasing positive charge on the nucleus and it makes possible the determination of atomic numbers from a study of X-ray line spectra (p. 401).

(c) Electron collision experiments

Direct evidence for the existence of energy levels is provided by experiments, first successfully performed in 1914 by Franck and G. Hertz (nephew of the radio pioneer H. Hertz), in which electrons have collisions with the atoms of a gas at low pressure. While a free electron completely detached from an atom can be accelerated to any energy, these experiments lead to the conclusion that the electrons *in an atom* can have only *certain* values, i.e. those permitted by its energy levels.

When an electron has an encounter with a gas atom one of three things can happen.

(i) An elastic collision occurs in which the bombarding electron, being much less massive than the gas atom, suffers only a slight loss of kinetic energy (see p. 491, question 11).

(ii) An inelastic collision occurs in which an electron in a gas atom gains *exactly* the amount of energy it requires to reach a higher energy level. Excitation has then occurred and the gas atom is in an excited state. The collision is inelastic since the kinetic energy lost by the bombarding electron does not reappear as kinetic energy of the gas atom but is emitted as electromagnetic radiation when the gas atom returns to a lower state or the ground state, usually after about 10^{-9} s. The p.d. through which the bombarding electron has to be accelerated from rest to cause excitation is called the **excitation potential** and since every atom has many energy levels, there are numerous excitation potentials characteristic of a particular atom.

(*iii*) An inelastic collision occurs in which an electron in a gas atom gains enough energy to escape from the atom, so causing ionization. The accelerating p.d. is then the **ionization potential** of the atom.

The principle of a Franck–Hertz type of experiment is illustrated in Fig. 22.27*a*. Electrons emitted by the hot cathode C in a tube containing gas at a low pressure are accelerated by a positive potential V_1 on the wire mesh G, called the grid. When V_1 just exceeds the small negative potential V_2 which the anode A has with respect to G, electrons reach A and a small current is indicated on the galvanometer. As V_1 is increased the current increases. During this phase, PQ in Fig. 22.27*b*, the bombarding electron energies are small and all collisions between electrons and gas atoms are elastic. At a certain value of V_1, called the **first excitation potential**, some electrons have inelastic collisions with gas atoms near G and raise them to their first excited energy level. The negative potential V_2 ensures that the electrons, having lost their kinetic energy, are carried back to G. The anode current therefore falls abruptly, shown by QR. Further increase of V_1 allows even those electrons which have inelastic collisions to overcome V_2 and reach A. The galvanometer reading increases again, RS.

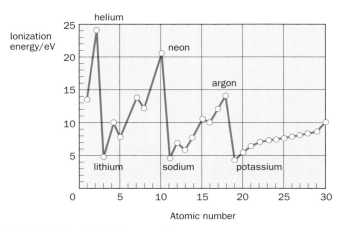

Fig. 22.28 Periodicity of ionization energies

Other effects may be obtained at higher values of V_1 depending on the tube and circuit conditions. It is possible to find the ionization potential of the gas, i.e. the potential at which the bombarding electrons have enough energy to cause gas molecules to be ionized.

Collision experiments, as just outlined, indicate that an atom cannot absorb any amount of energy but only certain definite amounts, determined by its energy levels.

Franck–Hertz type tubes that give reliable results are difficult to make and so are expensive. An experiment with a commercial tube containing the inert gas xenon is described in Appendix 5. It gives rough values of the ionization potential and one excitation potential.

The ionization potentials and energies of many gases and vapours have been investigated. The graph of Fig. 22.28 shows the variation of ionization energy with atomic number for the first 30 elements in the periodic table. A periodicity, like that displayed by other properties, is evident. The small ionization energies of the alkali metals suggests they have a loosely held electron in a higher energy level. By contrast, the inert gases must have very stable electronic structures since they require most energy for ionization.

(d) Spectrum of mercury vapour

Electron collisions experiments with mercury vapour show that there is an energy difference of 4.9 eV (7.8×10^{-19} J) between two of the energy levels in an isolated mercury atom. We might therefore expect photons with this energy to be emitted by a mercury vapour lamp. The wavelength of such radiation is 2.5×10^{-7} m, i.e. 0.25 μm; it is in the ultraviolet region as may be shown using the arrangement of Fig. 22.29*a*, **in which the lamp should be well screened from observers**. The ultraviolet lines A and B in Fig. 22.29*b* appear only on the fluorescent paper (the visible lines are also on the white screen) and are cut off if a piece of glass is held in front of the lamp. Assuming the wavelength of the green line C is 5.5×10^{-7} m (it can be measured readily using a transmission diffraction grating of known spacing), the wavelength λ of the ultraviolet line A is given approximately by

$$\frac{\lambda}{5.5 \times 10^{-7}\,\text{m}} = \frac{\text{AA}}{\text{CC}}$$

If *cold* mercury vapour is puffed *gently* into the air from a polythene bottle held in front of the reflection grating, both A and B fade or disap-

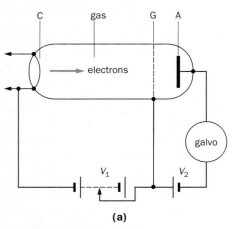

Fig. 22.27 Franck–Hertz experiment

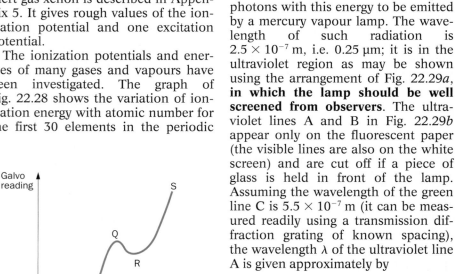

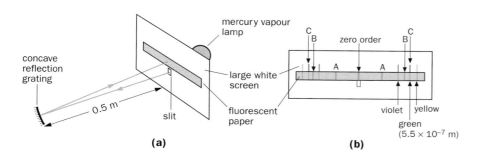

Fig. 22.29 Investigating the spectrum of mercury vapour

pear. What does this suggest about B? *Note.* **Mercury vapour is highly poisonous and on no account must the mercury be warmed.**

(e) Absorption spectrum of iodine vapour

An arrangement for viewing lines in the absorption spectrum of iodine vapour is given in Fig. 22.30a and offers further evidence for the existence of energy levels; Fig. 22.30b shows what is seen.

A test-tube (hard glass) containing one or two small iodine crystals, after being warmed along its length and lightly corked, is heated at the bottom until the iodine vaporizes and colours the tube strongly. As it cools it is observed in front of the straight filament lamp through a fine diffraction grating (about 300 lines per millimetre).

The iodine molecules absorb from the light those frequencies whose quanta have the correct amount of energy to enable them to jump from one energy level to another. Certain frequencies are therefore missing in the continuous spectrum of the light from the lamp and show up as dark lines. The presence of so many of these lines indicates that a whole 'ladder' of energy levels exists. In a gas or vapour the molecules are sufficiently far apart not to affect each other's energy levels to any extent.

WAVE–PARTICLE DUALITY OF MATTER

The wave–particle nature of electromagnetic radiation (discussed on p. 391) led de Broglie (pronounced 'de Broy') to suggest in 1923 that matter might also exhibit this duality and have wave properties. His ideas can be expressed quantitatively by first considering radiation. A photon of frequency f and wavelength λ, has, according to the quantum theory, energy $E = hf = hc/\lambda$ where c is the speed of light and h is the Planck constant. Then, by Einstein's energy–mass relation (p. 480) the equivalent mass m of the photon is given by $E = mc^2$. So

$$\frac{hc}{\lambda} = mc^2$$

$$\therefore \quad \lambda = \frac{h}{mc}$$

By analogy, de Broglie suggested that a particle of mass m moving with speed v behaves in some ways like waves of wavelength λ given by

$$\lambda = \frac{h}{mv} = \frac{h}{p}$$

where p is the momentum of the particle.

Calculation shows that electrons accelerated through a p.d. of about 100 V should be associated with **de Broglie** or **matter waves**, as they are called, having a wavelength of the order of 10^{-10} m. This is about the same as for X-rays and it was suggested that the conditions required to reveal the wave nature of X-rays might also lead to the detection of electron waves. At first de Broglie's proposal was no more than speculation, but within a few years a variety of experiments proved beyond dispute that moving particles of matter had wave-like properties associated with them.

Interference and diffraction patterns can be obtained with electrons. Figure 22.31, p. 406, shows interference fringes produced by Young's double-slit type experiments with (a) light and (b) a stream of electrons. An arrangement for producing electron interference is given in Fig. 22.32a. The stream of electrons is split into two by a very thin wire between two metal plates. When the wire is made a few volts positive with respect to the plates, the electron streams passing on each side of the wire are brought closer together and overlap on the photographic film as if they had come from two sources, Fig. 22.32b. The interference pattern is so small that it has to be magnified by an electron microscope and then by an optical enlarger to give fringes of the width shown in Fig. 22.31b.

Electron diffraction may be shown using the arrangement of Fig. 22.33, in which a beam of electrons strikes a

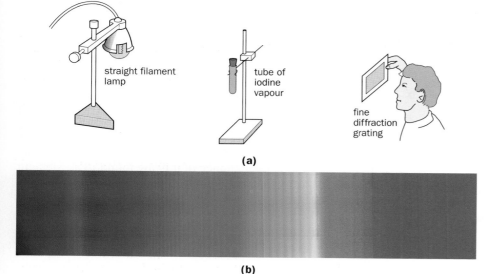

Fig. 22.30 Investigating the absorption spectrum of iodine vapour

thin film of graphite on a metal grid just beyond a hole in the anode. Diffraction effects have been obtained with streams of protons, neutrons and alpha particles, but it is evident from de Broglie's equation that the greater the mass of the moving particle, the smaller the associated wavelength and the more difficult detection becomes.

The idea of matter waves is useful when we consider the various kinds of microscope. The resolving power of any microscope increases as the wavelength of whatever is used to 'illuminate' the object decreases. Electron waves have a smaller wavelength than light waves and so an electron microscope reveals much more detail.

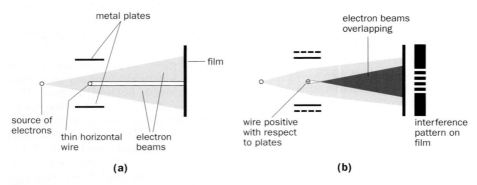

Fig. 22.31 Interference fringes with (*a*) light and (*b*) electrons

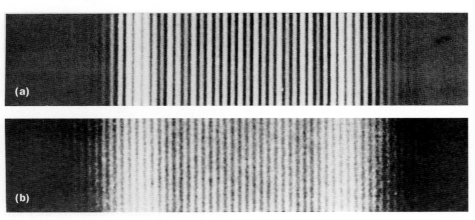

Fig. 22.32 Producing electron interference

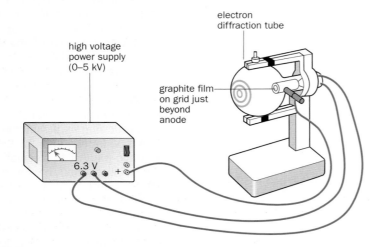

Fig. 22.33 Viewing electron diffraction

WAVE MECHANICS

(a) Wave–particle duality

In the early decades of the twentieth century physicists faced an apparently puzzling situation. How could matter and radiation have *both* wave-like and particle-like properties? It seemed to be a complete contradiction.

However, to some extent the dilemma was of the scientist's own making, for originally an attempt was made to explain this dual behaviour in terms of the real waves and real particles of everyday life. This was a perfectly reasonable thing to do when faced with phenomena that appeared to have similarities. But perhaps it was not altogether surprising that ideas which helped to deal with the large-scale world of water waves and tennis balls were not completely suited to the atomic world of photons and electrons. At any rate, it came to be realised that part of the dilemma, at least, arose from the use of models based on our experience of the macroscopic and that such analogies might not be valid when pushed too far.

A scientific model is an aid to understanding and not necessarily a true description. In the same way a map showing only towns and roads is no more a complete representation of a country than is one showing only physical features; each describes one aspect and neither is wrong. Both models are necessary for an adequate description of the behaviour of matter and radiation: they are complementary not contradictory. The sensible thing to do is to use one or other model when it is appropriate and helpful.

Such arguments help to make the duality dilemma more acceptable but they do not explain the nature of the 'particles' and 'waves'. The present view is that the 'waves' are not waves of moving matter or varying fields but are 'probability waves'. The wave-like behaviour of a moving electron is considered to be due to the fact that the chance of it being found in an element of volume depends on the intensity (i.e. the square of the amplitude) at the point of the electron wave associated with it. Where this is high there is likely to be a high electron density. In the same way an electromagnetic wave, although undoubtedly connected with electric and magnetic fields, is regarded as consisting of photons (i.e. bundles of energy) whose probable locations are given by the intensity of the wave. This view forms the basis of wave mechanics which, as the name implies, considers that the motion of a 'particle' is determined by a wave equation. It deals in probabilities and not certainties. For 'particles' such as electrons and photons it predicts quite a different behaviour from that given by the particle mechanics of Newton, but the two theories agree for macroscopic bodies.

(b) Wave-mechanical model of the atom

The real objection to the Bohr model of the atom is that it pinpoints an electron in a definite orbit and takes no account of its wave-like aspect. The wave-mechanical model developed by Erwin Schrödinger requires advanced mathematics but we can attempt to explain it by making analogies with real waves.

In a vibrating string fixed at both ends, waves are reflected to and fro and a stationary wave system is set up (p. 318) having nodes and antinodes. In its simplest mode of vibration the string has a node at each end and an antinode in the centre but it can vibrate with two, three or any integral number of loops (Fig. 19.16). The vibration is restricted to those modes having a complete number of loops and each mode produces a note of different frequency.

Similarly we can imagine that an electron, confined to an atom by the attractive force exerted on it by the nucleus, is associated with electron waves. Figure 22.34 shows how a stationary electron wave could be fitted into a circular orbit in an atom; modes of vibration possible are those in which a whole number of wavelengths λ fit smoothly around the circle of radius R. For the different modes $2\pi R = n\lambda$, where $n = 1, 2, 3, \ldots$, correspond to the allowed orbits in the Bohr model of the atom. In the wave-mechanical model, the circular stationary waves do not represent the actual path of an electron but give the amplitude of the electron's probability wave in the atom. Each mode of vibration corresponds to a particular frequency and, therefore, to a particular energy level of the atom. When an

electron wave changes its mode of vibration, the energy difference between the levels is emitted (or absorbed) as radiation. This argument enables us to see how wave mechanics can account for the existence of energy levels.

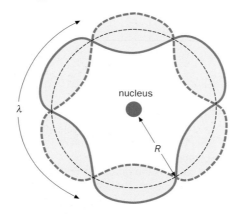

Fig. 22.34 Stationary electron wave ($n = 3$) in an atom

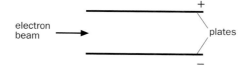

QUESTIONS

Electrons

1. Figure 22.35 shows a beam of electrons in a vacuum entering the region between two parallel plates. The electric field strength E between the plates is 2.4×10^4 V m^{-1}.

Fig. 22.35

a) Copy the diagram and draw a possible path for one electron in the beam as it passes between the two plates. At a point on your line, draw an arrow to show the direction of the electric force which acts on the electron at that point.
b) Calculate
    i) the force on the electron due to the electric field,
    ii) the acceleration of the electron due to the electric field.
c) Suggest how this apparatus could be used to separate
    i) positively charged particles from negatively charged particles,

   ii) particles of the same sign, speed and mass but with different magnitudes of charge.
   (UCLES, Nuclear Physics, June 1998)

2. How can cathode rays be produced? What are their main properties?
   An electron starts from rest and moves freely in an electric field whose intensity is 2.4×10^3 V m^{-1}. Find
   a) the force on the electron,
   b) its acceleration, and
   c) the velocity acquired in moving through a potential difference of 90 V. (The charge on an electron $= 1.6 \times 10^{-19}$ C and the mass of an electron $= 9.1 \times 10^{-31}$ kg.)

3. What do you understand by an **electron**?
   Electrons in a cathode ray tube are accelerated through a potential difference of 2.0 kV between the cathode and the screen. Calculate the velocity with which they strike the screen. Assuming they lose all their energy on impact and given that 10^{12} electrons pass per second, calculate the power dissipation.
   (Charge on an electron $= 1.6 \times 10^{-19}$ C; mass of an electron $= 9.1 \times 10^{-31}$ kg.)

4. Describe a method for measuring the ratio of the charge to mass (e/m) for an electron.
   Calculate
   a) the speed achieved by an electron accelerated in a vacuum through a p.d. of 2.00×10^3 V, and
   b) the magnetic flux density required to make an electron travelling with speed 8.00×10^6 m s^{-1} traverse a circular path of diameter 10.0×10^{-2} m.
   Take e/m for an electron as 1.76×10^{11} C kg^{-1}.

5. Give an account of a method by which the charge associated with an electron has been measured.
   Taking this electronic charge to be -1.60×10^{-19} C, calculate the potential difference in volts necessary to be maintained between two horizontal conducting plates, one 5.00×10^{-3} m above the other, so that a small oil drop, of mass 1.31×10^{-14} kg with two electrons attached to it, remains in equilibrium between them. Which plate would be at the positive potential? ($g = 9.81$ N kg^{-1})

6. a) Electrons in a cathode-ray tube leave the cathode with negligible speed at a potential of -9000 V and are accelerated to an anode at a potential of -200 V. For an electron in this tube

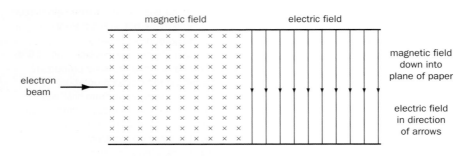

magnetic field · electric field

electron beam →

magnetic field down into plane of paper

electric field in direction of arrows

Fig. 22.36

calculate
  i) the gain in electrical potential,
  ii) the loss in potential energy,
  iii) the gain in kinetic energy,
  iv) the speed on reaching the anode.
b) Explain why a) i) is a gain but a) ii) is a loss.
c) While travelling between the anode and the screen of a cathode-ray tube, electrons move through adjacent electric and magnetic fields, as illustrated in Fig. 22.36.
  Copy Fig. 22.36 and on it sketch a possible path of an electron through both fields.
(*OCR, 9244/2, June 1999*)

Photoelectric effect

Speed of light = 3.0×10^8 m s^{-1}
Planck constant = 6.6×10^{-34} J s
Mass of electron = 9.1×10^{-31} kg
Electronic charge = 1.6×10^{-19} C

7. The Einstein photoelectric equation is

$$hf = \Phi + E_k$$

a) State the meaning of each of the terms in the equation.
b) In a laboratory demonstration of the photoelectric effect, a metal plate is given an electric charge and light of various wavelengths is shone on to the surface of the plate in turn. It is found that the plate loses its electric charge when the plate is given a negative charge *and* when ultraviolet light is shone on to the plate.
  Explain why the plate does not lose its charge when
  i) the plate is given a positive charge and illuminated by ultraviolet light,
  ii) the plate is given a negative charge and illuminated by visible light.
(*NEAB, AS/A PH02, June 1998*)

8. a) If a surface has a work function of 3.0 eV, find the longest wavelength light that will cause the emission of photoelectrons from it.
b) What is the maximum velocity of the photoelectrons liberated from a surface having a work function of 4.0 eV by ultraviolet radiation of wavelength 0.20 μm?

9. When light of wavelength 0.50 μm falls on a surface it ejects photoelectrons with a maximum velocity of 6.0×10^5 m s^{-1}. Calculate
a) the work function in electronvolts, and
b) the threshold frequency for the surface.

10. Figure 22.37 shows monochromatic light falling on a photocell.

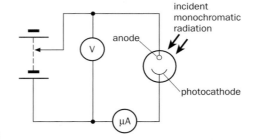

incident monochromatic radiation

anode

photocathode

Fig. 22.37

As the reverse potential difference between the anode and cathode is increased, the current measured by the microammeter decreases. When the potential difference reaches a value V_s, called the stopping potential, the current is zero. Explain these observations.
  What would be the effect on the stopping potential of
  i) increasing only the intensity of the incident radiation,
  ii) increasing only the frequency of the incident radiation?
(*L, June AS/A PH2, June 1998*)

11. The graph in Fig. 22.38 shows how the maximum kinetic energy T of photoelectrons emitted from the surface of sodium metal varies with the frequency f of the incident radiation.

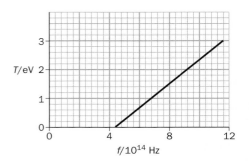

Fig. 22.38

Why are no photoelectrons emitted at frequencies below 4.4×10^{14} Hz?
  Calculate the work function Φ of sodium in eV.
  Explain how the graph supports the photoelectric equation $hf = T + \Phi$.
  How could the graph be used to find a value for the Planck constant?
  Sketch the graph and add a line to show the maximum kinetic energy of the photoelectrons emitted from a metal which has a greater work function than sodium.
(*L, AS/A PH2, Jan 1996*)

12. a) In the photoelectric effect equation

$$hf = \Phi + E_k$$

explain what is meant by hf, Φ and E_k.
b) Monochromatic light of wavelength 500 nm falls on a metal cathode of area 2000 mm^2 and produces photoelectrons. The light intensity at the surface of the metal is 1.0×10^{-2} W m^{-2}.

  charge of electron
    = -1.6×10^{-19} C
  the Planck constant
    = 6.63×10^{-34} J s
  speed of light in vacuo
    = 3.0×10^8 m s^{-1}

Calculate
  i) the frequency of the light,
  ii) the energy of a single photon of the light,
  iii) the number of photons falling on one square millimetre of the metal in one second,
  iv) the total photoelectric current, assuming that each photon releases one photoelectron.
(*NEAB, AS/A PH02, Feb 1997*)

X-rays

13. If the p.d. applied to an X-ray tube is 20 kV, what is the speed with which electrons strike the target? Take the specific charge of the electron as 1.8×10^{11} C kg^{-1}.

14. What is the minimum wavelength of the X-rays produced when electrons are accelerated through a potential difference of 1.0×10^5 V in an X-ray tube? Why is there a minimum wavelength? ($e = 1.6 \times 10^{-19}$ C; $h = 6.6 \times 10^{-34}$ J s; $c = 3.0 \times 10^8$ m s^{-1}.)

15. a) i) Describe, with the aid of a labelled diagram of an X-ray tube, how X-rays are produced.
ii) Sketch a typical X-ray spectrum and explain the origin of the continuous spectrum and the characteristic spectrum.
iii) Electrons are accelerated in a vacuum tube through a potential difference of 2.00 kV to strike a heavy metal target. Determine the minimum wavelength of the X-rays produced.
iv) The minimum wavelength X-rays from part iii) are directed at the surface of a crystal and a first maximum scattering intensity is observed at an angle of 30.0°. Describe why this happens and determine the spacing between the crystal planes.

b) When X-rays are produced only about 10% of the initial input energy appears as X-ray energy. Explain what has happened to the other 90% of the energy.

(*IB*, Higher Paper 2, Nov 1997)

16. An X-ray tube operates with a potential difference of 100 kV between the anode and cathode. The tube current is 20 mA.
a) Calculate
i) the rate at which energy is transformed in the target of the X-ray tube,
ii) the number of electrons which reach the target each second,
iii) the maximum energy of an X-ray photon produced.

b) Figure 22.39 is a sketch of the X-ray spectrum produced by this tube for a particular metal target. The tube voltage is 100 kV and the current is 20 mA.
i) Copy Fig. 22.39 and sketch on it a spectrum for X-rays from the tube if the tube voltage is reduced to

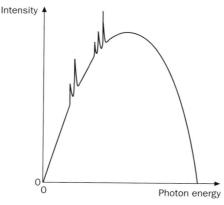

Fig. 22.39

50 kV, the current remaining at 20 mA. Label this spectrum A.
ii) Also on the same axes, sketch a spectrum for X-rays from the tube if the tube current is increased to 30 mA, the tube voltage remaining at 100 kV. Label this spectrum B.

c) An X-ray beam is used to produce the image of an object. State two factors which may reduce the sharpness of the image. In each case, suggest how the sharpness may be improved.

(*UCLES*, Nuclear Physics, June 1998)

Structure of the atom

17. a) Describe the principal features of the **nuclear model of the atom** suggested by Rutherford.
b) When gold foil is bombarded by alpha particles it is found that most of the particles pass through the foil without significant change of direction or loss of energy. A few particles are deviated from their original direction by more than 90°. Explain, in terms of the nuclear model of the atom and by considering the nature of the forces acting,
i) why some alpha particles are deflected through large angles,
ii) why most of the alpha particles pass through the foil without any significant change in direction or significant loss of energy.

(*NEAB*, AS/A PHO2, March 1998)

18. a) When electrons collide with atoms, the atoms may be excited or may be ionised. Explain what is meant by
i) excitation by collision,
ii) ionisation by collision.
b) Explain, in terms of what happens to the atom, how the lines in an atomic line spectrum are produced.

(*NEAB*, AS/A PHO2, March 1998)

19. a) Describe briefly how you could demonstrate in a school laboratory that different elements can be identified by means of their optical spectra.
b) Figure 22.40 is a simplified energy level diagram for atomic hydrogen.

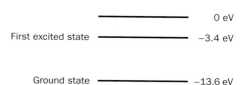

Fig. 22.40

A free electron with kinetic energy 12 eV collides with an atom of hydrogen and causes it to be raised to its first excited state.
Calculate the kinetic energy of the free electron (in eV) after the collision.
Calculate the wavelength of the photon emitted when the atom returns to its ground state.

(*L*, AS/A PH2, Jan 1996)

20. a) i) State the two postulates of the Bohr model of the hydrogen atom.
ii) Describe two limitations of the Bohr model.
b) The three lowest energy levels of a fictitious atom are shown in Fig. 22.41.

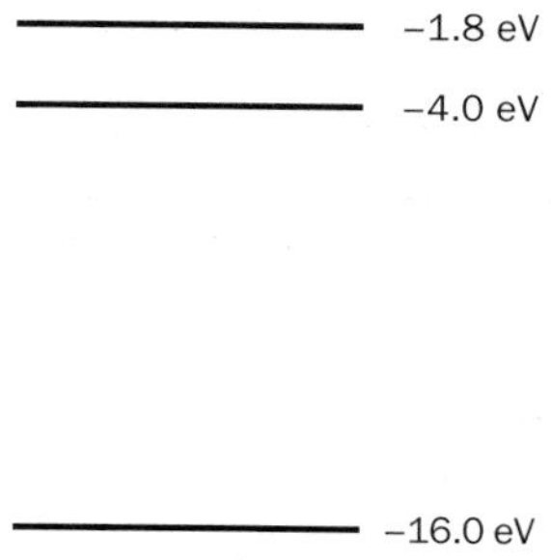

—————————————— −1.8 eV

—————————————— −4.0 eV

—————————————— −16.0 eV

Fig. 22.41

i) Determine the minimum energy required in joules to eject an electron in the lowest state from the atom.
ii) Assuming that energy level n has energy k/n^2 determine the energy of level $n = 4$ in electronvolts.
iii) Determine the wavelength of the radiation associated with a transition from level $n = 2$ to level $n = 3$.
iv) Name the region of the electromagnetic spectrum in which this radiation is found.

v) A tube containing this fictitious element in gaseous form is placed between a beam of white light and a prism. The spectrum produced is crossed by dark lines. Explain the existence of these lines.

(*IB, Subsidiary/Standard Paper 2, Nov 1997*)

21. Some of the energy levels of the mercury atom are shown in Fig. 22.42.

a) How much energy in electronvolts is required to raise an electron from the ground state to each of the levels shown?

b) What is the ionization energy of mercury?

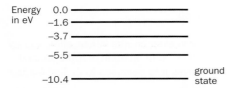

Fig. 22.42

c) If a mercury vapour atom in the ground state has a collision with an electron of energy

i) 5 eV, and

ii) 10 eV,

how much energy might be retained by the electron in each case?

d) What would happen to a photon of energy

i) 4.9 eV, and

ii) 8 eV,

which has a collision with a mercury atom?

22. a) Calculate the frequency of electromagnetic radiation emitted by a hydrogen atom which undergoes a transition between energy levels of -1.36×10^{-19} J and -5.45×10^{-19} J.

b) A different emission from a hydrogen atom has a wavelength of 4.34×10^{-7} m.

Calculate the angle of the first order diffraction maximum for this radiation when using a diffraction grating with a spacing of 2.00×10^{-6} m.

(*AEB, 0635/7, Jan 1998*)

Wave–particle duality

---

23. a) A beam of electrons is accelerated through a voltage V in an evacuated tube and hits a graphite film to produce a system of concentric rings on a fluorescent screen behind the film, Fig. 22.43.

i) What are the rings evidence of?

ii) If V is increased, what happens to the diameter of the rings?

b) What is the ratio of the de Broglie wavelength of electron X to that of electron Y if X has three times the kinetic energy of Y?

Fig. 22.43

24. The spacing of atoms in a crystal is 1.0×10^{-10} m.

mass of the electron = 9.1×10^{-31} kg
the Planck constant = 6.6×10^{-34} J s

a) Estimate the speed of electrons which would give detectable diffraction effects with such crystals.

b) State and explain how the speed of electrons would have to be different in an experiment to observe their diffraction by atomic nuclei.

c) Give *two* pieces of evidence to demonstrate that electrons have particle properties.

(*NEAB, AS/A PH02, June 1997*)

25. Quantum physics is a branch of physics which has developed over the last century. It commenced as a theory, put forward by Planck, to explain certain phenomena of electromagnetic radiation which could not be explained by wave theory. It was used subsequently to explain the photoelectric effect, line spectra and particle diffraction.

a) Describe the photoelectric effect, making reference to the significance of threshold frequency, work function energy, maximum energy of photoelectrons and intensity of incident radiation.

b) Describe how the quantum theory has been used to explain the existence of line spectra.

c) The phenomenon of particle diffraction showed that a wave nature could be assigned to moving particles. Calculate the wavelength of an electron travelling with speed 4.58×10^6 m s^{-1}.

(*UCLES, Basic 1, March 1998*)

26. This question is about models of the atom. Two models of the atom are based on Bohr's and Schrödinger's theories. Such models are devised to explain observed atomic phenomena.

a) In what ways do the Bohr and Schrödinger models of the hydrogen atom differ?

b) How does the existence of discrete atomic energy levels arise in each model? Give explanations in physical terms rather than mathematical.

c) Both models can at least explain the observed spectra of hydrogen. Why then is one model preferred to the other? In what respects is the model arising from Schrödinger theory deemed superior to Bohr's model?

(*IB, Higher Paper 3, Nov 1998*)

27. a) Figure 22.44 at the foot of the page shows the envelope of the vibrations of a stretched string that is emitting a note at its fundamental frequency.

i) Draw the shape of the envelope of the vibrations when the string is emitting a note at three times its fundamental frequency.

ii) Explain briefly how a stationary wave, such as that shown in Fig. 22.44, is produced.

b) A simple model of the hydrogen atom assumes that the wave associated with an electron in the atom is a stationary wave as shown in Fig. 22.45. The nodes correspond to opposite edges of the atom and the centre of the atom.

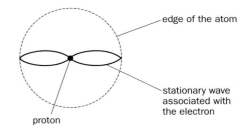

Fig. 22.45

Radius of a hydrogen atom = 1.0×10^{-10} m
The Planck constant, h = 6.6×10^{-34} J s

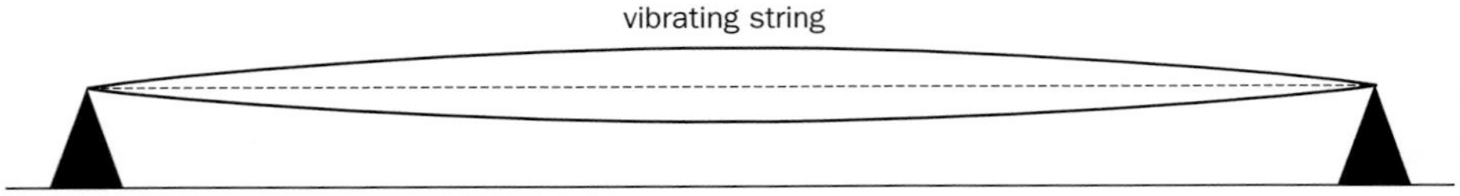

vibrating string

Fig. 22.44

i) State the de Broglie wavelength of the electron in Fig. 22.45.

ii) Calculate the momentum of an electron in a hydrogen atom.

iii) The mass of an electron is 9.1×10^{-31} kg. Calculate the kinetic energy of an electron in a hydrogen atom.

iv) State and explain briefly where, using this model, the electron in a hydrogen atom is most likely to be found.

(*AEB*, 0635/1, Summer 1997)

28. a) i) Define the **electronvolt**.

ii) Show that the speed of an electron accelerated through a potential difference of 6.0 kV is $4.6 \times 10^7 \, \text{m s}^{-1}$.

b) State what is meant by the duality of the nature of electrons.

c) In a demonstration of electron diffraction, a narrow beam of electrons is accelerated through 6.0 kV and passes normally through a thin film of graphite mounted in a vacuum tube. Concentric rings appear on a fluorescent screen at the end of the tube.

i) Calculate the wavelength associated with the electrons.

ii) What information does your answer to part c) i) suggest about the spacing of carbon atoms in graphite?

(*AQA: NEAB*, AS/A PHO2, March 1999)

29. A muon is a particle which has the *same charge* as an electron but its *mass* is 207 times the mass of an electron.

An unusual atom similar to hydrogen has been created, consisting of a muon orbiting a single proton. An energy level diagram for this atom is shown in Fig. 22.46.

State the ionisation energy of this atom.

Calculate the maximum possible wavelength of a photon which, when absorbed, would be able to ionise this atom.

To which part of the electromagnetic spectrum does this photon belong?

Calculate the de Broglie wavelength of a muon travelling at 11% of the speed of light.

(*L*, AS/A PH2, June 1999)

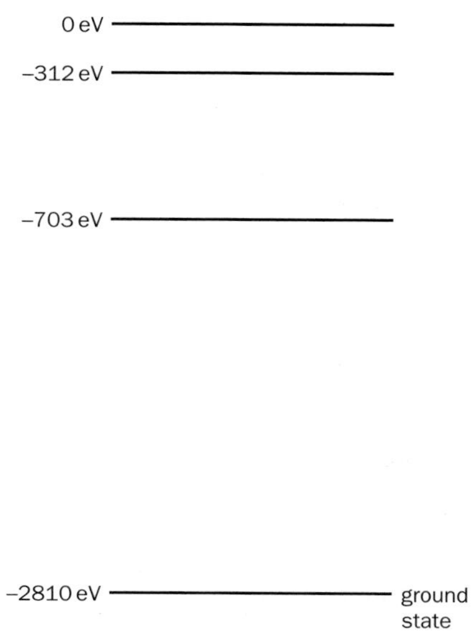

0 eV ────────────────

−312 eV ────────────────

−703 eV ────────────────

−2810 eV ──────────────── ground state

Fig. 22.46

23
Electronics

INTRODUCTION

The striking advances in electronics during recent years have been the result of the development of semiconductor devices such as junction diodes, transistors, operational amplifiers and integrated circuits. Their small size, low power requirements and very long life make them ideal for use in modern electronic equipment, but they can be damaged by too high temperatures and by p.ds that are too great or of the wrong polarity.

CATHODE RAY OSCILLOSCOPE

The cathode ray oscilloscope (CRO) is an important tool in electronics; it consists of a cathode ray tube and associated circuits. It has four main parts — the **electron gun**, the **deflecting system**, the **fluorescent screen** and the **time base** — described in the following sections (*a*) to (*d*).

(a) Electron gun

This is an electrode assembly for producing a narrow beam of cathode rays. A typical CRO tube is shown in Fig. 23.1 connected to a potential divider which supplies appropriate voltages from a high voltage power supply.

The gun comprises an indirectly heated cathode C, a cylinder G called the **grid** and two anodes A_1 and A_2. The grid G has a negative potential (or bias) with respect to C and controls the number of electrons reaching A_1 from C. The resistor R_1 determines the potential of G; making it less negative increases the brightness of the spot produced by the electron beam on the fluorescent screen S at the end of the tube. R_1 is the **brightness** control.

A_1 and A_2 are metal discs or cylinders with central holes and both have positive potentials relative to C, A_2 more so than A_1. They accelerate the electrons to a high speed down the tube (which is highly evacuated) and their shapes and potentials are such that the electric fields between them

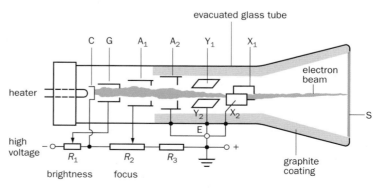

Fig. 23.1   Cathode ray oscilloscope tube

focus the beam to a small spot on the fluorescent screen S. Adjustment of the potential of A_1 by R_2 provides a **focus** control. A_1 and A_2 are an 'electron lens' system. Typical potentials for a small tube are 1000 V for A_2, 200 to 300 V for A_1 and -50 V to zero for G.

There must be a return path to A_2 (and hence the positive supply) for electrons reaching S. It is provided by a coating of graphite on the inside of the tube which is connected to A_2. The coating also shields the beam from external electric fields. A Mumetal screen round the tube protects it from stray magnetic fields.

To prevent earthed objects (e.g. the experimenter) near the screen affecting the beam, A_2 (and the coating) are usually earthed. This means that the positive supply terminal is also at earth potential, Fig. 23.1, and the other electrodes become negative with respect to A_2. The electrons are still accelerated through the same p.d. between A_2 and C.

(b) Deflecting system

The beam from A_2 passes first between a pair of horizontal metal plates, the **Y-plates**, whose electric field causes vertical deflection, and then between two vertical plates, the **X-plates**, which cause horizontal deflections when a p.d. is applied to them.

To minimize the effect of electric fields between the deflector plates and other parts of the tube (which might cause defocusing), one of each pair of plates is at the same (earth) potential as A_2. In practice X_2, Y_2 and A_2 are connected internally and brought out to a single terminal marked E (for earth). Deflecting p.ds are then applied to Y_1 (marked **Y** or **input**) and E or to X_1 (marked **X**) and E.

The deflection sensitivity for p.ds applied directly to the deflector plates is typically 50 V cm^{-1}, which is not high. X- and Y-deflection amplifiers are usually built into the CRO to amplify p.ds that are too small to give measurable deflections before they are connected to the plates.

X and Y **position** controls are used to move the spot 'manually' in the X or Y directions. They apply a positive or negative voltage to one of the deflecting plates according to the shift required.

(c) Fluorescent screen

The inside of the wide end of the tube is coated with a **phosphor** which emits light when struck by fast-moving particles. This may occur as fluorescence, i.e. the emission stops with the bombardment, and phosphorescence,

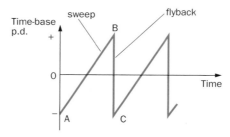

Fig. 23.2

when there is 'afterglow', possibly for a few seconds. Zinc sulphide is commonly used in cathode ray tubes; it emits blue light and has no afterglow.

(d) Time base

When the variation of a quantity with time is to be studied, an alternating p.d. representing the quantity is applied to the Y-plates via the vertical or Y-amplifier and the X-plates are connected via the horizontal or X-amplifier to a circuit in the CRO, called the **time base**, which generates a sawtooth p.d. like that in Fig. 23.2. The 'sweep' AB must be linear so that the deflection of the beam is proportional to time and causes the spot to travel across the screen from left to

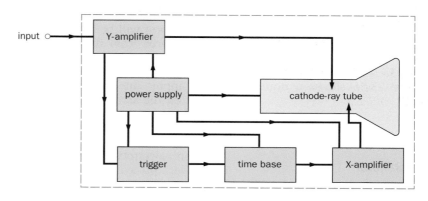

(a) Block diagram of a CRO

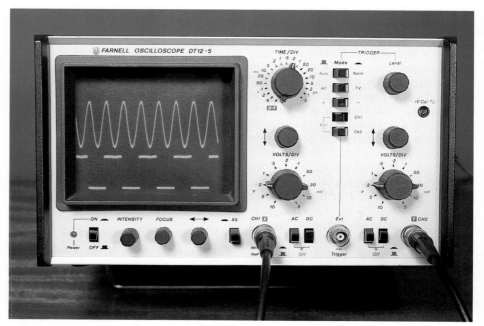

(b) Double-trace CRO

Fig. 23.3

right at a speed that is steady but which can be varied by the time-base control(s) to suit the frequency of the alternating input p.d.

To maintain a stable trace on the screen each horizontal sweep must start at the same point on the waveform being displayed. This is done by feeding part of the input signal to a **trigger** circuit that gives a pulse (at a chosen point on the input signal selected by the **trig level** control) which is used to start the sweep of the time-base sawtooth, i.e. it 'triggers' the time base and initiates the horizontal motion of the spot at the left-hand side of the screen. Automatic triggering by the input is obtained by selecting the **auto** position on the **trig level** control.

A trace appears only during the sweep part of the sawtooth p.d. since when the time base is not triggered by an input signal, the electron beam is suppressed so that it does not reach the screen.

Figure 23.3a is a block diagram of a CRO. Figure 23.3b shows a double-trace CRO.

SOME USES OF THE CRO

(a) As a voltmeter

The CRO can be used as a d.c./a.c. voltmeter by connecting the p.d. to be measured across the Y-plates, with the time base off and the X-plates earthed. D.c. deflects the spot; a.c. causes it to move up and down, and if the motion is fast enough (e.g. at 50 Hz) a vertical line is observed.

In many CROs the 'gain' control (marked **volts/div**) of the Y-deflection amplifier is calibrated. If it is on the '50' setting a deflection of 1 division of the screen graticule would be given by a 50 V d.c. input; a line 1 division long would be produced by an a.c. input of 50 V peak-to-peak (i.e. peak voltage $V_0 = 25$ V and r.m.s. voltage $V_{r.m.s.} = V_0/\sqrt{2} = 18$ V). On the '0.1' setting the gain of the amplifier is greatest and small p.ds can be measured, 0.1 V causing a deflection of 1 division. (The calibration can be checked by applying known input p.ds.) The use of a CRO as a voltmeter is shown in Fig. 16.6b (p. 251) and Fig. 16.30b (p. 259).

Advantages of the CRO as a voltmeter include the following.

- The electron beam behaves as a pointer of negligible inertia, responding instantaneously and having a perfect 'dead-beat' action.
- Direct and alternating p.ds can be measured, the latter at frequencies of several megahertz.
- It has an almost infinite resistance to d.c. and a very high impedance to a.c., so the circuit to which connection is made is little affected.
- It is not damaged by overloading.

(b) Displaying waveforms

The signal to be examined is connected to the Y-plates and the time base to the X-plates. As the spot is drawn horizontally across the screen by the time base, it is also deflected vertically by the alternating signal p.d. which is therefore 'spread out' on a time axis and its waveform displayed. The waveform will be a faithful representation, free from distortion, *if* the time base has a linear sweep. When the time base has the same frequency as the input, one complete wave is formed on the screen; if it is half that of the input, two waves are displayed. The CRO is used to display waveforms in the circuit of Fig. 17.32b (p. 282).

In an electrocardiograph (ECG), electrodes placed on the patient's body are connected to a CRO which is used to monitor small changes in the p.d. occurring in heart muscles during a heartbeat (Fig. 23.4).

Fig. 23.4 Electrocardiogram of a normal heartbeat.

(c) Measuring time intervals

This may be done if the CRO has a calibrated time base, as many CROs do. The time-base control is marked **time/div** and gives a range of pre-set sweep speeds.

If the time/div control is on 10 ms, the spot takes 10 ms to move 1 division and so travels 10 divisions of the screen graticule in 100 ms (0.1 s). How could the calibration of the time base be checked using a 50 Hz a.c. mains signal?

A CRO was used to estimate a very small time interval in the measurement of the speed of sound in air and in a metal rod (pp. 320 and 321). A CRO is also used to measure time delays in a number of pulse-reflection techniques such as radar, the ultrasonic testing of materials and medical imaging (pp. 306–8).

(d) Measurement of phase relationships

If two sinusoidal p.ds of the same frequency and amplitude are applied simultaneously to the X- and Y-plates (time base off), the electron beam is subjected to two mutually perpendicular simple harmonic motions and the trace produced on the screen by the spot depends on the phase difference between the p.ds. Figure 23.5 shows the traces obtained using 50 Hz supplies (of a few volts) from, say, a signal generator and a step-down transformer. In general, the resultant motion of the spot is an ellipse except when the phase difference is 0°, 90° or 180°.

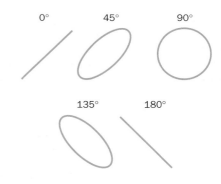

Fig. 23.5

The method can be used to find the phase relationships between p.ds and currents in a.c. circuits if a double-beam CRO is not available.

(e) Comparison of frequencies

When two p.ds of different frequencies f_x and f_y are applied to the X- and Y-plates (time base off), more complex figures than those in Fig. 23.5 are obtained, known as **Lissajous' figures**. Some are shown in Fig. 23.6.

In any particular case the frequency ratio can be found from inspection by imagining a horizontal and a vertical line to be drawn at the top and side of the trace. It is given by

$$\frac{f_y}{f_x} = \frac{\text{no. of loops touching horizontal line}}{\text{no. of loops touching vertical line}}$$

$f_y : f_x = 2:1 \qquad f_y : f_x = 3:1 \qquad f_y : f_x = 3:2$

Fig. 23.6 Lissajous' figures

The pattern is stationary only when f_y/f_x is a ratio of whole numbers and the phase difference is constant.

When f_y/f_x is large, comparison by Lissajous' figures is difficult and in such cases a 'circular time base' method can be used in which one alternating p.d. of known frequency (say 50 Hz) is applied to a 'phase-splitting' *RC* series circuit and produces a circular trace on the CRO. (For a 2 V, 50 Hz input and a CRO gain setting of 1 V, convenient values of *R* and *C* are 10 kΩ and 0.47 µF respectively.) Three connections from the *RC* circuit to the Y-, E- and X-terminals are necessary, Fig. 23.7a. If an unknown, much higher frequency (from a signal generator) is applied to the Z- and E-terminals on the CRO the brightness of the trace varies and a broken circle is obtained, Fig. 23.7b. Here the frequency ratio is 8:1. (The Z-input on the CRO is joined to the grid of the tube via an isolating capacitor, and an external p.d. applied to it brightens or blacks-out the trace, depending on whether it is positive or negative. The beam is thus 'intensity modulated'.)

The method can be used to check the calibration of a signal generator at frequencies up to 2 kHz against the 50 Hz mains supply. If the generator has a square-wave output it should be used in preference to the sine-wave output. Why?

BAND THEORY OF SOLIDS

An electric current arises from a movement of charge. In an insulator electrons are firmly bound to their atoms and no current flows. In a conductor some electrons are loosely bound and can move from atom to atom so that a current will flow if a p.d. is applied to a circuit. In the 'free electron' model of conduction these electrons could take any energy; applying quantum mechanics we expect that the electrons can only have certain allowed energies.

For an isolated atom in a gas we saw that the electrons occupy well-defined energy levels (p. 402). As atoms are brought closer together, as in a solid, electron waves from different atoms overlap and each single energy level broadens into a band of allowed energy levels which is shared by all the atoms; the bands are separated by gaps of forbidden energies. Electrical conductivity of the solid depends on the spacing and occupancy of the energy bands. In a conductor, such as metal, Fig. 23.8a, the lower energy bands are full, but the topmost band, the **conduction–valence band**, is only half-full of electrons. Electrons in this band can easily accept energy from an electric field and move to an unoccupied energy level in the conduction–valence band, i.e. current flows.

A semiconductor is a material whose conductivity lies between that of a good conductor and an insulator. For a semiconductor the topmost band, the **conduction band**, is empty; the **valence band** below it is full at absolute zero and no conduction occurs. As the temperature rises, electrons can gain enough energy to jump the small forbidden gap from the valence band into the conduction band; 'holes' are left in the valence band (Fig. 23.8b). Both the electrons in the conduction band and the holes in the valence band contribute to current flow when a p.d. is applied; the holes behave like positive charge carriers. In contrast to a metal, the conductivity of a semiconductor increases (and its resistance decrease) as the temperature rises, since more electrons are able to jump from the valence band into the conduction band.

An electron can also be promoted to the conduction band by the absorption of a photon of energy greater than the size of the energy gap E_g. In a semiconductor, infrared photons of frequency f will have sufficient energy to lift electrons to the conduction band if $hf > E_g$; absorption of a photon leads to the creation of an electron–hole pair and an increase in conductivity of the semiconductor. The semiconductor appears to be opaque at frequencies at which photons are absorbed.

In an insulator, Fig. 23.8c, the valence band is full and the conduction band is empty so that there are no

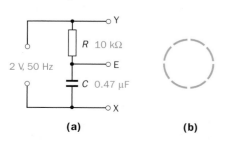

2 V, 50 Hz — Y — *R* 10 kΩ — E — *C* 0.47 µF — X

(a)   **(b)**

Fig. 23.7

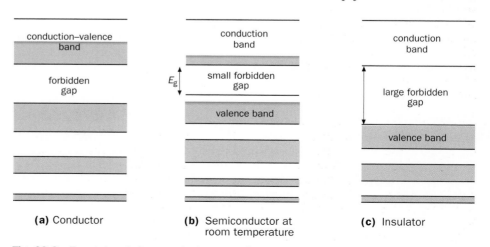

(a) Conductor   **(b)** Semiconductor at room temperature   **(c)** Insulator

Fig. 23.8 Energy bands in a conductor, a semiconductor and an insulator

electrons or holes available for current flow; the forbidden gap is generally too big for electrons to gain sufficient energy through thermal or photon absorption to jump from the valence band to the conduction band and so conduction does not occur; the resistance of insulators is very high.

SEMICONDUCTORS

Silicon and germanium are the best-known semiconductors because they are used to make the most common devices, i.e. transistors and diodes, the present trend being towards silicon devices. Other semiconductor materials are also in common use, such as cadmium sulphide, lead sulphide, gallium arsenide. These tend to be used in specialist devices such as photoelectric cells (p. 393). A selection of semiconductor devices is shown in Fig. 23.9.

Back row: Fast-recovery rectifier diode, high current silicon diode, 3 A bridge rectifier, 5 V voltage regulator
Middle row: General purpose silicon diodes
Front row: EPROM integrated circuit (type 2732), low power p-n-p transistor, temperature sensor, low power n-p-n transistor, red LED, green LED, lower voltage power transistor, photodiode with integral amplifier, dual-timer integrated circuit, precision instrumentation amplifier integrated circuit

The use of semiconducting devices depends on increasing the conductivity of the extremely pure material, which is very small at room temperatures, by introducing tiny but controlled amounts of certain 'impurities'. The process is known as **doping** and the material obtained is called an **extrinsic** semiconductor because the impurity introduces charge carriers additional to the **intrinsic** ones, i.e. to those already present.

Two types of semiconductor are obtained in this way. In **n-type**, the main charge carriers are **n**egative electrons. In **p-type**, conduction occurs chiefly due to **p**ositive charge carriers called holes. In n-type semiconductors the impurity atoms have more valence electrons than are needed for bonding to adjacent semiconductor atoms; they introduce 'extra' electrons in the crystal lattice which are only loosely bound to the impurity atoms and so can easily be thermally excited to the conduction band where they will increase the conductivity of the semiconductor. In p-type semiconductors the impurity atoms are deficient in valence electrons compared with the pure semiconductor and they introduce extra holes into the valence band.

In the manufacture of n- and p-type semiconductors, the semiconducting material is first purified to 1 part in 10^{10} and then impurity atoms added to produce the required conductivity.

JUNCTION DIODE

A junction diode consists of a junction of p- and n-type silicon or germanium formed in the same continuous crystal lattice. One connection, the **anode** A, is taken from the p-side and another, the **cathode** K, from the n-side, Fig. 23.10.

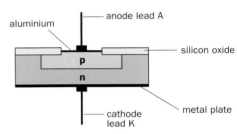

Fig. 23.10 Construction of a junction diode (not to scale)

(a) Action

A diode allows current to pass through it in one direction only. One is shown in Fig. 23.11 with its symbol; the end nearest the band is connected to the cathode.

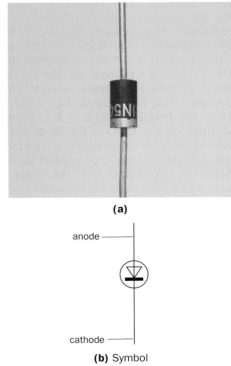

(a)

anode ———

cathode ———

(b) Symbol

Fig. 23.11 A junction diode

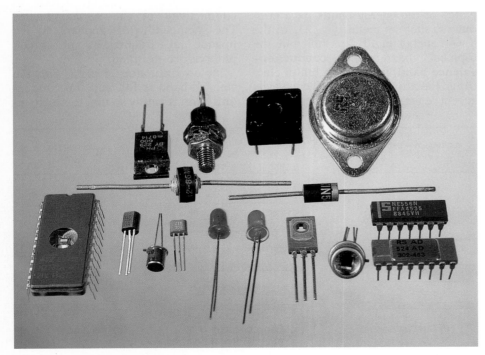

Fig. 23.9 Semiconductor devices

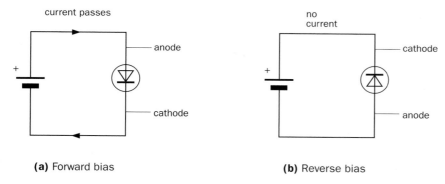

(a) Forward bias   **(b)** Reverse bias

Fig. 23.12

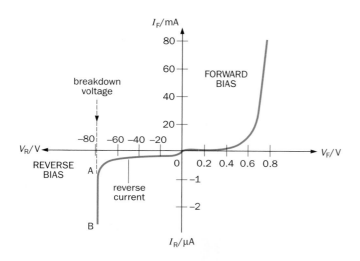

Fig. 23.13   Characteristic of a typical silicon junction diode

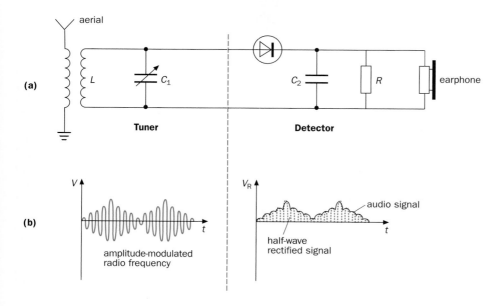

Fig. 23.14   (*a*) AM radio receiver and (*b*) action of the detector circuit

Conduction occurs when the anode goes to the positive terminal of the voltage supply and the cathode to the negative terminal, Fig. 23.12*a*. The diode is then **forward biased** and has a very low resistance. Conventional current passes in the direction of the arrow on its symbol, i.e. from anode to cathode. If the connections are the other way round, it has a very high resistance and does not conduct. It is then **reverse biased**, Fig. 23.12*b*.

(b) Characteristic

A typical characteristic for a silicon diode is shown in Fig. 23.13. You can see from it that the forward current I_F is small until the forward voltage V_F is about 0.6 V, then a very small change in V_F causes a large increase in I_F. The reverse current I_R is negligible (note that the scales on the graph have been changed to show it) and remains so as the reverse voltage V_R is increased. At a certain reverse voltage called the **breakdown voltage**, the operation of the diode breaks down — I_R increases suddenly and rapidly and damage may occur by overheating. Breakdown can occur at any value of the reverse voltage, from a few volts up to 1000 V, depending on the construction of the diode and the level of doping.

For every diode there is also a maximum forward current, called the **forward current rating**, which should not be exceeded or overheating and destruction occurs. For the 1N4001 silicon diode it is 1 A.

The characteristic of a germanium junction diode is similar to the one in Fig. 23.11 but I_F 'turns on' when V_F is about 0.2 V, I_R is greater and the breakdown voltage has a maximum value of about 100 V.

As rectifiers, silicon junction diodes are preferred to germanium types because their much lower reverse current makes them more efficient (i.e. more complete conversion of a.c. to d.c. occurs). Silicon can also have a higher breakdown voltage and can work at higher temperatures (150 °C as compared with 100 °C).

(c) Use in radio receiver

A diode is used in the detector circuit of a simple AM radio receiver (chapter 24), shown in Fig. 23.14*a*. An incoming radio frequency is selected in the *LC* tuner circuit (p. 279) by changing the variable capacitor C_1. In the detector circuit a diode half-wave rectifies (p. 283) the signal voltage, V, as shown in Fig. 23.14*b*. During each positive half-cycle the capacitor C_2 becomes charged, but in the rest of the cycle, when the diode is non-conducting, C_2 partially discharges through R as in a smoothing circuit (p. 284). The resulting voltage developed across R, V_R, follows the amplitude of the radio frequency signal which is the audio frequency heard in the earphone; the detector circuit has acted to separate the audio signal from the radio frequency carrier wave (p. 452).

OTHER SEMICONDUCTOR DIODES

(a) Light-emitting diode (LED)

An LED, shown in Fig. 23.15 with its symbol, is a junction diode made from the semiconductor gallium arsenide phosphide. When forward biased it emits red, yellow or green light depending on its exact composition. The forward bias injects electrons into the n-type material and holes into the p-type material when current flows; the increased density of electrons and holes leads to the combination of electron and holes at the junction and the creation of photons. The energy of a photon emitted from the junction will equal the energy lost by an electron in falling from the conduction to the valence band.

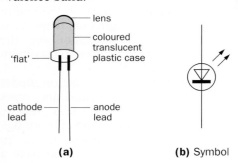

(a)   **(b)** Symbol

Fig. 23.15 An LED

No conduction or light emission occurs for reverse bias which, if it exceeds 5 V, may damage the LED.

To limit the current through an LED, a resistor R must be connected in series with it, Fig. 23.16, otherwise it could be destroyed. The forward voltage drop V_{LED} across an LED is

about 2 V (compared with 1 V for a silicon diode) and a typical forward current $I = 10$ mA $= 0.01$ A. The value of R depends on the supply voltage V_s and is given by

$$R = \frac{V_R}{I} = \frac{V_s - V_{LED}}{I}$$

since $V_s = V_R + V_{LED}$. For example, if $V_s = 5$ V then $R = (5$ V $- 2$ V$)/0.01$ A $= 300\ \Omega$.

LEDs are used as indicator (or signal) lamps in radios and other electronic systems. Their advantages over filament lamps are their small size, long life, reliability, fast response and small current needs. They are also used as optical fibre transmitters (p. 460) and in digital electronic circuits to show whether outputs are 'high' or 'low'.

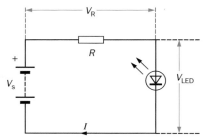

Fig. 23.16

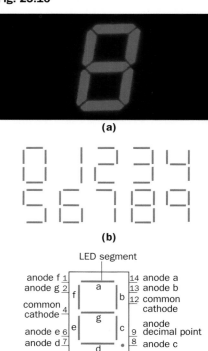

(a)

(b)

LED segment

anode f 1   14 anode a
anode g 2   13 anode b
common cathode 4   12 common cathode
anode e 6   9 anode decimal point
anode d 7   8 anode c

(c) Common cathode 7-segment LED display (typical pin connections)

Fig. 23.17 An LED numerical display

Many electronic calculators, clocks, cash registers and measuring instruments show their 'output' on several seven-segment red or green numerical displays, like the one in Fig. 23.17*a*. Each segment is an LED and depending on which have a p.d. across them, the display lights up the numbers 0 to 9, as in Fig. 23.17*b*. Every segment needs a separate current-limiting resistor and all the cathodes (or anodes) are joined together to form a common connection, Fig. 23.17*c*.

(b) Zener diode

This is used to regulate or stabilize (i.e. keep steady) the voltage output of a power supply (since this falls if the output current rises, due to the increase in 'lost' volts caused by the internal resistance of the supply, p. 51). It looks like an ordinary junction diode but its symbol is slightly different, Fig. 23.18.

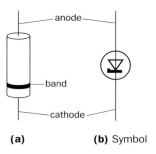

(a)   **(b)** Symbol

Fig. 23.18 A Zener diode

A Zener diode is a type of silicon junction diode, used in *reverse bias*. When the reverse p.d. reaches a certain value V_Z, called the **Zener**, **reference** or **breakdown voltage**, the reverse current increases suddenly and rapidly, as it does with an ordinary diode (part AB in Fig. 23.13). No breakdown occurs however if a resistor is connected in series with the diode to limit the reverse current. The p.d. across the diode then remains constant at V_Z over a wide range of reverse current values. It is this property of a Zener diode which makes it useful in stabilizing power supplies.

Zener diodes with specified values of V_Z between 2.0 and 200 V are made by controlling the doping.

In the stabilizing circuit of Fig. 23.19, the Zener diode is reverse biased and is in parallel with the load. The input voltage V_{in} is taken from an unstabilized supply, e.g. a battery, and is greater than the output voltage V_{out} developed across the load. If I_L and I_Z are the currents in the load and diode respectively, the total current I supplied is given by

$$I = I_L + I_Z$$

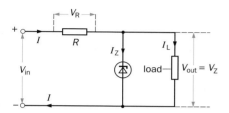

Fig. 23.19

If I_L rises (or falls), I_Z is found to fall (or rise) by the same amount, so keeping I and therefore V_{out} ($= V_Z$) constant.

The resistance R of the current-limiting resistor is given by

$$R = \frac{V_R}{I} = \frac{V_{in} - V_Z}{I}$$

For example, if $V_{in} = 9$ V, $V_{out} = 5$ V, $I_L = 40$ mA and $I_Z = 10$ mA, then $I = 50$ mA $= 0.05$ A. Therefore

$$R = \frac{9\text{ V} - 5\text{ V}}{0.05\text{ A}} = \frac{4\text{ V}}{0.05\text{ V}} = 80\ \Omega$$

The power P_R dissipated in R is

$$P_R = V_R \times I = 4\text{ V} \times 50\text{ mA}$$
$$= 200\text{ mW}$$

The power P_Z dissipated in the diode is

$$P_Z = V_Z \times I_Z = 5\text{ V} \times 10\text{ mA}$$
$$= 50\text{ mW}$$

If the load is disconnected, $I_L = 0$ and $I_Z = 50$ mA, making $P_Z = 250$ mW which the diode must be able to dissipate.

A circuit for demonstrating the action of a Zener diode is given in Fig. 23.20. When the input p.d. V_{in} is gradually increased from zero we see that:

(*i*) if $V_{in} < 6.2$ V (the Zener voltage), $I = 0$ and the output p.d. $V_{out} = V_{in}$ since there is no current through R and so no p.d. across it;

(*ii*) if $V_{in} > 6.2$ V, the diode conducts, even though it is reverse biased, and $V_{out} = 6.2$ V, with the excess p.d. appearing across R.

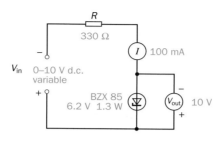

Fig. 23.20

(c) Photodiode

This consists of a normal p-n junction in a case with a transparent 'window' through which light can enter. Reverse bias is applied and the reverse current increases in proportion to the amount of light falling on the junction. The effect is due to the absorption of photons allowing electrons to gain sufficient energy to jump from the valence to the conduction band of the semiconductor, so increasing the number of electrons and holes available for conduction.

Photodiodes are used as optical fibre receivers (p. 460) and as fast 'counters' which generate a pulse of current every time a beam of light is interrupted. One is shown with its symbol in Fig. 23.21.

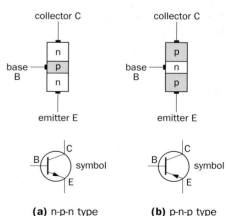
(a)

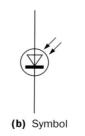
(b) Symbol

Fig. 23.21   A photodiode

TRANSISTORS

Transistors are the most important devices in electronics today. As well as being made as discrete (separate) components, integrated circuits (ICs) may contain several thousand transistors on a tiny slice of silicon.

Transistors are three-terminal devices used as **amplifiers** and as **switches**. There are two basic types:

- the **bipolar** or **junction transistor** (usually called *the* transistor) which consists of two p-n junctions; and
- the **unipolar** or **field effect transistor** (called the FET) in which there is just one p-n junction.

We will consider only the first, the junction transistor.

(a) Construction

A junction transistor consists of two p-n junctions in the same crystal. A very thin wafer of lightly doped p- or n-type semiconductor (the **base** B) is sandwiched between two thicker, heavily doped materials of the opposite type (the **collector** C and **emitter** E). The two possible arrangements are shown diagrammatically in Fig. 23.22. The arrows on the symbols give the direction in which conventional (positive) current flows; in the n-p-n type it points from B to E and in the p-n-p type from E to B.

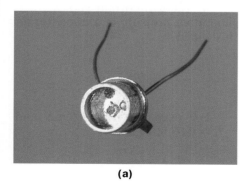

(a) n-p-n type   **(b)** p-n-p type

Fig. 23.22   The junction transistor

As with diodes, silicon transistors are in general preferred to germanium ones because they withstand higher temperatures and voltages. Silicon n-p-n types are more easily mass-produced than p-n-ps; the opposite is true of germanium. A simplified cross-section of an n-p-n silicon transistor is shown in Fig. 23.23a; Fig. 23.23b shows a transistor complete with case (called the encapsulation) and three wire leads.

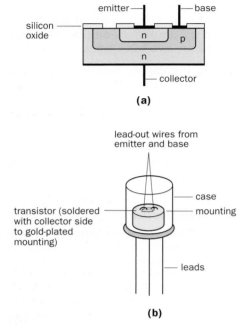

(a)

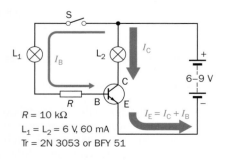

(b)

Fig. 23.23 Construction of a transistor

(b) Action

The switching and amplifying action can be shown using the circuit of Fig. 23.24. In an n-p-n transistor the collector C and the base B must both have *positive voltages* with respect to the emitter E, here obtained from the same battery.

There are two current paths through the transistor. One is taken by the **base current** I_B which enters by B and the other is followed by the **collector current** I_C with entry at C. Both leave by E to form the **emitter current** I_E.

$R = 10\ k\Omega$
$L_1 = L_2 = 6\ V,\ 60\ mA$
Tr = 2N 3053 or BFY 51

Fig. 23.24

When S is open, I_B is zero and neither of the lamps L_1 or L_2 lights up. When S is closed, B is connected through R to the battery + and L_2 lights up, but not L_1. This shows that I_C is now passing through L_2 and that it is much greater than I_B, which passes through L_1 but is too small to light it. With S open again, I_B becomes zero as does I_C and L_2 goes out.

To sum up: I_B **switches on and controls the much greater** I_C. Measurements using meters show that the switch-on does not occur until B is at least 0.6 V positive with respect to E. That is, the **turn-on voltage** for a transistor is about 0.6 V or so. The base–emitter behaves like a diode and must have a forward bias of about 0.6 V before I_B flows and turns on I_C.

Two further points to note are:

(i) R must be in the circuit to limit I_B, otherwise it causes an excessive I_C which destroys the transistor by overheating; and

(ii) the current paths for I_C and I_B have a common connection at E and the transistor is said to be in **common-emitter connection**.

(c) d.c. current gain

If we consider I_B as the input to a transistor and I_C as the output from it, then the transistor acts as a **current amplifier** since I_C is greater than I_B. Typically I_C is 10 to 1000 times greater than I_B depending on the type of transistor. The **d.c. current gain** h_{FE} is an important property of a transistor and is defined by

$$h_{FE} = \frac{I_C}{I_B}$$

For example, if $I_C = 5\ mA$ and $I_B = 0.05\ mA\ (50\ \mu A)$

$$h_{FE} = \frac{5\ mA}{0.05\ mA} = 100$$

Although h_{FE} is approximately constant for one transistor over a limited range of I_C values, it varies between transistors of the same type due to manufacturing tolerances.

Also note that, since the current leaving a transistor equals that entering,

$$I_E = I_B + I_C$$

In the above example, $I_E = 5.05\ mA$.

In the symbol h_{FE}, F indicates that *forward* currents are being considered and E that the transistor is connected in the *common-emitter* mode.

TRANSISTOR CHARACTERISTICS

These are graphs, found by experiment, which show the relationships between various currents and voltages and enable us to see how best to use a transistor. A module[1] and circuit for investigating an n-p-n transistor (e.g. BFY 51 or 2N3053) in common-emitter connection are shown in Fig. 23.25a and b opposite.

The voltmeter for measuring V_{BE} must have a very high resistance (e.g. an electronic type with a resistance of the order of 1 MΩ or more); if it is a moving-coil type, allowance has to be made in **(a)** below for the current it takes. The resistor R_2 protects the transistor from damage by excessive base currents. Three characteristics are important.

(a) Input (base) characteristic (I_B–V_{BE})

The collector–emitter voltage V_{CE} is kept constant (e.g. at the battery voltage of 6 V) and the base–emitter voltage V_{BE} measured for different values of the base current I_B, obtained by varying R_1. (The procedure when using a 'low-resistance' voltmeter for V_{BE} is to open switch S, adjust R_1 to make $I_B = 5\ \mu A$ say, record I_C, close S, readjust R_1, until I_C is the same as before, record V_{BE}.)

A typical graph for a silicon transistor is given in Fig. 23.26a. Note that I_B is negligibly small until V_{BE} exceeds about 0.6 V and thereafter small changes in V_{BE} cause large changes in I_B. The **input resistance** r_i is defined as the ratio $\Delta V_{BE}/\Delta I_B$ where ΔI_B is the change in I_B due to a change of ΔV_{BE} in V_{BE}. Since the input characteristic is non-linear, r_i varies but is of the order of 1 to 5 kΩ.

(b) Output (collector) characteristic (I_C–V_{CE})

I_B is fixed at a low value, e.g. 10 μA, and I_C measured as V_{CE} is increased in stages by varying R_3. This is repeated

[1] Unilab 'Blue Chip': 511.019

for different values of I_B to give a family of curves as in Fig. 23.26*b*.

You can see that I_C depends almost entirely on I_B and hardly at all on V_{CE} (except when V_{CE} is less than about 0.5 V). As an **amplifier**, a transistor operates well to the right of the sharp bend or 'knee' of the characteristic, i.e. where I_C varies linearly with V_{CE} for a given I_B. The small slope of this part of the characteristic shows that the **output resistance** r_o of the transistor is fairly high, of the order of 10 to 50 kΩ; it is given by $r_o = \Delta V_{CE}/\Delta I_C$ where ΔI_C is the change in I_C caused by a change of ΔV_{CE} in V_{CE}. As a **switch**, a transistor operates in the shaded parts of Fig. 23.26*b* and changes over rapidly from the 'off' state in which $I_C = 0$ (cut-off) to the 'on' state in which I_C is a maximum (saturation).

(c) Transfer characteristic (I_C–I_B)

V_{CE} is kept fixed and I_C measured for different values of I_B by varying R_1. The graph, Fig. 23.26*c*, is almost a straight line showing that I_C is directly proportional to I_B, i.e. the relation between I_C and I_B is linear.

The **a.c. current gain** h_{fe} is defined by

$$h_{fe} = \frac{\Delta I_C}{\Delta I_B}$$

where ΔI_C is the change in I_C produced by a change of ΔI_B in I_B. For most purposes h_{fe} and h_{FE} (the d.c. current gain = I_C/I_B) can be considered equal.

The characteristic also shows that when I_B is zero, I_C has a small value (about 0.01 μA for silicon and 2 μA for germanium at 15 °C). This is called the **leakage current** I_{CEO}; it increases with temperature rise (much more so for germanium than silicon) and upsets the working of the transistor.

Summing up, in a silicon junction transistor

- I_C is zero until I_B flows;
- I_B is zero until V_{BE} is about 0.5 V; and
- V_{BE} remains close to 0.6 V for a wide range of values of I_B.

(a)

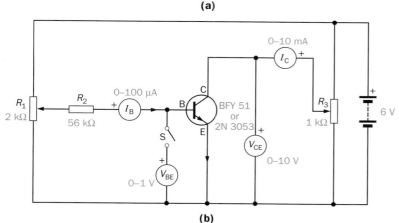

(b)

Fig. 23.25   Module and circuit for investigating an n-p-n transistor

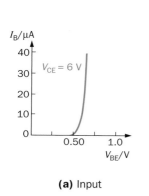

(a) Input

(b) Output

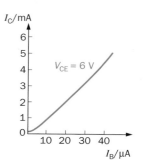

(c) Transfer

Fig. 23.26   Transistor characteristics

TRANSISTOR AS A VOLTAGE AMPLIFIER

Amplifiers are necessary in many types of electronic equipment such as oscilloscopes, radios and other audio systems. Often it is a small alternating voltage that has to be amplified. A junction transistor in the common-emitter mode can act as a voltage amplifier if a suitable resistor, called the **load**, is connected in the collector circuit.

The small alternating voltage, the **input** v_i, is applied to the base–emitter circuit and causes small changes of base current which produce large changes in the collector current flowing through the load. The load converts these current changes into voltage changes which form the alternating **output** voltage v_o, v_o being much greater than v_i. (Note the use of small italic letters to represent instantaneous values of alternating quantities.)

A simple circuit is shown in Fig. 23.27. To see why voltage amplification occurs, consider first the situation when there is no input, i.e. $v_i = 0$, called the 'quiescent' (quiet) state.

(a) Quiescent state

For transistor action to take place the base–emitter junction must always be forward biased (even when v_i is applied and goes negative). A simple way of ensuring this is to connect a resistor R_B, called the **base-bias resistor**, as shown. A steady (d.c.) base current I_B flows from battery +,

through R_B into the base and back to battery −, via the emitter. The value of R_B can be calculated (see the *Worked example* below) once the value of I_B for the best amplifier performance has been decided.

If V_{CC} is the battery voltage and V_{BE} is the base–emitter junction voltage (always about +0.6 V for an n-p-n silicon transistor), then for the *base–emitter* circuit, since d.c. voltages add up, we can write

$$V_{CC} = (I_B \times R_B) + V_{BE} \qquad (1)$$

I_B causes a much larger collector current I_C which produces a voltage drop $I_C \times R_L$ across the load R_L. The voltage at the end of R_L joined to battery + is fixed and so the voltage drop must be at the end connected to the collector. For the *collector–emitter* circuit, if V_{CE} is the collector–emitter voltage, then

$$V_{CC} = (I_C \times R_L) + V_{CE} \qquad (2)$$

Component values are chosen so that the quiescent collector–emitter voltage V_{CE} is about *half the battery voltage* V_{CC}. This ensures the best working conditions for the amplifier (p. 424).

(b) Input applied

When v_i is applied and goes positive it increases the base–emitter voltage slightly (e.g. from +0.60 V to +0.61 V). When v_i swings negative the base–emitter voltage decreases slightly (e.g. from +0.60 V to +0.59 V). As a result a small alternating current is superimposed on the quiescent base current I_B, which in effect becomes a varying d.c.

When the base current *increases*, large proportionate increases occur in the collector current. From equation (2) it follows that there is a corresponding large *decrease* in the collector–emitter voltage (since V_{CC} is fixed). A decrease of base current causes a large increase of collector–emitter voltage. In practice positive and negative swings of a few millivolts in v_i can result in a fall or rise of several volts in the voltage across R_L and so also in the collector–emitter voltage.

The collector–emitter voltage is a varying direct voltage and may be regarded as an alternating voltage superimposed on a steady direct voltage, i.e. on the quiescent value of V_{CE}. Only the alternating part is wanted and capacitor C blocks the direct part but allows the alternating part, i.e. the output v_o, to pass.

(c) Further points

The transistor *and* load together bring about voltage amplification.

The output v_o is 180° out of phase with the input v_i, i.e. when v_i has its maximum positive value, v_o has its maximum negative value (see the graphs in Fig. 23.27).

The emitter is common to the input, output and battery circuits and is usually taken as the reference point for all voltages, i.e. 0 V. It is called 'common' or 'ground' (symbol ⏚) or 'earth' (symbol ⏚) if connected to earth.

WORKED EXAMPLE

A silicon transistor in the simple voltage-amplifier circuit of Fig. 23.27 operates satisfactorily on a quiescent (no a.c. input) collector current (I_C) of 3 mA. If the battery supply (V_{CC}) is 6 V, what must be the value of (a) the load resistor (R_L) and (b) the base-bias resistor (R_B), for the quiescent collector–emitter voltage (V_{CE}) to be half the battery voltage? The transistor d.c. current gain (h_{FE}) is 100.

(a) The collector–emitter circuit equation is

$$V_{CC} = (I_C \times R_L) + V_{CE}$$

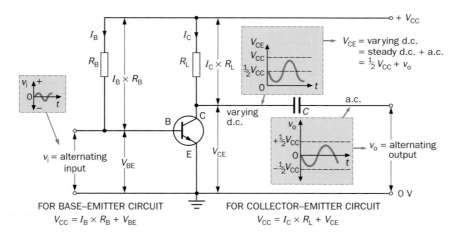

FOR BASE–EMITTER CIRCUIT
$V_{CC} = I_B \times R_B + V_{BE}$

FOR COLLECTOR–EMITTER CIRCUIT
$V_{CC} = I_C \times R_L + V_{CE}$

Fig. 23.27 Voltage amplification

Rearranging, we get

$$I_C \times R_L = V_{CC} - V_{CE}$$

That is

$$R_L = (V_{CC} - V_{CE})/I_C$$

Substituting
$V_{CC} = 6\,V$, $V_{CE} = \frac{1}{2}V_{CC} = 3\,V$ and
$I_C = 3\,mA$ gives

$$R_L = (6-3)V/3\,mA = 1\,k\Omega$$

(b) The d.c. current gain is given by

$$h_{FE} = \frac{I_C}{I_B}$$

where I_B is the quiescent base current to produce the quiescent collector current I_C.

Rearranging,

$$I_B = \frac{I_C}{h_{FE}}$$

Substituting
$I_C = 3\,mA$ and $h_{FE} = 100$, we get

$$I_B = 3\,mA/100 = 0.03\,mA\ (30\,\mu A)$$

The base–emitter circuit equation is

$$V_{CC} = (I_B \times R_B) + V_{BE}$$

where V_{BE} is the base–emitter voltage.

Rearranging gives

$$R_B = (V_{CC} - V_{BE})/I_B$$

Substituting
$V_{CC} = 6\,V$, $V_{BE} = 0.6\,V$ (for a silicon transistor) and $I_B = 0.03\,mA$ gives

$$\begin{aligned}R_B &= (6-0.6)V/0.03\,mA \\ &= 5.4\,V/0.03\,mA \\ &= 540/3\,k\Omega = 180\,k\Omega\end{aligned}$$

VOLTAGE GAIN OF AN AMPLIFIER

The **voltage gain** A of an amplifier is the ratio of the output voltage v_o to the input voltage v_i:

$$A = \frac{v_o}{v_i}$$

(a) Measurement

A circuit for finding A experimentally using the module of Fig. 23.25a is shown in Fig. 23.28. Variable resistor

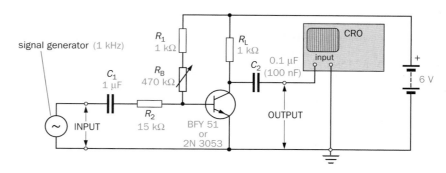

Fig. 23.28

R_B controls the base current and initially is turned fully anticlockwise to its maximum resistance. This sets the quiescent collector voltage at 2.5 to 3.0 V, i.e. about half the d.c. supply voltage. R_1 prevents damage to the transistor by limiting the base current should R_B be made zero.

A sine-wave input of several millivolts at 1 kHz is applied via C_1 and R_2 from a signal generator. C_1 stops the resistance of the generator upsetting the base–emitter bias. R_2 protects the transistor in the input circuit.

The peak-to-peak value of v_o is noted from the CRO (in volts if it is calibrated, otherwise in scale divisions). To measure v_i the CRO input connection is transferred to the input terminals on the amplifier module.

The gain (v_o/v_i) may be only in the range 5 to 10. This is due to most of the input being 'lost' across R_2 and just a fraction of it applied to the base–emitter. If R_2 is short-circuited (but the input still applied to C_1) and v_o and v_i are remeasured, the value obtained for A is much greater, e.g. 70 to 100.

(b) Theoretical value

Suppose that the input voltage v_i causes a change ΔI_B in I_B, then $\Delta I_B = v_i/(\text{resistance of input circuit})$, that is

$$\Delta I_B = \frac{v_i}{R_2 + r_i}$$

where r_i is the input resistance of the transistor, the resistance of the signal generator being neglected. If the resulting change of I_C is ΔI_C, the voltage change across R_L, i.e. v_o, is given by

$$v_o = \Delta I_C \times R_L$$

But

$$\Delta I_C = h_{fe} \times \Delta I_B = h_{fe} \times \frac{v_i}{(R_2 + r_i)}$$

So

$$v_o = h_{fe} \times \frac{R_L}{(R_2 + r_i)} \times v_i$$

Therefore

$$A = \frac{v_o}{v_i} = h_{fe} \times \frac{R_L}{(R_2 + r_i)}$$

In the circuit of Fig. 23.28, $R_L = 1\,k\Omega$ and $R_2 = 15\,k\Omega$. Transistor characteristics like those in Fig. 23.26 give typical values for r_i and h_{fe} of 1 kΩ and 100 respectively. Substituting in the equation for A we get

$$A = \frac{100 \times 1\,k\Omega}{(15+1)\,k\Omega} = \frac{100}{16} = 6\ (\text{approx.})$$

If $R_2 = 0$ then $A = 100$. These values for A are approximate since h_{fe} has been taken from characteristics obtained at a constant collector–emitter voltage. In practice, in a voltage amplifier this voltage changes when the input is applied. A better way of calculating A is given in the next section.

LOAD LINES: OPERATING POINT

When designing a voltage amplifier the aim is to obtain

- the desired voltage gain;
- minimum distortion of the output so that it is a good copy of the input; and
- operation within the current, voltage and power limits for the transistor.

The choice of the quiescent (d.c.) operating point, i.e. the values of I_C and V_{CE}, determines whether these requirements will be met. This is made by constructing a **load line**.

The output characteristics of a transistor (p. 421, Fig. 23.26b) show the relation between V_{CE} and I_C with *no load* in the collector circuit. With a load R_L, the equation connecting them is (p. 422)

$$V_{CC} = (I_C \times R_L) + V_{CE}$$

where V_{CC} is the battery voltage. Rearranging, we get

$$V_{CE} = V_{CC} - (I_C \times R_L) \qquad (3)$$

Knowing V_{CC} and R_L this equation enables us to calculate V_{CE} for different values of I_C. If a graph of I_C (as the y-axis) is plotted against V_{CE} (as the x-axis) we get a straight line, called a load line. If we accept that (3) is the equation of a straight line — you can see this by rewriting (3) as

$$I_C = -\left(\frac{1}{R_L} \times V_{CE}\right) + \frac{V_{CC}}{R_L}$$

which is of the form $y = mx + c$, the line can be drawn if we know just two points. The easiest to find are the end points A and B where the line cuts the V_{CE} and I_C axes respectively.

For A we put $I_C = 0$ in (3) and get $V_{CE} = V_{CC} = 6$ V (say).

For B we put $V_{CE} = 0$ in (3) and get $I_C = V_{CC}/R_L$. If $R_L = 1$ kΩ, then $I_C = 6$ V/1 k$\Omega = 6$ mA.

In Fig. 23.29, AB is the load line for $V_{CC} = 6$ V and $R_L = 1$ kΩ. It is shown superimposed on the output characteristics of the transistor used in the circuit of Fig. 23.27. We can regard a load line as the output characteristic

of the *transistor and load* for particular values of V_{CC} and R_L. Different values of either give a different load line; for example, a smaller value of R_L gives a steeper line.

The choice of load line and d.c. operating point affects the shape and size of the output waveform. Choosing a line that cuts the characteristics where they are not linear (straight) or where they are not equally spaced can cause distortion. Selecting an operating point too near either the I_C or the V_{CE} axis can have the same effect. The best position for the d.c. operating point is *near the middle of the chosen load line*, e.g. at Q on Fig. 23.29. The 'swing' capability of the output in both directions is then a maximum (from near V_{CC} to near 0 V) and distortion a minimum.

Having chosen Q, the quiescent (d.c.) values of V_{CE} and I_C can be read off. In Fig. 23.29, they are $V_{CE} = 3$ V (i.e. half V_{CC}) and $I_C = 3$ mA. The value of I_B that gives these values is obtained from the transistor output characteristics passing through Q (since I_C and V_{CE} have to satisfy *at the same time* both the characteristic and the load line). Here it is the 30 μA characteristic. R_B can then be calculated as on p. 425.

The voltage gain can be obtained from the load line by noting that when the input causes I_B to vary from 10 to 50 μA (from R to P), V_{CE} varies from 4.5 to 1.5 V (from Y to X). From the input characteristic of the transistor (Fig. 23.26a, p. 421) we can find the change in V_{BE} to cause this change of 40 μA in I_B. If it is, say, 40 mV (0.04 V), the voltage gain $A = (4.5 - 1.5)$ V/0.04 V = 3.0/0.04 = 75.

STABILITY AND BIAS

(a) Thermal runaway

If the temperature of a transistor rises, there is greater vibration of the semiconductor atoms resulting in the production of more free electrons and holes. The collector current increases, causing further heating of the transistor and so on until it is damaged or destroyed. The initial temperature rise may be due to slight overloading of the transistor, to an increase in the surrounding temperature or to the replacement of one transistor by another of greater h_{FE}.

To stop this 'thermal runaway' effect and to stabilize the d.c. operating point, special **bias circuits** have been designed which automatically compensate for variations of collector current. The simplest of these will now be considered.

(b) Collector-to-base bias

The basic circuit of Fig. 23.27 (p. 422) for a voltage amplifier can be adequately stabilized for many applications by *halving* the value of R_B and connecting it between the collector and base as in Fig. 23.30, rather than between battery + and base.

For this circuit we can write

$$V_{CC} = (I_C \times R_L) + V_{CE} \qquad (4)$$

where

$$V_{CE} = (I_B \times R_B) + V_{BE} \qquad (5)$$

From (4) you can see that if I_C increases for any reason, V_{CE} decreases since V_{CC} is fixed. From (5) it therefore follows that since V_{BE} is

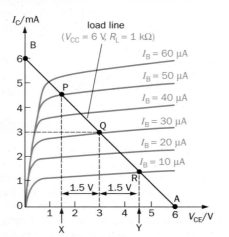

Fig. 23.29 Constructing a load line

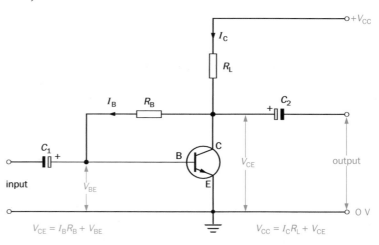

Fig. 23.30 Amplifying circuit with collector-to-base bias

constant (0.6 V or so), I_B must also decrease and in so doing tends to bring back I_C to its original value.

Taking the quiescent conditions to be (as in the *Worked example* on pp. 422–3) $V_{CE} = \frac{1}{2}V_{CC} = 3.0$ V, $I_C = 3.0$ mA and $I_B = 0.03$ mA, the value of R_B in Fig. 23.30 is found by rearranging equation (5) to be

$$R_B = \frac{V_{CE} - V_{BE}}{I_B} = \frac{(3.0 - 0.6)\ \text{V}}{0.03\ \text{mA}}$$

$$= \frac{2.4\ \text{V}}{0.03\ \text{mA}} = 80\ \text{k}\Omega$$

This is about half the value of 180 kΩ for R_B in the unstabilized circuit of Fig. 23.27, which was calculated in the *Worked example*.

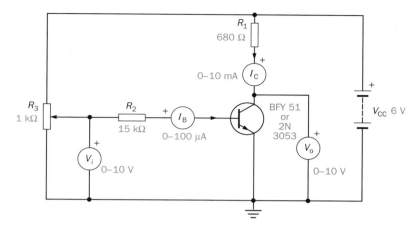

Fig. 23.31

TRANSISTOR AS A SWITCH

Transistors are used as switches in many important electronic circuits.

(a) Action

The switching action can be investigated with the circuit of Fig. 23.31 (using the module of Fig. 23.25a) in which an n-p-n silicon transistor (e.g. BFY 51 or 2N3053) is connected in common-emitter mode with a resistor R_1 in its collector (output) circuit. By adjusting R_3 the input voltage V_i can be varied and we find that

(*i*) when V_i increases from 0 to about 0.6 V, the base current I_B and the collector current I_C are both zero and the output voltage V_o remains almost equal to the battery voltage V_{CC} (6 V);

(*ii*) when V_i increases from about 0.6 V to 1.4 V, I_B and I_C increase rapidly from zero and V_o falls rapidly;

(*iii*) when V_i increases from 1.4 V to 6 V, I_B goes on increasing but soon I_C reaches a maximum and V_o falls to nearly zero.

The graph of V_o against V_i is shown in Fig. 23.32a; the dotted curve shows the effect of increasing the current-limiting resistor R_2. Table 23.1 summarizes the behaviour of V_i and V_o. Depending on the value of V_i, V_o has one of two 'levels' — either a 'high' positive voltage V_{CC} (here + 6 V) or a 'low' voltage (almost zero).

Figure 23.32b shows the graph of I_C against I_B.

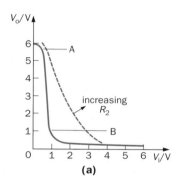

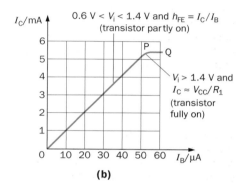

Fig. 23.32

Table 23.1

V_i	V_o
'low' (< 6.0 V)	'high' (6 V)
'high' (> 1.4 V)	'low' (0 V)

(b) Explanation

The basic transistor switching circuit is shown in Fig. 23.33, p. 426. R_2 is a current-limiting resistor to keep I_B below the maximum allowed value.

V_{CC} is applied across R_1 and the transistor in series, therefore

$$V_{CC} = (I_C \times R_1) + V_o$$

or

$$V_o = V_{CC} - (I_C \times R_1) \qquad (6)$$

When V_i is 'low' (e.g. by connecting the input terminal to 0 V), $I_C = 0$ and, from (6), $V_o = V_{CC}$, i.e. all the battery voltage is dropped across the transistor which behaves like a very large resistor and is said to be **cut off**.

When V_i is 'high' (e.g. by connecting the input terminal to $+V_{CC}$), it makes I_C and therefore $I_C \times R_1$ large. From (6), the greatest value $I_C \times R_1$ can have is V_{CC}. Then $V_o = 0$ and all the battery voltage is dropped across R_1 and none across the transistor, which behaves as if it had zero resistance. I_C has its maximum possible value, given by $I_C = V_{CC}/R_1$, and any further increase in I_B does not increase I_C. In this case the transistor is said to be **saturated** or **bottomed**.

Note that although V_i varies from 0 to 6 V, V_{BE} remains almost steady at 0.6 V. ($V_i - 0.6$ V) is dropped across the protective resistor R_2 which is essential to limit I_B when V_i exceeds 0.6 V.

As an amplifier a transistor acts *linearly* (i.e. the output is more or less directly proportional to the input) because it is biased to work on the straight part AB of Fig. 23.32a. As a switch it operates between cut-off and saturation; its behaviour is *non-linear*, the output having one of two values.

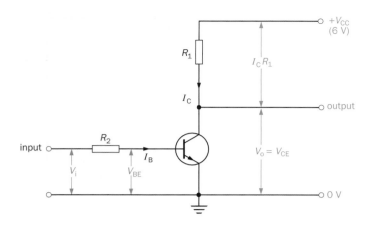

Fig. 23.33 The transistor as a switch

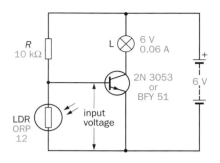

Fig. 23.34 Light-operated switch

(c) Power considerations

The power P used by a transistor as a switch should be as small as possible. It is given by $P = V_o \times I_C$. At cut-off $I_C = 0$ and at saturation $V_o \approx 0$ (typically 0.2 V; for the ideal switching transistor it should be zero), therefore $P \approx 0$ in both cases. To ensure saturation the transistor must be driven 'hard' by increasing I_B so that I_C/I_B is about *five* times less than h_{FE}.

For example, if $V_{CC} = 6$ V, $R_1 = 1$ kΩ and $h_{FE} = 100$, $I_C(\text{max}) = 6$ V/1 kΩ $= 6$ mA. Since 20 is one-fifth of 100, if $I_C/I_B = 20$, i.e. $I_B = I_C/20 = 6$ mA/20 $= 0.3$ mA, then 'hard' bottoming (satisfactory saturation) occurs.

Power is also used if the switch-over from cut-off to saturation takes place slowly. The larger R_2, the slower the switching rate and the greater the value of V_i at which it happens (as the dotted curve in Fig. 23.32a, p. 425, shows). Fast switching is desirable to avoid over-heating and damage to the transistor and for this reason n-p-n types are preferred because their charge carriers (electrons) travel faster than the charge carriers (holes) in p-n-p types.

(d) Advantages of transistor switching

Transistor switching occurs electrically and may be done by taking the input from another circuit. A further important advantage is that, it can occur millions of times a second.

(e) Comparison of transistor as an amplifier and as a switch

In both cases a 'load' (e.g. a resistor) is required in the collector circuit but the base bias is different. As an amplifier, the bias has to cause a collector current that makes the quiescent value of $V_{CE} \approx \frac{1}{2}V_{CC}$. As a switch, the bias has to make the collector current either zero or a maximum. In the first case the transistor is cut off and $V_{CE} = V_{CC}$; in the second case the transistor is saturated and $V_{CE} \approx 0$. Table 23.2 sums up these facts.

Table 23.2

Use	V_{CE}
Amplifier	$\frac{1}{2}V_{CC}$
Switch— —on	0
—off	V_{CC}

TRANSISTOR ALARM CIRCUITS

The two alarm circuits to be described use transistors as switches. The first is controlled by a light-dependent resistor (LDR) (p. 393) and the second by a thermistor (p. 50).

(a) Light-operated

A simple circuit which switches on a lamp L when it gets dark is shown in Fig. 23.34. The resistor R and the LDR form a potential divider across the supply. The input, applied between the base and emitter of the transistor, is the voltage across the LDR and depends on its resistance.

In bright light, the resistance of the LDR is low (e.g. 1 kΩ) compared with that of R (10 kΩ). Most of the supply voltage is dropped across R and the input is too small to switch on the transistor (for silicon about 0.6 V is needed). In the dark the LDR has a much greater resistance (e.g. 10 MΩ) and more of the supply voltage is dropped across it and less across R. The voltage across the LDR is now enough to switch on the transistor and produce a collector current sufficient to light L. If R is replaced by a variable resistor the light level at which L comes on can be adjusted.

When R and the LDR are interchanged, L is on in the light and off in the dark.

(b) Temperature-operated

In the circuit of Fig. 23.35, a thermistor and a resistor R form a potential divider across the supply and the input to the transistor is the voltage across R. When the temperature of the thermistor rises (e.g. by heating with a match), its resistance decreases, causing less of the supply voltage to be dropped across the thermistor and more across R. If the latter exceeds 0.6 V (for silicon), the transistor is switched on and collector current (too small to ring the bell directly) flows through the coil of the relay (p. 244). As a result the 'normally open' contacts close, enabling the bell to obtain the larger current it requires directly from the supply.

The diode across the relay coil protects the transistor from damage by the large e.m.f. induced in the coil when the current through it is switched off (p. 262). It is connected in reverse bias and offers an easy path for the current that the e.m.f. produces.

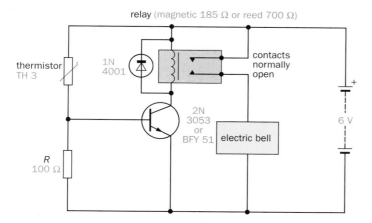

Fig. 23.35 Temperature-operated switch

How would you modify the circuit in Fig. 23.35 to make it a low-temperature alarm?

MULTIVIBRATORS

Multivibrators are two-stage transistor switching circuits in which the output of each stage is fed to the input of the other by coupling resistors or capacitors. As a result the transistors are driven alternately into saturation and cut-off, and when the output from one is 'high' the other is 'low'; we say their outputs are **complementary**.

The switch-over in each transistor from one output level or state to the other is so rapid that the collector voltage waveforms are almost square. The term 'multivibrator' refers to this since a square wave can be analysed into a large number of sine waves with frequencies that are multiples (harmonics) of the fundamental.

Multivibrators are of three types.

(a) The bistable or 'flip-flop'

This has two stable states. In one, the output of the first transistor is 'high' and of the second 'low'. In the other state the opposite is the case. It will remain in either state until an external trigger pulse makes it switch. Bistables are used as electrical **memories** or registers in computers to store the binary digits 0 ('low' output) and 1 ('high' output). They are also at the heart of **binary counters** (p. 436).

(b) The astable or 'free-running multivibrator'

This has no stable states. It switches from one state to the other automatically at a rate determined by the circuit components. Consequently it generates a continuous stream of almost square-wave pulses, i.e. it is a square-wave oscillator or pulse generator. It has many uses, including producing timing pulses to keep the various parts of a computer in step (it is then known as the 'clock') and generating musical notes in an electronic keyboard.

(c) The monostable or 'one-shot'

This has one stable state and one unstable state. Normally it rests in its stable state but can be switched to the other state by applying an external trigger pulse, where it stays for a certain time before returning to its stable state. Monostables are used to produce a square pulse of a certain height (voltage) and length (time) which may act as a 'gate' (high-speed switch) for another circuit and allow a number of timing pulses to pass for a certain time.

Multivibrators can be constructed using junction or field effect transistors, logic gates (p. 430) and operational amplifiers (p. 438). The astable will be considered in detail next and the bistable later.

ASTABLE

The basic circuit is shown in Fig. 23.36. The collector of each transistor is coupled to the base of the other by a capacitor C_1 or C_2.

(a) Action

When the supply is first connected both transistors draw base current but, because of slight differences (e.g. in h_{FE}), one conducts more than the other. A cumulative effect occurs and as a result one, say Tr_1, rapidly saturates while the other, Tr_2, is driven to cut-off. Each then switches automatically to its other state, then back to its first state and so on. As a result, the output voltage, which can be taken from the collector of either transistor, is alternately 'high' ($+6$ V) and 'low' (near 0 V) and is a series of almost square pulses.

To see how these are produced, suppose that Tr_2 was saturated (i.e. on) and has just cut off, while Tr_1 was off and has just saturated. Plate L of C_1 was at $+6$ V, i.e. the collector voltage of Tr_1 when it was off; plate M was at $+0.6$ V, i.e. the base voltage of

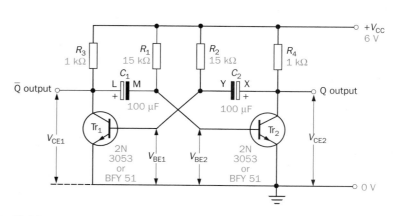

Fig. 23.36 Astable circuit

Tr$_2$ when it was saturated. C_1 was therefore charged with a p.d. between its plates of $(6\text{ V} - 0.6\text{ V}) = +5.4$ V. At the instant when Tr$_1$ suddenly saturates, the voltage of the collector of Tr$_1$ and so also of plate L, falls to 0 V (nearly). But since C_1 has not had time to discharge, there is still $+5.4$ V between its plates and therefore the voltage of plate M must fall to -5.4 V. This negative voltage is applied to the base of Tr$_2$ and turns it off.

C_1 starts to charge up via R_1 (the p.d. across which is now 11.4 V), aiming to get to $+6$ V but when it reaches $+0.6$ V it turns on Tr$_2$. Meanwhile plate X of C_2 has been at $+6$ V and plate Y at $+0.6$ V (i.e. the p.d. across it is 5.4 V); therefore when Tr$_2$ turns on, X falls to 0 V and Y to -5.4 V and so turns Tr$_1$ off. C_2 now starts to charge through R_2 and when Y reaches $+0.6$ V, Tr$_1$ turns on again. The circuit switches continuously between its two states.

(b) Voltage waveforms

The voltage changes at the base and collector of each transistor are shown graphically in Fig. 23.37. We see that V_{CE1} (collector voltage of Tr$_1$) and V_{CE2} (collector voltage of Tr$_2$) are complementary (i.e. when one is 'high' the

other is 'low'). Also, they are not quite square but have a rounded rising edge. This happens because as each transistor switches off and V_{CE} rises from near 0 V to $+V_{CC}$, the capacitor (C_1 or C_2) has to be charged (via the collector load R_3 or R_4). In doing so it draws current and causes a small, temporary voltage drop across the collector load which prevents V_{CE} rising 'vertically'.

(c) Frequency of the square wave

The time t_1 for which Tr$_1$ is *on* (i.e. saturated with $V_{CE1} \approx 0$) depends on how long C_1 takes to charge up through R_1 from -5.4 V to $+0.6$ V (i.e. 6.0 V $= V_{CC}$) and switch on Tr$_2$. It can be shown that

$$t_1 = 0.7C_1R_1$$

That is, it depends on the time constant C_1R_1 (p. 221). If C_1R_1 is reduced, Tr$_1$ is on for a shorter time.

Similarly the time t_2 for which Tr$_2$ is *on* (i.e. the time for which Tr$_1$ is off) is given by

$$t_2 = 0.7C_2R_2$$

Fig. 23.38 Transistor unit

The frequency f of the square wave is therefore

$$f = \frac{1}{t_1 + t_2} = \frac{1}{0.7(C_1R_1 + C_2R_2)}$$

If the circuit is symmetrical $C_1 = C_2$ and $R_1 = R_2$ and $t_1 = t_2$, i.e. the transistors are on and off for equal times. Their **mark-to-space ratio** is said to be 1. The frequency f in hertz (Hz) is therefore

$$f = \frac{1}{1.4C_1R_1} = \frac{0.7}{C_1R_1}$$

where C_1 is in farads (F) and R_1 in ohms (Ω). For example, if $C_1 = C_2 = 100 \ \mu\text{F} = 100 \times 10^{-6}$ F $= 10^{-4}$ F and $R_1 = R_2 = 15 \text{ k}\Omega = 1.5 \times 10^4 \ \Omega$ then

$$f = \frac{0.7}{(10^{-4}\text{ F} \times 1.5 \times 10^4\ \Omega)}$$

$$= \left(\frac{0.7}{1.5}\right)\text{Hz} \approx 0.5\text{ Hz}$$

The mark-to-space ratio of V_{CE} (the output from either transistor) can be varied by choosing different values for the time constants C_1R_1 and C_2R_2. However, the base resistors R_1 and R_2 should be low enough to saturate the transistors. The condition for satisfactory saturation and cut-off is R_1/R_3 and $R_2/R_4 < h_{FE}$. In Fig. 23.36, the transistors must have $h_{FE} > 15$.

(d) Demonstration

The 'oscillations' of an astable can be shown in slow motion by flashing lamps, using two transistor units, Fig. 23.38, and two indicator units[2] (each with a 6 V, 60 mA lamp), connected as in Fig. 23.39. The mark-to-space ratio and frequency of flashing can be changed by the variable resistors R_1 and R_2.

[2]Unilab 'Basic Units'

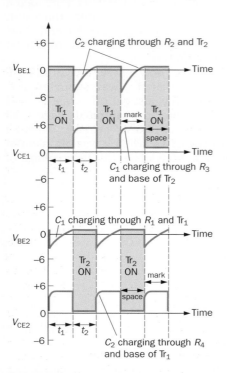

Fig. 23.37 Voltage waveforms of the astable

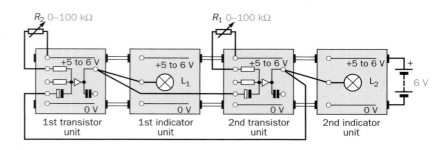

Fig. 23.39 Demonstrating oscillations of an astable

Fig. 23.40a Integrated circuit on a silicon 'chip' (×8)

Fig. 23.40b The casing and connections of a 'chip'

Fig. 23.41 Electrical signals

INTEGRATED CIRCUITS (ICs)

Integrated circuits are densely populated electronic circuits which are much smaller, much cheaper to make and much less likely to fail than circuits built from separate components. They may contain hundreds of thousands of transistors and often diodes, resistors and small capacitors. All are made on a tiny 'chip' of silicon, no more than 5 mm square and 0.5 mm thick and connected together by thin aluminium strips. They are the building blocks of microelectronics.

A micrograph of one is shown in Fig. 23.40a. In Fig. 23.40b a 'chip' is shown in its protective plastic case, partly removed to reveal the connections radiating from the 'chip' to the pins that enable it to make external connections. This type of package is the dual-in-line (d.i.l.) arrangement with the pins (from 6 to 64 in number but often 14 or 16) 0.1 inch (2.5 mm) apart, in two lines on each side of the case.

There are two broad groups.

(a) Linear or analogue ICs

These include **amplifier-type circuits** of many kinds. They handle signals that are often electrical representations, i.e. analogues, of physical quantities such as sound intensity, which change smoothly and continuously over a range of values between a maximum and a minimum, Fig. 23.41a. Their output is more or less directly proportional to the input, i.e. it varies *linearly* with the input. **Analogue electronics** is the branch of electronics dealing with such signals.

One of the most versatile linear ICs is the operational amplifier or **op amp**, to be considered on p. 438.

(b) Digital ICs

These contain **switching-type circuits** which process electrical signals that have only one of two values, Fig. 23.41b. Their inputs and outputs are either 'high' (e.g. near the supply voltage, often 5 V) or 'low' (e.g. near 0 V). The output jumps suddenly from one value to another when the input changes, and not smoothly as in a linear IC. They contain circuits such as logic gates (p. 430), astables (p. 431), bistables (p. 433), adders (p. 432), counters (p. 436), memories and microprocessors, which are the subject matter of **digital electronics**, the other main branch of electronics.

There are two main families of digital ICs. TTL (standing for Transistor-Transistor-Logic) uses junction transistors; CMOS (pronounced 'see-moss' and standing for Complementary Metal Oxide Semiconductor) is based on field effect transistors.

Each family has its own advantages and disadvantages. TTL requires a stabilized power supply of 5 V ± 0.25 V while CMOS works from any unstabilized voltage between 3 and 15 V. TTL gives faster switching than CMOS but it requires much larger currents when the IC is inactive. Greater component density is possible with CMOS, so making large-scale integration (LSI) easier, but it is damaged by static electricity.

The 74LS series is a *low-power* version of TTL and the 74HC series a *high-speed* version of CMOS.

LOGIC GATES

Logic gates are circuits in which transistors act as high-speed switches. They are used in calculators, digital watches, computers, robots, industrial control systems, and in telecommunications.

Their outputs and inputs involve only *two levels* of voltage, referred to as 'high' and 'low'. 'High' is near the supply voltage V_{CC}, e.g. +5 V, and 'low' is near 0 V. In the scheme known as **positive logic**, 'high' is referred to as logic level 1 (or just 1) and 'low' as logic level 0 (or just 0). The output depends on the inputs (of which there are usually more than one). Each of the various types of gate only 'opens' and gives a 'high' output for certain input combinations, i.e. they use combinational logic. Their behaviour is summed up in a **truth table** which shows in terms of 1s ('high') and 0s ('low') what the output is (1 or 0) for all possible inputs.

(a) NOT gate or inverter

This is the most basic gate, with one input and one output. The simple transistor switch of Fig. 23.33 behaves as one. It produces a 'high' output if the input is 'low' and vice versa. That is, the output is 'high' if the input is **not** 'high' and, whatever the input, the gate 'inverts' it. The commonly used symbol is given in Fig. 23.42, together with the truth table.

input	output
0	1
1	0

Fig. 23.42   NOT gate

(b) NOR gate

This is a NOT gate with two (or more) inputs, Fig. 23.43. Its output F is 'high' only when both inputs A and B are 'low', i.e. neither one input **nor** the other is 'high'.

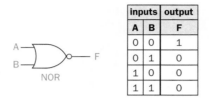

inputs		output
A	**B**	**F**
0	0	1
0	1	0
1	0	0
1	1	0

Fig. 23.43   NOR gate (2-input)

(c) OR gate

This is a NOR gate followed by a NOT gate; its symbol and truth table are shown in Fig. 23.44. The truth table can be worked out from that of the NOR gate by changing 0s to 1s and 1s to 0s in the output of the NOR gate. The output F is 1 when either input A **or** input B **or** both are 1s, i.e. if any of the inputs is 'high', the output is 'high'.

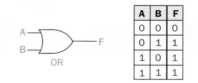

A	B	F
0	0	0
0	1	1
1	0	1
1	1	1

Fig. 23.44   OR gate (2-input)

(d) AND gate

The symbol and truth table are given in Fig. 23.45. The output F is 1 only when input A **and** input B are both 1.

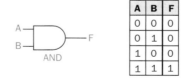

A	B	F
0	0	0
0	1	0
1	0	0
1	1	1

Fig. 23.45   AND gate (2-input)

(e) NAND gate

This is an AND gate followed by a NOT gate, Fig. 23.46. If inputs A **and** B are both 'high', the output F is **not** 'high'. The gate gets its name from this AND NOT behaviour.

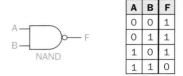

A	B	F
0	0	1
0	1	1
1	0	1
1	1	0

Fig. 23.46   NAND gate (2-input)

(f) Exclusive-OR gate

Unlike the ordinary OR gate (sometimes called the *inclusive*-OR gate), this excludes the case of both inputs being 'high' for a 'high' output, Fig. 23.47. It is also called the **difference** gate because the output is 'high' when the inputs are different.

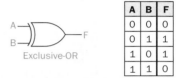

A	B	F
0	0	0
0	1	1
1	0	1
1	1	0

Fig. 23.47   Exclusive-OR gate

(g) Exclusive-NOR gate

This gate gives a 'high' output when its inputs are equal, i.e. both 0 or both 1, Fig. 23.48, and for this reason it is called a **parity** or **equivalence** gate. In effect it is an exclusive-OR gate followed by a NOT gate.

A	B	F
0	0	1
0	1	0
1	0	0
1	1	1

Fig. 23.48   Exclusive-NOR gate

(h) Summary

NOT: output always *opposite* of input

NOR: output 1 *only* when all inputs 0

OR: output 1 *unless* all inputs 0

AND: output 1 *only* when all inputs 1

NAND: output 1 *unless* all inputs 1

excl-OR: output 1 when inputs *different*

excl-NOR: output 1 when inputs *equal*

(i) Experiment

The truth tables for all the above logic gates can be derived or checked using one or more transistor units (see Fig. 23.38, p. 428) and an indicator unit for finding the level of the output voltage. If the indicator lamp lights, the output is 1, if it does not, the output is 0. Connecting an input to the '+5 to 6 V' socket makes it 'high' (1); a 'low' input (0) is obtained from the 0 V socket.

(j) Logic gate ICs

These are available as 14- or 16-pin d.i.l. packages with several gates on the same chip.

A quad 2-input NAND gate has four identical NAND gates, with two inputs and one output per gate. Every gate therefore has three pins, making 14 pins altogether on the package, including one each for the positive and negative power supply connections which are common to all four gates.

The TTL quad 2-input NAND IC 7400 is shown in Fig. 23.49a and its CMOS equivalent, the 4011B, in Fig. 23.49b. There are similar IC packages for other logic gates but, as we will see in the next section, NAND gates are especially useful.

Note that when using CMOS ICs, any unused inputs should be connected to supply positive (V_{DD}) or 0 V (V_{SS}) and not left 'floating'. Unused TTL inputs behave as if they were connected to 0 V (Gnd). It is good practice to join them to supply positive (V_{CC} = +5 V).

OTHER CIRCUITS FROM NAND GATES

All logic gates and other circuits can be made by combining only NAND (or NOR) gates.

(a) Logic gates

Figure 23.50 shows the NAND gate equivalents of different 2-input logic gates. For example, if the inputs of a NAND gate are joined together, it behaves as a NOT gate since a NAND gives a 0 output if both its inputs are 1 and a 1 output if either input is 0. You can check that the others give the correct output for the various input combinations by constructing a stage-by-stage truth table, as has been done for the NOR gate made from four NAND gates. Note the following boxed summary.

NOT = single-input NAND (or NOR), by joining all inputs together

AND = NAND followed by NOT

OR = NOT of each input followed by NAND

NOR = OR followed by NOT

These facts can often be used to minimize the number of ICs required for a system, using only one type of gate (e.g. NANDs).

(b) Astable multivibrator

The transistor version has been considered earlier (p. 427). The circuit in Fig. 23.51 uses two single-input NANDs as NOTs. The output from B is the input to A and, if at switch-on it is 'high', the output from A, and so also the astable output V_o, will be 'low'. Since the input to B must be 'low' (if its output is 'high'), C will start to *charge up* through R.

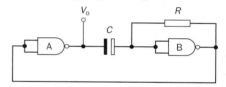

Fig. 23.51

Top views

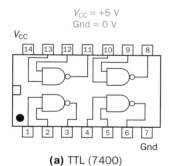

(a) TTL (7400)

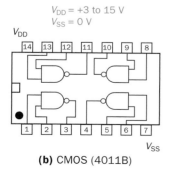

(b) CMOS (4011B)

Fig. 23.49   Quad 2-input NAND gate ICs

(a) NOT = single-input NAND

(b) AND = NAND + NOT

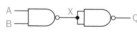

(c) OR = NOT of each input + NAND

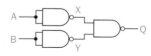

(d) NOR = OR + NOT

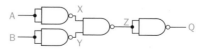

(e) Exclusive-OR

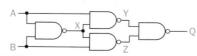

Fig. 23.50   NAND equivalents of different logic gates

NOR					
A	B	X	Y	Z	Q
0	0	1	1	0	1
0	1	1	0	1	0
1	0	0	1	1	0
1	1	0	0	1	0

When the right side of C is positive enough, i.e. 'high', the output from B begins to go 'low', driving the input to A 'low' and making V_o 'high'. As a result the voltage on the right side (as well as on the left) of C rises suddenly (due to the charge on it not having time to change). The resulting **positive feedback** reinforces the direction the input to B was going and confirms the 'high' state of V_o.

Because the input to B is now 'high' and its output 'low', C starts to *discharge* through R and when the input voltage to B becomes 'low', B switches back to its first state with a 'high' output. This in turn makes V_o 'low'. The process is repeated to give a **square-wave output** V_o, at a frequency determined by the values of C and R.

CONTROL SYSTEMS USING LOGIC GATES

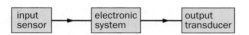

Fig. 23.52 Basic control system

Logic gates are decision-making devices. They can be used to control an electronic system, a block diagram for which is given in Fig. 23.52. The **input sensor** (e.g. a thermistor or an LDR) converts external physical effects (e.g. changes of temperature or light intensity) to **electronic information** (e.g. voltages) for processing by the system. The **output transducer** converts the processed electrical information from the system into some useful action (e.g. operates an electric motor or heater or sounds an alarm). The input and output devices enable the system to communicate with the outside world. Three simple examples will be outlined.

(a) Cooling fan

The block diagram in Fig. 23.53 is for a system that uses an AND gate to switch on a fan automatically when it gets too hot in a room during daylight

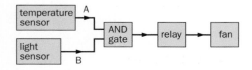

Fig. 23.53

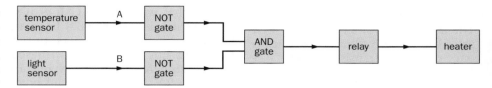

Fig. 23.54

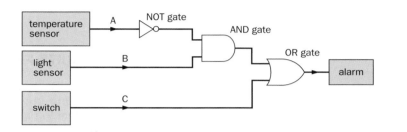

Fig. 23.55

hours. When inputs A and B are *both* 1 (i.e. when the temperature sensor is 'hot' and the light sensor is in daylight), the output from the AND gate is a 1 and this 'high' voltage operates the relay which in turn activates the fan. The relay is necessary because the output current from the gate is too small to drive the fan directly.

(b) Greenhouse heater

The system represented in Fig. 23.54 switches on a greenhouse heater when it is cold at night. Under these conditions inputs A and B are both 0 but become 1 by the action of the NOT gates. This produces a 1 output from the AND gate which turns on the relay and heater.

(c) Radiator alarm

The logic circuit diagram in Fig. 23.55 is for a system that activates an alarm if the temperature of a hot radiator falls during the day. The switch allows the operation of the alarm to be tested at any time. The output from the OR gate is 1 if input C is 1 (i.e. switch on) *or* if the output from the AND gate is 1. The latter happens only when input A is 0 (i.e. temperature sensor 'cold') and input B is 1 (i.e. light sensor in daylight).

If the radiator is 'hot', input A is 1 but becomes a 0 input to the AND gate (due to the NOT gate), so making its output 0. In that case, if the switch is off (i.e. C is 0), both inputs to the OR gates are 0, as is its output and so the alarm is not activated.

BINARY ADDERS USING LOGIC GATES

(a) Half-adder

This adds two **bits** (**b**inary dig**its**, i.e. 0 and 1) at a time and has to deal with four cases (essentially three since two are the same). They are:

$$\begin{array}{cccc} 0 & 0 & 1 & 1 \\ +0 & +1 & +0 & +1 \\ \hline 0 & 1 & 1 & 10 \end{array}$$

In the fourth case, 1 plus 1 equals 2, which in binary is 10, i.e. the right column is 0 and 1 is carried to the next column on the left. The circuit for a half-adder must therefore have two inputs, i.e. one for each bit to be added, and two outputs, i.e. one for the 'sum' and one for the 'carry'.

One way of building a half-adder from logic gates uses an exclusive-OR gate and an AND gate. From their truth tables in Figs 23.56a and b, you can see that the output of the exclusive-OR gate is always the 'sum' of the addition of two bits (being 0 for

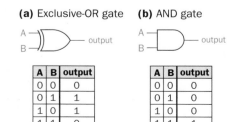

(a) Exclusive-OR gate

A	B	output
0	0	0
0	1	1
1	0	1
1	1	0

(b) AND gate

A	B	output
0	0	0
0	1	0
1	0	0
1	1	1

Fig. 23.56

1 + 1), while the output of the AND gate equals the 'carry' of the two-bit addition (being 1 only for 1 + 1). Therefore if both bits are applied to the inputs of both gates at the same time, binary addition occurs. The circuit is shown with its inputs A and B in Fig. 23.57, along with the half-adder truth table.

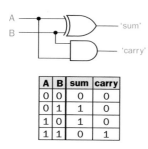

A	B	sum	carry
0	0	0	0
0	1	1	0
1	0	1	0
1	1	0	1

Fig. 23.57   Half-adder

(b) Full-adder

This adds *three* bits at a time and is a necessary operation when two multi-bit numbers are added. For example, to add 3 (11 in binary) to 3 we write:

$$\begin{array}{r} 11 \\ +11 \\ \hline 110 \end{array}$$

The answer 110 (6 in decimal) is obtained as follows. In the least significant (right-hand) column we have

$$1 + 1 = \text{sum } 0 + \text{carry } 1$$

In the next column three bits have to be added because of the carry from the first column, giving

$$1 + 1 + 1 = \text{sum } 1 + \text{carry } 1$$

A full-adder circuit therefore needs three inputs A, B, C (two for the digits and one for the carry) and two outputs (one for the sum and the other for the carry). It is realised by connecting two half-adders (HA) and an OR gate, as in Fig. 23.58a. We can check that it produces the correct answer by putting A = 1, B = 1 and C = 1, as in Fig. 23.58b. The first half-adder HA_1 has both inputs 1 and so gives a sum of 0 and a carry of 1. HA_2 has inputs 1 and 0 and gives a sum of 1 (i.e. the sum output of the full-adder) and a carry of 0. The inputs to the OR gate are 1 and 0 and, since one of the

inputs is 1, the output (i.e. the carry output of the full-adder) is 1. The addition of 1 + 1 + 1 is therefore sum 1 and carry 1.

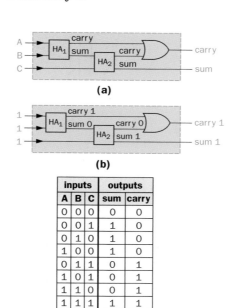

(a)

(b)

inputs			outputs	
A	B	C	sum	carry
0	0	0	0	0
0	0	1	1	0
0	1	0	1	0
1	0	0	1	0
0	1	1	0	1
1	0	1	0	1
1	1	0	0	1
1	1	1	1	1

(c)

Fig. 23.58   Full-adder

The truth table with the other 3-bit input combinations is given in Fig. 23.58c. You may like to check that the circuit does produce the correct outputs for them.

The block diagram for an adder which adds two 2-bit numbers, i.e. $A_2A_1 + B_2B_1$ where $A_2A_1 = 11$ (3 in decimal) $= B_2B_1$ is shown in Fig. 23.59. It consists of two full-adders (FA) in parallel. The sum is $C_2S_2S_1$ (S_1, like A_1 and B_1, being the least significant bit), i.e. 110 (6 in decimal).

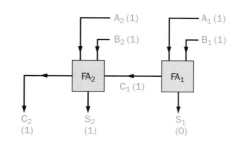

Fig. 23.59

To add two 4-bit numbers requires four full-adders.

RS BISTABLE

Bistable multivibrators or **flip-flops** (p. 427) are switching circuits with two outputs, one of which is 'high' when the other is 'low' and vice versa. The outputs, denoted by Q and $\overline{Q}$ (pronounced 'not Q'), are complementary. In one state Q = 1 and $\overline{Q} = 0$ and in the other Q = 0 and $\overline{Q} = 1$. Each of the two states is a *stable* one (hence bistable) which is retained (i.e. stored or remembered) until the bistable is 'triggered', causing one output to 'flop' from 1 to 0 and the other to 'flip' from 0 to 1.

There are several different types but the RS bistable is the basic one from which the others are developed. It has two inputs, R (for RESET) and S (for SET), Fig. 23.60a, and can be made from two 2-input NAND gates, connected as in Fig. 23.60b so that there is feedback from each output to one input (not R or S) of the other NAND gate.

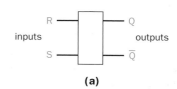

(a)

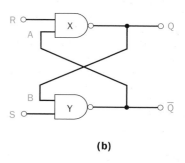

(b)

State	R	S	Q	$\overline{Q}$
Set	0	1	1	0
	1	1	1	0
Reset	1	0	0	1
	1	1	0	1

(c)

Fig. 23.60   RS bistable

(a) Set state

If R = 0 and S = 1, NAND gate X has at least one of its inputs at logic 0, therefore its output Q must be at logic 1 (since the output of a NAND is always 1 unless all its inputs are 1s). Q is fed back to input B and so both inputs to NAND gate Y are 1s, i.e. S = 1 and B = 1; hence output $\overline{Q}$ = 0. This is the stable set state with Q = 1 and $\overline{Q}$ = 0.

If R becomes 1 with S still 1, gate X inputs are now R = 1 and A = 0 (since $\overline{Q}$ = 0), i.e. one input is 0 and so Q stays at 1. The circuit has thus **latched** the state Q = 1 and $\overline{Q}$ = 0.

(b) Reset state

In this second stable state Q = 0 and $\overline{Q}$ = 1. It is given by R = 1 and S = 0. Since gate Y has one of its inputs at 0, its output $\overline{Q}$ = 1. $\overline{Q}$ is fed back to input A and so both inputs to X are 1s, i.e R = 1 and A = 1; hence output Q = 0.

If S becomes 1 with R still 1, Q stays at 0, showing that the reset state has been **latched**.

The truth table for the RS bistable is shown in Fig. 23.60c.

Note. When R = 1 *and* S = 1, Q and $\overline{Q}$ can be either 1 or 0 depending on the state before this input condition existed. The previous output state is retained as shown by the second and fourth rows of the truth table.

RS bistable ICs are not common; other more versatile types of bistable (see below) can do the same job.

DEBOUNCING A SWITCH

If a mechanical switch, for example on a keyboard, is used to change the state of a bistable (or other logic system), more than one electrical pulse may be produced due to the metal contacts of the switch not staying together at first but bouncing against each other rapidly and creating extra unwanted pulses, as in Fig. 23.61a. The effect of this 'contact bounce' can be eliminated by using an RS bistable in the debouncing circuit of Fig. 23.61b to 'clean up' the switch action.

When the switch is in the position shown, R is connected to 0 V and is at logic 0 while S is at logic 1 due to its connection via R_2 to supply positive.

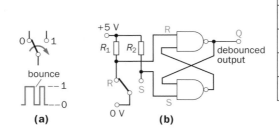

Fig. 23.61

Hence Q = 1 (from the truth table for the RS bistable). If the switch is then operated to make the circuit at the other contact and 'bounces' once, the logic levels at R and S for different positions are given in Fig. 23.61c. The last three lines show that Q stays at logic 0 despite the bounce.

D-TYPE BISTABLE (DATA LATCH)

The symbol for a D-type bistable is given in Fig. 23.62.

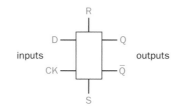

Fig. 23.62 D-type bistable

(a) Action

The D input is the one to which a bit of data (i.e. a 0 or 1) is applied for 'processing' by the bistable. The CK (clock) input has **clocking** or **triggering** pulses applied to it and these control exactly when the D input is processed. S is the set input allowing Q to be set to 1 and R is the reset input by which Q can be made 0.

Triggering is necessary in large digital systems where there are hundreds of interconnected bistables and we have to ensure that they all change state at the same time, i.e. are synchronized. As a result, the outputs Q and $\overline{Q}$ do not respond immediately to

changes at the input. They wait until a clock pulse is received at the CK input.

In most bistables triggering occurs during the rise of the clock pulse from 0 to 1. For a D-type bistable this means that the data (a 0 or a 1) at the D input passes to the Q output on the *rising edge* of the clock pulse. For example, in Fig. 23.63, if initially D = 1 and Q = 0 (therefore $\overline{Q}$ = 1), then on the rising edge AB of the first clock pulse the D input, i.e. a 1, is transferred to the Q output making Q = 1 (and $\overline{Q}$ = 0). When the D input changes to 0, Q does not become 0 until another clock-pulse rising edge occurs, i.e. along CD.

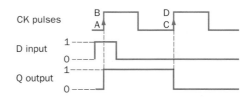

Fig. 23.63

To sum up:

> The Q output 'latches' on to the D input (and stores it as a 1 or 0) *at the instant a clock pulse changes its level from 0 to 1*. During the rest of the clock pulse, if the D input changes, it does not affect the Q output. **A D-type bistable is a one-bit data latch.**

(b) Use as a latch

A latching circuit consisting of interconnected D-type bistables is used to obtain a steady reading on what would

Switch position		R	S	Q
	Making contact at R	0	1	1
	Moving to make contact at S	1	1	1
	Making contact at S	1	0	0
	Bounces back from contact at S	1	1	0
	Remakes contact at S	1	0	0

(c)

otherwise be a rapidly changing numerical display. For example, pulses coming at a high rate from a counter would, without a latch, be seen as a blur.

A simple circuit for a three-light display is shown in Fig. 23.64. When the latch receives an appropriate clock pulse, the outputs from the three-bit binary counter (p. 436) are passed from the D inputs on the bistables to their respective Q outputs. They are held there until the next count enters the latch. In the meantime, the counter can carry on counting.

A latch is used in a digital voltmeter (p. 445) and is basically a store or 'memory'.

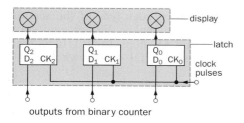

Fig. 23.64

CLOCKS AND TRIGGERING

(a) Clocks

Clock pulses are supplied by some sort of pulse generator. It may be a crystal-controlled oscillator with a very steady repetition frequency, e.g. 10 MHz, or an astable multivibrator or a mechanical switch turning a d.c. supply on and off. The pulses should have fast rise and fall times, i.e. be 'good' square waves, and any switches should be debounced (see p. 434).

(b) Triggering

There are two types of triggering or clocking operation. In **level** triggering, bistables change state when the logic level of the clock pulse is 1 or 0. The RS bistable described earlier is level triggered.

In **edge** triggering a *change* in voltage level causes switching, usually on the rise of the clock pulse from logic level 0 to logic level 1. The D-type bistable considered above is rising-edge triggered.

In general, edge triggering is more satisfactory than level triggering because in the former the output changes occur at an exact instant during the clock pulse. Any further changes do not affect the output until the next clock pulse rise. In level triggering, output changes can occur at any time while CK is 1, whereas in edge triggering the precise time of data capture is known.

T-TYPE BISTABLE

(a) Action

T-type bistables, also called **triggered** or **toggling flip-flops**, are the building blocks of binary counters. They are modified RS bistables with extra components that enable successive pulses, applied to an input called the **trigger** T, Fig. 23.65*a*, to make it switch to and fro (or 'toggle') from one stable state (e.g. Q = 0 and $\overline{Q}$ = 1) to the other (Q = 1 and $\overline{Q}$ = 0).

The truth table for a T-type bistable triggered by the *rising edge* of an input pulse is given in Fig. 23.65*b*. Two trigger pulses are required to give one

(a)

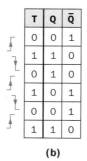

T	Q	$\overline{Q}$
0	0	1
1	1	0
0	1	0
1	0	1
0	0	1
1	1	0

(b)

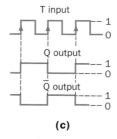

(c)

Fig. 23.65   T-type bistable

output pulse at either Q or $\overline{Q}$. The frequency of the output pulses is therefore half that of the input pulses, as the waveforms of the timing diagram in Fig. 23.65*c* show. A T-type flip-flop *divides the input frequency by two*. It can be built from transistors or NAND gates but is not available as an IC since other types of flip-flop, e.g. D-type, can be made to toggle, as we will now see.

(b) D-type as a T-type

If the $\overline{Q}$ output on a D-type bistable is connected to the D-input as in Fig. 23.66, successive clock (trigger) pulses applied to CK make it 'toggle'. If the first *rising edge* leaves Q = 1 and $\overline{Q}$ = 0 (because D = 1), then feedback (from $\overline{Q}$ to D) makes D = 0 and during the second *rising edge* D is transferred to Q, so now Q = 0 and $\overline{Q}$ = 1. D now becomes 1 and the third clock-pulse *rising edge* makes Q = 1 and $\overline{Q}$ = 0 again and so on, i.e. toggling occurs.

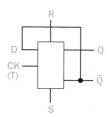

Fig. 23.66   D-type as a T-type

So Q = 1 *once* every two clock-pulse rising edges, that is, the output pulses have *half* the frequency of the input pulses.

> The T-flip-flop is a divide-by-2 circuit.

(c) Demonstration

The divide-by-2 action can be shown using the modules connected as in Fig. 23.67. Pressing and releasing the switch on the switch unit once applies to the trigger input a pulse which goes from 0 V to +6 V to 0 V. When it falls from +6 V to 0 V (i.e. when the switch is released) the bistable switches, i.e. it is *falling-edge* triggered. L_1 and L_2 show that (*i*) the Q and $\overline{Q}$ outputs are *complementary* (i.e. when L_1 is on, L_2 is off and vice versa), and (*ii*) *half* as many pulses appear at Q (or $\overline{Q}$) as are applied to the trigger.

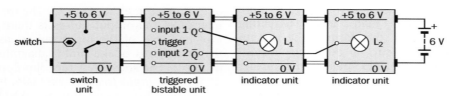

Fig. 23.67

J-K BISTABLE

The clocked J-K bistable is as versatile as a bistable can be and has many uses. Its symbol and truth table are given in Fig. 23.68. There are two inputs, called J and K (for no obvious reason), a clock input CK, two outputs Q and $\bar{Q}$ and set S and reset R inputs.

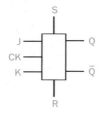

J	K	**Q** after CK pulse
0	0	stays at 0 or 1
0	1	resets to 0
1	0	sets to 1
1	1	toggles

Fig. 23.68 J-K bistable

When clock pulses are applied to CK:

1. It retains its present state if $J = K = 0$.
2. It acts as a D-type flip-flop if J and K are different (e.g. $J = 1$ and $K = 0$).
3. It acts as a T-type flip-flop if $J = K = 1$.

BINARY COUNTERS

Electronic counters consist of bistables connected so that they toggle, i.e. behave as T-type flip-flops, when the pulses to be counted are applied to their clock (CK) input. Counting is done in binary code, the binary digits (bits) 1 and 0 being represented by the 'high' and 'low' states respectively of the bistable's outputs.

There are two main types of counter — **ripple** and **synchronous**.

(a) Ripple counter

In Fig. 23.69a a 3-bit binary ripple (or asynchronous) counter is shown, consisting of three *falling-edge* triggered toggling flip-flops FF_0, FF_1 and FF_2 in which the Q output of each one feeds the CK (T) input of the next. To explain the action, suppose that the outputs Q_0, Q_1 and Q_2, which give the total count at any time, have all been reset to zero.

On the falling edge [ab] of the first clock pulse applied to CK_0 of FF_0, shown in Fig. 23.69b, Q_0 switches from 0 to 1. The resulting *rising* edge [AB] of Q_0 is applied to CK_1 of FF_1 which does not change state. So $Q_2 = 0$, $Q_1 = 0$ and $Q_0 = 1$, giving a binary count of 001 after one pulse.

The falling edge [cd] of the second clock pulse makes FF_0 change state again and Q_0 flops from 1 to 0. The falling edge [CD] of Q_0 switches FF_1

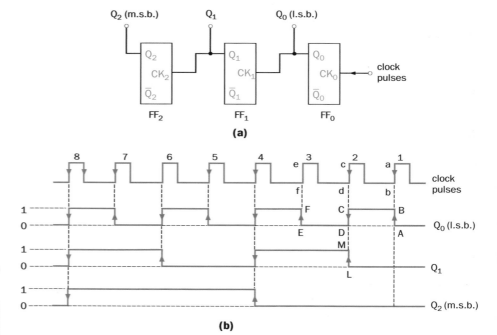

(a)

(b)

Number of clock pulse	Outputs		
	$Q_2 = 2^2 = 4$ (m.s.b.)	$Q_1 = 2^1 = 2$	$Q_0 = 2^0 = 1$ (l.s.b.)
0	0	0	0
1	0	0	1
2	0	1	0
3	0	1	1
4	1	0	0
5	1	0	1
6	1	1	0
7	1	1	1

(c)

Fig. 23.69 Ripple counter

this time, making $Q_1 = 1$. The rising edge [LM] of Q_1 leaves FF$_2$ unchanged. The count is now $Q_2 = 0$, $Q_1 = 1$ and $Q_0 = 0$, i.e. 010 in binary (2 in decimal).

The falling edge [ef] of the third clock pulse to CK$_0$ of FF$_0$ changes Q_0 from 0 to 1 again but the rising edge [EF] does not switch FF$_1$, leaving $Q_1 = 1$, $Q_2 = 0$ and the count at 011 (3 in decimal). The action thus 'ripples' along the flip-flops, each one waiting for the previous one to supply a falling edge at its CK input before changing state.

The table in Fig. 23.69c shows how the states of Q_0, Q_1 and Q_2 give the count in binary of the clock pulses fed to CK$_0$, Q_0 being the least significant bit (l.s.b.) and Q_2 the most significant bit (m.s.b.).

(b) Synchronous counter

In this type all flip-flops are triggered simultaneously and the *propagation delay time* (i.e. the time lapse between the clock pulse being applied and the output of the counter changing) is much less than for a ripple counter with a large number of flip-flops. Synchronous counters are used for high-speed counting but their circuits are more complex than those of ripple counters.

MORE ABOUT COUNTERS

(a) Modulo

The modulo of a counter is the number of output states it goes through before resetting to zero. A counter with three flip-flops counts from 0 to $(2^3 - 1) = 8 - 1 = 7$; it has eight different output states representing the decimal numbers 0 to 7 and is a modulo-8 counter. A counter with n flip-flops counts from 0 to $(2^n - 1)$ and has 2^n states, i.e. it is a modulo-2^n counter. If $n = 12$, 0 to 4095 can be counted.

(b) Frequency division

In a modulo-8 counter one output pulse appears at Q_2 every eighth clock pulse. That is, if f is the frequency of a regular train of clock pulses, the frequency of the pulses from Q_2 is $f/8$. Q_2 is 0 for four clock pulses and 1 for the next four pulses, as Fig. 23.69c shows.

This follows because each flip-flop produces pulses at its Q output which have *half* the frequency of those at its CK input. Therefore in a modulo-8 counter with three flip-flops, the frequency of the pulses at successive Q outputs are (if the clock pulses have frequency f), $f/2$ at the first, $\frac{1}{2} \times f/2 = f/4$ at the second and $\frac{1}{2} \times f/4 = f/8$ at the third.

A modulo-8 counter is also a 'divide-by-8' circuit. Counters are used as dividers in digital watches.

(c) Rising-edge triggered counters

If flip-flops that toggle on a *rising edge* (e.g. a D-type with D connected to output $\overline{Q}$) are used to make a counter, the $\overline{Q}$ outputs must be used to trigger the CK inputs, not the Q outputs as in Fig. 23.69a. (This can be checked by drawing a timing diagram like that in Fig. 23.69b for falling-edge triggered flip-flops and including the $\overline{Q}$ output waveforms).

A 3-bit rising-edge triggered ripple counter is shown in Fig. 23.70.

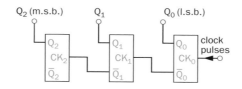

Fig. 23.70

(d) Up/down counters

The two counters considered so far have been up-counters in which the count increases with each clock pulse. In a down-counter the count decreases by one for each clock pulse. To convert an up-counter into a down-counter, the $\overline{Q}$ outputs are used to indicate the total count.

(e) Demonstration

A 3-bit binary up-counter to count to 7 can be made using three bistable units each with an indicator unit connected to its Q output, Fig. 23.71. The counter is first set to zero by briefly connecting to +6 V the input 1 of any bistable whose Q output is 'high', i.e. lamp lit.

Pulses are fed in by pressing and releasing the switch on the switch unit. It will be seen that the changes of state occur on 'release' when the input pulse *falls* from +6 V to 0 V. L$_1$ gives the least significant digit.

A down-counter is made by connecting the indicator units to $\overline{Q}$, not Q; the bistables still go from Q to trigger.

DECADE (BCD) COUNTER

A decade or 'binary-coded decimal' (BCD) counter counts from 0 to 9 in binary before resetting. It is a modulo-10 counter, made by modifying a 4-bit binary up-counter as in Fig. 23.72.

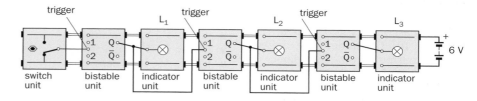

Fig. 23.71

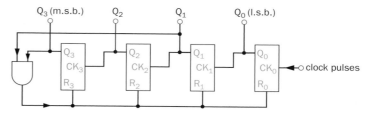

Fig. 23.72 Decade (BCD) counter

When the count is 1010 (decimal 10), $Q_3 = 1$, $Q_2 = 0$, $Q_1 = 1$ and $Q_0 = 0$ and since both inputs to the AND gate are 1s (i.e. Q_3 and Q_1), its output is 1 and this resets all the flip-flops to 0 (otherwise it would be a modulo-16 counter counting from 0 to 15).

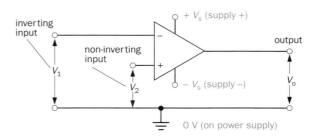

Fig. 23.73 Op amp

OPERATIONAL AMPLIFIER

An operational amplifier, or **op amp**, is so called because it can perform electronically mathematical operations such as addition, multiplication and integration. These operations form the basis of analogue computing (in which mathematical equations representing physical systems, e.g. the forces on a bridge, are solved) and it was for this that the op amp was designed originally. Nowadays it is also used widely as a high-gain amplifier of d.c. and a.c. voltages and as a switch.

The first op amps were made from discrete components, but they are now available in IC form and belong to the linear (analogue) group, although they can perform non-linear (i.e. digital) operations. One might contain twenty or so transistors as well as resistors and small capacitors, all on the same tiny silicon chip.

(a) Properties

The chief properties of an op amp are:

(*i*) *a very high voltage gain*, called the **open-loop gain** A_o, which typically is 10^5 for d.c. and low frequencies but decreases with frequency;

(*ii*) *a very high input resistance* r_i, typically $10^{12}\,\Omega$: it therefore draws a minute current from the device or circuit supplying its input (or, putting it another way, it does not alter the value of the voltage applied to its input, i.e. it behaves like a high-resistance voltmeter); and

(*iii*) *a very low output resistance* r_o, typically $100\,\Omega$, which means that its output voltage can be transferred with little loss to a load greater than a few kilohms.

(b) Description

An op amp has one output and two inputs. The non-inverting input is marked + and the inverting input is marked −, as shown on the amplifier

symbol in Fig. 23.73. Operation is most convenient from a dual-balanced d.c. power supply giving equal positive and negative voltages $\pm V_s$ (i.e. $+V_s$, 0 and $-V_s$) in the range $\pm 5\,V$ to $\pm 15\,V$. The centre point of the power supply, i.e. $0\,V$, is common to the input and output circuits and is taken as their voltage reference level.

Do not confuse the input signs with those for the power supply polarities, which, for clarity, are often left off circuit diagrams.

(c) Action

If the voltage V_2 applied to the non-inverting (+) input is positive relative to the other input, the output voltage V_o is positive; similarly if V_2 is negative, V_o is negative, i.e. V_2 and V_o are in phase. At the inverting (−) input, a positive voltage V_1 relative to the other input causes a negative output voltage V_o and vice versa, i.e. V_1 and V_o are in antiphase.

Basically an op amp is a *differential* voltage amplifier, i.e. it amplifies the difference between the voltages V_1 and V_2 at its inputs. There are three cases:

- if $V_2 > V_1$, V_o is positive;
- if $V_2 < V_1$, V_o is negative;
- if $V_2 = V_1$, V_o is zero.

In general, the output V_o is given by

$$V_o = A_o(V_2 - V_1)$$

where A_o is the open-loop voltage gain.

(d) Transfer characteristic

A typical voltage characteristic showing how the output V_o in volts (V) varies with the input $(V_2 - V_1)$ in microvolts (μV) is given in Fig. 23.74*a*. It reveals that it is only within the very small input range AOB that the output is directly proportional to the input,

i.e. when the op amp behaves more or less *linearly* and there is minimum distortion of the amplifier output. Inputs outside the linear range cause *saturation* and the output is then close to the maximum value it can have, i.e. $+V_s$ or $-V_s$, Fig. 23.74*b*.

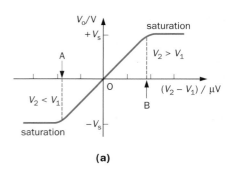

(a)

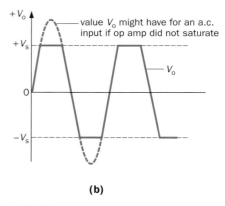

(b)

Fig. 23.74

The limited linear behaviour is due to the very high open-loop gain A_o, and the higher it is the greater the limitation. For example, on a $\pm 9\,V$ supply, since the output voltages can never exceed these values, and have maximum values close to either $+9\,V$ or $-9\,V$, the maximum input voltage swing (for linear amplification) is, if $A_o = 10^5$, $\pm 9\,V/10^5 = \pm 90\,\mu V$. A smaller value of A_o would allow a greater input.

OP AMP VOLTAGE AMPLIFIERS

(a) Negative feedback

The very high open-loop voltage gain of an op amp makes it unsuitable for most purposes as an amplifier. It either tends to saturate permanently or the gain varies violently with the frequency of the input signal, causing instability.

In practice an op amp is almost always used with negative feedback, obtained by feeding back a certain fraction of the output to the *inverting* input. This reduces the input and also the output, because the feedback voltage is in antiphase with the input (due to the output being 180° out of phase with the inverting input), as shown in Fig. 23.75. The gain with negative feedback, called the **closed-loop gain** A, is less than the open-loop gain A_o, but the following considerable advantages result:

(*i*) the gain is predictable and more or less independent of the characteristics of the op amp;

(*ii*) stability is greater;

(*iii*) distortion of the output is less, i.e. the amplification is more linear;

(*iv*) the gain is constant over a wider band of frequencies, as shown by the frequency response curve in Fig. 23.76. Note that the greater the negative feedback, the less the gain but the greater the **bandwidth**. The loss of gain is easily restored by using two or more op amp stages.

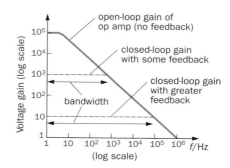

Fig. 23.76 Frequency response of op amp

(b) Inverting amplifier

The input voltage V_i (a.c. or d.c.) to be amplified is applied via resistor R_i to the inverting (−) terminal, Fig. 23.77. The output voltage V_o is therefore in antiphase with the input. The non-inverting (+) terminal is held at 0 V. Negative feedback is provided by R_f, the feedback resistor, feeding back a certain fraction of the output voltage to the inverting (−) terminal.

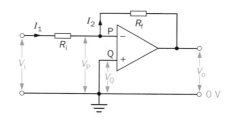

Fig. 23.77 Op amp inverting amplifier

Two assumptions are made to simplify the theoretical derivation of an expression for the gain of the amplifier, i.e. we consider an *ideal* op amp. The assumptions are:

(*i*) each input draws zero current from the signal source (it is less than 0.1 µA for the 741 op amp), i.e. their input impedances are infinite; and

(*ii*) the inputs are both at the same potential if the op amp is not saturated, i.e. $V_P = V_Q$. (The greater the open-loop gain A_o, the truer this will be, because, as we saw on p. 438, if the supply voltage $V_s = \pm 9$ V and $A_o = 10^5$, then maximum $V_o \approx \pm 9$ V and maximum $V_i \approx \pm 9$ V$/10^5 \approx \pm 90$ µV.)

If A_o is made infinite, we are making the 'infinite gain approximation'.

In the circuit of Fig. 23.77 $V_Q = 0$, therefore $V_P = 0$ and P is called a **virtual earth** (or ground) point, though of course it is not connected to ground. So

$$I_1 = (V_i - 0)/R_i \quad \text{and} \quad I_2 = (0 - V_o)/R_f$$

But $I_1 = I_2$, since by assumption (*i*) the input takes no current, therefore

$$\frac{V_i}{R_i} = -\frac{V_o}{R_f}$$

The negative sign shows that V_o is negative when V_i is positive and vice versa. The closed-loop gain A is given by

$$A = \frac{V_o}{V_i} = -\frac{R_f}{R_i} \qquad (1)$$

For example, if $R_f = 100$ kΩ and $R_i = 10$ kΩ, $A = -10$ exactly and an input of 0.1 V will cause an output change of 1.0 V.

Equation (1) shows that the gain of the amplifier depends only on the two resistors (which can be made with precise values) and not on the characteristics of the op amp (which vary from sample to sample).

If $R_f = R_i = 100$ kΩ, $A = -1$ and the circuit acts as a NOT gate or inverter.

The other versatile feature of this circuit is the way its input impedance can be controlled. Since point P is a virtual earth (i.e. at 0 V), R_i may be considered to be connected between the inverting (−) input terminal and 0 V. The input impedance of the *circuit* is therefore R_i in parallel with the much greater input impedance of the op amp (assumed infinite in the

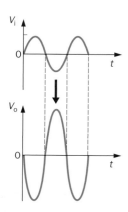

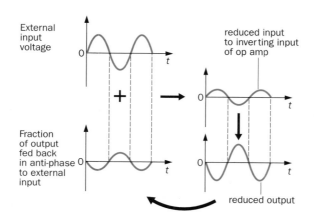

(a) No negative feedback

(b) Negative feedback

Fig. 23.75

previous theory), i.e. effectively R_i, whose value can be changed.

(c) Non-inverting amplifier

The input voltage V_i (a.c. or d.c.) is applied to the non-inverting (+) terminal of the op amp, Fig. 23.78. This produces an output V_o that is in phase with the input. Negative feedback is obtained by feeding back to the inverting (−) terminal the fraction of V_o developed across R_i in the voltage divider formed by R_f and R_i across V_o. This fraction, called the **feedback factor** β, is given by

$$\beta = \frac{R_i}{R_i + R_f}$$

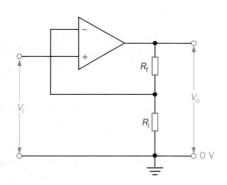

Fig. 23.78   Non-inverting op amp amplifier

It can be shown (p. 444) that if the open-loop voltage gain A_o is large (say 10^5), then for the closed-loop voltage gain A we can write

$$A = \frac{1}{\beta}$$

Therefore

$$A = \frac{R_i + R_f}{R_i} = 1 + \frac{R_f}{R_i} \qquad (2)$$

For example, if $R_f = 100\ \text{k}\Omega$ and $R_i = 10\ \text{k}\Omega$, then $A = 110/10 = 11$. As with the inverting amplifier, the gain depends only on the values of R_f and R_i and is independent of the open-loop gain A_o of the op amp.

Since there is no virtual earth at the non-inverting (+) input terminal, the input impedance is much higher (typically 50 MΩ) than that of the inverting amplifier. Also it is unaffected if the gain is altered by changing R_f and/or R_i. This circuit gives good

matching when the input is supplied by a high-impedance source such as a crystal microphone.

(d) Voltage follower

This is a special case of the non-inverting amplifier in which 100% negative feedback is obtained by connecting the output directly to the inverting (−) terminal, as shown in Fig. 23.79. Thus $R_f = 0$ and R_i is infinite. Because all of the output is fed back $\beta = 1$ and, since $A = 1/\beta$ (when A_o is very large), and then $A \approx 1$. The voltage gain is nearly 1 and V_o is the same as V_i to within a few millivolts.

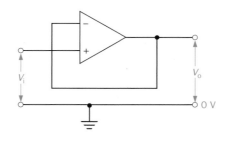

Fig. 23.79   Voltage follower

The circuit is referred to as a 'voltage follower' (because V_o follows V_i) or alternatively a 'unity gain follower' (because the gain is unity). It has an extremely high input impedance and a low output impedance. Its main use is as a **buffer amplifier** (giving current amplification) to match a high-impedance source to a low-impedance load. For example, it is used as the input stage of an analogue voltmeter where the highest possible input impedance is required (so as not to disturb the circuit under test), and the output voltage is measured by a relatively low-impedance moving-coil meter. It is used in this way in the d.c. amplifier shown in Fig. 14.24 (p. 222), which can be used as a 'coulomb-meter' to measure the charge on a capacitor.

(e) Summing amplifier

When connected as a multi-input inverting amplifier, an op amp can be used to add a number of voltages (d.c. or a.c.) because of the existence of the virtual earth point. This in turn is a consequence of the high value of A_o. Such circuits are employed as 'mixers' in audio applications to combine

outputs of microphones, electric guitars, pick-ups, special effects, etc. They are also used to perform the mathematical process of addition in analogue computing.

In the circuit of Fig. 23.80 three input voltages V_1, V_2 and V_3 are applied via input resistors R_1, R_2 and R_3 respectively. Assuming that the inverting terminal of the op amp draws no input current, all of it passing through R_f, then

$$I = I_1 + I_2 + I_3$$

Since P is a virtual earth, i.e. at 0 V,

$$\frac{-V_o}{R_f} = \frac{V_1}{R_1} + \frac{V_2}{R_2} + \frac{V_3}{R_3}$$

Therefore

$$V_o = -\left(\frac{R_f}{R_1} \cdot V_1 + \frac{R_f}{R_2} \cdot V_2 + \frac{R_f}{R_3} \cdot V_3\right)$$

The three input voltages are added and amplified if R_f is greater than each of the input resistors, i.e. 'weighted' summation occurs. Alternatively, the input voltages are added and attenuated if R_f is less than every input resistor.

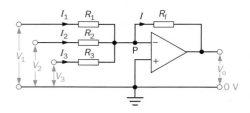

Fig. 23.80   Summing amplifier

If $R_1 = R_2 = R_3 = R_i$, the input voltages are amplified or attenuated equally and

$$V_o = \frac{-R_f}{R_i}(V_1 + V_2 + V_3)$$

Further, if $R_f = R_i$, then

$$V_o = -(V_1 + V_2 + V_3)$$

In this case the output voltage is the sum of the input voltages but is of opposite polarity.

Point P is also called the **summing point** of the amplifier. It isolates the inputs from one another so that each behaves as if none of the others existed and none feeds any of the other inputs, even though all the resistors are connected at the inverting input. Also, we can select the resistors to produce the best impedance matching with the transducer supplying the input, but compromise may be necessary to obtain both the gain required and the correct input impedance.

(f) Demonstration

A module[3] and circuit for demonstrating the amplification of an alternating voltage by an op amp (the TL081C, which has the same pin connections as the 741 but has greatly improved characteristics) in inverting mode are shown in Figs 23.81a and b. The signal generator, op amp module and CRO are set up with the values indicated. The amplitude of the input from the generator is slowly increased until the sine-wave output waveform on the CRO starts to be 'clipped' (i.e. the crests and troughs are 'squared'). The input is then slightly reduced to give a pure sine-wave output and the Y-gain on the CRO adjusted until the sine wave just occupies ten vertical divisions of the screen.

[3]Unilab 'Blue Chip': 511.006

(a) Op amp module

If the Y-input 'high' lead from the op amp output (point A) is transferred to the inverting input on the op amp (point B), the input waveforms should occupy about one vertical division, confirming that $A = V_o/V_i = R_f/R_i = 100/10 = 10$ (numerically). Other values of R_f and R_i can be tried.

The same arrangement may be used to show that the gain falls off rapidly above a certain frequency. The signal generator is set to 10 kHz and 25 mV (e.g. 2.5 V with −40 dB attenuation) and the CRO to 1 V/division and 0.1 ms/division. The Y-gain on the CRO is adjusted until the trace just occupies 10 vertical divisions. If the generator frequency is now raised to 100 kHz, the amplitude of the output waveform decreases appreciably.

The pin connections for the TL081C (or 741) are shown in Fig. 23.81c. The offset connections are used in analogue computing applications to ensure that the output is zero when the inputs are zero.

OP AMP VOLTAGE COMPARATOR

(a) Action

If both inputs of an op amp are used simultaneously then, as we saw earlier (p. 438), the output voltage V_o is given by

$$V_o = A_o(V_2 - V_1)$$

where V_1 is the inverting (−) input, V_2 the non-inverting (+) input and A_o the open loop gain. The voltage difference between the inputs, i.e. $(V_2 - V_1)$, is amplified and appears at the output, Fig. 23.82.

When $V_2 > V_1$, V_o is *positive*, its maximum value being the positive supply voltage $+V_s$, which it has when $(V_2 - V_1) \geq V_s/A_o$. The op amp is then saturated. If $V_s = +15$ V and $A_o = 10^5$, saturation occurs when $(V_2 - V_1) \geq 15$ V/10^5, i.e. when V_2 exceeds V_1 by 150 µV and $V_o \approx 15$ V.

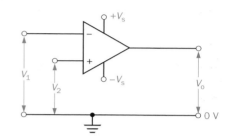

Fig. 23.82

When $V_1 > V_2$, V_o is *negative* and saturation occurs if V_1 exceeds V_2 by V_s/A_o, typically by 150 µV. But in this case $V_o \approx -V_s \approx -15$ V.

A small change in $(V_2 - V_1)$ thus causes V_o to switch between near $+V_s$ and near $-V_s$ and enables the op amp to indicate when V_2 is greater or less than V_1, i.e. to act as a **differential amplifier** and compare two voltages. It does this in an electronic digital voltmeter (p. 445).

(b) Some examples

The waveforms in Fig. 23.83 show what happens if V_2 is an alternating voltage (sufficient to saturate the op amp).

In Fig. 23.83a, $V_1 = 0$ and $V_o = +V_s$ when $V_2 > V_1$ (positive half-cycle) and $V_o = -V_s$ when $V_2 < V_1$ (negative half-cycle). V_o is a 'square' wave with a mark-to-space ratio of 1.

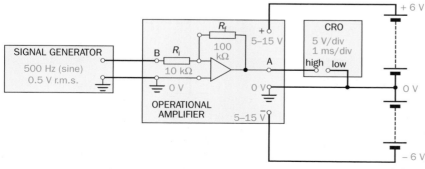

(b) Circuit

Fig. 23.81 Demonstrating amplification by an op amp

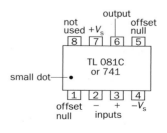

(c) Pin connections of op amp

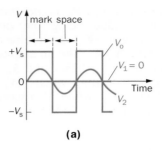

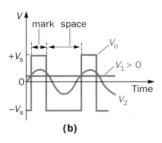

Fig. 23.83

In Fig. 23.83b, $V_1 > 0$ and switching of V_o occurs when $V_2 \approx V_1$, the mark-to-space ratio now being less than 1 and the output a series of 'pulses'.

In effect, the op amp in its saturated condition converts a continuously varying analogue signal (V_2) to a two-state (i.e. 'high'–'low') digital output (V_o); it is a **one-bit analogue-to-digital converter**.

(c) Alarm circuit

In the light-operated alarm circuit of Fig. 23.84, the inputs V_1 and V_2 are supplied by potential dividers and the op amp compares them. It is more sensitive than the transistor version described earlier (p. 426), i.e. switching occurs for smaller changes in the conditions that set off the alarm, and it is also easier to change the reference voltage at which switching occurs.

In the dark, the resistance of the LDR is much greater than R_1, making V_2 (the voltage across R_1) less than V_1, the reference voltage (set by the ratio of R_2 to R_3), the difference being sufficient to saturate the op amp. Since V_1 (at the inverting input) is positive, V_{out} will be negative and close to -9 V.

If light falls on the LDR, its resistance decreases and causes V_2 to increase. When V_2 (at the non-inverting input) exceeds V_1, the op amp switches to its other saturated state with V_{out} being close to $+9$ V. This positive voltage lights the LED (i.e. the alarm).

OP AMP OSCILLATORS

An op amp oscillator is an amplifier with a feedback loop from output to input which ensures the feedback is

- in phase with the input, i.e. is *positive*; and
- sufficient to make good the inevitable energy losses in the circuit, so that the amplified output consists of *undamped* electrical oscillations.

In effect, an oscillator supplies its own input and converts d.c. from the power supply into a.c. The alternating voltage generated can have a sine, square or some other waveform, depending on the circuit.

Audio frequency (a.f.) oscillators produce alternating voltages in the range 20 Hz to 20 kHz and are used as a.f. signal generators to make music or for use in sound experiments. Radio frequency (r.f.) oscillators generate higher frequency a.c., from about 100 kHz upwards, and are used in radio and television transmitters and receivers.

(a) Wien oscillator

Sine waves in the audio frequency range can be generated by an op amp using a Wien network circuit containing resistors and capacitors.

We saw earlier (Fig. 17.16, p. 277) that if a.c. is applied to a resistor and a capacitor in series, the voltages developed across them are 90° out of phase. In the Wien circuit, a network of two resistors R_1, R_2 (usually equal) and two capacitors C_1, C_2 (also usually equal), arranged as in Fig. 23.85a, acts as the *positive feedback* circuit. The network is an a.c. voltage divider and theory shows that the output voltage V_o is *in phase* with the input voltage V_i, i.e the phase shift is zero, at one frequency f, in hertz, given by

$$f = \frac{1}{2\pi RC}$$

where R is in ohms and C in farads. At all other frequencies there is a phase shift.

To obtain oscillations the network must therefore be used with a non-inverting amplifier which gives an output to the Wien network that is in phase with its input.

In the circuit shown in Fig. 23.85b, the frequency-selective Wien network R_1C_1–R_2C_2 applies positive feedback to the non-inverting ($+$) input. Negative feedback is supplied via R_3 and R_4 to the inverting ($-$) input. It can be proved that so long as the voltage gain of the amplifier exceeds 3, oscillations will be maintained at the desired frequency f.

A variable-frequency output, as in an a.f. signal generator, can be obtained if R_1 and R_2 are variable, 'ganged' resistors (i.e. mounted on the

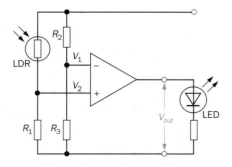

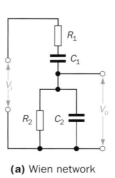

(a) Wien network

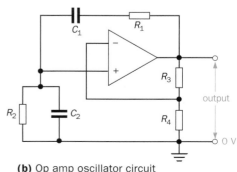

(b) Op amp oscillator circuit

Fig. 23.84

Fig. 23.85

same spindle so that they can be altered simultaneously by one control). C_1 and C_2 would be pairs of capacitors that are switched in for different frequency ranges.

(b) Astable multivibrator

An op amp voltage comparator (p. 441) in its saturated condition can operate as an oscillator if suitable external components are connected and *positive feedback* used.

The circuit and various voltage waveforms are shown in Figs 23.86a, b. Suppose the output voltage V_o is positive at a particular time. A certain fraction β of V_o is fed back as the non-inverting input voltage V_2, which equals βV_o where $\beta = R_2/(R_2 + R_3)$. V_o is also fed back via R_1 to the inverting terminal and V_1 rises (exponentially) as C_1 is charged.

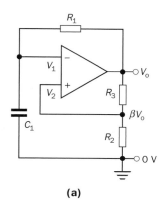

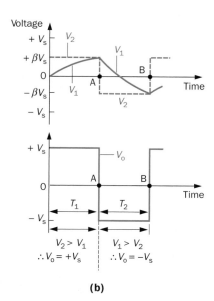

(a)

(b)

Fig 23.86 Op amp astable multivibrator

After a time T_1 which depends on the time constant C_1R_1, V_1 exceeds V_2 and the op amp switches into negative saturation, i.e. $V_o = -V_s$ (point A on the graphs). Also, the positive feedback makes V_2 go negative (where $V_2 = -\beta V_o = -\beta V_s$).

C_1 now starts to charge up in the opposite direction, making V_1 fall rapidly during time T_2 (AB on the graphs) and eventually become more negative than V_2. The op amp therefore switches again to its positive saturated state with $V_o = +V_s$ (point B on the graphs). This action continues indefinitely at frequency f given by $f = 1/(T_1 + T_2)$.

The output voltage V_o provides *square* waves and the voltage across C_1 is a source of *triangular* waves. The latter have 'exponential' sides, but they are 'straightened' by extra circuitry which ensures C_1 is charged by a constant current, rather than by the exponential one supplied through R_1.

OP AMP INTEGRATOR

The circuit is the same as for the op amp inverting amplifier (Fig. 23.77) but feedback occurs via a capacitor C, as in Fig. 23.87, rather than via a resistor. If the input voltage V_i, applied through the input resistor R, is constant, and $CR = 1$ s (e.g. $C = 1$ μF, $R = 1$ MΩ), the output voltage V_o after time t (in seconds) is given by

$$V_o = -V_i t$$

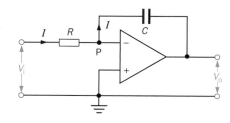

Fig. 23.87 Op amp integrator

The negative sign is inserted because, when the inverting input is used, V_o is negative if V_i is positive and vice versa.

For example, if $V_i = -3$ V as in Fig. 23.88a, V_o rises steadily by $+3$ V/s and, if the power supply is ±15 V, V_o reaches about $+15$ V after 5 s, when

the op amp saturates, Fig. 23.88b. V_i is effectively 'added up' or integrated over a time t to give V_o, a ramp voltage waveform whose slope is proportional to V_i.

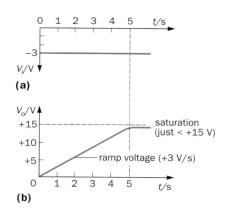

Fig. 23.88 Producing a ramp output voltage by integration

The integrating action of the circuit on V_i can be compared to that of a petrol pump, which 'operates' on the rate of flow (in litres per second) and the delivery time (in seconds) to produce as its output the volume of petrol supplied (in litres).

MORE ABOUT AMPLIFIERS

(a) The feedback equation

Suppose the amplifier in Fig. 23.89 has an open-loop (no feedback) gain A_o. If a fraction β of the output voltage V_o is fed back so as to add to the *actual* input signal to be amplified, V_i, i.e. there is *positive feedback*, the *effective* input to the amplifier becomes $V_i + \beta V_o$.

The gain of the amplifier itself is still A_o and so

$$V_o = A_o(V_i + \beta V_o)$$

That is

$$V_o(1 - \beta A_o) = A_o V_i$$

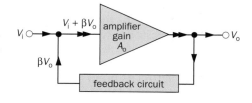

Fig. 23.89

Therefore

$$\frac{V_o}{V_i} = \frac{A_o}{1 - \beta A_o}$$

The voltage gain of the amplifier *with* feedback, i.e. the **closed-loop gain** *A*, is given by

$$A = \frac{V_o}{V_i} = \frac{A_o}{1 - \beta A_o}$$

This is the general equation for an amplifier with feedback. Three cases are important.

(*i*) *Positive feedback*, i.e. β is a positive fraction and $\beta A_o > 0$ but < 1, then $A > A_o$ and net amplification occurs.

(*ii*) *Oscillation occurs* if β is positive and $\beta A = 1$, i.e. $1 - \beta A_o = 0$, therefore $A = A_o/0 = $ infinite gain. This is the principle of many oscillators.

(*iii*) *Negative feedback*, i.e. if β is negative and $\beta A_o < 0$ making $1 - \beta A_o > A_o$ and $A < A_o$. If $\beta A_o \gg 1$, then

$$A = \frac{A_o}{\beta A_o} = \frac{1}{\beta}$$

As we saw earlier (p. 440), this result shows that with negative feedback the gain of an amplifier can be less dependent on its characteristics (e.g. A_o).

(b) Input and output impedance

An amplifier has to be 'matched' to the transducer or circuit supplying its input or receiving its output. Usually this means ensuring that either the maximum voltage or the maximum power is transferred to or from the amplifier. In all cases, the input or output impedance (since we are usually concerned with a.c.) of the amplifier is an important factor.

The **input impedance** Z_i equals V_i/I_i where I_i is the a.c. flowing into the amplifier when voltage V_i is applied to the input. It depends not only on the input resistance r_i (p. 420) of the amplifying device (e.g. transistor or op amp) but also on the presence of capacitors, resistors, etc., in the amplifier circuit. In effect, the amplifier behaves as if it had an impedance Z_i connected across its input terminals.

The **output impedance** Z_o is the a.c. equivalent of the internal or source resistance of a battery (p. 51). It causes

a 'loss' of voltage at the output terminals when the amplifier is supplying current. We can think of an amplifier as an a.c. generator of voltage *V* which on open circuit equals the voltage V_o at the output terminals. On closed circuit, when an output current flows, V_o is less than *V* by the voltage dropped across Z_o, as in the d.c. case.

The diagram in Fig. 23.90, called the **equivalent circuit** of an amplifier, is useful when considering matching problems. Z_i and Z_o can be measured (in ohms) by suitable meters.

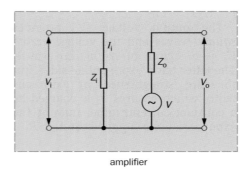

Fig. 23.90 Equivalent circuit

(c) Matching to signal

The source supplying the a.c. signal to the *input* of an amplifier can also be regarded as an a.c. generator, of voltage V_s, having an output impedance Z_s, as shown by its equivalent circuit in Fig. 23.91. If the input current is I_i, we can write

$$V_s = I_i(Z_s + Z_i)$$

where $V_i = I_i Z_i$. It follows that

$$V_i = V_s \left(\frac{Z_i}{Z_s + Z_i} \right)$$

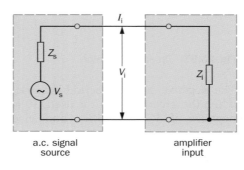

a.c. signal   amplifier
source   input

Fig. 23.91

The commonest requirement is for *maximum voltage transfer* from the signal source to the input of the amplifier. That is, V_i should be as large as possible. The above equation shows that if Z_i is large compared with Z_s (say, ten times larger), then $V_i \approx V_s$ and very little of V_s is 'lost' across Z_s.

In practice, things may be very different if, for example, a common-emitter transistor amplifier with an input impedance Z_i of 1 kΩ is supplied with an a.c. input signal from a crystal microphone having an output impedance Z_o of 1 MΩ. Only about 1/1000 of the voltage V_s generated by the microphone would be available at the amplifier input (since $Z_i/(Z_s + Z_i) \approx 1/1000$). An op amp with its appreciably larger Z_i (p. 438) would give much better matching.

(d) Matching to load

If the output of the amplifier supplies a load of impedance Z_L with current I_o, then from Fig. 23.92 we have

$$V = I_o(Z_o + Z_L)$$

where $V_o = I_o Z_L$. Therefore

$$V_o = V \left(\frac{Z_L}{Z_o + Z_L} \right)$$

For *maximum voltage transfer* this equation shows that we need $Z_L \gg Z_o$ if the voltage across Z_o is to be small.

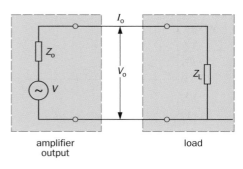

amplifier   load
output

Fig. 23.92

For *maximum power transfer* the maximum power theorem (p. 53) states that this occurs when $Z_L = Z_o$. In the case of an amplifier supplying a loudspeaker with a typical impedance Z_L of 8 Ω, a common-emitter transistor amplifier would not deliver maximum power on account of its much greater output impedance Z_o. Steps have to be taken to improve the matching.

(e) Worked example

Calculate (i) the voltage across, and (ii) the power developed in the load Z_L in the circuit of Fig. 23.93 if $V_s = 15$ mV, $Z_s = 500\,\Omega$ and $Z_L = 8.0\,\Omega$. The amplifier characteristics are $Z_i = 1000\,\Omega$, $A_o = 100$ and $Z_o = 8.0\,\Omega$.

(i) Voltage applied to input =

$$V_i = V_s\left(\frac{Z_i}{Z_i + Z_s}\right)$$

(see above)

Therefore

$$V_i = 15\text{ mV}\left(\frac{1000\,\Omega}{500\,\Omega + 1000\,\Omega}\right)$$
$$= 10\text{ mV}$$

The amplifier output voltage V is given by

$$V = A_o \times V_i = 100 \times 10\text{ mV} = 1.0\text{ V}$$

Voltage across Z_L =

$$V_o = V\left(\frac{Z_L}{Z_o + Z_L}\right)$$

(see above)

Therefore

$$V_o = 1.0\text{ V}\left(\frac{8.0\,\Omega}{8.0\,\Omega + 8.0\,\Omega}\right) = 0.50\text{ V}$$

(ii) Power in load =

$$V_o^2/Z_L = (0.50)^2\text{ V}^2/8.0\,\Omega$$
$$= 0.03\text{ W}$$

(f) Voltage gain in decibels (dB)

If the power output of an amplifier increases from P_1 to P_2, the power gain in decibels (see p. 325) is given by

$$\text{power gain} = 10\log_{10}\left(\frac{P_2}{P_1}\right)$$

Sometimes it is more convenient to express the gain in terms of the ratio of the corresponding r.m.s. voltages. If these are V_1 and V_2 when the powers are P_1 and P_2 then, since power is directly proportional to the square of the voltage (assuming V_1 and V_2 are developed across the same impedance), we can say $P_2/P_1 = (V_2/V_1)^2$ and power gain $= 10\log_{10}(V_2/V_1)^2$.

Therefore

$$\text{gain in dB} = 20\log_{10}\left(\frac{V_2}{V_1}\right)$$

since $\log_{10}x^2 = 2\log_{10}x$.

ELECTRONIC SYSTEMS

No matter how complex an electronic system such as a radio receiver, a digital voltmeter, a CRO or a computer may seem, it can be regarded as consisting of a number of basic building blocks or **modules** (often in the form of ICs), each performing a certain task. This may be to amplify, switch, count or store a signal.

The job of the electronics engineer is to know what different modules can and cannot do and how to assemble the minimum number to achieve a particular end. An understanding of the system as a whole is required and not necessarily a detailed knowledge of how individual circuits work.

In this so-called **systems approach**, the block diagram rather than the circuit diagram is used to simplify matters and give a broad but working appreciation of the system overall. Its use will be illustrated for a digital voltmeter.

DIGITAL VOLTMETER

An electronic digital voltmeter gives a reading on a numerical display (e.g. LED). It eliminates errors due to parallax that can arise in instruments requiring the position of a pointer on a scale to be estimated and also has a very high input resistance (for example, 10 MΩ).

The simplified block diagram and the waveforms in Figs 23.94a and b (p. 446) help us to follow the action. The d.c. voltage to be measured is fed to one input of a voltage **comparator** (p. 441). The other input of the comparator is supplied by a **ramp generator** which produces a repeating sawtooth waveform. The output from the comparator is 'high' (1) until the ramp voltage equals the input voltage when it goes 'low' (0).

The comparator output is applied to one input of an **AND gate**, the other input of the gate being fed by a steady train of pulses from a **pulse generator**. When both these inputs are 'high' the gate opens (p. 430) and gives a 'high' output, i.e. a pulse. The number of output pulses so obtained from the AND gate depends on the length of the comparator output pulse, i.e. on the time taken by the ramp voltage to reach the value of the input voltage. If the ramp is linear, this time is proportional to the input voltage.

The output pulses from the AND gate are recorded by a **binary counter**, then fed to a **latch** which passes them to a **decoder** for conversion to decimal before they reach the **display**, where the count is held until the next one enters the latch. The voltmeter is thus sampling the input voltage at regular intervals.

The whole process commences when the voltmeter is switched on and a pulse from a trigger circuit starts the ramp generator and sets the counter to

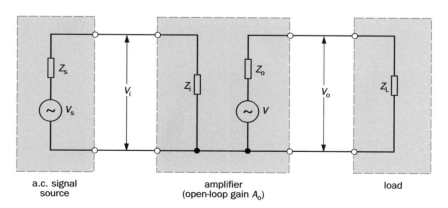

a.c. signal source   amplifier (open-loop gain A_o)   load

Fig. 23.93

zero. When the input voltage is of the order of millivolts it is amplified before being measured. With some additional circuitry the voltmeter can be adapted for use as a **multimeter**, Fig 23.95, to measure a.c. voltages, current and also resistance.

CHARGE-COUPLED DEVICES

A charge-coupled device (CCD) is a semiconductor device in which 'packets' of charge can be moved in an ordered sequence. It contains a large number of regions (pixels) arranged in rows and columns; when a photon falls on a pixel, charge is generated. The amount of charge that accumulates over time is proportional to the intensity of the light falling on the pixel. If the charge is moved electronically in an ordered way which preserves the arrangement of the pixels, a digital electronic image of the light intensity can be recorded. Highly sensitive CCD cameras are increasingly being used in astronomy to build up an image from a faint source and store it digitally. They are linear, allowing accurate intensity measurements to be made, can be used over a larger range of wavelengths and have a higher quantum efficiency (QE) than other types of detectors. Defining QE as follows,

$$QE = \frac{\text{number of photons detected}}{\text{number of photons incident}} \times 100\%$$

then the QE for a CCD is 50–100% for wavelengths in the range 400–1000 nm; this compares with 1% for a photographic plate over a 300–700 nm range. (The eye has a maximum QE of 1% at 550 nm.)

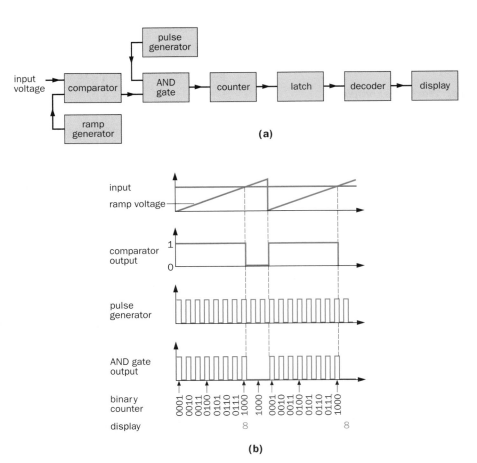

(a)

(b)

Fig. 23.94   Block diagram and voltage waveforms for a digital voltmeter

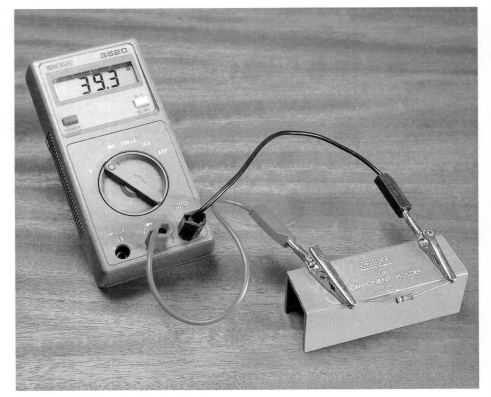

Fig. 23.95   Electronic digital multimeter

QUESTIONS

Cathode ray oscilloscope

1. An oscilloscope is connected across a signal generator as shown in Fig. 23.96a.

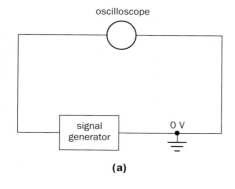

(a)

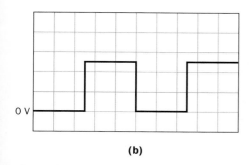

(b)

Fig. 23.96

a) Figure 23.96b shows the oscilloscope display. The y sensitivity is 2 V div^{-1} and the timebase is set to 50 µs div^{-1}. Calculate
   i) the maximum voltage of the signal,
   ii) the time period of the signal,
   iii) the frequency of the signal.
b) A 2.0 V cell is now included in the circuit as shown in Fig. 23.97.

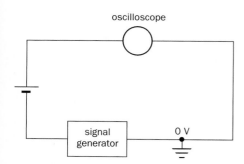

Fig. 23.97

On a grid like that in Fig. 23.96b draw the display you would expect to see.
(*NEAB*, AS/A PHO1, June 1997)

2. a) Draw a block diagram for an oscilloscope.
b) Sketch and explain the forms of the traces seen on an oscilloscope screen when a p.d. alternating at 50 Hz is connected across the Y-plates if the time base is linear and has a frequency of
   i) 10 Hz, and
   ii) 100 Hz.
c) What is the frequency of an alternating p.d. which is applied to the Y-plates of an oscilloscope and produces five complete waves on a 10 cm length of the screen when the time-base setting is 10 ms cm^{-1}?

3. The gain control of an oscilloscope is set on 1 V cm^{-1}. What is
a) the peak value, and
b) the r.m.s. value of an alternating p.d. that produces a vertical line trace 2 cm long when the time base is off?

Semiconductors; diodes; transistors; multivibrators

4. The circuit of Fig. 23.98 contains an ideal diode.

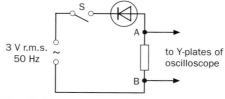

Fig. 23.98

a) The power supply produces a sinusoidal alternating voltage with an r.m.s. of value 3.0 V and a frequency of 50 Hz. For this supply, calculate
   i) the peak voltage,
   ii) the period.
b) Figure 23.99 shows the trace on the oscilloscope screen when the switch S is open.

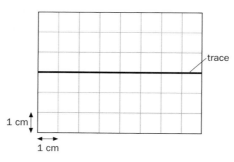

Fig. 23.99

i) The Y-plate sensitivity and the timebase are set at 2.0 V cm^{-1} and 5.0 ms cm^{-1} respectively. The switch S is now closed. Sketch the new trace on the oscilloscope screen.
ii) Indicate the direction of conventional current in the resistor in Fig. 23.98 when the diode is conducting.
iii) State which of the points, A or B, must be at 0 V in order to obtain the trace you have drawn in part i).
c) The trace you have drawn in part b) i) can be smoothed by including a capacitor in the circuit.
   i) Show how the capacitor should be included in the circuit of Fig. 23.98.
   ii) Explain how the capacitor smoothes the voltage.
(*UCLES*, Further Physics, March 1998)

5. a) Explain the terms **p-type** and **n-type** semiconductors.
b) Describe a p-n junction diode and draw a graph to show how the current through it varies with the p.d. across it.

6. In the junction-transistor voltage amplifier circuit of Fig. 23.100, if $R_1 = 100$ kΩ, $R_2 = 1$ kΩ, $V_{CC} = 6.0$ V and $V_{BE} = 0.6$ V, calculate
a) the voltage across R_1,
b) I_B,
c) I_C if $h_{FE} = 60$,
d) the voltage across R_2, and
e) the voltage across the collector–emitter.

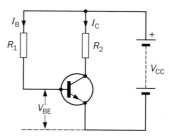

Fig. 23.100

7. The output characteristics of a junction transistor in common-emitter connection are shown in Fig. 23.101a. The transistor is used as an amplifier with a 9 V supply and a load resistor of 1.8 kΩ.
a) Copy the characteristics onto graph paper and draw the load line.
b) Choose a suitable d.c. operating point and read off the quiescent values of I_C, I_B and V_{CE}.
c) What is the quiescent power consumption of the amplifier?

d) If an alternating input voltage varies the base current by ± 20 μA about its quiescent value, what is
  i) the variation in the collector–emitter voltage, and
  ii) the peak output voltage?

e) An input characteristic of the transistor is given in Fig. 23.101*b*. Use it to find the base–emitter voltage variation which causes a change of ± 20 μA in the base current.

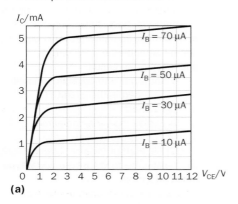

(a)

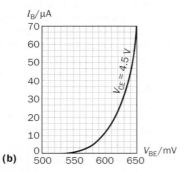

(b)

Fig. 23.101

f) Using your answers from d) and e), find the voltage gain of the amplifier.

g) If the amplifier uses the collector-to-base bias circuit of Fig. 23.30 (p. 424), calculate the value of R_B to give the quiescent value of I_B. (Assume $V_{BE} = 0.6$ V.)

8. a) In the circuit of Fig. 23.102, what will be the approximate reading on the voltmeter when the input is
  i) $+6$ V,
  ii) 0 V?

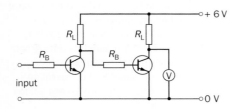

Fig. 23.102

b) In the voltage amplifier circuit of Fig. 23.103*a*, the alternating input v_i produces a varying d.c. voltage waveform V_{CE} on a CRO screen like that in Fig. 23.103*b*. If R_L is made slightly larger, what happens to
  i) the value of the steady d.c. component V_{DC} of V_{CE}, and
  ii) the amplitude of the alternating output voltage component v_o?

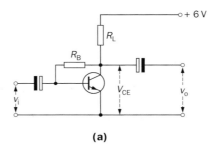

(a)

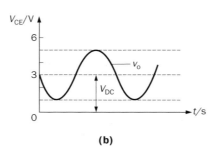

(b)

Fig. 23.103

9. State what each of the *three* types of multivibrator does and give one use for each.

Logic gates

10. A NOR gate 'opens' and gives an output only if *both* its inputs are 'low', but an OR gate 'closes'. An AND gate 'opens' only if both inputs are 'high', but a NAND gate 'closes'.

Copy and complete the truth table in Fig. 23.104 for each gate. 'High' is represented by 1 and 'low' by 0.

Input 1	Input 2	NOR output	OR output	AND output	NAND output
0	0				
1	1				
1	0				
0	1				

Fig. 23.104

11. Write a truth table for the circuit in Fig. 23.105, including the states at C, D, E, F and G.

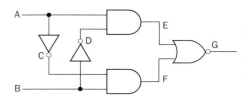

Fig. 23.105

12. The system in Fig. 23.106*a* makes three lamps flash in the order of British traffic signals. The amber lamp is connected to the output of the slow

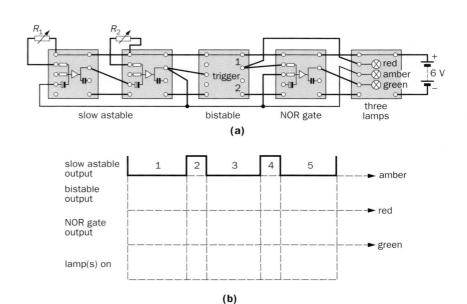

Fig. 23.106

astable, the red one to one of the outputs of the bistable and the green one to the output of the NOR gate.

In Fig. 23.106*b* the output pulses from the astable are shown. Copy it and add the corresponding outputs from the bistable and the NOR gate. The bistable switches only when its input goes from 'high' to 'low' and a NOR gate 'opens' only when both its inputs are 'low'.

Also show which lamp(s) is (are) on during intervals 1, 2, 3, 4 and 5. Write out the truth table for the system.

Op amps; systems

13. a) State *three* important properties of an op amp and say why they are important.

b) Explain the term **negative feedback** and state *four* advantages of using it in an op amp (or any type of) voltage amplifier.

c) Define **closed-loop gain** *A* and derive an expression for it for an inverting op amp voltage amplifier with an input resistor R_1 and a feedback resistor R_2. Use the expression to calculate *A* if $R_1 = 10 \text{ k}\Omega$ and $R_2 = 100 \text{ k}\Omega$.

14. a) If the **open-loop voltage gain** A_0 of an op amp is 10^5, calculate the maximum input voltage swing that can be applied for linear operation on a ±15 V supply.

b) Calculate the **closed-loop voltage gain** *A* for a non-inverting op amp voltage amplifier (Fig. 23.78, p. 440), in which $R_f = 220 \text{ k}\Omega$ and $R_i = 10 \text{ k}\Omega$.

15. If light falls on the LDR in the circuit of Fig. 23.107, current flows through R_1 and causes a voltage at the + input of the op amp. What happens when this voltage equals the 'reference' voltage (set by the values of R_2 and R_3) at the − input?

16. a) Figure 23.108 shows a non-inverting amplifier.

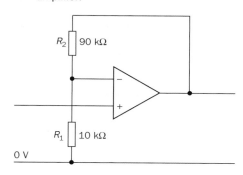

Fig. 23.108

Explain the term **non-inverting amplifier**.

The closed loop voltage gain of a non-inverting amplifier is given by the formula

$$1 + \frac{R_2}{R_1}$$

Calculate the gain of the amplifier shown. Show how this expression for the closed loop voltage gain is derived.

b) Op-amps can be used as comparators. Describe two differences in function between a comparator circuit and an amplifying circuit.

c) Draw a labelled circuit diagram of a simple a.m. radio receiver. Indicate on your diagram the parts which
   receive the frequency signal
   select the desired signal
   detect the signal
   smooth the signal.

(*L*, A PH3, June 1997)

17. a) The op-amp is a high gain differential amplifier with open loop gain A_{OL} and negligible input currents. Explain the following terms:
   high gain
   differential amplifier
   open loop gain
   negligible input current.

b) Figure 23.109 shows a switched-range voltmeter. The voltmeter is set to the 10 V range. Show that the output voltage from the op-amp is −3.0 V when the input is 10 V. You are not required to derive any formula you use.

The indicating meter has a resistance of 150 Ω. It needs a current of 2.0 mA for full scale deflection. Calculate the value of R_1 if the meter is to indicate full scale deflection when the voltmeter input is 10 V.

Calculate the feedback resistances R_2 and R_3 for the 1 V and 100 mV ranges.

c) Sketch the current–voltage characteristics for a semiconductor diode.

Draw waveform diagrams which show an amplitude modulated radio input signal and its audio frequency component as the output signal.

With reference to these waveforms, explain how a diode–capacitor arrangement produces the audio frequency output signal from the a.m. radio input signal.

(*L*, A PH3, Jan 1998)

18. a) State the properties of an ideal operational amplifier (op-amp).

b) i) Explain the meaning of the term **feedback** when applied to an op-amp circuit.
ii) State the effect of negative feedback on
   1. the gain,
   2. the bandwidth of an op-amp circuit.

c) Describe the use of an op-amp *either* as a summing amplifier *or* as a comparator.

d) Figure 23.110 (p. 450) shows a simple combination lock system. When contact is made by the 'push to operate' switch, the solenoid is energised and the lock opened only if

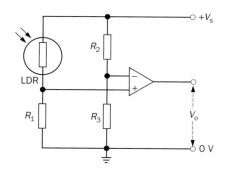

Fig. 23.107

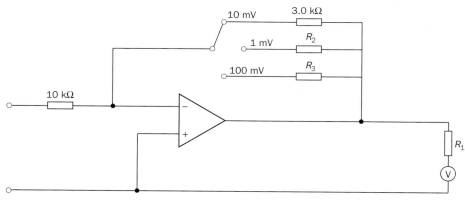

Fig. 23.109

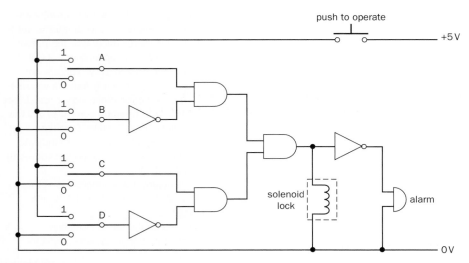

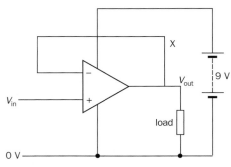

Fig. 23.110

the switches A, B, C and D are correctly set. For any other settings of A, B, C and D an alarm will sound.

Explain how the circuit functions. In your answer give the code required for the switches A, B, C and D to open the lock.

(OCR, 9244/3, Nov 1999)

19. The light-operated switch in Fig. 23.111 uses an op amp as a voltage comparator.

a) How must V_1 and V_2 compare if the op amp output is to be negative in daylight?

b) In darkness what happens to
   i) the LDR,
   ii) V_2 compared with V_1,
   iii) the output of the op amp,
   iv) Tr,
   v) the relay?

c) How would you alter the circuit to make the relay be off in the dark and switch on in daylight?

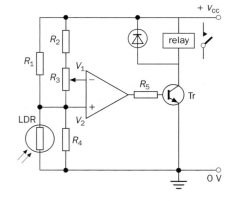

Fig. 23.111

20. a) Figure 23.112 shows a unity gain follower.

The input voltage is 4.5 V and the load resistance is 6.8 kΩ. What is
   i) the input current,
   ii) the current through the load,
   iii) the current in the wire at point X?

Fig. 23.112

Explain how the circuit keeps V_{out} equal to V_{in}.

What is the point of a circuit like this where the output voltage is equal to the input voltage?

The circuit as shown has a disadvantage which can be solved by a dual-ended power supply. Explain the disadvantage and draw a circuit diagram for a unity-gain follower with a dual-ended supply.

b) Figure 23.113 shows a coulombmeter.

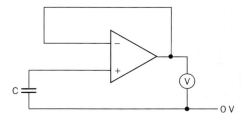

Fig. 23.113

The meter has a full-scale deflection of 100 mV. Calculate the capacitance of C if the coulombmeter has a full-scale deflection of 1000 nC.

c) Draw a circuit diagram of a comparator that will switch a lamp on automatically when it is dark. Explain how your circuit works.

(L, A PH3, June 1998)

INTRODUCTION

Telecommunications is concerned with the transmission and reception of **information** over a distance. In the broadest sense, information can be supplied in the form of spoken, written or keyed words, diagrams, numbers, music, pictures, or encoded computer data. Today information is seen, like minerals and energy, as a basic resource which is becoming more easily and widely available.

REPRESENTING INFORMATION

Information, or **data**, can be represented electrically in two ways.

(a) Digital method

In this method, which is the concern of digital electronics, the information is converted to electrical **pulses** produced by switching electricity on and off in **switching-type** circuits, e.g. logic gates (see chapter 23). For example, in the simple circuit of Fig. 24.1*a*, data can be sent by the 'dots' and 'dashes' of the Morse code by closing the switch for a short or a longer time. In Fig. 24.1*b* the letter A (· –) is shown.

Digital systems such as computers handle all information, not just numbers, in the binary code with the digits 1 and 0 being represented by 'high' and 'low' voltages respectively. It follows that if text is to be processed, a pattern of 1s and 0s has to be agreed for the 26 letters of the alphabet. A 5-bit code has 2^5 (32) variations, i.e. from 00000 to 11111, which would be enough. However, many digital systems use 8-bit 'words', or **bytes**. The *American Standard Code for Information Interchange* (ASCII) is an 8-bit code that allows $2^8 = 256$ characters to be coded in binary. This is adequate for all letters of the alphabet (capitals and small), the numbers 0 to 9, punctuation marks and other symbols. Figure 24.1*c* shows one 8-bit word, which represents the letter capital S in ASCII.

(b) Analogue method

In this case the information is changed to a voltage or current that *varies continuously* and smoothly over a range of values. The waveform of the voltage or current, i.e. its variation with time, represents the information and is an analogue of it, Fig. 24.2. For example, the loudness and pitch of a sound determines the amplitude and frequency of the waveform of the voltage produced by a microphone on which the sound falls.

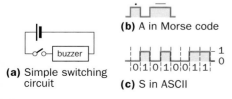

(a) Simple switching circuit
(b) A in Morse code
(c) S in ASCII

Fig. 24.1 Digital information

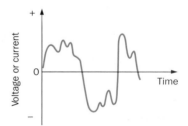

Fig. 24.2 Analogue information

Analogue electronics deals with the processing of analogue signals, e.g. in audio and television systems, using **amplifier-type** circuits.

(c) Advantages of digital signals

Information in digital form has certain important advantages over that in analogue form, despite the fact that many input devices (e.g. microphones, thermistors) produce analogue signals. There are two main reasons for this.

(*i*) Digital signals can be transmitted over long distances without error because of their ability to cope with 'noise'. All signals are weakened as they travel and also pick up electrical 'noise', i.e. stray, unwanted voltages or currents which distort the waveform and cause 'hiss' and 'hum' in loudspeakers. (For example, the random motion of the carbon granules in a telephone mouthpiece generates noise, as does the sparking of a car ignition system.)

Analogue signals require amplification (and correction) at suitable intervals, but the noise is amplified as well and may 'drown' the signal if it is weak. Digital signals, on the other hand, can be regenerated as 'clean' pulses, free from noise, Figs 24.3a, b, c, since it is only necessary to detect the presence or absence of a pulse (i.e. whether it is a 1 or a 0) and not its shape.

(*ii*) Digital signals fit in with modern technology and can be used with both telecommunications and data-processing equipment. Digital systems are easier to design and build (mostly from logic gates) and can be integrated on a single silicon chip.

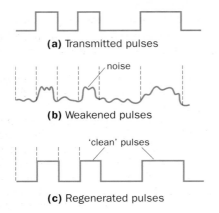

(a) Transmitted pulses

(b) Weakened pulses

(c) Regenerated pulses

Fig. 24.3

TRANSMISSION OF INFORMATION

Electrical signals representing information from a microphone, a TV camera, a computer, etc., can be sent from place to place using either radio waves or cables of copper or glass. Although audio-frequency signals may be transmitted directly by cable, in general, and certainly in radio and TV, a 'carrier' wave is required. This has a higher frequency than the information signal, its amplitude is constant and its waveform sinusoidal.

The general plan of any communication system is shown in Fig. 24.4. Signals from the **information source** are added to the carrier in the **modulator**. The modulated signal is sent along a 'channel' in the 'propagating medium' (i.e. cable or radio wave) by the **transmitter**.

At the receiving end, the **receiver** may have to select (and perhaps amplify) the modulated signal before the **demodulator** extracts from it the information signal for delivery to the **receptor of information**.

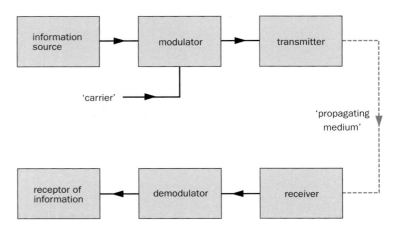

Fig. 24.4   Communication system

AMPLITUDE MODULATION (AM)

In amplitude modulation, the information signal is used to *vary the amplitude* of the carrier so that it follows the wave shape of the information signal, Fig. 24.5.

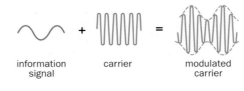

information signal   carrier   modulated carrier

Fig 24.5   Amplitude modulation

The modulated signal contains other frequencies, called **side frequencies**, which are created on either side of the carrier (a single frequency) in the process of modulation. If the carrier frequency is f_c and the modulating frequency of the information signal is f_m, two new frequencies of $f_c - f_m$ and $f_c + f_m$ are produced, one below f_c and the other above it, Fig. 24.6a.

If, as usually occurs in practice, the carrier is modulated by a range of audio frequencies, each a.f. gives rise to

a pair of side frequencies. The result is a band of frequencies, called the **lower** and **upper sidebands**, stretching below and above the carrier by the value of the highest modulating frequency. For example, if $f_c = 1$ MHz and the highest $f_m = 5$ kHz = 0.005 MHz, then $f_c - f_m$ = 0.995 MHz and $f_c + f_m$ = 1.005 MHz, Fig. 24.6b.

The **bandwidth** of an analogue signal is the range of frequencies the signal occupies. For intelligible speech it is about 3 kHz (e.g. 300 Hz to 3400 Hz as in the telephone system), for high-quality music it is 16 kHz or so and for television signals around 8 MHz. The bandwidth of the carrier sidebands in Fig. 24.6b is 10 kHz.

FREQUENCY MODULATION (FM)

In this case the *frequency* of the r.f. carrier, not the amplitude, is changed by the a.f. signal. The change or 'deviation' is proportional to the amplitude of the a.f. at any instant.

For example, if a 100 MHz carrier is modulated by a 1 V 1 kHz sine wave, the carrier frequency might swing

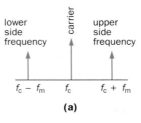

lower side frequency   carrier   upper side frequency

$f_c - f_m$ f_c $f_c + f_m$

(a)

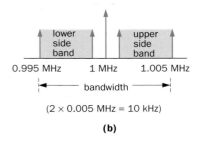

lower side band   upper side band

0.995 MHz   1 MHz   1.005 MHz

bandwidth

(2×0.005 MHz = 10 kHz)

(b)

Fig. 24.6

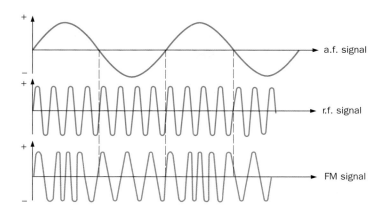

Fig. 24.7 Frequency modulation

15 kHz either side of 100 MHz, i.e. from 100.015 MHz to 99.985 MHz, and this would happen 1000 times a second. A 2 V 1 kHz signal would cause a swing of ± 30 kHz at the same rate; for a 2 V 2 kHz signal the swing remains at ± 30 kHz but it occurs 2000 times a second. By international agreement, the maximum deviation allowed is ± 75 kHz. Figure 24.7 shows frequency modulation; note that when the a.f. signal is positive, the carrier frequency increases but it decreases when the a.f. signal is negative.

In FM, each a.f. modulating frequency produces a large number of side frequencies (not two as in AM) but their amplitudes decrease the more they differ from the carrier. In theory, therefore, the bandwidth of an FM system should be extremely wide but in practice the 'outside' side frequencies can be omitted without noticeable distortion. The bandwidth may be taken as roughly $\pm(\Delta f_c + f_m)$ where Δf_c is the deviation and f_m the highest modulating frequency. The BBC uses a 250 kHz bandwidth which is readily accommodated in the v.h.f.

radio broadcasting band and also allows f_m to have the full range of audio frequencies. This accounts for the better sound quality of FM radio.

'Quiet' reception is another good feature of FM since unwanted 'noise' from, for example, lightning flashes ('static') is limited by the FM receiver. FM is also used for u.h.f. TV sound signals.

PULSE CODE MODULATION (PCM)

This is the process by which an analogue signal is changed into a digital one before it is transmitted by cable or radio wave.

The amplitude of the analogue signal, Fig. 24.8*a*, is 'sampled' at regular time intervals to find its value and a **pulse amplitude modulated** (PAM) signal is obtained, Fig. 24.8*b*. The values are measured on a scale of equally spaced voltage levels, six in Fig. 24.9*a*. Each level is represented in binary code by the appropriate pattern of electrical pulses, i.e. by a certain bit-pattern. A 3-bit code, Fig. 24.9*b*, can represent up to eight levels (0 to 7); a 4-bit code would allow sixteen levels to be coded. The bit-pattern is sent as a series of pulses, called a **pulse code modulated** (PCM) signal, Fig. 24.10.

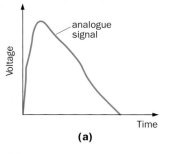

(a)

Fig. 24.8

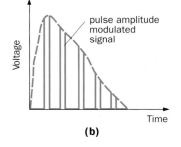

(b)

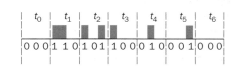

Fig. 24.10

The accuracy of the representation increases with the number of voltage levels and with the sampling frequency. The sampling frequency should be at least *twice* as great as the highest frequency of the analogue signal since it is sampled least often, Fig. 24.11*a, b*.

The highest frequency of intelligible speech in the telephone system is 3400 Hz and a sampling frequency of 8000 Hz is chosen, i.e. samples are taken at 125 μs intervals, each sample lasting for 2 to 3 μs. An 8-bit code (giving $2^8 = 256$ levels, represented by 00000000 to 11111111) is used and so

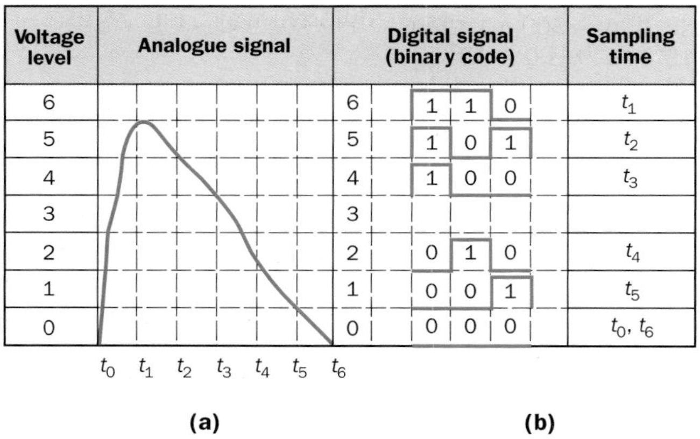

Voltage level	Analogue signal		Digital signal (binary code)			Sampling time
6		6	1	1	0	t_1
5		5	1	0	1	t_2
4		4	1	0	0	t_3
3		3				
2		2	0	1	0	t_4
1		1	0	0	1	t_5
0		0	0	0	0	t_0, t_6

$t_0 \; t_1 \; t_2 \; t_3 \; t_4 \; t_5 \; t_6$

(a)   **(b)**

Fig. 24.9

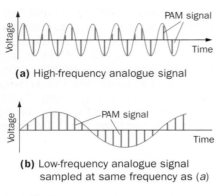

(a) High-frequency analogue signal

(b) Low-frequency analogue signal
sampled at same frequency as (a)

Fig. 24.11

the number of bits that have to be transmitted per second, called the **bit-rate**, is $8 \times 8000 = 64\,000 = 64$ kilobits/s.

In general, we can write

bit-rate = no. of bits × sampling frequency

For good quality music where frequencies up to about 16 kHz must be transmitted, the sampling frequency is 32 kHz and a 16-bit code (i.e. $2^{16} = 65\,236$ levels) is used. The bit-rate is therefore $16 \times 32 = 512$ kbits/s. For television signals, which carry much more information, a bit-rate of $70\,000\,000 = 70$ Mbits/s is needed.

When several signals are to be sent along the same communication channel, **multiplexing** (electronic switching) is used to prevent them interfering. In 'time-division' multiplexing each signal is assigned its own time allocation and the signals are sent in sequence; Fig. 24.12 shows the process for three signals. The multiplexer samples each signal in turn and sends each to the encoder for conversion into a set of pulses which are transmitted in the same order. The process is reversed at the receiving end by the decoder and demultiplexer which are synchronized with the multiplexer and encoder.

Table 24.1

Frequency band	Some uses
Low (l.f.) 30 kHz–300 kHz	Long-wave radio and communication over large distances
Medium (m.f.) 300 kHz–3 MHz	Medium-wave, local and distant radio
High (h.f.) 3 MHz–30 MHz	Short-wave radio and communication, amateur and CB radio
Very high (v.h.f.) 30 MHz–300 MHz	FM radio, police, meteorology devices
Ultra high (u.h.f.) 300 MHz–3 GHz	TV (bands 4 and 5), aircraft landing systems
Microwave Above 3 GHz	Radar, communication satellites, mobile telephones and TV links

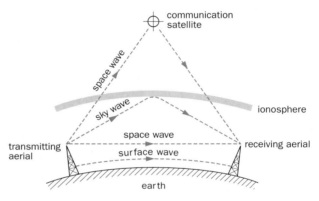

Fig. 24.13 Transmission of radio waves

RADIO WAVES

Radio waves are a member of the electromagnetic family of waves considered in chapter 20. They are energy-carriers which travel at the speed of light c, their frequency f and wavelength λ being related, as for any wave motion, by the equation

$$v = f\lambda$$

where $v = c = 3.0 \times 10^8\ \mathrm{m\ s^{-1}}$ in a vacuum (or air). If $\lambda = 300$ m, then $f = v/\lambda = 3.0 \times 10^8\ \mathrm{m\ s^{-1}}/(3.0 \times 10^2\ \mathrm{m}) = 10^6$ Hz = 1 MHz. The smaller λ is, the larger f.

Radio waves can be described either by their frequency or their wavelength, but the former is more fundamental since, unlike λ (and v), f does not change when the waves travel from one medium to another. They have frequencies extending from about 30 kHz upwards and are grouped into bands, as in Table 24.1 (1 GHz = 10^9 Hz).

Radio waves can travel from a transmitting aerial in one or more of three different ways, Fig. 24.13.

(a) Surface or ground wave

This travels along the ground, following the curvature of the earth's surface. Its range is limited mainly by the extent to which energy is absorbed from it by the ground. Poor conductors such as sand absorb more strongly than water, and the higher the frequency the greater the absorption. The range may be about 1500 km at low frequencies (long waves) but much less for v.h.f.

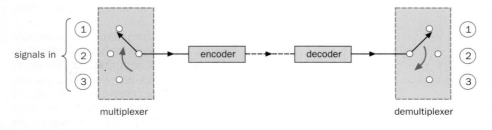

Fig. 24.12

(b) Sky wave

This travels skywards and, if it is below a certain **critical frequency** (typically 30 MHz), is returned to earth by the **ionosphere**. This consists of layers of air molecules (the D, E and F layers), stretching from about 80 km above the earth to 500 km, which have become positively charged through the removal of electrons by the sun's ultraviolet radiation. On striking the earth the sky wave bounces back to the ionosphere where it is again gradually refracted and returned earthwards as if by 'reflection'. This continues until it is completely attenuated.

The critical frequency varies with the time of day and the seasons. Sky waves of low, medium and high frequencies can travel thousands of kilometres but at v.h.f. and above they usually pass through the ionosphere into outer space.

If both the surface wave and the sky wave from a transmitter are received at the same place, interference can occur if the two waves are out of phase. When the phase difference varies, the signal 'fades', i.e. goes weaker and stronger. If the range of the surface wave for a signal is less than the distance to the point where the sky wave first reaches the earth, there is a zone which receives no signal.

(c) Space wave

For v.h.f., u.h.f. and microwave signals, only the space wave, giving line-of-sight transmission, is effective. A range of up to 150 km is possible on earth if the transmitting aerial is on high ground and there are no intervening obstacles such as hills, buildings or trees. Transmission via communication satellites is considered later.

AERIALS (ANTENNAE)

An aerial, or antenna, radiates or receives radio waves. Any conductor can act as one but proper design is necessary for maximum efficiency.

(a) Transmitting aerials

When a.c. from a transmitter flows in a transmitting aerial, radio waves of the same frequency, f, as the a.c. are emitted if the length of the aerial is comparable to the wavelength λ of the waves. For example, if $f = 100$ MHz $= 10^8$ Hz, $\lambda = c/f = 3 \times 10^8/10^8 = 3$ m; but if $f = 1$ kHz, $\lambda = 300\,000$ m. Therefore if aerials are not to be too long they must be supplied with r.f. currents from the transmitter.

An electromagnetic wave is regarded as an alternating electric field accompanied by an alternating magnetic field of the same frequency and phase, the fields being at right angles to each other and to the direction of travel of the wave, Fig. 20.48, p. 346. Consideration of the way in which a transmitting aerial produces radio waves shows that the electric field emerges parallel to the aerial and the magnetic field at right angles to it. If the aerial is vertical, the waves are said to be vertically polarized, the direction of polarization being given by the direction of the electric field, so following the practice adopted with light waves (p. 347).

Fig. 24.14 Microwave aerial tower

The **dipole aerial** consists of two vertical or horizontal conducting rods or wires, each of length one-quarter of the wavelength of the wave to be emitted, and is centre-fed, Fig. 24.15a. It behaves as a series LC circuit (p. 278) whose resonant frequency depends on its length, since this determines its inductance L (which all conductors have) and its capacitance C (which is distributed along it and arises from the capacitor formed by each conductor and the air between them). The radiation pattern of Fig. 24.15b shows that a vertical dipole emits equally in all horizontal directions, but not in all vertical directions, Fig. 24.15c.

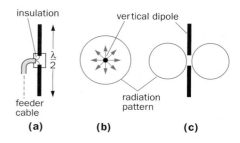

Fig. 24.15 Dipole aerial

When required (e.g. in radio telephony but not in most sound or television broadcasting systems), concentration of much of the radiation in one direction can be achieved by placing another slightly longer conductor, a **reflector**, about $\lambda/4$ behind the dipole and a slightly shorter one, a **director**, a similar distance in front, Fig. 24.16a, p. 456. The radiation that these extra elements emit arises from the voltage and current induced in them by the radiation from the dipole. It adds to that from the dipole in the wanted direction and subtracts from it in the opposite direction. Figure 24.16b shows the radiation pattern given by the array, called a **Yagi aerial**. The extra elements reduce the input impedance of the dipole and cause a mismatch with the feeder cable unless a **folded dipole** is used, Fig. 24.16c, to restore it to its original value.

In the microwave band, **parabolic dish aerials** in the shape of large metal 'saucers' are used (see Fig. 24.14). The radio waves fall on them from a small dipole at their focus, which is fed from the transmitter, and are reflected as a highly directed beam, nearly parallel, Fig. 24.17a. The radiation pattern has *one* very narrow loop or lobe and a number of much smaller side lobes, Fig. 24.17b; it resembles the diffraction pattern given by light falling on a single slit (Fig. 20.26, p. 340). The relative inten-

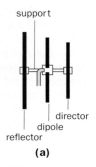

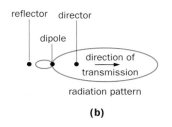

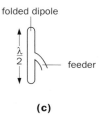

Fig. 24.16 Yagi aerial

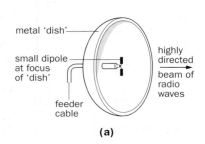

(a)

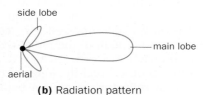

(b) Radiation pattern

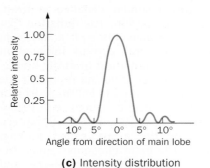

(c) Intensity distribution

Fig. 24.17 Parabolic dish aerial

sity distribution curve, Fig. 24.17c, is similar to that for light (Fig. 20.29); in this case the first minimum on either side makes an angle θ with the direction of the main lobe given by

$$\sin \theta = \frac{1.22\lambda}{a}$$

where λ is the wavelength of the microwaves and a is the diameter of the dish. This equation shows that the beam emitted is most directional (i.e. the main lobe is narrowest) when λ is small and a is large, thus making θ small.

(b) Receiving aerials

Whereas transmitting aerials may handle many kilowatts of r.f. power, receiving aerials, though basically similar, deal with only a few milliwatts due to the voltages (of a few microvolts) and currents induced in them by passing radio waves.

Dipole arrays are used to receive TV (where the reflector is often a slotted metal plate) and v.h.f. radio signals. They give maximum output when they are (*i*) lined up on the transmitter, and (*ii*) vertical or horizontal if the transmission is vertically or horizontally polarized.

RADIO SYSTEMS

(a) Transmitter

An aerial must be fed with r.f. power if it is to emit radio waves effectively, but speech and music produce a.f. voltages and currents. The transmission of sound by radio therefore involves modulating r.f. so that it 'carries' the a.f. information (p. 452). Amplitude modulation (AM) is used in medium-, long- and short-wave broadcasting.

A block diagram for an AM transmitter is shown in Fig. 24.18. In the

modulator the amplitude of the r.f. carrier from the **r.f. oscillator** is varied at the frequency of the a.f. signal from the **microphone**.

The modulated signal consists of the carrier and the upper and lower sidebands (p. 452, Fig. 24.6b). The bandwidth it requires to transmit a.fs up to 5 kHz is 10 kHz. In practice, in the medium wave band, which extends from about 500 kHz to 1.5 MHz, 'space' is limited if interference between stations is to be avoided and the bandwidth is restricted to 9 kHz.

(b) Simple receiver

The various parts of a **straight** or **tuned radio frequency** (TRF) receiver are shown in Fig. 24.19. The wanted signal from the **aerial** is selected and amplified by the **r.f. amplifier**, which is a voltage amplifier with a **tuned circuit** (p. 279) as its load, and should have a bandwidth of 9 kHz to accept the sidebands.

The a.f. is next separated from the modulated r.f. by the **detector** or **demodulator** and amplified by the **a.f. pre-** and **power amplifiers** to operate the **loudspeaker**. The bandwidths of the a.f. amplifiers need not exceed 4.5 kHz, i.e. the highest a.f. allowed.

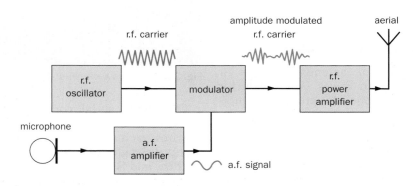

Fig. 24.18 AM transmitting system

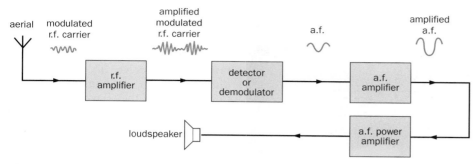

Fig. 24.19 TRF receiving system

(c) Other receivers

A much superior, but more complex, type of receiver is the 'superheterodyne' type, used in commercial AM radios. Frequency modulated (FM) receivers also use the superheterodyne principle at v.h.f. frequencies; they give better sound quality and are freer from interference.

COMMUNICATION SATELLITES

Many intercontinental telephone calls, as well as television broadcasts giving worldwide coverage of major events, use the satellite communication network.

British television viewers saw their first live broadcast from America in 1962. The microwave signals were beamed from a large steerable horn antenna at Andover, Maine (USA) to the satellite *Telstar*, which then amplified and re-transmitted the signals back to earth where they were received by a 30 m steerable dish aerial at Goonhilly Downs in Cornwall (UK). *Telstar* orbited the earth every $2\frac{1}{2}$ hours at a height varying from 320 to 480 km. Signals could only be transmitted for about 20 minutes when the satellite was 'visible' from both sides of the Atlantic.

(a) Geosynchronous (geostationary) satellites

The idea proposed by the space-science writer Arthur C. Clarke as long ago as 1945 was realised with the advent of powerful rockets which enabled satellites to be placed in orbit 36 000 km above the equator. At this height they orbit in 24 hours and appear to be stationary from the earth. Such satellites 'see' 120° of longitude,

so that the whole of the populated earth's surface can be covered by three. *Early Bird*, launched in 1965 into orbit over the Atlantic, was the first geosynchronous satellite to give round-the-clock use.

Throughout the world there are now thousands of earth stations with large steerable dish aerials like that shown in Fig. 24.20. The International Telecommunications Satellite Organ-ization (INTELSAT), a commercial cooperative of 143 member countries, is a leading provider of global satellite telecommunications services. With its constellation of 17 geostationary satellites, INTELSAT connects with its customers in more than 200 countries and territories. Over 50 INTELSAT satellites have been successfully launched by American and European launch companies, in over 35 years. The communication satellite INTELSAT VIII is shown in Fig. 24.21, p. 458. It has a capacity of 22 500 two-way telephone circuits plus three TV channels, or up to 112 500 two-way telephone circuits with the use of digital circuit multiplication equipment (DCME). It has a number of aerials which receive signals from earth, amplify them and re-transmit them back. Six INTELSAT VIII satellites were launched during 1997 and 1998; they are expected to have a life of 14–17 years.

Fig. 24.20 Earth station satellite dish at Whitehill Satellite Communications Centre, Oxfordshire

Fig. 24.21 The communication satellite INTELSAT VIII

(c) Direct broadcasting by satellite (DBS)

This enables homes in any part of a country to become low-cost 'earth stations' and receive TV programmes directly from a national geosynchronous satellite if they have a small rooftop dish aerial pointing towards the satellite. In this way one satellite, having had its programme beamed from an earth station, gives country-wide coverage. The dish needs to be steerable to obtain reception from the satellites of other countries.

Alternatively, programmes may be received via cable from an earth station with dish aerials.

OPTICAL FIBRE SYSTEMS

The suggestion that information could be carried using light sent over long distances in thin fibres of very pure (optical) glass, was first made in the 1960s. In 1977 the world's first fibre-optics telephone link was working in the UK. Now copper cables in almost all long-distance connections between exchanges in the UK have been replaced by optical fibre cables, and local lines from homes and offices to exchanges are being replaced. Undersea optical cables are also in operation, between Britain and the Continent and the USA.

(a) Outline of system

A simplified block diagram of an optical fibre communication system is shown in Fig. 24.23. The electrical signals representing the **information** (e.g. speech, television pictures, computer data) are pulse code modulated in a **coder** and then changed into the equivalent digital 'light' signals by the **optical transmitter**. This is either a miniature laser or an LED bonded on to the end of the fibre. The 'light' used is infrared radiation in the region just beyond the red end of the visible spectrum (with wavelength 0.85, 1.3 or 1.5 μm) because it is attenuated less by absorption in glass than 'visible' light.

The **optical fibre** is 125 μm (0.125 mm) in diameter and has a glass core of higher refractive index than the glass cladding around it. As a result the infrared beam is trapped in the core by total internal reflection

(b) Satellite networks

A typical satellite telephone call goes from the caller to a local exchange, to an international exchange and then by either cable or terrestrial microwave link to an earth station. There it is beamed up to the satellite by microwaves, Fig. 24.22, and re-transmitted down to another country.

INMARSAT (International Maritime Satellite Organization) uses satellites to provide direct communication with ships at sea and to other regions lacking cable links.

Satellite systems are frequently used for continuous transmission of data, especially from remote locations such as the Antarctic.

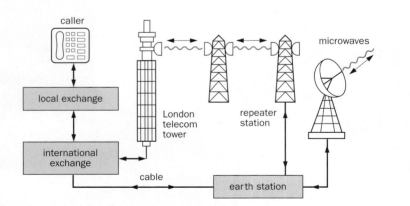

Fig. 24.22 Satellite telephone network

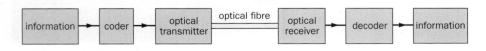

Fig. 24.23 Optical fibre communication system

(p. 84) at the core–cladding boundary and bounces off it in zig-zag fashion along the length of the fibre, Fig. 24.24. The photograph in Fig. 24.25 shows fibres grouped around the strength member in a multi-core optical fibre cable.

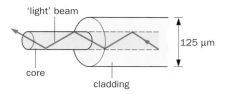

Fig. 24.24 Optical fibre

The **optical receiver** is a photodiode which converts the incoming infrared signals into the corresponding electrical signals before they are processed by the **decoder** for conversion back into **information**.

Fig. 24.25 Optical fibre cable

(b) Advantages

A digital optical fibre system has important advantages over other communication systems.

- Optical fibres have a very high information-carrying capacity. A typical optical fibre cable can carry about 9000 telephone channels, or over 1000 music channels, or 8 television channels, which is over five times the capacity of the best copper cable.
- It is free from 'noise' due to electrical interference.
- Greater distances (e.g. 50 km) can be worked without regenerators. (Copper cables require boosters to be much closer.)
- An optical fibre cable is lighter, smaller and easier to handle than a copper cable.
- Crosstalk between adjacent channels is negligible.
- It offers greater security to the user.

OPTICAL FIBRES

(a) Types

There are two main types – **multimode** and **monomode** fibres.

In the **step-index multimode** type, the core has the relatively large diameter of 50 μm and the refractive index changes abruptly at the cladding, Fig. 24.26a. The wide core allows the infrared to travel by several different paths or modes. Paths that cross the core more often are longer, Fig. 24.26b, and signals in those modes take longer to travel along the fibre. Arrival times at the receiver are therefore different for radiation from the same pulse, 30 ns km⁻¹ being a typical maximum difference. The pulse is said to suffer **dispersion**, i.e. it is spread out, Fig. 24.27. In a long fibre separate pulses may overlap and errors and loss of information will occur at the receiving end.

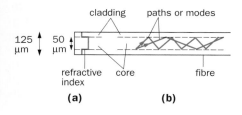

Fig. 24.26 Step-index multimode fibre

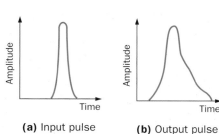

Fig. 24.27 Dispersion of a pulse

In the **graded-index multimode** type, the refractive index of the glass varies continuously from a high value at the centre of the fibre to a low value at the outside, Fig. 24.28a, so making the boundary between core and cladding indistinct. Radiation following longer paths, Fig. 24.28b, travels faster on average, since the speed of light is inversely proportional to the refractive index. The arrival times for different modes are then about the same (to within 1 ns km⁻¹) and all arrive more or less together at the receiving end. Dispersion is thereby much reduced.

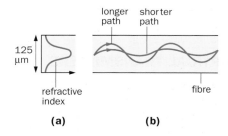

Fig. 24.28 Graded-index multimode fibre

In the **monomode** fibre the core is only 5 μm in diameter, Fig. 24.29a (p. 460), and only the straight-through transmission path is possible, i.e. one mode, Fig. 24.29b. This type, although more difficult and expensive to make, is being used increasingly. For short distances and low bit-rates, multimode fibres are quite satisfactory.

(b) Attenuation

Absorption of infrared occurs when it travels through glass, but less for longer wavelengths than shorter ones. The **optical power** decays exponentially with fibre length x according to the equation

$$P = P_0 e^{-\alpha x}$$

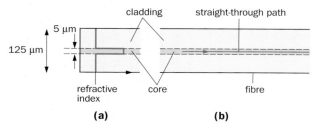

Fig. 24.29 Monomode fibre

where P and P_0 are the output and input powers respectively and α is a constant called the **attenuation coefficient** of the fibre. It is expressed in km^{-1} and is given by

$$\ln\left(\frac{P}{P_0}\right) = -\alpha x$$

As well as loss due to absorption by impurity atoms in the glass, scattering of the radiation at imperfect joints in the fibre also adds to the attenuation.

Attenuated pulses have their size and shape restored by the use of **regenerators** at intervals along the optical cable.

OPTICAL TRANSMITTERS

(a) Light-emitting diode (LED)

An LED is a junction diode made from the semiconducting compound gallium arsenide phosphide; its action was considered earlier (p. 418). Those used as optical fibre transmitters emit infrared radiation at a wavelength of about 850 nm (0.85 μm). Pulse code modulated signals from the coder supply input current to the LED which produces an equivalent stream of infrared pulses for transmission along the fibre. The spectral spread of wavelengths in the output is 30 to 40 nm.

LEDs are a cheap, convenient 'light' source. They are generally used only with multimode fibres because of their low output intensity, and in low bit-rate digital systems (up to 30 Mbits/s or so) where the 'spreading' of output pulses due to dispersion is less of a problem. A lens between the LED and the fibre helps to improve the transfer of 'light' energy between them.

(b) Laser

A laser, named from the first letters of **l**ight **a**mplification by the **s**timulated **e**mission of **r**adiation, produces a very intense beam of light or infrared radiation which is

- **monochromatic**, i.e. consists of one wavelength;
- **coherent**, i.e. all parts are in phase;
- **collimated**, i.e. all parts travel in the same direction.

Those used in optical fibre systems are made from the semiconductor gallium arsenide phosphide. One the size of a grain of sand can produce a continuous power output of around 10 mW. The speed at which a laser can be switched on and off by the digital pulses of input current is much faster than for an LED. Spectral spreading of the radiation emitted is also smaller (1 to 2 nm or less) and so dispersion is not such a problem (since refractive index and speed depend on wavelength). Lasers are therefore more suitable for use with monomode, high bit-rate fibre systems.

OPTICAL RECEIVER

The receiver converts 'light' signals into the equivalent stream of electrical pulses. A reverse-biased photodiode (p. 419) is used to do this both at the end of the system and in regenerators along the cable.

The p-i-n photodiode has a low-doped intrinsic (i) 'depletion layer' between the p- and n-regions, Fig. 24.30. When 'light' photons are absorbed in this i-region, the resulting electrons and holes then move in opposite directions under the applied voltage to form the current through the external circuit of the diode. Reverse bias depletes the i-region completely and produces an electric field high enough to cause rapid motion of the charge carriers. This ensures that the current responds rapidly to changes of 'light' intensity on the photodiode.

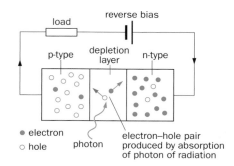

Fig. 24.30 Photodiode

LASERS

(a) Action

The action of a laser can be explained in terms of energy levels.

A material whose atoms are excited emits radiation when electrons in higher energy levels return to lower levels. Normally this occurs randomly, i.e. **spontaneous emission** occurs, Fig. 24.31a, and the radiation is emitted in all directions and is incoherent. The emission of light from ordinary sources is due to this process. However, if a photon of exactly the correct energy approaches an excited atom, an electron in a higher energy level may be induced to fall to a lower level and emit another photon. The remarkable fact is that this photon has the same phase, frequency and direction of travel as the stimulating photon which is itself unaffected. This phenomenon was predicted by Einstein and is called **stimulated emission**; it is illustrated in Fig. 24.31b.

In a laser it is arranged that light emission by stimulated emission exceeds that by spontaneous emission. To achieve this it is necessary to have more electrons in an upper than a lower level. Such a condition, called an 'inverted population', is the reverse of the normal state of affairs but it is essential for light amplification, i.e. for a beam of light to increase in intensity as it passes through a material rather than to decrease as is usually the case.

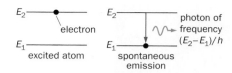

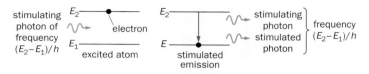

(a) Spontaneous emission

(b) Stimulated emission

Fig. 24.31

One method of creating an inverted population is known as 'optical pumping' and consists of illuminating the laser material with light. Consider two levels of energy E_1 and E_2, where $E_2 > E_1$. If the pumping radiation contains photons of frequency $(E_2 - E_1)/h$, electrons will be raised from level 1 to level 2 by photon absorption. Unfortunately, however, as soon as the electron population in level 2 starts to increase, the pumping radiation induces stimulated emission from level 2 to level 1, since it is of the correct frequency, and so no build-up occurs.

In a three-level system, Fig. 24.32, the pumping radiation of frequency $(E_3 - E_1)/h$ raises electrons from level 1 to level 3, from which they fall by spontaneous emission to level 2. An inverted population can arise between levels 2 and 1 if electrons remain long enough in level 2. The spontaneous emission of a photon due to an electron fall from level 2 to level 1 may subsequently cause the stimulated emission of a photon, which in turn releases more photons from other atoms. The laser action thus occurs between levels 2 and 1 and the pumping radiation has a different frequency from that of the stimulated radiation.

(b) Ruby laser

Many materials can be used in lasers. The ruby rod laser consists of a synthetic crystal of aluminium oxide containing a small amount of chromium as the laser material. It is a type of three-level laser in which 'level' 3 consists of a band of very close energy levels. The pumping radiation, produced by intense flashes of yellow-green light from a flash tube, Fig. 24.33, raises electrons from level 1 (the ground level) into one of the levels of the band. From there they fall spontaneously to the metastable level 2 where they can remain for approximately 1 millisecond, as compared with 10^{-8} second in the energy band. Red laser light is emitted when they are stimulated to fall to level 1 from level 2.

One end of the ruby rod is silvered to act as a complete reflector, while the other is thinly silvered and allows partial transmission. Stimulated light photons are reflected to and fro along the rod producing an intense beam, part of which emerges from the partially silvered end as the useful output of the laser.

(c) Helium–neon laser

This uses a gaseous mixture of helium and neon, and whereas the ruby laser emits short pulses of light, it works continuously and produces a less divergent beam. In one form the gas is in a long quartz tube with an optically flat mirror at each end. Pumping is done by a 28 MHz r.f. generator instead of a flash tube. An electric discharge in the gas pumps the helium atoms to a higher energy level. They then excite the neon atoms to a higher level by collision and produce an inverted population of neon atoms which emit radiation when they are stimulated to fall to a lower level.

(d) Semiconductor laser

A semiconductor laser operates on the same principle as an LED (p. 418) but the p- and n-type materials either side of the p-n junction are more heavily doped, enabling laser action to occur. A forward bias on the p-n junction injects electrons into the n-type material and holes into the p-type material when current flows so that an inverted population results — extra electrons exist in the conduction band. When an electron falls from the conducting band to the valence band a photon is created (of energy equal to the band gap) which can stimulate another electron and hole to combine and so stimulate the emission of another photon; laser light is generated when a sequence of stimulated emissions occurs. The opposite faces of the semiconductor are carefully cut and polished so that light is reflected back and forth within the crystal to encourage laser action.

(e) Uses

Semiconductor lasers are used in optical fibre communication systems, as described above, and in other information-handling systems; for example they read digitally coded music on CDs (Fig. 24.34) and retrieve data from computer disks. They are also used in high-speed computer printers.

Ruby lasers are used for range-finding, welding, cutting, drilling, and micro-circuit fabrication and repair. Helium–neon lasers are used for the precision measurement of length,

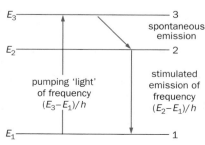

Fig. 24.32 Optical pumping

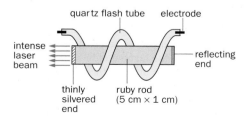

Fig. 24.33 Ruby laser

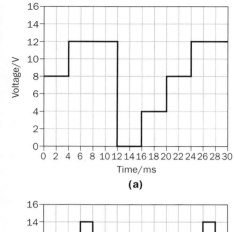

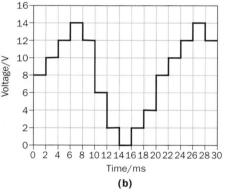

Fig. 24.36

Fig. 24.34   Semiconductor laser beam used in a CD player

surveying, for example in the construction of tunnels, and holography (p. 354).

Lasers are being used increasingly in surgical procedures.

QUESTIONS

1. The audio signal shown in Fig. 24.35 is composed of two sinusoidal components, a fundamental and a harmonic of frequency 400 Hz.

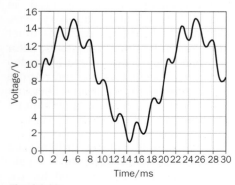

Fig. 24.35

a) Calculate the fundamental frequency of this audio signal.

b) This signal is passed into two separate analogue-to-digital converters (ADC) before being recovered by two matched digital-to-analogue converters (DAC). The outputs from the two ADCs are shown in Figs 24.36a and b.

   i) For the signal in Fig. 24.36a, calculate the sampling frequency and the number of bits for the ADC used to produce this waveform.

   ii) Explain how you arrived at your answer in i).

   iii) Explain which, if any, of the frequency components of the signal in Fig. 24.35 could be recovered from the output shown in Fig. 24.36a.

   iv) For the signal in Fig. 24.36b, calculate the sampling frequency and the number of bits for the ADC used to produce this waveform.

c) The more detailed of the two digitised signals, i.e. that of Fig. 24.36b, does not give a faithful reproduction of the signal of Fig. 24.35. Explain why this is so and suggest minimum values for the sampling frequency and the number of bits of the ADC necessary to recover all the important features in the original signal.

(*UCLES*, Telecommunications, March 1998)

2. **a)** Explain the terms **amplitude modulation** and **bandwidth**.

   b) What is the bandwidth when frequencies in the range 500 Hz to 3500 Hz amplitude-modulate a carrier?

   c) A carrier of frequency 800 kHz is amplitude modulated by frequencies ranging from 1 kHz to 10 kHz. What frequency range does each sideband cover?

3. **a)** Explain the terms **pulse code modulation** and **bit-rate**.

   b) If a TV picture is made up of 350 000 dots (picture elements or pixels) and an 8-bit binary code is needed to describe their many different colours and brightness levels, how many bits per second must be sent when digital transmission is used? (*Note.* 25 complete pictures occur on the screen per second.)

4. In 1887 Heinrich Hertz discovered how to generate electromagnetic waves and demonstrated coded radio communication over a short distance by simply switching a transmitter on and

off. Since that time there has been considerable development in the generation and use of radio waves.

a) Describe two processes by which radio waves can be made to transmit audio information. You should illustrate your answer with suitable diagrams.

b) Describe three different paths by which radio waves can be propagated from a transmitter to a receiver over short and long distances. Include typical frequency and transmission ranges in your answer.

c) Explain why the development of radio wave communications has essentially involved a progression to higher and higher frequencies.

(*UCLES*, Telecommunications, March 1998)

5. a) Sketch the radiation patterns of a horizontal $\lambda/2$ dipole in
  i) the horizontal plane,
  ii) the vertical plane.

b) At what frequency is a dipole of length 1.0 m half a wavelength long?

6. Draw a block diagram for
a) a radio transmitter,
b) a simple radio receiver.

7. Figure 24.37 is a diagram of a very simple radio receiver which can be used for broadcasts from one station only.

a) The tuning circuit selects one station (one frequency). Is the output of energy of the selected signal coming from the tuning circuit greater than the energy of the signal collected by the aerial? Explain your answer briefly.

b) For which of the other three boxes is the energy of the output signal of the box larger than the energy of the input signal to the box? Give the name of the box or boxes for which there is energy increase.

c) How could the tuning circuit be altered so that it could select other stations?

d) Which box or boxes is/are transducers?

e) The frequency of the signal received at the aerial is about 10^6 Hz, and the frequency of the speaker output is about 10^3 Hz. Would it matter if the amplifier could only amplify signals of frequency less than 10^5 Hz and so failed to amplify signals of 10^6 Hz? Explain your answer.

8. a) What is a geosynchronous satellite?
b) Give *two* reasons why microwaves are used for satellite communication.

9. a) Draw a block diagram for an optical fibre communication system.
b) State *six* advantages that optical fibre cables have over copper cables.
c) Explain the terms **pulse spreading** and **attenuation** and state their cause in optical fibres.
d) Distinguish between
  i) multimode and monomode fibres,
  ii) step-index and graded-index fibres.

10. Figure 24.38 shows light entering a straight length of step-index optic fibre.

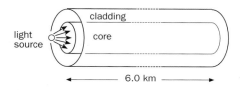

Fig. 24.38

A pulse of light is emitted by switching on the source for 1.0 μs. The light is emitted in many directions as shown. The critical angle at the core/cladding boundary is 59° and the length of the fibre is 6.0 km.

a) Explain what is meant by a **step-index** optic fibre.

b) State what happens to light which is incident on the core/cladding boundary at angles which are
  i) less than 59°,
  ii) greater than 59°.

c) Some of the light in the pulse will travel straight down the centre of the fibre for its entire length. The refractive index of the core is 1.5. Calculate how long this light takes to be transmitted along the full length of the fibre.

d) Light incident on the core/cladding boundary at the critical angle undergoes the greatest number of reflections down the fibre. Calculate how long this light takes to be transmitted.

e) Hence calculate the duration of the pulse as it leaves the fibre.

f) Figure 24.39 is a sketch graph of the variation with time of the power of the light pulse as it enters the fibre.

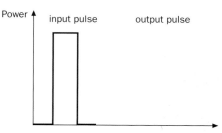

Fig. 24.39

Copy Fig. 24.39 and, at the position indicated, draw a sketch graph of the variation with time of the power of the light pulse as it leaves the fibre.

g) Name the effect illustrated in f).

h) Calculate the maximum frequency at which 1 μs pulses could enter this fibre and be distinguished at the other end.

i) Describe and explain how modern optic fibres have been modified to eliminate the effect in f).

(*UCLES*, Telecommunications, March 1998)

11. The graph in Fig. 24.40 shows the variation of refractive index across the diameter of a step index multimode optical fibre.

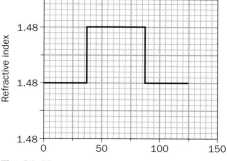

Fig. 24.40

Calculate the critical angle for light incident on the boundary between the core and the cladding of the fibre.

Copy the graph and add to it the variation of refractive index for a step index monomode optical fibre made from materials of refractive index 1.45 and 1.47.

Calculate the time taken for a light pulse to travel 5.00 km along this monomode optical fibre.

(*L*, AS/A PH2, Jan 1998)

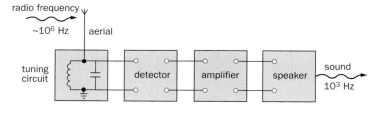

Fig. 24.37

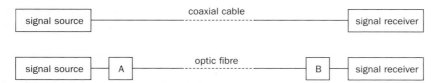

Fig. 24.41

12. In the transmission of an electrical analogue signal along a coaxial copper cable, the signal source can simply be connected to the cable and the signal recovered at the other end. When the same signal is transmitted as an analogue signal by optic fibre, energy transformations must be performed at both ends of the fibre. This is illustrated in Fig. 24.41.

a) Name the two transducers A and B which can perform these energy transformations.

b) The maximum unbroken length of cable or fibre is governed by the minimum value of the signal-to-noise ratio which is acceptable at the signal receiver.

  i) Explain what is meant by **noise**.
  ii) Figure 24.42 shows a signal from the signal source. Copy it, and show the effect of noise on this signal.

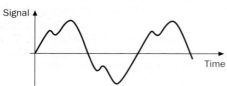

Fig. 24.42

  iii) Explain why the signal-to-noise ratio does not remain constant along the cable or fibre.

c) In the systems shown in Fig. 24.41, the output power of both signal sources is 480 mW. The minimum acceptable signal power at both receivers is 0.12 nW. The efficiency of each of the energy transducers A and B is 10%.

The ratio of two powers P_1 and P_2 is expressed as a number of decibels (dB) according to

$$\text{number of dB} = 10\,\lg\!\left(\frac{P_1}{P_2}\right)$$

  i) For the coaxial cable, calculate the reduction in signal power.
  ii) The coaxial cable has a loss per unit length of 8.0 dB km^{-1}. Calculate the maximum unbroken length of this cable.
  iii) For the optic fibre, calculate the reduction in signal power.

  iv) The optic fibre has a loss per unit length of 0.76 dB km^{-1}. Calculate the maximum unbroken length of this fibre.
    (UCLES, Telecommunications, March 1998)

13. a) Compare the spectral output and radiation characteristics of two types of optical transmitter.

b) Describe the structure and operation of a p-i-n photodiode.

14. a) Two amplitude modulated radio waves of the same frequency, carrying audio signals, are represented by Figs 24.43*a* and *b*. The correct numbers of oscillations are not shown because of the problem of scaling but time intervals of 1 µs and 10 ms are indicated. The vertical scales are the same.

  i) Calculate the radio frequency of the carrier wave.
  ii) Calculate the audio frequency of each of the waves.
  iii) Describe in what respects the audio signals being carried
    1. sound the same,
    2. sound different.
  iv) Calculate the frequencies involved in the transmission of these waves.

b) Outline the steps which would have to be taken if an audio signal, similar to one of those in a), were to be transmitted using an optic fibre system in place of a radio wave.
    (OCR, 9244/3, Nov 1999)

15. Radio stations which broadcast on the Long-Wave (LW) region of the electromagnetic spectrum must use a carrier frequency in the range 140 kHz to 280 kHz. They must limit the *sidebands* to within ±4.5 kHz of the *carrier frequency*.

a) i) Sketch a graph to show the variation with frequency of the power of a typical LW radio transmission with a carrier frequency of 200 kHz.
  ii) On your graph, label the carrier frequency and the sidebands.

b) State the bandwidth of each station in the LW waveband.

c) Calculate the maximum number of radio stations which could transmit on this waveband.
    (OCR, Telecommunications 4838, Nov 1999)

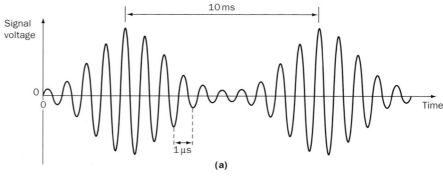

(a)

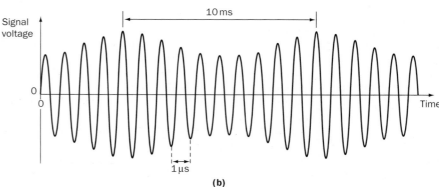

(b)

Fig. 24.43

25
Nuclear physics

- ■ Radioactivity
- ■ Nuclear radiation detectors
- ■ Radioactive decay
- ■ Measuring half-lives
- ■ Absorption of alpha, beta and gamma rays

- ■ Radioisotopes and their uses
- ■ Radiation hazards
- ■ Nuclear reactions
- ■ Mass and energy
- ■ Nuclear energy
- ■ The particles of physics

- ■ Quarks
- ■ The forces of nature
- ■ The exchange nature of forces
- ■ Chaos theory

RADIOACTIVITY

In 1896 the French scientist Becquerel found that uranium compounds emitted radiation which affected a photographic plate wrapped in black paper and, like X-rays, ionized a gas. The search for other such **radioactive** substances was taken up by Marie Curie, who extracted from the ore pitchblende two new radioactive elements which she named polonium and radium.

(a) Alpha, beta and gamma rays

One or more of three types of radiation may be emitted which can be identified by their different penetrating power, ionizing ability and behaviour in a magnetic field.

Alpha (α) rays have a range of a few centimetres in air (at s.t.p.) and are stopped by a thick sheet of paper. They produce intense ionization in a gas and are deflected by a *strong* magnetic field in a direction which suggests they are relatively heavy, positively charged *particles*.

Beta (β⁻) rays are usually more penetrating, having ranges which can be as high as several metres of air at s.t.p., or a few millimetres of aluminium. They cause much less intense ionization than alpha particles but are more easily deviated by a magnetic field, in a direction which indicates they are negatively charged *particles* of small mass. The magnetic deflection

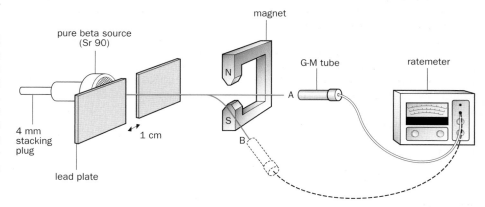

Fig. 25.1 Investigating beta particles

of beta particles may be demonstrated using the apparatus of Fig. 25.1. With the Geiger–Müller (G-M) tube in position A and without the magnet, the count-rate produced by the beam of beta particles is observed on the ratemeter. When the magnet is inserted, the count-rate decreases but rises again when the G-M tube is moved to some position such as B.

Gamma (γ) rays of high energy can penetrate several centimetres of lead. They ionize a gas weakly and are not deflected in a magnetic field. Their behaviour is *not* that of charged particles.

Alpha, beta and gamma rays are termed **nuclear radiation**, since, as we shall see later, they originate in atomic nuclei.

(b) Nature of the 'rays'

The specific charges of alpha and beta particles can be deduced from measurements of their deflections in electric and magnetic fields and give information about their nature.

Alpha particles have a specific charge that suggests they might be helium atoms which have lost two electrons, i.e. helium ions with a double positive charge – a **helium nucleus** $_2^4$He. In 1909 Rutherford and Royds confirmed this by compressing some of the radioactive gas radon in a tube A, Fig. 25.2, whose walls were thin enough to allow the alpha particles emitted by radon to escape into the evacuated space enclosed by the thicker-walled tube B. After a week the mercury level was raised so that gas which had collected in B was forced into C. On passing a current through C the line spectrum of helium was observed. Each alpha particle penetrating A had collected two electrons and changed into a helium atom.

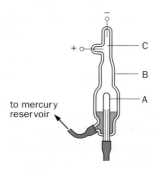

Fig. 25.2 Rutherford–Royds experiment

Specific-charge measurements for β⁻ **particles** show they are **high-speed electrons**. (Allowance has to be made for the increase of mass with speed as predicted by special relativity, see p. 506, since in some cases beta particles are emitted with speeds very close to that of light. If the relativistic mass is not used, the specific charge is found to decrease with increasing speed.)

Gamma rays were the subject of controversy until they were shown to be diffracted by a crystal, thus establishing their wave-like nature. Their wavelengths are those of very short X-rays, and like X-rays they are a form of **electromagnetic radiation** travelling with the speed of light (since among other things they give the photoelectric effect). While diffraction gives the most accurate method of finding gamma ray wavelengths, it is difficult and rarely done. The rays are usually absorbed by a solid-state detector calibrated by rays of known energy.

(c) Energy and speed of emission

The energies of alpha and beta particles are found from measurements of their paths in magnetic fields and the energy of gamma rays is found as explained above.

(*i*) *Alpha particles.* In many cases the alpha particles emitted by a particular nuclide all have the same energy and are said to be monoenergetic. Energies vary from 4 to 10 MeV, corresponding to emission speeds of 5–7% of the speed of light.

(*ii*) *Beta particles.* These exhibit quite a different behaviour. Their energy spectrum is a continuous one in which all energies are present, from quite small values up to a certain maximum, as shown by Fig. 25.3. The maximum energy is characteristic of the nuclide and varies from 0.025 to 3.2 MeV for natural radioactive sources. The highest energy represents a beta particle emission speed of 99% of the speed of light.

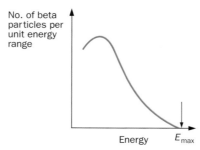

Fig. 25.3 Beta emission spectrum

(*iii*) *Gamma rays.* These fall into several distinct monoenergetic groups, giving a 'line' spectrum. The gamma rays from cobalt-60 (see below) have two different energies of 1.2 and 1.3 MeV.

(d) Sources

Sources used for instructional purposes are usually supplied mounted in a holder with a 4 mm plug. The active material is sealed in metal foil which is protected by a wire gauze cover, Fig. 25.4a. When not in use they are stored in a small lead 'castle', Fig. 25.4b, inside a wooden box.

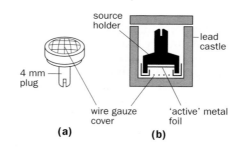

Fig. 25.4

The health hazard is negligible when using the weak closed sources listed shortly, provided they are

- **always lifted with forceps;**
- **held so that the open window is directed away from the body; and**

- **never brought close to the eyes for inspection.**

Approved *closed* sources have low activities (see p. 471) and include the following.

(*i*) Radium-226 for α, β⁻, and γ-rays.

(*ii*) Americium-241 for α-particles only (but it also emits some low-energy γ-rays).

(*iii*) Strontium-90 for β⁻-particles only.

(*iv*) Cobalt-60 for γ-rays. (An aluminium cover disc absorbs the beta particles also emitted.)

(*v*) Some other very weak sources are not completely enclosed but the radioactive material is secured to a support. Such sources are used in school cloud chambers.

Sources (*ii*), (*iii*) and (*iv*) are prepared in nuclear reactors.

NUCLEAR RADIATION DETECTORS

In a nuclear radiation detector *energy* is transferred from the radiation to atoms of the detector and may cause

- ionization of a gas as in an ionization chamber, a Geiger–Müller tube, a cloud or bubble chamber (see below);
- exposure of a photographic emulsion;
- fluorescence of a phosphor as in a scintillation counter; or
- mobile charge carriers in a semiconducting solid-state detector.

The radiation is thus detected by the effects it produces.

(a) Ionization chamber

In its simplest form this comprises two electrodes between which ion-pairs, i.e. electrons and positive ions, can be produced from neutral gas atoms and molecules by ionizing radiation from a source inside or outside the chamber (p. 397). One electrode of the chamber is often a cylindrical can and the other a metal rod along the axis of the cylinder. Under the influence of an electric field between the electrodes, electrons move to the anode and positive gas ions to the cathode to form an ionization current. Some means of detecting the current is necessary. Figure 25.5a shows the basic arrangement required.

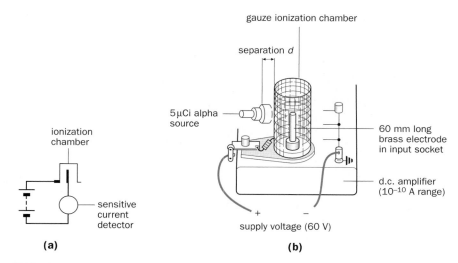

Fig. 25.5   Detecting radiation with an ionization chamber

The current depends on the nature of the radiation and the volume of the chamber. An approved alpha source creates a current of the order of 10^{-10} A in a small chamber; beta particles and particularly gamma rays cause very much smaller currents. A school-type d.c. amplifier (p. 222) can detect the currents due to alpha particles but is insensitive to beta and gamma radiation.

To measure the ionization current the d.c. amplifier is first calibrated (p. 223) then arranged as in Fig. 25.5b. The supply voltage is increased until the output meter reading reaches a maximum value which is recorded. All ions produced are then being collected by the electrodes of the chamber and the ionization current has its saturation value I. This is a measure of the intensity of the radiation from the alpha source. It is calculated from $I = V/R$ where V is the p.d. across the input resistor R (10^{10} Ω) at saturation; V is obtained from the calibration measurement. A very rough estimate of the energy of an alpha particle may also be made (see question 1, p. 489).

Other experiments with an ionization chamber and d.c. amplifier are given on pp. 472 and 474.

(b)   Geiger–Müller (G-M) tube

The G-M tube is a very sensitive type of ionization chamber which can detect single ionizing events. It consists of a cylindrical metal cathode (the wall of the tube) and a coaxial wire anode, containing argon at low pressure, Fig. 25.6a. The very thin mica end-window allows beta and gamma radiation to enter, as well as more energetic alpha particles; gamma rays can also penetrate the wall.

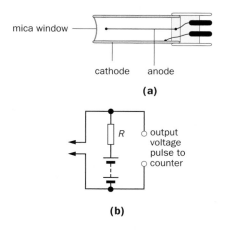

Fig. 25.6   G-M tube and its connection to a counter

A p.d. of about 450 V is maintained between anode and cathode and, since the anode is very thin, an intense electric field is created near it. If an ionizing 'particle' passing through the tube produces an ion-pair from an argon atom, the resulting electron is rapidly accelerated towards the anode and when close to it has sufficient energy itself to produce ion-pairs in encounters. The electrons thus freed produce additional ionization and an avalanche of electrons spreads along the whole length of the wire, which absorbs them to produce a large pulse of anode current. In this way, one electron freed in one ionizing event can lead to the release, in a few tenths of a microsecond, of as many as 10^8 electrons.

During the electron avalanche the comparatively heavy positive ion members of the ion-pairs have been almost stationary round the anode. After the avalanche has occurred they move towards the cathode under the action of the electric field, taking about 100 microseconds to reach it. They now have appreciable energy and would cause the emission of electrons from the cathode by bombardment. A second avalanche would follow, maintaining the discharge and creating confusion with the effect of a later ionizing particle entering the tube. The presence in the tube of a small amount of a **quenching agent** such as bromine tends to prevent this, since the positive ion energy is used to decompose the molecules of the quencher. In a halogen-quenched tube these subsequently recombine and are available for further quenching.

A G-M tube has a **dead time** of about 200 microseconds due to the time taken by the positive ions to travel towards the cathode. Ionizing particles arriving within this period will not give separate pulses, i.e. are not resolved. If radioactive substances emitted particles at regular intervals a maximum of 5000 pulses per second could be detected, but this is not so and in practice the counting rate is less. Almost every beta particle entering a G-M tube is detected. By contrast, the detection efficiency for gamma rays is less than 1%. Gamma photons produce ion-pairs indirectly in the gas of the tube as a result of the secondary electrons they create when absorbed by the tube wall (cathode). The number of such electrons is small since gamma rays interact weakly with matter, and this accounts for the low detection efficiency.

If a resistor R is connected in series with a G-M tube as in Fig. 25.6b, a voltage pulse (of about 1 V) is created which can be applied to an electronic counter such as a scaler or a ratemeter. A **scaler**, Fig. 25.7a, counts the pulses and indicates on a digital display the total received in a certain time. A **ratemeter**, Fig. 25.7b, has a meter marked in 'counts per second' from which the average pulse rate can

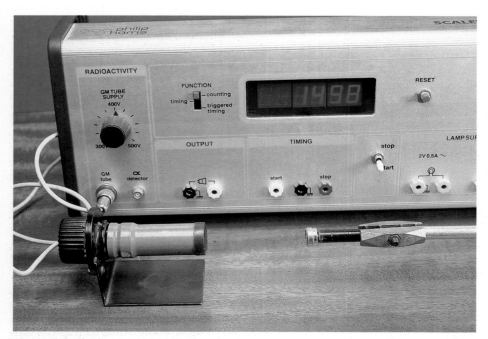

Fig. 25.7a Scaler

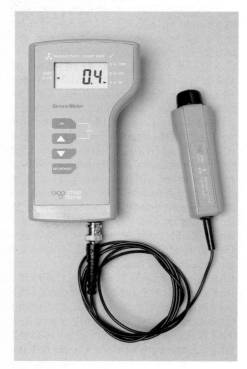

Fig. 25.7b Ratemeter

be read directly. It also has a loud-speaker which gives a 'click' for every pulse. For quantitative work the scaler is more accurate.

The characteristic curve of a G-M tube shows that its response depends on the applied p.d., and allows the selection of an appropriate working p.d. The curve is obtained by placing a

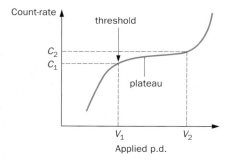

Fig. 25.8 Characteristic of G-M tube

beta source at a short distance from the tube and noting the count-rates on a scaler (or ratemeter) as the p.d. is increased. A typical curve is given in Fig. 25.8. When the p.d. reaches the **threshold** value, the count-rate remains almost constant over a range

called the **plateau** (of 100 V or more). Here a full avalanche is obtained along the entire length of the anode and all particles whatever their energy (and nature) produce the same output pulse. Beyond the plateau the count-rate for the same intensity of radiation increases rapidly with p.d. and a continuous discharge occurs which damages the tube.

Normally a tube is operated at about the middle of the plateau where the count-rate is unaffected by p.d. fluctuations. The plateau usually has a slight upward slope, calculated from

$$\text{percentage slope} = \frac{C_2 - C_1}{V_2 - V_1} \times \frac{100}{C_M} \text{ \% per volt}$$

where $C_M = (C_1 + C_2)/2$. As the condition of a tube deteriorates, the length of the plateau decreases and the slope increases. The slope should be less than 0.15% per volt.

Other experiments with a G-M tube are outlined on pp. 473 and 474.

(c) Cloud chambers

If air containing saturated vapour (e.g. alcohol) is cooled and ionizing radiation passes through it, condensation of the vapour occurs on air ions created by the radiation. The resulting white line of tiny liquid drops shows up as a track in the chamber when suitably illuminated. Note that what we see is the track, not the ionizing radiation.

Figure 25.9 shows a simple **diffusion cloud chamber**. The upper compartment contains air which is at room temperature at the top and at about −78 °C at the bottom due to the layer of 'dry ice' (solid carbon dioxide) in the lower compartment. The felt

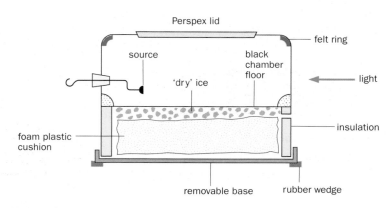

Fig. 25.9 Diffusion cloud chamber

(a) Alpha particles

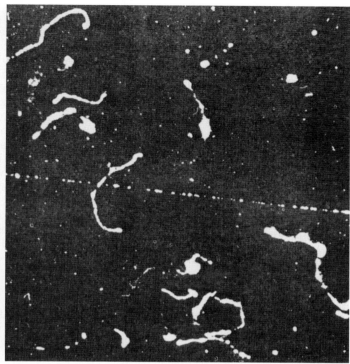

(b) Slow beta particles

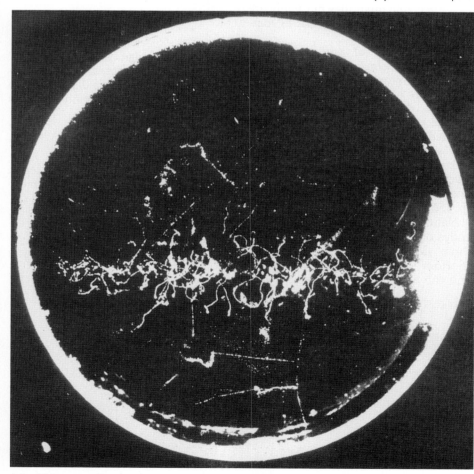

(c) High-energy X-rays

Fig. 25.10 Tracks of ionizing radiation in a cloud chamber

ring at the top of the chamber is soaked with alcohol, which vaporizes in the warm upper region, diffuses downwards and is cooled. About 1 cm from the floor of the chamber, the air contains a layer of saturated alcohol vapour where conditions are suitable for the condensation of droplets on air ions produced along the path of the radiation from the source. Tracks are seen in this sensitive layer which are well defined if an electric field is created by frequently rubbing the Perspex lid of the chamber with a cloth.

Alpha particles give bright, straight tracks like those in Fig. 25.10*a*. Very fast beta particles produce thin, straight tracks while those travelling more slowly give short, thicker, tortuous ones, Fig. 25.10*b*. Gamma rays (like X-rays) cause electrons to be ejected from air molecules and give tracks similar to those in Fig. 25.10*c*, which are due to an intense beam of X-rays.

(d) Bubble chamber

In a bubble chamber charged particles leave a trail of bubbles in superheated liquid hydrogen. (Uncharged particles or radiation such as γ-rays do not leave tracks.) It has now replaced the cloud chamber for studying collisions in nuclear physics, because the higher density of atoms in the liquid gives better defined tracks. A magnetic field is generally applied across the bubble chamber which causes charged particles to move in circular paths; from the curvature of these paths the sign of the charge and the momentum of the particle can be deduced.

RADIOACTIVE DECAY

(a) Disintegration theory

The changes accompanying the emission of radiation from radioactive substances are unlike ordinary chemical changes in certain fundamental respects. They are spontaneous, they cannot be controlled and they are unaffected by chemical combination and physical conditions such as temperature and pressure. Energy considerations too suggest that they are different. The energy released by a radioactive substance emitting alpha particles is several million electronvolts per atom, compared with a few electronvolts per atom in any chemical change.

We believe that radioactivity involves the nucleus of an atom — not its extra-nuclear electrons as in chemical changes — and is an attempt by an *unstable* nucleus to become more stable. The emission of an alpha or beta particle from the nucleus of a radioactive atom produces the nucleus of a different atom which may itself be unstable. The disintegration process proceeds at a definite rate through a certain number of stages until a stable end-product is formed. When first advanced by Rutherford and others the theory conflicted with existing ideas about the permanency of atoms but it is a view that accounts satisfactorily for the known facts.

An alpha particle is a helium nucleus consisting of 2 protons and 2 neutrons and when an atom decays by alpha emission, its nucleon number A decreases by 4 and its proton number Z by 2. It becomes the atom of an element two places lower in the periodic table. For example, when radium of nucleon number 226 and proton number 88 emits an alpha particle, it decays to radon of nucleon number 222 and proton number 86. The change may be written as follows.

$$^{226}_{88}\text{Ra} \rightarrow {}^{222}_{86}\text{Rn} + {}^{4}_{2}\text{He}$$

The values of A and Z must balance on both sides of the equation since nucleons and charge are conserved.

Radon in turn disintegrates and, after seven more disintegrations, a stable isotope of lead is formed.

When β^{-} decay occurs a neutron changes into a proton and an electron. The proton remains in the nucleus and the electron is emitted as a beta particle. The new nucleus has the same nucleon number, but its proton number increases by one since it has one more proton. It becomes the atom of an element one place higher in the periodic table. Radioactive carbon, called carbon-14, decays by beta emission to nitrogen:

$$^{14}_{6}\text{C} \rightarrow {}^{14}_{7}\text{N} + {}^{0}_{-1}\text{e} + \bar{\nu}_{e}$$

Note that an uncharged, massless *antineutrino* ($\bar{\nu}_e$) is also emitted in β^{-} decay (see p. 484).

β^{-} emitters have a higher proportion of neutrons; few occur naturally in appreciable quantities but many are obtained artificially by irradiating matter with neutrons.

The emission of gamma rays is explained by considering that nuclei (as well as atoms) have energy levels and if an alpha or beta particle is emitted, the nucleus is left in an excited state. Gamma ray photons are emitted when the nucleus returns to the ground state (Fig. 25.11). The existence of different nuclear energy levels accounts for the 'line-type' energy spectrum of gamma rays (p. 466).

Naturally occurring radioactive nuclides of high proton number fall into three decay series, known as the **thorium** series, the **uranium** series and the **actinium** series. Radium is a member of the uranium series and at any time all the decay nuclides will be present – which explains why a radium source emits alpha particles and gamma rays of several energies, as well as beta particles.

(b) Decay law

Radioactive decay is a completely haphazard or random process in which nuclei disintegrate quite independently. Since there is always a very large number of active nuclei in a given amount of radioactive material, we can apply the methods of statistics to the process and obtain an expression for the certain fraction of the nuclei originally present that will have decayed on average in a given time interval.

We assume that the **rate of disintegration of a given nuclide at any time is directly proportional to the number of nuclei N of the nuclide** present at that time; that is, in calculus notation,

$$-\frac{dN}{dt} \propto N$$

The negative sign indicates that N decreases as t increases. The **radioactive decay constant** λ is defined as the constant of proportionality in this expression, giving

$$-\frac{dN}{dt} = \lambda N \qquad (1)$$

If there are N_0 undecayed nuclei at some time $t = 0$ and a smaller number N at a later time t, then integrating (1),

$$\int_{N_0}^{N} \frac{dN}{N} = -\lambda \int_{0}^{t} dt$$

$$\therefore \quad [\ln N]_{N_0}^{N} = -\lambda t \qquad (\ln = \log_e)$$

$$\ln N - \ln N_0 = \ln(N/N_0) = -\lambda t \quad (2)$$

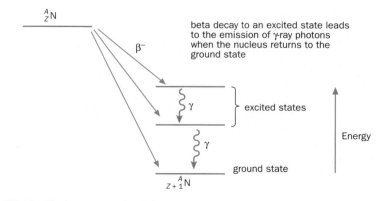

Fig. 25.11   Nuclear energy levels

Therefore

$$N = N_0 e^{-\lambda t} \qquad (3)$$

This is the decay law and states that a **radioactive substance decays exponentially with time** — a fact confirmed by experiment. The law is a statistical one; it does not tell us when a particular nucleus will decay but only that after a certain time a certain fraction will have decayed. On the microscopic scale the process is purely random, as is evident from the variations that occur when particles from a source are counted. On the macroscopic scale, however, where large numbers of particles are concerned, a definite law is followed.

The rate of decay $-\,dN/dt$ is called the **activity** A of a source; it is one **becquerel** (1 Bq) if the average rate of decay is one disintegration per second, i.e. one atom decays in one second. This is the unit used to measure the amount of radioactivity in substances such as milk and water. Previously the **curie** (Ci) was the unit of activity (1 Ci = 3.7×10^{10} Bq).

Approved sources used in school laboratories have activities of about 2×10^5 Bq which is very small compared with some medical sources.

From equation (1) we see that

$$\lambda = -\frac{dN}{N\,dt}$$

and so the decay constant is the fraction of the total number of nuclei present that decays in unit time, provided the unit of time is small.

(c) Half-life

An alternative but more convenient term to the decay constant λ is the **half-life** $t_{\frac{1}{2}}$.

The half-life is the time for the number of active nuclei present in a source at a given time to fall to half its value.

Whereas it is often difficult to know when a substance has lost practically all its activity, it is less difficult to find out how long it takes for the activity to fall to half the value it has at some instant.

Half-lives vary from millionths of a second to thousands of millions of years. Radium-226 has a half-life of 1622 years, therefore starting with 1 g of pure radium, $\frac{1}{2}$g remains as radium after 1622 years, $\frac{1}{4}$g after 3244 years and so on. An exponential decay curve, like that given on p. 221 for the discharge of a capacitor through a high resistor, is shown in Fig. 25.12 and illustrates the idea of half-lives.

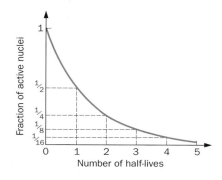

Fig. 25.12 Radioactive decay curve

The relationship between λ and $t_{\frac{1}{2}}$ can be derived from equation (3) for the decay law.

$$N = N_0 e^{-\lambda t}$$

$$\therefore \quad \frac{N_0}{N} = e^{\lambda t}$$

When $N = N_0/2$, $t = t_{\frac{1}{2}}$,

$$\therefore \quad 2 = e^{\lambda t_{\frac{1}{2}}}$$

Taking logs to base e,

$$\ln 2 = \lambda t_{\frac{1}{2}}$$

Therefore

$$t_{\frac{1}{2}} = \frac{0.693}{\lambda}$$

The half-life $t_{\frac{1}{2}}$ (and λ) is characteristic for each radionuclide and is an important means of identification.

In general it is found that as half-life decreases, the energy of an emitted α-particle increases. This would be expected if the α-particle maintained the same energy inside and outside the nucleus; higher energy α-particles would escape more easily, resulting in shorter half-lives. Quantum mechanics explains the escape of the α-particle with no energy loss in terms of 'tunnelling' through the surface of the nucleus.

(d) Nuclear stability

While the chemical properties of an atom are governed entirely by the number of protons in the nucleus (i.e. the proton number Z), the stability of an atom appears to depend on both the number of protons and the number of neutrons. In Fig. 25.13 the number of neutrons $(A - Z)$ has been plotted against the number of protons for all known nuclides, stable and unstable, natural and artificial. A continuous line has been drawn approximately through the stable nuclides (only a few are labelled) and the shading on either side of this line shows the region of unstable nuclides.

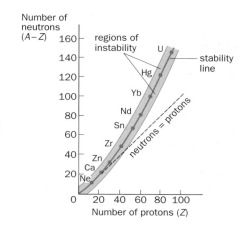

Fig. 25.13 Line of nuclear stability

For *stable* nuclides the following points emerge.

(i) The lightest nuclides have almost equal numbers of protons and neutrons.

(ii) The heavier nuclides require more neutrons than protons, the heaviest having about 50% more.

(iii) Most nuclides have both an even number of protons and an even number of neutrons. The implication is that two protons and two neutrons, i.e. an alpha particle, form a particularly stable combination and in this connection it is worth noting that oxygen ($^{16}_{8}O$), silicon ($^{28}_{14}Si$) and iron ($^{56}_{28}Fe$) together account for over three-quarters of the earth's crust.

For *unstable* nuclides the following points can be made.

(i) Distintegrations tend to produce new nuclides nearer the 'stability' line and continue until a stable nuclide is formed.

(*ii*) A nuclide above the line decays so as to give an increase of proton number, i.e. by β⁻ emission (in which a neutron changes to a proton and an electron). Its neutron-to-proton ratio is thereby decreased.

(*iii*) A nuclide below the line disintegrates in such a way that its proton number decreases and its neutron-to-proton ratio increases. In heavy nuclides this can occur by alpha emission.

(e) Radioactive decay analogue using dice

Radioactive atoms are represented by small cubes having one face marked in some way. If 100 of these are placed in a large can or jar and then thrown on to the bench so that they are in a single layer, the cubes with the marked face uppermost are considered to have 'decayed' and are removed and counted. The remaining 'undecayed' cubes are returned to the can, thrown again, the 'decayed' cubes removed and counted as before. The process is repeated at least 15 times until only a few cubes have not 'decayed'.

The whole procedure is repeated *another four times*, always starting with 100 cubes and making the same number of throws, so that the effect is the same as if 500 cubes had been thrown initially. The results can be recorded as shown in Table 25.1, the total number N of surviving cubes being obtained for each throw t.

A graph of N against t is plotted and a smooth curve drawn through the points. An estimate of the half-life for cube decay, i.e. the number of throws necessary for half the original number of cubes to decay, can be made.

Assuming the decay law is exponential we have

$$N = N_0 e^{-\lambda t}$$

So

$$\ln N = \ln N_0 - \lambda t \quad (\ln = \log_e)$$

But $\ln N = 2.303 \log_{10} N$,

$$\therefore \quad \log_{10} N = \log_{10} N_0 - \frac{\lambda}{2.303} \cdot t$$

The graph of $\log_{10} N$ against t should be a straight line.

Table 25.1

Number of throw	Number of 'decayed' cubes						Number of 'surviving' cubes					
t	(1)	(2)	(3)	(4)	(5)	Total	(1)	(2)	(3)	(4)	(5)	Total N
0	0	0	0	0	0	0	100	100	100	100	100	500
1												
2												

MEASURING HALF-LIVES

The next two experiments can be performed in a school laboratory.

(a) Half-life of radon-220 ($^{220}_{86}Rn$)

This is an alpha-emitting radioactive gas, usually called *thoron*, which collects above its solid, parent nuclides in an enclosed space and is readily separated from them. If a polythene bottle containing thorium hydroxide is connected to an ionization chamber mounted on a d.c. amplifier, Fig. 25.14, a small quantity of thoron enters the chamber when the bottle is squeezed.

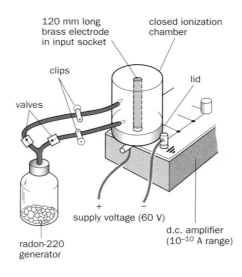

120 mm long brass electrode in input socket

closed ionization chamber

clips

lid

valves

+ − supply voltage (60 V)

d.c. amplifier (10⁻¹⁰ A range)

radon-220 generator

Fig. 25.14 Measuring the half-life of radon-220

As the thoron decays the ionization current decreases and is always a measure of the number of alpha particles present (i.e. the quantity of thoron remaining) so long as the p.d. across the chamber produces the saturation value of the current.

If the time is determined for the reading of the output meter on the d.c. amplifier to fall to a half (or better still, a quarter) of some previous value, the half-life can be found. Alternatively the meter may be read every ten seconds and the half-life found from a graph of meter reading against time. A continuous plot of the decrease in ionization current with time could be recorded using a datalogger and computer.

The chief constituent of the thoron source is thorium-232 and part of its decay series is shown in Fig. 25.15.

The decay products following radon-220 do not affect the result appreciably since their half-lives are so very different from that of radon-220.

If some thoron is puffed into a closed diffusion cloud chamber, two tracks can be observed which are due to the two alpha particles emitted almost simultaneously (why?) when radon-220 decays to polonium-216 and then to lead-212. V-shaped tracks are obtained.

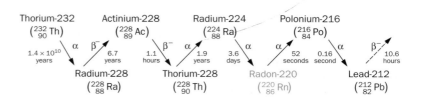

Thorium-232 ($^{232}_{90}Th$) Actinium-228 ($^{228}_{89}Ac$) Radium-224 ($^{224}_{88}Ra$) Polonium-216 ($^{216}_{84}Po$)

1.4 × 10¹⁰ years α β⁻ 6.7 years 1.1 hours α 1.9 years 3.6 days α 52 seconds 0.16 second α β⁻ 10.6 hours

Radium-228 ($^{228}_{88}Ra$) Thorium-228 ($^{228}_{90}Th$) Radon-220 ($^{220}_{86}Rn$) Lead-212 ($^{212}_{82}Pb$)

Fig. 25.15 Decay series of thorium-232

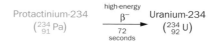

Uranium-238 $\xrightarrow[\substack{4.5 \times 10^9 \\ \text{years}}]{\alpha}$ Thorium-234 $\xrightarrow[\substack{24 \\ \text{days}}]{\substack{\text{low-energy} \\ \beta^-}}$ Protactinium-234 $\xrightarrow[\substack{72 \\ \text{seconds}}]{\substack{\text{high-energy} \\ \beta^-}}$ Uranium-234
$(^{238}_{92}U)$ $(^{234}_{90}Th)$ $(^{234}_{91}Pa)$ $(^{234}_{92}U)$

Fig. 25.16 Decay series of uranium-238

(b) Half-life of protactinium-234 ($^{234}_{91}$Pa)

This is a beta-emitting decay product in the decay of uranium-238; part of the series is given in Fig. 25.16.

The protactinium-234 is extracted almost completely by an organic solvent, such as amyl acetate, from an acidified aqueous solution of uranyl nitrate (proportions — 1 g uranyl nitrate, 3 cm³ water, 7 cm³ concentrated hydrochloric acid). Only the high-energy beta particles from the decay of protactinium-234 are detected by a G-M tube.

Equal volumes of the organic and aqueous solutions are contained in a thin-walled, stoppered, polythene bottle so that each layer is as deep as the width of the G-M tube window. The bottle is well shaken and then arranged as in Fig. 25.17 with the G-M tube opposite the top half of the bottle. As soon as the layers have separated, the scaler or ratemeter connected to the G-M tube is started and counts taken at ten-second intervals without stopping the counter. After allowing for the background count (p. 477), a graph of count-rate against time can be plotted and the half-life found.

The thorium from which the protactinium-234 has been extracted is left in the aqueous solution and immediately starts to generate the protactinium again as it decays. A second experiment may be carried out with the G-M tube opposite the lower half of the bottle to observe the growth of $^{234}_{91}$Pa.

ABSORPTION OF ALPHA, BETA AND GAMMA RAYS

(a) Absorption processes

In their passage through matter alpha and beta particles are 'absorbed' by losing kinetic energy in ionizing encounters with atoms of the absorbing medium, i.e. an electron is 'knocked out' of an atom to form an ion-pair. After travelling a certain distance, called the **range**, they have insufficient energy to produce any more ion-pairs and are then considered to have been absorbed. An alpha particle can also have an encounter with a nucleus (p. 400). Beta particles, because of their small mass, are readily 'back-scattered' by atoms of the medium to emerge from the incident surface.

A measure of the intensity of the ionization produced when a charged particle passes through a gas is given by the **specific ionization**. This is defined as the number of ion-pairs formed per centimetre of path. It increases with the size of the charge on the particle and decreases as the speed of the particle increases. A fast-moving particle spends less time near an atom of the gas through which it is travelling and so there is less chance of an ion-pair being formed. An alpha particle may produce 10^5 ion-pairs per centimetre in air at atmospheric pressure while a beta particle of similar energy produces about 10^3 on account of its higher speed (on average about ten times greater) and smaller charge.

It should, however, be noted that for alpha and beta particles of the same energy the *total* number of ion-pairs formed would be of the same order since a beta particle travels about one hundred times farther in air than an alpha particle. For each ion-pair produced in the track of any type of charged particle in the air the *average* energy loss is 34 eV.

Gamma rays are usually most strongly absorbed by elements of high atomic number such as lead, and the absorption process is complex and differs from that occurring with charged particles. Whereas the alpha or beta particle gradually loses kinetic energy by a series of ionizing encounters with electrons belonging to atoms of the absorber, a gamma ray photon may interact with either an electron or, if it has enough energy, with a nucleus in several ways. The energy given up by the photon produces one or more high-speed 'secondary' electrons and it is these that are responsible for the ionization created in a gas by gamma rays. They enable gamma rays to be detected by a G-M tube. The specific ionization of gamma rays therefore depends on the energy of the secondary electrons.

(b) Range of alpha particles in air

An ionization chamber with a d.c. amplifier can be used with the source (e.g. americium-241) supported just outside the wire gauze ionization chamber which is connected to the d.c. amplifier, Fig. 25.18a, p. 474. The ionization current (saturation) in the chamber is large when the source is close to the gauze and decreases as d increases by, say, 5 mm steps. The corresponding readings of the d.c. amplifier output meter are noted and from a graph like that in Fig. 25.18b the range in air can be estimated.

(c) Range of beta particles in aluminium

Beta particles have a continuous energy spectrum and the number emerging from an absorber falls off gradually as the thickness of the absorber increases; this behaviour contrasts with the fairly sharp cut-off given by alpha particles. For practical purposes the range of beta particles is defined as the thickness of aluminium

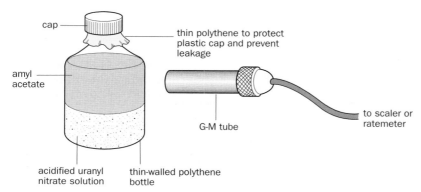

cap

thin polythene to protect plastic cap and prevent leakage

amyl acetate

acidified uranyl nitrate solution

thin-walled polythene bottle

G-M tube

to scaler or ratemeter

Fig. 25.17 Measuring the half-life of protactinium-234

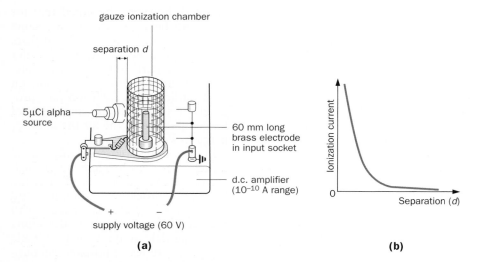

(a)

(b)

Fig. 25.18 Investigating the range of α-particles in air

beyond which very few particles can be detected.

The range may be determined by inserting an increasing number of aluminium sheets between a pure beta source (e.g. strontium-90) and a G-M tube connected to a scaler or ratemeter, Fig. 25.19a. Count-rates are measured and an absorption curve is drawn.

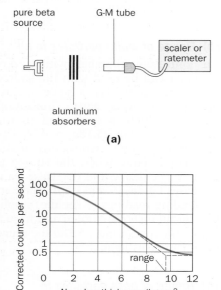

(a)

(b)

Fig. 25.19 Investigating the range of β-particles in aluminium

A typical curve is given in Fig. 25.19b, in which the wide range of count-rates is accommodated by plotting the log of the count-rate. This is most conveniently done by plotting

the count-rate (after subtracting the background count-rate) directly on the log scale of semi-log graph paper. The absorber thickness can be expressed either in millimetres of aluminium or, as is more common, in terms of the **surface density** of the absorber. This is the mass per unit area and equals the product of the density and the thickness of the absorber. For example, a sheet of aluminium 2.0 mm (2.0×10^{-3} m) thick and of density 2.7×10^3 kg m^{-3} has a surface density of 5.4 kg m^{-2}. The range in kg m^{-2} is found by extrapolation from the absorption curve, as shown.

(d) Inverse square law for gamma rays

Gamma rays are highly penetrating on account of their small interaction with matter. In air they suffer very little absorption or scattering and, like other forms of electromagnetic radiation, their intensity falls off with distance according to an inverse square law. This states that the intensity of radiation I is inversely proportional to the square of the distance d from a *point* source. That is, $I = k/d^2$ where k is a constant.

The law may be investigated by placing a G-M tube at various distances from a pure gamma source (e.g. cobalt-60) and measuring the corresponding count-rates, Fig. 25.20a. If, after correction for background, C is the count-rate, then C is directly proportional to I and we can write $C = k_1/d^2$, where k_1 is another constant. A graph of C against $1/d^2$

should be a straight line through the origin but small errors occur in the measurement of d and become important as d decreases. Non-linearity of the graph results. The errors can be eliminated by an alternative procedure.

Let $d = D + x$, where D = distance measured from the source to any point on the G-M tube and x = an unknown correction term which gives the true distance d. The law may then be written

$$C = k_1/(D + x)^2$$
$$\therefore \quad D + x = (k_1/C)^{\frac{1}{2}}$$

and so

$$D = (k_1/C)^{\frac{1}{2}} - x$$

A graph of D against $1/C^{\frac{1}{2}}$ should be a straight line of slope $k_1^{\frac{1}{2}}$ and intercept $-x$ on the D-axis, as shown in Fig. 25.20b.

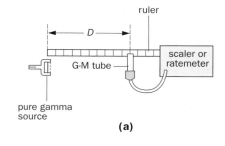

(a)

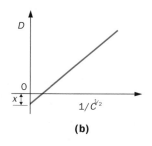

(b)

Fig. 25.20 Investigating the inverse square law for γ-rays

In practice, departure from the law arises because of (*i*) the finite size of the source, and (*ii*) counting losses due to the G-M tube having a 'dead-time' when the source is close to the tube and the count-rate is high.

(e) Half-thickness of lead for gamma rays

The 'half-thickness', denoted by $x_{\frac{1}{2}}$, is a convenient term for dealing with the absorption of gamma rays. It is defined as the thickness of absorber, usually lead, that reduces the intensity of the gamma radiation to half its incident value.

The half-thickness can be determined by inserting sheets of lead between a gamma source (e.g. cobalt-60) and a G-M tube and counter in a similar way to that described for beta absorption.

The count-rate C, corrected for background, is directly proportional to the intensity of the radiation, and it is again convenient to plot C directly on semi-log graph paper. The thickness of lead is usually expressed in the surface density unit of $kg\,m^{-2}$. An absorption curve for a parallel beam of monoenergetic gamma rays is shown in Fig. 25.21; by considering two (or more) half-thicknesses, i.e. the thickness required to half the count-rate twice, a more accurate result is obtained for $x_{\frac{1}{2}}$.

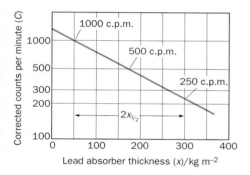

Fig. 25.21 Absorption curve for monoenergetic γ-rays

The graph of log C against absorber thickness is seen to be a *straight line*. This implies that the intensity of the gamma radiation decreases *exponentially* with increasing absorber thickness. If I_0 is the intensity of the gamma rays from the source in the absence of absorbing material and I is the intensity after passing through a thickness x of absorber, then

$$I = I_0 e^{-\mu x}$$

where μ is a constant, called the **linear absorption coefficient**. Taking count-rate as a measure of intensity,

$$C = C_0 e^{-\mu x}$$
$$\therefore \ \ln C = \ln C_0 - \mu x$$

and

$$\log C = \log C_0 - \frac{\mu}{2.303} \cdot x$$

The graph of log C against x should therefore be a straight line if the absorption is exponential. This is only true for a parallel beam of monochromatic (i.e. of one wavelength and energy) gamma rays. The value of μ, which is an alternative term to $x_{\frac{1}{2}}$, can be found from the slope of the line and is smaller for high-energy (more penetrating) rays than for low-energy rays.

The linear absorption coefficient μ can be shown to be equal to $(\ln 2)/x_{\frac{1}{2}}$.

When $x = x_{\frac{1}{2}}$, the intensity I has fallen to $I_0/2$, and so

$$\tfrac{1}{2} = e^{-\mu x_{\frac{1}{2}}}$$

Taking natural logarithms:

$$\ln \tfrac{1}{2} = -\mu x_{\frac{1}{2}}$$

or $\qquad \ln 2 = \mu x_{\frac{1}{2}}$

and so $\qquad \mu = (\ln 2)/x_{\frac{1}{2}}$

RADIOISOTOPES AND THEIR USES

(a) Artificial radioactivity

The first artificial radioactive substance was produced in 1934 by Irène Joliot-Curie (daughter of the discoverer of radium) and her husband. They bombarded aluminium with alpha particles and, as a result of a nuclear reaction (p. 477), obtained an unstable isotope of phosphorus:

$$^{27}_{13}Al + ^{4}_{2}He \rightarrow ^{30}_{15}P + ^{1}_{0}n \qquad [^{27}_{13}Al(\alpha, n)^{30}_{15}P]$$

The expression in brackets is the symbolic way of writing a nuclear reaction, i.e.

[initial nuclide (incoming particles,
      outgoing particles) final nuclide]

Since then artificial radioisotopes (i.e. radioactive isotopes) of every element have been produced and today at least 1500 are known. They are made by bombarding a stable element with neutrons in a nuclear reactor (p. 481) or with charged particles in a particle accelerator. Their use in industry, research and medicine is growing.

(b) Some uses of radioisotopes

Some applications use the fact that the extent to which radiation is absorbed when passing through matter depends on the material's thickness and density. For example, in the manufacture of paper the thickness can be checked by having a beta source below the paper and a G-M tube and counter above it. A thickness gauge of this type may be adapted for automatic control of the manufacturing process. Level indicators also depend on absorption and are used to check the filling of toothpaste tubes and packets of detergents.

Leaks can be detected in underground pipe-lines carrying water, oil, etc., by adding a little radioactive solution to the liquid being pumped. Temporary activity gathers in the soil around the leak which can be detected from the ground above. This technique is now used widely when studying river pollution, sand-wave movement in river estuaries and in the accurate measurement of fluid flow. In research into wear in machinery, a small amount of radioactive iron is introduced into the bearings and the rate of wear found from the resulting radioactivity of the lubricating oil.

The use of radioactive 'tracers' in medicine, agriculture and biological research depends on the fact that the radioisotope of an element can take part in the same processes as its nonactive isotope since it is chemically identical (and only slightly different physically). Gamma-ray scans of the body, such as the brain scan in Fig. 25.22, p. 476, map the concentration of a radioisotope given earlier to the patient. Information on the activity of different parts of the brain can be revealed, and tumours can be detected. The use of radioactive phosphorus as a tracer in agriculture has provided information about the best type of phosphate fertilizer to supply to a particular crop and soil.

Gamma rays from high-activity cobalt-60 sources have many applications. In radiotherapy they are replacing X-rays from expensive X-ray machines in the treatment of cancer. The rapidly growing cells of the diseased tissue that cause cancer are more affected by the radiation than are healthy cells, but there is a danger

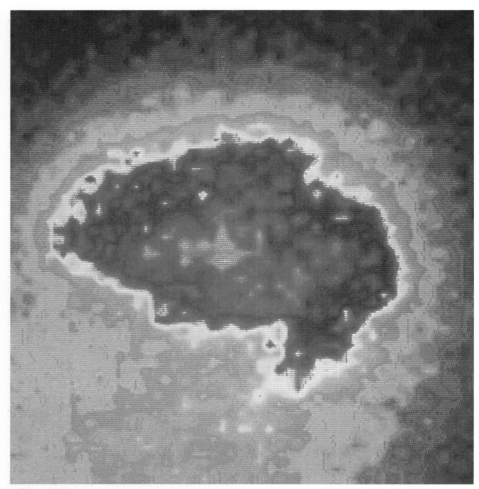

Fig. 25.22 Scan of a normal brain

of damage to healthy cells. In a recent development 'monoclonal antibodies' are used, which find and stick to cancer cells in the body. Minute radioactive particles are attached to the antibodies before they are given to the patient and the radioactivity reveals the cancer cells. Further tiny doses of radioactivity fixed to the antibodies can then destroy the harmful cancer cells with minimum damage to the surrounding healthy cells.

Medical instruments and bandages are sterilized after packing by brief exposure to gamma rays. Food may be similarly treated and meat made to stay fresh for fifteen days instead of three. This is perfectly safe since no radioactivity is produced in the material irradiated by gamma rays.

A smoke detector contains a tiny radioactive source which continually checks the air for ionized particles of smoke from a fire. An increase sets off an alarm.

(c) Carbon-14 dating

A natural radioisotope which is interesting from an archaeological point of view is carbon-14, formed in a nuclear reaction when neutrons ejected from nuclei in the atmosphere by cosmic rays collide with atmospheric nitrogen:

$$^{14}_{7}N + ^{1}_{0}n \rightarrow ^{14}_{6}C + ^{1}_{1}H \qquad [^{14}_{7}N(n, p)^{14}_{6}C]$$

Subsequently carbon-14 forms radioactive carbon dioxide which is taken up by plants and trees for the manufacture of carbohydrates by photosynthesis. The normal activity of living carbonaceous material is 15.3 counts per minute per gram of carbon, but when the organism dies (or when part of it ceases to have any interaction with the atmosphere, as in the heartwood of trees) then no fresh carbon is taken in and carbon-14 starts to decay by beta emission with a half-life of 5.70×10^3 years. By measuring the residual activity, the age of any

ancient carbon-containing material such as wood, linen or charcoal may be estimated within the range 1000 to 50 000 years. This has been done with the Dead Sea Scrolls (about 2000 years old) and charcoal from Stonehenge (about 4000 years old).

The ages of radioactive rocks have been similarly estimated.

(d) Transuranic elements

Uranium has the highest atomic number (92) of any naturally occurring element, but traces of elements with higher atomic numbers have been made artificially and, like other artificial isotopes made from stable atoms in particle accelerators or nuclear reactors, they are radioactive; they are called the **transuranic** elements.

RADIATION HAZARDS

The radiation hazards to human beings arise from

- exposure of the body to external radiation; and
- ingestion or inhalation of radioactive matter.

The effect of radiation depends on the nature of the radiation, the part of the body irradiated and the dose received. The hazard from alpha particles is slight (unless the source enters the body) since they cannot penetrate the outer layers of skin. Beta particles are more penetrating in general, but most of their energy is absorbed by surface tissues and adequate protection is given by a sheet of Perspex or aluminium a few millimetres thick. Gamma rays present the greatest external radiation hazard since they penetrate deeply into the body; substantial lead or concrete shielding may be required.

Radiation can cause immediate damage to tissue and, according to the dose, is accompanied by radiation burns (i.e. redness of the skin followed by blistering and sores which are slow to heal), radiation sickness, loss of hair and, in extremely severe cases, by death. Damage to body cells is due to the creation of ions which upset or destroy them. The most susceptible parts are the reproductive organs, blood-forming organs such as the liver, and to a smaller extent the eyes.

The hands, forearms, feet and ankles are less vulnerable. Delayed effects such as cancer, leukaemia and eye cataracts may appear many years later. Hereditary defects may also occur in succeeding generations due to genetic damage.

The **absorbed dose** D is the energy absorbed (in J kg^{-1}) by unit mass of the irradiated material. The unit of D is the **gray** (Gy) where 1 Gy = 1 J kg^{-1}. Equal doses of different ionizing radiations provide the same amount of energy in a given absorber but they do not have the same biological effect on the human body. To allow for this the **dose equivalent** H is used as a measure of the effect that a certain dose of a particular kind of ionizing radiation has on a person. It takes into account the *type* of radiation as well as the *amount of energy* absorbed, and is obtained by multiplying D by a number called the **relative biological effectiveness** (r.b.e.). The unit of H is the **sievert** (Sv).

$$\begin{array}{ccc} \text{dose} \\ \text{equivalent} \end{array} = \begin{array}{c} \text{absorbed} \\ \text{dose} \end{array} \times \text{r.b.e.}$$
$$(H \text{ in sievert}) \qquad (D \text{ in gray})$$

For beta particles, X- and gamma rays, the r.b.e. is about 1; for alpha particles, protons and fast neutrons it is about 20.

We all unavoidably absorb background radiation due to cosmic rays, radioactive minerals, radon in the atmosphere, potassium-40 in the body, X-rays from television and computer screens. A year's dose from this natural background radiation is about 0.0015 Sv (1.5 mSv); the dose from a chest X-ray is roughly one-fifth of this. The dose rate from one of the weak sources used for experimental work in school physics is very small. In industry and research strong sources have to be handled by a 'master–slave' manipulator and protection obtained for the operator from concrete and lead walls. The dose equivalent for radiation workers must not exceed 0.05 Sv (50 mSv) in a year. A dose of 5 Sv to every part of the body would kill at least 50% of those receiving it in two to three months.

NUCLEAR REACTIONS

Much information concerning the nucleus has been obtained by studying the effects of bombarding atomic nuclei with fast-moving particles. Three historic nuclear reactions will be considered.

(b) Bombardment of nitrogen by alpha particles

In 1919 Rutherford bombarded gases with alpha particles. His apparatus is shown in Fig. 25.23. One end of the metal cylinder was closed by thin silver foil capable of stopping most of the alpha particles from a radioactive source near the other end. Any that penetrated fell on a fluorescent screen and their scintillations were observed through a microscope. Various gases were admitted: with nitrogen, scintillations were obtained which indicated that more penetrating particles were produced. Further experiments involving the deflection of these longer-range particles in a magnetic field showed they were protons of high energy.

Two explanations seem plausible. Either an alpha particle 'chips' a proton off a nitrogen nucleus, in which case the alpha survives the encounter; or the alpha particle enters the nitrogen nucleus and the latter immediately ejects a proton. In either case the proton is energetic enough to cause a scintillation on the screen. Since only one alpha in about a million caused a disintegration the chance of identifying the residual nucleus seemed small.

In 1925, however, about 20 000 photographs of the tracks of alpha particles passing through nitrogen in a cloud chamber were taken. In eight of these, a forked track like that of Fig. 25.24 was obtained. The alpha source is at the bottom and the collision occurs near the top. The short thick track at one o'clock is due to the residual nucleus and the long thin one at eight o'clock is that of a proton. There is nothing to indicate the existence of the alpha particle after the disintegration.

Fig. 25.24 The first nuclear reaction observed

This nuclear reaction may therefore be represented by the following equation (assuming neutrons and protons are conserved).

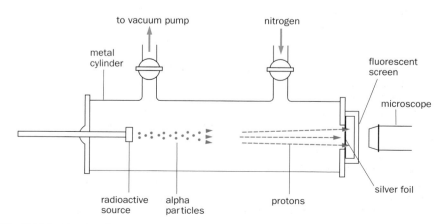

Fig. 25.23 Bombarding nitrogen with alpha particles

$$^{14}_{7}\text{N} + {}^{4}_{2}\text{He} \rightarrow {}^{17}_{8}\text{O} + {}^{1}_{1}\text{H}$$

nitrogen   alpha   residual   proton
nucleus   particle   nucleus

$$[^{14}_{7}\text{N}(\alpha, \text{p})^{17}_{8}\text{O}]$$

The residual nucleus is an isotope of oxygen. This was the first occasion on which one element (nitrogen) was changed into another (oxygen) although the number of atoms transformed was small. The event is sometimes referred to as the 'splitting of the atom'. One other point requires comment.

If the inverse square law for electrostatic force holds right up to the nucleus the repulsive force would become infinitely large as the distance approaches zero. A charged particle would never enter a nucleus, however great its energy. Nuclear reactions do occur, however, as Rutherford demonstrated and so the repulsive force must be replaced by an extremely powerful attractive force acting over a range of about 10^{-15} m. It is in fact this force that keeps nucleons together in a nucleus and is responsible for the existence of matter; it is called the **strong force**.

The tracks produced by protons in a diffusion cloud chamber (p. 468) may be observed by making a small pinhole in a thin sheet of polythene and placing it over an americium source, Fig. 25.25*a*, in the chamber. Alpha particles emerging from the pinhole produce short tracks all about four centimetres long, but occasionally a much longer track appears, due to the ejection of a proton from the hydrogen-rich polythene as a result of alpha-particle bombardment, Fig. 25.25*b*.

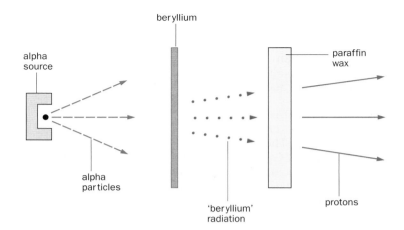

Fig. 25.26   Discovery of the neutron

(b) Discovery of the neutron

After the transmutation of nitrogen many other light elements were found to emit protons when bombarded by alpha particles. Beryllium, however, behaved unexpectedly and was found to emit radiation capable of penetrating many centimetres of lead. It could also cause protons of high energy to be shot out from materials such as paraffin wax which contain hydrogen, Fig. 25.26.

At first it was suggested that 'beryllium radiation' might be very energetic gamma rays until it was realised that if this were so then energy and momentum were not conserved in the collision producing it. However, in 1932 Chadwick suggested that there would be conservation if the radiation consisted of uncharged particles with a mass almost the same as a proton's. This neutral particle was called a **neutron** ($^{1}_{0}\text{n}$). Chadwick was able to account for all the observed effects and established that the action of alpha particles on beryllium was represented by

$$^{9}_{4}\text{B} + {}^{4}_{2}\text{He} \rightarrow {}^{12}_{6}\text{C} + {}^{1}_{0}\text{n} \qquad [^{9}_{4}\text{Be}(\alpha, \text{n})^{12}_{6}\text{C}]$$

Because of its lack of charge the neutron penetrates matter easily and leaves no tracks in a cloud chamber unless it has a collision with a nucleus, when charged particles can be released that do give tracks. Although a neutron decays outside the nucleus with a half-life of 13 minutes to a proton, an electron and an antineutrino, inside it is perfectly stable and an entity in its own right.

In general, nuclear reactions induced by alpha particles cause the emission of either protons or neutrons depending on the energy of the alpha particles and the nucleus under attack.

(c) Cockcroft and Walton's experiment

Alpha particles from radioactive sources have limitations as atomic projectiles. This is due partly to the fact that, because their energy is limited, they cannot overcome the powerful repulsion of the large positive nuclear charge of a heavy atom. In addition only a very small proportion of particles are generally able to cause disintegration, and so radioactive sources of high activity which emit a large number of particles per second would be necessary. These are both difficult to obtain and to manipulate. Consequently, in the early 1930s attention was given to the problem of building machines, called **particle accelerators**, to accelerate charged particles such as

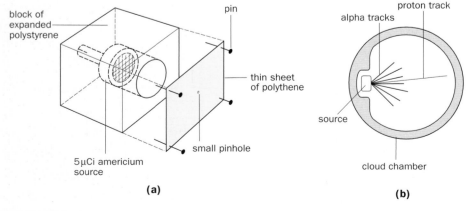

(a)   **(b)**

Fig. 25.25   Observing tracks of ejected protons

protons to high speeds by means of p.ds of hundreds of thousands of volts.

The first nuclear reaction to be induced by artificially accelerated particles was achieved in 1932 by Cockcroft and Walton. They bombarded lithium with protons and obtained alpha particles. The reaction is represented by the equation

$$^{7}_{3}\text{Li} + ^{1}_{1}\text{H} \rightarrow ^{4}_{2}\text{He} + ^{4}_{2}\text{He} \quad [^{7}_{3}\text{Li}(p, \alpha)^{4}_{2}\text{He}]$$

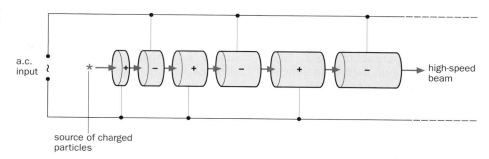

Fig. 25.27 Linear accelerator

(d) Particle accelerators

Several types of accelerator are currently in use. **Van de Graaff** generators produce high p.ds which can be used to accelerate charged particles to about 30 MeV before they impact on a fixed target (p. 226). A **linear accelerator**, such as that shown in Fig. 25.27, successively accelerates electrons as they pass through a sequence of cylindrical electrodes. An alternating p.d. is applied to each electrode so that it is positive as electrons approach it but negative as they leave it; the electrons are thus accelerated at each gap. As the electrons speed up they travel further in a given time and so successive electrodes need to be longer. The Stanford linear accelerator in California is about 3 km long and can accelerate electrons to around 50 GeV.

The size of an accelerator can be reduced if instead of using a linear path the charged particles follow a circular path as in the **cyclotron** shown in Fig. 25.28.

A charge Q moving with velocity v at right angles to a magnetic field B moves in a circle of radius R given by $BQv = mv^2/R$ (see p. 387). In the cyclotron, the magnetic field is applied perpendicularly to two D-shaped cavities in which the charges, often protons, circulate; an alternating p.d. is applied between the two 'Dees' so that the proton is accelerated each time it crosses the gap. The angular velocity of particles in the magnetic field $\omega_c = v/R = BQ/m = 2\pi f_c$ where f_c will be the required frequency of the alternating p.d. (the cyclotron frequency); then the electric field reverses direction every half-revolution of the proton so that it is always accelerated at a gap. As the proton moves faster, the radius of its path increases and it spirals outwards from its source at the centre of the cyclotron. At the outer

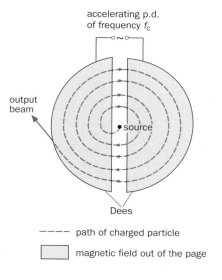

Fig. 25.28 Cyclotron

edge of the cyclotron, the proton has maximum velocity and is directed onto a target.

Accelerating charges lose energy by emitting radiation and this sets a limit to the maximum speed attainable; the energy loss is less severe in a linear accelerator than in a cyclotron where centripetal acceleration is always present. Also, relativistic effects cause the mass of a particle to increase as it approaches the speed of light (p. 506); since the cyclotron frequency f_c depends on m, at high speeds synchronization with the accelerating p.d. is lost and the proton no longer accelerates. The new breed of **synchrotron** accelerators overcomes this limitation by varying B and f_c during the accelerating cycle so that synchronization is retained. The protons then follow a circular path and the magnetic field can be restricted to a ring, rather than the large area of the Dees. Figure 25.29 shows the European Synchro-

Fig. 25.29 ESRF synchrotron, Grenoble

tron Radiation Facility (ESRF) at Grenoble, France, where the high-energy radiation emitted is used in a variety of experimental fields including medicine and materials science.

Protons can be accelerated to energies of 500 GeV in the accelerator at the European Laboratory for Particle Physics (CERN) where the radius of the ring is about 1 km. Even higher energies can be achieved if two beams of particles travelling in opposite directions are made to collide; however a wider range of interactions can be studied when an ion beam is incident on a stationary target. Analysis of the debris from such high-energy collisions has led to the discovery of many new sub-atomic particles and allowed the internal structure of protons and neutrons to be probed.

MASS AND ENERGY

(a) Einstein's relation

In 1905, while developing his special theory of relativity, Einstein made the startling suggestion that energy and mass are equivalent. He predicted that if the energy of a body changes by an amount E, its mass changes by an amount m given by the equation

$$E = mc^2$$

where c is the speed of light. Everyday examples of energy gain are much too small to produce detectable changes of mass.

For example, when 1.0 kg of water absorbs 4.2×10^3 J of energy to produce a temperature rise of 1 K, according to $E = mc^2$ the increase in mass of the water is

$$m = \frac{E}{c^2} = \frac{4.2 \times 10^3 \text{ J}}{(3.0 \times 10^8)^2 \text{ m}^2 \text{ s}^{-2}}$$
$$= 4.7 \times 10^{-14} \text{ kg}$$

Check that $\text{J m}^{-2} \text{ s}^2$ reduces to kg.

The changes of mass accompanying energy changes in chemical reactions are not much greater and cannot be used to prove Einstein's equation.

However, radioactive decay, which is a spontaneous nuclear reaction, is more helpful. For a radium atom, the combined mass of the alpha particle it emits and the radon atom to which it decays is, by atomic standards, appreciably less than the mass of the original radium atom. Atomic masses can now be measured to a very high degree of accuracy by mass spectrographs.

The mass decrease m for the decay of one radium atom is 8.6×10^{-30} kg. The energy equivalent E is given by

$$E = mc^2$$
$$= 8.6 \times 10^{-30} \times (3.0 \times 10^8)^2 \text{ J}$$
$$= 7.74 \times 10^{-13} \text{ J}$$

But 1 eV $= 1.6 \times 10^{-19}$ J

$$\therefore \quad E = \frac{7.74 \times 10^{-13}}{1.6 \times 10^{-19}} \text{ eV}$$
$$= 4.84 \times 10^6 \text{ eV}$$
$$= 4.84 \text{ MeV}$$

The alpha particle carries off 4.80 MeV of this energy; some appears as k.e. of the recoiling nucleus, and the rest is emitted soon afterwards as a γ-ray photon.

Mass appears as energy and the two can be regarded as equivalent. In nuclear physics mass is measured in **unified atomic mass units** (u), 1 u being one-twelfth of the mass of the carbon-12 atom and equal to 1.66×10^{-27} kg. It can be shown using $E = mc^2$ that 931 MeV has an equivalent mass of 1 u.

A unit of energy may therefore be considered to be a unit of mass, and in tables of physical constants the masses of various atomic particles are often given in MeV/c^2 as well as in kg and u. For example, the electron has a rest mass of about 0.5 MeV/c^2.

If the principle of conservation of energy is to hold for nuclear reactions it is clear that mass and energy must be regarded as equivalent. The implication of $E = mc^2$ is that any reaction producing an appreciable mass decrease is a possible source of energy.

The energy change occurring in a nuclear process is termed the **Q value**. For the disintegration of a radium atom considered above, $Q = 4.84$ MeV. Q can be calculated more easily if the atomic masses are expressed in u, as follows.

$$^{226}_{88}\text{Ra} = 226.0254 \text{ u}$$
$$^{222}_{86}\text{Rn} = 222.0176 \text{ u}$$
$$^{4}_{2}\text{He} = 4.002602 \text{ u}$$

For the disintegration reaction

$$^{226}_{88}\text{Ra} \rightarrow \; ^{222}_{86}\text{Rn} + \; ^{4}_{2}\text{He}$$

the mass difference is then

$$m = 226.0254 - 222.0176 - 4.0026$$
$$= 0.0052 \text{ u}$$

and so

$$Q = 0.0052 \times 931 = 4.84 \text{ MeV}$$

as already found.

(b) Binding energy

The mass of a nucleus is found to be less than the sum of the masses of the constituent protons and neutrons. This is explained as being due to the binding of the nucleons together into a nucleus, and the **mass defect** represents the energy that would be released in forming the nucleus from its component particles. The energy equivalent is called the **binding energy** of the nucleus; it is equal to the energy needed to split the nucleus into its individual nucleons if this were possible.

Consider an example. The helium atom, $^{4}_{2}\text{He}$, has an atomic mass of 4.0026 u; this includes the mass of its two electrons. Its constituents comprise two neutrons each of mass 1.0087 u and two hydrogen atoms (i.e. two protons and two electrons) each of mass 1.0078 u; the total mass of the particles is therefore

$$(2 \times 1.0087) + (2 \times 1.0078)$$
$$= 4.0330 \text{ u}$$

The mass defect is

$$(4.0330 - 4.0026) = 0.0304 \text{ u}$$

and since 1 u is equivalent to 931 MeV, it follows that the binding energy of helium is 28.3 MeV.

The binding energy, derived in a similar manner for other nuclides, is found to increase as the mass (nucleon) number increases. For neon, $^{20}_{10}\text{Ne}$, it is 160 MeV. If the binding energy of a nucleus is divided by its mass number, the **binding**

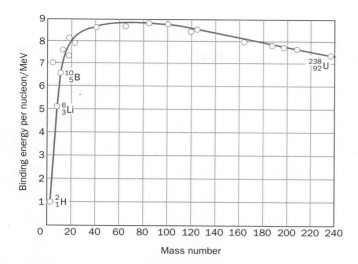

Fig. 25.30

neutrons is available for new fissions since some are lost by escaping from the surface of the uranium before colliding with another nucleus. The ratio of neutrons escaping to those causing fission decreases as the size the piece of uranium-235 increases and there is a critical size (about the size of a cricket ball) that must be attained before a chain reaction can start.

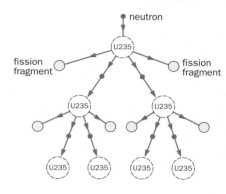

Fig. 25.31   A chain reaction

energy per nucleon is obtained. The graph of Fig. 25.30 shows how this quantity varies with mass number; in most cases it is about 8 MeV. Nuclides in the middle of the graph have the highest binding energy per nucleon and are thus the most stable since they need most energy to disintegrate. The smaller values for higher and lower mass numbers imply that potential sources of nuclear energy are reactions involving the disintegration of a heavy nucleus or the fusing of particles to form a nucleus of high nucleon number. In both cases nuclei are produced having a greater binding energy per nucleon and there is consequently a mass transfer during their formation.

NUCLEAR ENERGY

(a) Fission

The discovery of the neutron provided nuclear physicists with an important new projectile. Being uncharged, it is not repelled on approaching the large positive charge on the nucleus of a heavy atom as are protons and alpha particles. In 1939, following some work by Fermi in Italy and Hahn and Strassmann in Germany, it was found that the bombardment of uranium by neutrons may split the uranium nucleus into two large nuclei — for example, those of barium and krypton.

This nuclear reaction, called **nuclear fission**, differs from earlier nuclear reactions in three respects.

- The nucleus is deeply divided into two large **fission fragments** of roughly equal mass.
- The **mass decrease**, or Q value, is appreciable.
- Other neutrons, called **fission neutrons**, are emitted in the process.

Fission occurs in several heavy nuclides as well as in the two main isotopes of uranium, $^{238}_{92}U$ and $^{235}_{92}U$. The latter is the more useful and one reaction that occurs is

$$^{235}_{92}U + ^{1}_{0}n \rightarrow ^{144}_{56}Ba + ^{90}_{36}Kr + 2^{1}_{0}n$$

$$[^{235}_{92}U(n, 2n)^{144}_{56}Ba, ^{90}_{36}Kr]$$

The Q value of this reaction is about 200 MeV per fission, which is 45 million times greater per atom of fuel (uranium-235) than in a chemical reaction where the energy change is no more than a few electronvolts per atom. The uranium-235 nucleus is evidently a vast storehouse of energy. The fission energy appears mostly as kinetic energy of the fission fragments (e.g. barium and krypton nuclei) which fly apart at great speed. The kinetic energy of the fission neutrons also makes a contribution. In addition, one or both of the large fragments are highly radioactive and a small amount of energy takes the form of beta and gamma radiation.

The production of fission neutrons (two in the above case) raised the possibility of creating a **chain reaction** in which fission neutrons cause further fission of uranium-235 and the reaction keeps going, Fig. 25.31. In practice only a proportion of the fission

In the 'atomic bomb' an increasing uncontrolled chain reaction occurs in a very short time when two pieces of uranium-235 (or plutonium-239) are rapidly brought together to form a mass greater than the critical size. The reaction is started either by having a small neutron source in the bomb or by stray neutrons from the occasional spontaneous fission of a uranium-235 nucleus.

(b) Nuclear reactors

In a nuclear reactor the chain reaction is steady and controlled so that on average only one neutron from each fission produces another fission. The reaction rate is adjusted by inserting neutron-absorbing rods of boron steel into the uranium-235. The basic arrangement is shown in Fig. 25.32.

The graphite core is called the **moderator** and is needed to maintain a chain reaction when the fuel is not fairly pure uranium-235. Natural uranium contains over 99% of uranium-238 and less than 1% of uranium-235 and unfortunately the uranium-238 captures the medium-speed fission neutrons without fissioning. The reaction that occurs is

$$^{238}_{92}U + ^{1}_{0}n \rightarrow ^{239}_{92}U \rightarrow ^{239}_{93}Np + ^{0}_{-1}e + \bar{\nu}_e$$

$$[^{238}_{92}U(n, e)^{239}_{93}Np]$$

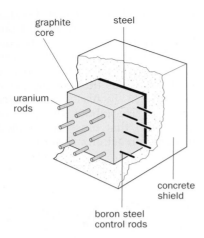

graphite
core

steel

uranium
rods

concrete
shield

boron steel
control rods

Fig. 25.32 Nuclear reactor

Fig. 25.33 The pressurized water reactor (lower centre) at Sizewell B

Neptunium-239 is the first transuranic element; it decays, also by beta emission, to plutonium, the second transuranic element:

$$^{239}_{93}\text{Np} \rightarrow {}^{239}_{94}\text{Pu} + {}^{0}_{-1}\text{e} + \bar{\nu}_e$$

Uranium-238 fissions only with very fast neutrons. On the other hand, uranium-235 (and plutonium-239) fissions with **slow (thermal) neutrons** and the job of the moderator is to slow down the fission neutrons very quickly so that most escape capture by uranium-238 and then cause the fission of uranium-235. A bombarding particle gives up most energy when it has an elastic collision with a particle of similar mass. For neutrons, hydrogen atoms would be most effective but unfortunately absorption occurs. However, deuterium (in heavy water) and carbon (as graphite) are both suitable.

In a nuclear power station a nuclear reactor provides the heat required to produce steam, instead of a coal-, oil- or gas-burning furnace. In the reactor core the large fission fragments share their kinetic energy with surrounding atoms as a result of collisions and raise the internal energy of the system. In a gas-cooled reactor a current of high-pressure carbon dioxide gas is pumped through the core of the reactor and transfers heat to the heat exchanger where water is converted to steam which is then used as in a conventional power station to drive a turbo-generator (see p. 162). A thick concrete safety shield gives protection from neutrons and gamma rays.

Nuclear reactors like that just described are called **thermal reactors** because fission is caused by slow neutrons with thermal energies, i.e. energies equal to the average kinetic energy of the surrounding atoms. The first thermal reactors were called **magnox reactors** after the magnesium alloy cans containing the natural uranium metal fuel. From these were developed **advanced gas-cooled reactors** (AGR) using enriched uranium (containing 2.3% of uranium-235 compared with 0.7% in natural uranium) and having the carbon dioxide coolant gas at a higher temperature and pressure (650 °C and 40 atmospheres instead of 400 °C and 20 atmospheres). A third type of thermal reactor is the **pressurized water reactor** (PWR), Fig. 25.33, which uses even more-highly enriched uranium (3.2% uranium-235) and high-pressure water (to stop it boiling) as both moderator and coolant.

The **fast breeder reactor** has a core made of highly enriched fuel (uranium and plutonium); fission occurs by *fast* neutrons and no moderator is required. The core is surrounded by a 'breeder assembly' of natural uranium which is converted to plutonium-239 by fission neutrons that escape from the core and would otherwise be lost. New fissionable material is thus 'bred' for the reactor. Liquid sodium metal is used as the coolant.

Important by-products of nuclear reactors are artificial radioisotopes; they are made when stable nuclides, inserted in the core of the reactor, are bombarded by neutrons.

Radioactive waste products pose problems for the nuclear energy industry since both reprocessing and 'safe' disposal (see p. 164) are expensive and many people are not convinced that complete safety is possible. A proportion of the public are also against nuclear energy stations because of the risk of accidents such as that in Chernobyl in 1986, and potential health hazards to those working in the power stations or living in their vicinity. However, some believe that producing electricity from nuclear power rather than from burning coal, oil or gas helps to reduce the longer-term dangers to life of the greenhouse effect (see p. 3).

(c) Fusion

The union of light nuclei into heavier nuclei can also lead to a transfer of mass and a consequent liberation of energy. Such a reaction has been achieved in the 'hydrogen bomb' and it is believed to be the principal source of the sun's energy.

A reaction with heavy hydrogen, or deuterium, which yields 3.3 MeV per fusion is

$$^2_1\text{H} + ^2_1\text{H} \rightarrow ^3_2\text{He} + ^1_0\text{n} \qquad [^2_1\text{H}(^2_1\text{H}, \text{n})^3_2\text{He}]$$

By comparison with the 200 MeV per fission of uranium-235 this seems small, but per unit mass of material it is not. Fusion of the two deuterium nuclei will only occur if they overcome their mutual electrostatic repulsion. This may happen if they collide at very high speed when, for example, they are raised to a very high temperature. If fusion occurs, enough energy is released to keep the reaction going and since heat is required, the process is called **thermonuclear fusion**.

Temperatures between 10^8 and 10^9 K are required and in the uncontrolled thermonuclear reaction occurring in the hydrogen bomb, the high initial temperature is obtained by using an atomic (fission) bomb to trigger the fusion.

If a controlled fusion reaction can be achieved an almost unlimited supply of energy will become available from deuterium in the water of the oceans. The problem is, first, how to achieve the high temperature necessary and, second, how to keep the hot reacting gases (called the plasma) from touching the vessel holding it. Research continues on a worldwide scale into this challenging task and the largest project is the JET (Joint European Torus) Joint Undertaking at Abingdon, Oxfordshire — a collaborative venture involving fourteen European countries. Current experiments there involve deuterium–tritium plasmas, confined by magnetic fields in a toroidal vessel (Fig. 25.34), a set-up called a 'Tokamak'. The reaction is

$$^2_1\text{H} + ^3_1\text{H} \rightarrow ^4_2\text{He} + ^1_0\text{n}$$

yielding 17.6 MeV per fusion. Construction of an experimental fusion reactor — the ITER (International Thermonuclear Experimental Reactor) — is now in progress. This is a multinational project involving Europe, Russia and Japan.

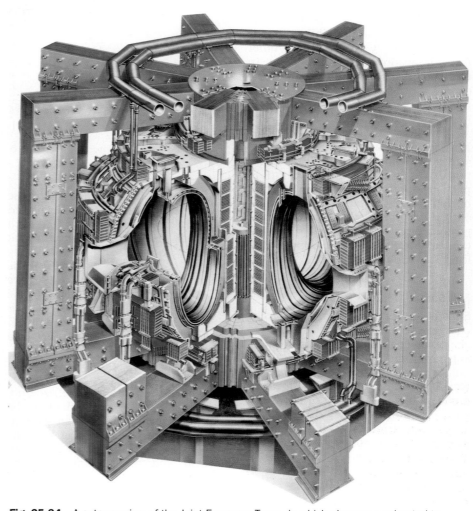

Fig. 25.34 A cutaway view of the Joint European Torus, in which plasmas are heated to temperatures up to 300 million degrees

THE PARTICLES OF PHYSICS

Until about 1930 the nucleus was thought to contain (or emit) only protons (p), neutrons (n) and electrons (e$^-$ or β^-). Since then, mainly as a result of the study of cosmic rays (i.e. high-energy ionizing radiation from outer space) and the bombardment of targets by beams of high-speed charged particles in accelerators, many hundreds more, mostly short-lived (e.g. 10^{-10} s) sub-nuclear particles have been discovered.

We will now look at the concepts and classifications that particle physicists have introduced in order to place all these sub-atomic particles in an ordered scheme in which the fundamental constituents of matter can be discerned.

(a) Antiparticles

The idea of antiparticles was first proposed by Dirac in 1928 when considering the electron in terms of wave mechanics and relativity. He called the electron's antiparticle the **positron** (e$^+$) or 'positive electron' because it had the same mass as the electron and a charge of equal size but of positive sign. It was identified from cloud chamber studies of cosmic rays by Anderson in 1932. It is now thought that every particle has an antiparticle counterpart which has the same mass but opposite charge to the particle; the antiproton ($\bar{\text{p}}$), detected in 1955, has the same mass as a proton but unit negative charge. In general, particle and antiparticle have opposite signs for **charge**, **lepton number**, **baryon number**, **spin** and **strangeness**, terms that will shortly be explained.

Antiparticles are difficult to detect because when a particle meets its antiparticle both are annihilated; their combined energy is released in other forms in accordance with $E = mc^2$. In **electron–positron annihilation** the energy is transferred to two gamma rays:

$$e^+ + e^- \rightarrow \gamma + \gamma \qquad (Q = 1.02 \text{ MeV})$$

If the particles are initially at rest, the γ-rays are emitted in opposite directions so as to conserve momentum. In **pair production** the reverse reaction occurs when radiation of sufficient energy collides with a nucleus and produces an electron–positron pair; the radiation must have energy of at least 1.02 MeV.

(b) Spin

Spin is an additional property beside charge and mass, which is used to characterize particles; it is a form of angular momentum (but it has features unlike anything in the macroscopic world) and is quantified by allocating numbers $0, \frac{1}{2}, 1, \frac{3}{2}, 2, \ldots$ to the various particles. For example, the neutrino and antineutrino both have spin of value $\frac{1}{2}$ but the spins are oppositely directed. Spin must be conserved in particle reactions.

(c) Leptons

These form the lightest group of particles and contain electrons, **muons**, **taus** and their corresponding **neutrinos** and antiparticles; their symbols and some of their properties are summarized in Table 25.2.

All lepton particles have a **lepton number** (L) of $+1$ and a spin of $\pm\frac{1}{2}$. Their antiparticles have lepton number -1 and opposite spin and charge to the corresponding particle.

The twelve leptons are thought to be fundamental particles which cannot be subdivided further; they are acted on by the **weak force** (p. 486) but not the strong force.

Neutrinos were postulated by Pauli in 1933 to account for the continuous spectrum of beta particles (p. 466) and were first detected experimentally in a nuclear reactor in 1953. Neutrinos have no charge and appear to have a rest mass equal or very close to zero. They interact only weakly with matter so are difficult to detect, but are stable

Table 25.2 Leptons

Particle	Symbol	Charge	Antiparticle	Mass (MeV/c^2)
electron	e^-	-1	e^+	0.5
electron neutrino	ν_e	0	$\bar{\nu}_e$	0
muon	μ^-	-1	μ^+	106
muon neutrino	ν_μ	0	$\bar{\nu}_\mu$	0
tau	τ^-	-1	τ^+	1780
tau neutrino	ν_τ	0	$\bar{\nu}_\tau$	0

particles with long lifetimes. Because of their weak interaction with matter, the millions of neutrinos emitted each day from the sun pass right through ourselves and the earth without being detected or leaving a trace; those generated in the big bang (p. 496) are still the most abundant particles in the universe. Three different types of neutrino occur, each associated with an electron, muon or tau lepton (see Table 25.2).

Charge, baryon number (see next section), lepton number and spin must all be conserved in a particle reaction; lepton numbers for μ, e and τ must be conserved separately. We can see that this occurs in the following example where a neutron decays to a proton, electron and antineutrino in a β^- decay process:

		n	$\rightarrow$	p	+	e^-	+	$\bar{\nu}_e$
charge		0	=	$+1$		-1		0
baryon number		$+1$	=	$+1$		0		0
lepton number (L_e for electrons)		0	=	0		$+1$		-1
spin		$\frac{1}{2}$	=	$\frac{1}{2}$		$+\frac{1}{2}$		$-\frac{1}{2}$

Check that charge, baryon number, lepton number and spin are all conserved in the alternative mode of β^+ decay when a proton changes to a neutron with the emission of a positron and neutrino:

$$p \rightarrow n + e^+ + \nu_e$$

Beta particles emitted from nuclei have a range of energies (see Fig. 25.3) since the energy released is shared between the emitted leptons — the maximum energy of the electron or positron results when no energy is carried away by the antineutrino or neutrino and this maximum energy is equal to the Q value of the disintegration.

A reaction related to beta decay is that of **electron capture**, in which a nucleus can combine with one of its own orbiting electrons. An example is that of beryllium in the reaction

$$^7_4\text{Be} + e^- \rightarrow ^7_3\text{Li} + \nu_e$$

where a proton in the nucleus captures an inner electron and changes to a neutron; a neutrino is emitted in the process. An X-ray is emitted soon afterwards when an electron falls from a higher energy level to take the place of the missing electron.

(d) Baryons

The particles of this group have relatively large masses (> 938 MeV/c^2) and include the proton, neutron, **omega** (Ω), **lambda** (Λ), **sigma** (Σ) and **Xi** (Ξ) particles. Baryons all have half-integral spin ($\pm\frac{1}{2}$ or $\pm\frac{3}{2}$); some properties of those with spin $\pm\frac{1}{2}$ are summarized in Table 25.3 opposite, where it can be seen that some exist in both charged and uncharged forms. An additional property, **strangeness**, is assigned to the baryons to classify them — see section (**f**). Each particle again has a corresponding antiparticle.

The **baryon number** (B) assigned to each baryon is $+1$ while the corresponding antibaryon is assigned -1; all other particles have $B = 0$ and baryon number must be conserved in a reaction. For example, the following reaction cannot occur since baryon number (and other quantities) is not conserved:

		Σ^+	$\rightarrow$	e^+	+	ν_e
baryon number		$+1$	$\neq$	0	+	0

The proton is stable outside the nucleus (with a lifetime of 10^{32} years) but the neutron can only exist for about 13 minutes outside the nucleus

Table 25.3 Baryons

Particle	Symbol	Charge	Strangeness	Antiparticle
proton	p	+1	0	$\bar{\text{p}}$
neutron	n	0	0	$\bar{\text{n}}$
lambda	Λ^0	0	−1	$\bar{\Lambda}^0$
sigma	Σ^+	+1	−1	$\bar{\Sigma}^-$
sigma	Σ^0	0	−1	$\bar{\Sigma}^0$
sigma	Σ^-	−1	−1	$\bar{\Sigma}^+$
xi	Ξ^0	0	−2	$\bar{\Xi}^0$
xi	Ξ^-	−1	−2	Ξ^+

Table 25.4 Mesons

Particle	Symbol	Charge	Strangeness	Antiparticle
pion	π^+	+1	0	π^-
pion	π^0	0	0	self
kaon	K^+	+1	+1	K^-
kaon	K^0	0	+1	$\bar{K}^0$
eta	η	0	0	self

before it decays by β^- emission; other particles have much shorter lifetimes (for example, 10^{-10} s).

Baryons are acted upon by both the strong and the weak force and are not fundamental particles but, as we will soon see, are built up from **quarks**.

(e) Mesons

These particles have masses intermediate between leptons and baryons (< 960 MeV/c^2); the π-meson or **pion** has a mass of 140 MeV/c^2 (about 270 times heavier than the electron). Mesons are very short-lived and differ from baryons in that they all have integral spin (and $B = 0$, $L = 0$). Some properties of mesons with zero spin are shown in Table 25.4 above.

Like baryons, mesons are not fundamental particles but are built up from quarks and are acted upon by both the strong and the weak force.

(f) Strangeness; conservation laws

The property called **strangeness** (S) is needed to explain why certain reactions never occur. All leptons, the proton and neutron are considered to have a strangeness of zero, but many baryons, mesons and their antiparticles have strangeness values represented by digits from −3 to +3; particles and antiparticles have strangeness of equal value but opposite sign.

Strangeness must be conserved if the strong force (p. 486) is involved in a reaction or decay process; energy, momentum, charge, spin, lepton number and baryon number must be conserved in *all* reactions and decays. For example, the following reaction is possible since all the conservation laws are satisfied:

$$\pi^- + p \rightarrow \Lambda^0 + K^0$$

charge	−1	+1 =	0	0
spin	0	$+\frac{1}{2}$ =	$+\frac{1}{2}$	0
lepton number	0	0 =	0	0
baryon number	0	+1 =	+1	0
strangeness	0	0 =	−1	+1

QUARKS

By the early 1960s so many new short-lived particles had been discovered that the notion of the existence of a few fundamental particles, from which all others were built, was in doubt. However, the following twenty years saw considerable progress towards clarifying the situation, due mainly to the quark theory.

In 1963 Gell-Mann and Zweig proposed independently that baryons and mesons, collectively called **hadrons**, consisted of smaller particles, which became known as **quarks** (from James Joyce's novel *Finnegan's Wake*).

Originally three different types of quark were suggested, called **up** (u), **down** (d) and **strange** (s). Later it was found another three were required; they were known as **charm** (c), **top** (t) and **bottom** (b). Some of their properties are listed in Table 25.5. For charge and baryon number to work out correctly when they are combined as hadrons they must have fractional charges (Q) of $\pm\frac{1}{3}$e or $\pm\frac{2}{3}$e and baryon number (B) of $\frac{1}{3}$; quarks all have spin $\frac{1}{2}$. All six have antiparticles (denoted by a bar above the symbol) whose signs of spin, Q, B and strangeness S are opposite to the corresponding quark; for example $\bar{u}$ has $Q = -\frac{2}{3}$e, $B = -\frac{1}{3}$ and $S = 0$. It is believed that quarks cannot be subdivided, so are fundamental particles and the building blocks of the hadrons.

The up, down and strange quarks and their corresponding antiquarks occur in the more commonly encountered particles; they have lower masses than charm, top and bottom. The high masses of the latter mean they are only identified in very high-energy particle accelerators; they have the additional characteristics of *charm*, *topness* and *bottomness* which must be conserved

Table 25.5 Quarks

Quark type	Symbol	Charge (*Q*)	Baryon no. (*B*)	Strangeness (*S*)	Mass (MeV/c^2)
up	u	$+\frac{2}{3}$e	$\frac{1}{3}$	0	5
down	d	$-\frac{1}{3}$e	$\frac{1}{3}$	0	10
strange	s	$-\frac{1}{3}$e	$\frac{1}{3}$	−1	200
charm	c	$+\frac{2}{3}$e	$\frac{1}{3}$	0	1500
top	t	$+\frac{2}{3}$e	$\frac{1}{3}$	0	180 000
bottom	b	$-\frac{1}{3}$e	$\frac{1}{3}$	0	4300

Table 25.6

Particle	Symbol	Charge	Strangeness	Structure
neutron	n	0	0	u d d
proton	p	+1	0	u u d
sigma +	Σ^+	+1	−1	u u s
kaon 0	K^0	0	+1	d $\bar{s}$

in strong interactions. The identification of the top–antitop quark pair (t $\bar{t}$) in 1995 gave strong support to the quark theory. The six quarks and six leptons together with their antiparticles are now believed to be the fundamental constituents of matter.

A baryon is considered to be formed from three quarks and a meson from one quark and one antiquark. Table 25.6 shows the quark structure of some sub-nuclear particles. Using Table 25.5 on p. 485, check that it gives the correct charge and strangeness in each case. What will be the quark structure of the antiproton?

It is interesting to look at the beta decay process in terms of quarks. β^- decay occurs when a neutron changes to a proton within a nucleus, causing the emission of an electron and antineutrino:

$$n \rightarrow p + e^- + \bar{\nu}_e$$

In terms of quarks, a neutron (u d d) changes to a proton (u u d) and the reaction can be written:

$$d \rightarrow u + e^- + \bar{\nu}_e$$

The quark theory has been able to explain a wide range of sub-nuclear effects; but a single quark has not yet been isolated. However, progress is being made and early in 2000 it was reported that 'quark–gluon matter' had been observed with the new atom smasher, the Large Hadron Collider (LHC) at the European Laboratory for Particle Physics (CERN).

THE FORCES OF NATURE

Four different kinds of force seem to be at work in nature.

(a) The gravitational force

This force between masses is the weakest of the four and while it can be ignored inside atoms, it dominates everyday life due to the proximity of the huge mass of the earth. Its range is infinite, and it is always attractive.

(b) The electromagnetic (electric) force

This force acts between all charged particles, e.g. electrons in atoms, and produces interatomic and intermolecular attractions and repulsions that bind atoms and molecules together (p. 18). It is also a force of infinite range.

(c) The weak force

This force acts on all particles — both leptons and quarks — and is intermediate in strength between the gravitational and electromagnetic forces. It has a range of less than 10^{-2} fermi (1 fermi = 1 fm = 10^{-15} m) and is responsible for radioactive decay when beta particles are emitted. The electromagnetic and weak forces are now thought to be different aspects of the same **electroweak force**.

(d) The strong force

This force holds the protons and neutrons together in the nucleus. It must be a very short-range force for, if its effects went much outside the nuclear surface, Rutherford scattering (p. 400) could not be explained by electrostatic repulsion alone. It must be strong enough to overcome the repulsion that protons (with their positive charge) exert on one another. Further, at very small distances, it must be sufficiently repulsive to prevent the nucleus collapsing.

From scattering experiments it is found that the radius R of a spherical nucleus is given approximately by $R = R_0 A^{1/3}$ where $R_0 = 1.2$ fm and A is the mass number. Since the density of a nucleus is $\rho = \text{mass/volume} = 3A/(4\pi R^3)$ and substituting for A this becomes $3/(4\pi R_0^3)$, we expect nuclear material to have a constant density, independent of the size of the nucleus. For internucleon separations of about 1.2 to 3 fm the strong force is attractive but for distances less than 1 fm it must be repulsive.

(e) Unification of forces

Theoretical physicists are working towards a connection of the electroweak force and the strong force in a **Grand Unified Theory (GUT)**. The next step forward would be to include gravitation and have a fully unified theory — the 'Theory of Everything' (TOE) — in which all four forces are shown to have a common origin in some single 'superforce'.

THE EXCHANGE NATURE OF FORCES

The present explanation of how it is thought the forces of nature arise, is based on three of the most difficult topics of modern physics — quantum theory, special relativity and field theory — and is highly mathematical. While the theory is complex and has some unsatisfactory aspects, it has successfully explained a great deal of sub-atomic particle activity. The account given here is much simplified.

The basic idea is that force is due to certain particles, called **field quanta**, which are 'exchanged' between the interacting particles. The field quanta are considered to form clouds surrounding the particles (e.g. electrons), creating a field and acting as 'messengers' that convey the force. The implication is that the particles of matter can no longer be distinguished from the fields that connect them.

For the electromagnetic field and force, the field quanta are photons, called **virtual photons** to identify them from 'ordinary' photons. For the weak force, which has a very short range of less than 10^{-2} fm, the field quanta are the **W-particle** with a mass of about 80 protons and the **Z-particle** with a mass of around 90 protons. The electromagnetic and weak forces can be regarded as resulting from the same electroweak force, in which different messenger particles are used. The W-particle may be positively or negatively charged so can transfer charge, for example in beta decay.

For the strong force the messenger particles are called **gluons** and like photons they are expected to be massless. Exchange of gluons is thought to be responsible for binding quarks

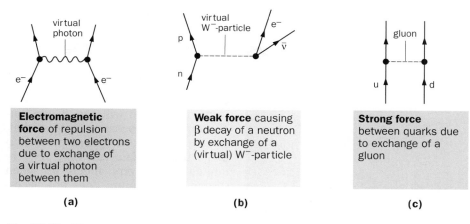

Fig. 25.35 Diagrammatic representations of the exchange nature of forces

together into baryons and mesons and also for holding neutrons and protons together within a nucleus.

The diagrams in Figs 25.35*a*, *b* and *c* give pictorial representations of the three exchange processes described above but it should be remembered that they do not represent reality nor are they a substitute for mathematics; they are aids to understanding.

Direct evidence for the existence of virtual particle clouds has been obtained from scattering experiments involving electrons.

CHAOS THEORY

(a) Introduction

The traditional approach to science has been to find the simplest laws that enable sense to be made of the world around us. Newton's laws of motion are a good example. However, they have been derived for an idealized situation in which practical complications such as friction and air resistance are largely ignored. In mathematical terms, they deal with **linear** systems in which the equations expressing their behaviour involve directly proportional relationships between quantities. Such equations are readily solved and so enable predictions about future behaviour to be made.

This is the Newtonian 'mechanistic' or 'deterministic' view in which the universe is regarded as a piece of clockwork (see p. 142) whose future is determined by the past. It held sway for 200 years before it was challenged early in the twentieth century by two

developments. The first was the theory of **relativity** (p. 504), which shattered the idea of absolute space and time. The second was **quantum mechanics**, with its 'uncertainty principle' which stated that it was impossible to measure both the position and the velocity of a particle simultaneously.

For most of the last 100 years the exploration of the building blocks of matter and the forces of nature has been spear-headed by particle physicists. They have attempted to reduce our understanding of the physical world to explanations in terms of its constituent entities such as quarks and gluons. While this 'reductionist' approach has had its successes, it has been accompanied by narrow specialization. Some scientists have felt that broader questions relating to everyday problems and systems have been by-passed and left unanswered. Many of these are undoubtedly complex and apparently lacking in any kind of order. They involve **non-linear** equations that are notoriously difficult to solve. For these problems it would appear that a quite different approach is needed.

(b) The chaologist's approach

It is against the above background that chaos theory emerged in the 1960s and 70s. It developed for three reasons:

(*i*) the availability of fast computers for repeating thousands of calculations again and again (and so taking a system through a great many steps);

(*ii*) the existence of a new type of geometry called **topology** (based on

work done by the French mathematician Poincaré at the end of the nineteenth century); and

(*iii*) an increase of scientific interest in non-linear systems where *disorder* and *unpredictability*, i.e. **chaos** (in a scientific sense, not 'mess' as the everyday meaning of the word suggests), have been shown to be common characteristics.

The chaologist's approach involves creating a mathematical model of the system under study (e.g. the atmosphere) in the form of equations (usually non-linear), containing the variable quantities that affect its behaviour (e.g. temperature, pressure, humidity, wind speed and direction, etc.). The equations and data are then fed into a computer which is programmed to perform a large number of similar operations, i.e. to **iterate**, resulting in an output that represents the final *predicted* state of the system (e.g. a weather forecast). How close the prediction is to reality depends on how good a likeness the model is to the actual system.

The output is commonly in the form of computer graphics that are infinitely complicated colourful shapes which appear to repeat themselves on a smaller and smaller scale (Fig. 25.36) — an effect known as 'self-similarity'. In 1975 Mandelbrot, an American mathematician and physicist, who played a major role in developing chaos theory, proposed the word **fractal** (from the Latin for broken) for such shapes. A tree is a natural fractal. The formation of a fractal from a triangle is shown in Fig. 25.37. The process can be repeated indefinitely.

For those skilled at the new fractal geometry, conclusions can be drawn from the shape and repeating pattern of the computer graphics, enabling limited predictions to be made about the behaviour of the system.

One feature of chaotic systems revealed by this approach is that many changing systems are shown to be greatly affected by their initial state. Even if the states are very similar to start with, they rapidly diverge. This is called the 'butterfly effect'. In the context of weather forecasting, the statement about the flapping of a butterfly's wings in the Amazonian jungle causing a tornado in Texas, derives from this 'sensitivity to initial conditions' of chaotic systems.

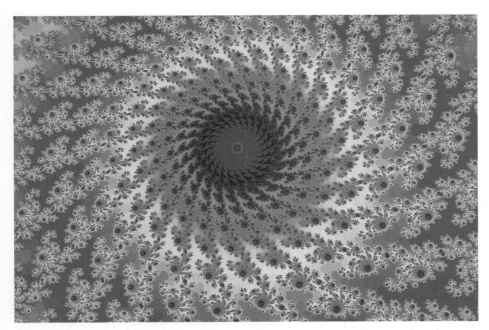

Fig. 25.36 The *Wheel of Dreams* — a fractal image of part of the Mandelbrot Set of complex numbers

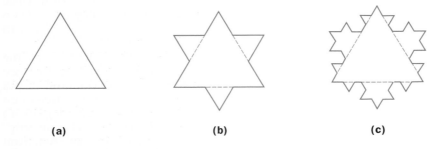

(a)   (b)   (c)

Fig. 25.37 Formation of a fractal

(c) Applications of chaos theory

Chaos theory now permeates almost every branch of science and has highlighted the interdisciplinary aspect of modern research. It has brought together the computer, one of today's chief research tools, and abstract mathematics. There now exists a way of probing complex dynamical problems of the real world and finding patterns where disorder and unpredictability, i.e. chaos, formerly appeared to reign. It attempts to bridge the gap between knowing what one item (e.g. atom, cell) will do and what millions will do.

Initially chaos theory emerged from the work on **meteorology** in the early 1960s by Lorenz at the Massachusetts Institute of Technology, where he discovered the butterfly effect. Today the study of the atmosphere's movements, which determines our weather, is an important application.

The theory is used by **biologists** dealing with changing bird and insect populations and with the action of cells. **Medical researchers** use it to help estimate the spread of diseases such as AIDS. **Chemists** use it to study certain catalytic reactions that can behave oddly, while **physicists** employ it to gain insight into the motion of electrons in atoms and the loss of particles in accelerators. Not all electrical circuits behave predictably and **electronic engineers** find the theory helps in understanding their unusual behaviour.

The way nuclear power stations rock during earthquakes, the reaction of offshore oil platforms to large waves and the sudden disappearance of ships in high seas are all investigated by **engineers** using chaos theory. Such events usually involve non-linear dynamical systems.

The complex fluctuations of financial and commodity markets are studied by **economists** using the theory.

Astronomers and **cosmologists** try, on the one hand, to model the early stages of the birth of the universe using chaos theory and, on the other, to predict how it will end. Their conclusions so far seem to suggest that, since it is a chaotic system, prediction of its ultimate fate can only be very limited. It is very much an open system whose future does not necessarily depend on its earlier states, as did the 'deterministic' Newtonian universe.

(d) Pendulums, order and chaos

Pendulums have long been regarded as providing classical examples of simple, orderly motion that can be used to control clocks because of their regularity. However, a closer look reveals that this is not always so.

(*i*) *Simple pendulum.* The period of a simple pendulum of a certain length is constant for all angles of swing so long as they are small (p. 153). If the angle becomes too large, the period depends on it and varies, due to the gravitational force pulling the bob to the vertical position no longer being directly proportional to the angle of swing. The equation describing its motion is then non-linear and its period becomes unpredictable, i.e. it exhibits the **chaotic** behaviour characteristic of non-linear systems.

(*ii*) *Driven conical pendulum.* One arrangement is shown in principle in Fig. 25.38*a*. The point of suspension of the string oscillates horizontally in a straight line when the motor (whose speed has to be precisely controlled) sets the eccentric into rotation. The amplitude of the oscillation is small compared with the length of the pendulum. Two very different types of motion of the bob can be observed.

The first type occurs if the driving frequency is *just greater* than the pendulum's natural frequency, when it swings freely; it is **non-chaotic**. Initially the bob moves parallel to the drive, gaining in amplitude since near-resonance occurs due to the closeness of the driving and natural frequencies. Subsequently the swings acquire a component perpendicular to the drive, causing the bob to settle into an almost circular path. It makes one

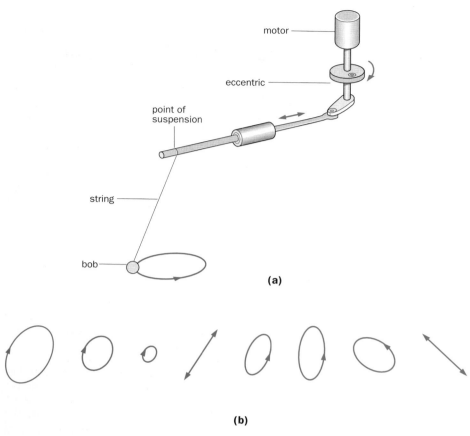

Fig. 25.38 A driven conical pendulum and its chaotic behaviour

complete orbit during one period of the drive. This motion, which is maintained for as long as the system is driven, is orderly and predictable except that it is impossible to say beforehand whether the circulation will be clockwise or anticlockwise.

The second type of motion occurs when the driving frequency is *just less* than the natural frequency; it is **chaotic**. Figure 25.38*b* shows some of the many possible paths described by the bob; in general, each one is an ellipse but the length, width, orientation and direction of rotation vary with time. There is no observable order to the changes. Furthermore, if the experiment is repeated, the sequence of orbits followed is quite different and unpredictable.

If we wished to study the unstable chaotic behaviour shown in the second case using chaos theory, we would have to make detailed observa-

tions of the bob's motion over, say, a period and feed these into a computer programmed with the system's equations of motion. The predictions about future motion would agree with the actual motion initially but eventually it would fail. The more exact the observations fed to the computer, the longer the agreement would last but in the end the observed motion would differ from that predicted. The discrepancy escalates very rapidly as time passes and militates against reliable prediction.

Unfortunately we can never know precisely how the bob is moving at a certain time since a slight puff of air or an external vibration could alter it. And, as the butterfly effect tells us, the smallest change in a chaotic system affects its behaviour profoundly. The 'sensitivity to initial conditions' is the feature that distinguishes such systems from well-ordered ones.

QUESTIONS

Radioactivity; nuclear reactions

Atomic masses:
$^{238}_{92}$U   238.0508 u
$^{234}_{90}$Th   234.0436 u
$^{212}_{84}$Po   211.9890 u
$^{208}_{82}$Pb   207.9767 u
$^{4}_{2}\alpha$ 4.0026 u

1. Estimate the energy in MeV of an alpha particle from a source of activity 1.0 μCi (3.7×10^4 Bq) which creates a saturation current of 1.0×10^{-9} A in an ionization chamber. Assume $e = 1.6 \times 10^{-19}$ C, and 30 eV is needed to produce one ion-pair.

2. Part of the uranium decay series is shown below.

    $$\begin{array}{ccccc} (1) & (2) & (3) & (4) & (5) \end{array}$$
    $$^{238}_{92}\text{U} \rightarrow {}^{234}_{90}\text{Th} \rightarrow {}^{234}_{91}\text{Pa} \rightarrow {}^{234}_{92}\text{U} \rightarrow {}^{230}_{90}\text{Th} \rightarrow {}^{226}_{88}\text{Ra}$$

    a) What particle is emitted at each decay?

    b) How many pairs of isotopes are there?

    c) If the stable end-product of the complete uranium series is lead-206, how many alpha particles are emitted between radium-226 and the end of the series?

    d) What is the Q value of the first decay? (1 u = 931 MeV)

3. A radioisotope of silver has a half-life of 20 minutes.

    a) How many half-lives does it have in one hour?

    b) What fraction of the original mass would *remain* after one hour?

    c) What fraction would have *decayed* after two hours?

4. **a)** *Ionising radiation* is emitted as a consequence of *spontaneous* and *random* nuclear decay. Different types of radiation have different *penetrating properties*.

    i) What is meant by the four terms in italics?

    ii) Give an account of the types and nature of the ionising radiation from naturally occurring radioactive sources.

    iii) Describe the different penetrating properties of the radiations in ii).

    b) A source of ionising radiation contains 3.87×10^{24} radioactive atoms of a certain nuclide and has an activity of 4.05×10^{13} Bq. Calculate

    i) the decay constant and half-life of the nuclide,

ii) the time taken for the activity of the source to fall to 8.53×10^{11} Bq.
(*UCLES*, Basic 1, June 1998)

5. Radiocarbon dating is possible because of the presence of radioactive carbon-14 ($^{14}_{6}$C) caused by the collision of neutrons with nitrogen-14 ($^{14}_{7}$N) in the upper atmosphere. The equation for the reaction is:

$$^{14}_{7}N + ^{1}_{0}n = ^{14}_{6}C + X$$

The half-life of carbon-14 is 5.7×10^3 years.

a) i) Write out the above equation, adding the proton and nucleon numbers of the particle X.
ii) Identify the particle X.

b) The mass of carbon-14 produced by this reaction in one year is 7.5 kg. 14 g of carbon-14 contains 6.0×10^{23} atoms.

i) Show that the number of carbon-14 atoms produced each year is approximately 3.0×10^{26}.
ii) Calculate the decay constant of carbon-14 in year^{-1}.
iii) Assuming that the number of carbon-14 atoms in the Earth and its atmosphere is constant, then 3.0×10^{26} carbon-14 atoms must decay each year. Use this fact and your answer to ii) to calculate the number of carbon-14 atoms in the Earth and its atmosphere.

c) A sample of wood containing carbon-14 from a tree which had recently been shopped down had an activity of 0.80 Bq. A sample of similar size from an ancient boat had an activity of 0.30 Bq.

i) Sketch, on axes like those in Fig. 25.39, a graph to show how the activity of carbon-14 in the sample having an initial activity of 0.80 Bq will vary with time over a period of three half-lives.

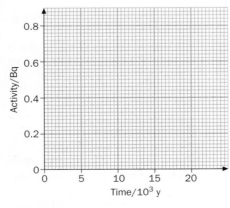

Fig. 25.39

ii) Use the graph to estimate the age of the boat.
iii) Explain briefly why an activity of 0.80 Bq would be hard to measure in a typical A Level Physics laboratory.
(*AEB*, 0635/1, Summer 1998)

6. Describe, with the aid of a diagram, the structure of a Geiger–Müller tube. Why does the tube contain a small quantity of a halogen or an organic gas?

Explain in what important respect a Geiger–Müller tube suitable for detecting beta particles differs from a tube used for detecting gamma radiation. State, giving your reasons, whether either tube would be suitable for detecting alpha particles.

A Geiger–Müller tube in conjunction with a scaler was used to investigate the rate of decay of a radioactive isotope of protactinium. Counts were made over periods of ten seconds, each count starting 30 seconds after the previous count was completed. The results, corrected for background radiation, were recorded as follows:

Time interval from start:	0–10	40–50	80–90 seconds
Count:	3410	2310	1620

Time interval from start:	120–130	160–170	200–210 seconds
Count:	1110	770	550

Determine, graphically or otherwise, the half-life of the isotope.

7. a) Figure 25.40 shows α particle tracks from a radioactive source in a cloud chamber.

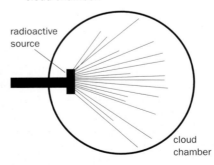

Fig. 25.40

i) Explain how a track is produced by an α particle from the source.
ii) What feature of the diagram shows that α particles are emitted

with *two* distinct values of kinetic energy?

b) The unstable isotope $^{212}_{84}$Po is formed when $^{212}_{83}$Bi decays. Write out and complete the equation below representing this decay.

$$^{212}_{83}Bi \rightarrow ^{212}_{84}Po +$$

c) i) The $^{212}_{84}$Po nucleus emits an α particle when it decays. Write out and complete the equation below representing this decay.

$$^{212}_{84}Po \rightarrow ^{4}_{2}\alpha +$$

ii) Show that the Q value for this change is 9.0 MeV.
iii) A source containing $^{212}_{83}$Bi and its daughter $^{212}_{84}$Po emits α particles of kinetic energies 8.9 MeV and 9.7 MeV. Suggest an explanation for the emission of the α particle of kinetic energy 9.7 MeV.
(*NEAB*, AS/A PHO6, June 1998)

8. Two elements A and B have the following properties.

Element	Atomic mass	Total mass of the separate protons and neutrons
A	15.9994 u	16.1320 u
B	17.0049 u	17.1407 u

Calculate the binding energy in joules of a nucleus of element A and the binding energy in joules of a nucleus of element B.

Which of these two elements would you expect to be the more stable? Explain your answer.
(*L*, AS/A PH2, June 1998)

9. a) Define the following terms:
i) 'atomic number';
ii) 'mass number';
iii) 'isotope';
iv) 'radioactive half-life'.

b) A radioactive isotope has a half-life of 6 hours. A sample of the isotope has an initial activity of 1000 disintegrations per second.
i) Draw a graph to show how the activity varies with time over four half-lives.
ii) Does the probability of decay of a single atom decrease with time in this way? Explain.

c) The isotope ^{215}Po can decay by emitting either an alpha or a beta particle.
i) Write down the equation for each of these possible decay sequences.

ii) It is known that an alpha particle from this decay can ionise about a million atoms. Estimate the energy of the alpha particle.

d) Describe briefly one use of radioactive isotopes.

(*IB*, Subsidiary/Standard Paper 2, Nov 1997)

10. a) Sketch a graph showing how the binding energy per nucleon varies with mass number. Explain the relevance of the graph to nuclear fission and nuclear fusion.

b) What purpose do the moderator, control rods and coolant each serve in a nuclear reactor? How many fissions must occur per second to generate a power of 1.0 MW if each fission of uranium-235 liberates 200 meV?

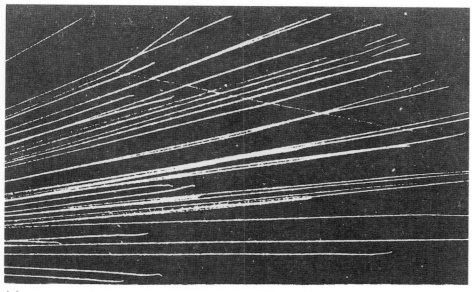

(a)

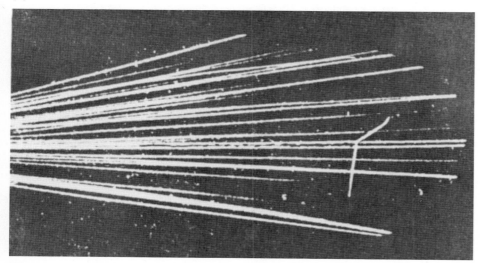

(b)

Fig. 25.41

Particle physics

11. A particle of mass m and speed v has a head-on collision with a stationary particle of mass M. Assuming the collision is elastic, derive an expression for the velocity of each particle after impact.

Hence determine what happens if the moving particle is an alpha particle and the stationary particle is

a) an electron,

b) a helium atom, and

c) a gold atom.

Under what conditions is there maximum energy transfer?

12. Two collisions between atomic particles are shown in the cloud chamber photographs of Figs 25.41*a, b*. In both photographs most of the tracks are due to alpha particles travelling from left to right. In one case there is hydrogen in the chamber and in the other nitrogen and in each an elastic collision has occurred giving a 'split' track.

a) Which photograph shows the nitrogen collision? Give reasons for your choice.

b) Copy the hydrogen collision tracks (three parts) and label each part to show the tracks of the alpha particle before and after the collision and of the hydrogen atom set into motion.

c) Draw the collision tracks for an alpha particle having an oblique elastic collision with a helium atom. Mark the size of any angle which you think is important.

13. Choose one type of particle accelerator and describe how it works.

A proton is accelerated from rest by a p.d. of 30 GV. Calculate the energy gained by the proton and hence its increase in mass during the acceleration. (1 eV = 1.60×10^{-19} J)

14. a) Describe how the idea of particle exchange can be used to explain

   i) the repulsion between two electrons,

   ii) the binding of a neutron and a proton in the nucleus.

b) State *two* differences between the exchange particles involved in a) i) and ii).

15. a) Use the conservation laws of charge, spin, strangeness, baryon and lepton number to decide whether the following reactions occur:

   i) $p + \bar{p} \rightarrow 4\pi^+ + 4\pi^-$

   ii) $K^+ + \bar{p} \rightarrow K^- + p$

   iii) $p + \pi^- \rightarrow \pi^0 + n$

   iv) $\mu^- \rightarrow e^- + \nu_e + \nu_\mu$

b) Write down the quark composition of the following particles:

   i) neutron,

   ii) proton,

   iii) antiproton,

   iv) kaon K^+.

State whether each is a meson, baryon or other type of particle.

16. a) Explain why a bubble chamber is more efficient at detecting high-energy particles than a cloud chamber.

b) Two pions π^+ and π^0 travelling at the same speed enter a bubble chamber. State and explain which pion is more likely to be detected, giving *two* reasons for your choice.

lifetime of π^+ is 2.6×10^{-8} s
lifetime of π^0 is 0.8×10^{-16} s

c) An electron produces a track of initial radius 0.55 m in a magnetic field of flux 3.5 T in a bubble chamber. Calculate the momentum of the electron.

d) Explain why the tracks produced in a bubble chamber by electrons are spiral in form.

(AQA: NEAB, AS/A PHO6, June 1999)

17. a) In a head-on collision between two protons of equal kinetic energy the following interaction was observed:

$$p + p \rightarrow p + 7\pi^+ + 7\pi^- + K^+ + \Lambda$$

Data:

mass of p	$= 938 \text{ MeV}/c^2$
mass of π^+ or π^-	$= 140 \text{ MeV}/c^2$
mass of K^+	$= 494 \text{ MeV}/c^2$
mass of Λ	$= 1115 \text{ MeV}/c^2$

Calculate the minimum kinetic energy in MeV of each proton for this interaction to occur. Explain why this is the minimum possible value.

Why would this interaction not be observed if one of the protons were stationary and the other had twice your calculated minimum kinetic energy?

b) Sketch a graph showing the energy spectrum of β^- particles emitted during β^- decay. Explain in detail why the shape of this graph led to the prediction of the existence of the neutrino.

c) An electron and a positron can annihilate by either of the mechanisms in Fig. 25.42.

Which of the fundamental interactions is represented by each figure?

Which figure illustrates the most frequent mechanism for electron–positron annihilation?

Why is the interaction shown in Fig. 25.42*b* short range compared to that shown in Fig. 25.42*a*?

(L, A PH4, June 1996)

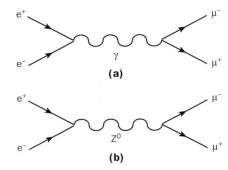

Fig. 25.42

18. a) Define **magnetic flux density**.

b) In Fig. 25.43 the arrowed circle is the path of a proton in an evacuated circular tube at a nuclear research centre. The circular path is maintained by a uniform magnetic field acting in a direction perpendicular to the plane of the circle.

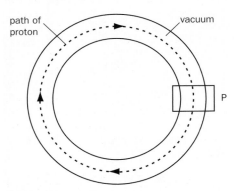

Fig. 25.43

i) At a certain instant, the speed of the proton is $2.5 \times 10^7 \text{ m s}^{-1}$ and the radius of its path is 120 m. The mass and charge of a proton are 1.7×10^{-27} kg and 1.6×10^{-19} C respectively. Calculate the flux density of the uniform magnetic field.

ii) Each time the proton passes through the region labelled P in the tube, its speed is boosted by the application of an appropriate electric field. Without further calculation, describe and explain what action must be taken to maintain the proton in its original circular path.

(OCR, 6843, June 1999)

19. a) Describe *two* experiments which can be carried out in a school or college laboratory to distinguish β radiation from γ radiation.

b) A carbon-14 nucleus undergoes β^- decay, forming a new nucleus, releasing a β^- particle and one other particle which is difficult to detect.

i) Write down the proton number and the nucleon number of the new nucleus.

ii) Name the particle which is difficult to detect.

iii) Name the baryons and leptons involved in the decay.

iv) Give the quark structure for the neutron and the proton.

Hence state the quark transformation that occurs during β^- decay.

(AQA: NEAB, AS/A PHO6, June 1999)

26

Cosmology and astrophysics

- ■ Introduction
- ■ Astronomical measurements
- ■ Hubble's law
- ■ The big bang theory

- ■ History of the universe
- ■ Stellar structure and evolution
- ■ Formation of the solar system
- ■ Astronomy

- ■ Dark matter
- ■ Future of the universe
- ■ Special relativity
- ■ General relativity

INTRODUCTION

Where did our universe come from? When did it emerge? Will it end? What is its structure? Such are the questions asked by cosmologists trying to provide a theoretical understanding of the birth, life and death of the universe. Some of the answers may be provided by astrophysicists who use the techniques and concepts of physics to study the stars.

Optical and radio telescopes which probe the heavens show us that millions of stars grouped together in clusters and nebulae within numerous galaxies make up our universe. The stars range in size from a few kilometres across for neutron stars, to over a million times larger than our sun for red giants. Our own galaxy, the Milky Way, contains about 100 billion stars; Fig. 26.1 shows a spiral galaxy similar to our own.

Fig. 26.1 The *Grand Spiral* galaxy NGC2997

ASTRONOMICAL MEASUREMENTS

(a) Distance measurements

Viewed from earth it is difficult to gauge the scale of the universe but astrophysicists have developed techniques to help to do this. Stars and galaxies are so far away that a new unit of distance measurement, the **light-year** (**ly**), is often used. For light travelling at 3×10^8 m s^{-1}, the distance travelled in one year is

$$1 \text{ ly} = (3 \times 10^8 \text{ m s}^{-1}) \times (365 \times 24 \times 60 \times 60 \text{ s}) = 9.46 \times 10^{15} \text{ m}$$

The Milky Way is about 100 000 ly across; our own sun is located on one of the spiral arms of the galaxy at a distance of 28 000 ly from the galactic centre.

How can the distance of a star or galaxy from earth be measured? For nearby stars (less than 100 ly away) **trigonometric parallax** is used; the angle of the line of sight of a star relative to the plane of the earth's orbit around the sun, θ, is measured at intervals of six months from opposite sides of the earth's orbit, Fig. 26.2.

The parallax angle is $\phi = (90 - \theta)$, and

$$\phi \approx \tan \phi = R/d$$

for small parallax angles, so the distance of the star from earth is given by

$$d = R/\phi$$

where ϕ is in radians and $R = 1.5 \times 10^{11}$ m (the earth–sun distance).

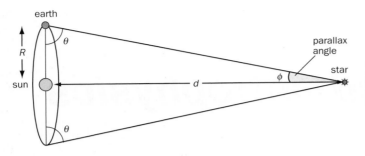

Fig. 26.2   Parallax measurement

Astrophysicists often quote distance in **parsecs** (**pc**). One parsec is defined as $1/\phi$, with ϕ in seconds of arc (rather than radians as above); for a parallax angle $\phi = 1"$, the distance from earth $d = 1/\phi = 1$ pc. To obtain the distance from earth in metres we convert the parallax angle to radians and substitute in the equation $d = R/\phi$.

$$1" = 2\pi/(360 \times 60 \times 60) \text{ rad}$$

so

$$d = \frac{R}{\phi} = \frac{(1.5 \times 10^{11} \text{ m})}{2\pi/(360 \times 60 \times 60)}$$
$$= 3.084 \times 10^{16} \text{ m}$$

i.e.

$$1 \text{ pc} = 3.084 \times 10^{16} \text{ m}$$
$$= \frac{3.084 \times 10^{16} \text{ m}}{9.46 \times 10^{15} \text{ m ly}^{-1}} = 3.26 \text{ ly}$$

The **astronomical unit** (**AU**), defined as the mean earth–sun distance, i.e. R, is another useful astronomical measurement:

$$1 \text{ AU} = 1.496 \times 10^{11} \text{ m}$$

Saturn is about 10 AU from the sun, and Pluto 57 AU; the nearest star, Alpha Centauri, is 300 000 AU or 4.3 ly away. The distance to our nearest galaxy, the Andromeda Nebula, is about 200 000 ly, while the furthest galaxies visible with the Hubble space telescope are 10^{10} ly distant.

For stars further than 100 ly away the parallax angle becomes too small to measure accurately and estimates of distances have to be made instead from brightness measurements.

(b) Brightness measurements

The brightness of a star depends on the rate at which it radiates energy into space. Stars behave like black bodies and the energy radiated per unit surface area per second is given by Stefan's law (p. 353) as $E = \sigma T^4$, where T is the absolute temperature of the star and Stefan's constant $\sigma = 5.7 \times 10^{-8} \text{ W m}^{-2} \text{ K}^{-4}$. The total power radiated by a star is termed the **absolute luminosity**, L, so

$$L = A\sigma T^4 = 4\pi r^2 \sigma T^4$$

(where the surface area $A = 4\pi r^2$ for a spherical star of radius r). Knowledge of the absolute luminosity and temperature of a star allows its size to be calculated.

Example. Estimate the radius of the sun which has a surface temperature of 5700 K and a luminosity of 3.9×10^{26} W.

From Stefan's law,

$$L = 4\pi r^2 \sigma T^4$$

so

$$r^2 = \frac{L}{4\pi \sigma T^4}$$

$$= \frac{(3.9 \times 10^{26} \text{ W})}{4\pi \times (5.7 \times 10^{-8} \text{ W m}^{-2} \text{ K}^{-4}) \times (5700 \text{ K})^4}$$

$$= 5.16 \times 10^{17} \text{ m}^2$$

and therefore the radius of the sun

$$r = 7.2 \times 10^8 \text{ m}$$

For most stars the temperature is proportional to the mass of the star; this is known as the **mass–luminosity relationship** — the more massive the star the hotter and the brighter it is. At a distance d from the star, the power is spread over the surface of a sphere of radius d and so when the star is viewed from the earth the **apparent brightness** or **luminosity**, l, is given by

$$l = \frac{L}{4\pi d^2}$$

l falls off as the inverse square of the distance and is equal to the energy *received* per square metre per second from the star.

Since the brightness of the stars varies enormously it is convenient to use a logarithmic scale of brightness in which the **apparent magnitude**, m, of two stars is defined to differ by 5 when the ratio of their luminosities is 100; this results in the definition

$$m = \text{constant} - 2.5 \log l$$

The **absolute magnitude**, M, of a star is then defined as the apparent magnitude it would have at a distance of 10 pc; this gives

$$M = m - 5 \log\left(\frac{d}{10}\right) \qquad \text{where } d \text{ is in parsecs}$$

The brighter the star the *lower* the value of its apparent magnitude; our sun has an apparent magnitude of -27; for very faint stars just visible with the naked eye, $m = +6$.

For certain types of stars known as **Cepheid variables**, the brightness varies in a periodic way over time; a well defined relationship exists between the time period of oscillations and the absolute luminosity of the star so that by measuring the time period, the absolute luminosity can be determined (see question 3, p. 507). By comparison with apparent luminosity measurements, a reliable value of the distance of the Cepheid variable can be found and hence the distance of the galaxy in which it lies; this method is very useful for finding the distance of nearby galaxies.

(c) Hertzsprung–Russell diagram

The Hertzsprung–Russell (H-R) diagram, Fig. 26.3, shows the absolute luminosities (or absolute magnitudes) of stars plotted against their surface temperature; by convention, a logarithmic scale is used for the luminosity, and increasing temperature is plotted from right to left. The colour of a star is used to gauge its surface temperature; cool stars look red while very hot stars shine blue. If a star is considered to radiate as a black body, the peak of the black body radiation curve will be in the red region of the spectrum for cool stars and in the blue for hot ones. Wien's displacement law (p. 353) enables us to estimate the surface temperature of the star from the wavelength corresponding to the peak intensity λ_{max}:

$$\lambda_{max}T = 2.9 \times 10^{-3} \text{ m K}$$

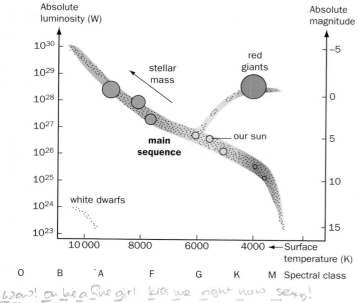

Fig. 26.3 Hertzsprung–Russell (H-R) diagram

The horizontal temperature scale on the H-R diagram is sometimes replaced by the sequence of **stellar spectral classes**, O B A F G K M; the sequence is a temperature sequence and the letters refer to the strongest absorption lines (see p. 344) that appear superimposed on the continuous black body radiation spectrum. In very hot O-stars (30 000–40 000 K) the helium absorption lines dominate the radiation spectrum, while in B- and A-stars (20 000–10 000 K) the hydrogen Balmer lines are strongest. Metal absorption lines dominate in F- and G-stars (7500–5500 K) and molecular bands are strongest for the coolest K- and M-stars (4500–3000 K).

On the H-R diagram most stars lie in a diagonal band termed the **main sequence**; cool, red M-stars with low mass lie in the low luminosity region, while hotter, blue A-stars with high mass appear in the high luminosity section. Our own yellow-hot sun is a G-star with a surface temperature of about 5700 K and it lies on the main sequence (see Fig. 26.3). Some stars lie away from the main sequence; for example the **red giants** — massive (high luminosity) but cool (red) stars which lie on the 'giant branch' — and the low luminosity (small) but hot stars, known as **white dwarfs**, which lie in the bottom left-hand corner of the H-R diagram. These positions correspond to different stages in the life cycle of a star which last for relatively short times. A star spends most of its life on the main sequence.

The absolute luminosities of a large number of stars in a cluster a known distance away are determined from their apparent luminosities; if it is assumed that all the stars in the cluster are of the same age, their position on the H-R diagram indicates their mass. Conversely, if the temperature of a main sequence star is known, its absolute luminosity can be read off from the H-R diagram; knowledge of both the absolute and apparent luminosity then allows its distance from earth to be calculated.

Example. How far from earth is a main sequence star if the wavelength at the peak of the intensity spectrum is 450 nm and its apparent brightness is 1.0×10^{-11} W m^{-2}?

First calculate the temperature of the star using Wien's displacement law:

$$\lambda_{max}T = 2.9 \times 10^{-3} \text{ m K}$$

so

$$T = \frac{2.9 \times 10^{-3} \text{ m K}}{450 \times 10^{-9} \text{ m}} = 6640 \text{ K}$$

From the H-R diagram, a main sequence star of this temperature has estimated absolute luminosity $L = 10^{27}$ W.

The apparent brightness is $l = L/4\pi d^2$, where d is the distance of the star from Earth, so

$$d^2 = \frac{L}{4\pi l} = \frac{10^{27} \text{ W}}{4\pi \times (1 \times 10^{-11} \text{ W m}^{-2})}$$

$$= 7.96 \times 10^{36} \text{ m}^2$$

and therefore

$$d = 2.82 \times 10^{18} \text{ m}$$

$$= \frac{2.82 \times 10^{18} \text{ m}}{(9.46 \times 10^{15} \text{ m ly}^{-1})} = 298 \text{ ly}$$

$$= \frac{2.82 \times 10^{18} \text{ m}}{(3.084 \times 10^{16} \text{ m pc}^{-1})} = 91.4 \text{ pc}$$

(d) Red shifts

The chemical elements present in distant stars are identified by their characteristic emission and absorption lines, observed in the radiation received from the star. Measurements from stars in neighbouring galaxies show that spectral lines from known elements are shifted in wavelength compared with the values measured in earth-based laboratories. A shift of $\Delta\lambda$ towards the red (longer wavelength) end of the spectrum is known as a **red shift** and it indicates that the galaxy is receding from the earth.

The red shift is due to the Doppler effect (p. 322) and for low relative velocities, v along the line of sight ($v \ll c$), the shift for electromagnetic waves of wavelength λ is given by

$$\Delta\lambda = \frac{\lambda v}{c}$$

So for a galaxy receding from earth with velocity v,

$$v = \frac{c\,\Delta\lambda}{\lambda}$$

(At high relative velocities, near the speed of light, this equation is no longer valid and a relationship derived from the concepts of Einstein's special theory of relativity must be used.)

Measurement of the size of the red shift enables the speed of recession of a galaxy to be calculated; very distant galaxies are found to be moving away from the earth at about one-third the speed of light.

HUBBLE'S LAW

(a) The Hubble constant

Working at the Mount Wilson Observatory in California in the 1920s, Edwin Hubble analysed data from a large number of galaxies and made the unexpected discovery that the recession speed v of a galaxy is proportional to its distance d from earth. Hubble's law is written as

$$v = H_0 d$$

where the constant of proportionality, H_0, is the **Hubble constant**, which represents the rate at which the universe is expanding at the present time. Hubble's law tells us that the more distant the galaxy, the greater its recession speed. Observations of distant galaxies suggest that they are also moving away from all other galaxies, so there is no preferred direction of expansion and all observers would see galaxies at the same distance away recede at the same speed.

Hubble's law is one of the cornerstones of modern cosmology and great effort has gone into finding the best value for H_0. Recent data from the Hubble space telescope gives

$$H_0 = 72 \pm 10 \text{ km s}^{-1}\,\text{Mpc}^{-1} \quad \text{(Mpc = megaparsec)}$$

or

$$H_0 = 23 \pm 3 \text{ km s}^{-1}\,\text{Mly}^{-1}$$

The uncertainty in H_0 is high since it is hard to obtain reliable distance measurements for galaxies.

(b) The age of the universe

Let the size of the universe be R and the age of the universe be T_0. For a galaxy at the farthest region of the universe receding from us with velocity v, Hubble's law gives

$$R = \frac{v}{H_0}$$

If we assume that the universe has always been expanding at the same rate, we have

$$R = vT_0$$

So

$$T_0 = \frac{1}{H_0}$$
$$= \frac{(3.084 \times 10^{19} \text{ km Mpc}^{-1})}{(72 \text{ km s}^{-1}\,\text{Mpc}^{-1})(3.16 \times 10^7 \text{ s year}^{-1})}$$
$$= 14 \times 10^9 \text{ years} = 14 \text{ billion years}$$

The reciprocal of the Hubble constant thus enables an estimate to be made of the age of the universe; it is only an approximation since the recession speed of galaxies and hence the Hubble constant are thought to have changed with time. However, the value obtained is consistent with other data; the oldest rocks on earth have been dated from their radioactivity to be about 4.5 billion years old, while calculations of stellar evolution suggest the age of the oldest stars to be 10–15 billion years.

THE BIG BANG THEORY

The universe is believed to have originated some 10–20 billion years ago in an event known as the **big bang**. This event is regarded as the beginning of space–time — to talk about events 'before' the big bang is meaningless! According to the big bang theory, the universe expanded tremendously rapidly from an initial, extremely compact, high density state and it has continued to expand ever since. In the expansion, the density and temperature of the universe have continually decreased, and radiation and matter have evolved into the forms in which we see them today. We will now see what evidence there is for the big bang theory of the origin of the universe.

(a) Expansion of the universe

Measurements on the spectra from distant galaxies show that they are *all* red-shifted. This surprising result means that the galaxies are all moving away from us and therefore the universe must be expanding. Electromagnetic signals from distant galaxies take a long time to reach us and so tell us about the galaxy not as it is now, but as it was millions of years ago when the signals left them; they allow us to 'look back' in time. Hubble's observations show that more distant galaxies (those nearer the big bang in time) are moving faster than nearby galaxies; it seems then that the expansion of the universe is slowing down. Latest results from galaxies at the furthest reaches of observation have however raised some doubts about this conclusion.

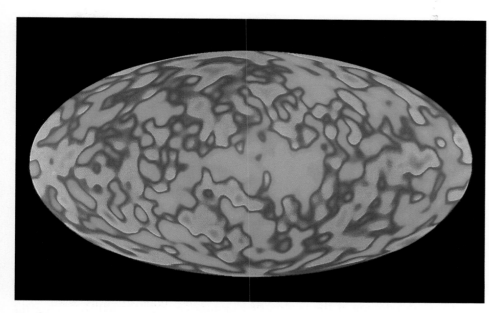

Fig. 26.4 Ripples in the cosmic microwave background radiation recorded by COBE

(b) Cosmic microwave background radiation

The big bang produced radiation energy which still exists in the universe today in the form of **cosmic microwave background** radiation, or **CMB**. It was first detected in 1964 by Arno Penzias and Robert Wilson as a background hiss in the microwave signals detected by their radio antenna, and its discovery gave strong support to the big bang theory. CMB fills the whole universe with an intensity that is nearly the same in every direction; it has a peak wavelength of 1.1 mm corresponding to a black body temperature of 2.7 K. NASA's Cosmic Background Explorer satellite (COBE) has mapped the CMB in detail. Ripples in the 2.7 K value at 'hot' and 'cold' spots have been found, as shown in Fig. 26.4; these may act as 'seeding' sites for galaxy formation.

(c) Light element abundance

Further evidence for the big bang theory comes from the quantity of light nuclei such as the isotopes of hydrogen, helium and lithium present in the universe. These are thought to have formed a few minutes after the big bang when protons and neutrons combined to form nuclei in nucleosynthesis; for example

$$^1H + {^2H} \rightarrow {^3He} + \gamma$$

The relative abundances of the light nuclei measured in the stars and interstellar gas clouds of today agree well with those predicted by the big bang theory.

HISTORY OF THE UNIVERSE

In recent years many of the concepts of particle physics have been employed in cosmology in order to understand both how the early universe evolved and the processes that continue today. It seems that the physics of the very small is linked to the physics of the very large.

(a) The first three minutes

In the standard model of the universe (the big bang) a number of changes occurred in a very short time as the universe underwent an explosive expansion and cooling. According to the **Grand Unified Theories (GUTs)** the forces of nature (p. 486) were initially unified but 'condensed' out very rapidly after the big bang in a series of **phase transitions**. The gravitational force was the first to separate at 10^{-43} s after the big bang, when the temperature of the universe was about 10^{32} K. The universe then expanded rapidly causing the temperature to drop quickly; in the **inflationary scenario** the expansion was exponential around

10^{-35} s and 10^{27} K when leptons and quarks were formed from radiation energy generated in the big bang. Such a transformation between energy and matter was predicted by Einstein and quantified by his equation $E = mc^2$ (see p. 480).

A short time later (10^{-12} s) further cooling allowed the strong force to separate from the electroweak forces and the quarks combined to form protons, neutrons and their antiparticles in what is known as the **quark era**. Radiation in the form of photons no longer had sufficient energy to break up the new particles; matter and antimatter collided, annihilating each other, but leaving a residual excess of matter which currently makes up our universe. Further expansion and cooling led to the **lepton era** at about 10^{-4} s after the big bang when the electroweak forces separated into the electromagnetic and weak forces; the lighter particles (such as electrons, muons, taus and neutrinos) dominated.

From a few seconds after the big bang the universe was **radiation-dominated**. After the initial balance between radiation and matter when there were roughly equal numbers of particles, antiparticles and photons, now antimatter had been eliminated and the universe consisted mainly of photons and neutrinos — there was more energy in the photons than in the matter.

At about three minutes after the big bang when the temperature was around 10^9 K **nucleosynthesis** became possible; the average kinetic energy of particles had fallen to a value insufficient to break up stable nuclei: $kT = (1.4 \times 10^{-23} \, \text{J K}^{-1}) \times (10^9 \, \text{K}) / (1.6 \times 10^{-19} \, \text{J eV}^{-1}) = 10^5 \, \text{eV} = 0.1 \, \text{MeV}$. When protons and neutrons collided they remained together and light nuclei such as deuterium, helium and lithium could be formed. As the universe continued to expand and cool, collisions became more infrequent and the kinetic energy of the colliding particles was not enough to overcome Coulomb repulsive forces, so that nucleosynthesis was not sustained for more than a few minutes. At the end of this time the hydrogen:helium ratio was about 3:1, a ratio that remains in the universe to this day. In this period only light nuclei were made — it was at a much later stage in the evolution of the universe that heavier elements could be manufactured in the interior of stars.

(b) After 300 000 years

It was not until around 300 000 years after the big bang that the universe was cool enough for electrons to remain attached to nuclei and for atoms to form — at 3000 K, $kT = 0.3$ eV, which is less than the energy needed to ionize atoms. The light emitted in the formation of these atoms is the source of the cosmic microwave background radiation observable today; its directional uniformity suggests that at this time all matter in the universe was uniformly distributed.

Prior to atom formation, matter existed in a plasma of free nuclei and electrons; photons interacted strongly with the plasma and were not able to travel far without being absorbed; they were 'coupled' with matter so that radiation and matter had the same effective temperature. However photons do not interact strongly with atoms and so could now travel further without collision; groups of nuclei which would previously have been scattered by interaction with radiation could now survive and grow as clumps of atoms — matter and energy 'decoupled' and could go their separate ways. Atoms collected together under the influence of gravitation and the process of accretion into stars began. The energy in the background radiation had been decreasing as the universe expanded and around this time became less than the energy in the matter — the universe became **matter-dominated**. It is now about

15 billion years since stars began to form and in this time the background radiation has cooled from 3000 K to 3 K. The history of the universe is summarized in Fig. 26.5.

STELLAR STRUCTURE AND EVOLUTION

(a) The birth of stars

Gravitational attraction between the atoms of the early universe such as hydrogen led them to collect first into clouds of gas and then into stars grouped in clusters and galaxies. Gravitational contraction of a young star leads to high pressures and temperatures developing at the centre, or core, of the star. When the mass and temperature are sufficient, hydrogen atoms reach speeds high enough for colliding nuclei to overcome Coulomb repulsion and to fuse, forming helium in a three-stage thermonuclear fusion process known as 'hydrogen burning':

Fusion reaction	*Energy released*
$^1_1H + ^1_1H \rightarrow ^2_1H + e^+ + \nu$	0.4 MeV
$^1_1H + ^2_1H \rightarrow ^3_2He + \gamma$	5.5 MeV
$^3_2He + ^3_2He \rightarrow ^4_2He + ^1_1H + ^1_1H$	12.9 MeV

The net result of the sequence is that four protons are converted to one helium nucleus; a central core of helium builds up in this process, which is also known as the **proton-proton (PP) cycle** or **PP chain**. The Q value for the cycle is given by

$$Q = (2 \times 0.4 + 2 \times 5.5 + 12.9) \text{ MeV}$$
$$= 24.7 \text{ MeV}$$

A small amount of this energy is carried off by the neutrinos, but most remains to heat up the core of the star.

Hydrogen can be converted to helium in an alternative cycle, the **carbon (CNO) cycle**, if the temperature is higher and if carbon is available as a catalyst:

$$^{12}_6C + ^1_1H \rightarrow ^{13}_7N + \gamma$$
$$^{13}_7N \rightarrow ^{13}_6C + e^+ + \nu$$
$$^{13}_6C + ^1_1H \rightarrow ^{14}_7N + \gamma$$
$$^{14}_7N + ^1_1H \rightarrow ^{15}_8O + \gamma$$
$$^{15}_8O \rightarrow ^{15}_7N + e^+ + \nu$$
$$^{15}_7N + ^1_1H \rightarrow ^{12}_6C + ^4_2He$$

Again four protons are converted to one helium nucleus in the sequence; a carbon atom is needed to initiate the cycle but is released at the end, so only acts as a catalyst. Carbon is produced in the later stages of stellar evolution and so hydrogen burning by this mechanism did not occur in the stars that first formed in our universe. However second-generation stars such as our own sun — which may have formed from the gas and dust remnants of supernovae explosions — do contain carbon. The CNO cycle dominates in stars with temperatures above 2×10^7 K, if carbon is available.

Energy is released at each step of the cycles so that tremendous amounts of energy are generated in the core, the nuclear 'furnace' of the star. This energy increases the kinetic energy of nuclei and is eventually transferred to the outer hydrogen layers by collisions. In these outer regions the mechanism of energy transfer changes, as convection currents are set up in the less dense surface layers or 'envelope'; on finally reaching the surface the energy is radiated into space as heat and light. The kinetic energy of nuclei in the interior (the thermal pressure) counteracts the gravitational collapse of material towards the core so that a balance is established and a stable star results. It takes the relatively short time of about 30 million years for a star to contract and progress to stability on the main sequence of the Hertzsprung-Russell diagram; it will then stay there, burning hydrogen, for around 10^{10} years.

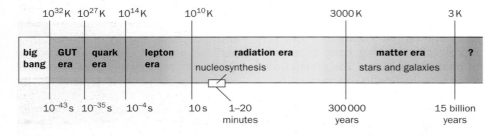

Fig. 26.5 History of the universe

(b) Stellar fuels and life cycles

Individual stars have very long lives but eventually the hydrogen fuel reserves are used up in thermonuclear fusion; the bigger the star the faster this occurs since the temperature in the core is higher and so the rate of fusion reactions is greater. The balance between gravitational and thermal pressures is upset and gravity again dominates, causing the core to contract. As the core contracts, gravitational potential energy is transferred to kinetic energy so that the core becomes hotter; there is a fast burn-up of the remaining hydrogen envelope (in contrast to the core) and a huge expansion and cooling of the outer surface gases; the spectrum of radiation that originates from the surface region shifts towards a lower temperature and the star moves off the main sequence of the H-R diagram, becoming a luminous **red giant**. As the temperature of the core increases 'helium burning' begins, in which helium fuses to carbon via the following reactions:

$$^4_2He + {^4_2}He \leftrightarrow {^8_4}Be + \gamma$$

$$^8_4Be + {^4_2}He \rightarrow {^{12}_6}C + \gamma$$

Ejection of material from the hydrogen envelope occurs as the core heats up.

Further evolution of the star depends on its mass. For small stars (less than 1.4 times the mass of our sun) the central core gradually burns to carbon. The core continues to contract, since the energy released in fusion is not enough to overcome gravitational attraction; material lost from the outer envelop forms a **planetary nebula**, while the core of the star shrinks to a **white dwarf** (a star of high temperature but low luminosity) which radiates heat until it has cooled to a **black dwarf**. It is expected that our own sun will eventually 'go out' in this way. Matter is thought to be more closely packed than normal in a white dwarf; the gravitational pressure is balanced by the quantum mechanical force that prevents electrons occupying the same quantum state — the 'electron degeneracy pressure'.

In more massive stars thermonuclear fusion can continue and a different scenario emerges. The temperature in the core increases so much that carbon fuses to produce heavier nuclei such as oxygen, silicon and eventually, in stars greater than 8 solar masses, iron. The core takes on a layered 'onion' structure as the product of one fusion process acts as fuel for the next and a sequence of collapses and 'burnings' occur. Energy cannot be extracted from the fusion of elements heavier than iron so such reactions do not fuel the nuclear furnace and the core of massive stars is gradually converted to iron. As contraction under gravitational forces progresses, the density and temperature in the core become so enormous that electrons can actually combine with protons in nuclei to form neutrons. The star then collapses catastrophically in a very short time until the density of neutrons is so high that further contraction is resisted; then the core 'bounces back' and a shock wave is generated which blows off the outer layers of the star in a giant **supernova** explosion. Many different elements may be formed and dispersed into the expanding shell of material ejected in the explosion; the large variety of elements found on earth are thought to originate from supernovae of the past. Huge amounts of energy are released as radiation, kinetic energy and neutrinos; a supernova may shine more brightly than a whole galaxy of stars. The most recent supernova visible to the naked eye was SN1987a; it occurred when the Blue Giant Sanduleak in the Large Magellanic Cloud exploded. First visible in 1987, it is still providing astrophysicists with data.

The supernova explosion observed by the Chinese in AD1054 in the Crab nebula (Fig. 26.6) is still visible today; at its centre is a **pulsar** — a rapidly rotating **neutron star** which emits intense radio waves in a cone about its magnetic axis. As a result of the rotation of the star about a different axis, radio telescopes detect short pulses of radiation occurring at regular intervals (like the beam from a lighthouse). In a neutron star it is 'neutron degeneracy pressure' that counteracts further gravitational collapse — such stars have a radius of only about 10 km but a density of 10^{17} kg m^{-3}.

Sometimes a neutron star occurs in a **binary star** configuration, in which two stars move in elliptical orbits about their common centre of mass. The stars are held together by mutual gravitational attraction and some are so close that material is actually transferred from one to the other; energy is radiated in the mass transfer process, in this case as ultraviolet and X-ray photons.

If a star is sufficiently massive, contraction may continue until the density is so high and gravity predominates to such an extent that a **black hole** results (see p. 507). Objects believed to be massive black holes have been found at the centre of some galaxies — they appear to power very luminous objects such as **quasars** and **active galactic nuclei** (**AGNs**). Matter and radiation cannot escape from a black hole due to its immense gravitational field, but as material falls towards it intense radiation is emitted.

Fig. 26.6　The Crab nebula

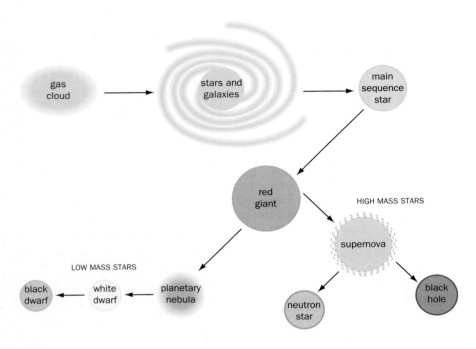

Fig. 26.7   Birth, life and death of stars

The life cycles of low-mass and high-mass stars are summarized in Fig. 26.7. The development of a star like our own sun is traced on the H-R diagram shown in Fig. 26.8. It should be realised that a star spends most of its life as a main sequence star — its birth and death occupy relatively short times in the evolution process.

FORMATION OF THE SOLAR SYSTEM

Our own sun condensed around 4.5 billion years ago from a cloud of gas and 'dust' (see p. 502). Since elements heavier than iron are present in the sun's black body radiation spectrum (and are also found on the earth) it is clear that the sun is a second-generation star, probably formed from the remnants of an earlier supernova explosion.

The planets of the solar system all rotate around the sun in the same direction and in roughly the same plane. Two hundred years ago, the French physicist LaPlace suggested that this motion indicated the sun and planets were formed from a rotating gas cloud. As the infant sun drew in surrounding material by gravitational attraction, the rotating gas cloud flattened into a disc shape. As more material collected at its centre the cloud rotated faster, as a consequence of conservation of angular momentum (just as a ballet dancer rotates faster when she draws in her arms). Large particles in the cloud captured nearby smaller ones and eventually most of the material was swept into the sun or into smaller entities which became planets, asteroids and comets orbiting the sun. The motion of these entities can be described by Kepler's third law (see p. 139). Asteroids are small rocky bodies which today lie mainly in the asteroid belt between Mars and Jupiter; if their orbits around the sun cross the earth's path, they may fall to earth as meteors or shooting stars. Comets have much larger orbits (generally outside that of Uranus) and consist of dust embedded in ice formed from water and methane; they are often referred to as 'dirty snow-balls'. Halley's comet, last seen in 1986, returns to the vicinity of earth every 76 years.

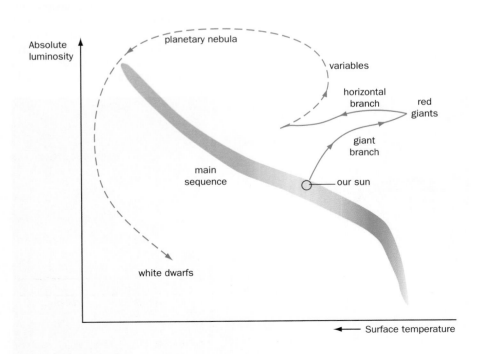

Fig. 26.8   H-R diagram showing the evolution of a star of one solar mass

Like asteroids and comets, the inner and outer planets are found to be very different from each other both in size and chemical composition; the inner planets such as Venus, Mars and the earth are small, rocky worlds made of high density material compared with the large, gaseous, low density outer planets such as Jupiter and Saturn. When the sun grew sufficiently large that nuclear fusion became possible, it must have been rocked by violent nuclear explosions in which nearby material was blown outwards against the gravitational forces; light elements such as hydrogen and helium would have been blown to the outer regions of the solar system while the heavier elements remained in the inner regions — hence the difference in density between the inner and outer planets.

As the sun began to radiate energy and a 'solar wind' developed, any gas or small particles left between the planets was swept to the outer regions of the solar system. The solar wind consists of a stream of charged particles, mainly hydrogen and helium ions, continually emitted by the sun; it distorts the magnetic fields of the planets. Sudden surges in the sun's activity are in evidence today in the form of solar flares and sunspots.

The planets were too small for nuclear fusion to commence and so since their formation they have gradually cooled; they are only visible through reflected light from the sun. Any original atmospheres of the inner planets were lost when thermonuclear fusion began in the sun; the secondary atmospheres now present result from the release of gases from the interiors of the planets; on earth the composition of the atmosphere has been modified by biological processes.

ASTRONOMY

(a) Telescopes

Our present-day knowledge of the universe comes from a variety of astronomical telescopes. Optical telescopes (p. 98) give us information about bodies in the temperature range 3000–10 000 K — our own sun has a surface temperature of 5700 K. Infrared and radio telescopes (p. 103) tell us about cooler regions of space, less than 3000 K, and are particularly useful for determining the properties of cool interstellar gas and dust clouds. At the other end of the spectrum, ultraviolet, X-ray and γ-ray astronomy can map the very hot bodies (10^4 to 10^9 K) in our universe. Radio waves are produced when high-energy electrons are accelerated by magnetic fields, while X-rays and γ-rays result when energetic electrons and protons collide with interstellar material; each type of radiation can tell us about the physical conditions at their source.

Ground-based telescopes suffer from blurring of images due to fluctuations in the refractive index of the

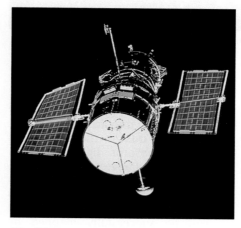

Fig. 26.9 The Hubble space telescope

earth's atmosphere; this causes scattering of incoming light and the apparent 'twinkling' of stars. Sharper images can be obtained from telescopes mounted in satellites above the atmosphere — NASA's Hubble space telescope (Fig. 26.9) has sent back exceptionally clear optical images. Ultraviolet, X-ray, γ-ray and some infrared wavelengths are absorbed by the earth's atmosphere (Fig. 26.10) and so telescopes based on these wavelengths must also be operated above the atmosphere. IRAS (the Infrared Astronomical Satellite) has mapped the whole of the sky in the infrared wavebands while the COBE (Cosmic Background Explorer) satellite is providing detailed astronomical data in the microwave region of the electromagnetic spectrum. At short wavelengths, the Einstein X-ray satellite, the HEAO (High Energy Astrophysical Observatory) and the CGRO (Compton γ-ray Observatory) are collecting information on supernova explosions and neutron stars. Recently γ-ray bursts brighter than supernovae, but lasting for only a few seconds, have been detected from all regions of the sky — they are thought to be due to the collision of two neutron stars in a binary system.

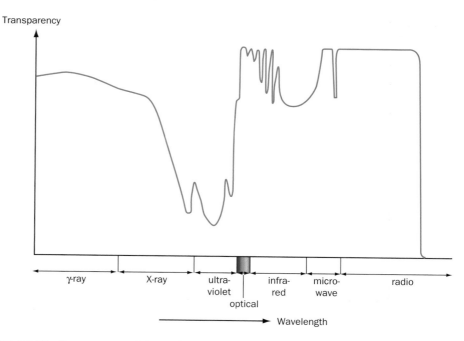

Fig. 26.10 Transparency of the earth's atmosphere to electromagnetic radiation of different wavelengths

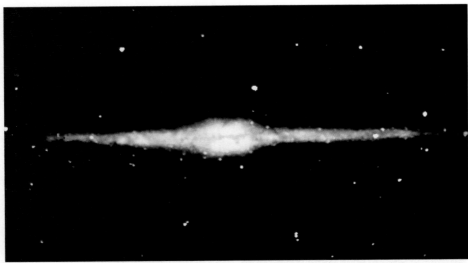

Fig. 26.11   View of the Milky Way galaxy at infrared wavelengths (by COBE)

(b) Galaxies

Galaxies occur in a variety of forms; the main types have an elliptical or spiral shape. Our own galaxy, the Milky Way, is spiral with a central bulge and a disc of stars and interstellar clouds of dust and gas, which rotate about the galactic centre (Fig. 26.11). New stars form in the arms of spiral galaxies, giving them a bluish appearance. In elliptical galaxies the disc is absent; light comes from old redder stars and there is little sign of interstellar dust, gas or new star formation. Some galaxies have a very luminous object such as a quasar or active galactic nucleus (AGN) at their centre.

It is possible to estimate the mass of a galaxy by equating the centripetal force on a star of mass m, orbiting the galactic centre at radius R, with the gravitational force of the mass M inside the orbit:

$$\frac{mv^2}{R} = \frac{GmM}{R^2}$$

so

$$M = \frac{v^2R}{G}$$

For our sun $v = 2.5 \times 10^5 \text{ m s}^{-1}$ and $R = 28\,000$ ly, which gives

$$M = 2.7 \times 10^{41} \text{ kg}$$
$$= 1.3 \times 10^{11} \text{ solar masses}$$

The calculation assumes most of the mass is concentrated in the central region of our galaxy which may not be true if 'dark matter' (see the next section) is taken into account.

Interstellar dust in the spiral arms of galaxies absorbs light and may obscure optical images of distant stars. Absorption of radiation leads to heating of the dust particles which then re-radiate in the infrared and so can be successfully mapped by infrared and radio telescopes to give information about the temperature, composition and motion of the gas cloud; rotation of our galaxy was deduced from red-shift measurements on the 21 cm spectral line emitted from cool interstellar hydrogen. If a large explosion occurs nearby the dust may be heated sufficiently to emit in the visible range and give colour images in optical telescopes. Interstellar dust is mainly composed of silicates and carbon particles about 1 μm in diameter which produce scattering of radiation of wavelength of comparable size to the dust particles. When we wish to map the stars at the centre of our own galaxy we need to use radio waves, since the dust is transparent to radio wavelengths. **Interstellar gas clouds** may consist of hydrogen, carbon monoxide and many other molecular species and they can vary greatly in temperature. Young hot blue stars can be seen condensing in the spiral arms of galaxies from the gas and dust left from older exploded stars; the matter in the universe is constantly being recycled.

DARK MATTER

The present range of telescopes — radio, optical through to X- and γ-ray — gives us information about stars and galaxies which emit radiation over the whole electromagnetic spectrum. But is there matter in the universe that remains undetected because it does not emit electromagnetic radiation? It is now thought that our galaxy and others contain such 'dark matter'. Observations of the rotation of stars in a galaxy show that their speed remains constant with distance from the galactic centre, suggesting there is more mass associated with the galaxy than is visible. A knowledge of the amount of dark matter is essential if cosmologists are to predict the future of the universe. Dark matter could be present in a variety of forms, from cold clouds of interstellar dust to the more exotic 'MACHOs' and 'WIMPs'.

(a) MACHOs and black holes

Possible candidates for MACHOs (massive compact halo objects) are planets, brown dwarfs ('failed' stars which were not large enough to initiate thermonuclear fusion), black dwarfs and black holes.

As we have seen, black holes arise from the collapse of a massive star into a very small high-density object in which gravitational forces are so great that no matter or radiation (and hence no electromagnetic signature) can escape. An estimate can be made of the radius, known as the **Schwarzchild radius**, R_S, to which a body of mass M must shrink in order for a black hole to result. The escape speed from mass M of radius r was found (see p. 146) to be $v_e = \sqrt{(2GM/r)}$. If the radius shrinks so that the escape speed becomes c, even light could not escape and

$$c^2 = \frac{2GM}{R_S}$$

or

$$R_S = \frac{2GM}{c^2}$$

A black hole with the mass of the earth would have a radius of 1 cm!

General relativity tells us we cannot see events at distances from a mass within the 'event horizon'; for a spherical black hole the event horizon is a sphere with the Schwarzchild radius.

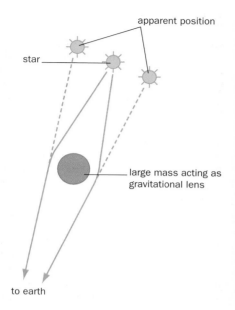

(a) Gravitational lensing

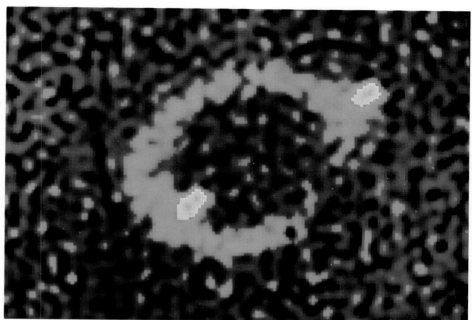

(b) 'Einstein ring' image of the radio source MG1131+0456

Fig. 26.12

Einstein's general theory of relativity predicts the bending of light by massive objects as shown in Fig. 26.12*a*; such 'gravitational lensing' is observed when an invisible MACHO passes in front of a more distant star and enables an estimate of the mass of the MACHO to be determined. The 'Einstein ring' image of a distant radio galaxy shown in Fig. 26.12*b* is due to the presence of a massive ellipsoidal galaxy situated between the source and earth. Astrophysicists also use gravitational lensing as a tool to obtain information about distant galaxies.

(b) WIMPs

Calculations suggest that so many neutrinos were produced in the initial big bang that they may now be the most common particle; the universe may consist of an ocean of neutrinos with relatively few islands of atoms. If they prove to have a mass, even of only 30 eV, neutrino WIMPs (weakly interacting massive particles) could provide a large amount of the dark matter. Perhaps the fate of the universe lies in the hands of the ubiquitous neutrino which passes through everything without leaving a visiting card.

FUTURE OF THE UNIVERSE

Will the universe continue to expand? Calculations of the total matter in the universe, including both light and dark matter, may give us the answer. If the density of matter is greater than a critical value, the **critical density** ρ_c, gravitational forces will be able to reverse the expansion and cause the eventual collapse of the universe in the 'big crunch'.

A rough estimate of ρ_c can be made by considering a large galaxy of mass m, at a distance d from the earth, receding with velocity v. If the volume of space around the earth, $4\pi d^3/3$, contains mass M and the galaxy has just enough kinetic energy to escape the gravitational field associated with this mass then

$$\frac{mv^2}{2} = \frac{GMm}{d} \quad \text{(see p. 146)}$$

(see p. 146)

and the universe would be expected to be 'open', i.e. continue to expand. Using Hubble's law and substituting $v = H_0 d$ gives

$$(H_0 d)^2 = \frac{2GM}{d}$$

Taking $M = \rho_c 4\pi d^3/3$ gives

$$H_0^2 d^3 = \frac{2G \cdot 4\pi d^3 \cdot \rho_c}{3}$$

and

$$\rho_c = \frac{3H_0^2}{8\pi G}$$

A more rigorous derivation using special relativity gives the same result.

Using the best value of the Hubble constant, the critical density is found to be about 10^{-26} kg m^{-3}; current estimates of the total mass density ρ of the universe give a value less than this (about $0.3\rho_c$) so we can expect the universe to continue to expand.

Defining the **density parameter** $\Omega = \rho/\rho_c$, then if $\Omega < 1$ we have an 'open' universe which will continue to expand. If $\Omega > 1$, gravity will eventually dominate and the universe is 'closed', i.e. it will eventually collapse. For $\Omega = 1$, the universe is 'flat' and will expand until the recession speed of galaxies is zero. Figure 26.13 depicts the three possible scenarios for the evolution of the universe since the big bang. Many theorists believe that the universe is flat and that the amount of matter in the universe has been underestimated. In this scenario the critical density and hence Hubble's constant must decrease with time; our estimate of the age of the universe (from H_0) and its ultimate fate both depend on the value of ρ_c.

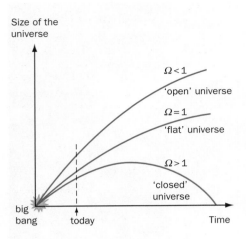

Fig. 26.13 How the value of the density parameter (Ω) affects the expansion of the universe

One of the assumptions of modern cosmological theory is that our galaxy is not located at any special position in the universe; the universe would be expected to look much the same to an observer in another galaxy as it does to us. This assumption is known as the **cosmological principle**. The non-directional nature of the cosmic microwave background radiation lends support to this idea that the universe is isotropic and homogeneous.

Why is the sky dark at night? If it is assumed that the universe is uniform, static and infinite, then light should eventually reach us from every part of the sky so it should be bright at night; however in an expanding universe with a finite age, light may not yet have reached us from some regions of the sky. This obvious fact, known as **Olber's paradox**, lends support to our current model of the universe.

SPECIAL RELATIVITY

Einstein proposed his **special theory** in 1905 to explain the relative motion of bodies travelling at speeds near that of light. While the special theory predicts the same behaviour as Newton's laws at low velocities (much less than the speed of light) it gives very different results for bodies moving at high speeds — such as stars, galaxies and elementary particles.

(a) Frames of reference

The driver of a car travelling with *uniform* velocity has *zero* velocity relative to a passenger in the car but a *constant* velocity relative to a pedestrian standing at the roadside. The car is in one **frame of reference** and the roadside in another. The observed motion of a body clearly depends on the motion of the observer as well as the body — a well established statement of classical physics.

The laws of motion hold in each frame of reference since in each the resultant force on the car is zero and the observed velocity is constant or zero (Newton's first law). Frames for which this is true are termed **inertial frames**.

(b) Postulates

By definition the laws of motion are constant in all inertial frames; Einstein extended this idea to include all the laws of physics. The postulates of special relativity are

> **1.** The laws of physics are the same for all observers in all inertial reference frames.
> **2.** The measured velocity of light in a vacuum, c, is the same in all inertial frames of reference and is independent of the motion of the light source or the observer.

The second postulate contradicts the previous classical physics statement about the relative velocity of bodies in motion. It was derived by rejecting the 'common sense' view of the constancy of space, time and mass measurements and allows for a change of value in different inertial frames; the difference is negligible except when the relative velocity of the frames is near the speed of light.

Evidence for the second postulate comes from the null result of the experiments of Michelson and Morley in the 1890s. At that time it was thought that electromagnetic waves required a medium (the ether) in which to propagate. Michelson and Morley used the Michelson interferometer (p. 337) to try to measure a change in the optical path lengths in the two arms of the interferometer when the apparatus was rotated through 90°. Such a change would be expected if the speed of light was affected by the motion of the earth through the ether, but no change in the interference pattern could be detected despite the high sensitivity of the equipment. This was a very puzzling result until it was realised that no medium is required for electromagnetic waves to propagate and light can travel in a vacuum — where its speed is always c, regardless of the relative motion between source and detector.

(c) Time dilation

A consequence of the second postulate is that an observer who is moving uniformly relative to a pair of events measures a longer time interval between the two events than does an observer who is at rest relative to the events. Time dilation may be illustrated by the thought experiment shown in Fig. 26.14, in which two 'events' are viewed from two different frames of reference.

On the spaceship a flash of light sent out from the light clock (event 1) is reflected from a mirror and returned to the clock (event 2). To Zoe on the spaceship the time taken for the light to bounce off the mirror is the 'proper time'

$$T_{\mathrm{p}} = \frac{2D}{c}$$

so

$$4D^2 = (cT_{\mathrm{p}})^2$$

If the spaceship travels in a direction perpendicular to D, at speed v relative to Adam on the earth, he measures a time interval

$$T = \frac{2L}{c}$$

between the two events occurring on the spaceship. He will need two synchronized clocks in different places to measure T, since the spaceship moves a distance vT in the time interval. Since

$$L = \sqrt{[D^2 + (vT/2)^2]}$$

$$T = \frac{2L}{c}$$

$$= \frac{2\sqrt{[D^2 + (vT/2)^2]}}{c}$$

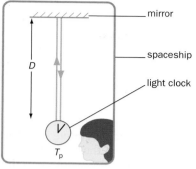

(a) Zoe, on a spaceship, measures the time between two events on the spaceship

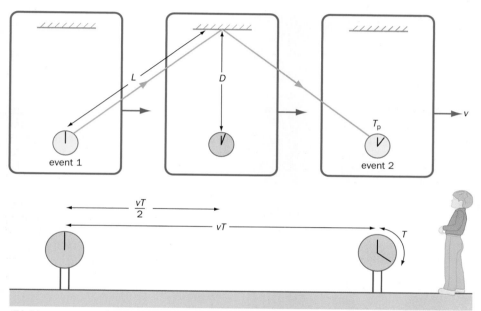

(b) Adam measures the time between the same two events, as seen from the earth

Fig. 26.14 Time dilation thought experiment

and so

$$(cT)^2 = 4D^2 + (vT)^2$$

Substituting for $4D^2$,

$$(cT)^2 = (cT_p)^2 + (vT)^2$$

or

$$T = \frac{T_p}{\sqrt{\left(1 - \dfrac{v^2}{c^2}\right)}}$$

The time measured from the earth is longer than the proper time measured by Zoe. Proper time requires only one clock to measure the time interval between two events that occur at the same place in that frame of reference.

Example. A stationary muon decays in 2.2 μs. What is its lifetime if it is moving at $0.97c$ relative to laboratory clocks? How far does it travel before it decays?

The proper time $T_p = 2.2 \times 10^{-6}$ s since the lifetime measurement on a muon at rest requires only one clock. The lifetime of the moving muon is then

$$T = \frac{T_p}{\sqrt{[1 - (v/c)^2]}}$$

$$= \frac{2.2 \times 10^{-6} \text{ s}}{\sqrt{(1 - 0.97^2)}}$$

$$= 9 \times 10^{-6} \text{ s} = 9 \text{ μs}$$

During this time it travels a distance

$$vt = (0.97 \times 3 \times 10^8 \text{ m s}^{-1}) \times (9 \times 10^{-6} \text{ s})$$

$$= 2.6 \times 10^3 \text{ m}$$

Measurements made on the lifetimes of muons produced when cosmic rays enter the earth's atmosphere confirm these results and so provide evidence for time dilation. Further confirmation is obtained using atomic clocks; when flown around the world in aeroplanes they record a shorter time for the trip than do ground-based clocks.

(d) Length contraction

An observer who is moving uniformly relative to an object (parallel to its length) measures a shorter length than does an observer who is at rest relative to the object. Length contraction can also be illustrated by our thought experiment. In Fig. 26.14 Adam measures the distance between his two clocks to be the 'proper length'

$$L_p = vT$$

To Zoe on the spaceship the clocks on the earth appear to pass her at a velocity v in a time interval T_p and so she believes the distance between them to be

$$L' = vT_p$$

From time dilation

$$T_p = T\sqrt{(1 - v^2/c^2)}$$

so

$$L' = vT\sqrt{(1 - v^2/c^2)}$$

or

$$L' = L_p \sqrt{\left(1 - \frac{v^2}{c^2}\right)}$$

Zoe measures a shorter length than Adam as a result of their relative motion. Note that in this case the proper length L_p is measured at rest and *length = velocity × time* holds in both frames of reference.

The special theory of relativity links space and time closely; when space shrinks, time lengthens. Space has three dimensions and time one, so **space–time** has four dimensions.

Figure 26.15 is a computer simulation showing how the appearance of a tram might change when it is moving with a speed close to c.

(a)

(b)

Fig. 26.15 View of a tram (a) at rest, (b) moving at a relativistic speed

(e) Relativistic mass

In Einstein's special theory of relativity the **relativistic mass** m is given by

$$m = \frac{m_0}{\sqrt{\left(1 - \dfrac{v^2}{c^2}\right)}}$$

where v is the speed of the moving frame of reference relative to the observer's frame, c is the speed of light and m_0 is the mass of the object in the observer's (stationary) frame of reference, called its **rest mass** (the mass obtained when $v = 0$). This result is confirmed by observations from particle accelerators where high-speed particles appear to have a higher mass than when measured at rest.

It is clear that nothing can be accelerated to the speed of light. At $v = c$, the relativistic mass would become infinite and infinite energy would then be needed to accelerate the mass, an impossible scenario. For $v > c$ the mass would be imaginary — again impossible!

The most far-reaching consequence of special relativity is Einstein's famous mass–energy equation

$$E = mc^2 \qquad \text{(see p. 480)}$$

(see p. 480)

Writing the momentum of the body as

$$p = mv = Ev/c^2$$

and

$$E^2 = m^2c^4 = \frac{m_0{}^2c^4}{(1 - v^2/c^2)}$$

then

$$E^2 - E^2v^2/c^2 = m_0{}^2c^4$$

and the relativistic relation between energy and momentum becomes

$$E^2 - p^2c^2 = m_0{}^2c^4$$

or

$$E^2 = m_0{}^2c^4 + p^2c^2$$

GENERAL RELATIVITY AND GRAVITATION

In the special theory of relativity Einstein considered frames of reference moving relative to each other with uniform velocity. In the **general theory** of relativity, published in 1915, he turned his attention to the situation where one frame of reference is accelerating relative to another, as is the case when gravitational forces are acting.

Consider a falling body. To an observer on the ground in an inertial frame of reference, the body is accelerating under the action, according to Newton, of a gravitational force. However, to an observer falling with the body, the body remains at rest relative to his accelerating frame of reference and there is no reason to suppose a gravitational force is acting. Therefore by choosing the correct frame of reference, we can eliminate the need for gravitation. In other words, the effects of acceleration and a gravitational field are equivalent.

The essence of Einstein's general theory, developed from the above ideas, is that gravitation is better regarded, not as a force, but as a *curvature in the geometry of space–time*. Put crudely, it states that the earth orbits the sun in a curved path because the sun distorts space–time in its vicinity and causes the earth to take the 'straightest' route through the distortion. It is, however, more than a complex piece of abstract mathematics since it can account for certain observable effects, inexplicable by gravitation, including

- the slight difference between the observed orbit of Mercury and that predicted by Newton's theory;
- the bending of stellar light as it passes close to the sun, resulting in the apparent displacement of the position of the star (gravitational lensing); and
- the slowing down of the vibrations within an atom near a star, causing the light from it to have a slightly longer wavelength, i.e. be redder, than the corresponding light from an atom on earth (the gravitational red shift).

While Einstein's theory reduces to Newton's under most conditions, there are two areas where major differences arise. First, general relativity predicts that the collapse of stars can cause great distortion of space–time, resulting in **black holes** with such intense gravitational fields that even light cannot escape. Second, general relativity predicts that heavy moving objects (like collapsing stars) emit gravitational *waves* in the form of massless **gravitons** travelling at the speed of light — just like photons. Their detection is one of the big challenges of experimental physics today. If evidence for their existence were found it would enable the gravitational force to be described in terms of an exchange mechanism (p. 486) and hence to confirm the unified theory of all force.

QUESTIONS

Cosmology and astrophysics

$H_0 = 72$ km s^{-1} Mpc$^{-1} = 23$ km s^{-1} Mly^{-1}
1 ly $= 9.46 \times 10^{15}$ m; 1 pc $= 3.084 \times 10^{16}$ m

1. **a)** The parallax angle for a star measured from the earth is 0.00005°. How far away is the star in
   i) parsecs,
   ii) light-years,
   iii) astronomical units?
   b) What is the parallax angle in seconds of arc for a star 20 parsecs distant?
   c) The apparent magnitude of a star is 0.2. If it is 7.2 pc distant from earth, what is its absolute magnitude?

2. **a)** The apparent magnitudes m_1 and m_2 of two stars differ by 5; the ratio of their luminosities l_1 to l_2 is 100.
   i) By writing

   $$l_1/l_2 = 100^{(m_2 - m_1)/5}$$

   show that

   $$m_2 - m_1 = 2.5 \log l_1/l_2$$

   ii) Hence show that, in general,

   $$m = \text{constant} - 2.5 \log l$$

   b) Two stars have the same apparent brightness when viewed from earth but one is 10 times further away than the other. What is the ratio of their absolute luminosities?

3. **a)** The Hubble Space Telescope was used to study the spiral galaxy M100. The resulting images allowed astronomers to measure the light curves of the most distant Cepheid variable stars ever observed. Figure 26.16 shows the variation of the apparent magnitude with time for such a star.

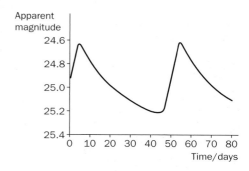

Fig. 26.16

Use Fig. 26.16 to estimate
   i) the period of the variable star,
   ii) the mean apparent magnitude of the star.
   b) Figure 26.17 shows the absolute magnitude as a function of the period for a group of such variable stars.

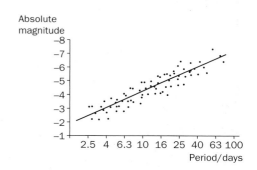

Fig. 26.17

   i) Use Fig. 26.17 to estimate the absolute magnitude of the star referred to in part a).
   ii) Hence calculate the approximate distance to this star.
   (NEAB, AS/A PHO4, March 1997)

4. Measurements of the apparent brightness of a star from earth give a value of 5×10^{-11} W m^{-2} and the peak intensity in the radiation spectrum occurs at 400 nm.
   a) What is the surface temperature of the star?
   b) If the star lies on the main sequence of the H–R diagram, what would you expect the absolute luminosity of the star to be?
   c) How far from earth is the star?

5. **a)** Astronomers have discovered from their observations of the star Capella that:

- its surface temperature T is 5200 K
- its distance from the Earth is 4.3×10^{17} m
- at the Earth's surface the intensity of the radiation received from Capella is 1.2×10^{-8} W m^{-2}.

Explain briefly how the surface temperature is determined.

Describe, in outline only, the parallax method for finding the distances of the star from the Earth.

Calculate the radius r of Capella, given that its luminosity L can be found by using

$$L = 4\pi r^2 \sigma T^4$$

where σ is the Stefan-Boltzmann constant which is 5.7×10^{-8} W m^{-2} K^{-4}.

b) Sketch and label an HR diagram showing the **main sequence** and the regions occupied by **red giants** and **white dwarfs**.

Explain why main sequence stars of large mass have higher luminosities and shorter lives than main sequence stars of low mass.

Describe the processes which occur within a star similar in mass to the Sun when it leaves the main sequence.

(L, A PH3, Jan 1997)

6. **a)** Write an account of the Hertzsprung–Russell diagram. In your account, you should include a diagram and refer to

i) the quantities which need to be known to be able to construct the diagram,
ii) how these quantities may be determined,
iii) the significance of the diagram to our understanding of stars.

b) A recent discovery is the neutron star GROJ 1744-28. It is believed to be the most powerful stellar X-ray source yet found. Measurements suggest that it has a radius of about 8 km and that its surface gravitational field is some 10^{11} times greater than that of the Earth.

Deduce a value for the mass of the star and suggest a mechanism for the emission of X-rays.
(mass of Earth = 6.0×10^{24} kg; radius of Earth = 6.4×10^6 m)

(UCLES, Cosmology, June 1998)

7. The red shift of a galaxy indicates it is receding at a speed of 2000 km s^{-1}.
a) Assuming Hubble's law holds, how far away is the galaxy from earth in
i) parsecs, and
ii) light-years?
b) What will be the measured wavelength of the hydrogen spectral line which occurs at 486 nm on earth?
c) Estimate the age of the universe for a Hubble constant of 70 km s^{-1} Mpc^{-1}.

8. **a)** Two radiant sources of equal size have different kelvin temperatures T_1 and T_2 where $T_1 = 2T_2$. Sketch two curves on a single graph comparing the radiant energy distributions against wavelength for the two radiant sources.

Explain which features of your curves illustrate the Stefan-Boltzmann law and Wien's law.

Explain which of these two laws you would expect to be of most help in estimating the temperature of a distant star.

b) List the steps in the pp chain by which energy is released in stars such as the Sun. Explain why the process takes a long time to complete.

Describe the physical processes which transport the energy released within the central core of a star such as the Sun to the surface of the star and away into space.

(L, A PH3, June 1997)

9. **a)** Outline the standard model of the universe giving the sequence of events from the big bang through to the formation of atoms. What evidence is there to support the big bang theory?
b) Derive the expression for the critical density of matter in the universe:

$$\rho_c = \frac{3H_0^2}{8\pi G}$$

Evaluate ρ_c numerically and explain what relevance it has to the future of the universe.

10. A spectrum of light from a distant galaxy shows the spectral lines of a certain element. The spectral lines are compared with those produced in the laboratory from the same element. Figure 26.18 shows the two sets of lines. (The dark lines represent the galaxy lines; the dotted lines represent the laboratory lines.)
a) i) Name the effect which causes the shift in the position of the spectral lines.

Fig. 26.18

ii) State which end (left or right) in Fig. 26.18 represents the long-wavelength end of the spectrum. Explain your answer.

b) Explain how the shifts of spectral lines, such as those shown in Fig. 26.18, have been used to support a theory about the evolution of the Universe.

(OCR, 6844, June 1999)

11. **a)** The mass of the Sun is approximately 1000 times greater than that of Jupiter. The radius of Jupiter's orbit (assumed circular) is 7.8×10^{11} m. The distance of their common centre of mass along the line of their centres may be calculated by using the principle of moments.

Show that x, the distance between the Sun's centre and the common centre of mass of the Sun–Jupiter system, is approximately 7.8×10^8 m.
b) Two bodies in orbit about each other rotate about their common centre of mass.

There is evidence that many stars may have planetary systems. Imagine that the star α Centauri, which is 4.3 light-years from the Sun, has a planet from which observations of the Solar System can be made.

Use the value of x given in a) to calculate the angle through which the Sun would appear to oscillate when viewed from this planet of α Centauri. (The effect of the planet's orbital motion may be ignored.)
c) Figure 26.19 is a sketch of the variation of planetary orbital speeds with distance from the central body.
i) Copy Fig. 26.19 and sketch on it another curve to show how the orbital speed of a star about the centre of the Milky Way Galaxy varies with its distance from the centre of the Galaxy.
ii) Explain the shapes of these two curves.

(OCR, Cosmology (4837), Nov 1999)

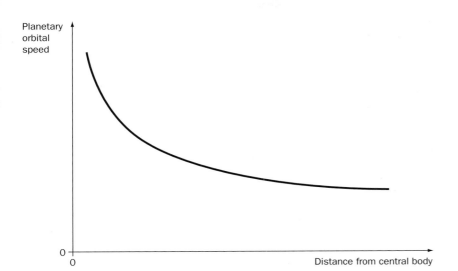

Planetary orbital speed

0

0   Distance from central body

Fig. 26.19

Relativity

12. The mean lifetime of a pion at rest is measured as 26 ns. For a pion travelling at 0.95c relative to a laboratory detector, determine
a) the laboratory lifetime of the pion,
b) the distance the pion travels in the laboratory.

13. a) A spaceship of proper length 100 m passes close to a space station. An observer on the station determines the speed of the spaceship to be 2.5×10^8 m s^{-1}. What is the length of the spaceship according to:
  i) the pilot of the spaceship,
  ii) the observer on the space station?
b) Use your answer in a) to calculate how much time it takes for the spaceship to pass the observer on the space station, according to:
  i) the pilot of the spaceship,
  ii) the observer on the space station.
c) The observer continues to track the spaceship, and notes that it travels at a constant speed in a straight line for 3 years, then turns around and returns at the same speed, taking a further 3 years to complete the journey. The total time for the journey according to the space station observer is 6 years.
  i) According to the travellers on the spaceship, is the journey time less than, equal to or greater than 6 years? Explain your answer.
  ii) Whose version of the time taken is 'correct', or are both correct or neither correct?
(*IB*, Subsidiary/Standard Paper 3, Nov 1998)

14. The mass m of an electron moving with a speed v is given, according to special relativity, by

$$m = \frac{m_0}{\sqrt{(1 - v^2/c^2)}}$$

where m_0 is the rest mass of the electron and c is the speed of light in vacuo.
a) What does the equation tell us about the effect on the electron's mass if it is accelerated to speeds near c?
b) Name one other relativistic effect shown by high-speed electrons.

15. Relativistic effects start becoming noticeable at about 10% of the speed of light. Sometimes, speeds greater than this can occur in everyday objects. Suppose an electron has been accelerated through a potential difference of 20 kV in an ordinary television. Calculate, using relativity,
a) the speed of the electron, and
b) the momentum of the electron.
(*IB*, Higher Paper 3, May 1998)

16. According to Einstein's theory, mass, length and time are quantities that are subject to relativistic effects.
a) Two observers in different inertial frames of reference each measure, in their own and the other frame, the quantities listed in the table below.

Quantity	Measured value
mass	larger/smaller/the same
length	larger/smaller/the same
time	larger/smaller/the same

State for each quantity whether the relative speed of the two frames results in the quantity in the *other* frame being measured as larger, smaller or the same as in that observer's own frame.
b) Use your answers to a) to suggest how the observer in one frame considers the density of an object which is stationary in the other frame to be affected.
(*OCR*, Cosmology (4837), June 1999)

17. In recent years, multiple images of a very distant quasar have been observed as a consequence of 'gravitational lensing' by a large galaxy lying between Earth and the quasar, as illustrated in Fig. 26.20.
Use Einstein's relativity ideas to suggest an explanation of why photons of light emitted from the quasar may be deviated as shown in Fig. 26.20.
(*OCR*, Cosmology (4837), June 1999)

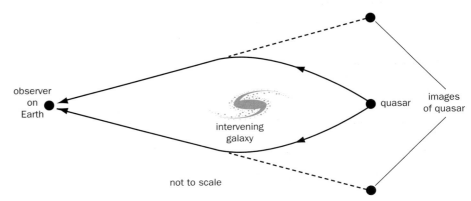

observer on Earth   intervening galaxy   quasar   images of quasar

not to scale

Fig. 26.20

Objective-type revision questions 5

The first figure of a question number gives the relevant chapter, e.g. **21.2** is the second question for chapter 21.

MULTIPLE CHOICE

Select the response that you think is correct

21.1 At pressure P and absolute temperature T a mass M of an ideal gas fills a closed container of volume V. An *additional* mass $2M$ of the same gas is introduced into the container and the volume is then reduced to $V/3$ and the temperature to $T/3$. The pressure of the gas will now be

   A $\dfrac{P}{3}$ **B** P **C** $3P$ **D** $9P$ **E** $27P$

21.2 If, at a pressure of 10^5 Pa (N m^{-2}), the density of oxygen is 1.4 kg m^{-3}, it follows that the root-mean-square velocity of oxygen molecules in m s^{-1} is

   A 5   **B** 18   **C** 120   **D** 270   **E** 460

21.3 An ideal gas at 300 K is adiabatically expanded to twice its original volume and then heated until the pressure is restored to its initial value. What is the final temperature?

   A 300 K   **B** 400 K   **C** 450 K   **D** 600 K

21.4 A Carnot engine operates between temperatures of 600 K and 300 K and accepts a heat input of 1000 joules. The work output is

   A 300 J   **B** 400 J   **C** 500 J   **D** 600 J

21.5 A heat pump is to be installed to heat a greenhouse; the heat is to be extracted from a neighbouring stream at a maximum rate of 20 kW. The maximum temperature of the greenhouse is 300 K.

   When the maximum temperature difference between the greenhouse and the stream is 40 K, assuming a Carnot cycle the input power required for the pump will be of the order of

   A 1 kW   **B** 3 kW   **C** 10 kW   **D** 30 kW

21.6 When the temperature of a fixed volume of an ideal gas is doubled, the root-mean-square speed of the gas molecules changes from c to

   A $\dfrac{c}{2}$ **B** c **C** $\sqrt{2}c$ **D** $2c$

21.7 An ideal gas of volume 1×10^{-3} m^3 is heated slowly so that it expands at a constant pressure of 3×10^5 Pa until the volume is 3.2×10^{-3} m^3. If 2400 J of energy are supplied to the gas during heating, by how much does its internal energy change?

   A 1440 J   **B** 1740 J   **C** 2100 J   **D** 2400 J

22.1 Which one of the following phenomena cannot be explained by the wave theory of light?

   A refraction   **B** interference   **C** diffraction
   D polarization   **E** photoelectric effect

22.2 An electrostatic field E and a magnetic flux density B act over the same region, and an electron enters the region. Which one of the combinations of E and B in Fig. R36 can be made to cause the electron to pass undeflected?

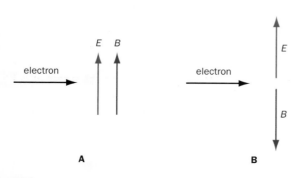

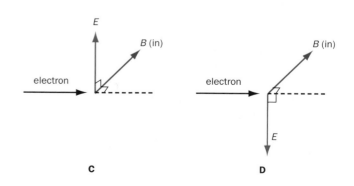

Fig. R36

22.3 A photocell is illuminated with ultraviolet. The intensity of the illumination is reduced, resulting in
A a reduction in the average kinetic energy of the electrons but no change in their rate of emission
B no change in either the rate at which electrons are emitted or in their average kinetic energy
C a reduction in the rate at which electrons are emitted but no change in their average kinetic energy
D a reduction in both the rate at which they are emitted and their average kinetic energy.

22.4 In a Rutherford α-particle scattering experiment, a beam of α-particles is incident normally on a thin gold foil. Which of the following statements is **incorrect**?
A Most of the α-particles are deflected back from the foil.
B A few of the α-particles are deflected back from the foil.
C Most of the α-particles pass undeviated through the foil.
D Some of the α-particles are deviated as they pass through the foil.

22.5 Figure R37 shows some of the electron energy levels in an atom. The transition labelled V results in the emission of visible light. What could be emitted when the transition labelled R occurs?
A infrared radiation   **B** ultraviolet radiation
C an electron   **D** microwave radiation

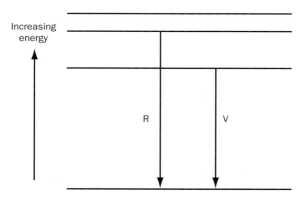

Fig. R37

22.6 An electron accelerated through a p.d. of 100 kV strikes a metal plate and causes the emission of an X-ray photon. What is the maximum frequency the X-ray photon can have? (The Planck constant $= 6.63 \times 10^{-34}$ J s; electronic charge $= 1.60 \times 10^{-19}$ C.)
A 1.6×10^{14} Hz   **B** 2.4×10^{19} Hz
C 2.4×10^{20} Hz   **D** 1.6×10^{24} Hz

23.1 An a.c. voltage of 50 Hz is connected to a CRO. If four complete cycles are to be displayed on the screen which is 8 cm wide, what should be the time-base setting?
A 100 µs/cm   **B** 1 ms/cm   **C** 10 ms/cm
D 0.1 s/cm

23.2 An LED is made from semiconducting material having a band gap of 1.1 eV. What type of radiation does the LED emit?
A ultraviolet   **B** blue light   **C** red light   **D** infrared

24.1 Figure R38 shows the voltage variation of an amplitude-modulated carrier wave. The frequency of the carrier wave is
A 125 kHz   **B** 250 kHz   **C** 2 MHz   **D** 4 MHz

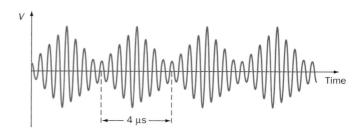

Fig. R38

25.1 Two radioactive elements X and Y have half-lives of 50 minutes and 100 minutes respectively. Samples of A and B initially contain equal numbers of atoms. After 200 minutes what is the value of the following fraction?

$$\frac{\text{number of atoms of X unchanged}}{\text{number of atoms of Y unchanged}}$$

A 4 **B** 2 **C** 1 **D** $\frac{1}{2}$ **E** $\frac{1}{4}$

25.2 The graph in Fig. R39 shows how the count-rate A of a radioactive source as measured by a Geiger counter varies with time t. The relationship between A and t is
A $A = 2.5\,e^{-10t}$ **B** $A = 12\,e^{10t}$ **C** $A = 2.5\,e^{-0.1t}$
D $A = 12\,e^{-0.1t}$ **E** $A = 12\,e^{0.1t}$

(Assume ln 12 = 2.5)

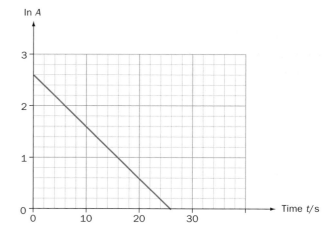

Fig. R39

25.3 When a lithium nucleus ($^{7}_{3}$Li) is bombarded with certain particles, two alpha particles only are produced. The bombarding particles are
A electrons **B** protons **C** deuterons **D** neutrons
E photons

25.4 Which of the following is not considered to be a fundamental constituent of matter?
A lepton **B** meson **C** quark **D** antineutrino

26.1 The following red shifts in the wavelength of a particular spectral line were observed in the spectra of radiation received from some receding galaxies. Which would have come from the galaxy most distant from earth?
A 5 nm **B** 50 nm **C** 70 nm **D** 0.1 µm

MULTIPLE SELECTION

In the questions below one or more of the responses may be correct. Choose one letter from the answer code given.

*Answer **A** if (i), (ii) and (iii) are correct.*
*Answer **B** if only (i) and (ii) are correct.*
*Answer **C** if only (ii) and (iii) are correct.*
*Answer **D** if (i) only is correct.*
*Answer **E** if (ii) only is correct.*

22.7 If the accelerating voltage across an X-ray tube is doubled
i) the wavelengths of the characteristic lines are halved
ii) the minimum wavelength of the X-rays is halved
iii) the X-rays are most probably more penetrating.

25.5 Which of the following statements about a nuclear fission reactor are true?
i) The moderator promotes the capture of neutrons by uranium-238.
ii) The control rods prevent the reactor overheating.
iii) The moderator reduces the speed of neutrons released on fission.

Additional questions

MECHANICAL PROPERTIES

1. Calculate the strain energy stored in a wire of unstretched length 2.0 m when it is stretched by 5.0 mm by a force of 60 kg. ($g = 10$ N kg^{-1})

2. In a railway bridge, a vertical steel girder of length 5.0 m, cross-section area 4.5×10^{-3} m^2, supports a load that increases by 30 tonnes (30 000 kg) when a train passes. The base of the girder is rigidly supported in concrete (see Fig. AQ1). The Young modulus for this steel is 2.0×10^{11} Pa.

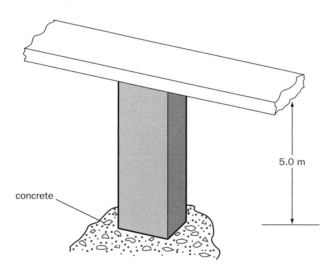

Fig. AQ1

a) Calculate the depression of the top of the girder resulting from the passage of a train.
b) Such girders have to be 'elastic'. What is meant by 'elasticity' and why is it a necessary property?
c) Calculate the elastic strain energy stored in the girder when compressed elastically by 0.05% of its unloaded length.

3. The force (F) – extension (e) graph of Fig. AQ2 shows that the strain energy stored in the material under test for an extension of 4 m is greater than which of the following values?
A 100 J **B** 80 J **C** 60 J **D** 40 J

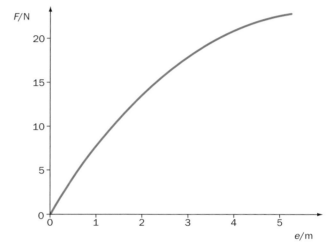

Fig. AQ2

4. Figure AQ3 shows curves of the force, F, against separation, r, for pairs of atoms of two crystalline solids X and Y. The solids have the same crystal structure.

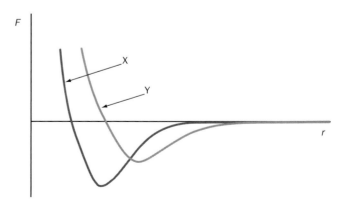

Fig. AQ3

State and explain which solid has the
i) larger atoms,
ii) greater Young modulus,
iii) greater breaking stress.

(*NEAB*, AS/A PHO5, June 1997)

ELECTRICAL PROPERTIES

5. If each battery in Fig. AQ4 has negligible internal resistance, what should be the e.m.f. of X for there to be no deflection on the galvanometer?

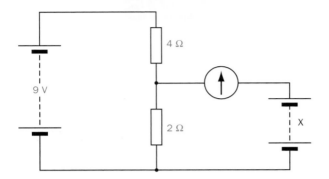

Fig. AQ4

6. If the high-resistance voltmeter connected across the battery in Fig. AQ5 reads 9 V when S is open and 6 V when it is closed, the internal resistance of the battery is

A 2.0 Ω **B** 1.5 Ω **C** 1.0 Ω **d** 0.50 Ω

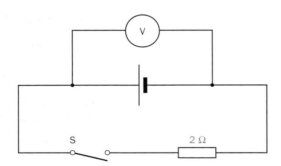

Fig. AQ5

7. A 240 volt 1.0 kilowatt electric fire element is to be manufactured using resistance wire of diameter 0.70 mm and resistivity (at its working temperature) of 15×10^{-6} Ω m.
a) Calculate
i) the resistance of the element at its working temperature, and
ii) the length of resistance wire required.
b) In terms of the flow of electrons, explain why there is a significantly greater current when the fire is first switched on than when the element has reached its working temperature.

8. **a)** State Kirchhoff's second law.
b) i) In the circuit shown in Fig. AQ6, *E* is a cell of source (internal) resistance *r* and the resistance of *R* is 4.0 Ω. With the switch S open, the high resistance voltmeter reads 10.0 V and with S closed the voltmeter reads 8.0 V. Show that $r = 1.0$ Ω.

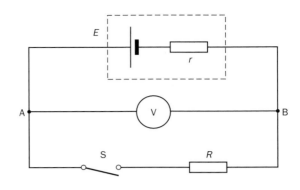

Fig. AQ6

ii) If *R* were replaced by a cell of e.m.f. 4.0 V and source resistance 1.0 Ω with its negative terminal connected to B, what would be the reading of the voltmeter with S closed?

9. Suppose you have been asked to investigate the variation of current with potential difference for a device with a resistance which is thought to vary slightly about a value of 10 Ω. The only additional apparatus which is available is:
■ a battery of e.m.f. 9.0 V and internal resistance 5.0 Ω
■ three resistors each of resistance 18 Ω
■ a 500 mA ammeter with negligible resistance
■ a 5 V voltmeter with a very high resistance
■ a switch and connecting leads.
a) Draw a diagram showing the circuit you would use.
b) State briefly the steps you would take to obtain a range of values of p.d. and current.
c) Estimate the maximum and minimum values of current and p.d. you would expect to obtain with the procedure you use. Show your working clearly.

10. Figure AQ7 shows the way in which eight heating elements of the rear window heater in a car are connected.

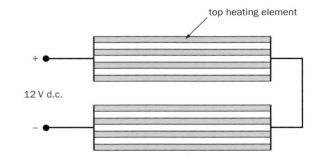

Fig. AQ7

Each of the elements of the heater has a resistance of 8.0 Ω and the heater is connected to a 12 V d.c. supply.
a) Calculate
i) the resistance of the heater,
ii) the potential difference across each of the elements,
iii) the current through each of the elements.

b) The top heating element, marked with the arrow in the diagram, is damaged in use and stops conducting. Calculate

i) the new resistance of the heater,

ii) the current flowing in each of the three top conducting elements.

(NEAB, AS/A PHO1, June 1998)

THERMAL PROPERTIES

11. The Celsius temperature equivalent to 608 K is

A 881 °C   **B** 335 °C   **C** 235 °C   **d** 781 °C

12. A simple thermocouple thermometer (assumed to have a linear response) consists of copper and constantan wires connected to a sensitive digital voltmeter (Fig. AQ8). When the thermojunction J is placed in melting ice at 0 °C the voltmeter gives a reading of −0.56 mV; when J is in steam at 100 °C the reading is 3.44 mV.

a) What was room temperature when these measurements were made?

b) What will be the reading of the voltmeter when J is grasped firmly between finger and thumb so that J is at blood temperature (37 °C)?

c) State one advantage of a thermocouple thermometer in relation to thermometers of other types.

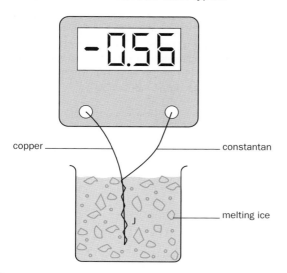

copper ——————   —————— constantan

—— melting ice

J

Fig. AQ8

13. Hot water from a central heating boiler suffers a fall in temperature of 10 °C when it passes through a radiator. If the radiator supplies heat to the surroundings at a rate of 2.0 kW, what is the rate at which water flows through the radiator? (Specific heat capacity of water = 4200 J kg^{-1} K^{-1})

14. The graph in Fig. AQ9 shows how the temperature of 2.0 kg of a substance, initially solid, rises when it is heated uniformly at a rate of 2000 J min^{-1}.

What is

a) the melting-point of the substance,

b) the specific heat capacity of the substance when solid, and

c) the specific latent heat of fusion?

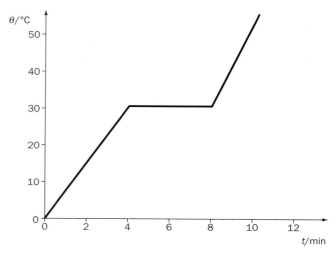

Fig. AQ9

15. a) Explain (in terms of atomic structure) why metals are better conductors of heat than non-metals.

b) The base of an aluminium saucepan has a cross-section area of 0.050 m^2 and a thickness of 2.5 mm. The saucepan is partly filled with water and placed on an electric hotplate at a temperature of 140 °C. When the water is boiling it is found that the water vaporizes at a rate of 2.0 gram per minute.

(Take the specific latent heat of vaporization of water as 2.2×10^6 J kg^{-1}.)

i) Calculate the rate at which heat is used in vaporizing the water.

ii) Calculate a value for the thermal conductivity of aluminium.

iii) Suggest one reason why the value calculated in ii) is far less than the accepted value for aluminium.

OPTICAL PROPERTIES

16. The graph in Fig. AQ10 shows the variation of refractive index n with wavelength λ for light travelling in water.

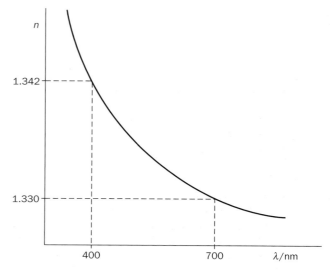

Fig. AQ10

Figure AQ11 shows a mixture of red and violet light incident on an air/water interface.

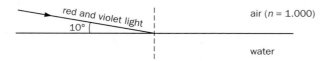

Fig. AQ11

Calculate the angles of refraction for red and violet light. Copy Fig. AQ11 and draw in the approximate paths of the refracted rays.

If refractive index and wavelength were related as shown in Fig. AQ12, what changes would you need to make to your diagram?

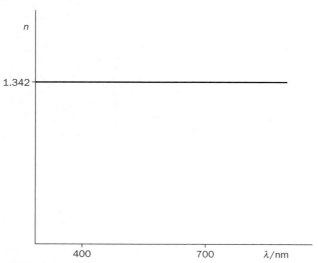

Fig. AQ12

(L, AS/A PH2, Jan 1997)

17. A swimming pool 2.0 m deep is illuminated at night by a small underwater lamp U placed on the tiles at the bottom of the pool. Figure AQ13a shows two rays of light (1) and (2) from U incident on the water/air surface.
a) Calculate the critical angle for total internal reflection of light at a water/air interface. (Take the refractive index for air/water as 1.33.)
b) Without detailed calculation, show accurately on a copy of Fig. AQ13a all the subsequent paths of the light from rays (1) and (2) immediately after the rays meet the water/air surface.
c) Figure AQ13b shows an aerial view of the pool. Explain why the light from U emerges from the water surface over a circular area and calculate its radius r.

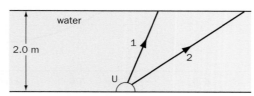

Fig. AQ13a

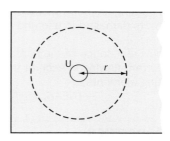

Fig. AQ13b

STATICS AND DYNAMICS

18. Figure AQ14 shows a heavy uniform beam. The ratio of the supporting forces F_1 to F_2 is
A 2:3 **B** 4:3 **C** 2:5 **D** 3:2

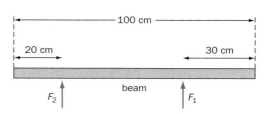

Fig. AQ14

19. A uniform bar of mass 1.0 kg is pivoted as in Fig. AQ15 with a mass of weight 4.0 N hung at one end. What is the value in N m of the resultant moment about the pivot when the bar is horizontal?

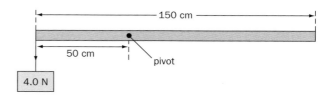

Fig. AQ15

20. Figure AQ16a shows the bone and muscle structure of a person's arm supporting a 5.0 kg mass in equilibrium. The forearm is horizontal and is at right angles to the upper arm. Figure AQ16b shows the equivalent mechanical system. F_M is the force exerted by the biceps muscle and F_J is the force at the elbow joint.
a) i) Explain why the 20 N force has been included.
   ii) State the conditions which must be met by the forces when the arm is in equilibrium.
b) i) Calculate the magnitude of the force F_M.
   ii) Show that F_M has the same magnitude when the forearm is at 45° to the horizontal with the upper arm still vertical.
   (Acceleration of free fall $g = 10.0$ m s^{-2})
c) In many athletes the distance between the elbow joint, E, and the muscle attachment, P, is greater than 5.0 cm. Explain how this is an advantage in lifting and throwing events.

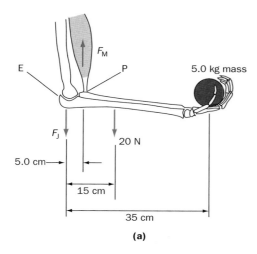

(a)

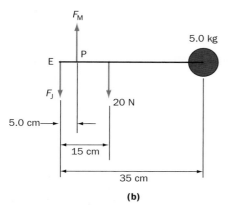

(b)

Fig. AQ16

21. A public house sign is fixed to a vertical wall as shown in Fig. AQ17.

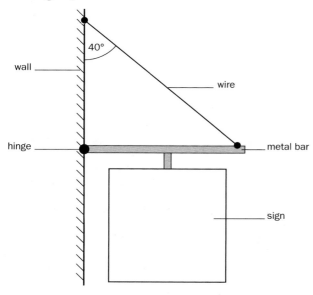

Fig. AQ17

A uniform metal bar 0.75 m long is fixed to the wall by a hinged joint that allows free movement in the vertical plane only. The wire is fixed to the wall directly above

the hinge and to the free end of the horizontal metal bar. The wire makes an angle of 40° with the wall. A single support holds the sign and is mounted at the mid-point of the metal bar so that the weight of the sign acts through that point.

a) i) Copy the diagram and draw three arrows showing the forces acting on the metal bar, given that the system is in equilibrium. Label the arrows A, B and C.

ii) State the origin of the forces A, B and C.

b) The combined mass of the metal bar and sign is 12 kg and the mass of the wire is negligible. By taking moments about the hinged end of the bar, or otherwise, calculate the tension in the wire.

(AQA: NEAB, AS/A PHO1, March 1999)

22. a) Explain what is meant by a **couple**.

b) Two small masses A and B (weighing 1.0 N and 4.0 N respectively) are fixed at the end of a rod of length 1.0 m and weight 2.0 N, Fig. AQ18. By taking moments (or otherwise) show that there is a single point X on the rod at which the system may be supported in equilibrium by a single upward force. Find the position of X and the magnitude of this force.

c) The system could also be supported by a single force applied at the centre of the rod and a couple. Find the magnitude of this force and the torque (moment) of the couple.

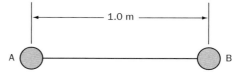

Fig. AQ18

23. Three forces P, Q and R hold an object in equilibrium. Draw

a) a space diagram,

b) a force diagram,

and hence find

c) the angle between the directions of P and R if $P = 13$ N, $Q = 12$ N and $R = 5.0$ N.

24. If two spheres of weights 1 N and 2 N are joined by a string as in Fig. AQ19 and allowed to fall, what is the tension in the string, neglecting air resistance? How would air resistance affect their behaviour?

Fig. AQ19

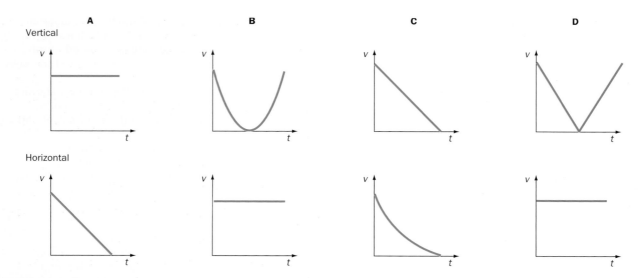

A
Vertical

B

C

D

Horizontal

Fig. AQ20

25. Which pair of graphs in Fig. AQ20 represents the vertical and horizontal speeds *v* with time *t* of a projectile?

26. Small pebbles are dropped from ground level down a well 5.0 m deep at intervals of 0.50 s. How far below ground level is the next pebble when the previous one reaches the bottom? ($g = 10$ m s^{-2})

27. To keep the wagon of weight *W* moving up the slope in Fig. AQ21 with a steady speed, what is the magnitude of the force *F* that must be applied parallel to the slope, neglecting air resistance and friction?
A $W \sin \theta$ **B** $W \tan \theta$ **C** $W \cos \theta$ **D** $W/\tan \theta$

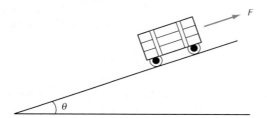

Fig. AQ21

28. A car driver suddenly brakes the vehicle to avoid a collision. The car (mass 1500 kg), originally moving at 20 m s^{-1}, comes to a standstill with uniform deceleration in 4.0 s.
a) Calculate:
  i) the deceleration of the car,
  ii) the braking force acting on the car,
  iii) the distance travelled before the car comes to rest.
b) Describe the horizontal forces acting on the driver during the braking. Explain how the wearing of a seat belt helps reduce the risk of injury.

29. A block of wood resting on a fixed horizontal platform covered in felt, as shown in Fig. AQ22*a*, can, when placed in a car, be used to show the driver when the acceleration is greater than a certain value.

a) Given that the mass of the block is 50 g, calculate the minimum frictional force between the block and the felt which is required to prevent the block sliding backwards when the car is accelerating at a rate of 1.5 m s^{-2}.
b) When the acceleration of the car exceeds a certain value, the block topples backwards, rotating about its rear lower edge, Fig. AQ22*b*. Draw a side view of the block as the block is just starting to rotate about its rear lower edge. On your diagram mark and name the forces acting on the block.
c) i) What kind of changes would you make to the dimensions of the block to make it topple over at a smaller value of the acceleration? (A qualitative answer only is required.)
  ii) State what would happen if, with the block standing vertically on the platform, the car were to brake sharply.

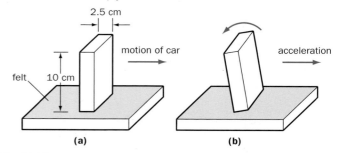

(a)   (b)

Fig. AQ22

30. A force *F* causes two blocks P and Q, of masses 3*m* and *m* respectively, Fig. AQ23, to accelerate along a smooth horizontal surface. The magnitude of the force exerted by Q on P during this acceleration is:
A *F* **B** 3*F*/4 **C** *F*/3 **D** 0

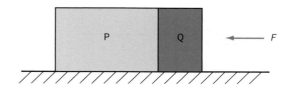

Fig. AQ23

31. In a fairground side-show a prize (a can of drink) is won by knocking it off a shelf by firing a wooden ball from a spring-loaded gun. The can is made to slide along the shelf by the impact of the ball. The dimensions of the prize and the shelf are shown in Fig. AQ24.
The can has a mass of 0.40 kg.
The acceleration of free fall, $g = 9.8$ m s^{-2}.
a) When the speed of the can immediately after the collision is 0.90 m s^{-1}, the can just falls from the rear of the shelf.
   i) Calculate the kinetic energy of the can immediately after the collision.
   ii) Calculate the average frictional force between the can and the surface.

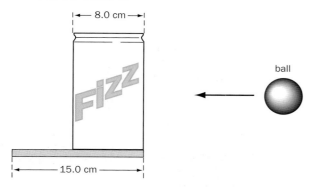

Fig. AQ24

   iii) Sketch Fig. AQ24 and draw the forces acting on the can as it slides along the shelf. Include the magnitude of the forces on the diagram.
b) i) If the collision were perfectly elastic, the can would just fall from the shelf when the impact is 'head on' and the speed of the ball of mass 0.020 kg is 9.5 m s^{-1}. Determine the velocity of the ball immediately after impact with the can assuming an elastic collision.
   ii) The stall-holder thinks that there would be less chance of a prize being won if a way could be found of making the ball, which is travelling at 9.5 m s^{-1}, stick to the can. Suggest whether this idea is worth following up and briefly justify your answer.

(AEB, 0635/6, Jan 1998)

32. Figure AQ25a shows a body K of mass 0.20 kg and moving at a speed of 6.0 m s^{-1}, colliding with a stationary body L of mass 1.0 kg. K rebounds from the collision with a speed of 4.0 m s^{-1} and L is driven forwards with a speed of 2.0 m s^{-1}, Fig. AQ25b.

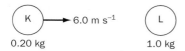

(a) Before collision

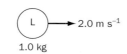

(b) After collision

Fig. AQ25

Show that the collision is elastic.
Calculate the impulse applied to L by K.
If the collision were inelastic, would the impulse be smaller, the same, or greater?

(L, AS/A PH1, Jan 1998)

33. A firm of industrial cleaners uses high-pressure water to clean the exterior wall of an old building. Water is sprayed normally onto the wall at a speed of 15 m s^{-1}. The cross-section area of the water jet is 2.0×10^{-4} m^2. Calculate:
a) the mass of water used per second,
b) the linear momentum of this water,
c) the force exerted by the incoming jet on the surface of the wall, assuming that the water runs down the wall after impact, and
d) the pressure exerted by the jet on the wall.

34. Bales of wool are sometimes raised to the upper floor of a storehouse up an inclined ramp, as shown in Fig. AQ26.

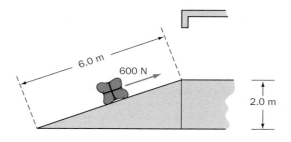

Fig. AQ26

a) At one storehouse, bales, each of mass 120 kg, are raised through a height of 2.0 m up a ramp 6.0 m long. The force needed to slide each bale along the ramp is 600 N. Calculate, for each bale,
   i) the gain in potential energy,
   ii) the work done in dragging it up the ramp,
   iii) the work done against frictional forces.
b) Suggest *one* way of reducing the energy wasted in moving the bales up the ramp.

35. A bullet of mass 10 g travelling at 100 m s^{-1} is fired vertically from below into a block of wood of mass 990 g initially at rest. If the bullet remains in the block, to what height above its rest position does the block rise?

36. Figure AQ27a shows the masses and velocities of two particles A and B about to undergo an elastic collision. Figure AQ27b shows their velocities after the collision.

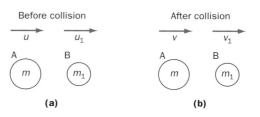

Fig. AQ27

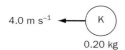

a) Write down the equations relating the following quantities for A and B before and after the collision:
  i) momentum,
  ii) kinetic energy,
  iii) relative velocity.
b) Use your equations to determine v and v_1 in terms of u for an elastic collision in which $u_1 = \frac{1}{2}u$ and $m_1 = \frac{1}{2}m$.

CIRCULAR MOTION AND GRAVITATION

37. **a)** What is meant by **angular velocity**? Illustrate your answer by a suitable diagram.
b) The minute hand (120 mm long) of a clock moves uniformly around the clock face. Calculate
  i) the linear speed of the tip of the hand,
  ii) its angular velocity.
c) The tip of the hand, although moving at uniform speed, changes its velocity continually and hence continually accelerates. Calculate the magnitude of this acceleration, and state its direction.

38. A gymnast of mass m swings from rest on a light rope of length l from a point P, Fig. AQ28. Considering resistances to forward motion to be negligible, derive expressions for
a) the gymnast's speed on passing point Q,
b) the angular velocity of the rope at this moment,
c) the tension in the rope as the gymnast passes Q.

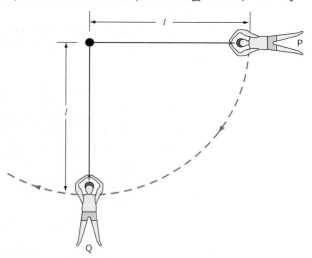

Fig. AQ28

39. **a)** The concept of **moment of inertia** is often used in rotational mechanics. Explain the usefulness of this concept.
b) A car wheel has a moment of inertia (about its central perpendicular axis) of 2.0 kg m². When the car starts from rest with a uniform acceleration, the wheel acquires an angular velocity of 50 rad s⁻¹ by the end of the first 5.0 s of motion.
  Calculate the following quantities associated with the wheel's rotation (in each case specifying an appropriate unit):
  i) the resultant torque experienced,
  ii) the rotational energy acquired,
  iii) the angular momentum acquired.

40. **a)** The conservation of linear momentum is an example of a conservation law. Explain what is meant by
  i) linear momentum,
  ii) a conservation law.
b) When a car accelerates from rest it gains momentum and when a car comes to rest it loses momentum. Explain how these changes are consistent with the law of the conservation of momentum.
c) The earth moves in an approximately circular orbit round the sun at a constant speed v.
  i) Given that r, the radius of the earth's orbit, is 150×10^6 km, calculate the value of v.
  ii) The force F, required to keep the earth moving in its circular orbit, is given by

$$F = \frac{mv^2}{r}$$

  Use your answer to i) and Newton's law of gravitation to find the mass of the sun, given that the universal gravitational constant $G = 6.7 \times 10^{-11}$ m³ kg⁻¹ s⁻².

MECHANICAL VIBRATIONS

41. **a)** A simple pendulum executes simple harmonic motion with amplitude A. Using axes like those in Fig. AQ29, sketch:
  i) a graph showing the variation of the potential energy of the pendulum bob with displacement. Label this graph **P**;
  ii) a graph showing the variation of the kinetic energy of the pendulum bob with displacement. Label this graph **K**.

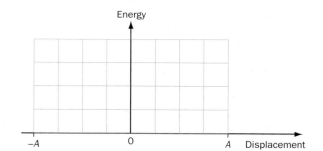

Fig. AQ29

b) A 1.5 m long simple pendulum oscillates with an amplitude of 0.042 m. The acceleration of free fall, g, is 9.8 m s⁻².
  i) Calculate the frequency of oscillation of the pendulum.
  ii) Calculate the maximum speed of the pendulum bob during an oscillation.
c) You are supplied with a set of light springs, each of a different stiffness. You also have access to normal laboratory equipment. Explain how you would investigate the relationship between the period of oscillation of a mass–spring system and the stiffness of the spring.
  (*AEB*, 0635/7, Jan 1998)

42. **a)** A body oscillates with simple harmonic motion back and forth along a horizontal line, Fig. AQ30. Copy the diagram and label the points on the line where the body has

  i) maximum speed,
  ii) maximum acceleration.

Fig. AQ30

b) Give *three* different examples of simple harmonic motion (s.h.m.). In one case indicate (with the aid of some mathematical detail) why the motion is s.h.m.
c) How is the energy associated with s.h.m. oscillations related to

  i) the amplitude,
  ii) the frequency of the s.h.m.?

ENERGY AND ITS USES

43. A car of mass 1.6 tonne (1.6×10^3 kg) consumes 0.050 kg of petrol in a 1.00 km test run at a constant speed of 20 m s^{-1} along a horizontal straight track. The heat of combustion of the fuel is 44 MJ kg^{-1}.

The efficiency of the engine is 25% and the efficiency of the power transmission (from engine to driving wheels) is 80%.

a) Calculate

  i) the energy input to the engine during the test run,
  ii) the external work done by the car during the test run.

b) Explain briefly the nature of the external forces against which the car performs work. State briefly what happens to that 75% of the input power to the engine which is lost.
c) Suppose you have been given the job of designing a simple device to measure the volume of petrol consumed in 1 km test runs at different speeds. Recalling that in the test described above 50 g of petrol was consumed,

  i) estimate the volumes you will be required to measure, stating any assumptions you have made,
  ii) outline, with the aid of a diagram, the apparatus you would use,
  iii) outline the method you would adopt, including any precautions you would take to make your results as reliable as possible.

d) Some motorized wheelchairs are not powered by an internal combustion engine but by lead–acid batteries. List the advantages and disadvantages of these two power sources for motorized wheelchairs.

44. **a)** i) State what is meant by the **efficiency** of a heat engine.
  ii) What factors determine the maximum value of this efficiency?
b) Use the following information to estimate the minimum rate of energy dissipation in the exhaust system of a car engine.

Mean cylinder temperature 1600 °C
Mean exhaust temperature 800 °C

Petrol consumption 1.0 g s^{-1}
Calorific value of petrol 40 MJ kg^{-1}
75% of the waste energy from the engine passes out through the exhaust

c) Tidal energy is classified as a **renewable** energy source.

  i) Explain the distinction between **renewable** and **non-renewable** energy sources.
  ii) What are the origins of tidal energy?
  iii) State and explain *one* environmental disadvantage of exploiting tidal energy.

d) Identify one other renewable energy source. Comment on its availability, its advantages and its drawbacks.

45. Figure AQ31 shows steam from a power station boiler passing through a turbine which is used to turn an electric generator. Once the steam has passed through the turbine, it is cooled by river water and condensed before being pumped back to the boiler for re-use.
a) The cooling water is removed from the river at 16 °C and returned to the river at 30 °C. The water flow used is 15 tonnes per second (15×10^3 kg s^{-1}). Calculate the power used to heat this water. (Specific heat capacity of water = 4200 J kg^{-1} K^{-1})
b) The generator turned by the turbine produces 560 MW of electrical power. Calculate the efficiency with which energy in the steam is actually converted into electrical energy.
c) If the temperature of the water taken from the river is much lower, suggest what effect this may have on the efficiency of the system.

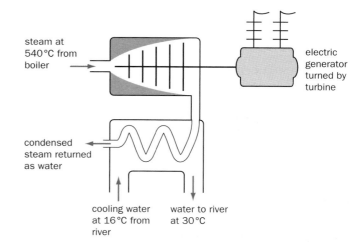

steam at 540 °C from boiler

electric generator turned by turbine

condensed steam returned as water

cooling water at 16 °C from river

water to river at 30 °C

Fig. AQ31

46. **a)** Write a simple explanation of how energy is transmitted by each of the processes of conduction, convection and radiation. Assume that the reader knows a little about elementary physics including the ideas of molecules, movement, density and waves.
b) When engineers are designing heating systems they assume that the rate of energy loss through any wall, door, window, etc., in a room is directly proportional to the difference in temperature between the room and the surroundings (i.e. rate of energy transfer = constant × temperature difference).

A firm of heating engineers, called in to advise on the heating of the small classroom shown in Fig. AQ32, draws up a table of values of the relevant constants.

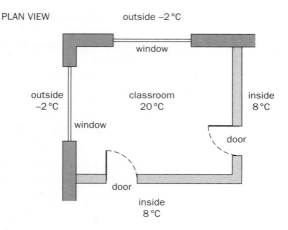

PLAN VIEW

Fig. AQ32

Structure	Rate of energy transfer/kJ h⁻¹ K⁻¹
outside walls	120
2 windows	300
inside walls	280
2 doors	40
ceiling	360
floor	320

i) Estimate the rate of energy loss when the classroom is maintained at a steady temperature of 20 °C when the outside temperature is −2 °C and the temperature of each of the other surrounding spaces is 8 °C.

ii) When the classroom is in use it is partially heated by the warm bodies of the 15 students in the room. Assuming that each student emits energy at the rate of 120 W, calculate the additional rate of energy input, in kW, required to make good the losses of b)i).

iii) It is necessary to change the air of the room every so often. The volume of the room is 150 m³. Assuming that there are two complete changes of air every hour and that the air is initially at −2 °C, estimate the additional rate of energy input required. (Density of air = 1.3 kg m⁻³, specific heat capacity of air = 1000 J kg⁻¹ K⁻¹)

iv) Outline *three* ways of reducing the rate of energy loss from rooms such as the classroom.

47. Figures AQ33*a*, *b* are of an evacuated heat-pipe solar collector. The heat-pipe contains a special fluid which evaporates when heated and rises to the condensing unit. Here it condenses, thus warming the water which flows through the heat exchanger. The fluid then falls back down the pipe and the cycle is repeated. These collectors are much more efficient than simple flat-plate collectors and supply hot water at significantly higher temperatures.

Identify two of the design features shown in Fig. AQ33*b* which account for the high efficiency. Explain in each case the physics principle employed.

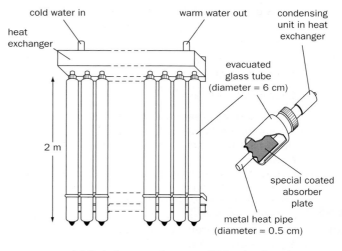

(a) Part of one panel   (b) Section through one tube

Fig. AQ33

Suppose that one of these collectors is 70% efficient. Estimate the area of collector required to produce hot water for an average household on a winter day. You may find the following data helpful, but you must state clearly any additional assumptions and approximations you have made.

Solar flux at the collector's surface at midday on a fine day = 0.9 kW m⁻²
Specific heat capacity of water = 4 kJ kg⁻¹ K⁻¹
Average hot water consumption per household per day = 80 litres
Mass of 1 litre of water = 1 kg

(*L*, A PH3, Jan 1996)

48. Figure AQ34 shows the variation of temperature with time for the contents of a freezer during a power cut. The contents warm up and start to defrost at time *T*.

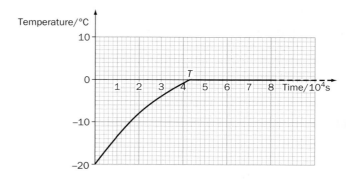

Fig. AQ34

a) i) Heat transfer through the walls of the freezer happens by thermal conduction. State *two* ways in which the rate of heat transfer can be reduced.

ii) Show that the rate of change of temperature for the curved part of the graph at the time T ($= 4.4 \times 10^4$ s) is approximately 2.3×10^{-4} K s^{-1}.
iii) Calculate the rate of transfer of energy into the freezer at time T.

Mass of freezer contents $= 30$ kg
Specific heat capacity of freezer contents $= 2.1$ kJ kg^{-1} K^{-1}

iv) Estimate the time taken, from the time T, for the contents of the freezer to defrost.

Specific latent heat of freezer contents $= 330$ kJ kg^{-1}

v) The surface area of the freezer is 2.3 m^2 and the U value of the freezer walls is 0.49 W m^{-2} K^{-1}. Determine the ambient temperature outside the freezer.

b) i) Energy is transferred by convection from the air in the room to the outer walls of the freezer. Briefly explain how this transfer occurs.
ii) Manufacturers usually make the outer surface of the walls of freezers smooth and shiny. Briefly explain how the nature of the surface minimises the transfer of energy into the freezer.

(*AEB*, 0635/2, Summer 1997)

FLUIDS AT REST

49. In an experiment to measure the density of atmospheric air, a standard flask of volume 2.00 litres ($= 2.00 \times 10^{-3}$ m^3) was weighed before and after the air in the flask had been removed by a vacuum pump. The readings on a laboratory balance were respectively 487.74 gram and 485.20 gram.

At the beginning of the experiment the laboratory mercury barometer showed a reading of 750 mm. (Take the density of mercury as 13 600 kg m^{-3}.)
a) Calculate
i) the atmospheric pressure (in pascal),
ii) the density of atmospheric air.
b) How would the result a)ii) be affected if, immediately after taking the barometer reading, there was a sharp drop in
i) atmospheric pressure,
ii) atmospheric temperature?

50. **a)** State the physical principle or law relevant to the production of each of the following forces. Explain in terms of these principles or laws how each force is produced:
i) the force from the ground propelling a bicycle,
ii) the force from the sea propelling a ship,
iii) the reverse thrust from the air decelerating an aircraft after landing.
b) A motor cruiser, shown in transverse section in Fig. AQ35*a*, has a total mass of 50 tonnes. The point G is the centroid (centre of gravity) of the cruiser, and point B is the centroid of the displaced water (termed the **centre of buoyancy**). The distance between G and B is the **metacentric height** (h).
i) Calculate the underwater volume of the cruiser when floating in sea-water of density 1030 kg m^{-3}.

ii) Figure AQ35*b* shows the cruiser heeling over. The distance GB is now 1.2 m and the line GB makes an angle of 25° with the vertical. Considering the buoyancy forces from the sea-water to be concentrated at B, calculate the moment of the restoring couple tending to return the cruiser to the upright position.
c) i) Explain why the position of the centre of buoyancy B changes when the boat heels over.
ii) Discuss how the value of h, the metacentric height, affects the stability of the cruiser.

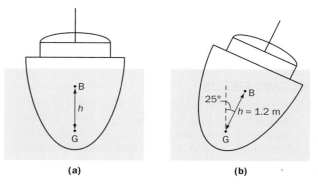

Fig. AQ35

FLUIDS IN MOTION

51. You have been asked to write an article suitable for the general public, e.g. in a newspaper or 'popular science' journal, on *The Bernoulli effect and its relevance to flying and sailing*. Write brief notes to indicate the points you would want to include in the article and indicate the relative importance that you would place on the points you identify.

52. Figure AQ36*a* shows a sailing dinghy, sailing at 40° to the wind. The wind produces a mean pressure difference of 400 Pa across the sail S which has an area of 5.0 m^2. C is the centre-board (or drop-keel) of the dinghy, and T is the tiller attached to the rudder as shown in Fig. AQ36*b*.

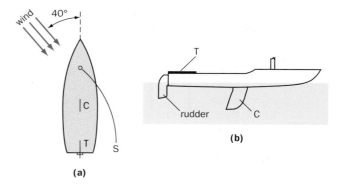

Fig. AQ36

a) Copy Fig. AQ36*a* and add to it:
i) streamlines showing the passage of air over the sail,
ii) a letter L indicating a region in which the air pressure is less than atmospheric pressure,

iii) arrows showing the direction of the principal horizontal forces experienced by the sail and the centre-board C due to the action of the wind. Hence explain how the dinghy can sail forward against the general direction of the wind.

b) Calculate:
   i) the total force exerted by the wind on the sail,
   ii) the force exerted by the sail in the forward direction.

c) i) State and explain how the rudder is used to change the direction of motion of the dinghy.
   ii) Suggest why the scoring of tiny grooves, sometimes called 'riblets', along the immersed surfaces of the hull in a fore-and-aft direction reduces the drag experienced by the dinghy.

ELECTRIC FIELDS

53. a) Estimate the magnitude of the electrical force between the nucleus of a hydrogen atom and the electron of that atom given that the charge on an electron is -1.6×10^{-19} C and that the numerical value of $1/(4\pi\epsilon_0)$ is 9×10^9. Assume that the separation of the nucleus and the electron is of the order of 10^{-10} m.
b) Why would your answer to a) have been the same if you had considered a deuterium ($^2_1$H) nucleus instead of a hydrogen ($^1_1$H) nucleus?

54. Fiona decides to investigate the movement of small charged bodies in an electric field. She suspends a small light conducting sphere at the bottom of a very long fine vertical nylon thread, placed between two large metal plates. The plates are connected to a high voltage d.c. supply, Fig. AQ37.

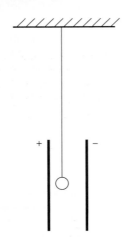

Fig. AQ37

a) She notices that, when the supply is switched on, the sphere oscillates between the two charged plates. Explain why the sphere acts in this way. You should include all the changes that occur in one complete oscillation.
b) Fiona estimates that, when the ball is not touching either plate, the force on the ball is 0.04 N. Assuming that she is correct, estimate the size of the charge on the sphere, given that the plates are 10 cm apart and the voltage between them is 1 kV.

55. Figure AQ38 shows a pair of horizontal, parallel conducting plates. Plate **A** is positively charged and the plate **B** is negatively charged. The separation of the plates is 0.05 m and the potential difference between the plates is 100 V.
a) Copy the diagram and sketch the electric field pattern produced by the plates.

A

B

Fig. AQ38

b) Calculate the electric field strength between the plates.
c) A small sphere carrying a -6.0 nC charge placed at a point midway between the plates remains stationary. Calculate the mass of the charge.

(*IB*, Subsidiary/Standard Paper 2, Nov 1998)

56. Two identical table tennis balls, A and B, each of mass 1.5 g, are attached to non-conducting threads. The balls are charged to the same positive value. When the threads are fastened to a point P the balls hang as shown in Fig. AQ39. The distance from P to the centre of A or B is 10.0 cm.

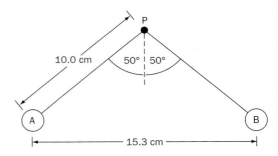

Fig. AQ39

Draw a labelled free-body force diagram for ball A.
Calculate the tension in one of the threads.
Show that the electrostatic force between the two balls is 1.8×10^{-2} N.
Calculate the charge on each ball.
How does the gravitational force between the two balls compare with the electrostatic force?

(*L*, A PH4, June 1997)

57. Figure AQ40 shows the variation of electric field strength E with distance r from the centre of a hydrogen nucleus.
a) i) Determine the value of the electric field strength at a distance of 5.0×10^{-11} m from the centre of the nucleus.
   ii) Use your value of the electric field strength to calculate the magnitude of the force between the

nucleus and an electron separated by a distance of 5.0×10^{-11} m.

Charge on an electron, $e = -1.6 \times 10^{-19}$ C

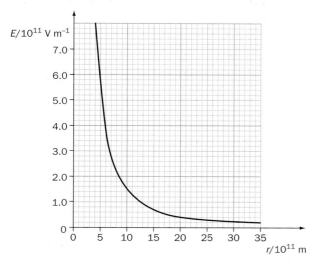

Fig. AQ40

b) Explain briefly how the graph could be used to estimate the amount of work required to remove the electron from the atom when the electron is at a distance of 5.0×10^{-11} m from the nucleus.

c) Calculate the magnitude of the gravitational force between an electron and a proton at a separation of 5.0×10^{-11} m and show that the gravitational attraction does not contribute significantly towards keeping the proton and electron together.

Universal gravitational constant	$= 6.7 \times 10^{-11}$ N m^2 kg^{-2}
Mass of a proton	$= 1.7 \times 10^{-27}$ kg
Mass of an electron	$= 9.1 \times 10^{-31}$ kg

(AEB, 0635/8, Jan 1998)

58. On a grid like that in Fig. AQ41, sketch a graph to show how the electric potential V varies with distance r from an isolated positive point charge.

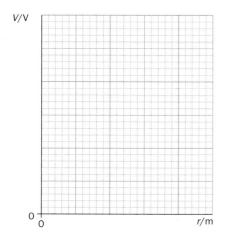

Fig. AQ41

How would your graph change if the isolated charge were negative?

Write down an expression for the work done in bringing an α-particle, $^4_2$He, from infinity up to a distance x from a single gold nucleus, $^{197}_{79}$Au.

Hence calculate the closest distance of approach of a 4 MeV α-particle travelling directly towards a stationary gold nucleus.

(L, A PH4, Jan 1998)

59. **a)** i) State Newton's law of gravitation. Give the meaning of each symbol you use.
ii) Define **gravitational field strength**.
iii) Use your answers to i) and ii) to show that the magnitude of the gravitational field strength at the earth's surface is GM/R^2 where M is the mass of the earth, R is the radius of the earth and G is the gravitational constant.

b) Define **gravitational potential**. Use the data below to show that its value at the earth's surface is approximately -63 MJ kg^{-1}.

c) A communications satellite occupies an orbit such that its period of revolution about the earth is 24 hours. Explain the significance of this period and show that the radius, R_0, of the orbit is given by $R_0 = \sqrt[3]{(GMT^2/4\pi^2)}$ where T is the period of revolution and G and M have the same meanings as in a)iii).

d) Calculate the least kinetic energy which must be given to a mass of 2000 kg at the earth's surface for the mass to reach a point a distance R_0 from the centre of the earth. Ignore the effect of the earth's rotation.

$G = 6.7 \times 10^{-11}$ N m^2 kg^{-2}
$M = 6.0 \times 10^{24}$ kg
$R = 6.4 \times 10^6$ m

CAPACITORS

60. The circuit shown in Fig. AQ42 can be used for measuring the charge stored on a parallel plate capacitor. The capacitor is formed from two square metal plates, each 0.40 m $\times$ 0.40 m, separated by small insulating spacers of thickness d.

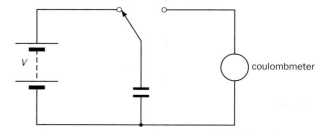

Fig. AQ42

The capacitance C of such a capacitor is given by

$$C = \frac{\epsilon_0 A}{d}$$

Write down an expression for the charge Q stored on the capacitor when the p.d. between its plates is V.

Calculate the value of the electric field between the plates which would produce a reading of 1.0 µC when the capacitor is discharged. Use your answer to suggest suitable values for *V* and *d*.

(L, A PH4, Jan 1996)

61. What is the current in the circuit of Fig. AQ43 *immediately after* S is closed if the charge stored by the 2 µF capacitor is 10 µC?

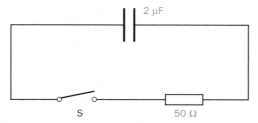

Fig. AQ43

62. Figure AQ44*a* shows a series circuit containing a battery of e.m.f. 6.0 V and internal resistance 2.5 Ω, a resistor of resistance 12.5 Ω and an uncharged capacitor.

The graph in Fig. AQ44*b* shows the variation of current *I* in the circuit with time *t* after the switch is closed.

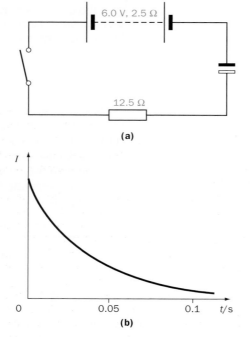

(a)

(b)

Fig. AQ44

a) Calculate the value of the current in the circuit immediately after the switch is closed.
b) Determine the potential differences across
  i) the 12.5 Ω resistor,
  ii) the battery, and
  iii) the capacitor,
immediately after the switch is closed.
c) What are the corresponding values of potential difference 100 s after the switch is closed?

MAGNETIC FIELDS

63. What is the direction of the magnetic field at O due to current in a straight conductor at A, directed out of the plane of the diagram, Fig. AQ45?

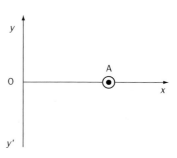

Fig. AQ45

64. The three long straight wires A, B and C in Fig. AQ46 are normal to the plane of the paper and each carries the same current. In A the current is directed out of the plane of the paper and in B it is into it. If the resultant force on B is as shown by the arrow, what is the direction of the current in C?

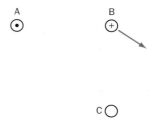

Fig. AQ46

65. The magnitude of the force on a current-carrying conductor in a magnetic field is directly proportional to the magnitude of the current in the conductor. With the aid of a diagram describe how you could demonstrate this in a school laboratory.

At a certain point on the Earth's surface the horizontal component of the Earth's magnetic field is 1.8×10^{-5} T. A straight piece of conducting wire 2.0 m long, of mass 1.5 g, lies on a horizontal wooden bench in an East–West direction. When a very large current flows momentarily in the wire it is just sufficient to cause the wire to lift up off the surface of the bench.
State the direction of the current in the wire.
Calculate the current.
What other noticeable effect will this current produce?

(L, A PH4, June 1997)

66. PQRS in Fig. AQ47 is a section of a circuit, with right-angled corners at Q and at R. It is mounted so that it lies in the same horizontal plane as a uniform magnetic field of flux density *B*, which is parallel to the sides PQ and RS. When a current, *I*, flows in the direction shown in the diagram, the wire QR experiences a force which acts vertically upwards.

Fig. AQ47

a) i) Sketch the diagram (Fig. AQ47) and draw an arrow to show the direction of the magnetic field.
ii) Complete the diagram in Fig. AQ48 by sketching the resultant magnetic field pattern around QR, ignoring the effects of the fields due to PQ and RS.

wire QR, carrying
current out of
plane of paper

Fig. AQ48

iii) If the length of QR is 0.15 m, and B is 0.20 T, the force on QR is 0.18 N. Calculate I.
iv) Explain why the magnetic forces on sides PQ and RS need not be considered when investigating the equilibrium of PQRS.
b) An experiment based on the arrangement described in a) may be used to determine the magnitude of the flux density of a magnetic field. A current is passed through PQRS, which is a wire frame pivoted about a horizontal axis through PS. The upwards magnetic force on QR is balanced by the weight of a length of ticker tape placed over it. The experiment is repeated for other lengths of tape.
The following results were obtained in such an experiment.

Length of tape x/mm	41	83	120	162	208	249
Current I/A	0.51	0.98	1.53	2.02	2.55	3.05

The mass of a 2.00 m length of the uniform ticker tape was 1.20×10^{-3} kg.
i) Plot a graph of I against x.
ii) Show that I and x are related by the equation

$$I = \left(\frac{\mu g}{Bl}\right) x$$

where μ is the mass per unit length of the tape, l is the length of QR and g is the acceleration due to gravity.
iii) Hence calculate the magnitude of B from the gradient of the graph.

(NEAB, AS/A PHO3, Feb 1997)

67. **a)** i) The force on a current-carrying conductor in a magnetic field is given by $F = BIl$. State the condition for which this equation is valid.

ii) State *two* necessary conditions for a charged particle to be subject to a magnetic force in a magnetic field.
b) An alpha particle is travelling at 8.5×10^6 m s^{-1} and a beta particle is travelling at 3.0×10^7 m s^{-1}. They both enter a magnetic field of flux density 2.5 T, at right angles to the field. Given that the magnitude of the charge of an alpha particle is double that of a beta particle, calculate the ratio

$$\frac{\text{magnetic force on alpha particle}}{\text{magnetic force on beta particle}}$$

(NEAB, AS/A PHO3, June 1997)

68. Figure AQ49 shows a current flowing through a piece of semiconductor material. The charge carriers are electrons. A magnetic field acts perpendicular to the current in the semiconductor.

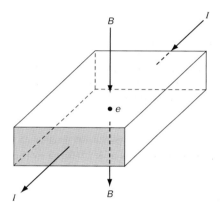

Fig. AQ49

a) i) State the magnitude of the force experienced by the electron e, due to its movement with velocity v at right angles to a magnetic field of flux density B.
ii) Copy Fig. AQ49 and draw an arrow, labelled **F**, to indicate the direction of the force experienced by the electron.
b) Electrons build up on one side of the semiconductor, leaving an equal positive charge on the opposite side. This leads to a potential difference, known as the Hall voltage, V_H, across the piece of semiconductor.
i) State the magnitude of the electric field strength set up across the semiconductor, in terms of V_H and the width d of the semiconductor.
ii) Draw onto another copy of Fig. AQ49 lines of equipotential, labelled **P**, and electric field lines, labelled **E**, for the field set up by the Hall voltage.
iii) Electrons entering the semiconductor will experience a force due to the electric field. State the magnitude of this force in terms of V_H, d, and the charge e on an electron.
iv) The Hall voltage builds up until the forces on the electrons, caused by the magnetic field and by the electric field, are equal and opposite. Show that the Hall voltage is given by:

$$V_H = Bvd$$

where v = the drift velocity of the electrons.

v) The semiconductor has a width of 0.83 mm. It is placed in a magnetic field of flux density 1.6 T and the drift velocity of the electrons is 0.048 m s^{-1}. By performing a calculation, suggest a suitable range for a voltmeter to measure the Hall voltage.

(*AEB*, 0635/8, Jan 1998)

ELECTROMAGNETIC INDUCTION

69. Sketch a graph to show how the deflection θ of the centre-zero galvanometer in Fig. AQ50 varies with time t when the short bar magnet passes at a steady speed right through the solenoid.

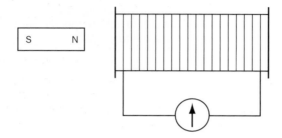

Fig. AQ50

70. Figure AQ51*a* shows in outline a small a.c. generator in which permanent magnets are driven with angular velocity ω around a stationary coil C. Figure AQ51*b* shows how the generated e.m.f. E varies with time t as the magnets rotate through one revolution.

a) Will the plane of the coil C be perpendicular or parallel to the direction of the field of the magnets at the moment when the e.m.f. is zero? Justify your answer.

b) Copy Fig. AQ51*b* and draw a graph on the same axes to show how the e.m.f. will change over the same time interval if the angular velocity ω is doubled.

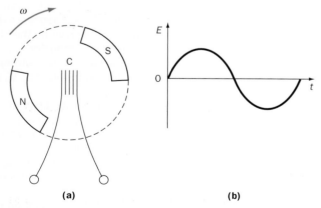

Fig. AQ51

71. A student investigates the behaviour of a magnetic pendulum. A magnet swings above a coil attached to a counter, Fig. AQ52. A count is made every time the voltage across the terminals changes polarity.

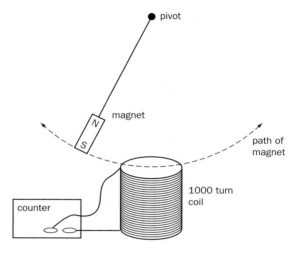

Fig. AQ52

a) Explain why a voltage is induced in the coil.

b) Sketch the diagram and label with an X *every* position of the magnet where the induced voltage changes polarity.

c) The counter is replaced with one that counts each time the induced voltage exceeds 2.0 mV. Calculate the rate of change of flux needed to induce this voltage in the coil, which has 1000 turns.

d) Write down *three* changes that could be made to increase the maximum voltage induced in the coil.

(*NEAB*, AS/A PHO3, June 1998)

72. **a)** i) A power station turbine rotates at 3.0×10^3 revolutions per minute. The torque produced by the turbine is 1.1×10^5 N m. Calculate the power output of the turbine.

ii) The moment of inertia of the turbine and alternator rotor to which it is connected is 3.2×10^4 kg m^2. You may assume that there is no torque resisting the motion. Determine the angular acceleration of the rotor.

iii) Assuming that the angular acceleration remains constant, calculate the time taken for the rotor to achieve its operational speed of 3.0×10^3 revolutions per minute.

b) The alternators at a power station are based on the same principles as the simple alternator, except that the e.m.f. is induced in the static coils (stator) shown in Fig. AQ53 opposite. A small direct current is passed into the rotating coil (rotor). Turbines cause the rotor to rotate between the static coils (stator).

Explain briefly, using the principles of electromagnetic induction, how electricity is produced in the static coils.

c) The alternator produces an r.m.s. voltage of 11 kV. The current of 2.7×10^3 A is fed into a transformer which produces an r.m.s. voltage of 132 kV for the national grid.

i) Calculate the turns ratio of the transformer and state which of the transformer coils has the greater number of turns.

ii) Calculate the current output of the transformer, assuming that there are no energy losses.

(*AEB*, 0635/2, Summer 1997)

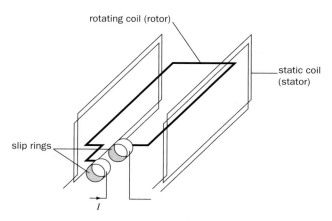

Fig. AQ53

73. Figure AQ54 shows a solenoid with a small coil placed inside. There is an alternating current in the solenoid and the small coil is connected to the Y plates of an oscilloscope, so that the induced e.m.f. can be measured.

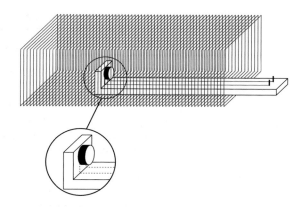

Fig. AQ54

a) i) Figure AQ55 shows the variation of the current I in the solenoid with time t. Sketch the graph and mark on it a time **T** when the e.m.f. induced in the small coil is at its maximum value.

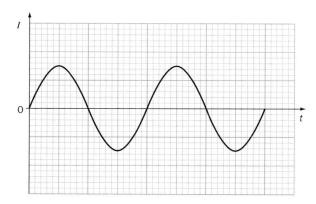

Fig. AQ55

ii) Explain briefly, in terms of electromagnetic principles, why the induced e.m.f. has a maximum value at this time.

iii) The maximum e.m.f. induced in the small coil is 0.54 V when the frequency of the alternating current is 750 Hz. The oscilloscope screen is 10 cm × 10 cm. Give suitable time-base *and* Y amplification settings for the oscilloscope in order to make accurate measurements of the period and amplitude of the e.m.f. Justify your answers.

iv) State *two* factors, apart from the frequency of the supply, that will affect the size of the induced e.m.f. For each factor, state what effect the factor will have on the induced e.m.f.

b) A solenoid has 500 turns and a length of 0.25 m. Calculate the magnetic flux density at its centre when the current is 4.0 A.

Permeability of free space, $\mu_0 = 4\pi \times 10^{-7}$ H m^{-1}

(AEB, 0635/8, Jan 1998)

74. **a)** Describe the structure of a simple transformer.

b) Explain the principles of the operation of an ideal transformer. State the equations relating input and output currents and voltages to the structure of the transformer.

c) Discuss the use of transformers to improve the efficiency of transfer of electrical energy over large distances by reducing power loss in the cables. Support your answer by calculations.

(UCLES, Further Physics, March 1998)

75. **a)** Define the **self inductance** of an inductor.

b) Figure AQ56 shows an inductor L of inductance 10 H, which has a negligible resistance, connected in series with a lamp, a switch and a supply of e.m.f. 6.0 V. After the switch is closed, the lamp takes a short time to develop its full brightness. When at its full brightness its resistance is 3.0 Ω.

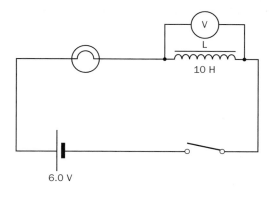

Fig. AQ56

i) When would the voltmeter connected across the inductor display its largest reading? Explain your answer.

ii) When does the energy stored by the inductor have its greatest value, E? Calculate this energy E.

(AQA: NEAB, AS/A PH03, March 1999)

76. **a)** Explain why the current in a d.c. motor has a greater steady value when it is lifting a load.
b) If the final steady current is 3.0 A when a certain 12 V motor is loaded and 1.0 A when unloaded, calculate
  i) the power input to the loaded motor,
  ii) the power input to the unloaded motor, and
  iii) the useful power output of the motor.

77. **a)** Describe briefly how the demand for electrical power is likely to vary
  i) throughout a working day,
  ii) over the course of a year.
b) How do the electricity-generating companies plan their operations in order to be able to cope with the variations in demand referred to in part a)?
c) For a school's Open Day, some AS level students set up a small exhibition to demonstrate the electricity distribution system. They use the two circuits shown in Fig. AQ57. Each of the resistors is supposed to represent the resistance of one of the two conductors in the transmission line. In circuit 2 the step-up transformer connected to the power supply has a turns ratio of 1:5 and the step-down transformer connected to the lamp has a turns ratio of 5:1.
  i) Each lamp is rated as 6 V, 12 W. Calculate the current through and the resistance of each lamp when the voltage across it is 6.0 V.
  ii) Calculate the voltage across and the current through the lamp in circuit 1, assuming that the resistance of the lamp does not change.
  iii) Assuming that, in circuit 2, each transformer is 100% efficient and that the resistance of the lamp does not change, write down expressions for

V_3 in terms of I_3
V_2 in terms of V_3
V_2 in terms of V_1 and I_2
I_2 in terms of I_3

Calculate the value of V_1 and hence, or otherwise, determine the voltage across and the current through the lamp in circuit 2.
d) Practical transformers are not ideal. Outline briefly three sources of energy loss in transformers and indicate how each may be kept small.

ALTERNATING CURRENT

78. The graph in Fig. AQ58 shows the variation of potential difference with time across a 2 kW electric kettle connected to an a.c. supply. The water in the kettle is boiling.

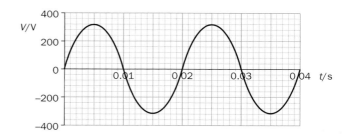

Fig. AQ58

What is the frequency of the a.c. supply?
On axes like those in Fig. AQ59, sketch the current–time graph which would be obtained for the same kettle over the same time interval. Add a scale to the current axis.

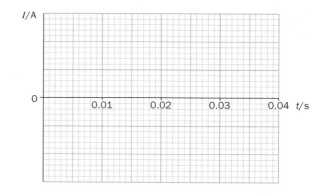

Fig. AQ59

Calculate the resistance of the kettle.

(*L*, A PH4, June 1998)

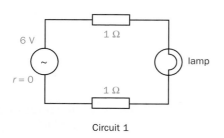

Circuit 1

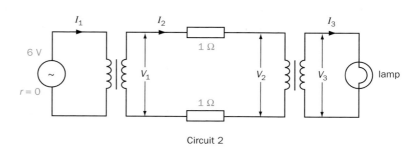

Circuit 2

Fig. AQ57

79. **a)** What is meant by the **reactance** of an electrical component? Name *two* components that possess reactance.

b) Figure AQ60 shows a circuit for investigating the frequency characteristics of electrical components.

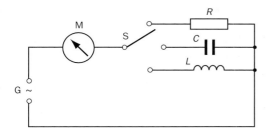

Fig. AQ60

G is a constant-voltage generator producing sinusoidal signals of variable frequency *f*.

M is a low-resistance a.c. milliammeter.

S is a switch connecting to a resistor *R*, *or* a capacitor *C*, *or* an inductor *L*.

Draw three sketch graphs showing how the current *I* recorded by the ammeter varies with frequency *f* when S is switched in succession to *R*, *C* and *L*.

c) The electronic circuitry of a music centre requires a steady d.c. voltage of 30 V for its operation. A mains transformer with a secondary winding (output) marked '22.5 V r.m.s.' is available.

　i) Draw the circuit of a full-wave rectifier that could be connected to the transformer secondary winding to give a smoothed d.c. output.

　ii) Estimate the d.c. voltage available at the output of your rectifier circuit.

d) When the music centre is switched on, a faint low note (or 'mains hum') is heard.

　i) What is the frequency of this note?

　ii) How would you modify the circuit to reduce the volume of the mains hum?

80. **a)** Draw a labelled diagram of a simple a.c. generator. Explain how the generator produces an alternating e.m.f.

b) A sinusoidally varying p.d. of frequency 250 Hz and r.m.s. value 12.0 V is connected to an oscilloscope giving the trace shown in Fig. AQ61.

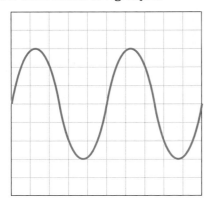

Fig. AQ61

Estimate

　i) the p.d. represented by 1.0 division in the Y direction,

　ii) the time represented by 1.0 division in the X direction.

WAVE MOTION AND SOUND

81. The same progressive wave is represented by the two graphs in Fig. AQ62. Graph 1 shows how the displacement *y* varies with the distance *s* along the wave at a given time. Graph 2 shows how *y* varies with time *t* at a given point on the wave.

a) What does AB represent on graph 1?

b) What does CD represent on graph 2?

c) What does AB/CD represent?

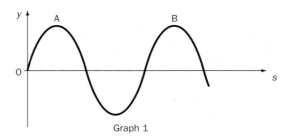

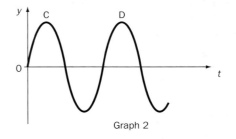

Fig. AQ62

82. The three waves A, B and C in Fig. AQ63 are travelling to the right. What is the phase difference between

a) A and B,

b) A and C?

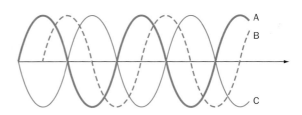

Fig. AQ63

83. Figure AQ64 is a simplified diagram of an ultrasound transducer.

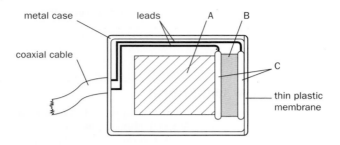

Fig. AQ64

a) Name parts A, B and C in Fig. AQ64 and explain the function of each.

b) Figure AQ65 shows the media into which an ultrasound signal is passed.

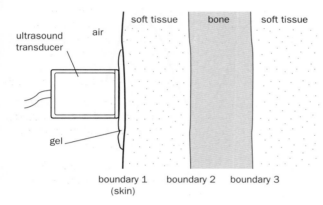

Fig. AQ65

The table shows the speed c of ultrasound in different media and the acoustic impedance Z of these media. Acoustic impedance is a measure of the opposition of a medium to ultrasound.

Medium	c/m s^{-1}	Z/kg m^{-2} s^{-1}
air	3.3×10^2	4.3×10^2
soft tissue	1.5×10^3	1.4×10^6
bone	4.1×10^3	7.8×10^6
gel	1.5×10^3	1.5×10^6

i) The time interval, as detected by the transducer, between pulses of ultrasound reflected from boundaries 2 and 3 is 4.9 μs. Calculate the thickness of the bone.

ii) In general, the fraction F of the ultrasound signal reflected at the boundary between medium 1 and medium 2 is given by

$$F = \frac{(Z_2 - Z_1)^2}{(Z_2 + Z_1)^2}$$

where Z_1 = the acoustic impedance of medium 1 and Z_2 = the acoustic impedance of medium 2.

Use the above equation and the data in the table to calculate the fraction F for an air–soft tissue boundary.

With the aid of a second calculation, suggest an explanation for the use of gel between the ultrasound transducer and the skin.

(*UCLES*, Health Physics, March 1998)

84. If the limits of audibility for a certain person are 16.5 Hz and 16 500 Hz and if the speed of sound in air is 330 m s^{-1}, what is the longest wavelength of sound in air that the person can hear?

85. S_1 and S_2 in Fig. AQ66 are two point sources of sound of the same frequency which produce consecutive maxima of sound at A and B and consecutive minima at C and D. What are represented by
a) $S_2B - S_1B$,
b) $S_2C - S_1C$?

Fig. AQ66

86. a) A microwave transmitter T and a suitable receiver R are placed side by side as shown in Fig. AQ67. A movable sheet of hardboard B and a fixed metal sheet M are set perpendicular to the microwave beam. The hardboard sheet reflects 30% of the microwaves incident on it.

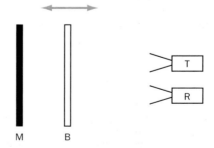

Fig. AQ67

i) When B is moved backwards and forwards, the reading on R is high for some positions of B and for other positions it is low. Explain these observations.

ii) It is found that when B is moved 140 mm towards T and R, the reading on R goes from a high reading through nine high readings and then to a final high reading. Calculate the wavelength of these microwaves.

iii) The speed at which B is moving can be found by measuring the frequency of the high–low readings for microwaves of known wavelength.

Describe another way of measuring the average speed of B when it moves through a metre in about two seconds.

Discuss any possible errors in your measurements. (Assume that any instruments you use are accurately calibrated.)

b) It is difficult to explain the experiment in a)i) using the photon model for microwaves.

i) In what way does the photon model make the experiment difficult to explain?

ii) Calculate the energy of photons of wavelength 30 mm and express this energy in electronvolts.

iii) How does the value you have calculated compare to the energy of a typical photon of visible radiation (light)?

(L, AS PH5, Jan 1998)

87. In passing from radio waves to X-rays in the electromagnetic spectrum how do the following change:

a) the frequency,

b) the wavelength, and

c) the speed in a vacuum?

PHYSICAL OPTICS

88. a) i) Describe and explain, with the aid of a labelled diagram, the double slit arrangement which you would set up in order to produce and observe Young's interference fringes using monochromatic light.

ii) Suggest approximate values for the slit separation and the distance from slits to screen.

b) i) Describe *two* features of the interference fringes.

ii) Describe and explain *one* change in the appearance of the fringes if the slit separation is reduced.

c) i) State the measurements you would make in order to determine the wavelength of the light used.

ii) Calculate the fringe separation which would result, using the approximate values which you gave in part a)ii), assuming that the wavelength of light used is 600 nm.

(NEAB, AS/A PHO2, March 1998)

89. a) Figure AQ68 shows a plan view of part of a diffraction grating of spacing *d*. Plane monochromatic waves of wavelength λ are incident normally on the diffraction grating. The diffracted waves make an angle θ with the normal. Derive from first principles the diffraction equation

$$n\lambda = d \sin \theta$$

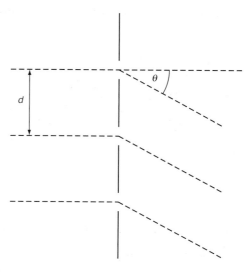

Fig. AQ68

b) A diffraction grating is illuminated with light of wavelength 577 nm. The second order maximum is formed at an angle of 41.25° to the normal. Calculate the total number of lines on the grating if it is 30 mm wide.

(NEAB, AS/A PHO2, March 1998)

90. Monochromatic light of wavelength 500 nm is incident normally on a diffraction grating with 5.0×10^5 lines per metre in a spectrometer experiment. At what angle to the normal is the second-order line seen?

91. In Fig. AQ69, A and B are 'crossed' Polaroids. If A is kept fixed, through what angle must B be rotated to produce 'darkness' again?

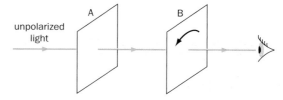

Fig. AQ69

KINETIC THEORY AND THERMODYNAMICS

92. Figure AQ70 shows a hot-air balloon tethered to the ground. It is ready for release and the tension in the rope connecting it to the ground is 400 N. The total weight of the balloon (including the hot air in it) is 16 500 N.

a) i) Draw a free-body force diagram of the balloon and hence calculate the upthrust force *U* on the balloon.

ii) Explain how the upthrust force arises. What factors affect its magnitude during the flight of the hot-air balloon?

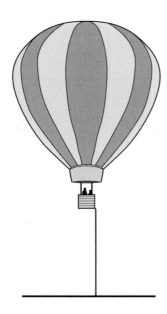

Fig. AQ70

b) Use the equation

$$pV = \frac{m}{M} RT$$

where the symbols have their usual meanings, to find a relationship between the density ρ of the air in the balloon and its kelvin temperature T.

Under what circumstances will ρ be inversely proportional to T?

Using the data below
i) show that the hot air in the balloon has a temperature of 81 °C,
ii) calculate the average power of the gas burner.

Density of cold air outside balloon = 1.29 kg m^{-3}
Density of hot air inside balloon = 1.05 kg m^{-3}
Temperature of cold air outside
balloon = 15 °C
Radius of balloon, assumed
spherical = 5.5 m
U-value of balloon material in air = 3.2 W m^{-2} K^{-1}

c) Suggest how you would demonstrate that air has mass, i.e. that the density of air is not zero.

(L, A PH6, Jan 1998)

93. When a gas expands isothermally which *one* of the following statements is **true**?
A No heat is supplied to the gas from the surroundings.
B The mean kinetic energy of the gas molecules is unchanged.
C The expansion must take place quickly.
D The volume and temperature of the gas both change.

94. Which one of the statements in question 93 is **false** for the adiabatic expansion of a gas?

95. The first law of thermodynamics can be expressed by the equation

$$\delta Q = \delta U + \delta W$$

a) State what each term represents.
b) Rewrite the equation for
i) an isothermal change,
ii) an adiabatic change.

96. **a)** A heat engine working in a cycle consisting of two constant-pressure and two constant-volume changes has an idealized indicator diagram as shown in Fig. AQ71.

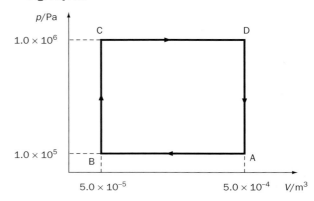

Fig. AQ71

Using the values taken from the diagram where necessary, find:
i) the change in internal energy round a whole cycle, A → B → C → D → A,
ii) the work done on the gas from A → B,
iii) the work done on the gas from B → C,
iv) the work done by the gas from C → D,
v) the work done by the gas from D → A,
vi) the net work done by the gas during the whole cycle.
vii) If 730 J of heat are supplied during the B → C → D part of the cycle, find how much heat is lost during the D → A → B part of the cycle, and calculate the efficiency of the engine.
b) Some of the wasted heat from a car engine may be used to heat the car interior. However the waste heat from a power station is not usually used despite the fact that the operating efficiency of a power station is typically less than 40%.
Explain
i) why the wasted heat from the car engine can conveniently be used in the car heater,
ii) why the efficiency of a power station is as low as 40%.

97. The diagram in Fig. AQ72 shows the variation of pressure, p, and volume, V, for one complete cycle of an ideal hot-air heat engine. The two curved lines represent isothermal changes, i.e. changes at constant temperature. Explain why
i) the order of events is ADCB rather than ABCD,
ii) between A and D, energy must be given to the air in the engine,
iii) between D and C, energy must be extracted from the air in the engine,
iv) the same amount of energy must be returned to the air in the engine between B and A as is extracted between D and C.

Fig. AQ72

ATOMIC PHYSICS

98. An electron of mass m and charge e moves in a circular path in a magnetic field of flux density B. How long does it take to complete one orbit?
A $2me/B\pi$ **B** $2B/me\pi$ **C** $2\pi m/Be$
D $Be/2\pi m$

99. If an electron of mass m and velocity v has kinetic energy 0.10 keV, what is the kinetic energy in keV of a proton of mass $1840m$ and velocity $v/2$?

100. a) The incomplete diagram, Fig. AQ73, represents (not to scale) the main regions of the electromagnetic spectrum.

Fig. AQ73

i) Copy it and label the four regions where labels have been omitted.
ii) State *two* features that all electromagnetic waves have in common.
b) The following equation represents the photoelectric effect.

$$\tfrac{1}{2}mv^2_{\text{max}} = hf - e\Phi$$

i) Explain what is meant by the photoelectric effect.
ii) Identify each of the three terms in this equation.

101. A stream of electrons, accelerated through a p.d. of 12 kV, is directed against the target of an X-ray tube. Taking the charge on an electron to be -1.6×10^{-19} C, and the Planck constant to be 6.6×10^{-34} J s, estimate
a) the kinetic energy of each electron,
b) the minimum wavelength of the X-rays which could be emitted by the target. (Speed of light in vacuum, $c = 3.0 \times 10^8$ m s^{-1})

102. a) A student carried out an experiment to determine the grating spacing, d, of a diffraction grating by determining the diffraction angle in the second order for several spectral lines of known wavelength. These results are given in the table.

Wavelength/nm	435	521	589	652
Second order angle/degrees	20.4	24.6	28.1	31.4

Use these results to obtain the values needed to plot a straight line graph. Draw the graph and use it to determine a value for d.
b) Three diffraction gratings, illustrated in Fig. AQ74, are available for observing line spectra. The grating width and the *total* number of vertical rulings on the grating are given for each one. Determine which of the three gratings will give the largest diffraction angle in the first order with a given spectral line.

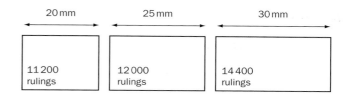

Fig. AQ74

c) The first of the spectral lines in part a) (435 nm) corresponds to one of the lines in the spectrum of atomic hydrogen. The energy level diagram in Fig. AQ75 represents the first five energy levels for this spectrum. Determine which of the energy level transitions will give this spectral line.

level 4	—————————	−0.54 eV
level 3	—————————	−0.85 eV
level 2	—————————	−1.51 eV
level 1	—————————	−3.4 eV
ground state	—————————	−13.6 eV

Fig. AQ75

(*NEAB*, AS/A PHO2, June 1998)

103. Electrons appear to possess two very different sets of properties. Sometimes their behaviour suggests they act like particles and sometimes it suggests they act like waves.

Describe briefly *one* particle-like property and *one* wave-like property of electrons.

What is the wavelength associated with a stream of electrons moving at one tenth of the speed of light?

(The Planck constant, $h = 6.6 \times 10^{-34}$ J s, the mass of an electron, $m = 9.1 \times 10^{-31}$ kg, the speed of light, $c = 3.0 \times 10^8$ m s^{-1}.)

ELECTRONICS AND TELECOMMUNICATIONS

104. An a.c. signal of frequency 100 Hz is applied to the Y-plates of a CRO with its time base set at 2 ms cm^{-1}. How many cycles will be seen on the screen if it is 15 cm wide?

105. Figure AQ76 shows a trace on the screen of an oscilloscope. The Y-sensitivity of the oscilloscope is set at 5.0 V per division and the time base is set at 0.50 ms per division.

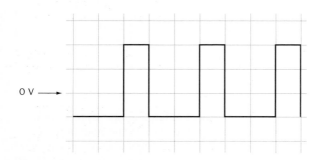

Fig. AQ76

a) For the trace, determine
i) the maximum positive value of potential difference,
ii) the maximum negative value of potential difference,
iii) the frequency of the signal.
b) The trace shows the variation in the potential difference across a 100 Ω resistor. Calculate the energy dissipated in the resistor
i) for the first 1.00 ms,
ii) between 1.00 ms and 1.50 ms,
iii) in one cycle,
iv) in one second.

(*NEAB*, AS/A PHO1, March 1999)

106. The amorphous semiconductor cadmium sulphide is sometimes used in the construction of light dependent resistors (LDRs). The resistance of such devices is typically about 10 MΩ in the dark and less than 1 kΩ in bright daylight.

The energy band diagram of an amorphous semiconductor may be taken as being identical to that of an intrinsic semiconductor for the purposes of this question. See Fig. AQ77.

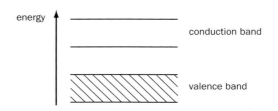

Fig. AQ77

a) By referring to Fig. AQ77, discuss the operating principle of the LDR.
b) Sketch a graph showing how the resistance R of an LDR varies with the intensity I of incident light.

(*NEAB*, AS/A PHO5, June 1997)

107. a) Draw up a truth table for a two-input NAND gate, and use it to show that a NAND gate will behave as a NOT gate if its input terminals are joined together.
b) Show that the arrangement of NAND gates in Fig. AQ78 has the same truth table as an OR gate.

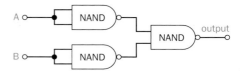

Fig. AQ78

c) The logic system shown in Fig. AQ79 acts as a 'majority voting system' for three judges A, B and C. Each judge presses a switch for a YES vote causing the respective input to rise to logic state 1: if a switch is not pressed the input stays at logic 0. Logic 1 at the output represents a majority YES vote.
i) Construct a truth table for the system showing the states of the inputs A, B, C and the output.
ii) Redesign the system (and redraw it) so that NAND gates only are used.

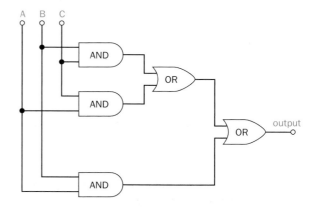

Fig. AQ79

108. a) An operational amplifier, Fig. AQ80*a*, with a (+15 V)–0–(−15 V) supply can be used in circuits giving low-frequency voltage amplifications of
  i) −10,
  ii) −20,
  iii) −100.
  By means of graphs (drawn on the same axes) show how the amplification given by these circuits will vary at higher frequencies.
b) Draw the circuit giving amplification of −20, indicating suitable component values.
c) A low-frequency a.c. sinusoidal voltage signal of amplitude 1.0 V (Fig. AQ80*b*) is applied to the inputs of each of the three circuits.
  Draw three diagrams (with the same time scale as the input signal) showing the corresponding output waveforms.

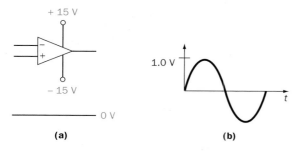

(a)   (b)

Fig. AQ80

109. Figure AQ81 shows the aerial and tuning circuit of a simple radio receiver which can be tuned to two frequency bands. With the switch in position A and the capacitor C at the middle of its range, the circuit is tuned to a frequency *f*.
a) State whether each of the following changes, on its own, would increase or decrease the value of *f*:
  i) increasing the capacitance of C,
  ii) switching from position A to position B.

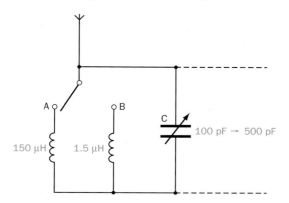

Fig. AQ81

b) i) State two effects caused by the addition of resistance to a tuned circuit.
  ii) Explain why it is more difficult to tune a domestic receiver precisely to a station on the short-wave band than to one on the medium-wave band.

c) Figure AQ82 shows a block diagram of a simple radio receiver for amplitude-modulated (AM) signals.

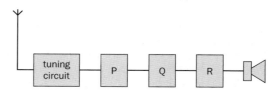

Fig. AQ82

  i) Identify the stages P, Q and R, and explain the function of each.
  ii) Draw diagrams to show the waveform at the output of each of these stages.
d) Broadcast radio transmissions are either amplitude modulated (AM) or frequency modulated (FM). Describe the principal features of a frequency-modulated (FM) radio wave. Refer particularly to the differences in the modulated waveforms of
  i) high- and low-frequency audio notes of equal amplitude,
  ii) loud and soft audio notes of equal frequency.
  Explain why FM transmissions are confined to the v.h.f. radio band.

110. a) The output signal from a microphone is in analogue form. However, there are advantages in transmitting the signal in digital form.
  Explain how a digital coding represents
  i) changes in amplitude of an analogue waveform,
  ii) the frequency of an analogue signal.
b) Why is the quality of speech or music transmitted in digital form generally superior to that transmitted in analogue form?
c) Data can be transmitted in digital form through electric cable links or through optic fibre links.
  i) State and explain why the transit time for pulses along optic fibre links differs from that for pulses along electric cable links of the same length.
  ii) Explain why long-distance optic fibre links between the centres of communication can be laid alongside electrified railway tracks, whereas this method is seldom used with electric cable data links.
  iii) State briefly *three* further advantages of optic fibre transmission compared with electric cable transmission.
d) Figure AQ83 shows the waveform of digital light pulses as they enter an optic fibre communications link.

Fig. AQ83

  i) Draw a diagram to the same scale as Fig. AQ83 to show the waveform of the pulses after passing through several kilometres of optic fibre.
  ii) Explain the consequence of not reshaping the pulses at intervals along the fibre link.

NUCLEAR PHYSICS, COSMOLOGY AND ASTROPHYSICS

111. Which type of radiation from radioactive materials
  a) has a positive charge?
  b) is the most penetrating?
  c) is easily deflected by a magnetic field?
  d) consists of waves?
  e) causes the most intense ionization?
  f) has the shortest range in air?
  g) has a negative charge?
  h) is not deflected by an electric field?

112. As a result of four successive disintegrations the radioactive isotope $^{232}_{90}$Th emits two alpha particles and two beta particles. If the daughter product is ^A_ZX, what are the values of A and Z?
   Are $^{232}_{90}$Th and ^A_ZX isotopes? Explain your answer.

113. What is the half-life of the radioactive substance whose decay curve is shown in Fig. AQ84?

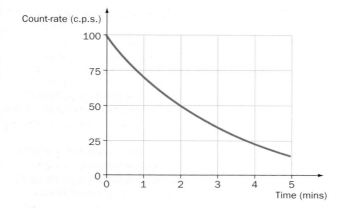

Fig. AQ84

114. What is the half-life of a radioactive material if its activity falls to one-eighth of its initial value in 60 minutes?

115. The energy level diagram for the decay of $^{130}_{53}$I into $^{130}_{54}$Xe is shown in Fig. AQ85.

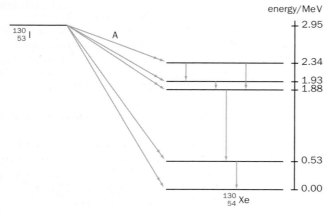

Fig. AQ85

a) i) State what emission is represented by the vertical arrows.
  ii) State the Q value for the decay of $^{130}_{53}$I.
b) i) Sketch a graph on axes like those in Fig. AQ86 to show the energy spectrum of β radiation represented by transition A in the energy level diagram.

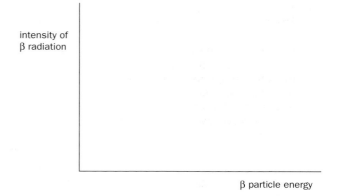

Fig. AQ86

  ii) Give values on the β particle energy axis.
  iii) Explain why the β particles represented by transition A in the energy level diagram are *not* monoenergetic.

(*NEAB, AS/A PHO6SP, March 1999*)

116. Calculate the energy available when two nuclei of deuterium fuse together as shown:

$$^2_1H + ^2_1H \rightarrow ^3_2He + ^1_0n$$

Given that 2 kg of deuterium contains 6×10^{26} deuterium nuclei, calculate the energy released per kilogram by this fusion reaction.
   In a fission reactor, 1.9×10^{16} J of energy are released when 6×10^{26} uranium nuclei (mass 240 kg) disintegrate.
   Compare the energy released per kilogram by the two processes and comment on your answer.

Mass of a deuterium nucleus = 3.345×10^{-27} kg
Mass of a helium-3 nucleus = 5.008×10^{-27} kg
Mass of a neutron = 1.675×10^{-27} kg
Speed of light = 3.00×10^8 m s^{-1}

117. State
  a) *two* similarities, and
  b) *two* differences between the electromagnetic force and the gravitational force.

118. Using the information in Table 25.5 (p. 485), determine the quark structure of each of the fundamental particles in the table below, given that each is composed of two or three quarks.

Particle	Symbol	Charge	Strangeness
pion +	π^+	+1	0
sigma −	Σ^-	−1	−1

119. The results of electron scattering experiments using different target elements show that the radius, R, of the nuclei varies with the nucleon number, A, of the target nuclei according to the equation

$$R = r_0 A^{1/3} \qquad \text{where } r_0 = 1.3 \times 10^{-15} \text{ m}$$

a) Use this equation to show that the density of nuclear matter is independent of its mass.

b) The binding energy per nucleon of a large nucleus is about 8 MeV.

   i) Calculate the work done, in J, to remove a nucleon from a large nucleus.

   ii) State an approximate value for the range of the strong nuclear force.

   iii) Estimate the magnitude of the strong nuclear force between two neighbouring nucleons in the nucleus.

(NEAB, AS/A PHO6, June 1998)

120. a) A negative pion, π^-, quark composition $\overline{u}d$, with momentum 1100 MeV/c collides with a stationary proton. The only products of this collision are two different new particles.

Mass of proton $= 938$ MeV/c^2
Mass of pion $\quad = 139$ MeV/c^2

   Calculate
   i) the energy of the incident pion, given that $E^2 = m_0^2 c^4 + p^2 c^2$,
   ii) the total energy before the collision.

State the total momentum of the two new particles after the collision.

One of the particles created in the collision is a neutral K meson, K^0, quark composition $d\overline{s}$. Explain why this particle will not leave a track in the particle detector.

What can be deduced about the properties of the other new particle produced in the collision?

b) At one stage in its expansion the Universe resembled a giant nuclear fusion reactor with ${}^1_1\text{H}$, ${}^2_1\text{H}$ and ${}^3_1\text{H}$ nuclei fusing to form helium nuclei. This era is shaded in on the graph in Fig. AQ87, which shows the relationship between temperature and time in the expanding Universe.

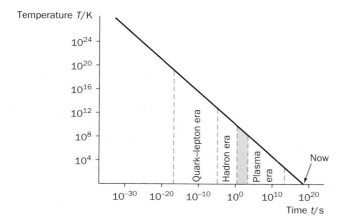

Fig. AQ87

Use the graph to estimate the time at which helium formation began, and the temperature of the Universe at this time.

Explain in terms of the appropriate fundamental interactions why the fusion reactions only take place at very high temperatures.

In an accelerator at CERN, electrons and positrons with energies of the order of 100 GeV are made to collide resulting in the interaction shown in Fig. AQ88.

Fig. AQ88

In what era of the expanding Universe did such interactions take place? What exchange particle is involved in this interaction?

(L, A PH4, June 1998)

121. a) Calculate the force of repulsion between two protons separated by a distance of 1.0×10^{-14} m in an atomic nucleus.

The Rutherford scattering experiment was used to investigate the structure of the atom. Deep inelastic scattering provided evidence about the structure of protons and neutrons. Outline the similarities and differences between these two experiments. What conclusion was drawn from the deep inelastic scattering experiment?

The quarks in a nucleon are held together by the strong interaction. What exchange particle mediates this interaction? State the mass and charge of this exchange particle. What evidence is there that the range of this exchange particle is less than 10^{-14} m?

b) State one piece of evidence that supports the Big Bang Theory and explain how it does so.

The very early Universe was made up of a collection of quarks, leptons and exchange particles. Through what stages has it passed in order to reach its present state?

Explain why atoms were unstable when the Universe was at a temperature greater than about 4000 K.

(L, A PH4, Jan 1997)

122. a) Sketch the Hertzsprung–Russell diagram. On the diagram locate the following: pre-main sequence stars, main sequence stars, red giants, white dwarfs.

b) Changes occur in a star similar to the Sun as it evolves from the main sequence to the next stage of its life cycle. Describe the changes taking place in
   i) core composition and size, and
   ii) surface temperature, size, brightness, and type of star.

c) Name three possible final equilibrium states for a star at the end of its life and describe any initial stellar mass dependencies of these states.

d) How do the spectra observed from very massive stars compare with the spectrum from the Sun? Explain.

(IB, Higher Paper 2, Nov 1997)

123. Write a few sentences stating why the fate of the universe may depend on the mass and abundance of the neutrino.

124. Suppose the speed of light was 75 mph.
a) How would the length of a car travelling at 50 mph seem to change to someone standing at the roadside?
b) Would you seem older or younger to the friends you saw off one morning on a 50 mph coach trip when you met them again on their return in the evening?
c) If you walked forward at 4 mph in a train travelling at 50 mph your speed relative to the ground would be 54 mph according to Newtonian mechanics. Would Einsteinian mechanics give a larger or a smaller value? Explain your answer.

Practical investigations

MATERIALS

1. Strength of paper 'pillars'

Apply loads centrally to a hardboard square placed on top of a hollow cylinder of paper (the 'pillar') until it collapses. Repeat with pillars of different dimensions.

Comment on the results. Do they help you to make a rough prediction of the load that a particular pillar will support? What affects the strength of each pillar? Try to improve the design of the pillars so that their strength increases appreciably.

2. Forces in a roof truss

Fit up a model truss as in Fig. P1 with the laths bolted together so that they pivot freely and there is room between them to hang the 1 kg mass. Unscrew a wheel from each trolley and pass the screw from the wheel through a hole at the lower end of the lath before loosely screwing the wheel back on the trolley. Attach pieces of string to each end of the spring balance and also to the dowel rods in the trolleys so that they form a tie rod with the balance.

Make and record the observations which enable you to work out the forces in the struts.

Investigate how lowering the 'pitch of the roof' (by lengthening the tie rod) affects the forces in the struts.

Also investigate what happens to the force in the tie rod if it is connected across the mid-points of the laths, keeping the angle between the laths the same.

In your report discuss the pros and cons of high- and low-pitched roofs.

3. Model bridge

Design and construct a model road bridge using, for example, drinking straws or Meccano strips. Its minimum length should be at least twice the length of the longest piece of strip or straw used. It must be able to bear a load 50 times its own weight placed anywhere on it.

4. Fusing currents

Before starting read the section on *Fuses* on p. 53. Draw the circuit you propose to use for finding the values of the currents at which wires of different gauges of a given material fuse and have it approved by your teacher.

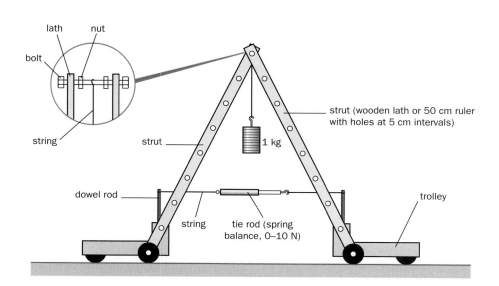

Fig. P1

Assuming that the relationship between the fusing current I and the diameter d of a wire of the material is

$$I = kd^n$$

find the values of the constants k and n.

5. Thermistors

The aim of this experiment is to find how the resistance of both n.t.c. and p.t.c. thermistors (see p. 50) varies with temperature.

Devise a circuit to measure the currents through the thermistor and the p.d. across it at temperature intervals between 0 °C and 100 °C.

Display your results graphically for each type of thermistor.

Comment on the results and try to explain what is occurring as the temperature changes in terms of the equation $I = nAev$ (see p. 41).

Note. Suitable thermistors are Radiospares types RS 151-079 (n.t.c.) and RS 158-272 (p.t.c.).

6. Strain gauge

In this experiment you have to design an electrical bridge circuit containing a strain gauge (see p. 49) which will show on a suitable output device (e.g. a microammeter or a CRO) that a steel hacksaw blade, on which the gauge is mounted, is under strain when it vibrates with one end clamped to the bench.

Try to roughly calibrate your gauge using one of the output devices.

Find and investigate other applications for your strain-detecting system.

Note. A suitable strain gauge is Radiospares type RS 308-102 which has a resistance of 120 Ω.

MECHANICS

7. Accelerometer

Design and construct a device that will indicate the acceleration of a dynamics trolley.

Investigate how it might be calibrated.

8. Vibrating cantilever

A lath clamped securely at one end to the bench and with a mass fixed to the other, free, end can be set into vibration at the free end; it is called a vibrating cantilever.

Set up such an arrangement with a suitable lath and measure the period of vibration T for:

(*a*) several different vibrating lengths l with a fixed mass on the free end; and

(*b*) several different masses m with a fixed vibrating length.

Plot graphs to show any relationships between T, l and m.

FIELDS

9. Speed of an electric motor

Devise *two* different ways of measuring the speed of rotation of a small electric motor which has a pulley wheel on its shaft.

In your report comment on the accuracy attainable by each method and the range of speeds for which it is suitable.

10. Efficiency of an electric motor

The electrical power input to the motor and its mechanical power output have to be measured. Devise and perform methods for doing both using a small d.c. motor which has a pulley wheel on its shaft.

Determine how much power is dissipated in heating up the motor, and suggest how other losses might occur.

Investigate the effect of a greater load on (*i*) the current taken by the motor, (*ii*) the speed of the motor, (*iii*) the power output, and (*iv*) the efficiency of the motor, at its rated p.d. Discuss your findings.

11. d.c. generator (dynamo)

Use a 12 V fractional horsepower motor (with separate field coil and rotor connections) as a dynamo. Couple a hand drill to its shaft using rubber pressure tubing and investigate how the e.m.f. generated depends on:

(*a*) the speed of rotation of the drill; and

(*b*) the field current.

Comment on your findings.

12. CR coupled circuits

Different parts of an electronic circuit are often coupled (joined) by a capacitor and a resistor. The time constant (see p. 221) of the coupling circuit determines the shape of the output waveform.

To investigate this, design and assemble a circuit in which a **square-wave input** is applied to a capacitor and resistor in series in which

(*a*) the capacitor couples and the output is taken from across the resistor; and

(*b*) the resistor couples and the output is taken from across the capacitor.

Record and explain the shape of the output waveforms for time constants that are (*i*) much smaller and (*ii*) much greater than the period of the input waveform.

WAVES

13. Photoelastic stress analysis

Before starting read the section on *Stress analysis* on p. 348. Observe the sequence of colour changes that occur in a transparent film held between crossed Polaroids, placed in front of a white light source, when there is:

(*a*) an increase of stress in the film;

(*b*) an increase of thickness of the film;

(*c*) a small V-shaped notch in the film.

For (*a*) use a piece of polythene that you stretch gradually; for (*b*) build up on a microscope slide staggered layers of transparent adhesive tape that has been stretched previously until it looks pale yellow between the crossed Polaroids; and for (*c*) use a stretched strip of transparent adhesive tape fixed on a slide viewed in both white and red light.

Illustrate your report with sketches whenever possible to show the isochromatic (same colour) fringes, i.e. the regions of equal stress.

14. Light-dependent resistor (LDR)

Design and construct a circuit in which an LDR (e.g. ORP 12) can be used as a light-meter.

Use your light-meter to investigate

(*a*) how the light intensity of a 12 V lamp varies with the power applied to it; and

(*b*) how the LDR responds to light of different wavelengths.

ATOMS

15. Electronic systems I: flasher and bleeper

Use either the modules from a commercial electronics kit or chips and discrete components wired up on a prototype board.

(*a*) Design and assemble a system which makes a lamp flash six times in a row when suitably pulsed.

(*b*) Modify the system made in (*a*) so that it causes a loudspeaker to emit six audible pips one after the other.

16. Electronic systems II: temperature and light-level alarms

Use either the modules from a commercial electronics kit or chips and discrete components wired up on a prototype board.

(*a*) Design and construct a system which emits a warning note when the temperature of a thermistor becomes high.

Modify the system to act as a low-temperature warning device.

(*b*) Design and construct a system using an LDR which produces a warning note when the light level is low.

Mathematics for physics

Modify the system to act as a high light-level alarm. Mathematics is an essential tool of physics. Many laws are written as mathematical equations which may have to be manipulated. Arithmetic, algebra and trigonometry are also required at times and graphs are frequently used to represent results and relationships visually.

EQUATIONS — TYPE 1

An equation has sometimes to be changed round before it can be solved. In the equation $x = a/b$, the **subject** is x. To change the subject we **multiply or divide both sides** of the equation **by the same quantity**.

(a) To change the subject to a

We have

$$x = \frac{a}{b}$$

If we multiply both sides by b the equation is still true:

$$x \times b = \frac{a}{b} \times b$$

The bs on the right-hand side cancel:

$$x \times b = \frac{a}{\cancel{b}} \times \cancel{b} = a$$

$$\therefore \quad a = b \times x$$

(b) To change the subject to b

We have

$$x = \frac{a}{b}$$

Multiplying both sides by b as before, we get

$$a = b \times x$$

Divide both sides by x:

$$\frac{a}{x} = \frac{b \times x}{x} = b \times \frac{\cancel{x}}{\cancel{x}} = b$$

$$\therefore \quad b = \frac{a}{x}$$

Questions

1. Find the value of x if
   a) $3x = 15$ **b)** $x/2 = 10$
   c) $4/x = 2$ **d)** $x/6 = \frac{4}{3}$
2. Change the subject to
   a) s in $v = s/t$ **b)** a in $F = ma$
   c) I in $P = I^2R$ **d)** p_1 in $p_1V_1/T_1 = p_2V_2/T_2$
   e) v in $\frac{1}{2}mv^2 = mgh$

EQUATIONS — TYPE 2

To change the subject in the equation $x = a + by$ we **add or subtract the same quantity from each side**. We may also have to divide or multiply as in a type 1 equation. Suppose we wish to change the subject to y in

$$x = a + by$$

Subtract a from both sides:

$$x - a = a + by - a = by$$

Divide both sides by b:

$$\frac{x - a}{b} = \frac{by}{b} = y$$

$$\therefore \quad y = \frac{x - a}{b}$$

Questions

3. What is the value of x if
   a) $2x + 3 = 7$ **b)** $2(x - 3) = 10$
   c) $x/3 + \frac{1}{4} = 0$ **d)** $2x + \frac{5}{3} = 6$
   e) $3/x + 2 = 5$
4. Change the subject to
   a) u in $v = u + at$ **b)** m in $y = mx + c$
   c) a in $v^2 = u^2 + 2as$

PROPORTION (OR VARIATION)

One of the most important mathematical operations in physics is finding the relation between two sets of measurements.

(a) Direct proportion

Suppose that in an experiment two sets of readings are obtained for the quantities x and y as in the table below (units omitted):

x	1	2	3	4
y	2	4	6	8

We see that when x is doubled, y doubles; when x is trebled, y trebles; when x is halved, y halves and so on. There is a one-to-one relationship between each value of x and the corresponding value of y.

We say that y is **directly proportional** to x, or y **varies directly** as x. In symbols,

$$y \propto x$$

Also, the ratio of one to the other, say of y to x, is always the same, that is, it has a constant value which in this case is 2. So we can write

$$\frac{y}{x} = \text{a constant}$$

The constant, called the **constant of proportionality** or **variation**, is given a symbol, such as k, and the relation between y and x is then summed up by the equation

$$\frac{y}{x} = k \qquad \text{or} \qquad y = kx$$

Note that:

- In practice, because of inevitable experimental errors, the readings seldom show the relation so clearly as here.
- If instead of using numerical values for sets of readings of x and y we use letters, such as $x_1, x_2, x_3 \ldots$ and $y_1, y_2, y_3 \ldots$ then we can also say

$$\frac{y_1}{x_1} = \frac{y_2}{x_2} = \frac{y_3}{x_3} \ldots = k$$

or $\qquad y_1 = kx_1, \quad y_2 = kx_2, \quad y_3 = kx_3 \ldots$

(b) Inverse proportion

Two sets of readings for the quantities p and V are given in the table below.

p	3	4	6	12
v	4	3	2	1

There is again a one-to-one relationship between each value of p and the corresponding value of V, but here when p is doubled, V is halved; when p is trebled, V has one-third its previous value, and so on.

We say that V is **inversely proportional** to p, or V **varies inversely** as p, that is

$$V \propto \frac{1}{p}$$

Also, the product $p \times V$ is always the same (here 12) and we can write

$$V = \frac{k}{p} \qquad \text{or} \qquad pV = k$$

where k is the constant of proportionality or variation and equals 12 in this case.

Using letters for values of p and V, we can also say

$$p_1 V_1 = p_2 V_2 = p_3 V_3 \ldots = k$$

MATHEMATICAL NOTATION IN PHYSICS

(a) Standard form or scientific notation

This is a neat way of writing numbers, especially those that are very large or very small. Table M1 shows how it works.

Table M1

Number	Standard form
300 000 000	3×10^8 (stated as 'three times ten to the power eight')
100 000	1×10^5 or 10^5
420	4.2×10^2
9.81	9.81×10^0 or 9.81 (since $10^0 = 1$)
0.75	7.5×10^{-1} (stated as 'seven point five times ten to the power minus one')
0.1	1×10^{-1} or 10^{-1}
0.055	5.5×10^{-2}
0.000 16	1.6×10^{-4}

The small figures showing the powers of ten are called **exponents** and specify how many times the number before the multiplication sign has to be multiplied by 10 if the number concerned is greater than 1, and divided by 10 if it is less than 1. For numbers less than 1, the power has a negative sign. The number 1 itself is written as 10^0.

Numbers in standard form are less likely to be copied incorrectly than numbers consisting of long strings of digits.

In calculations involving the multiplication or division of numbers in standard form, certain rules have to be followed. They are given in the next section, p. 546.

(b) Significant figures

Every measurement of a quantity is an attempt to find its true value but is subject to errors arising from limitations of the apparatus and the experimenter. The number of figures, called significant figures, in the stated value of a quantity indicates how accurate we think it is, and more figures should not be given than the experiment justifies. For example, a value of 4.5 given for a measurement has two significant figures; 0.0385 has three significant figures, 3 being the most significant figure and 5 the least, that is, it is the one we are least sure about — it might be 4 or it might be 6 and perhaps had to be estimated by the experimenter because the reading was between two marks on a scale.

When doing a calculation your answer should contain the same number of significant figures as the measurements used in the calculation. For example, if your calculator gave an answer of 3.4185062, you should write this as 3.4 if the measurements had two significant figures and as 3.42 if they had three. In deciding what the least significant figure should be, you look at the number on its right. If this is less than 5 you leave the least significant figure as it is (hence 3.41 to two significant figures becomes 3.4) but if it is equal to or greater than 5 you increase the least significant figure by 1 (hence 3.418 to three significant figures becomes 3.42).

Where a number is expressed in standard form, the number of significant figures is the number of digits before the power of ten, for example, 2.73×10^3 has three significant figures.

If an experimental result is estimated to be accurate to say ±2%, i.e. to 1 part in 50, then giving the result to two significant figures is justified. For example, if a quantity Q is found to have a numerical value of 4.2, measured to ±2% accuracy, it can be written

$$Q = 4.2 \pm 2\% = 4.2 \pm \tfrac{1}{50} \times 4.2 = 4.2 \pm 0.1$$

The treatment of errors in practical work is discussed on p. 552.

(c) Mathematical symbols

The following symbols are often met.

=	is equal to
<	is less than
≪	is much less than
∝	is proportional to
≈	is approximately equal to
>	is greater than
≫	is much greater than

Certain other symbols are used as a short-hand way of indicating the result of particular mathematical operations, for example those in the following list.

Δx means the difference between two values of x or a change in the value of x (pronounced 'delta x')

δx means a *very small* change in the value of x (also pronounced 'delta x': δ and Δ are the small and capital letters 'd' and 'D' in the Greek alphabet)

$\bar{x}$ means the average or mean of several values of x

Σx means the sum of several values of x (pronounced 'sigma x')

Questions

5. Write the following in standard form:

4000	200 000	1 000 000
2500	186 000	0.1
0.05	0.29	0.0076
0.000 013		

6. How many significant figures are there in a length measurement of
a) 3.8 cm   **b)** 9.76 cm
c) 5.349 cm   **d)** 0.0062 cm?

7. A rectangular block measures 4.1 cm by 2.8 cm by 2.1 cm. Calculate its volume, giving your answer to an appropriate number of significant figures.

RULES FOR CALCULATIONS IN STANDARD FORM

There are three rules for performing calculations involving the multiplication and division of numbers that are in powers of ten form.

Rule 1

When the number is changed from numerator (top) to denominator (bottom) or vice versa, the sign of the power of ten is changed. For example:

$$10^3 = \frac{1}{10^{-3}} \qquad 10^{-6} = \frac{1}{10^6}$$

$$\frac{6}{2 \times 10^{-6}} = \frac{6 \times 10^6}{2} = 3 \times 10^6$$

Rule 2

When two numbers are multiplied, the powers of ten are added. For example:

$$10^2 \times 10^6 = 10^{2+6} = 10^8$$
$$10^5 \times 10^{-3} = 10^{5-3} = 10^2$$
$$10^4 \times 10^{-9} = 10^{4-9} = 10^{-5}$$
$$3 \times 10^{12} \times 2 \times 10^{-5} = 3 \times 2 \times 10^{12-5} = 6 \times 10^7$$
$$4 \times 10^6 \times 2 \times 10^{-6} = 4 \times 2 \times 10^{6-6} = 8 \times 10^0$$
$$= 8 \times 1 = 8$$

Rule 3

When one number is divided by another, their powers of ten are subtracted. For example:

$$\frac{10^8}{10^3} = 10^{8-3} = 10^5$$

$$\frac{8 \times 10^{-5}}{2 \times 10^{-3}} = \frac{8 \times 10^{-5+3}}{2} = 4 \times 10^{-2}$$

$$\frac{10^{-2}}{10^{-6}} = 10^{-2+6} = 10^4$$

Questions

8. Find the value of
a) $10^4 \times 10^5$ **b)** $10^{12} \times 10^{-9}$
c) $10^7/10^3$ **d)** $10^2/10^5$

e) $\dfrac{6 \times 10^3}{3 \times 10^2}$ **f)** $\dfrac{5.0 \times 10^5}{2.0 \times 10^{-3}}$

g) $2 \times 10^4 \times 4 \times 10^{-3}$ **h)** $\dfrac{8}{4 \times 10^6}$

i) $3.5 \times 10^{-4} \times 2 \times 10^4$ **j)** $5 \times 10^6/10^6$

9. Find the value of
a) $3 \times 10^3 + 2 \times 10^3$ **b)** $6 \times 10^5 + 3 \times 10^4$
c) $5 \times 10^3 - 4 \times 10^2$ **d)** $7 \times 10^6 - 9 \times 10^5$

SI SYSTEM OF UNITS

(a) Basic units

Before a physical quantity can be measured, a standard or **unit** must be chosen and defined. In the SI system (Système Internationale d'Unités) there are seven **basic units** and from these the units of all other quantities, called **derived units**, can be obtained by multiplying or dividing the basic units. The system is a convenient one because all the derived units are connected with as few basic units as possible.

We need only consider six of the basic units, those for measuring the quantities: **length**, **mass**, **time**, **temperature**, **electric current** and **amount of substance**. The quantities, units and symbols are given in Table M2.

The definitions of the basic units are agreed internationally and must not vary with time. Apart from the prototype kilogram (a piece of metal) all are defined in terms of reproducible measurements of physical phenomena.

Table M2

Quantity		Basic unit	
Name	*Symbol*	*Name*	*Symbol*
mass	m	kilogram	kg
length	l	metre	m
time	t	second	s
temperature	T	kelvin	K
electric current	I	ampere	A
amount of substance	n	mole	mol

(b) Derived units

Some derived units with *combined* names are given in Table M3.

Table M3   Derived units with combined names

Quantity		Derived unit	
Name	*Symbol*	*Name*	*Symbol*
area	A	square metre	m^2
volume	V	cubic metre	m^3
density	ρ	kilogram per cubic metre	$kg\,m^{-3}$
velocity	v	metre per second	$m\,s^{-1}$
acceleration	a	metre per second squared	$m\,s^{-2}$
momentum	p	kilogram metre per second	$kg\,m\,s^{-1}$

The unit of area is derived by multiplying the unit of length by itself, i.e. $m \times m = m^2$. Similarly, since a volume basically involves a length measurement three times, the unit of volume is $m \times m \times m = m^3$. Since density equals mass divided by volume, the unit of density is obtained by dividing the unit of mass by the unit of volume, i.e. $kg/m^3 = kg\,m^{-3}$. The unit of velocity, $m/s = m\,s^{-1}$ is derived from the units of length and time.

Derived units that are rather complex when expressed in terms of the basic units are given *special* names. They are named after famous scientists and their symbols are given capital letters, as shown in Table M4.

Table M4   Derived units with special names

Quantity		Derived unit	
Name	*Symbol*	*Name*	*Symbol*
force	F	newton	N
pressure	p	pascal	Pa
energy	E	joule	J
work	W	joule	J
power	P	watt	W
frequency	f	hertz	Hz
electric charge	Q	coulomb	C
resistance	R	ohm	Ω
electromotive force	E	volt	V
potential difference	V	volt	V

(c) Prefixes for SI units

Although there is only one unit, basic or derived, for each physical quantity, decimal multiples and submultiples with approved prefixes are sometimes required to give larger or smaller units. In general, prefixes involving powers of ten which are multiples of three are preferred but others are used, e.g. 10^{-2} (centi). Table M5 lists the prefixes for some commoner multiples and submultiples.

Table M5   Unit prefixes

Multiple	Prefix	Symbol	Submultiple	Prefix	Symbol
10^3	kilo	k	10^{-3}	milli	m
10^6	mega	M	10^{-6}	micro	µ
10^9	giga	G	10^{-9}	nano	n
10^{12}	tera	T	10^{-12}	pico	p

HOMOGENEITY OF EQUATIONS

A physical quantity consists of a **numerical magnitude** and a **unit** and is written as the product of a number and a unit abbreviation, e.g. 2 kg for a mass. When the quantity is represented by a symbol, e.g. m for mass, the symbol stands for a number *and* a unit. Unit abbreviations can, like numbers, be treated algebraically, so enabling the unit of the quantity required to be found as well as the number. If this is done but an incorrect mathematical equation is used, it shows up by the unit for the required quantity being wrong. This is called **quantity algebra**.

In a correct equation, both numbers and units can be equated for each term, i.e. the equation is **homogeneous** with respect to the units and provides a useful check. For example, suppose the density of a material is to be calculated. Using the correct equation

$$\rho = \frac{m}{V}$$

we have (from Tables M2 and M3)

$$\text{unit of } \rho = \text{kg m}^{-3}$$

$$\text{unit of } \frac{m}{V} = \frac{\text{kg}}{\text{m}^3} = \text{kg m}^{-3}$$

Therefore since the same units occur on both sides of the equation, it is homogeneous so far as units are concerned and is the correct one to use. Using the incorrect equation

$$\rho = m \times V$$

we have

$$\text{unit of } m \times V = \text{kg} \times \text{m}^3$$

which is not equal to the unit of ρ, i.e. kg m^{-3}, and so the equation, not being homogeneous, is incorrect.

Checking that the units on both sides of an equation are the same, i.e. that the equation is homogeneous, is a good way of guarding against using the wrong equation in a calculation.

The relationship of derived units to other units, including the basic units, can also be obtained by quantity algebra.

Example 1. Force is defined by the equation

$$F = ma$$

where a is the acceleration produced on a mass m by a force F. Substituting the values $m = 3$ kg and $a = 2$ m s^{-2}, we get

$$F = 3 \text{ kg} \times 2 \text{ m s}^{-2}$$
$$= 6 \text{ kg m s}^{-2}$$
$$= 6 \text{ N}$$

from which we see that the unit of force, the newton (1 N), is equivalent to 1 kilogram metre per second squared (1 kg m s^{-2}).

Example 2. Pressure is defined by the equation

$$p = \frac{F}{A}$$

where p is the pressure exerted by a force F acting at right angles to an area A. Substituting the values $F = 20$ N and $A = 2.0$ m^2, we get

$$p = \frac{20 \text{ N}}{2.0 \text{ m}^2}$$
$$= 10 \text{ N m}^{-2}$$
$$= 10 \text{ Pa}$$

The unit of pressure, the pascal (1 Pa) is therefore equivalent to 1 newton per square metre (1 N m^{-2}), or, since 1 N = 1 kg m s^{-2} (Example 1), to

$$1 \text{ N m}^{-2} = 1 \text{ kg m s}^{-2} \times \text{m}^{-2}$$
$$= 1 \text{ kg m}^{-1} \text{ s}^{-2}$$

(since m $\times$ m^{-2} = m$^{1-2}$ = m^{-1})

Example 3. Work is defined by the equation

$$W = F \times s$$

where W is the work done when a force F moves its point of application through a distance s in its own direction. Substituting the values $F = 4$ N and $s = 2$ m, we get

$$W = 4 \text{ N} \times 2 \text{ m}$$
$$= 8 \text{ N m}$$
$$= 8 \text{ J}$$

Therefore the unit of work, the joule (1 J) is equivalent to 1 newton metre (1 N m). In terms of basic units, since 1 N = 1 kg m s^{-2}, it is also given by 1 N m = 1 kg m s^{-2} $\times$ m = 1 kg m^2 s^{-2}.

Questions

10. Prove that the following equations are homogeneous with respect to units.
   a) Volume = mass/density   ($V = m/\rho$)
   b) Kinetic energy = $\frac{1}{2}$ mass $\times$ velocity^2 (k.e. = $\frac{1}{2}mv^2$)
   c) Centripetal acceleration = velocity^2/radius   ($a = v^2/r$)
   d) For a simple pendulum of length l, its period T is given by $T = 2\pi\sqrt{(l/g)}$ where g is the acceleration of free fall.

11. Use quantity algebra to obtain the basic unit equivalent of the unit of power, i.e. the watt (W), using the equation

   power = work done/time taken   ($P = W/t$)

SOLVING PHYSICS PROBLEMS

When tackling physics problems using mathematical equations it is sound technique *not to substitute numerical values until you have obtained the expression in symbols which gives the answer.* That is, work in symbols until you have solved the problem and only then insert the numbers and units in the expression to get the final result.

This has two advantages. First, it reduces the chance of arithmetical (and copying down) errors. Second, you write less since a symbol is usually a single letter whereas a numerical value is often a string of figures.

Adopting this 'symbolic' procedure frequently requires you to transpose (change round) an equation first. If you are not sure how to do this refer to the sections on *Equations*, p. 544.

Example 1 (Density). What is the mass of air in a room measuring $6.0 \text{ m} \times 5.0 \text{ m} \times 2.0 \text{ m}$ if the density of air is 1.3 kg m^{-3}?

The density equation is

$$\rho = \frac{m}{V}$$

Changing the subject to m gives

$$m = \rho \times V$$

Substituting numerical values and units, where $\rho = 1.3 \text{ kg m}^{-3}$ and $V = 6.0 \text{ m} \times 5.0 \text{ m} \times 2.0 \text{ m} = 60 \text{ m}^3$, gives

$$m = 1.3 \text{ kg m}^{-3} \times 60 \text{ m}^3$$
$$= (1.3 \times 60)(\text{kg m}^{-3} \times \text{m}^3)$$
$$= 78 \text{ kg}$$

Note that:

- The symbols in $m = \rho \times V$ stand for a number *and* a unit and *both* are substituted in the equation.
- The correct unit is obtained in the answer for m, i.e. kg, since the symbols m³ and m⁻³ cancel when the units are treated algebraically, so showing that the equation $m = \rho \times V$ is homogeneous and the correct one to use.
- Since the values for the measurements of ρ and V were given to two significant figures, so too is the answer for m.

Example 2 (Falling body). How long does an object take to reach the ground if it is dropped from a stationary hovercraft at a height of 45 m?

The equation for a body falling freely under gravity is

$$s = \tfrac{1}{2}gt^2 \qquad \text{(see p. 117)}$$

Changing the subject to t we get

$$t = \sqrt{\left(\frac{2s}{g}\right)}$$

Substituting numerical values and units, where $s = 45 \text{ m}$ and g = acceleration of free fall $\approx 10 \text{ m s}^{-2}$, gives

$$t = \sqrt{\left(\frac{2 \times 45 \text{ m}}{10 \text{ m s}^{-2}}\right)}$$
$$= \sqrt{\left(\frac{2 \times 45}{10}\right)} \times \sqrt{\left(\frac{\text{m}}{\text{m s}^{-2}}\right)}$$
$$= \sqrt{9} \times \sqrt{\text{s}^2}$$
$$= 3.0 \text{ s}$$

GRAPHS IN PHYSICS

A useful way of finding the relation between two quantities is by a graph.

When plotting a graph from experimental results, scales should be chosen so that as much of the paper as possible is used, points should be marked $\times$ or $\odot$ and a smooth curve or straight line drawn so that the points are distributed equally on either side of it.

The points plotted on a graph are pure numbers and have no units. Since the symbol for a physical quantity stands for a number and a unit, to obtain the label for an axis, we divide the quantity plotted on the axis by its unit to give just a number. For example, an axis used for time in seconds would be labelled t/s, for volume in cubic metres V/m^3. A velocity axis might be labelled $v/\text{m s}^{-1}$, a pressure axis p/Pa and a current axis I/A and so on. The same system is used for labelling the columns in a table of measurements.

(a) Straight-line graphs

If a straight line is obtained its equation is of the form

$$y = mx + c$$

where m is the slope or gradient of the line (AB/BC) in Fig. M1 and c is the intercept (OD) on the y-axis.

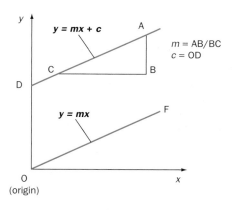

Fig. M1 Straight-line graphs

If a graph *passes through the origin* O, as does FO in Fig. M1, then $c = 0$ and $y = mx$, i.e. $y \propto x$ and y is directly proportional to x. A straight-line graph not passing through the origin only indicates a linear relationship between y and x, not proportionality.

If it is not possible to include on the graph the origins for one or both axes, c can be calculated by substituting the co-ordinates of a point on the graph, say (x_1, y_1) and the value of m, in the equation $y = mx + c$ and solving for c. Thus if $x_1 = 2$, $y_1 = 3$ and $m = +1$, then $3 = 1 \times 2 + c$, therefore $c = 1$ and the graph does not pass through the origin of the y-axis.

(b) Non-linear graphs

An **inverse proportion** graph, Fig. M2, is obtained if y against x is plotted using the readings in the following table. Its equation is $xy = $ a constant $= k$, i.e. $y \propto 1/x$.

y	3	4	6	12
x	4	3	2	1

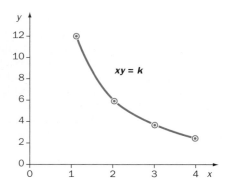

Fig. M2   Inverse proportion graph

A **square law** graph is obtained by plotting y against x for the equation $y = kx^2$; it is shown in Fig. M3.

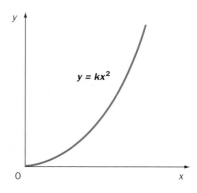

Fig. M3   Square law graph

An **inverse square law** graph with the equation $y = k/x^2$ is shown in Fig. M4 for a plot of y against x.

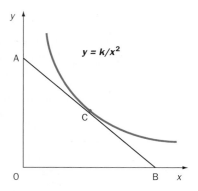

Fig. M4   Inverse square law graph

When possible, quantities should be plotted which give a straight-line graph since this is easier to interpret. For example, in the three previous cases, straight-line graphs can be obtained by plotting y against $1/x$, y against x^2 and y against $1/x^2$ respectively.

The gradient of a non-linear graph at a particular point is found by measuring the gradient of the tangent to the curve at that point. In Fig. M4 the gradient at C = AO/OB.

(c) Log graphs

Sometimes we want to find experimentally the relationship between two quantities x and y. If we assume that it is

$$y = kx^n$$

where k and n are constants, then taking logs:

$$\log y = \log k + n \log x$$

This is of the form $y = c + mx$ and so a graph of $\log y$ against $\log x$ should be a straight line of gradient n and intercept $\log k$ on the $\log y$ axis, Fig. M5. Hence k and n can be found.

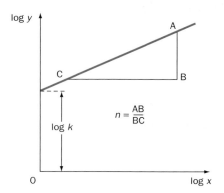

Fig. M5   Log graph

Note that:

$$\log (ab) = \log a + \log b$$
$$\log (a/b) = \log a - \log b$$
$$\log (x^n) = n \log x$$
$$\log_e (e^{kx}) = \ln (e^{kx}) = kx$$

Questions

12. In an experiment different masses m were hung from the end of a spring held in a stand, and the extensions Δl were as shown below.

m/g	$\Delta l/\text{cm}$
100	1.9
150	3.1
200	4.0
300	6.1
350	6.9
500	10.0
600	12.2

a) Plot a graph of Δl along the vertical axis against m along the horizontal axis.

b) What is the relation between Δl and m? Give a reason for your answer.

13. Pairs of readings of the quantities y and x are given below.

y	0.12	1.5	2.5	3.5
x	20	40	56	72

a) Plot a graph of y against x.

b) Is y directly proportional to x? Explain your answer.

c) Use the graph to find x when $y = 1$.

14. The distance s travelled by a car at various times t is shown below.

s/m	0	2	8	18	32	50
t/s	0	1	2	3	4	5

Draw graphs of

a) s against t,

b) s against t^2.

What can you conclude?

BASIC MEASUREMENTS

Some experiments require the accurate measurement of length. Two instruments that do this are considered below.

(a) Vernier scale

The simplest type enables a length to be measured to 0.01 cm. It comprises a small sliding scale which is 9 mm long and is divided into 10 equal divisions, Fig. M6a.

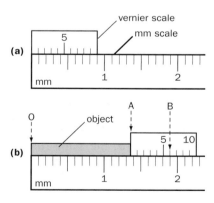

Fig. M6 Vernier scale

$$1 \text{ vernier division} = \tfrac{9}{10} \text{ mm}$$
$$= 0.9 \text{ mm}$$
$$= 0.09 \text{ cm}$$

One end of the length to be measured is made to coincide with the zero of the millimetre scale and the other end with the zero of the vernier scale. The length of the object in Fig. M6b is between 1.3 cm and 1.4 cm. The reading to the second place of decimals is obtained by finding the vernier mark which is exactly opposite (or nearest to) a mark on the millimetre scale. In this case it is the 6th mark and the length is 1.36 cm, since

$$OA = OB - AB$$
$$\therefore \quad OA = (1.90 \text{ cm}) - (6 \text{ vernier divisions})$$
$$= 1.90 \text{ cm} - 6(0.09) \text{ cm}$$
$$= (1.90 - 0.54) \text{ cm}$$
$$= 1.36 \text{ cm}$$

Vernier scales are often used on calipers, barometers, travelling microscopes and spectrometers.

(b) Micrometer screw gauge

This measures very small objects to 0.001 cm. One revolution of the drum opens the accurately plane, parallel jaws by 1 division on the scale on the shaft of the gauge; this is usually $\tfrac{1}{2}$ mm, i.e. 0.05 cm. If the drum has a scale of 50 divisions round it, then rotation of the drum by 1 division opens the jaws by $0.05/50 = 0.001$ cm, Fig. M7. A friction clutch ensures that the jaws exert the same force when the object is gripped.

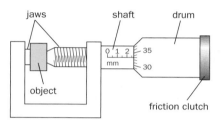

Fig. M7 Micrometer screw gauge

The object shown has a length of

2.5 mm on the shaft scale + 33 divisions on the drum scale
$$= 0.25 \text{ cm} + 33(0.001) \text{ cm}$$
$$= 0.283 \text{ cm}$$

Before making a measurement a check should be made to ensure that the reading is zero when the jaws are closed. Otherwise the zero error must be allowed for when the reading is taken.

Questions

15. What are the readings on the vernier scales in Figs M8a, b?

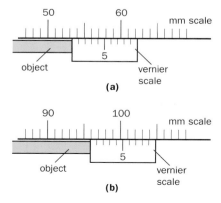

Fig. M8

16. What are the readings on the micrometer screw gauges in Figs M9a, b?

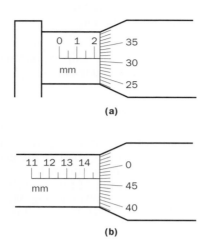

(a)

(b)

Fig. M9

TREATMENT OF ERRORS

(a) Types of error

Experimental errors cause a measurement to differ from its true value and are of two main types.

(*i*) A **systematic error** may be due to an incorrectly calibrated scale on, for example, a ruler or an ammeter. Repeating the observation does not help and the existence of the error may not be suspected until the final result is calculated and checked, say by a different experimental method. If the systematic error is small a measurement is accurate.

(*ii*) A **random error** arises in any measurement, usually when the observer has to estimate the last figure, possibly with an instrument that lacks sensitivity. Random errors are small for a good experimenter and taking the mean of a number of separate measurements reduces them in all cases. A measurement with a small random error is precise but it may not be accurate, as there may be a systematic error.

Fig. M10a shows random errors only in a meter reading, while in Fig. M10b there is a systematic error as well.

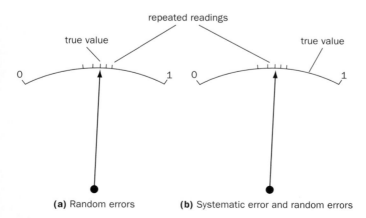

(a) Random errors   **(b)** Systematic error and random errors

Fig. M10

(b) Estimating errors in single measurements

In more advanced work, if systematic errors are not eliminated they can be corrected from the observations made. Here we shall assume they do not exist and then make a reasonable estimate of the likely random error.

> *Example 1.* Using a metre rule the length of an object is measured as 2.3 cm. At very worst the answer might be 2.2 or 2.4 cm, i.e. the maximum error it is possible to make using a ruler marked in mm is ±0.1 cm. The **possible error** (p.e.) is said to be ±0.1 cm and the length is written (2.3 ± 0.1) cm. The percentage possible error (p.p.e.) is (±0.1 × 100)/2.3 ≈ ±4%.
>
> *Example 2.* Using vernier calipers capable of measuring to 0.01 cm, the length of the same object might be read as (2.36 ± 0.01) cm. In this case the p.e. is ±0.01 cm and the p.p.e. is (±0.01 × 100)/2.36 ≈ ±0.4%.

If a large number of readings of one quantity are taken the mean value is likely to be close to the true value and statistical methods enable a **probable error** to be estimated. Here, we adopt the simpler procedure of estimating the maximum error likely, i.e. the **possible error**.

(c) Combining errors

The result of an experiment is usually calculated from an expression containing the different quantities measured. The combined effect of the errors in the various measurements has to be estimated. Three simple cases will be considered.

(*i*) *Sum.* Suppose the quantity Q we require is related to quantities a and b, which we have measured, by the equation

$$Q = a + b$$

Then total p.e. in Q = (p.e. in a) + (p.e. in b). Thus if $a = 5.1 ± 0.1$ cm and $b = 3.2 ± 0.1$ cm then $Q = 8.3 ± 0.2$ cm. That is, in the worst cases, if both a and b are read 0.1 cm too high $Q = (5.2 + 3.3) = 8.5$ cm, but if both are 0.1 cm too low then $Q = (5.0 + 3.1) = 8.1$ cm.

(*ii*) *Difference.* If $Q = a - b$, the same rule applies, i.e. the total p.e. in Q = (p.e. in a) + (p.e. in b).

(*iii*) *Product and quotient.* If the individual measurements have to be multiplied or divided it can be shown that the *total percentage possible error equals the sum of the separate percentage possible errors*. For example, if a, b and c are measurements made and

$$Q = \frac{ab^2}{c^{1/2}}$$

then if the p.p.e. in a is ±2%, that in b is ±1% and that in c is ±2%, then the p.p.e. in b^2 is $2(±1)\% = ±2\%$ and in $c^{1/2}$ is $\frac{1}{2}(±2)\% = ±1\%$. Then

total p.p.e. in Q = ±(p.p.e. in a + p.p.e. in b^2 + p.p.e. in $c^{1/2}$)
= ±(2 + 2 + 1) = ±5%

The answer for Q will therefore be accurate to 1 part in 20 and if the numerical result for Q is 1.8 then it is written

$$Q = 1.8 ± \tfrac{1}{20} × 1.8 = 1.8 ± 0.1$$

It would not be justifiable to write $Q = 1.842$ since this would be claiming an accuracy of four figures. According to our estimate this accuracy is not possible with the apparatus used.

It is instructive to estimate whenever possible the total p.p.e. for an experiment; it indicates (*i*) the number of significant figures that can be given in the result, (*ii*) the limits within which the result lies, and (*iii*) the measurements that require particular care. There is little point in making one measurement to a very high degree of accuracy if it is not possible with the others; a chain is only as strong as the weakest link.

It is useful to remember the following.

$$\sin 30° = \cos 60° = 0.5$$

$$\tan 45° = 1$$

$$\sin 0° = \cos 90° = 0$$

$$\sin 90° = \cos 0° = 1$$

$$\pi \approx 3, \quad \pi^2 \approx 10, \quad \pi \text{ radians} = 180°$$

TRIGONOMETRY

For the right-angled triangle ABC in Fig. M11:

$$\text{sine } a = \frac{\text{opposite side}}{\text{hypotenuse}} = \frac{BC}{AB}$$

$$\text{cosine } a = \frac{\text{adjacent side}}{\text{hypotenuse}} = \frac{AC}{AB}$$

$$\text{tangent } a = \frac{\text{opposite side}}{\text{adjacent side}} = \frac{BC}{AC}$$

These three terms are abbreviated to sin *a*, cos *a* and tan *a*. Note that:

$$\frac{\sin a}{\cos a} = \frac{BC/AB}{AC/AB} = \frac{BC}{AC} = \tan a$$

$$\sin^2 a + \cos^2 a = \frac{BC^2}{AB^2} + \frac{AC^2}{AB^2} = \frac{BC^2 + AC^2}{AB^2} = \frac{AB^2}{AB^2} = 1$$

(since by Pythagoras $AB^2 = BC^2 + AC^2$).

Question

17. In Fig. M11 what are the expressions for
   a) sin *b*   **b)** cos *b*   **c)** tan *b*?

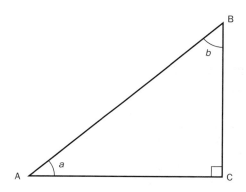

Fig. M11

Appendix 1

CONSTRUCTION OF MODEL CRYSTAL FOR MICROWAVE ANALOGUE DEMONSTRATION (p. 23)

The model has a face-centred cubic structure and is made from 190 5 cm diameter polystyrene balls glued together in seven hexagonal close-packed layers, Fig. A1.1.

Fig. A1.1 Model crystal

Start with **layer 4** which has 37 balls, Fig. A1.2a (the black dots represent the centres of the balls).

The black dots in Fig. A1.2b show the 36 balls in **layer 3** and their positions in the hollows of layer 4 which is shown by blue circles. Each ball should be glued to all those it touches.

The 27 balls of **layer 2** are the black dots in Fig. A1.2c (the blue circles are layer 3) and the layer 2 balls should be placed *over the hollows in layer 4* (not over the balls in layer 4) to give the ABCABC stacking of an FCC crystal.

Fig. A1.2d shows the 19 balls of **layer 1** as black dots and layer 2 as blue circles.

Half the model is now made and when set it can be turned over to build the other half.

In Fig. A1.2e the blue circles are layer 4 (now uppermost), the blue crosses represent layer 3. **Layer 5** is shown by 36 black dots.

The 23 black dots in Fig. A1.2f show **layer 6** and the blue circles layer 5. Care should again be taken to ensure that the balls in layer 6 are over *hollows* in layer 4.

Fig. A1.2g gives the position of the 12 balls in **layer 7**.

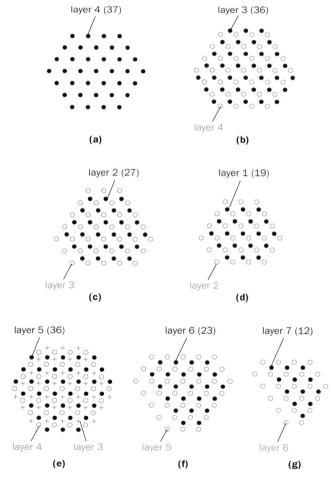

Fig. A1.2

Appendix 2

SPEED OF AN ELECTRICAL PULSE ALONG A CABLE (p.41)

The drift speed of the electrons forming the current in a conductor is about 1 mm s^{-1}, but the electric and magnetic fields which constitute the signal that sets them in motion almost simultaneously round the circuit travel at very great speeds. Basically these fields are the same as those of an electromagnetic wave in free space.

Using a double-beam CRO in a fast time-base speed (1 μs cm^{-1}), the time taken by an electrical pulse from a 200 kHz pulse generator to travel along a 200 m length of coaxial cable can be found, Fig. A2.1. The in-going pulse is applied to the Y_1 input (set at 0.2 V cm^{-1}) as well as to the near end of the cable, and the outcoming pulse from the far end of the cable is applied to the Y_2 input (also on 0.2 V cm^{-1}). The distance between the pulses is measured and the time it represents calculated. The speed is roughly 2×10^8 m s^{-1}. If air or a vacuum replaced the polythene insulation between the central and outer conductors of the cable and through which the fields travel, the speed would be that of light (3×10^8 m s^{-1}). Figure A2.2 shows the shapes of the electric and magnetic fields travelling along the cable; they are at right angles to each other and to the direction of the current.

Coaxial cable is used because it does not pick up unwanted interference if the outer conductor is connected to E on the CRO at both ends. The 68 Ω resistors should be connected to the CRO terminals directly. A full explanation of their action requires a more advanced treatment but without them the pulse would be reflected backwards and forwards along the cable, setting up a standing wave system on it. Instead, the resistors 'absorb' the pulses and ensure they are applied to the CRO.

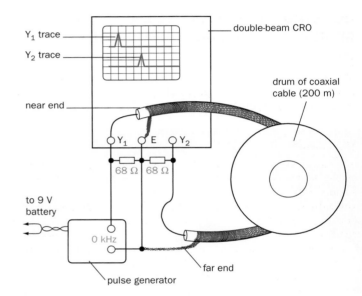

Fig. A2.1

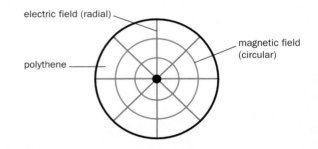

Fig. A2.2 Cross-section through coaxial cable showing electric and magnetic fields

Appendix 3

CONSTRUCTION AND ACTION OF A FLAME PROBE

(a) Construction

Details are shown in Fig. A3.1. In use the flame should be made as small as possible by *slowly* reducing the gas supply (a screw clip helps).

(b) Action

The deflection of the electroscope connected to the flame probe is a measure of the potential at the point where the probe is situated. Roughly, the action is as follows. When brought near to a positively charged body the metal probe (i.e. the needle) has a negative charge induced on it, while a positive charge appears on the electroscope movement. However, the flame is producing positive and negative ions; the positive ions are attracted to the probe and neutralize its charge while the negative ions are repelled. The electroscope remains positively charged and is at the same potential as the uncharged probe since they are connected. The closer the probe is to the charged body the higher is its potential and the greater is the positive charge induced in the electroscope, and so the greater the deflection.

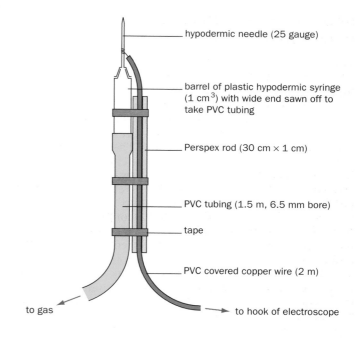

hypodermic needle (25 gauge)

barrel of plastic hypodermic syringe (1 cm^3) with wide end sawn off to take PVC tubing

Perspex rod (30 cm × 1 cm)

PVC tubing (1.5 m, 6.5 mm bore)

tape

PVC covered copper wire (2 m)

to gas

to hook of electroscope

Fig. A3.1

Appendix 4

PHOTOELECTRIC EFFECT AND ROUGH ESTIMATE OF THE PLANCK CONSTANT

The apparatus is shown in Fig. A4.1. Before the electrometer/d.c. amplifier is connected to the photoelectric unit any input resistor is removed so that it acts as a voltmeter whose resistance is as high as possible.

The photoelectric unit contains a photocell (see p. 393) with a photocathode of potassium (in a vacuum) which loses electrons when light falls on it. These travel to a collecting wire (of material that does not emit photoelectrons with incident light) which now becomes negatively charged and soon stops the arrival of further electrons. A steady p.d. is created between the cathode and the collecting wire and is measured by the electrometer connected across the photocell. If this is, say, 1.0 V then 1.0 eV must be the maximum k.e. of the electrons emitted by the cathode, otherwise they would continue to reach the collecting wire. The maximum k.e. of the photoelectrons is thus indicated by the electrometer reading (provided its input resistance is sufficiently high).

(a) Effect of colour

The spectrum is rotated *slowly* so that the colours from red to violet and beyond fall in turn on the slit in front of the photocell. The reading on the electrometer meter rises steadily indicating that the higher the frequency of the light, the greater the maximum k.e. of the photoelectrons.

The p.d. readings for red and blue light are noted.

(b) Effect of intensity

With blue light on the slit a stop is placed in front of the lamp to halve (approximately) the light intensity. The electrometer reading should remain *almost* constant showing that the maximum k.e. of the photoelectrons is more or less independent of the brightness of the light. (What is the effect of decreased intensity?)

(c) The Planck constant

An estimate of h may be obtained if we assume that for red light $f = 4.5 \times 10^{14}$ Hz and for blue light $f = 6.5 \times 10^{14}$ Hz. If the corresponding p.ds are, say, 0.25 V and 1.00 V then the maximum k.es are 0.25 eV (i.e. $0.25 \times 1.6 \times 10^{-19}$ J) and 1.00 eV (i.e. $1.00 \times 1.6 \times 10^{-19}$ J) respectively. Hence, from Einstein's photoelectric equation (p. 391), we have

$$hf - \Phi = \tfrac{1}{2}mv_{\text{max}}^2$$

For blue light

$$h(6.5 \times 10^{14}\ \text{s}^{-1}) - \Phi = 1.00 \times 1.6 \times 10^{-19}\ \text{J}$$

For red light

$$h(4.5 \times 10^{14}\ \text{s}^{-1}) - \Phi = 0.25 \times 1.6 \times 10^{-19}\ \text{J}$$

Subtracting,

$$h = \frac{(1.00 - 0.25) \times 1.6 \times 10^{-19}\ \text{J}}{(6.5 - 4.5) \times 10^{14}\ \text{s}^{-1}}$$

$$= 6.0 \times 10^{-34}\ \text{J s}$$

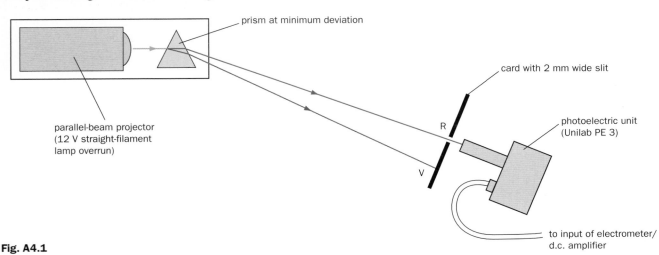

Fig. A4.1

Appendix 5

ELECTRON COLLISION EXPERIMENTS USING A COMMERCIAL XENON-FILLED THYRATRON VALVE (EN 91)

(a) Excitation potential

The circuit is shown in Fig. A5.1a. Electrons emitted by the cathode C of the thyratron are accelerated by the positive potential V_g on the grid G_1, and initially are able to reach the anode A since, for small values of V_g, any collisions they have with atoms of the xenon gas are elastic. The electron flow I_a through the thyratron is recorded by the galvanometer on its most sensitive range and maintains A at a small negative potential with respect to G_1.

As V_g is increased some electrons have just enough energy to cause excitation of xenon atoms and, having lost all their kinetic energy in the inelastic collision, are unable to overcome the retarding p.d. between A and G_1, causing I_a to fall. Further increase of V_g causes more electrons to lose their energy after excitation and I_a decreases further. When most of the electrons passing through G_1 produce excitation, I_a is a minimum and the corresponding value of V_g gives the **first excitation potential** of xenon. Increasing V_g beyond this value enables electrons, even after they have had inelastic collisions, to overcome the retarding p.d. and reach A, and I_a rises again. The form of the I_a–V_g graph is shown in Fig. A5.1b.

The 10 kΩ resistor in the grid circuit prevents the current exceeding 10 mA (and destroying the thyratron) in the event of the gas ionizing during the experiment. With some thyratrons it may be necessary to *apply* a small retarding p.d. to prevent electrons reaching A. This is done by connecting a 1.5 V cell between A and G_1, in series with the galvanometer, as shown by the dotted symbol in Fig. A5.1a.

(b) Ionization potential

The thyratron, used as a diode with A joined to G_1, is connected as in Fig. A5.2a overleaf. When V equals the ionization potential of xenon, the current recorded by the milliammeter increases due to electrons from C having inelastic collisions with xenon atoms and ionizing them. The positive xenon ions created act as a new source of current.

Figure A5.2b shows the form of the current–p.d. graph.

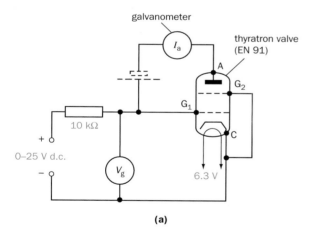

(a)

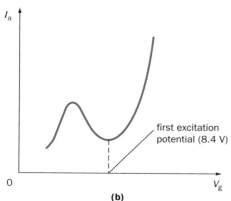

(b)

Fig. A5.1   Determining the excitation potential

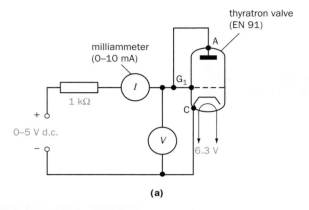

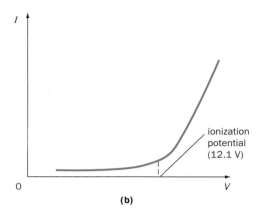

(a)

(b)

Fig. A5.2 Determining the ionization potential

Appendix 6

INTERNET WEBSITES

Interactive science centres around the world
http://www.exploratory.org.uk
http://www.exploratorium.edu
http://www.planet.eon.net/~essc/main/html
http://www.nasm.si.edu

Physics
Institute of Physics (information for schools)
http://www.iop.org/sc.html

The National Physical Laboratory
http://www.npl.co.uk

Particle Physics and Astronomy and Space Science
Research in the UK
http://www.pparc.ac.uk

Astronomical photographs
http://www.mso.anu.edu.au/~bessell/thumbnails

Special relativity
http://www.anu.edu.au/Physics/Searle

Science City
http://www.sciencecity.org.uk

Association for Women in Science
http://www.awise.org

American Institute of Physics
http://www.aip.org

Space
NASA
http://www.nasa.gov

Galileo
http://www.jpl.nasa.gov/galileo

Mars Global Surveyor
http://www.jpl.nasa.gov/mars

Hubble Space Telescope
http://www.stsci.edu

Goddard Space Flight Centre – Space Science Education
Home Page
http://www.gsfc.nasa.gov

Equipment
Pico Technology Ltd, 149–151 St Neots Road, Hardwick,
Cambs CB3 7QJ
http://www.picotech.co.uk

PEARLS software
http://www.aip.org/pas

ANSWERS

Answers

The answers given are the responsibility of the author alone and have not been approved by the various examining boards.

ABOUT PHYSICS

1. A quantity that can be measured and helps to 'explain' the behaviour of matter and/or energy. See text.
2. The relationship between concepts. See text.
3. To help us to think about and 'explain' things we cannot see directly. See text.
4. **a)** Washing machine, refrigerator, food mixer, vacuum cleaner.
   b) Starter motor, windscreen wiper motor, radiator cooling fan motor.
5. **a)** To do calculations quickly; to enable rapid communications and data transfer; to analyse medical information; to control industrial processes or robots in a factory.
   b) To control missiles in warfare; to store secret personal information and make it available to other people.
7. $34\,\Omega$

1 MATERIALS AND THEIR USES

2. **a)** AB, since it is in tension.
   b) i) AC and AD   ii) AB
3. **b)** The bottom of the bridge-deck is in tension and steel is strong in tension.
4. **a)** It uses the 'arch principle' and is made of material that is strong in compression.
   b) It is strong in compression.
5. **a)** AB is a tie; it is in tension due to the load pulling it at B and the wall pulling it at A; both pulls tend to stretch it.
   b) BC is a strut; it is in compression due to the load pushing it down at B and the wall resisting it at C; both pushes tend to squeeze it.
6. **a)** AD and BD
   b) AB, AE and BC

2 STRUCTURE OF MATERIALS

1. **a)** 4:3 **b)** 16 **c)** 16 g
2. **a)** 14 g **b)** 70 g
3. **a)** 1.5×10^{23} **b)** 18×10^{23} **c)** 3.0×10^{23}
4. **a)** 4.0×10^{23} g **b)** 1.2×10^{-21} g **c)** 3.0×10^{-23}
5. **a)** 6.0×10^{23} **b)** 2.0 g **c)** 2.2×10^{4} cm^3 **d)** 2.7×10^{17}
6. 6.1×10^{23} mol^{-1}
7. **a)** 3.0×10^{-23} g **b)** 3.9×10^{-10} m (0.39 nm)
8. **a)** 5.0×10^{3} J g^{-1} **b)** 5.3×10^{-19} J/atom
9. **a)** i) 4 ii) 6; closest in hexagonal packing
   b) i) hexagonal ii) square
10. **a)** $2\sqrt{2}\,r$ **b)** $16\sqrt{2}\,r^3$ **c)** $1/(16\sqrt{2}\,r^3)$
11. **a)** 8 **b)** 8 **c)** 8/8 = 1 **d)** 6 **e)** 2 **f)** 6/2 = 3
    g) 1 + 3 = 4 **h)** $1/(4\sqrt{2}\,r^3)$

12. 8.4×10^{22} atoms cm^{-3}
13. 1.3×10^{-10} m (0.13 nm)
14. If a is the separation between adjacent Na and Cl ions, a cube of side a and volume a^3 will contain 4 Na ions and 4 Cl ions, i.e. 4 NaCl molecules. But each corner (ion) of the cube is shared by 8 neighbouring cubes and so each volume a^3 contains $\frac{1}{8}$ of 4 molecules of NaCl, i.e. $\frac{1}{2}$ molecule of NaCl.

$$\therefore \text{ volume occupied by } \tfrac{1}{2}\text{ molecule} = a^3 = \frac{1}{2}\cdot\frac{M}{L\rho}$$

where M = formula weight (molecular mass) and ρ = density of NaCl and L = the Avogadro constant.

$$\therefore a = 2.82 \times 10^{-8} \text{ cm } (0.282 \text{ nm})$$

15. **a)** i) 10.6 nm ii) 0.3 nm iii) Oil molecules consist of several atoms; film may be thicker than a monolayer.
    b) Use the Avogadro constant.

3 MECHANICAL PROPERTIES

1. **b)** 4.0×10^{7} Pa **c)** yes
   d) i) $\frac{1}{2}$ ii) 50%
   e) 2.002 m
2. **c)** 1.0×10^{11} Pa; 0.040%; 4.0×10^{2} N; no; smaller
3. **a)** 1.1 (1.06) mm **b)** 1.0×10^{2} (101) N
4. **a)** i) Y ii) Z iii) X
   b) iii) copper, steel — ductile; glass — brittle
5. **a)** 1.86×10^{-3} m **b)** 7.75×10^{-2} J
6. **b)** 3.16×10^{3} N m^{-1} **c)** 2×10^{11} Pa
7. **a)** 110 °C
8. **a)** i) 0.009 ii) 0.044
   b) 1.8 J **c)** 17.55 J
9. rubber

4 ELECTRICAL PROPERTIES

1. 2.4×10^{2} C; 1.5×10^{21}
2. **a)** 8.5×10^{28} **b)** 8.5×10^{22} **c)** 1.4×10^{4} C **d)** 6.8×10^{3} s
   e) 0.15 mm s^{-1}
3. 8.0 A
4. **a)** 8.0 V; 2.5 A **b)** 240 V
5. **a)** 4.0 V **b)** 240 J
6. **a)** 2 A **b)** 2 A **c)** 6 V **d)** 10 V **e)** 5 Ω
7. **a)** $I = I_1 + I_2$ **b)** 6 V **c)** 3 A, 2 A, 1 A **d)** 2 Ω
8. **a)** i) 4 V ii) 2 V **b)** i) 4.8 ii) 1.2 V
9. **a)** 10 V **b)** 40 V
10. 5 Ω
12. **c)** 27 Ω; 90 Ω **d)** 70 mA
13. i) 95 Ω in series ii) 0.050 Ω in parallel
14. 3:2
15. 4.20×10^{-3} °C^{-1}

16. **a)** 1.00×10^3 J s^{-1} **b)** 798 °C
17. **a)** 12 J **b)** 36 J **c)** 240 J
18. **a)** $I_1 + I_2$
    b) i) $-10I_1 - 10(I_1 + I_2) - 10I_1 + 3 = 0$
    ii) $-10(I_1 + I_2) + 1.5 = 0$
19. **b)** 3×10^4 J **c)** 960 Ω
20. Internal resistance 5 Ω
22. 4.9 Ω; 8.9 Ω
23. 0.0020 °C^{-1}; 50.2 cm
24. 2.00 Ω

5 THERMAL PROPERTIES

2. True, true, true; 89.1 °C
3. 16.0 V
4. 3180 J K^{-1}; 0.073 °C s^{-1}; 233 W
5. 0.817 W; 35.6 g
6. 3.3×10^2 °C
7. True height is 76.46 *divisions* of scale. A division marked as 1 cm is only 1 cm at 0 °C. At 15 °C one division has length $1(1 + 1.9 \times 10^{-5} \times 15)$ cm, therefore 76.46 divisions at 15 °C have a true length of $76.46(1 + 1.9 \times 10^{-5} \times 15)$, i.e. 76.48 cm
8. 1.2 N
9. 1.9×10^3 W
10. Copper 3.0(2.98) °C cm^{-1}; aluminium 5.5 °C cm^{-1}
11. 4.4×10^{-3} W cm^{-2}
12. 18 °C

6 OPTICAL PROPERTIES

1. 15 cm; 5.0 cm
2. 20 cm; 10 cm behind mirror
3. Virtual, 13 (12.9) cm behind convex mirror
5. **a)** 40(.5)° **b)** 40(.5)° **c)** 35(.4)°
6. Angle in liquid must exceed 67.9°.
7. **a)** 37(.4)° **b)** 28(27.9)°
8. $a = 50$ mm; $b = 125$ mm; $r = 9.40$°; $i = 80.6$°; $n = 1.46$
9. Critical angle = 41.8°; $\alpha = 48.6$°
10. 55.6°; 39.6° to 90.0°
11. Critical angle glass/air = 41.8°; critical angle glass/liquid = 69.0°
12. **a)** 20 cm; +15 cm **b)** +4.0 cm
13. **a)** i) -60 cm
    b) converging — real, inverted, magnified; diverging — virtual, erect, diminished
14. 20 cm from first position of screen
15. **a)** +60 cm **b)** +2.4 m
17. **c)** i) 1.68 cm iii) -0.83 D
18. **a)** 15 cm
19. $f = 66.7$ cm gives near point at 25.0 cm. The new far point will be the object distance which gives a virtual image 200 cm from the spectacle lens, i.e. we have to find u when $v = -200$ cm and $f = 200/3$ cm. From $1/f = 1/v + 1/u$ we get $u = 50.0$ cm. The new range of vision is therefore 25.0 to 50.0 cm.
20. Magnifying power = 5.0; magnification = 6.0
21. **b)** 58(.3)
22. Separation of lenses 25 cm; $f = 6.0$ cm
23. 112(.4) cm; 14.7
24. **a)** 1.1 m to infinity **b)** ii) increase the f-number
25. i) f/diameter of aperture ii) 8/125 s
26. **a)** 15 (14.5) cm **b)** 0.051 **c)** 26 cm **d)** 24 cm
27. D
28. C
29. **a)** farther away **b)** closer to slide
30. **a)** ii) 6.25×10^{-3} rad

OBJECTIVE-TYPE REVISION QUESTIONS 1

2.1 B **2.2** A **2.3** B
3.1 C **3.2** C **3.3** B **3.4** A, E, D
4.1 A **4.2** D **4.3** C **4.4** E **4.5** D **4.6** B
4.7 C (this is a *balanced* Wheatstone bridge)
4.8 C **4.9** D **4.10** C **4.11** A **4.12** E
5.1 B **5.2** C **5.2** E
6.1 B **6.2** A **6.3** D **6.4** C **6.5** A

7 STATICS AND DYNAMICS

1. $F_A = 75$ kN; $F_B = 45$ kN
2. **a)** 1.2×10^3 N **b)** 1.1×10^3 N; 74° to horizontal
3. 4.3×10^2 N; 69° to horizontal
4. **b)** i) 3 N ii) 4 N **c)** 6 N
5. 1.59 m s^{-1}; 1.41 m s^{-2}
6. 2.17×10^4 m; 3.13×10^3 m; 25.0 s
7. $\sqrt{3}/3$; $\sqrt{3}$ g/3 m s^{-2}; $2\sqrt{3}$ mg/3 (m = mass of body)
9. $P/(5\,m)$; $P/5$; $2\,P/5$
10. i) 0.296 m s^{-2} ii) 7.69 m s^{-1} iii) 41.4 kN iv) 159 kJ s^{-1}
11. **a)** i) 1.62 kg m s^{-1} ii) 54 N **b)** 19.3 m
12. **a)** i) 33×10^6 N ii) 3×10^6 N
13. **a)** i) 5×10^3 kg m s^{-1} ii) 1.25×10^3 J **b)** 20 N
    c) i) constant ii) reduced
14. **a)** i) 39.6 m s^{-1} ii) 77.8 m s^{-1} at 30.6° to horizontal
    b) i) 2380 N
15. 8.00×10^3 N; 12.0×10^3 W
16. 30 N
17. $2mnu$
18. **a)** conserved **b)** conserved **c)** P: zero; Q: v
20. **c)** i) 10 125 J ii) 3600 J iii) 17.5 m s^{-1}
    d) i) 16 N
21. **a)** i) $m_1u_1 - m_2u_2 = m_1v_1 + m_2v_2$
    b) i) 1.1 m s^{-1} ii) 0.91 J
    c) i) $\rho A v$ ii) $\rho A v^2$
    d) 7.3 m s^{-1}

8 CIRCULAR MOTION AND GRAVITATION

1. **a)** Time $= \dfrac{\text{arc AB}}{\text{speed}} = \dfrac{\pi \times 5 \text{ m}}{11 \text{ m s}^{-1}} = \dfrac{22 \times 5 \text{ s}}{7 \times 11} = \dfrac{10 \text{ s}}{7}$

    b) Average velocity $= \dfrac{\text{displacement}}{\text{time}} = \dfrac{10 \text{ m}}{10/7 \text{ s}}$

    $= 7.0$ m s^{-1} *to the right*

    The displacement is diameter AB to the right

    c) Average acceleration $= \dfrac{\text{change of velocity}}{\text{time}}$

    $= \dfrac{22 \text{ m s}^{-1} \text{ downward}}{10/7 \text{ s}}$

    $= 154/10$ m s^{-2} *downwards*

2. **a)** 1.1π rad s^{-1} **b)** 0.13π m s^{-1}
3. **a)** i) 40π rad s^{-1} ii) 947.5 m s^{-2} iii) 0.095 N
    b) i) 645 rad s^{-1} ii) 38.7 m s^{-1} tangentially
4. **a)** 5.0 rad s^{-1}; 25 N **b)** 30 N; 20 N
5. **c)** 19.4 m s^{-2} **d)** friction
6. **c)** 28.4 m s^{-1}
7. **b)** 180 N m **c)** 4.05×10^6 J **d)** 10.1 km
8. 15.8 s
9. **b)** 4.0×10^{-4} kg m^2
10. 9.83 N
11. 365 days; 199×10^{-9} rad s^{-1}; 35.5×10^{21} N; gravitational attraction

12. 0.873×10^{-3} rad s^{-1}
13. 24 hours
14. **a)** 7.71×10^{22} kg **b)** 1.15×10^3 m
15. **b)** 65×10^6 m s^{-2} **c)** 12.5×10^6 m s^{-1}
16. 1.89×10^{27} kg
17. **d)** i) 62.5 AU ii) 1.7×10^{-6} rad

9 MECHANICAL OSCILLATIONS

1. **a)** 75 cm s^{-1} **b)** 1.4×10^4 cm s^{-2}
2. 0.5 Hz; 0.1π m s^{-1}; $0.1\pi^2$ m s^{-2}
3. 250 N m^{-1}; O; A to O; $T = 0.284$ s; potential to kinetic
4. **a)** i) 0.1π m s^{-1} ii) $0.002\pi^2$ J iii) $0.1\pi^2$ N m^{-1} **b)** 0.013 J
5. **c)** An s.h.m. of period $\pi/\sqrt{50}$ s; maximum velocity = 0.71 m s^{-1}; maximum acceleration = 10 m s^{-2}
6. 24(.5) N kg^{-1} m^{-1}
7. **c)** 1.3×10^2 cm s^{-1} **d)** 1.6 cm
8. **b)** i) 1250 Hz ii) 370×10^3 m s^{-2}

10 ENERGY AND ITS USES

2. **a)** about 22 MJ **b)** 180 MJ **c)** about 17×10^{12} J
d) i) about 67 kJ ii) about 250 kJ
e) 3.0×10^{20} J **f)** 110×10^6 J
3. Coal 240 years; oil 58 years; gas 54 years
4. 13 litres!
5. **a)** 120 W **b)** 2 kW
8. **a)** 200 kW **b)** 5000 kg s^{-1}
10. 3.1×10^{20} J (*Hint.* Regard the earth as a flat circular disc which always has the sun's rays falling on it.)
11. 3.15×10^9 J; 20.4×10^6 J
14. Maximum power P is obtained if the air approaching the turbine is brought to rest, i.e. loses all its k.e.
If m = mass of air stopped per second then

$$\text{k.e. converted per second} = P = \tfrac{1}{2}mv^2$$

If V = volume of air stopped per second then

$$V = \text{area swept out by blades} \times \text{speed of wind} = Av$$

But $m = \rho V$, $\therefore m = \rho Av$, and so

$$P = \tfrac{1}{2}(\rho Av)v^2 = \tfrac{1}{2}\rho Av^3$$

17. **a)** 560 W(J s^{-1}) **b)** 290 W
18. **c)** 3.6 W m^{-2} K^{-1}
19. **a)** 17×10^{-3} K W^{-1} **b)** 1.2 kW
20. **a)** 6.0×10^{-3} m^2 K W^{-1} **b)** 3.0×10^{-3} K W^{-1} **c)** 7.0 kW
22. **b)** i) 38.4 kW ii) 9×10^4 kg iii) 1.016×10^{-4} K s^{-1}
iv) 8.2 hours
23. 7.7 litres/minute

11 FLUIDS AT REST

1. 1.0×10^4 N; 5.0×10^2 N
2. **b)** 9 N
3. **b)** 8394 kg m^{-3}
4. 2.4 cm
5. 19.1×10^{-3} N; 1.43 g l^{-1}
6. 100.3 kPa
7. 7.1×10^{-2} N m^{-1}
8. 49 mm; the water does not overflow (and violate conservation of energy) but it remains at the top of the tube (i.e. $h = 30$ mm) with an angle of contact of 52.5°. The weight of the raised column of water is then supported by the vertical components of the surface tension forces.
9. 76.93 cm
10. 60 mm

12 FLUIDS IN MOTION

1. **b)** i) 9720 Pa ii) 95.8 cm
2. 8.7×10^{-4} m s^{-1}
3. **a)** i) 160 N ii) 0.167 **b)** 3200 W **c)** 4
5. **b)** $v_Y = A_X v_X / A_Y$
c) i) $m(v_Y^2 - v_X^2)/2$
6. **a)** i) 180 m^3 s^{-1} ii) 2.15 m s^{-1}
b) pressure reduced (conservation of energy)

OBJECTIVE-TYPE REVISION QUESTIONS 2

7.1 B **7.2** C **7.3** E **7.4** D **7.5** B **7.6** B
7.7 B **7.8** B **7.9** E **7.10** C
8.1 A **8.2** D **8.3** D **8.4** C **8.5** D **8.6** D **8.7** A
9.1 E **9.2** B **9.3** D **9.4** D **9.5** C
11.1 C
12.1 E **12.2** A

13 ELECTRIC FIELDS

2. i) A iii) 1500 V
3. **a)** 29 V **b)** 4.6×10^{-18} J **c)** 4.6×10^{-18} J
5. **b)** i) 1250 V m^{-1} ii) 1050 V m^{-1} iii) 1630 V m^{-1}; 40° (NE)
c) ii) 80 mm
6. **a)** 2.9×10^{11} N C^{-1}
b) 4.3×10^6 N C^{-1}; ratio (0.68×10^5):1
c) 1.4×10^{11} N C^{-1}
d) 5.8×10^7 N C^{-1}; ratio (0.25×10^4):1
7. negative; 200×10^3 V m^{-1}; 4.8×10^{-19} C; 3
8. **a)** 10^{-12} J **b)** 10^{-12} J **c)** 5×10^{-11} N **d)** 5×10^4 N C^{-1}
e) 5×10^4 V m^{-1}
9. 16 MJ
10. **b)** i) 1.3×10^{-17} C ii) 4.6×10^{19} V m^{-1}
iii) $E = -$(gradient at X)
c) i) 3.7×10^{-13} J ii) 3.7×10^{-13} J iii) 2.3×10^6 V
12. **b)** ii) 0.5 SX and $\sqrt{2}$ SX from S iii) 3.4×10^{10} J
13. **b)** positive **c)** plate is at a lower potential than the droplets
d) charge builds up on plate which will repel droplets

14 CAPACITORS

1. **a)** 10^{-12} F (1 pF) **b)** 10^{-9} C
c) $V = Q_1/C_1 = Q_2/C_2 \therefore Q_1/Q_2 = C_1/C_2 = 10^{-9}/10^{-12} = 10^3/1$
d) i) 10^{-9} C ii) zero
e) equally
2. **a)** 6 V; 18 µC; 3 µF
b) i) 4 µC ii) 4 µC iii) 4 V iv) 2 V v) 2 µF
vi) 2/3 µF; 4 µC
3. **b)** i) 3300 µF ii) 1.03 J
4. **a)** i) 3.7×10^{-8} C ii) 9.3×10^{-8} C
b) 2.3 m
5. **a)** i) 6.64×10^{-11} F ii) 1.66×10^{-7} C
b) charge increases
c) 0.833 mm
6. **a)** i) 10^{-3} C ii) 0.05 J
b) i) 50 V ii) 0.025 J iii) transformed to heat in wires when current flows
7. **a)** 2×10^6 V m^{-1} **b)** i) -1.2×10^{-7} C ii) 0.72 µA
8. ii) 580 V iii) 27×10^{-3} C iv) 13.5 A
9. **a)** 1.2×10^{-5} C **b)** 800 kV **c)** 20 cm
10. 75 kΩ
11. **a)** i) 4.4 V ii) 7.6 V iii) 16.7×10^{-6} C
b) 1.58×10^{-4} J
12. **a)** $4C$ **b)** $V/4$ **c)** $E/4$
13. **a)** 1.6 MΩ **b)** i) 72.13 s ii) 18 MΩ
c) i) 0.0128 C

14. c) i) C_F and C_V in parallel; $C_V = 20$ pF   ii) C_F and C_V in series; $C_V = 25$ pF
15. a) i) 1.7 J   ii) 3.8×10^{-3} F
16. 5.4×10^{-5} J; 2.0×10^{-5} J

15   MAGNETIC FIELDS

1. b) downwards   **c)** upwards
2. a) 1×10^{-2} N   **b)** 0.5×10^{-2} N
3. 10^{-5} T; it is alternating
4. 1.6 A $(5/\pi)$
5. 6.7×10^{-6} T
6. b) iii) 3.1 T   iv) 159.98 g
7. 3.6 A
8. 1.1 kV west to east
9. $Be/(2\pi m)$; 2.8×10^7 rev s^{-1}
10. upper plate +ve   i) 1.6×10^4 V m^{-1}
   ii) 2.56×10^{-15} N; $V = 728$ V
11. b) into page   **c)** 4×10^{-13} N   **d)** direction and magnitude
12. a) no
   c) $Ee = Bev$ $\therefore$ $E = Bv$. Also $E = V_H/d$ $\therefore$ $V_H = Bdv$
   d) 1 mm s^{-1}
13. ii) 1.5×10^{-6} T   iii) 4.5×10^{-6} N   iv) away from P   v) away from Q
14. a) into page   **b)** 0.005 N   **c)** 0 N   **d)** anticlockwise rotation
15. c) 1.1×10^{-3} A
17. b) i) 2.3×10^{-2} N
18. b) F **c)** $4I$, towards the observer

16   ELECTROMAGNETIC INDUCTION

2. $\mathscr{E} = Blv$; 1.5 mA
3. 0.77 V
6. b) i) N
7. 16 mV
8. b) i) 50 Hz   ii) 23.6 V
9. 0.31 mV
11. a) iii) WZYX   **b)** i) 0°, 0 Wb   ii) 90°, 1.2×10^{-6} Wb
   c) 19 mV
12. 1.6×10^{-6} Wb; 48×10^{-6} Wb; 6.5×10^{-3} m s^{-1}; reduced
13. a) i) 5.4×10^{-6} Wb   ii) 5.4×10^{-3} V
15. 237 V, 948 W; 195 V, 11.7×10^3 W, 329 r.p.m.
16. a) i) decreases   ii) decreases
   b) i) 18 W   ii) 75%
17. c) 4 A
19. ii) 6.25×10^{-3} J   iii) 0.83 A
20. b) 42 H   **c)** 20 Ω **d)** 1.78×10^{-2} T m^2
21. 0.50 T

17   ALTERNATING CURRENT

1. a) 12 V   **b)** 17 V   **c)** 2.8 cm
2. a) 99 kV, 0   **b)** 19.8 MW, 0; 141 kV
3. 230 V; $T/3$; 508 V
4. a) 7.0×10^{-4} C
   c) i) when rate of change of charge is a maximum, i.e. when $Q = 0$
   ii) when rate of change of charge is zero, i.e. when $Q = \pm 7.0 \times 10^{-4}$ C
   d) See Fig. 17.8, p. 275.   **e)** 1.6×10^{-1} A
5. 6.37 μF
6. a) 0   **b)** 10^{-4} A   **c)** 2 V s^{-1}
7. b) i) 94.2 Ω ii) 2.44 A   iii) coil has resistance
8. 120 Ω; 0.66 H; 15 μF
9. a) i) 8.84 V   ii) 0.13 A   **c)** 3.1 μF
10. a) 40 Ω **b)** 10 H   **c)** 1.6×10^3 V
11. b) i) halved   ii) doubled   **c)** 375 Hz

12. b) 0.707 V
13. a) 39.6 V   **b)** 0.4 W, 0.72 W
16. c) iii) 1592 Hz
17. a) 16 V   **b)** 12 W
18. a) increase ripple   **b)** decrease ripple   **c)** increase ripple
19. a) i) 10 V   ii) 100 W
   b) power varies at twice the frequency of the supply, i.e. 120 Hz; it has a peak value of 100 W and a time period of 8.3 ms

OBJECTIVE-TYPE REVISION QUESTIONS 3

13.1 C **13.2** C **13.3** E **13.4** A
14.1 A **14.2** i) E   ii) D   **14.3** C   **14.4** C
15.1 D **15.2** C **15.3** D **15.4** D
16.1 D **16.2** D **16.3** C **16.4** C **16.5** C **16.6** D **16.7** A
17.1 C **17.2** A **17.3** C **17.4** C

18   WAVE MOTION

1. 47°; 42° with the normal to the oil surface
2. a) $\lambda/2 = S_2Q - S_1Q$; $\lambda = S_2R - S_1R$
   b) the spacings PQ and QR i) decrease   ii) increase
3. b) 4.7 m s^{-1}
6. 20.0 m to 17.0 mm
7. a) 200 kHz   **b)** 60.0 cm; 59.4 cm   **c)** 5×10^{14} Hz
8 a) 3.3×10^2 m s^{-1} **b)** 6.6×10^{-4} m s^{-1}
9 e
10. b) 45 mm
11. a) 2.5 m   **b)** i) 5 m s^{-1} ii) 3.9 m s^{-1}
12. decrease; increase; stay the same

19   SOUND

1. a) 167 Hz   **b)** 1.5 ms   **c)** progressive longitudinal   **d)** A
   e) 2.04 m
2. a) 10 cm   **b)** i) maxima occur at 5 cm intervals (antinodes)
4. 215 Hz; 645 Hz; 1075 Hz
5. 306 m s^{-1}
6. true, false, true, false
7. $f = \dfrac{n}{2l} \sqrt{\dfrac{T}{\mu}}$ where $n = 1, 2, 3 \ldots$
8. 190 Hz; 202 N
9. Resonant vibration at 115 Hz
10. a) 3.2 m; 3.04×10^8 m s^{-1} **b)** 6 km
11. 320 Hz; 282 Hz; 1.0 s
12. 136 cm s^{-1}
13. 120 Hz
14. a) 4.35×10^8 m   **b)** ii) 9 km s^{-1} iii) towards   **c)** ii) 225 m
15. Stationary wave pattern formed of wavelength $2 \times (5.66 - 2.99)$ mm $= 1.34$ mm; 1.34×10^3 m s^{-1}
16. 0.266 m
17. 280, 350, 420 or 490 Hz
19. a) 2   **b)** 10
20. 7 dB
21. c) i) 34 Hz   ii) 102 Hz

20   PHYSICAL OPTICS

1. 3.0×10^8 m s^{-1}
2. a) 2.0×10^8 m s^{-1} **b)** 5.0×10^{14} Hz   **c)** 4.0×10^{-7} m
3. 5.0×10^{-7} m
4. b) 604 nm
6. 1.5; to side of covered slit
7. 100
8. 0.20 mm

9. **b)** i) 2.14×10^8 m s^{-1} ii) 376 nm   iii) 94 nm
10. 1.51 m; 1.33
11. 9°
12. 2.5×10^5 m^{-1}
13. 4.8×10^{-6} m
14. 20.1°
15. **b)** 24° 17′
17. **b)** i) 6×10^{-4} rad   ii) 2.2 mm
18. 198 kHz
20. **c)** i) yes, no, yes, yes
22. **b)** 7
23. 9.0×10^2 °C
24. 0.96 °C s^{-1}
25. 1.8×10^3 K
26. **b)** i) 812 m^2

OBJECTIVE-TYPE REVISION QUESTIONS 4

18.1 D **18.2** E **18.3** A **18.4** A **18.5** D
19.1 D **19.2** A **19.3** C
20.1 D **20.2** B **20.3** C **20.4** A **20.5** A **20.6** C

21   KINETIC THEORY AND THERMODYNAMICS

1. 8.20 J mol^{-1} K^{-1}
2. **b)** 2.66×10^{25}
3. **b)** i) 3740 J   ii) 1367 m s^{-1} **c)** 7480 J
4. **b)** ii) 478 m s^{-1}
   c) i) no change   ii) temperature and r.m.s. speed fall
   d) mercury
5. 663 m s^{-1}
8. **a)** 1.2×10^{16} **b)** 1.9×10^3 m s^{-1}
9. **b)** i) 1340 m s^{-1} ii) 6.0×10^{-21} J   iii) 80 kPa
10. **b)** i) 2.24×10^{-2} m^3; 2.60×10^{-2} m^3
    c) i) 7.0 m^3 ii) 7.35×10^5 J
11. **a)** 6.5 MJ   **b)** conservation of energy
    c) freezer at a lower temperature than room
    d) 52 W   **e)** 17.6×10^{-3} W m^{-1} K^{-1}
    f) 3.07 MJ
13. **d)** 200 cmHg   **e)** -205 °C (69 K); 273 °C (546 K)
    f) 136 J at constant pressure
14. **b)** i) 23 kJ   ii) 177 kJ   **c)** not quite (V/T varies by ~8%)
15. **b)** i) 5 J   ii) 440 K
    iii) Moves faster and travels further between collisions
16. **b)** i) Q ii) 66×10^5 Pa   iii) 20.6 J
17. **a)** ii) 900 K   iii) 600 kPa   **c)** 5983 J, decrease
18. 5.3×10^2 J kg^{-1} K^{-1}; $\frac{1}{2}$
19. 0.53 atmosphere, -130 °C (143 K); 6.3 litres
21. b, c, d

22   ATOMIC PHYSICS

1. **b)** i) 3.84×10^{-15} N   ii) 4.2×10^{15} m s^{-2}
2. **a)** 0.39×10^{-15} N   **b)** 4.2×10^{14} m s^{-2} **c)** 5.6×10^6 m s^{-1}
3. 2.7×10^7 m s^{-1}; 3.2×10^{-4} W
4. **a)** 2.65×10^7 m s^{-1} **b)** 9.09×10^{-4} T
5. 2.01×10^3 V, upper
6. **a)** i) 8800 V   ii) 1.41×10^{-15} J   iii) 1.41×10^{-15} J
   iv) 5.6×10^7 m s^{-1}
8. **a)** 4.1×10^{-7} m   **b)** 8.8×10^5 m s^{-1}
9. **a)** 1.5 eV   **b)** 3.5×10^{14} Hz
11. $\Phi = 1.8$ eV
12. **b)** i) 6×10^{14} Hz   ii) 4×10^{-19} J   iii) 2.5×10^{10} iv) 8 μA
13. 8.5×10^7 m s^{-1}
14. 1.2×10^{-11} m
15. **a)** iii) 6.2×10^{-10} m   iv) 1.2 nm
16. **a)** i) 2 kW   ii) 1.25×10^{17} iii) 10^5 eV

c) overlap of structures — use CAT scan; movement of the object — restrict movement of patient
19. **b)** 1.8 eV; 1.22×10^{-7} m
20. **b)** i) 25.6×10^{-19} J   ii) 1 eV   iii) 560 nm   iv) visible
21. **a)** 4.9 eV; 6.7 eV; 8.8 eV; 10.4 eV   **b)** 10.4 eV
    c) i) $5 - 4.9 = 0.1$ eV   ii) $10 - 4.9 = 5.1$ eV;
    $10 - 6.7 = 3.3$ eV
    d) i) absorbed and disappears   ii) scattered
22. **a)** 6.2×10^{14} Hz   **b)** 12.5°
23. **a)** i) wave nature of electrons   ii) decrease
    b) $1/\sqrt{3}$
24. **a)** 7.3×10^6 m s^{-1} **b)** higher
25. **c)** 0.158×10^{-9} m
27. **b)** i) 2×10^{-10} m   ii) 3.3×10^{-24} kg m s^{-1} iii) 37 eV
    iv) at radius 0.5×10^{-10} m
28. **c)** i) 1.6×10^{-11} m
29. 2810 eV; 4.4×10^{-10} m; X-ray; 1.07×10^{-13} m

23   ELECTRONICS

1. **a)** i) 5 V   ii) 250 μs   iii) 4 kHz
2. **b)** i) 5 waves   ii) $\frac{1}{2}$ wave   **c)** 50 Hz
3. **a)** 1 V   **b)** 0.7 V
4. **a)** i) 4.24 V   ii) 20 ms   **b)** ii) B to A
6. **a)** 5.4 V   **b)** 54 μA   **c)** 3.2 mA   **d)** 3.2 V   **e)** 2.8 V
7. **b)** $I_C = 2.5$ mA, $I_B = 30$ μA, $V_{CE} = 4.5$ V   **c)** 11.3 mW
   d) i) 4.5 V ± 2.0 V   ii) ± 2.0 V peak
   e) ± 20 mV   **f)** 100   **g)** 130 kΩ
8. **a)** i) $+6$ V   ii) 0 V
   b) i) decreases   ii) decreases
10. See Fig. 23.114.

Input 1	Input 2	NOR output	OR output	AND output	NAND output
0	0	1	0	0	1
1	1	0	1	1	0
1	0	0	1	0	1
0	1	0	1	0	1

Fig. 23.114

11. See Fig. 23.115.

A	B	C	D	E	F	G
0	0	1	1	0	0	1
0	1	1	0	0	1	0
1	0	0	1	1	0	0
1	1	0	0	0	0	1

Fig. 23.115

12. See Figs. 23.116a, b.

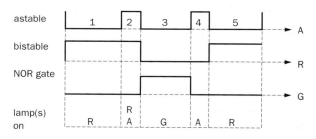

Fig. 23.116a

R	A	G
1	0	0
1	1	0
0	0	1
0	1	0

Fig. 23.116b

13. **c)** 10
14. **a)** $\pm 150\,\mu V$ **b)** 23
15. V_o switches from 'low' to 'high'
16. **a)** 10
17. **b)** 1350 Ω; 30 kΩ; 300 kΩ
18. **d)** 1010
19. **a)** $V_1 > V_2$
    b) i) resistance large ii) $V_2 > V_1$ iii) goes positive
    iv) switches on v) switches on
    c) interchange R_1 and LDR
20. **a)** i) 0 A ii) 0.66 mA iii) 0 A **b)** 10 μF

24 TELECOMMUNICATIONS

1. **a)** 50 Hz **b)** i) 250 Hz; 4 iii) 50 Hz iv) 500 Hz; 8
   c) 800 Hz; 16
2. **b)** 7000 Hz
   c) lower sideband 790–799 kHz; upper sideband
   801–810 kHz
3. **b)** $25 \times 8 \times 350\,000 = 70\,000\,000 = 70$ Mbits s^{-1}
5. **b)** 150 MHz
10. **b)** i) transmitted ii) total internal reflection
    c) 30 μs **d)** 35 μs **e)** 5 μs **g)** dispersion **h)** 100 kHz
    i) monomode or graded-index multimode optic fibres used
11. 80.6°; 24.5 μs
12. **c)** i) 96 dB ii) 12 km iii) 76 dB iv) 100 km
14. **a)** i) 1000 kHz ii) 100 Hz iii) same pitch (frequency) but
    different loudness iv) 999.9 kHz, 1000 kHz and 1000.1 kHz
15. **b)** 9 kHz **c)** 15

25 NUCLEAR PHYSICS

1. 5.1 MeV
2. **a)** alpha, beta, beta, alpha, alpha **b)** 2 **c)** 5 **d)** 4.3 MeV
3. **a)** 3 **b)** $\frac{1}{8}$ **c)** $\frac{63}{64}$
4. **b)** i) 1.05×10^{-11} s^{-1}; 2093 years ii) 11 690 years
5. **a)** i) 1, 1 ii) $^1_1$H **b)** ii) 1.2×10^{-4}/yr^{-1} iii) 2.47×10^{30}
   c) ii) 8000 years
6. 75 s
7. **a)** ii) two different length tracks **b)** $^0_{-1}$e **c)** i) $^{208}_{82}$Pb
8. A 19.8×10^{-12} J; B 20.2×10^{-12} J; B more stable as it has a
   higher binding energy
9. **c)** i) $^{215}_{84}$Po $\rightarrow$ $^4_2$He + $^{211}_{82}$Pb; $^{215}_{84}$Po $\rightarrow$ $^{215}_{85}$At + $^0_{-1}$e + $\bar{\nu}_e$
   ii) 10 MeV
10. **b)** 3.13×10^{16}
11. $v_m = (m - M)v/(m + M)$; $v_M = 2mv/(m + M)$
    a) alpha v; electron $2v$
    b) alpha zero; helium atom v
    c) alpha $-v$; gold atom zero
    There is maximum energy transfer when there is a head-on
    collision between a moving and a stationary particle of equal
    mass.
12. **a)** b **c)** 90°
13. 4.8×10^{-9} J; 5.3×10^{-26} kg
15. **a)** i) yes ii) no iii) yes iv) no

b) i) (udd) baryon ii) (uud) baryon iii) ($\overline{\text{u}}\,\overline{\text{u}}\,\overline{\text{d}}$) antibaryon
   iv) (u$\bar{\text{s}}$) meson
16. **a)** Liquid hydrogen has a higher density of atoms than the
    gas in a cloud chamber.
    b) π^+ has a longer lifetime and is charged.
    c) 3.1×10^{-19} kg m s^{-1}
    d) Particles do not necessarily enter bubble chamber at right
    angles to the magnetic field.
17. **a)** 1.18×10^{20} MeV
    c) a — electromagnetic, most frequent; b — weak
18. **b)** i) 2.2×10^{-3} T ii) increase magnetic flux density
19. **b)** i) 7 protons, 14 nucleons ii) antineutrino iii) carbon
    and nitrogen (baryons); electron and antineutrino (leptons)
    iv) (udd), (uud); d $\rightarrow$ u

26 COSMOLOGY AND ASTROPHYSICS

1. **a)** 5.56 pc, 18.1 ly, 1.14×10^6 AU **b)** 0.05" **c)** 0.91
2. **b)** 100
3. **a)** i) 50 days ii) 25
   b) i) -6 ii) 16 Mpc
4. **a)** 7250 K **b)** 10^{27} W **c)** 133 ly
5. **b)** 7.3×10^6 km
6. **b)** 9.4×10^{29} kg
7. **a)** i) 28 Mpc ii) 87 Mly
   b) 489 nm **c)** 14 billion years
10. **a)** i) Doppler ii) right
11. **b)** 3.8×10^{-8} rad
    c) i) orbital speed remains constant (due to presence of dark
    matter in galaxy)
12. **a)** 83 ns **b)** 23.7 m
13. **a)** i) 100 m ii) 55.3 m
    b) i) 0.4 μs ii) 0.22 μs
    c) i) 3.3 years
15. **a)** 8.1×10^7 m s^{-1} **b)** 7.7×10^{-23} kg m s^{-1}
16. **a)** larger, smaller, larger **b)** larger

OBJECTIVE-TYPE REVISION QUESTIONS 5

21.1 C **21.2** E **21.3** D **21.4** C **21.5** B **21.6** C **21.7** B
22.1 E **22.2** C **22.3** C **22.4** A **22.5** B **22.6** B **22.7** C
23.1 C **23.2** D
24.1 C
25.1 E **25.2** D **25.3** B **25.4** B **25.5** C
26.1 D

ADDITIONAL QUESTIONS

1. 1.5 J
2. **a)** 1.7×10^{-3} m **c)** 5.6×10^2 J
3. D
4. i) Y ii) X iii) X
5. 3 V
6. C
7. **a)** i) 58 Ω ii) 1.5 m
8. **b)** i) Apply Ohm's law to R: $I = V/R = 8.0$ V/4.0 $\Omega = 2.0$ A.
   For the complete circuit: $E = V + Ir$
   $\therefore r = (E - V)/I = (10$ V $- 8.0$ V$)/2.0$ A $= 1.0$ Ω
   ii) Apply Kirchoff's second law $\Sigma E = \Sigma IR$ to the 'outer'
   circuit EASBE: $\Sigma E = 10$ V $- 4.0$ V $= 6.0$ V (since the e.m.f.s
   oppose) and $\Sigma IR = I \times 1.0$ $\Omega + I \times 1.0$ $\Omega = I \times 2.0$ Ω (since
   the current I is anticlockwise and causes a drop of p.d. in the
   int. resist. of both cells): hence $I = 6.0$ V/2.0 $\Omega = 3.0$ A.
   For loop EAVBE: 10 V $- V = 3$ A $\times 1.0$ Ω $\therefore V = 7.0$ V
   For loop VASBV: $V - 4.0$ V $= 3$ A $\times 1.0$ Ω $\therefore V = 7.0$ V
10. **a)** i) 4 Ω ii) 6 V iii) 0.75 A
    b) i) 4.67 Ω ii) 0.857 A

11. B
12. **a)** 14 °C **b)** 1.48 mV
13. 48 g s^{-1}
14. **a)** 30 °C **b)** 133 J kg^{-1} K^{-1} **c)** 4000 J kg^{-1}
15. **b)** i) 73 W ii) 0.09 W m^{-1} K^{-1}
16. 47.8°, 47.2°
17. **a)** 49° **c)** 2.3 m
18. D
19. 0.50 N m clockwise
20. **a)** i) weight of arm ii) no resultant force in any direction: no resultant moment
b) i) 4.1 × 10^2 N
c) F_M has greater moment about E.
21. **b)** 7.8 N
22. **b)** 0.71 m from A: 7.0 N
c) 7.0 N and 7.0 N × 0.21 m
23. **c)** 113°
24. 0 N
25. D
26. 1.25 m
27. A
28. **a)** i) −5.0 m s^{-2} ii) 7.5 kN iii) 120 m
29. **a)** 0.075 N
30. B
31. **a)** i) 0.16 J ii) 1.5 N
b) i) 8.5 m s^{-1} ii) yes
32. 2 kg m s^{-1}; smaller
33. **a)** 3.0 kg **b)** 45 kg m s^{-1} **c)** 45 N
d) 225 kN m^{-2} = 225 kPa (kilopascals)
34. **a)** i) 2400 J ii) 3600 J iii) 1200 J
35. 0.050 m (5.0 cm)
36. **a)** i) $m(u - v) = m_1(v_1 - u_1)$
ii) $m(u^2 - v^2) = m_1(v_1^2 - u_1^2)$
iii) $u - u_1 = -(v - v_1)$
b) $v_1 = 7u/6$, $v = 2u/3$
37. **b)** i) $(\pi/15) \times 10^{-3}$ m s^{-1} ii) $\pi/1800$ rad s^{-1}
c $(\pi^2/27) \times 10^{-6}$ m s^{-2} towards centre of clock face
38. **a)** $\sqrt{(2gl)}$ **b)** $\sqrt{(2g/l)}$ **c)** $3mg$
39. **b)** i) 20 N m ii) 2500 J iii) 100 kg m^2 s^{-1}
40. **c)** i) 30 × 10^3 m s^{-1} ii) 2.0 × 10^{30} kg
41. **b)** i) 0.41 Hz ii) 0.11 m s^{-1}
43. **a)** i) 2.2 MJ ii) 0.44 MJ
44. 17 kW
45. **a)** 8.8 × 10^8 W = 880 MW **b)** 64%
46. **b)** i) 28 000 kJ h^{-1} (7.8 kW) ii) 6.0 kW iii) 8.6 kW
47. area of collector = 2.1 m^2 (assuming 4 h sunshine and 60 °C temperature rise needed)
48. iii) 14.5 J s^{-1} iv) 190 h v) 13 °C
49. **a)** i) 1.02 × 10^5 Pa ii) 1.27 kg m^{-3}
50. **b)** i) 49 m^3 ii) 2.5 × 10^5 N m
52. **b)** i) 2.0 × 10^3 N ii) 1.3 × 10^3 N
53. **a)** 2.3 × 10^{-8} N
54. **b)** 4 µC
55. **b)** 2000 V m^{-1} **c)** 1.2 × 10^{-6} kg
56. 0.023 N; 0.216 × 10^{-6} C; less
57. **a)** i) 0.54 × 10^{12} V m^{-1} ii) 86 × 10^{-9} N
b) area under curve **c)** 42 × 10^{-48} N (much less than **a)**ii))
58. 2 × 79e^2/(4$\pi\varepsilon_0 x$); 57 fm
59. **d)** 1.1 × 10^{11} J
60. $\varepsilon_0 AV/d$; 0.7 × 10^6 V m^{-1}; 70 V, 0.1 mm
61. 0.1 A
62. **a)** 400 mA **b)** i) 5.0 V ii) 5.0 V iii) 0 V **c)** i) 0 V
ii) 6.0 V iii) 6.0 V (all approx.)
63. Oy^1
64. Into the plane of the paper
65. west to east; 408 A; heat
66. **a)** iii) 6 A iv) no forces, since wires parallel to the magnetic field

b) iii) 3.2 mT
67. **a)** i) Direction of the magnetic field must be at right angles to the conductor.
ii) Charged particle must be moving and moving in a direction that is not parallel to the magnetic field.
b) 0.57
68. **b)** v) 64 µV
70. **a)** perpendicular
71. **a)** There is relative motion between the magnet and the conductor.
c) 2 × 10^{-6} Wb s^{-1}
d) increase the strength of the magnet, use a larger amplitude oscillation, use a coil with more turns
72. **a)** i) 34.6 MW ii) 3.44 rad s^{-2} iii) 91.4 s
c) i) 12:1 secondary ii) 225 A
73. **b)** 0.01 T
75. **b)** i) when switch first closed ii) when lamp brightest
76. **b)** i) 36 W ii) 12 W iii) 24 W
77. **c)** i) 2 A, 3 Ω ii) 3.6 V, 1.2 A
iii) $V_3 = 3I_3$, $V_2 = 5V_3$, $I_2 = I_3/5$, $V_2 = V_1 - 2I_2$; $V_1 = 30$ V, $I_3 = 1.95$ A, $V_3 = 5.85$ V
78. 50 Hz; 25.6 Ω
79. **c)** ii) 32 V **d)** i) 100 Hz
80. **b)** i) 5.6 V div^{-1} ii) 0.8 m s
81. **a)** wavelength λ = AB
b) period T = CD = 1/frequency f ∴ f = 1/CD
c) AB/CD = λf = speed v of wave
82. **a)** $\pi/2$ rad = 90° = $\lambda/4$ **b)** π rad = 180° = $\lambda/2$
83. **a)** A absorber, B quartz crystal, C electrical contacts
b) i) 1 cm ii) 0.9988; gel reduces reflection at skin surface
84. 20 m
85. **a)** wavelength **b)** half wavelength
86. **a)** i) interference occurs between waves reflected from B and M; a maximum will occur if the distance between M and B is an integral number of half wavelengths
ii) 28 mm iii) Doppler effect
b) ii) 4.1 × 10^{-5} eV iii) 2 eV
87. **a)** increases **b)** decreases **c)** stays constant
88. **c)** ii) 0.6 mm for a = 1 mm, d = 1.0 m
89. **b)** 17 142
90. 30°
91. π rad (180°)
92. **a)** U = 16 900 N **b)** $\rho = pM/RT$; inversely proportional if p constant ii) 80.3 kW
93. B
94. B
96. **a)** i) 0 ii) 45 J iii) 0 iv) 450 J v) 0 vi) 405 J
vii) W = 405 J, Q_1 = 730 J, Q_2 = 730 − 405 = 325 J = heat lost during DAB; $\eta = W/Q_1$ = 405 × 100/730 = 55%
98. C
99. 46 keV
101. i) 1.9 × 10^{-15} J ii) 1.0 × 10^{-10} m
102. **a)** 2.5 µm **b)** 20 mm grating **c)** level 4 to level 1
103. **e)** 0.24 × 10^{-10} m
104. 3
105. **a)** i) 10 V ii) −5 V iii) 667 Hz
b) i) 0.25 × 10^{-3} J ii) 0.5 × 10^{-3} J iii) 0.75 × 10^{-3} J
iv) 0.5 J
108. **a)** See Fig. 23.76.
109. **a)** i) decrease ii) increase
111. **a)** α **b)** γ **c)** β **d)** γ **e)** α **f)** α **g)** β **h)** γ
112. A = 224, Z = 88; no, since Zs are different
113. 2 min
114. 20 min
115. **a)** i) γ ii) 2.95 MeV
b) iii) antineutrinos take some energy
116. 6.3 × 10^{-13} J; 9.5 × 10^{13} J; 7.9 × 10^{13} J
118. pion+ ud̄; sigma− dds

119. **b)** i) 12.8×10^{-13} J   ii) 1–3 fm   iii) 600 N
120. **a)** i) 1108.75 MeV   ii) 2046.75 MeV; 1100 MeV/c; K^0 is
uncharged; the other particle will have $Q = 0$, $B = 1$, $S = -1$
b) 10 s; nuclei must overcome Coulomb repulsive forces;
quark–lepton era; photon
121. **a)** 2.3 N
124. **a)** Shorter since length decreases in a moving frame of
reference when viewed from one at rest.
b) Older since time passes more slowly in a moving frame
than for one at rest.
c) Smaller since time passes more slowly and a given
distance would be covered in a longer time.

MATHEMATICS FOR PHYSICS

1. **a)** 5 **b)** 20 **c)** 2 **d)** 8
2. **a)** $s = vt$ **b)** $a = F/m$ **c)** $I = \sqrt{(P/R)}$
d) $p_1 = p_2 V_2 T_1/(V_1 T_2)$ **e)** $v = \sqrt{(2gh)}$
3. **a)** 2 **b)** 8 **c)** $-3/4$ **d)** 13/6 **e)** 1
4. **a)** $u = v - at$ **b)** $m = (y - c)/m$ **c)** $a = (v^2 - u^2)/2s$
5. 4×10^3; 2×10^5; 1×10^6; 2.5×10^3; 1.86×10^5; 1×10^{-1};
5×10^{-2}; 2.9×10^{-1}; 7.6×10^{-3}; 1.3×10^{-6}
6. **a)** two **b)** three **c)** four **d)** two
7. 24 cm³

8. **a)** 10^9 **b)** 10^3 **c)** 10^4 **d)** 10^{-3} **e)** 20 **f)** 2.5×10^8 **g)** 80
h) 2×10^{-6} **i)** 7 **j)** 5
9. **a)** 5×10^3 **b)** 6.3×10^5 **c)** 4.6×10^3 **d)** 6.1×10^6
10. The units on both sides of each equation must be the same.
a) unit of $V = m^3$
unit of $m/\rho = $ kg/(kg m^{-3}) = kg m³/kg = m³
b) unit of k.e. = J = N m = kg m s^{-2} × m = kg m² s^{-2}
unit of mv^2 = kg (m s^{-1})² = kg m² s^{-2}
c) unit of acceleration = m s^{-2}
unit of v^2/r = (m s^{-1})²/m = m² s^{-2}/m = m²s^{-2} × m^{-1} = m s^{-2}
d) unit of $T = $ s
unit of $\sqrt{(l/g)} = \sqrt{[m/(m\ s^{-2})]} = \sqrt{(m\ s^2/m)} = \sqrt{s^2} = $ s
11. $P = W/t = $ J/s = J s^{-1} = N m s^{-1} = (kg m s^{-2}) × m × s^{-1}
= kg m² s^{-3}
12. **b)** Δl is directly proportional to m since the graph is a
straight line through the origin.
13. **a)** No; graph is a straight line but it does not pass through
the origin.
c) 32
14. **a)** is a curve **b)** is a straight line through the origin,
therefore $s \propto t^2$ or $s/t^2 = $ a constant = 2
15. **a)** 53.3 mm **b)** 95.8 mm
16. **a)** 2.31 mm **b)** 14.97 mm
17. **a)** $\sin b = $ AC/AB **b)** $\cos b = $ BC/AB **c)** $\tan b = $ AC/BC

INDEX

Index